International Symposium on Geohazards and Geomechanics 2015 (ISGG 2015)

IOP Conference Series: Earth and Environmental Science
Volume 26

Warwick, United Kingdom
10 - 11 September 2015

ISBN: 978-1-5108-1310-6
ISSN: 1755-1307

Printed from e-media with permission by:

Curran Associates, Inc.
57 Morehouse Lane
Red Hook, NY 12571

Some format issues inherent in the e-media version may also appear in this print version.

Printed by Curran Associates, Inc. (2016)

For permission requests, please contact the Institute of Physics
at the address below.

Institute of Physics
Dirac House, Temple Back
Bristol BS1 6BE UK

Phone: 44 1 17 929 7481
Fax: 44 1 17 920 0979

techtracking@iop.org

Additional copies of this publication are available from:

Curran Associates, Inc.
57 Morehouse Lane
Red Hook, NY 12571 USA
Phone: 845-758-0400
Fax: 845-758-2634
Email: curran@proceedings.com
Web: www.proceedings.com

Table of contents

Volume 26

International Symposium on Geohazards and Geomechanics (ISGG2015)
10–11 September 2015, Warwick, UK

Accepted papers received: 21 August 2015
Published online: 9 September 2015

Preface

Keynote Lectures

Keynote 1

Keynote 2

Keynote 3

Keynote 4

Constitutive modelling and thermo-mechanical characterisation of geomaterials

Field monitoring

Numerical and analytical modelling to predict landslide initiation and rockfalls

Susceptibility and hazard mapping, geohazard assessment

Numerical methods for landslide propagation and debris flows

Failure processes

IOP Conf. Series: Earth and Environmental Science **26** (2015) 011001 doi:10.1088/1755-1315/26/1/011001

FOREWORD

These Conference Proceedings contain the full papers in electronic format of the International Symposium on 'Geohazards and Geomechanics', held at University of Warwick, UK, on September 10-11, 2015.

The Symposium brings together the complementary expertise of world leading groups carrying out research on the engineering assessment, prevention and mitigation of geohazards. A total of 58 papers, including 8 keynote lectures cover phenomena such as landslide initiation and propagation, debris flow, rockfalls, soil liquefaction, ground improvement, hazard zonation, risk mapping, floods and gas and leachates. The techniques reported in the papers to investigate geohazards involve numerical modeling (finite element method, discrete element method, material point method, meshless methods and particle methods), experimentation (laboratory experiments, centrifuge tests and field monitoring) and analytical simplified techniques.

All the contributions in this volume have been peered reviewed according to rigorous international standards. However the authors take full responsibility for the content of their papers. Agreements are in place for the edition of a special issue dedicated to the Symposium in three international journals: *Engineering Geology*, *Computational Particle Mechanics* and *International Journal of Geohazards and Environment*. Authors of selected papers will be invited to submit an extended version of their work to these Journals that will independently assess the papers.

The Symposium is supported by the Technical Committee 'Geo-mechanics from Micro to Macro' (TC105) of the International Society for Soil Mechanics and Geotechnical Engineering (ISSMGE), 'Slope Stability in Engineering Practice' (TC208), 'Forensic Geotechnical Engineering' (TC302), the British Geotechnical Association and the EU FP7 IRSES project 'Geohazards and Geomechanics'. Also the organizers would like to thank all authors and their supporting institutions for their contributions.

For any further enquiries or information on the conference proceedings please contact the organizer, Dr Stefano Utili, University of Warwick, s.utili@warwick.ac.uk.

Warwick, September 2015

Peer review statement

All papers published in this volume of *IOP Conference Series: Earth and Environmental Science* have been peer reviewed through processes administered by the proceedings Editors. Reviews were conducted by expert referees to the professional and scientific standards expected of a proceedings journal published by IOP Publishing.

Physical modelling of rainfall-induced flow failures in loose granular soils

W A Take[1,3] and R A Beddoe[2]

[1] Department of Civil Engineering, Queen's University, Kingston, ON, K7L 3N6, Canada
[2] Department of Civil Engineering, York University, Toronto, ON, M3J 1P3, Canada

E-mail: andy.take@civil.queensu.ca

Abstract. The tragic consequences of the March 2014 Oso landslide in Washington, USA were particularly high due to the mobility of the landslide debris. Confusingly, a landslide occurred at that exact same location a number of years earlier, but simply slumped into the river at the toe of the slope. Why did these two events differ so drastically in their mobility? Considerable questions remain regarding the conditions required to generate flow failures in loose soils. Geotechnical centrifuge testing, in combination with high-speed cameras and advanced image analysis has now provided the landslides research community with a powerful new tool to experimentally investigate the complex mechanics leading to high mobility landslides. This paper highlights recent advances in our understanding of the process of static liquefaction in loose granular soil slopes achieved through observations of highly-instrumented physical models. In particular, the paper summarises experimental results aimed to identify the point of initiation of the chain-reaction required to trigger liquefaction flow failures, to assess the effect of slope inclination on the likelihood of a flowslide being triggered, and to quantify the effect of antecedent groundwater levels on the distal reach of landslide debris with the objective of beginning to explain why neighbouring slopes can exhibit such a wide variation in landslide travel distance upon rainfall-triggering.

1. Introduction

The southern California community of La Conchita is situated on a narrow strip of land situated between the Pacific Coast and the slopes of a 180 m high bluff inclined at approximately 35° [1]. The recent history of this community has been defined by two vastly different landslide events separated in time by only ten years. In March 1995, a 120 m wide, 330 m long, deep-seated coherent slump-flow [1] destroyed nine homes at the base of the bluff. Despite this significant damage to the community, the mobility of the landslide was not exceptional for its size. The term "mobility" in this context is often used by landslide researchers (e.g. Iverson et al., [2]) to collectively describe the speed and distal reach of landslide debris, as these quantities define the threat posed by the hazard. Ten years later, however, the landslide debris from the La Conchita event remobilised as a highly fluid debris flow. This much higher mobility January 2005 event had much higher consequences for the small community as the further distal reach and speed of the debris resulted in the deaths of 10 residents and the loss of 36 houses. As noted by Jibson [1], "the movement of the same landslide mass in 1995 and 2005 by two very different mechanisms and with markedly different results is difficult to explain";

[3] To whom any correspondence should be addressed.

however, as also noted by Jibson, the 2005 landslide occurred at the end of a 15-day period of near-record amounts of rainfall. This fact indicates that antecedent rainfall may play a significant role in controlling not only the triggering, but the subsequent mobility of a landslide event.

More recently, a similar scenario in which a pair of landslides of vastly different mobility and consequences occurred near Oso, Washington, USA. In this case, the community at risk was a small neighbourhood of houses known locally as Steelhead Haven situated along the bank of the North Fork Stillaguamish River at the base of a 180 m high slope inclined at approximately 20°. In 2006, a landslide occurred on the slope on the opposite side of the river from the neighbourhood which exhibited non-exceptional mobility. Although the debris from this event travelled approximately 100 m and partially blocked the river, the landslide mass came to rest before reaching the Steelhead Haven neighbourhood [3]. Less than ten years later, on 22 March 2014, this same slope remobilised as a high mobility flowslide (Figure 1) of a volume of approximately 8 x 10^6 m^3 which overran the full 1 km width of the floodplain, demolished the Steelhead Haven neighborhood, buried highway SR530, and resulted in 43 fatalities [2]. Again, antecedent rainfall was identified as a leading potential factor causing the vast difference in mobility between these two events. As noted by the GEER team [3], the cumulative precipitation for the three weeks prior to the high-mobility 22 March 2014 event corresponds to a return period of almost 100 years and the wettest early March on record for a nearby rain gauge. Seismic signals recorded during the landslide event have been interpreted to infer that failure occurred in two stages. However, the exact sequence of events through which a higher antecedent rainfall could lead to such an exceptional mobility landslide vent is currently unclear, with different mechanisms proposed by the GEER team [3] and Iverson et al. [2] to explain the cascading sequence of events which led to liquefaction of the base of the landslide.

Figure 1. Photograph of the Oso Landslide after the 2014 flowslide. Photo taken by Spc. Matthew Sissel, U.S. Navy.

The two pairs of tragic landslide events at La Conchita and Oso clearly underline the increased threat posed by high mobility landslides. In so doing, these events also demonstrate the importance of understanding how and why some landslides transition from slide to flow. In particular, research is needed to further investigate the sequence of triggering conditions needed to generate high mobility flows, the influence of antecedent groundwater on the likelihood of their triggering, and the range of slope inclinations susceptible to this type of failure. The objective of this paper is to present an overview of the findings of a 10 year research program using physical model experiments to begin to seek answers to these questions.

2. Physical modelling test program

Geotechnical physical modelling provides an avenue of experimental landslides research in which nominally identical slopes can be created and brought to failure under different highly controlled boundary conditions. This permits factors such as antecedent groundwater flow, or slope angle to be varied in the models whilst keeping all other factors constant. As a further advantage, slope geometry and soil properties within physical models can be sufficiently simplified to permit subsequent analysis of the observed landslide behaviour in the form of quantitative data capturing transient pore water pressures and landslide movement. Thus, rather than to explicitly attempt to model the more complex geological sequences at La Conchita, Oso or from another historical landslide, the physical modelling test program focused on capturing the transition from slide to flow for a simple idealised slope geometry to investigate how this much simpler system works as a first step.

The geometry used for the testing program consists of a 50 mm thick soil layer overlying bedrock, which when subjected to a centrifugal acceleration of 30 g in a geotechnical centrifuge, corresponds to a layer of 1.5 m in thickness (Figure 2). The decision to include an impermeable bedrock layer was made for a number of reasons, which have been fully described by Take [4], and as such will only be briefly summarised here. The primary reason for this permeability contrast layer relates to the difficulty in replicating static liquefaction events in physical models. Despite being observed in triaxial tests and field events, static flow liquefaction events within physical model tests has proved to be an surprisingly difficult experimental task. As described in the recent keynote lectures investigating the deformation and failure mechanisms in loose and dense granular slopes by Ng [5,6], static liquefaction has been seen in some physical model tests of homogeneous loose fill slopes (e.g. Zhang and Ng [7]) but not in others (e.g. Take et al. [8]). The primary difference between these two cases is the need for a high degree of saturation for liquefaction to occur. Whereas the saturated loose slope of Zhang and Ng [7] liquefied during inundation, the unsaturated homogeneous granular slope model reported by Take et al. [8] simply experienced minor volumetric collapse upon application of intense model rainfall. In other words, although the unsaturated slope experienced rainfall-induced instability, a flow failure was not triggered in this case through the mechanism of static liquefaction. Instead, the low degree of saturation of the granular slope prior to rainfall permitted the model slope to accommodate the volumetric strains associated with the wetting collapse through the compressibility and ease of migration of the pore air [8]. Although the slope material was contractile, and a triggering event was provided in the form of a rainfall-induced wetting collapse, this test result indicates the importance of a high degree of saturation to permit the development of shear-induced pore water pressures if the mechanism of static flow liquefaction is to be triggered [4]. The impermeable bedrock layer in the follow-on testing programme described here therefore permits the development of a water table within the potential slide mass, in order to create the pore network of contractile saturated voids filled with an incompressible fluid (water) which is prone to significant positive excess pore water pressure generation upon shearing. Thus, the presence of the impermeable bedrock layer is simply a strategy to encourage saturation of the potential slide mass.

The 50 mm thick layer of soil consisted of fine uniform sand (U.S. Silica F110 Ottawa Sand of d_{50} of 0.12 mm), mixed at a gravimetric moisture content of 7%, and rained through a sieve from an elevation of approximately 125 mm to create a very loose sand layer resting on a roughened impermeable base intended to model impermeable bedrock. A series of miniature PDCR-81 pore water pressure sensors were mounted into this base layer to capture both the static and transient pore water pressure response of the slope prior to and during failure. In this model, water was introduced as a seepage flux, Q_{in}, at the top of the model using a metering flow pump. The response of the slope was then monitored using a Phantom V9.0 high-speed digital camera recording at 1000 frames per second (fps) to capture the deformation of the soil at toe and base region of the slope. Readers interested in the full details of the physical model, soil properties, and instrumentation are referred to Take and Beddoe [9].

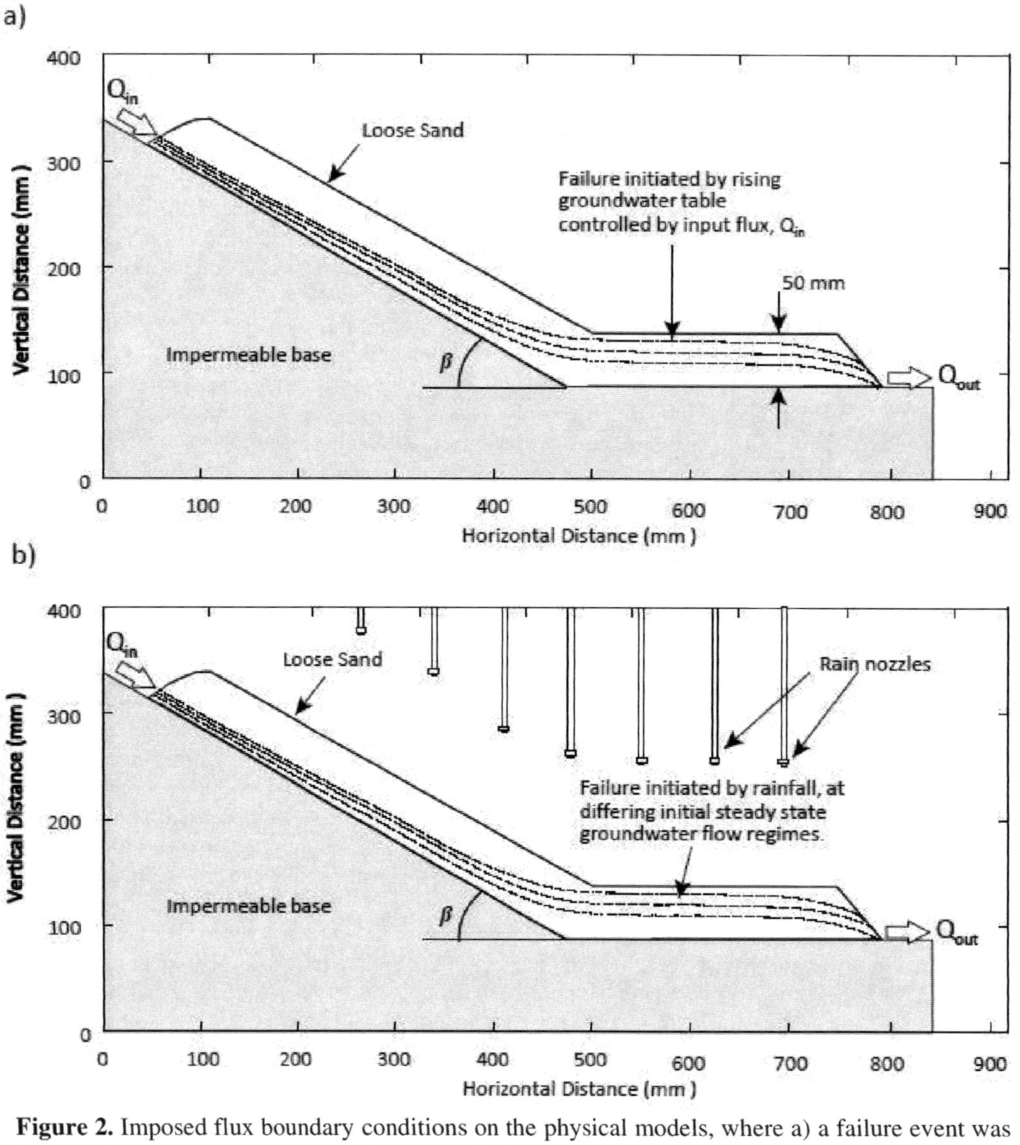

Figure 2. Imposed flux boundary conditions on the physical models, where a) a failure event was initiated by a rising groundwater table controlled by input flux, Q_{in}, and b) failure was initiated by rainfall at differing initial steady state groundwater flow regimes. Model scenarios were conducted at a slope angle, β, of 20 and 30 degrees.

Once constructed, the model was subjected to the test acceleration of 30 g on the C-CORE centrifuge in St John's, NL, Canada. The testing program aimed to investigate the sequence of triggering events leading to high mobility flows (Take and Beddoe [9]), the influence of antecedent groundwater on the likelihood of triggering high mobility flows (Take et al. [10]), and the range of slope inclinations susceptible to this type of failure (Beddoe and Take [11]). An overview of the findings from these investigations are presented in the following sections, with the objective of combining the observations of slope behaviour from these recently published papers to reach overarching conclusions from the entire combined dataset of physical model test results.

3. Triggering sequence

In order for the deviatoric strain-softening associated with static liquefaction to be generated, the soil must be contractile, be subjected to a monotonic loading trigger, and be sufficiently saturated to permit the generation of excess pore water pressures upon shearing. However, as proposed by Take and Beddoe [9], it is this third condition that is perhaps the most difficult to achieve for granular soil slopes. If one considers an infinite slope of granular material inclined at 30°, a significant seepage flow rate would be required to saturate even a small thickness of the layer. Instead, Take and Beddoe hypothesised that a more likely location of saturated soil could be located in regions where the inclination of the soil layer locally reduces, as this change in geometry can impose a lower hydraulic gradient resulting in a thicker zone of saturated soil for the same groundwater flux. One extreme case of a change in slope inclination is the toe of a slope. As shown in Figure 2a, water entering the top of the physical model slope with a flux Q_{in} will result in only a thin zone of saturated flow on the slope, but a significantly thicker zone at the base due to the reduction in hydraulic gradient. Thus, the change in soil layer inclination at the toe creates the hydraulic conditions that favor saturation at a lower flow rate than would be required to trigger liquefaction failure in an infinite slope of the same inclination. If this soil was sufficiently loose, and if a rainfall or seepage induced local toe failure were to provide a monotonic shearing trigger, it is theoretically possible that the base of the landslide may be at higher risk of experiencing static liquefaction than the better drained inclined portion of the slope [9]. To test this hypothesis, Take and Beddoe [9] carried out a physical model test of geometry described in Figure 2a in which the groundwater flux, Q_{in}, was incrementally increased until failure was triggered.

The sequence of failure observed by [9] in a rising groundwater experiment in a 30° physical model slope is shown in Figure 3. This figure contains displacement vectors, calculated using the image processing technique of digital image correlation [12] on image frames captured in the toe / base region of the slope every 0.001 s of the experiment. As the groundwater flux was increased in small increments, no significant landslide deformation was observed in the slope. However, once the groundwater flux was sufficient to saturate the sand layer at the base of the slope (i.e. note the total head values and measurement locations in Figure 3a), a small initial slip surface developed at the toe of the slope. This failure then rapidly sheared the highly saturated volume of soil in the base layer, causing the generation of positive excess pore water pressures (signified by the total head values significantly above the soil surface in Figure 3b) consistent with the undrained shearing of a contractile material. This then permits the failure surface to deepen, a larger volume of material to be entrained, and the debris to flow nearly horizontally as the toe region removes its buttressing support of the remainder of the material left on the inclined portion of the slope (Figure 3c). Although a highly idealised geometry, this failure sequence highlights the need for a small monotonic failure to be co-located with a zone of saturated, contractive soil for the system to be prone to liquefaction.

These observations illustrate that the sequence of failure illustrating the transition from a small localised slide to a more widespread flow can be captured in a physical model. But can physical models be used to explain why lower antecedent groundwater conditions led to lower mobility landslides? To answer this question, we present the results of a physical model test conducted using the same geometry as the previous experiment, but this time, brought to failure by the application of model rainfall using mist nozzles. This slightly revised testing configuration is shown in Figure 2b, and in this case that will be discussed in detail, no steady-state antecedent groundwater flow was induced in the model prior to rainfall application. Therefore, this test represents the extreme case of a landslide being brought to failure with the same rainfall event, but with very low initial groundwater levels. The sequence of failure observed by [10] for this case is presented in Figure 4. As the test aimed to have very low antecedent groundwater volumes, the pre-rainfall total head values presented in Figure 4a are consistent with the drainage of moisture from the slope during gravity turn-on (i.e.

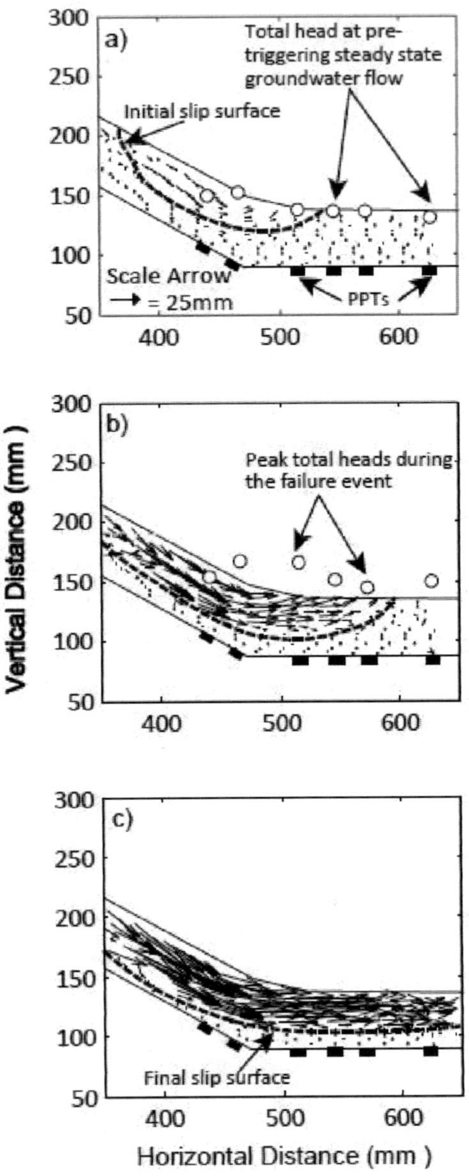

Figure 3. Evolution of the failure event for rising groundwater scenario, highlighting, a) pre-triggering total heads measured by the porewater pressure transducers (PPTs) super imposed on the initial failure event, b) mid failure event, when peak total heads are reached, and c) final vector displacement of the landslide.

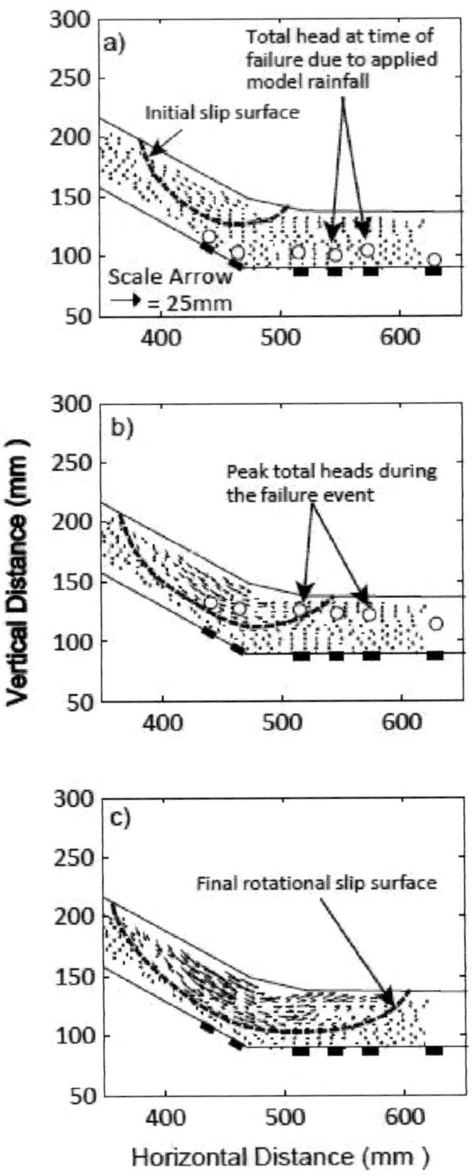

Figure 4. Evolution of the failure event for the rainfall only scenario, highlighting, a) pre-triggering total heads from rainfall event imposed on the initial failure event, b) mid failure event, when peak total heads are reached, and c) final vector displacement of the landslide.

recall that the soil was placed at a gravimetric water content of 7 percent). As model rainfall was applied, the reduction in suction associated with the infiltration front caused an initial localised failure surface at the toe, consistent with the first physical model. However, the consequences of this monotonic shearing event on the base layer are dramatically different in the second case. Although shear-induced pore water pressures are generated (Figure 4b), the lower saturation of this layer has the consequence of reducing the magnitude and duration of these total heads, resulting in a landslide of much lower mobility (i.e. The full extent of the difference between the landslide travel distance can be assessed by comparing Figures 3c and 4c as the displacement vectors are plotted to the same scale).

4. Effect of antecedent groundwater flow on mobility

The differences in mobility between the high and no antecedent groundwater flow cases in Figure 3 and 4 is drastic. But what is the nature of the relationship between antecedent groundwater flow and mobility? Is it linear? Or is there a threshold value of flux required to trigger the high mobility liquefaction landslides? To attempt to answer this question, Take et al. [10] performed additional experiments at intermediate antecedent groundwater levels prior to rainfall infiltration. As illustrated schematically in Figure 2b, this was achieved physically in the model, by controlling the metering flow pump to provide different groundwater fluxes to the top of the model, thereby creating different initial positions of the water table (i.e. thicknesses of saturated soil) along the base of the model. Frames captured every 0.001 s from each experiment were then analysed using digital image correlation to measure the total distal reach of the landslide debris and the maximum velocity. The results from the five experiments performed on a 30 degree slope are presented using solid markers in Figure 5 in model scale. The x-axis for this figure reports the antecedent groundwater condition of each experiment, as expressed by the steady-state flux applied prior to rainfall infiltration. Therefore the two tests already discussed in this paper correspond to 175 mm/min (i.e. the test failed by groundwater seepage alone) and 0 mm / min (i.e. the test with no pre-existing groundwater table prior to rainfall). Superficially, the results illustrate that the landslides that experienced the highest travel distance also experienced the highest peak velocity. This is both a logical and expected observation. More importantly, these results illustrate that the effect of antecedent groundwater flow on landslide travel distance and peak velocity is highly non-linear with seepage flux. For example, the increase in travel distance provided by a 75 mm/min increase in seepage flux from 0 – 75 mm/min is marginal; whereas an additional increase by the same 75 mm/min increment from 75 – 150 mm/min results in an enormous increase in mobility. Further, this data indicates that for the simple idealised geometry investigated in this research program, high antecedent groundwater conditions can increase the mobility of the landslide debris by a factor of eight.

5. Effect of slope angle

The average inclination of the slope at La Conchita was approximately 35°, whereas the inclination at Oso was only 20°. As noted by Iverson et al. [2], "landslides that transform into mobile, high-speed flows almost invariably begin on slopes >20°, and initiation sites steeper than 30° are typical [13,14]". In this context, the high mobility of the Oso landslide underscores the importance of defining the range of slope angles susceptible to high mobility liquefaction events. To this end, Beddoe and Take [11] performed a series of physical model tests at a lower inclination (20°) to enable a direct comparison to the test results already presented for 30° slope models. Intuitively, higher slope inclinations correspond to higher downslope components of gravity, which may lead to the conclusion that steeper slopes are more susceptible to large distal reach landsliding. However, as noted by Beddoe and Take [11] more gently inclined slopes require higher pore water pressures to trigger failure, which could possibly lead to higher mobility due to a larger region of saturated soil at the moment of triggering.

The sequence of failure observed in a rising groundwater experiment in a 20° physical model slope is shown in Figure 6. Once again, similar to the more steeply inclined soil layer, increasing

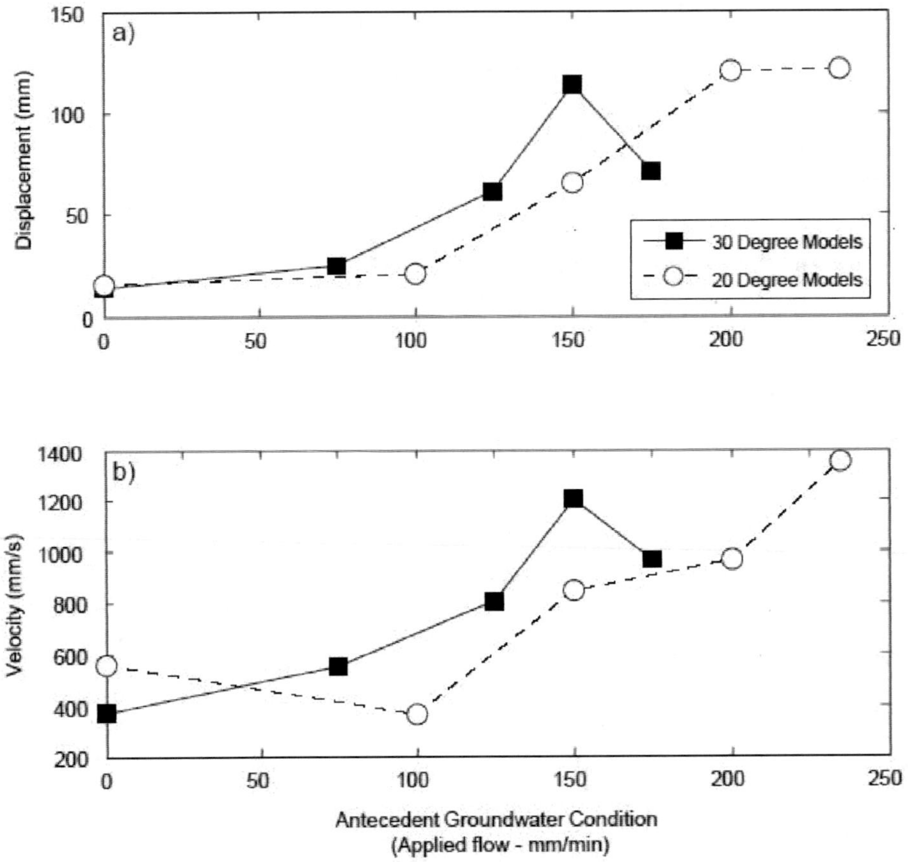

Figure 5. Comparison of landslide mobility observed in 30° [10] and 20° [11] slopes, where a) maximum displacement, and b) maximum velocity.

groundwater flow had the effect of saturating the base of the landslide prior to a small localised failure at the toe of the slope (Figure 5a). This monotonic shearing event then generated significant excess pore water pressures within the slope (Figure 5b) sufficient to liquefy the soil layer, and experience significant horizontal flow. These observations indicate that although more water was needed to trigger failure (230 mm/min rather than 175 mm/min flux), high mobility flow slides can still be initiated at slope angles as low as 20°.

The relationship between landslide mobility (i.e. maximum velocity and distal reach of debris) and antecedent groundwater conditions (expressed as a pre-rainfall imposed groundwater flux to the model) reported by Take et al. [10] for the 30° slopes and Beddoe and Take [11] for the 20° slopes are combined in Figure 5. This data clearly illustrates that a 20° slope can exhibit higher landslide mobility than a more steeply inclined soil layer subject to a lower groundwater seepage flux. Further, it is important to note that the relationship between mobility and groundwater seepage flux remains non-linear for the 20° slopes. Therefore, given the non-linearity of this relationship in both datasets, this data illustrates the enormity of the challenge posed by the prediction of the triggering of high mobility flows from meteorological data alone.

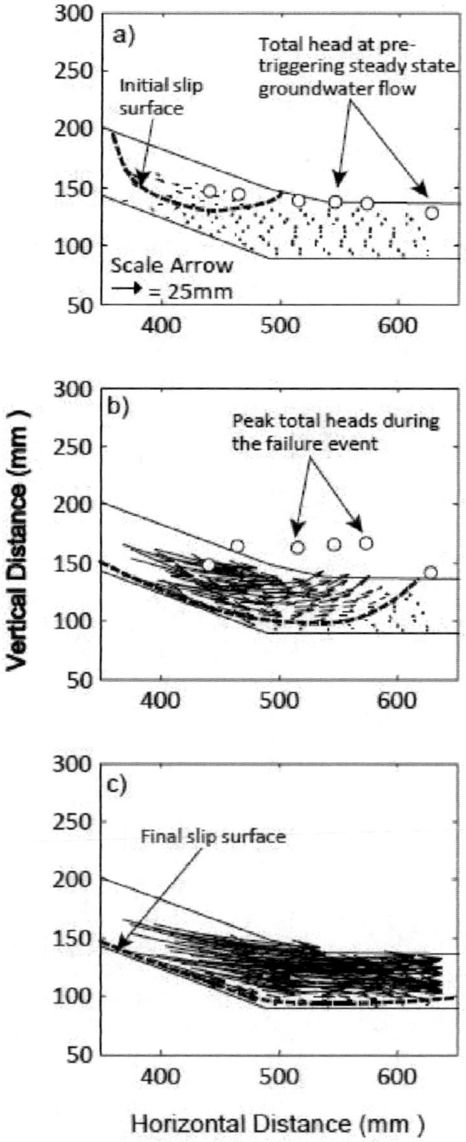

Figure 6. Evolution of the failure event for a 20 degree test, highlighting, a) pre-triggering total heads along with the initial failure event, b) mid failure, when peak total heads are reached, and c) total vector displacement of the landslide.

To further illustrate this point, the rainfall duration required to initiate failure as a function of antecedent groundwater condition for all tests are summarised in Figure 7. As all tests brought to failure using a constant intensity rainfall, the y-axis of the graph represents short-term rainfall (i.e. the rainfall duration required to trigger rainfall expressed in seconds), whereas the x-axis represents the antecedent groundwater seepage conditions prior to rainfall. This data underlines the metrological conditions required to trigger failure, with the lower inclination slope requiring more infiltration to initiate failure. However, what is completely lost in this treatment of the data is that not all of these landslide events have the same mobility, and therefore do not pose the same threat to life and livelihood.

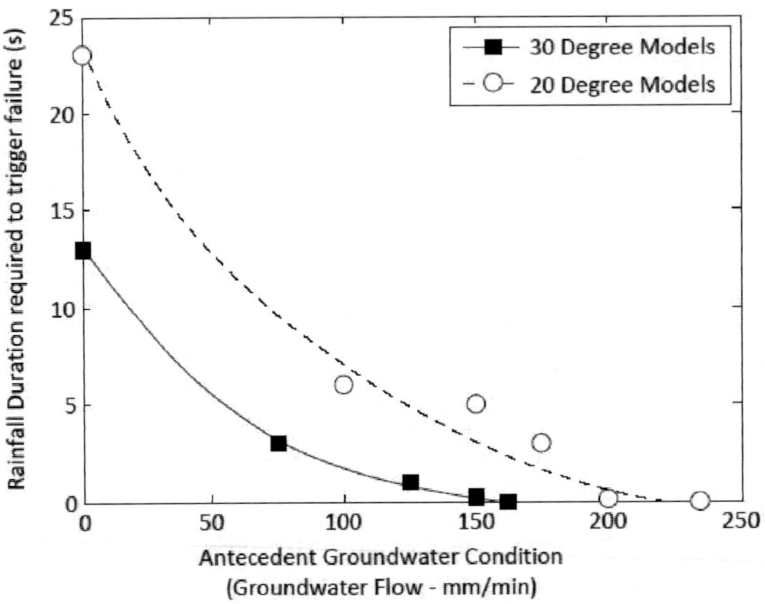

Figure 7. Comparison of the rainfall duration required to initiate failure for the 30° slope and 20° slope physical model tests. The curves emphasize a landslide warning rainfall threshold, however this type of intensity-duration curve does not capture the resulting landslide mobility.

6. Conclusions

The tragic landslide events in the communities of La Conchita and Oso clearly demonstrate the increased threat posed by high mobility landslides. Geotechnical centrifuge testing, in combination with high-speed cameras and advanced image analysis has now provided the landslides research community with a powerful new tool to experimentally investigate the complex mechanics leading to high mobility landslides. In this paper, an overview of the findings of a 10 year research program using physical model experiments aimed at better understanding these complex failure processes has been presented. These experiments have been successful in initiating deviatoric strain-softening associated with the cause of the extreme mobility (static liquefaction) in physical models under a wide range of different hydraulic boundary conditions.

The first contribution of these physical model tests to our understanding of these events it to illustrate one particularly highly-successful scenario in which high-mobility landslides can be generated. This scenario begins with a small initial slip surface developed at the toe of the slope. This failure then rapidly sheared the highly saturated volume of soil in the base layer, causing the generation of positive excess pore water pressures consistent with the undrained shearing of a contractile material. This then permits the failure surface to deepen, a larger volume of material to be entrained, and the debris to flow nearly horizontally as the toe region removes its buttressing support of the remainder of the material left on the inclined portion of the slope. Although a highly idealised geometry, this failure sequence highlights the need for a small monotonic failure to be co-located with a zone of saturated, contractive soil for the system to be prone to liquefaction. For the case of field events, heightened pore water pressures due to permeability contrasts associated with adverse layering or local bedrock exfiltration (e.g. Rüdlingen landslide experiment [15]) could potentially create other locations in addition to the toe of the slope where an initial localized failure event could trigger the onset of static liquefaction of a zone of saturated, contractive soil.

Secondly, the high mobility La Conchita and Oso landslide events were both hypothesized to be related to the extremely high antecedent rainfall that preceded both events. This hypothesis was tested using physical models of various antecedent groundwater levels prior to the application of model rainfall. The results from this testing program indicate that the effect of antecedent groundwater flow on landslide travel distance and peak velocity is highly non-linear with seepage flux. Further, this data indicates that for the simple idealised geometry investigated in this research program, high antecedent groundwater conditions can increase the distal reach of the landslide debris by a factor of eight. Therefore, these results confirm the enormous role that antecedent groundwater levels play in determining whether the landslide mass is sufficiently saturated to experience full liquefaction.

Finally, the influence of slope angle was investigated to contrast the evolution of the slide to flow mechanism at slope angles thought to be typical of flow slide generation (30°) and more gently inclined slopes (20°) consistent with the average slope at Oso. The physical model results indicate that liquefaction can indeed be triggered in more gently inclined slopes (20°), and in certain circumstances, can result in higher mobility landslides than those triggered on steeply inclined slopes (30°). However, higher groundwater flow seepage rates and durations of rainfall infiltration were required to achieve the pore water pressure required to trigger failure, consistent with expectations from a frictional effective stress framework to predict triggering.

It should be noted that all physical modelling experiments performed in the testing program were conducted at a void ratio that was very loose and exceedingly contractile under loading. As noted by Iverson et al [2] a critical void ratio exists that defines whether a landslide deposit is sufficiently contractile to generate the high excess positive pore water pressures required to experience liquefaction. The next step of the physical modelling research program is therefore to investigate the variation in mobility of landslides at different initial void ratio (and therefore state parameter).

References

[1] Jibson R W 2006 The 2005 La Conchita, California, landslide *Landslides* **3** 73-8

[2] Iverson R M, George D L, Allstadt K, Reid M E, Collins B D, Vallance J W, Schilling S P, Godt J W, Cannon C M, Magirl C S, Baum R L and Coe J A 2015 Landslide mobility and hazards: implications of the 2014 Oso disaster *Earth and Planetary Science Letters* **412** 197-208

[3] Keaton J R, Wartman J, Anderson S, Benoît J, deLaChapelle J, Gilbert R, and Montgomery D R 2014 The 22 March 2014 Oso Landslide, Snohomish County, Washington, *GEER report, NSF Geotechnical Extreme Events Reconnaissance*

[4] Take W A 2014 Keynote paper: Physical modelling of instability and flow in loose granular slopes. *8th International Conference on Physical Modelling in Geotechnics*, Perth, Australia.

[5] Ng C W W 2007 Keynote paper: Liquefied flow and non-liquefied slide of loose fill slopes. *Proc. 13th Asian Regional Conference on Soil Mechanics and Geotechnical Engineering*, 10-14 December, Kolkata. Vol. 2 (post-conference volume). 120-134. Allied Publishers Private Ltd, India.

[6] Ng C W W 2008 Invited special lecture: Deformation and failure mechanisms of loose and dense fill slopes with and without soil nails. *Proc. of 10th Int. Sym. On Landslides and Engineered Slopes*. 30 June – 4 July, 2008, Xi'an, China. Vol. 1 159-177

[7] Zhang M and Ng C W W 2003 Interim Factual Testing Report I—SG30 & SR30. Hong Kong University of Science & Technology.

[8] Take W A, Bolton M D, Wong P C P and Yeung F J 2004 Evaluation of landslide triggering mechanisms in model fill slopes *Landslides* **1** 173–184

[9] Take W A and Beddoe R A 2014 Base liquefaction: a mechanism for shear-induced failure of loose granular slopes *Can. Geotech. J.* **51** 496-507

[10] Take W A, Beddoe R A, Davoodi-Bilesavar R and Phillips R 2015 Effect of antecedent

groundwater conditions on the triggering of static liquefaction *Landslides* **12** 469-79

[11] Beddoe R A and Take W A 2015 Influence of slope inclination on the triggering and distal reach of hydraulically-induced flow slides *Engineering Geology* **187** 170-82

[12] White D J, Take W A and Bolton M D 2003 Soil deformation measurement using particle image velocimetry (PIV) and photogrammetry *Geotechnique* **53** 619-31

[13] Voight B (Ed.) 1978 Rockslides and Avalanches 1 Natural Phenomena Elsevier, Amsterdam

[14] Iverson R M, Reid M E and LaHusen R G 1997 Debris-flow mobilization from landslides *Annu. Rev. Earth Planet. Sci.* **25** 85–138

[15] Askarinejad A, Laue J, Zweidler A, Iten M, Bleiker E, Buschor H and Springman S M 2012 Physical modelling of rainfall induced landslides under controlled climatic conditions *Proc. of EuroFuge 2012* 23-24 April 2012 Delft Netherlands 1-10

Geotechnical hazards associated with closed municipal solid waste landfill sites

W Powrie[1], D Richards[1], R Beaven[1]

[1] Faculty of Engineering and the Environment, University of Southampton, UK.

E-mail: W.Powrie@soton.ac.uk

Abstract. As pressure for new infrastructure and development grows, it is inevitable that building projects will encounter some of the c20,000 closed former solid waste landfills in the UK, many of which will have accepted municipal solid wastes (MSW). Construction on or across these sites brings a special set of geohazards associated with the potential for large and difficult to predict settlements, gas (and odour) release or generation, contaminated leachate and the breach of containment systems and other environmental controls. The presentation will discuss these issues with reference to recent research into understanding and predicting settlements in municipal solid waste landfills; assessing the total, current and residual gas potential of biodegradable wastes; the role of the hydraulic regime in the flushing of contaminants from the waste and the quality of leachate; and the need or otherwise for the long term integrity of engineered barriers and controls.

1. Introduction

European waste management policy has for the past 20 years sought to move away from landfill in favour of waste avoidance and reduction, recycling and the recovery of energy from wastes. However, in Europe and much of the developed world, landfilling has been the primary means of municipal solid waste (MSW) disposal during more than 150 years of progressive industrialisation and population growth. As a result, there are estimated to be more than 20,000 closed former MSW landfills in the UK alone, containing ~1 billion tonnes of household waste and ~5 billion tonnes of Commercial and Industrial (C&I) type wastes.

The Environment Agency (EA) holds a GIS database of both current (authorised) and historic landfills that can be accessed via the "what's in your backyard" part of their website. The database is a collation of information held by local authorities and the Environment Agency. Authorised landfills are sites that have a permit issued under the Pollution Prevention and Control Act 1999, or closed sites that no longer take waste but still have a waste management licence issued under Part II of the Environmental Protection Act 1990. Some waste management licences date back to disposal licences issued under the Control of Pollution Act 1974.

Historic (closed) landfill sites are those where there is no PPC permit or waste management licence currently in force. This includes (most) sites that existed before the waste licensing regime, and previously licensed sites where the licence has been revoked, ceased to exist or surrendered and a certificate of completion issued. From 1948 until the introduction of the waste licensing regime, regulation of landfills was undertaken through planning law. There were no controls on waste disposal before the Planning Act of 1948, and no formalised means of recording the location of sites.

The National Planning Policy Framework (NPPF) published in 2012 sets out Government planning policy for England and how this is expected to be applied to development. Paragraphs 120 to 122 of Section 11 of the NPPF, *Conserving and enhancing the natural environment*, concern contaminated land and state the following:

120. *To prevent unacceptable risks from pollution and land instability, planning policies and decisions should ensure that new development is appropriate for its location. The effects (including cumulative effects) of pollution on health, the natural environment or general amenity, and the potential sensitivity of the area or proposed development to adverse effects from pollution, should be taken into account. Where a site is affected by contamination or land stability issues, responsibility for securing a safe development rests with the developer and/or landowner.*

Legislative controls over authorised landfills and contaminated land are set out in Part II and Part IIA respectively of the Environmental Protection Act (1990) and a raft of Statutory Instruments and amendments thereafter. Within this Legislation responsibility for the adverse effects of contaminated land (appropriate person) also lie with the original person(s) who caused or knowingly permitted contamination at the site – usually taken to be the site operator.

It follows that any development over an old MSW landfill will require the involvement of a variety of parties including the original landowner and operator, the local authority and the EA (including as statutory consultees on any planning application), as well as the developer.

Many old landfills are no longer actively managed or monitored, and their state of degradation is uncertain. Many are likely to pose a residual potential pollution threat through the leakage of leachate into ground or surface waters or the escape of landfill gas into the ground or the atmosphere. Ongoing degradation of the waste will result in continuing settlement. Even if degradation is complete, the potential for creep or further load-induced settlement may be high. These processes and features represent possible geohazards for the construction of new infrastructure and buildings across a former MSW landfill, as summarised in Table 1.

The presentation will discuss these issues with reference to:

- recent research into understanding and predicting settlements in municipal solid waste landfills
- the total, current and residual gas potential of biodegradable wastes
- the role of the hydraulic regime in the flushing of contaminants from the waste and the quality of leachate
- the point at which a landfill can be considered not to represent a potential environmental threat, hence the need or otherwise to maintain the long term integrity of engineered barriers and controls.

Table 1: Potential impacts of landfill processes on future site redevelopment

Landfill process	Mechanisms	Potential impacts on site redevelopment
Settlement	• Mechanical (self-weight stress) • Biodegradation • Creep	Unlikely – this is a short term phenomenon Could result in significant ongoing settlement if degradation is still ongoing Potentially an issue but decreases with log time
Settlement	• Mechanical (imposed additional stress)	Large settlements may occur as a result of structural loading owing to the generally low stiffness of wastes
Gas	• Biodegradation induced generation of CO_2 and CH_4.	Odours. Potential release of gases, including H_2S.
Leachate	• Washout of recalcitrant contaminants and products of biodegradation	Biochemical effects on sub-surface structures and foundations. Leakage of leachate into groundwater (liner failure) or surface water (drainage failure leading to overtopping)
Passive engineered control features	• Cap, liner, drains, gas vents etc	May need to avoid penetrating cap and liner to preserve integrity if biodegradation is still ongoing and/or contamination potential through leachate escape remains. May need to maintain drainage to avoid overtopping (see above)

2. Landfill processes

It is well established that the pollution potential of a waste is reduced over time through the processes of biodegradation and flushing (e.g. DoE, 1986; Knox 1990). Biodegradation generates gases, principally methane and carbon dioxide. It also causes a loss of mass and volume, resulting in settlement of the waste. Some contaminants are not readily biodegradable, and must be removed from the landfill by leaching or flushing, i.e. the continued flow-through of relatively clean water to wash these contaminants out.

In principle, an MSW landfill could be operated in such a way as to remove the contaminant load through biodegradation and flushing to achieve a state of practical completion (in which the landfill is in hydraulic equilibrium with its surroundings, and the waste is considered to be of "final storage quality", FSQ, in that it no longer represents a potential pollution threat) within a timescale of about a generation (~30 years; e.g. Harris et al., 1995). However, in the UK at least there is no financial or regulatory incentive to do this, and under current UK operational regimes it could take hundreds of years for a state of practical completion to be achieved (e.g. Knox, 1990; Hall et al., 2007). It follows that most closed landfills will still represent a potential pollution threat, which could be realised on redevelopment of the site through the disturbance of the waste and / or the engineered controls such as caps, liners and drains built to prevent the escape of pollutants from the waste during the operational and aftercare phases of the site.

Commonly accepted interpretations of the terms "Landfill completion" and "Final storage quality" are reflected in guidance published in the UK (Environment Agency; 2005, 2012), which requires that the flux of contaminants to the environment would still be acceptable assuming, inter alia:

- no active management;

- failure of all engineered containment;
- attainment of hydraulic equilibrium (i.e. water or leachate levels in the site have equilibrated with water fluxes into and out of the site in the absence of active management and failure of some or all of the engineered controls);
- no functioning gas or leachate management systems.

Biodegradation in landfills occurs primarily through the processes of anaerobic degradation, which produces landfill gas (LFG). The potential for aerobic degradation to occur is normally very limited. However, air injection (particularly in Germany) has been utilised as a final treatment stage in aged landfills. There is a close relationship between biodegradation and waste settlement.

3. Settlement

Settlement of biodegradable wastes may be considered in three stages, termed immediate, primary and secondary. Immediate settlement is due to the compression or expulsion of gas and/or the compression of particles. Primary settlement in a saturated waste is associated with the expulsion of liquid and the consolidation of the waste matrix. Secondary settlement results from mechanical creep and biodegradation.

The above was demonstrated by data from tests on raw MSW and mechanically biologically treated (MBT) wastes in consolidating anaerobic reactors (CARs) reported by Ivanova *et al.* (2008) and Siddiqui *et al.* (2012, 2013). Specimens of wastes were held under a constant vertical stress of 50 kPa or 150 kPa and the progress of degradation, gassing and settlement monitored. The data included control tests in which degradation was prevented, which enabled settlements due to mechanical creep to be separated from those due to biodegradation.

Terzaghi's classical theory of consolidation was successfully applied to the analysis of primary settlement by Siddiqui *et al.* (2013), although this mechanism may be of limited direct relevance to an unsaturated waste in the field.

If simple models for estimating medium and long term settlements are to be applied successfully, mechanical creep must be considered separately from biodegradation as the mechanisms are different and may proceed at different rates under different operational conditions or environmental conditions of (for example) water content, leachate recirculation or temperature. Also, the ultimate amount of biodegradation settlement (ε_{bt}) is finite, whereas creep continues in log time.

In simple terms, creep strains ε_c may be characterised by a linear (in log time) relationship of the form:

$$\varepsilon_c = C_{\alpha\varepsilon}.\log_{10}(t/t_1) \qquad (1)$$

where $C_{\alpha\varepsilon}$ is a coefficient of creep settlement, t is the time following the application of stress and t_1 is a reference time. The value of $C_{\alpha\varepsilon}$ reduces with increasing waste bulk density but is substantially insensitive to vertical stress, at least over the range 50-150 kPa.

Biodegradation settlements ε_b can be characterised using a relationship based on first order reaction kinetics,

$$\varepsilon_b = \varepsilon_{bt}.(1 - e^{-kb.t'}) \qquad (2)$$

IOP Conf. Series: Earth and Environmental Science **26** (2015) 012002 doi:10.1088/1755-1315/26/1/012002

where ε_{bt} is the total eventual settlement due to biodegradation, kb is a biodegradation rate constant and t' is the elapsed time since the start of degradation. This approach is similar in concept to the Tier 2 methods of the IPCC for estimating GHG emissions from landfills (IPCC 2000).

Alternatively, more fundamental but complex models that fully couple degradation, gas and liquid transport, settlement and mass transfer between phases such as LDAT (White et al., 2004, 2013, 2014; or McDougall, 2007) may be adopted. The advantage of such models is that they can take into account the potential impact of an infrastructure development on settlements not only in terms of new loadings but also in terms of waste degradation. This may be important, especially if any redevelopment results in the ingress of water, which could cause biodegradation in parts of the site that were previously dormant. If remediation of the landfill is a necessary part of redevelopment, this will probably involve further biological stabilisation of the landfill. The use of fully coupled models can then aid predictions of the impact of the remediation on settlement.

*"It is not possible to characterise the engineering properties of waste fully, because of its heterogeneous nature, but it is important that its basic behaviour is understood and that likely ranges of the key engineering properties are known." (*Dixon et al 2006). Evidence of the variability in the engineering properties of landfills is provided by results of cone penetration testing (CPT). For example, McKnight et al (2015) reported highly variable data compared to typical CPT use in soil.

For developments that involve the building of new infrastructure on top of old landfills there is a need to consider the impact of load and how the load will be carried, for example by shallow or piled foundations.

The Environment Agency have produced guidance on *Piling and Penetrative Ground Improvement Methods on Land Affected by Contamination: Guidance on Pollution Prevention* (EA, 2001). This is general guidance and does not cover MSW landfills in particular. However, the guidance discusses potential environmental hazards associated with piling through contaminated soils, and how sites may be improved. It focuses on the effects of piling on any controlled waters, especially groundwater, with brief consideration of gas and human health issues.

"*Where there is an option, concrete raft foundations are the preferred means of foundation on contaminated sites*" (EA, 2001) This may be a useful general principle for contaminated land, but in landfills the application of substantial additional surface loads may result in significant primary settlement. If the waste is already fully or partly saturated this may cause unacceptable rises in leachate levels.

Knowledge of the void ratio and the one dimensional constrained modulus of MSW waste allows the impact of this type of loading on leachate levels to be assessed.

There are limited data on the one dimensional (constrained) modulus of MSW waste (e.g. Hudson et al 2004, Gomes et al 2014 and McDougall 2008). However, in general constrained modulus values, lie within the range of compression ratios of 0.08 to 0.41. Compression ratio is the negative of the slope of a graph of vertical strain against the logarithm of vertical stress, and is a useful measure that normalises constrained moduli against vertical stress.

There is no obvious influence of age on compressibility; the largest influence relates to the nature of the landfilled material. Greater increases in leachate levels will be seen in landfills with a characteristic compression ratios at the top of this range.

EA (2001) considers various pollution scenarios associated with piling in contaminated lands within the framework of source–Pathway–Receptor risk assessments. Two of their scenarios are highly relevant to MSW landfills. The first covers the case where piling goes through the contaminated land and into underlying strata to provide the load bearing capacity required for the infrastructure development. A hydrogeological risk assessment would be needed, particularly if the landfill is located in a source protection zone or overlies a major or minor aquifer. Many historic landfills are unlikely to have any as-built engineered containment, and many sites in old sand and gravel exaction pits are in deposits that have the characteristics of minor aquifers. The geo-hazards associated with this situation are discussed in the section on Leachate and Groundwater Quality.

The second potential pollution scenario relates to the creation of preferential pathways through a low permeability surface layer, allowing migration of landfill gas and odours to the surface. The geo-hazards associated with this are covered in the section on Gassing.

4. Leachate and groundwater quality

The database associated with *What's in your backyard* holds general information (where known) on the waste types accepted at a site. For authorised sites, the nature of the licence held is recorded (e.g. Licence type A5 : Landfill taking non-biodegradeable wastes). For historic sites, information on whether a particular category of waste (e.g. Industrial waste, Commercial waste, Household waste, Liquid /sludge etc) was landfilled is held.

Historically, many landfills accepted a range of industrial/ chemical liquid and solid wastes that would not be licensed today. Consequently the development of an old landfill site would need to include a comprehensive organic and inorganic leachate analysis to characterise potential pollutants and toxins. Landfills that accepted predominantly MSW will, in comparison, contain or generate less aggressive leachate. However, leachates from many old MSW landfills will nevertheless contain species not suitable for direct discharge to controlled waters.

Many contaminants found within wastes do not biodegrade and can only be removed through the liquid phase. The mechanism of removal in landfills is invariably through the generation (and subsequent abstraction / treatment) of leachate. Landfill completion does not necessarily require that all contaminants are removed from the waste; the key point is rather that their rate of release can be attenuated by the receiving environment, at the point of completion and all times into the future.

There has been a considerable amount of research into the release of contaminants from wastes in the liquid phase. Various standard waste leaching protocols have been developed in Europe and America in support of waste acceptance criteria tests; and a large database of test results drawn up (e.g. LeachXS; van der Sloot and Kosson, 2012) including a detailed appraisal of the geochemistry of the leaching process. Leach test results are often expressed as a function of the liquid to solid (LS) ratio, which is the volume of liquid leached (in litres) from the mass of dry solids (in kg) of an element of waste. In general terms, the higher the LS ratio achieved, the lower the leached concentration of a contaminant.

A secondary control on the rate of release of contaminants from landfilled waste comes from the heterogeneity of the waste at a variety of scales. Although there appears to be a relatively simple relationship between the mass of contaminants leached and the LS ratio, an obvious inference from this is that if there has been no flow of water through waste there will have been no leaching of contaminants. Numerous studies of contaminant transport processes

IOP Conf. Series: Earth and Environmental Science **26** (2015) 012002 doi:10.1088/1755-1315/26/1/012002

in waste (e.g. Rosqvist and Destouni 2000; Beaven et al., 2003; Fellner and Brunner, 2010) have noted a dual porosity effect. This characterises flow in waste as occurring through a number of discrete mobile pathways around blocks of waste within which no advective flow occurs. Contaminants are removed from the blocks by a process of diffusion into the flow in the mobile pathways. The rate of diffusion of contaminants out of a block follows an inverse square law with the block size. Barker (1985) characterised dual porosity systems in terms of block geometry functions, with a characteristic diffusion time of a block related to its average size. The characteristic diffusion time for an 'immobile' 1 cm cube of waste is approximately 4 days, whereas for a 1 m cube of waste it is approximately 110 years (Beaven and Barker, 2010). The importance of this is that the size of the immobile blocks effectively controls the rate of release of contaminants. If in defining FSQ we are concerned about the rate of release, then a site containing effectively larger immobile blocks of waste may achieve FSQ more quickly than a site with the waste in smaller blocks.

Beaven et al. (2014) discussed the relevance of hydraulic equilibrium to the technical debate on aftercare. *"During aftercare, active management and the functioning of engineered controls will result in an imposed hydraulic equilibrium, which, in many sites, will mean the majority of the waste is unsaturated. As active management (e.g. leachate pumping) is discontinued and engineered controls (e.g. the cap and/or liner) deteriorate or fail, then a new hydraulic equilibrium will be established that in many cases may involve the slow filling of the site with leachate. For completion to occur, the regulator must be satisfied that future fluxes to the environment will be acceptable under a range of hydraulic equilibrium situations."* Extending this reasoning to the dual porosity concept, we could imagine a large block size that in unsaturated conditions has led to low fluxes of contaminants to leachate collection systems. If leachate pumping were then stopped and leachate levels rose, new saturated flow paths could access different parts of the site with a different block geometry, leading to different fluxes of contaminants.

5. Landfill gas and odours

The anaerobic degradation of organic material in landfill will yield landfill gas (LFG) comprising approximately 60% CH_4 and 40% CO_2. To be explosive, landfill gas has to be in a confined space and mixed with oxygen as methane is flammable only when the concentration in air is between 5 and 15 percent.

he lateral migration of gas from municipal solid waste landfill sites is recognised as a significant hazard (e.g. Christophersen et al 2001). It resulted in an LFG explosion at Loscoe in 1986, which destroyed a bungalow located within 20m of a landfill.

In landfills with LFG control systems overpumping is the primary mechanism by which air may be drawn into the landfill. However, most old landfills will not have LFG control measures installed, in which case barometric pumping is the primary mechanism of air ingress (Young, 1990).

The total gassing potential of an MSW landfill may be estimated from any of a variety of tests including loss on ignition, biological methane potential (BMP) and the ratio of cellulose and hemicellulose to lignin, (C+H)/N, in the waste as determined by the Fibrecap test (Zheng et al, 2007). Siddiqui et al. (2013) reported a close correlation between the biodegradation induced settlement and the cumulative volume of gas produced (at SPT) for both raw MSW and MBT waste, and hence between the potential ultimate biodegradation induced settlement and loss on ignition, BMP and (C+H)/L. Many old landfills retain a gas potential from cellulose and hemi-cellulose possibly as high as ~75m^3/tonne (Knox et al., 2011), but nevertheless have low gas generation rates. There are two main reasons for this. The first may relate to a lack of moisture in parts of the waste essential for biodegradation. The second is that lignin shielding restricts the access of bacterial exo-cellular enzymes to much of the degradable content (e.g. Van Soest, 1994). Lignin degradation occurs mainly under aerobic conditions.

Consequently, any development on an old landfill would need to consider the implications for landfill gas generation and control. Introducing more water and/or air could increase waste degradation leading to landfill settlement and potentially more LFG production. Any development that restricted the ability for landfill gas to be released to the atmosphere may increase the potential for off-site lateral migrations; this too would need to be accounted for.

Odours from landfill sites are predominantly associated with landfill gas, with 12 trace components contributing to odours (EA 2002). Hydrogen sulphide is a major contributor and has been detected in landfill gas in concentrations up to 12,000 ppm, although more typical concentrations in landfill gas are ~ 100 ppm. At these concentrations H_2S cause olfactory paralysis (desensitisation), which is a concern because H_2S is fatal to humans at concentrations of < 500 ppm. The HSE short and long term exposure limits for H_2S are 5 ppm and 10 ppm respectively. A main cause of high concentration of H_2S in landfills is linked to the disposal of gypsum (sulphate) based industrial wastes, which is now banned.

6. Summary and conclusions

1. Old MSW landfills can be considered as a sub-set of contaminated land, and in terms of their re-development the legislative controls are the same.
2. The main environmental geohazards arising from old MSW landfills relate to the potential for water pollution, and hazards linked to landfill gas including risk of explosions and the possibility of hazardous concentrations of H_2S. Any new development across an old landfill will need to assess its impact on the above, including the potential to reactivate degradation and gassing through a change in the landfill hydraulic regime.
3. While landfill settlement does not in itself constitute a hazard to the surrounding environment, its impact on the development infrastructure may be significant. Models exist to help predict the magnitude of this settlement which will be of help to designers.

7. Acknowledgments

The authors are grateful to the UK Engineering and Physical Sciences Research Council (EPSRC) for funding through grants EP/E00654X, EP/E041965 and EP/I012206.

8. References

[1] Barker, J.A. (1985). Block-geometry functions characterizing transport in densely fissured media. J. Hydrol., 77, 263–279.

[2] Beaven, R.P., Barker, J.A. and Hudson, A., (2003). Description of a tracer test through waste

and application of a double porosity model, Sardinia 2003: Proc. Ninth Int' Waste Manag' and Landfill Symp. S. Margherita di Pula, Sardinia, Italy.

[3] Beaven, R.P. and Barker, J.A. (2010) Scientific and management concepts in landfill flushing: pilot studies using a dual porosity model. Proc. Waste 2010: Waste and Resource Management – Putting Strategy into Practice Stratford-upon-Avon, UK

[4] Beaven, R.P., Knox, K., Gronow, J.R., Greedy, D, Hjelmar, O and Scharff, H (2014) A new economic instrument for financing accelerated landfill aftercare. Waste Management, 34, (7), 1191-1198. (doi:10.1016/j.wasman.2014.03.024).

[5] Christophersen, M., Kjeldesen, P., Holst, H and Chanton, J. (2001) Lateral gas transport in soil adjacent to an old landfill: factors governing emissions and methane oxidation. Waste Management & Research 19, 595-612

[6] DoE (1986) Landfilling Waste: A technical memorandum for the disposal of wastes on landfill sites. DoE Waste Management Paper 26. HMSO, London, 206pp. 1st pub 1986 ISBN 0 11 750891 3.

[7] Dixon, N., Whittle, R.W., Jones, D.R.V. and Ng'ambi, S. (2006) Pressuremeter tests in municipal solid waste: measurement of shear stiffness. Geotechnique, 56(3), pp. 211 - 222.

[8] Environment Agency (2001) Piling and Penetrative Ground Improvement Methods on Land Affected by Contamination: Guidance on Pollution Prevention. National Groundwater & Contaminated Land Centre report NC/99/73

[9] Environment Agency (2002) Investigation of the Composition and Emissions of Trace Components in Landfill Gas. R&D Technical Report P1 -438/TR.

[10] Environment Agency (2005) LFTGN 09: Guidance on Landfill Completion and Surrender. Bristol: The Environment Agency for England & Wales.

[11] Environment Agency (2012) How to surrender your environmental permit: The Landfill Sector (EPR 5.02) and other permanent deposits of waste. Bristol: The Environment Agency for England & Wales.

[12] Fellner, J. and Brunner, P., H., 2010. Modeling of leachate generation from MSW landfills by a 2-dimensional 2-domain approach. Waste Management, 30: 2084–2095.

[13] Gomes, C.C., Lopes M.L. and Oliveira, PJV (2014) Stiffness parameters of municipal solid waste. Bull Eng Geol Environ 73:1073–1087. DOI 10.1007/s10064-014-0621-9

[14] Hall, D.H., McDougall, J., Drury, D., Blakey, N., Braithwaite, P. and Rosevear, A. (2007) Sustainable landfill: the role of fail-safe engineering for landfill aftercare In: Proceedings Sardinia 2007, Eleventh International Waste Management and Landfill Symposium, Cagliari, Italy.

[15] Harris, R.C., Knox, K and Walker A.N. (1994). "A strategy for the development of sustainable landfill design and operation." Proceedings of the Institute of Waste Management. Jan-94: 26-29

[16] Hudson, A. P., White, J.K., Beaven, R. P. and Powrie, W. (2004) Modelling the compression behaviour of landfilled domestic waste. Waste Management, 24, (3), 259-269. (doi:10.1016/j.wasman.2003.12.003).

[17] IPCC (2000) Good Practice Guidance and Uncertainty Management in National Greenhouse Gas Inventories. http://www.ipcc-nggip.iges.or.jp/public/

[18] Ivanova, L. K., Richards, D. J. and Smallman, D. J. (2008). The long-term settlement of landfill waste. Proceedings of the Institution of Civil Engineers Waste and Resource Management, 161 (WR3): 121-133.

[19] Knox, K. (1990) The relationship between leachate and gas. pp367-386 in Proceedings, International Conference Landfill Gas: Energy and Environment Conference, Didcot: ETSU. ISBN 0-7058 1628-1.

[20] Knox, K., Cheng, K. and Hayward-Higham, S. (2011) Uncertainties in understanding and measuring the completion of waste degradation. Paper no 168 in Proceedings of Sardinia 2011, Thirteenth International Waste Management and Landfill Symposium, Cagliari: CISA

[21] McDougall, J. (2007). A hydro-bio-mechanical model for settlement and other behaviour in landfilled wastes. Computers and Geotechnics, 34 (4): 229-246.

[22] McDougall, J. (2008). Chapter 3 Settlement: the short and the long of it. In: Geotechnical characterization , field measurement and laboratory testing of MSW. ASCE Geotechnical Special Publication No. 209 Ed. Zekkos.

[23] McKnight, T., Cho, Y., Townsend, T., and Choate, A. (2015) Cone Penetration Testing for Characterizing Landfilled Municipal Solid Waste. Journal of Geotechnical and Geoenvironmental Engineering 141:3

[24] Rosqvist, H. and Destouni, G., 2000. Solute transport through preferential pathways in municipal solid waste. Journal of Contaminant Hydrology, 46: 39-60

[25] Siddiqui, A. A., Powrie, W. and Richards, D. J. (2012). Investigations into the landfill behaviour of pretreated wastes. Waste Management, 32: 1420-1426.

[26] Siddiqui, A. A., Powrie, W. and Richards, D. J. (2013). Settlement characteristics of mechanically biologically treated wastes wastes. ASCE Journal of Geotechnical and Geoenvironmental Engineering, 139 (10): 1676-1689.

[27] van der Sloot, H.A. and Kosson, D.S. (2012) Use of characterisation leaching tests and associated modelling tools in assessing the hazardous nature of wastes Journal of Hazardous

Materials 207–208 36–43

[28] Van Soest, P.J. (1994) Nutritional ecology of the ruminant. Cornell University Press. 476pp.

[29] White, J. K., Robinson, J. P. and Ren, Q. (2004). Modelling the biochemical degradation of solid waste in landfills. Waste Management, 24 (3): 227-240.

[30] White, J.K. and Beaven, R.P. (2013) Developments to a landfill processes model following its application to two landfill modelling challenges. Waste Management, 33, (10), 1969-1981. (doi:10.1016/j.wasman.2012.12.006).

[31] White, J.K., Nayagum, D. and Beaven, R.P. (2014) A multi-component two-phase flow algorithm for use in landfill processes modelling. Waste Management, 34, (9), 1644-1656. (doi:10.1016/j.wasman.2014.05.005).

[32] Young, A. (1990) Volumetric changes in landfill gas flux in response to variations in atmospheric pressure. Waste Management & Research 8, 379-385

[33] Zheng, B., Richards, D.J., Smallman, D.J. and Beaven, R.P. (2007) Assessing MSW degradation by BMP and fibre analysis. Proceedings of the Institution of Civil Engineers: Waste and Resource Management, 160, (4), 133-139. (doi:10.1680/warm.2007.160.4.133).

Recent developments of the Material Point Method for the simulation of landslides

E E Alonso[1], A Yerro and N M Pinyol

Department of Geotechnical Engineering & Geosciences, Universitat Politècnica de Catalunya-BarcelonaTech, Barcelona, Spain

E-mail: eduardo.alonso@upc.edu

Abstract. The paper describes first the theoretical framework of a "single layer" three phase material. The formulation is general and particular cases are dry and fully saturated soils. The formulation and discretization of the motion and balance equations is presented. Two constitutive equations are used in the applications described: A brittle model for saturated soils and a Mohr-Coulomb elastoplastic soil formulated in terms of net stress and suction. Three aspects of the behaviour of landslides are discussed: first time failures in over-consolidated clays; internal shearing in deep seated landslides and rain induced failures in unsaturated slopes. The discussion is supported by three real cases which are described and analyzed in detail.

1. Introduction

Material Point Methods (MPM) are rapidly evolving in the geotechnical field especially in slope stability problems. This is because of their capability to analyze the whole instability problem in a unified calculation including the transition from the pre-failure stage, characterized by the development of small strains, to the accelerated post-rupture phase in which large deformations of the sliding mass are involved.

Based on particle methods, MPM [1] discretizes the continuum by means of material points. Each such a point moves attached to the portion of media that represents, giving a Lagrangian description of the motion, and carrying all the information (strain, stresses, pressure, mass...). On the other hand, main governing equations are posed and solved at the nodes of a background mesh in a very similar way as it is done in finite element methods (FEM). However, the integration points are the material points themselves instead of the typical Gaussian quadrature. The mesh covers the whole computational domain and usually it is maintained fixed throughout the calculation. In this manner, most of the expertise learned from FEM can be applied to the MPM with the additional improvement that large deformations can be simulated without limitations of mesh tangling. Linear shape functions

[1] Department of Geotechnical Engineering and Geosciences.
Universitat Politècnica de Catalunya-BarcelonaTech, Campus Nord
C/ Jordi Girona 1-3, Edificio D-2. 08034 Barcelona
Phone: +34 93 401 6866; +34 93 401 7256
Fax: +34 93 401 7251
e-mail: eduardo.alonso@upc.edu

are typically used in order to transfer information between both discretization levels (nodes and material points). Moreover, another advantage of the MPM discretization is that it is well suited for history-dependent constitutive models. This is a very important capability, especially in landslide analysis where material properties can change during the instability process (degradation due to extensive shearing).

Three relevant aspects for the interpretation of Landslides are described in the paper. The first topic concerns the development of progressive failure and subsequent post-failure behavior in brittle materials. This is the case in first time failures in overconsolidated clays. The failure results in an accelerated motion. The case analyzed: The Selborne failure experiment illustrates this class of instabilities.

The second aspect analyzed focuses on the kinematics of deep seated landslides and, in particular, in the role played by internal shearing of rock masses. Compound landslides are possible if internal rock strength is exhausted along surfaces or shearing bands which make the motion kinematically admissible. The well-known case of Vaiont is examined from this perspective.

The final case deals with the stability of unsaturated slopes under rainfall. A case inspired in the surface ruptures experienced by some embankment slopes will be presented and discussed.

2. Theoretical framework

2.1. Basis of MPM

MPM was initially developed for solid mechanics [2]. However, the slope stability is closely related to changes in water table conditions due to climate effects, such as heavy rainfall or wetting-drying cycles. Therefore, in landslide prediction analysis coupled hydro-mechanical formulations are particularly useful because they are capable of modelling interaction between solid skeleton and fluid within the porous media or even with free water. During the last decade, the classic MPM formulation has been extended to solve multiphase problems and basically two approaches can be distinguished: the "single-layer" and the "multi-layer" one.

The "single-layer" strategy is the most extended formulation. It is based on representing the porous medium as a unique continuum by means a set of material points. Thus each point moves according to the solid skeleton motion, but also carries information of pore fluids -water and air in unsaturated media- as internal variables. This approach is appropriate to simulate seepage problems but it is not capable to model free liquid water. It has been adopted by several authors to solve problems in saturated conditions (two-phases) [3][4][5][6], and it has also been extended to unsaturated problems in Yerro et al. [6] (three-phases). The "multi-layer" approach represents each constituent -grains and water in saturated soils- separately by means of different sets of material points. Each Lagrangian point only carries information of the phase that represents. This MPM formulation has the capability of modeling both water within the pores and free water as a unique continuum which allows the simulation of fluid-structure interaction problems. However, the number of material points is much higher, requiring an additional computational cost. This approach has been recently adapted to the modelling of fully saturated soils [7][8].

Figure 1 shows a simplified scheme of both approaches depending on number of phases considered in a geotechnical problem. Each set of material points is represented by a circle and each color corresponds to a different phase (solid, liquid or gas).

2.2. Three-phase one-layer MPM formulation

In this lecture, the MPM formulation for unsaturated soils based on Yerro et al. [6] is outlined, as well as the explicit computational process. This is the most general "single-layer" approach because saturated and dry conditions are particular cases. Here the soil is understood as a unique medium integrated by three distinct phases (solid (s), liquid (l) and gas (g)) (figure 2). This is a dynamic formulation, in which all acceleration terms are considered.

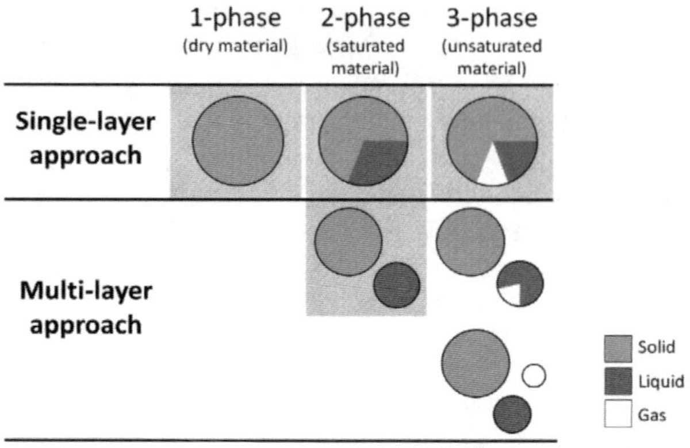

Figure 1. MPM approaches depending on the number of phases (solid, liquid and gas). Schemes colored in grey are those formulations that exist in the literature.

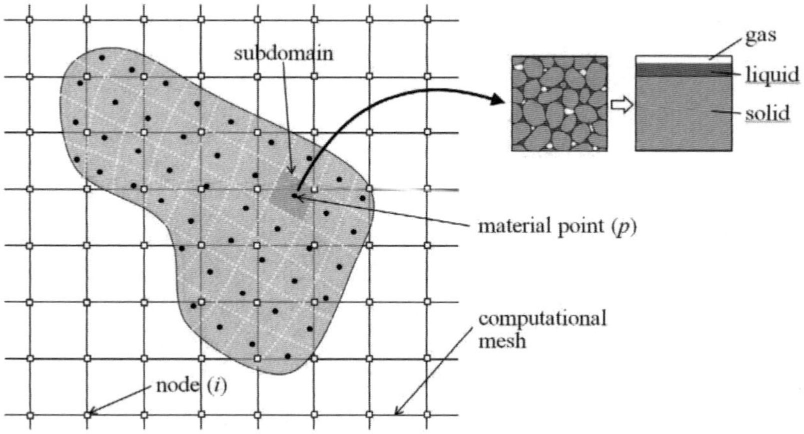

Figure 2. Spatial discretization scheme of the single-layer MPM approach for unsaturated porous material.

The main governing equations are the dynamic equilibrium of the gas, the liquid and the mixture. These are integrated into the volume and established at the nodes of the mesh at time t. At the beginning of each time step, information carried by material points is mapped to the computational mesh in order to calculate nodal mass matrixes ($\mathbf{M}_l$, $\mathbf{M}_g$ and $\mathbf{M}_s$); nodal velocities ($\mathbf{v}_l$, $\mathbf{v}_g$ and $\mathbf{v}_s$); internal and external forces of the liquid, gas and mixture ($\mathbf{F}_l^{\text{int}}, \mathbf{F}_g^{\text{int}}, \mathbf{F}^{\text{int}}, \mathbf{F}_l^{\text{ext}}, \mathbf{F}_g^{\text{ext}}$ and $\mathbf{F}^{\text{ext}}$); and dragging matrixes ($\mathbf{Q}_l$, $\mathbf{Q}_g$). It leads to the following system of equations (1), (2) and (3) in which liquid, gas and solid nodal accelerations ($\mathbf{a}_l$, $\mathbf{a}_g$, and $\mathbf{a}_s$) are taken as the unknowns of the problem.

$$\mathbf{M}_l \cdot \mathbf{a}_l = \mathbf{F}_l^{\text{ext}} - \mathbf{F}_l^{\text{int}} - \mathbf{Q}_l \cdot (\mathbf{v}_l - \mathbf{v}_s) \tag{1}$$

$$\mathbf{M}_g \cdot \mathbf{a}_g = \mathbf{F}_g^{\text{ext}} - \mathbf{F}_g^{\text{int}} - \mathbf{Q}_g \cdot (\mathbf{v}_g - \mathbf{v}_s) \tag{2}$$

$$\mathbf{M}_s \cdot \mathbf{a}_s + \mathbf{M}_l \cdot \mathbf{a}_l + \mathbf{M}_g \cdot \mathbf{a}_g = \mathbf{F}^{\text{ext}} - \mathbf{F}^{\text{int}} \tag{3}$$

Once the nodal accelerations are calculated, an explicit Euler scheme is used to update velocities, displacements and strains in the material points by means of shape functions.

Afterwards, the following mass balance equations for the solid, liquid and gas ((4), (5) and (6) respectively) are posed also in the material points; where n is the porosity, S_l is the degree of saturation and ρ_s, ρ_l and ρ_g are solid, liquid and gas densities.

$$\frac{\partial}{\partial t}\left((1-n)\rho_s\right) + \nabla \cdot \left((1-n)\rho_s \mathbf{v}_s\right) = 0 \tag{4}$$

$$\frac{\partial}{\partial t}\left(nS_l\rho_l\right)+\nabla\cdot\left(nS_l\rho_l\mathbf{v}_l\right)=0 \tag{5}$$

$$\frac{\partial}{\partial t}\left(n\left(1-S_l\right)\rho_g\right)+\nabla\cdot\left(n\left(1-S_l\right)\rho_g\mathbf{v}_l\right)=0 \tag{6}$$

Considering the solid grains to be incompressible, the mass balance of solid (4) becomes the material derivative of the porosity as follows.

$$\frac{Dn}{Dt}=\left(1-n\right)\nabla\cdot\mathbf{v}_s \tag{7}$$

Including (7) in (5) and (6) and assuming that liquid and gas pressures (P_l and P_g) are the state variables, the following expressions (8) and (9) are obtained. These provide the relationships to find liquid and gas pressure rates (dP_l and dP_g).

$$n\frac{\partial\left(\rho_l S_l\right)}{\partial P_l}dP_l+n\frac{\partial\left(\rho_l S_l\right)}{\partial P_g}dP_g=\nabla\cdot\left[nS_l\rho_l\left(\mathbf{v}_l-\mathbf{v}_s\right)\right]-nS_l\rho_l\nabla\cdot\mathbf{v}_s \tag{8}$$

$$n\frac{\partial\left(\rho_g\left(1-S_l\right)\right)}{\partial P_l}dP_l+n\frac{\partial\left(\rho_g\left(1-S_l\right)\right)}{\partial P_g}dP_g=\nabla\cdot\left[n\left(1-S_l\right)\rho_g\left(\mathbf{v}_g-\mathbf{v}_s\right)\right]-n\left(1-S_l\right)\rho_g\nabla\cdot\mathbf{v}_s \tag{9}$$

The general method is formulated in two stress variables: net stress tensor $\bar{\boldsymbol{\sigma}}$ and isotropic suction (s). These are written in the following convenient manner (10) and (11) respectively. $\boldsymbol{\sigma}$ is the total stress tensor and $\mathbf{I}$ is the identity matrix. It is important to highlight that when a saturated calculation is carried out, net stress defined as (10) coincides with Bishop's effective stress, which is consistent with the for saturated conditions.

$$\bar{\boldsymbol{\sigma}}=\boldsymbol{\sigma}-\max\left\{P_g,P_l\right\}\mathbf{I} \tag{10}$$

$$s=\max\left\{P_g-P_l,0\right\} \tag{11}$$

The general form of a suitable stress-strain relationship can be written incrementally as follows, where $\Delta\boldsymbol{\varepsilon}$ is the strain increment vector. $\mathbf{D}$ and $\mathbf{h}$ are, respectively, the tangent matrix and a constitutive vector that depends on the constitutive model used in the problem.

$$\Delta\bar{\boldsymbol{\sigma}}=\mathbf{D}\cdot\Delta\boldsymbol{\varepsilon}+\mathbf{h}\Delta s \tag{12}$$

At the end the computational cycle associated to a time step increment, the stress and also other soil properties are updated at material point; for instance, the degree of saturation -with the corresponding retention curve-, as well as intrinsic permeabilities and porosity.

Finally, the material points carry all the updated information, the computational grid is initialized and time is updated as

$$t=t+\Delta t \tag{13}$$

2.3. Suction-dependent constitutive model

In unsaturated conditions, the stability of a slope is essentially governed by the evolution of apparent strength with the evolving suction. Several authors examined the shear strength of unsaturated soils [9][10] and observed that friction angle and especially cohesion depend nonlinearly on suction.

A suction-dependent model is presented here [6]. The shear strength is written according to the well-known Mohr-Coulomb elastoplastic criterion. In order to include the suction effect, both strength parameters (cohesion c and friction angle φ) can be generally written as the sum of two terms as follows:

$$c = c' + c_s \tag{14}$$

$$\varphi = \varphi' + \varphi_s \tag{15}$$

where c' and φ' are the values for saturated conditions (effective cohesion and effective friction angle), whereas c_s and φ_s depend on the level of suction and provide an apparent strength.

It has been accepted that cohesion can increase from c' up to $c' + \Delta c_{\max}$ according to (16). P_{atm} is the atmospheric pressure and B is a constant parameter that controls the rate of apparent cohesion. Although changes in internal friction angle are typically less relevant, it is considered that it has linear dependence with suction depending on A.

$$c_s = \Delta c_{\max}\left(1 - e^{-B(s/P_{atm})}\right) \tag{16}$$

$$\varphi_s = A(s / P_{atm}) \tag{17}$$

2.4. Strain softening constitutive model

A strain weakening behavior is typically exhibited by rocks and overconsolidated clays. This is characterized by a reduction of the material strength from peak to residual conditions expressed as function of shear strains. The change of strength in the most loaded points of the slope causes a redistribution of the stresses. This fact may lead to more neighboring points reaching the maximum strength inducing a progression of the failure surface. Because of that, in this kind of materials a shear band does not occur simultaneously along the entire rupture surface, but shear strains localize gradually until a kinematically admissible mechanism is developed.

In order to simulate the complexity of the behavior of brittle materials, a strain softening Mohr-Coulomb model is implemented [11]. Softening behavior is accounted for by reducing the strength parameters (effective friction angle φ', and effective cohesion c') with accumulated deviatoric plastic stain invariant ε_d^p according to the following softening rules:

$$c' = c'_r + \left(c'_p - c'_r\right)e^{-\eta\varepsilon_{eq}^p} \tag{18}$$

$$\varphi' = \varphi'_r + \left(\varphi'_p - \varphi'_r\right)e^{-\eta\varepsilon_{eq}^p} \tag{19}$$

The model requires the specification of peak $\left(c'_p, \varphi'_p\right)$ and residual $\left(c'_r, \varphi'_r\right)$ strength parameters. The rate of strength decrease is essentially controlled by the plastic shear strain but an additional "shape factor" parameter η is also included in equations (18) and (19). The higher the shape factor η, the smaller the loss in strength.

3. First failure in brittle materials

First failures may be induced by different external actions. In man-made failures excavation of cuts (unloading) or the increase in loads may lead to failure. When examined in detail, some well-known cases of brittle failures (Aznalcóllar dam [12]; Selborne failure experiment [13]) are a combination of shear stressing and also a change in total stresses and pore pressures.

Soil brittleness leads to a progressive failure phenomenon. This process, conceptually independent of time, may be described as a static process in which slope deformations tend to be small because of the stiff nature of brittle clays. However, the final instability is associated with a non-equilibrated stress release and therefore with an accelerated motion. The subsequent motion is controlled by Newton law and the consequences of the instability depend on additional aspects, typically related to the geometry of the site at a larger scale. The case of Selborne will be analyzed here to illustrate these aspects.

The Selbourne cut was excavated in Gault clay (weathered in the upper 7–8 m). Figure 3a shows a representative cross section. Two lateral trenches, 25 m apart, were excavated to eliminate restrictions to a failure experiment conceived as a two-dimensional case. The instability was induced by increasing the pore water pressure in a series of boreholes drilled from the upper subhorizontal surface of the slope. In the simulation performed, the pore water pressure increase was simulated by means of a boundary increase in water pressure.

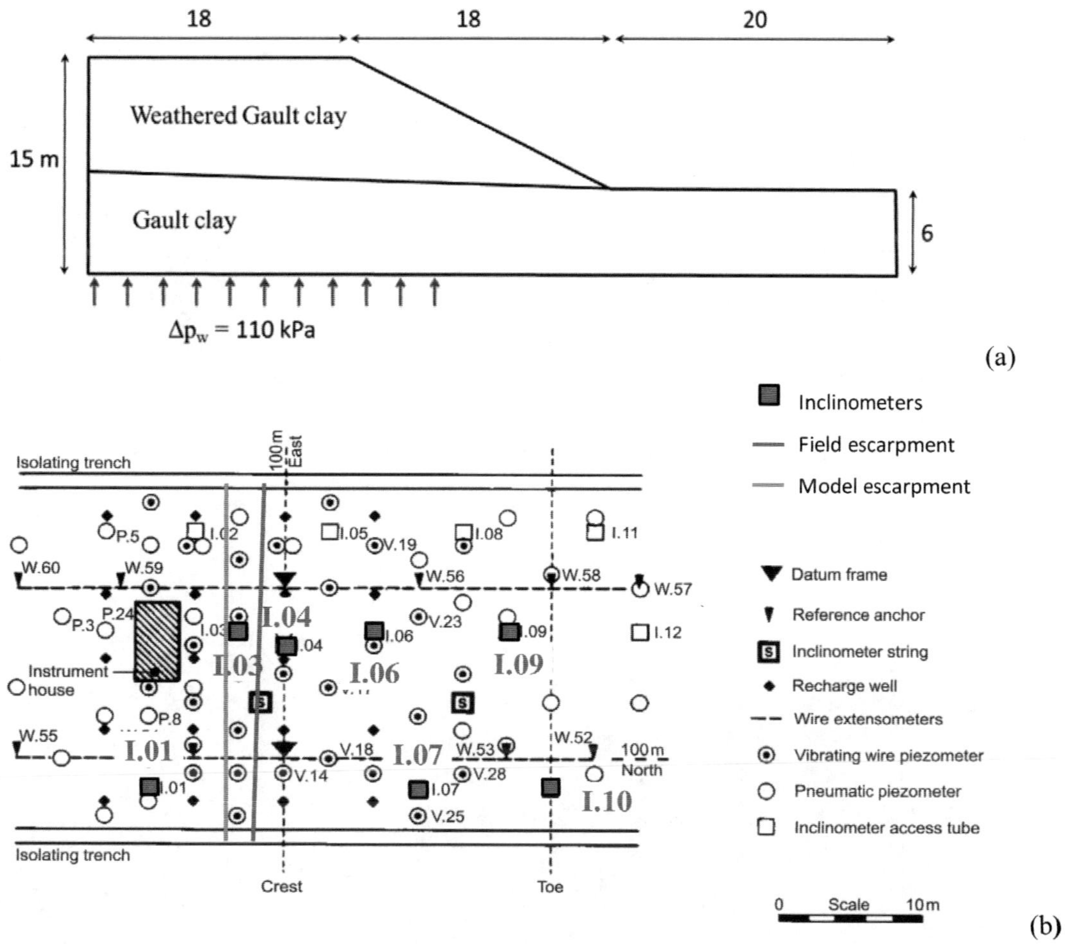

Figure 3. (a) Representative cross section of Selborne failure experiment; (b) Plan view of position of monitoring instruments [13]

The slope was monitored by means of surface topographic marks, piezometers and inclinometers. A detail of their position is given in plain view in figure 3b. The main Gault clay properties adopted in the analysis performed were collected in table 1. Porosity, unit weights, elastic parameters and effective strength parameters are derived from laboratory tests reported in Cooper et al. [13]. The clay permeability was artificially increased to reduce the computational time. In fact, the explicit formulation of the MPM analysis requires very small time increments. However, field and computational times may be compared in dimensionless terms: $t^* = t/T$, where t is the real time and T the time to failure.

Note in table 1 that the stiffness and strength properties of the weathered and unweathered clay are very similar.

The computational mesh as well as the initial location of material points is shown in figure 4. Elements are 3D tetrahedral elements. A plane strain "slice" of the slope was simulated. The density of material points was increased in a region when slope deformations are expected. The average size of tetrahedral elements was selected in connection with the strain softening "shape factor" parameter, η, of the clay constitutive model. Figure 5 shows the effect of varying η and the mesh size on the predicted displacement of the crest point of the slide. An element size of 0.7m and η = 400 was selected.

The initial stress state in the model was due to gravity effects. The excavation process was not simulated in the set of results described below. This is admittedly a limitation because of the relatively high K_o value expected in these overconsolidated clays ($K_o \cong 1.5$ to 2). However, this aspect is being currently investigated in more detail.

Table 1. Properties of Gault clay.

	Weathered Gault clay	Gault clay
Porosity	0.2	0.2
Permeability (m/s)	0.001	0.001
Dry unit weight (kN/m^3)	2,500	2,500
Young modulus (kPa)	20,000	20,000
Poisson's coefficient	0.33	0.33
Effective cohesion (peak/residual)(kPa)	13/4.7	25/1
Friction angle (peak/residual)(°)	24.5/13.5	26/15
Calibration parameter, η	500	500

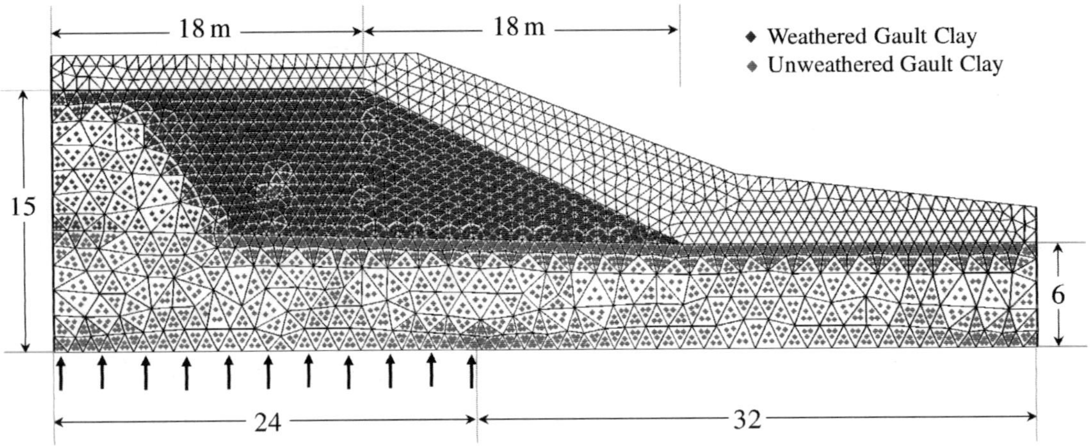

Figure 4. Computational mesh for the Selborne failure experiment

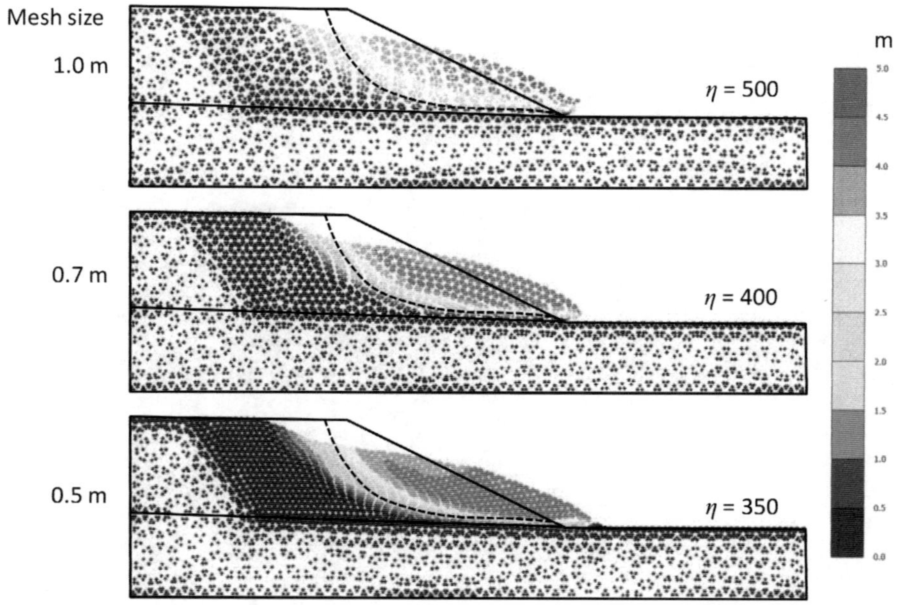

Figure 5. Effect of mesh size and the shape factor parameter on displacement pattern at failure.

3.1. Model calculations and slope response

Selborne is an interesting case because there is data available on the evolution of pore pressures and inclinometer readings during the hydraulic loading of the slope. In addition, the motion of the slide after failure was reported in Grant [14]. This information is particularly valuable to check the capability of the dynamic MPM model to capture the slide run-out, a piece of information that is generally absent in Limit Equilibrium and Finite Element analysis.

Pore pressures were recorded in a high number of sensors. Three records, selected for their proximity to the actual failure surface, are compared with model calculations in figure 6. The agreement is quite reasonable. Records for points B and C show a rapid decrease if pore pressure in the proximity of the final failure. This is probably due to the combined effect of positive shear-straining induced dilation and the unloading associated with the slope motion. However, a zero dilatancy angle was used in the analysis.

Post-failure displacements on two parallel lines to the sliding direction are shown in figure 7. The slide moved towards the East. Therefore, the two reference lines are indicated as North and South. The distance between them is around 17m. Calculated results are especially accurate for the Northern part. The maximum recorded motion of central points of the slope was about 4m. The MPM model reproduced correctly the kinematics of the motion. The smaller displacement of the Northern side was attributed by Cooper et al. [13] to an overall rotation of the slide, towards the North that introduced an additional restraining force in the Northern side of the slide. The figure shows also the MPM calculated deformed shape of the slope. It corresponds to an effective cohesion of 4.7kPa, which is a small value. It was found that the slide displacement was quite sensitive to the value of c', even if it is small (with reference to the usual understanding of small cohesion values in engineering practice). This sensitivity is illustrated in figure 8a. It shows the variation of the displacement of the crest point with residual cohesion. The value $c' = 4.7$kPa was justified because of the suction measured in a few piezometers located in the lower part of the slide. This suction varied between 0 and 40kPa. A mean value of 20kPa and a residual friction of 13.5° lead to a suction-induced apparent cohesion of 4.7kPa (the clay remains most probably saturated for the small range of measured suctions).

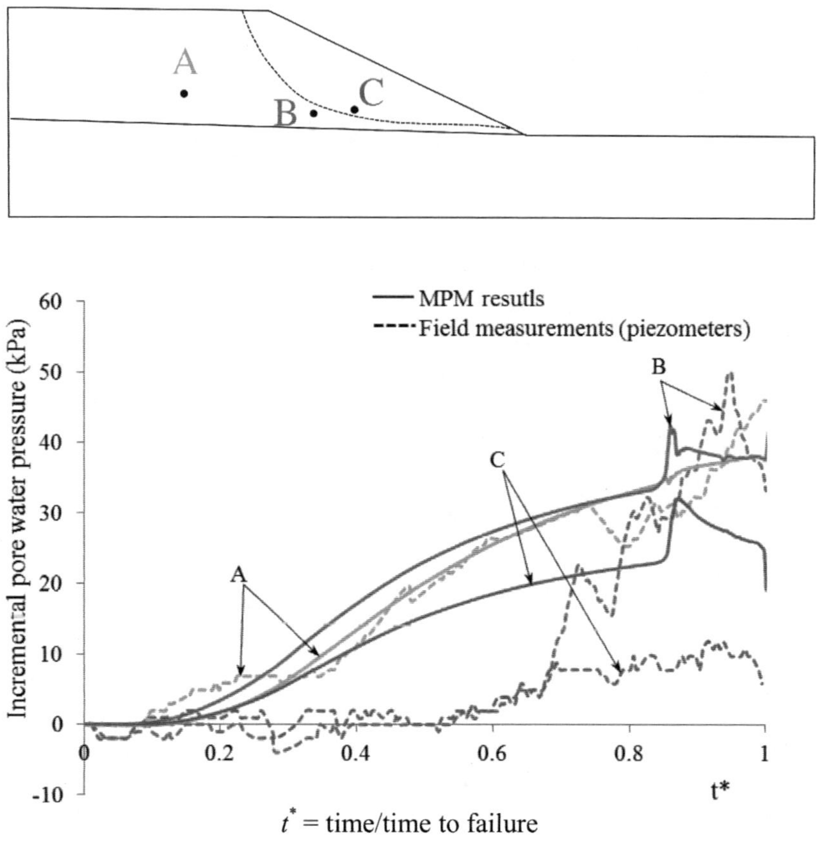

Figure 6. Piezometer records and model calculations.

Also indicated in figure 8b is the effect of time step in slide displacement. A “safe” value of $\Delta t = 1.3 \times 10^{-4}$s was used in calculations.

The geometry of the initial sliding surface was also recorded. It is compared with MPM calculations in figure 9. The predicted shape follows the maximum values of plastic deviatoric strains immediately before final failure. The model captures correctly the surface shape and the dominant effect of the contact between weathered and unweathered Gault clay. The actual failure surface was however somewhat shallower than the predicted one.

Inclinometer readings allow also a further set of data to check the validity of the model. Readings for inclinometers located on the excavated slope (I04, I06, I07 and I09; see figure 3.1b) are compared with model calculations in figure 10. Predicted displacements for two times close to the failure time are given. They approximate the time of inclinometer readings before the final run-out time. Note the scale of the horizontal displacements, which belong to the quasi-static, pre-failure stage of the slide. The calculation’s results bound the actual measurements. The patterns of displacements are also reasonably well captured by the model and show the position of the failure surface.

The progressive failure was qualitatively described by Cooper [15] by interpreting inclinometer data. It was accepted that the toe of the slope was in residual conditions at the beginning of the experiment as a result of the previous excavation.

The shearing band progressed simultaneously in opposite directions from the higher and lower levels of the final position of the sliding surface. The band propagated towards the center part.

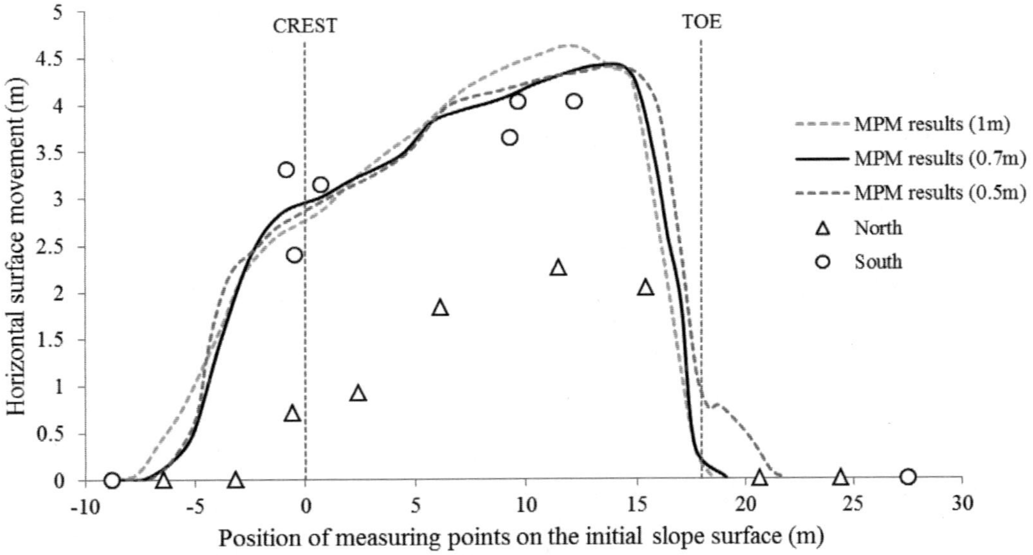

Figure 7. Measured and calculated displacements after failure.

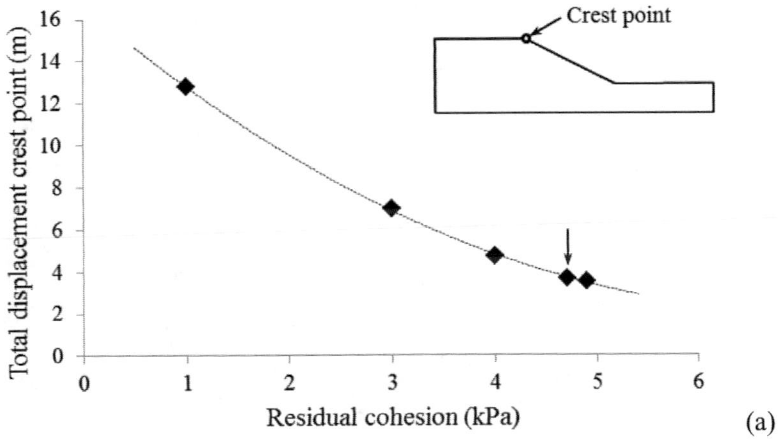

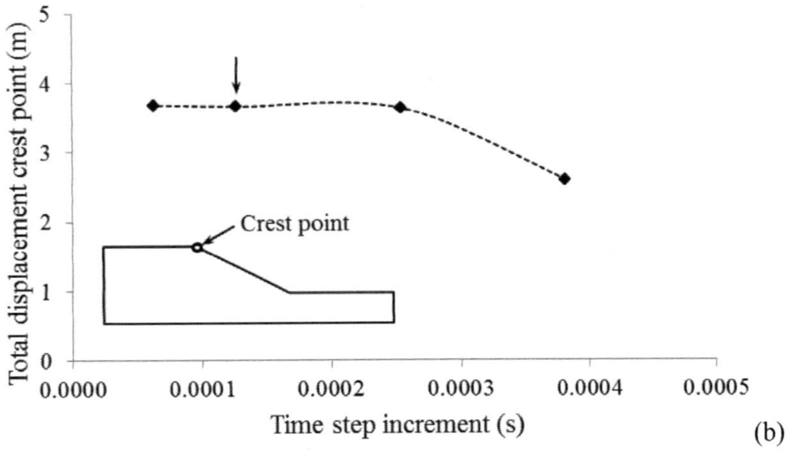

Figure 8. Total displacement of crest point after failure (a) Effect of residual cohesion strength; (b) Effect of calculation time step.

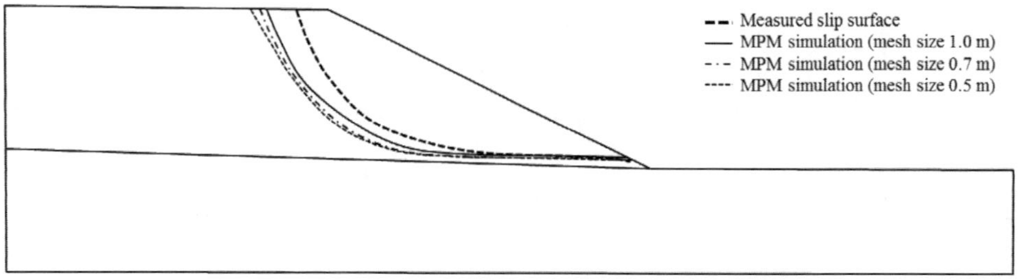

Figure 9. Comparison between measured and predicted shape of failure surface.

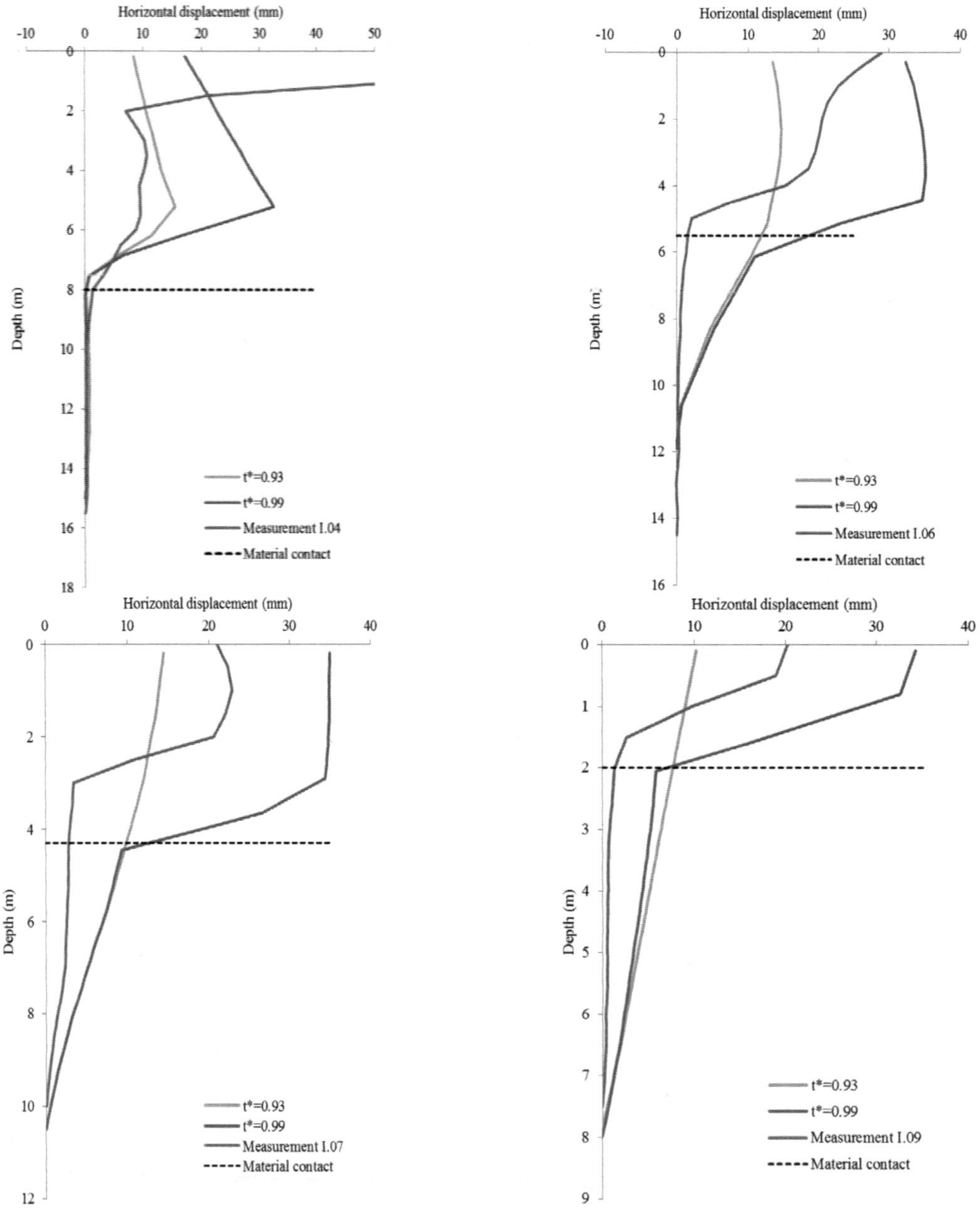

Figure 10. Measured and predicted inclinometer readings.

The model offers the opportunity to examine this result in some detail. A mobilized friction angle was defined as

$$\sin\hat{\varphi}' = \frac{q}{p^*} \quad \text{where } p^* = p' + \frac{c'}{\tan\varphi'} \tag{20}$$

q and p' are the deviatoric and mean effective stresses. Plots of sin $\hat{\varphi}'$ along the (final) shear surface for different dimensionless times are given in figure 11a,b,c.

Early times are given in figure 11a. Initially ($t^* = 0$) the slope remains in elastic conditions.

Maximum shear stresses are calculated at mid and upper heights (positions P2 and P3). As the water pressure increases the mobilization of shear strength continues in the upper and lower parts of the sliding surface. The first yielding (peak conditions) occur at $t^* = 0.83$ at P2. This is followed by a generalized plastification of the upper stretches of the failure surface which rapidly reach residual conditions at $t^* = 0.84$ and 0.845. Interestingly, this is immediately followed by the plastification of the toe of the slope (at $t^* = 0.85$). The evolution of shear stressing at later times, close to the failure is shown in figure 11c. The plastification advances in opposite directions from upper and lower parts, and the final point resisting under peak conditions is P4. This was also the inference based on inclinometer results.

The evolution of progressive failure in terms of an average friction angle can be followed in figure 12. The mean friction angle increases continuously until peak strength conditions are reached for the first time at a point around P2 for $t^* = 0.83$. Immediately afterwards there is a drop of available mobilized strength. Then the process of progressive failure progresses at essentially constant average mobilized strength. This process ends abruptly when the final point in the failure surface to reach peak conditions evolves towards a residual state. This sudden final drop of mobilized strength marks the beginning of an accelerated motion. The average (p', q) points at the three critical instants mentioned are plotted in figure 12b in a triaxial stress space.

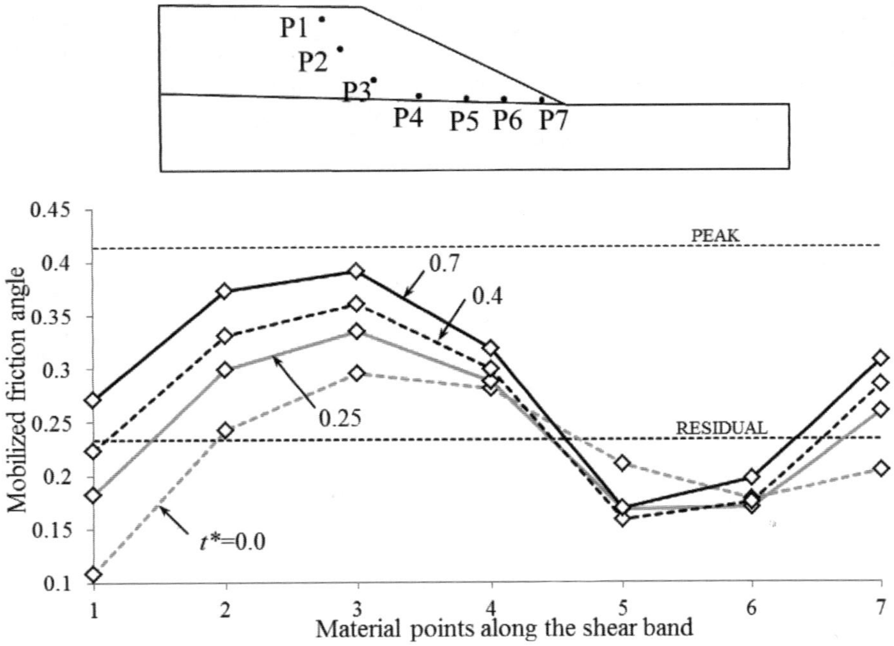

(a)

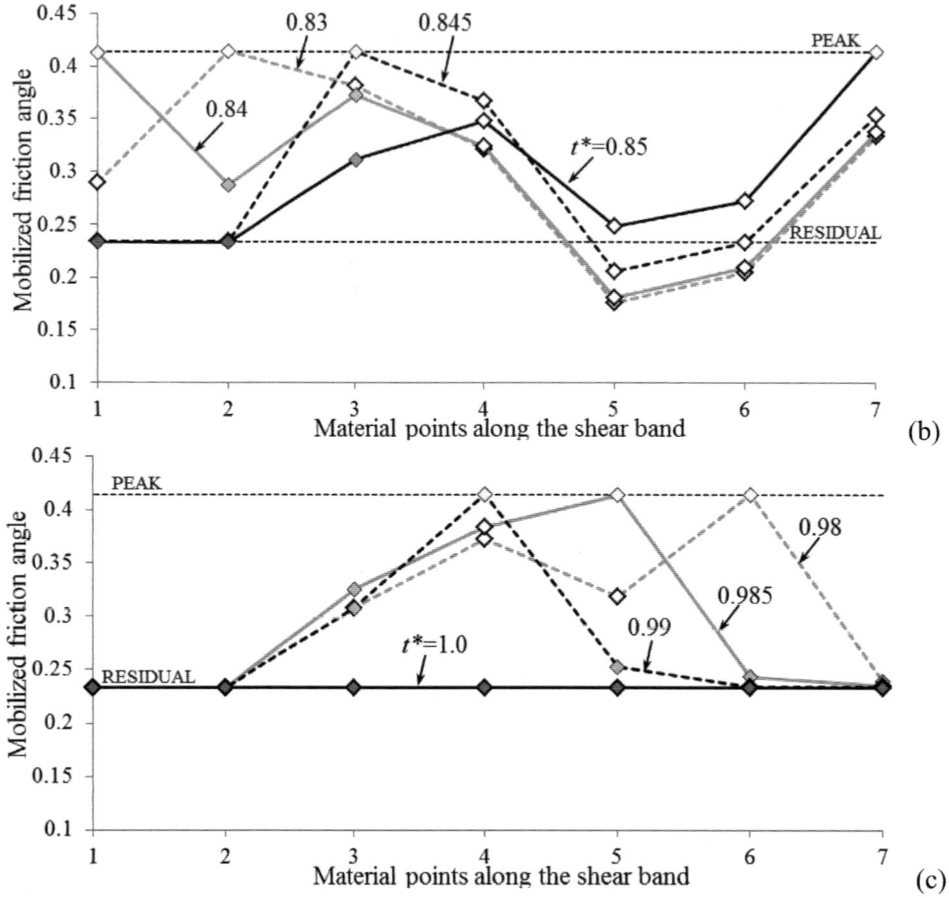

Figure 11. Development of progressive failure at (a): Early times; (b) Intermediate times; (c) Final times. The colour of the markers indicate the state of the point (white: elastic; yellow: peak conditions; orange: under softening; red: residual conditions).

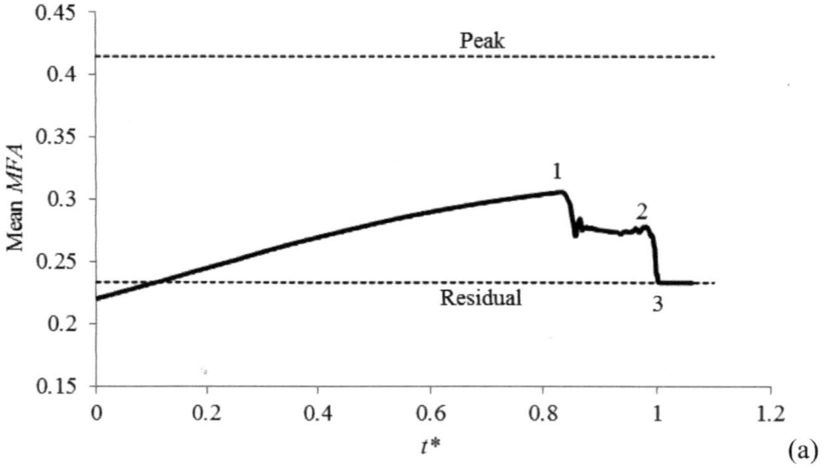

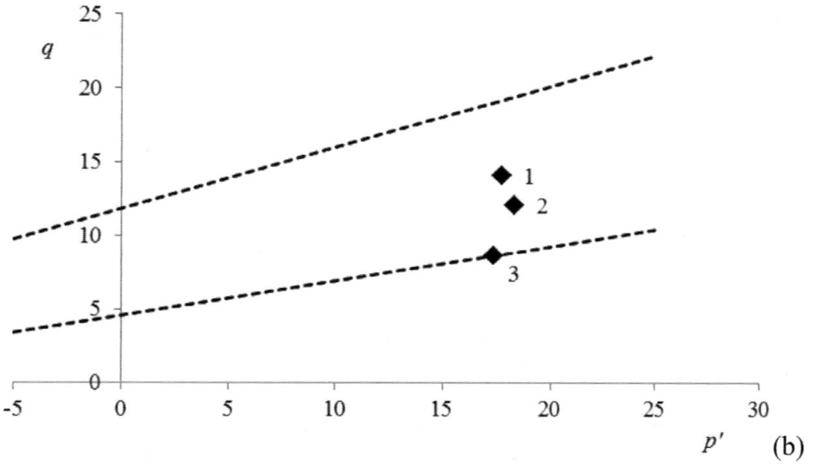

Figure 12. Evolution of (a) mobilized friction angle and (b) average (p', q) stress points.

The motion of the slide is summarized in figure 13. Velocity, displacement and acceleration of four points in the slope (C1 to C4) are plotted in terms of dimensionless time. Maximum velocities and displacements are calculated for points in the upper central part of the sliding mass. The maximum calculated velocity is close to 0.9m/s. The maximum calculated acceleration is a small fraction of gravity acceleration (0.036g).

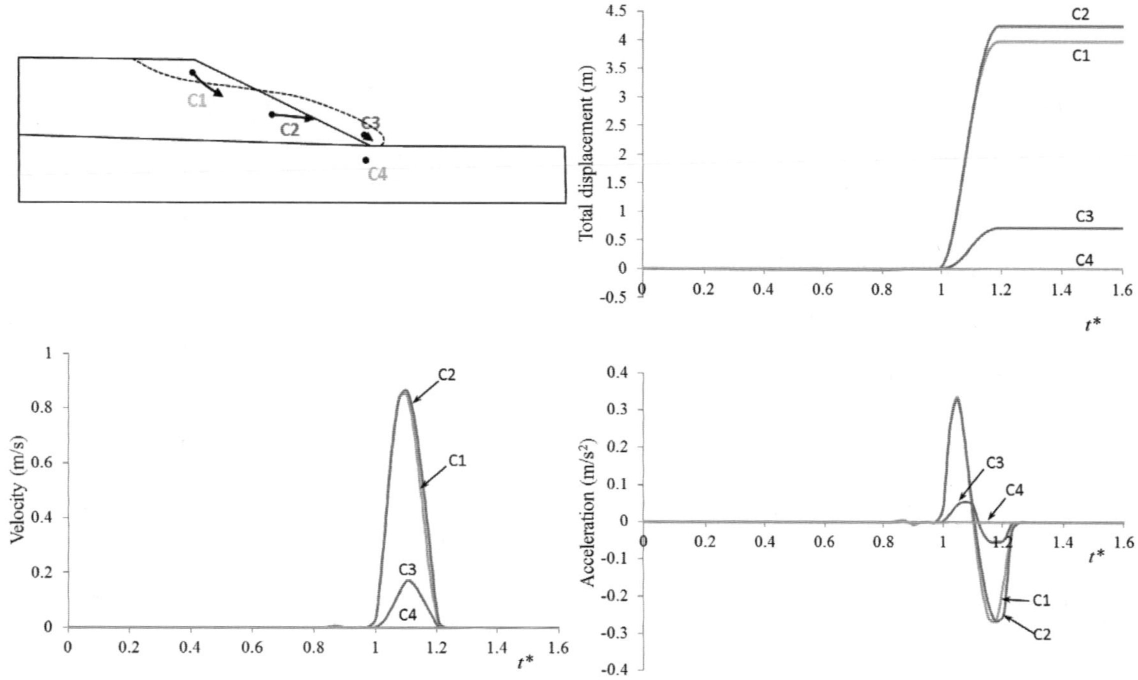

Figure 13. Kinematics of the motion: Total displacements, velocity and acceleration of four representative points of the slide.

Figure 14 relates the mobilized friction angle of three points on the sliding surface with the development of the accelerated motion. The sudden drop of (average) friction beyond the attainment of peak conditions for the first time is well identified as well as the final reduction of available mobilized strength and the associated acceleration of the sliding mass. Changes in geometry during sliding are reflected in small changes in mobilized friction due to some local unloading. The small arrows in figure 14 indicate the times selected in figure 11 to show the distribution of mobilized strength along the sliding surface.

A further illustration of the processes taking place inside the slope is given in figure 15. It shows the stress paths of two points: P2 in the upper part of the sliding surface and C4 at the toe of the slope, below the sliding surface. Point P2 is initially (A) in an elastic state. The progressive increase in pore pressure takes the point to peak strength conditions at a reduced mean effective stress. The small increase in q is associated with the effect of boundary water pressure applied only on part of the lower boundary. The sudden increase in q before reaching peak strength conditions is probably a result of internal stress changes due to the unloading of points reaching peak conditions. Beyond point B the continuing increase in pore pressure reduces the effective mean stress and, in parallel, the strength reduces progressively to residual conditions. The Mohr-Coulomb model used has not a cut-off, tensile strength and p' may become negative. Changes in pore pressure when the progressive failure begins to operate were also calculated and represented in figure 6 (beyond t^* = 0.83). The increase in pore pressure may be due to total stress changes induced by the sudden release of stresses along the failure surface. Reduction in pore pressure is explained by the flow/consolidation phenomenon and the boundary conditions. The end result is a cycle of increase/decrease of pore pressures, which is well identified in figure 6 but also in figure 15 for point P2 when the stress path moves along the residual strength envelope and reaches a final state at point C.

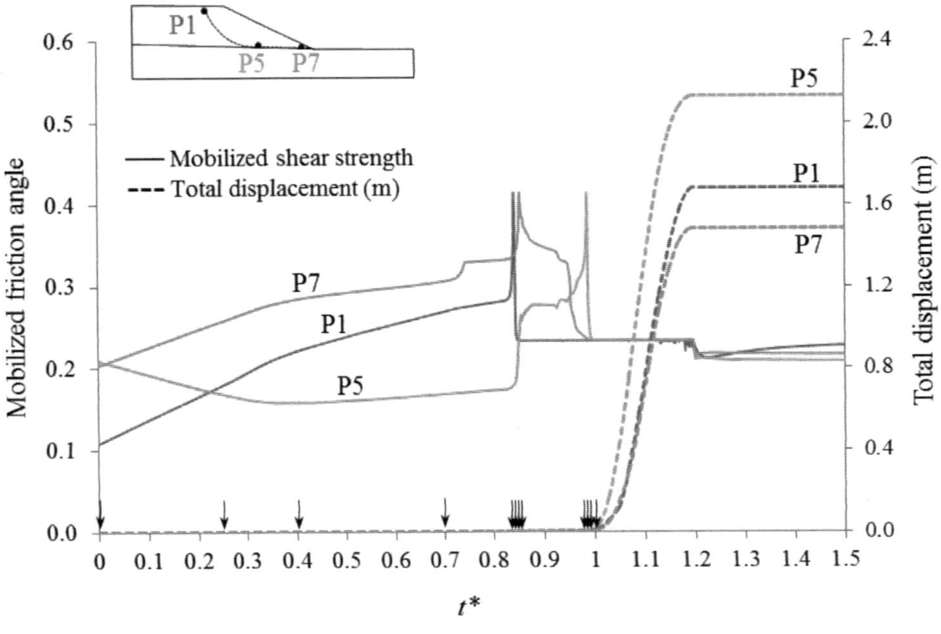

Figure 14. Evolution in time of mobilized friction angle and slope displacements before and after failure for points P1, P5 and P7.

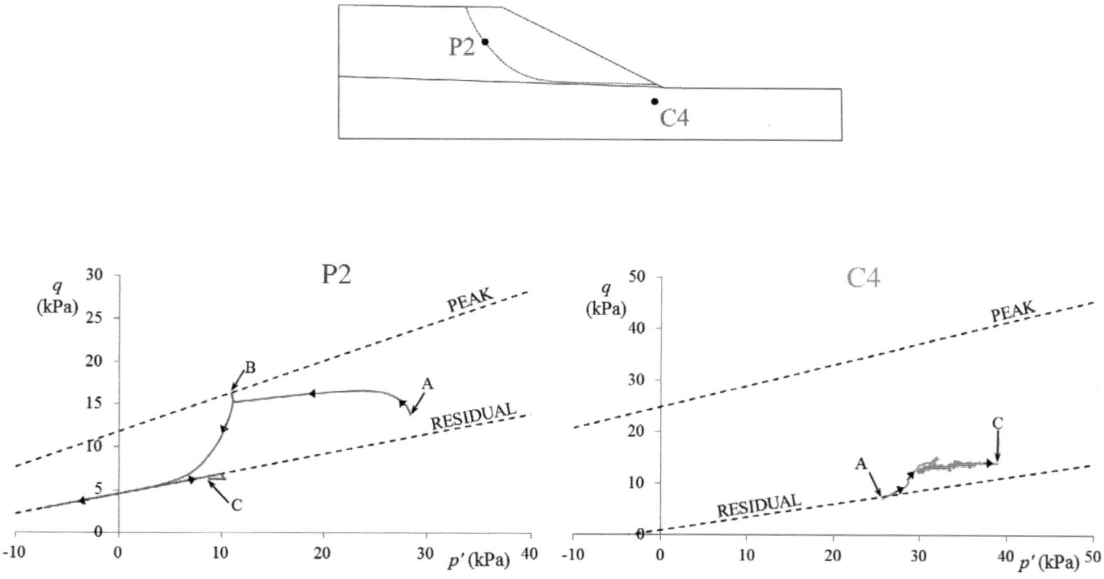

Figure 15. Stress paths in material points P2 and C4.

Point C4 behaves in a simpler way. The point, slightly affected by the instability, remains in an elastic condition throughout the test. The increase in mean effective stress is a consequence of the changes in slope geometry (mass of soil accumulating at the toe). Pore pressures are not expected to change in C4.

3.2. Lessons from the modelling of Selborne experiment

The Selborne failure experiment remains as one of the most informative field experiences to illustrate the nature of progressive failure in overconsolidated brittle, high plasticity clays. The interest of the case increases because of the available information on post-failure behavior. Therefore, an opportunity exists to perform an integrated analysis of pre and post-failure slope behavior. Validating such an experiment, which was preceded by a good identification of laboratory-based soil properties, provides an interesting support to the analysis performed. The "static" part of the analysis described is in itself a valuable addition to the interpretation of field measurements.

Some relevant points of the review of the pre-failure part of the experiment are:

- The identification of the progressive failure mechanism, which progresses from the two extremes of the failure surface towards the central part of the sliding surface.
- The quantitative evolution of the mobilized average shear strength that shows two distinct critical moments: the first arrival to peak strength and the last one. In between, the average mobilized strength is essentially constant. The final sudden drop in average mobilized strength triggers the accelerated slide motion.

The "availability" of field displacements after failure allowed a further check of the model. The calculated displacement pattern agreed well with actual measurements. However, the result is highly dependent on the adopted residual effective cohesion. Small changes of this strength parameter (justified in the Selborne case by the occurrence of "in situ" suctions) result in significant changes of run-out results.

One limitation of the analysis refers to the effect of the initial excavation, which was not considered. This is expected to modify to some degree the initial stress state. A second difficulty is the very long computational time for the explicit code in cases of low permeability. However, results plotted in terms of relative time with respect to the failure time seem to be consistent with field data.

4. Deep seated fast landslides

"Compound" landslides, a classification category included in the recent review of Hungr et al. [16], are a relatively common class of motions, which require internal shearing to accommodate the kinematics of the motion. In deep seated landslides internal shearing will affect long distances in rock masses that, as a rule, are brittle materials. Therefore, (a) the stability of the landslide will be, to a large extent, controlled by internal shearing; (b) progressive failure mechanisms are expected during the deformation of the slide and (c) a final accelerated motion will be a consequence of the release of energy once static equilibrium is lost.

The sliding surface, on the other hand, may have a completely different character, especially in the case of reactivated slides. Often the basal sliding plane follows a weak layer, typically a claystone in a sequence of stiffer levels. If high plasticity minerals dominate the weak clayey rock very low residual friction angles (and no cohesion) will be operating on the basal failure surface. In these cases, the internal shearing through competent but brittle rocks should play a major role to explain stability conditions but also to control the post-failure behaviour.

This section investigates the static and dynamic behaviour of a landslide characterized by the previous description. An archetypical case is Vaiont landslide, which has been simplified in some cases [25] as two interacting wedges through a common shearing plane, which is necessary to accommodate the kinematics of the motion if no gaps between the moving mass and the basal sliding plane are accepted. However, there are infinite possibilities to build an admissible deformation pattern. In fact, the kinematic mechanism is part of the solution of the problem and not a necessary previous assumption. One of the objectives of this work was to find the kinematic mechanism of a given sliding geometry knowing that the rock behaves as a brittle material.

4.1. The model

Consider in figure 16 the 2D model analysed. Geometry and dimensions correspond to "section 5" of Vaiont landslide defined in Hendron & Patton [18] and Pinyol & Alonso [25]. An equilibrium state will be found for a set of strength parameters of the rock, the basal sliding surface and a given position of the water level in the reservoir, h_w.

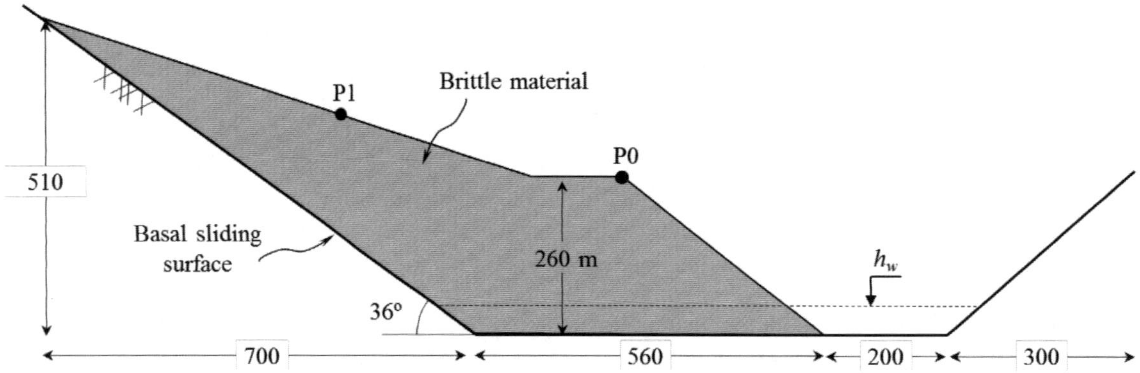

Figure 16. The analyzed landslide. Vaiont geometry.

Rock properties were derived from the analysis presented by Pinyol & Alonso [17]. A residual friction angle of 12° was adopted for the basal sliding surface following the data and discussion given by Hendron & Patton [18]. The set of material parameters is given in table 2.

Figure 17 shows the material and computational mesh. The computational mesh was extended to the right margin of Vaiont River to allow for a large displacement of the slide.

The basal sliding surface was simulated by means of a contact algorithm between the moving rock mass and a rigid lower layer, which acts as a lower boundary for the landslide.

IOP Conf. Series: Earth and Environmental Science **26** (2015) 012003 doi:10.1088/1755-1315/26/1/012003

The analysis was performed under drained conditions. At each time step, the current hydrostatic water pressure was imposed on material points. Points above the specified reservoir water level have no water pressure. Water in the reservoir was not modeled.

Table 2. Material parameters for the deep seated landslide.

Material parameter	Value
Rock porosity, n	0.2
Young Modulus, E [MPa]	5
Poisson ratio, ν	0.33
Peak effective cohesion, c'_r [kPa]	300
Peak effective friction angle, φ'_p	42°
Residual effective friction angle, φ'_r	36°
Shape factor, η	150
Basal residual friction, φ'_{res_0}	12°

4.2. Modelling results

The first step in the simulation was to introduce an initial stress state by activating gravity. The response of the slope to the imposed weight is shown in figure 18. The rock mass deforms slightly but the calculated deformation is enough to generate a band of plastic shearing deformations, which anticipates the kinematic mechanism in case of additional loading (increase in pore pressure). Deviatoric plastic strains reach a maximum at the kink between the two basal planes and extend upwards, almost vertically without reaching the slope surface. Strains are small but some points in the rock mass have already reached the peak strength.

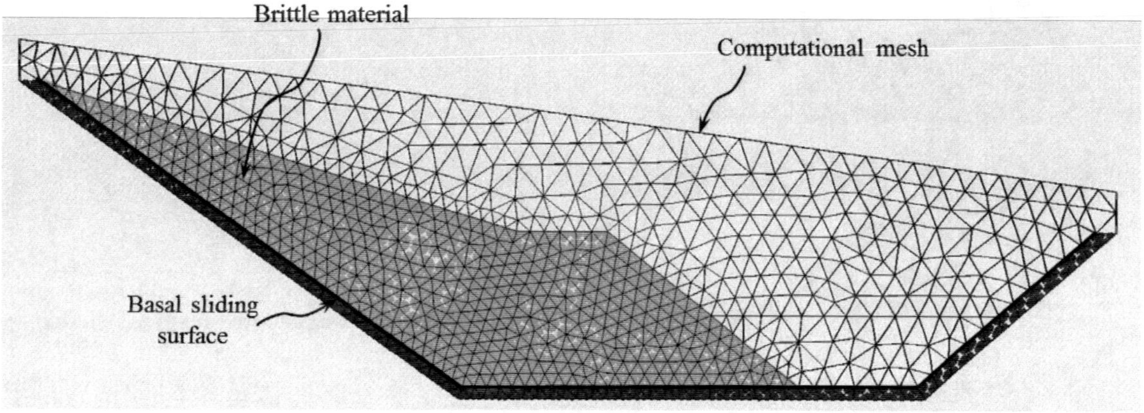

Figure 17. Computational mesh.

The next step is to increase the reservoir water level in the manner indicated in figure 19. The slide response is also shown in the figure. The internal shearing plane develops progressively (note the change in the scale of plastic shear strains). When the water level reaches 53m, a horizontal displacement of 8.5m is calculated for point P0 in figure 16.

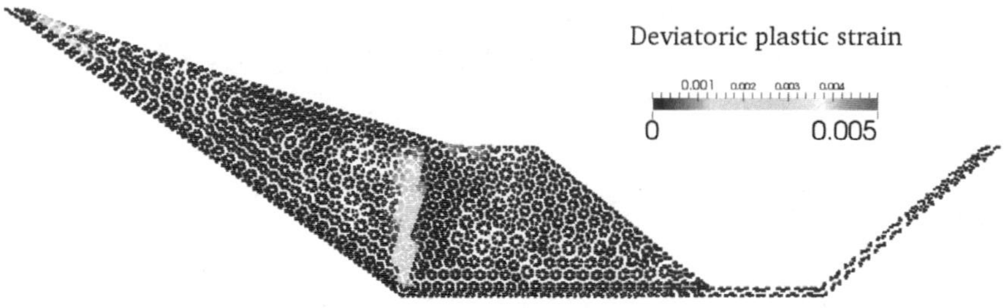

Figure 18. Deviatoric plastic strains after activation of gravity.

The progressive failure along the internal shear band starting at the kink point of the sliding surface is illustrated in figure 20. A mobilized friction was defined as in equation (20). Peak and residual thresholds are shown as well as the "available" strength for the initial time (t_0). This available strength is not the peak strength because it includes already the progressive failure taking place during the initial application of the gravity loading. Note that the lower points of the internal shearing plane had already suffered some amount of strength degradation. Increasing the water level mobilizes progressively upwards the friction angle. At time t_4 (h_w = 57 m) the entire plane is close to peak conditions and subsequent water level increments reduce the available friction to residual conditions. Note that the residual friction angle of the rock (36°) implies that a substantial shearing resistance is still being offered by the internal shearing of the rock.

The final stable geometry for h_w = 60m is shown in figure 21. The drop of rock friction to residual conditions leads to an increase of the plastic shearing strains but also to a change in the inclination of the shearing plane, which is now close to the bisection of the sliding surface dihedral angle.

The calculated displacement and velocity of the slide, once this internal plane has reached residual conditions is shown in figure 22. The maximum calculated velocity is around 0.5m/s very far from the estimated maximum velocity in Vaiont (25.30m/s).

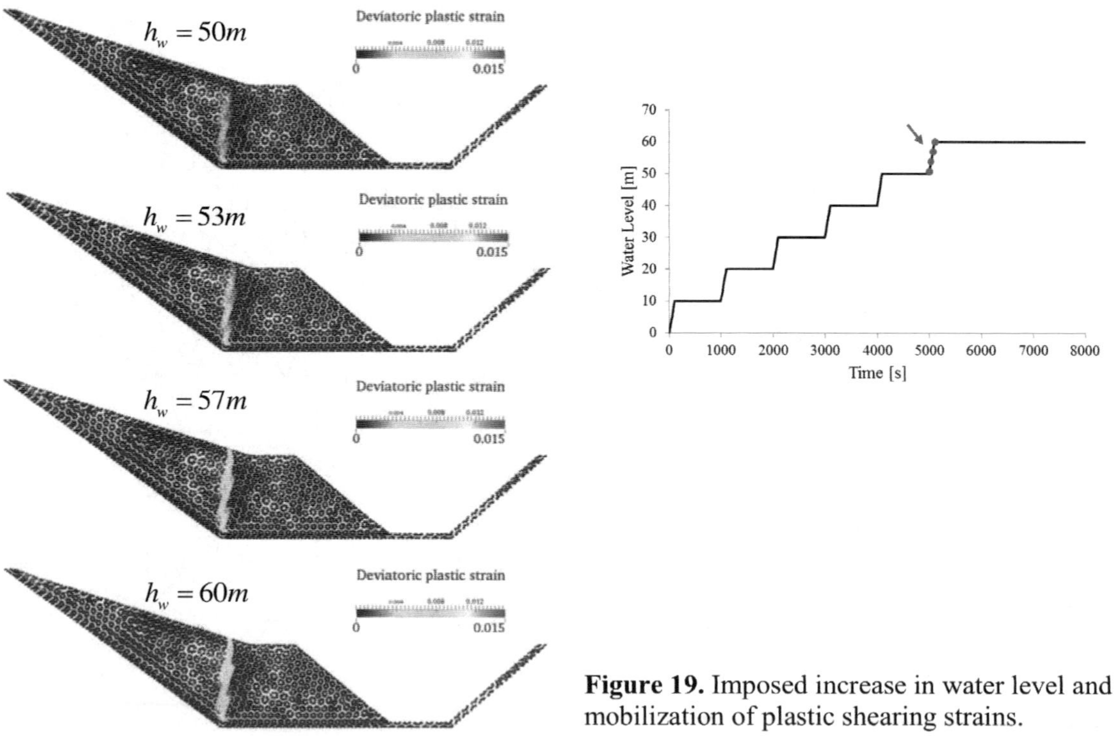

Figure 19. Imposed increase in water level and mobilization of plastic shearing strains.

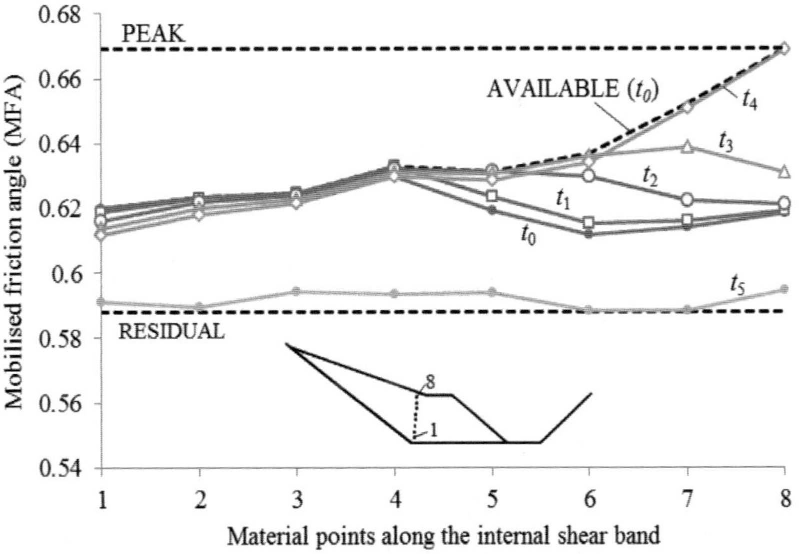

Figure 20. Mobilization of friction angle on internal shearing plane.

In order to increase significantly the run-out and the sliding velocity it is necessary to reduce the operating friction angle in the basal plane. This possibility is explained by the thermal pressurization of the basal clay layer and it has been proposed and justified by many authors [19][20][21][22][23][24]. Although some of the hypotheses change among the authors invoking the thermal pressurization, a plausible and simple explanation relies in the different thermal dilation coefficients of water and soil/rock skeleton. In practice this effect leads to a reduction in normal effective stress on the basal shearing surface. This reduction is fast and leads to an "explosive" reaction that reduces effective stress to zero (zero equivalent residual friction).

This phenomenon was simulated in the MPM by reducing the residual friction angle of the sliding surface. Two cases are graphically illustrated in figure 23: φ'_{res_b} = 5° and φ'_{res_b} =0°. The two sets of plots are similar. The slide accelerates and deforms and in the case of φ'_{res_b} = 0° it climbs up the opposite valley slope. The plot shows in a colour scale the calculated plastic deviatoric strains. Moving beyond the kink in the basal sliding surface results in a massive shearing of the rock. As new intact rock enters the position of a virtual shearing plane, plastic strains develop and the damaged rock moves forward to a new position. This process increases the volume of damaged rock in the direction of the motion. This is illustrated in figure 23. In the case of zero residual friction angle, most of the rock involved in the landslide has been damaged because of the two "kinks" of the basal surface at the two sides of the valley flat surface.

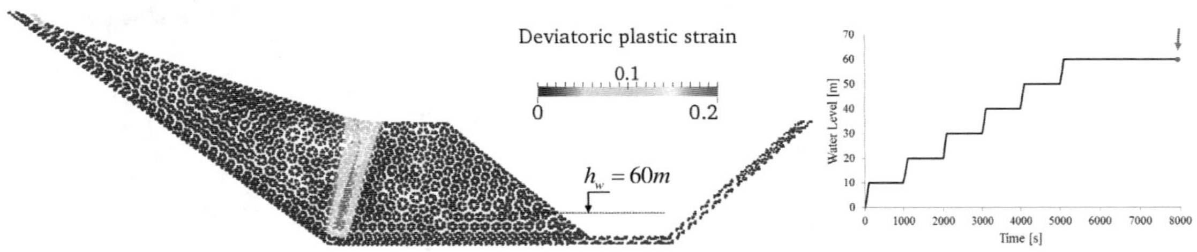

Figure 21. Stabilized geometry and development of shearing plane when h_w = 60m.

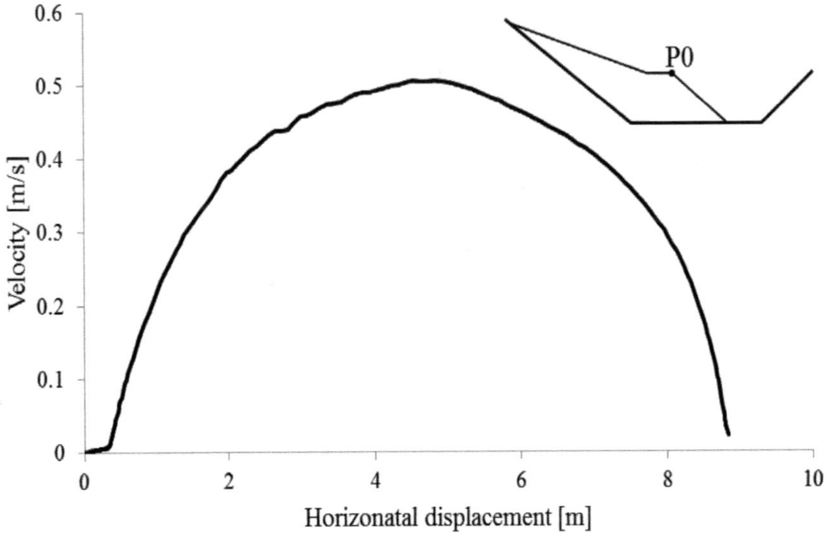

Figure 22. Velocity-displacement response of the slope for h_w = 60m (φ'_{res_0} =12°).

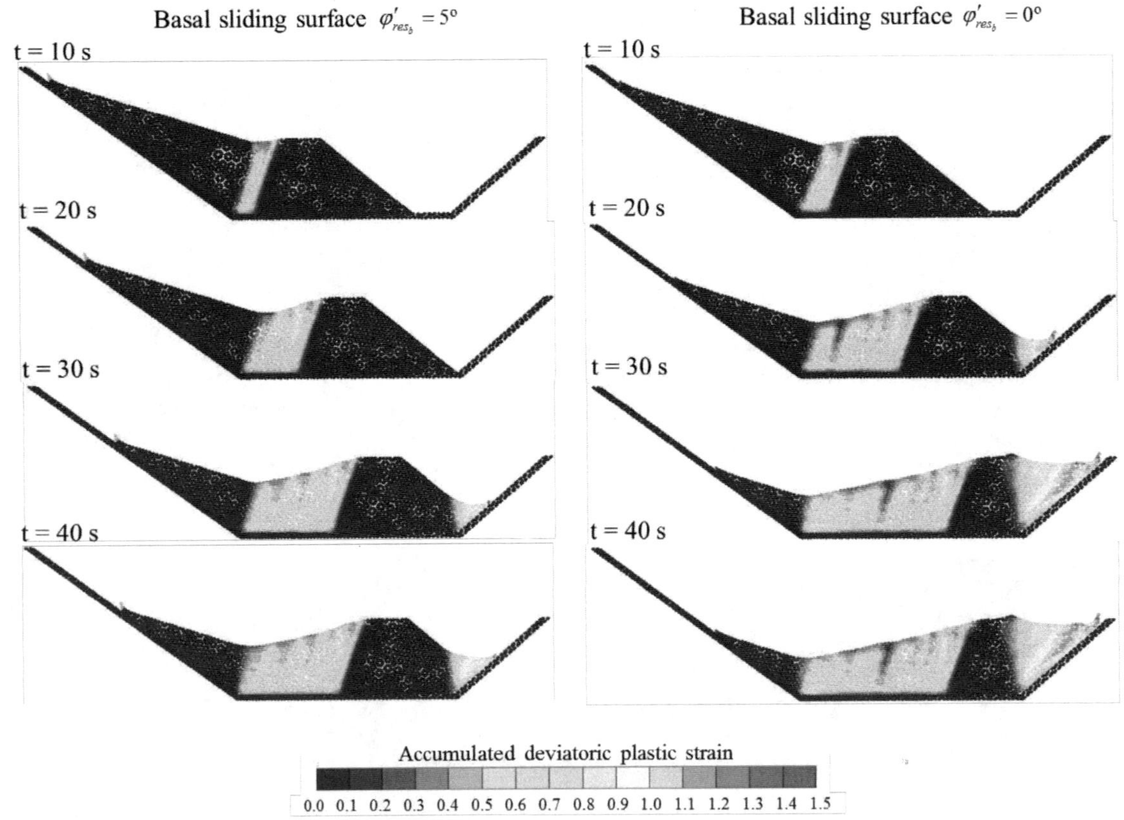

Figure 23. Evolution of landslide for two values of the basal residual friction angle.

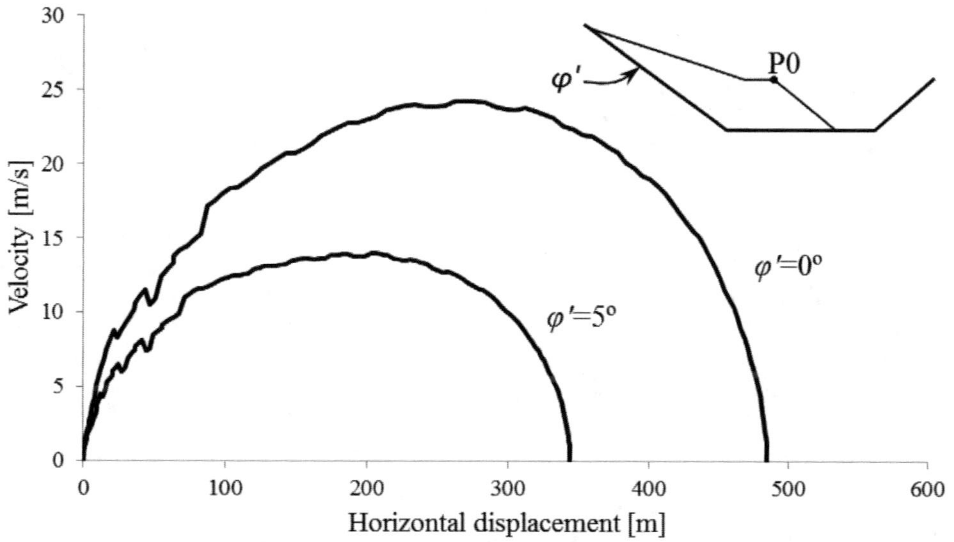

Figure 24. Velocity-displacement response when the basal residual friction is reduced.

Velocity and displacement of point P0 follow the diagram of figure 24. In the case $\varphi'_{res_b} = 0°$ the maximum speed is 25m/s, close to the estimations made for Vaiont. The slide reaches the opposite valley slope and has enough inertial energy to climb upwards. Both the maximum velocity and the run-out (490m) are consistent with field observations. The case $\varphi'_{res_b} = 5°$ does not explain equally well the actual failure.

The evolution of stresses inside the rock mass may be represented if some points are followed. Figure 25 shows the position of 13 material points located at mid-height at the initial and final states of deformation. The evolution of mobilized friction is shown in figure 26. The first 20s of motion are represented in figure 26a. Peak and residual friction angles provide a reference for the two figures.

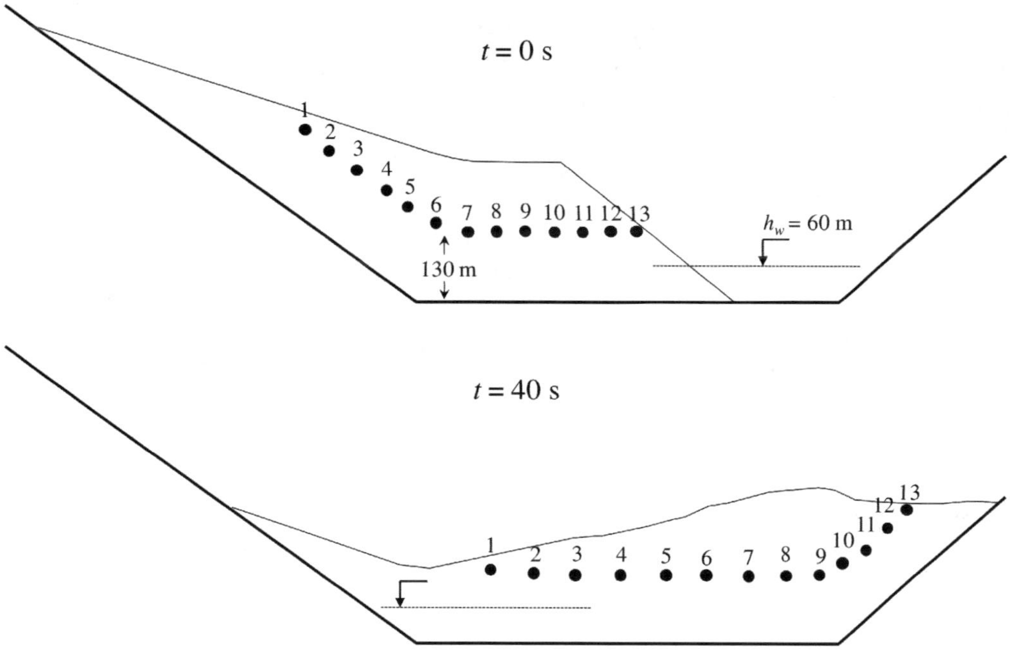

Figure 25. Position of material points selected in the landslide. Zero basal residual friction.

The effect of gravity loading is felt in the vicinity of the internal shearing zone. The available strength for $t = 0$ is lower than the peak strength, as shown in the figure. On both sides of this central zone, the rock remains elastic and the mobilized friction is below the limiting values (peak and residual). The initial stages of the motion damages the rock located on the upward side of the shearing plane. As time increases the process of progressive failure and the transition from elastic to peak and back to residual conditions is illustrated in figure 26. During this period ($t = 0$ to $t = 16$s), the rock on the right of the shearing plane remains elastic and the mobilized friction remains essentially unchanged because the unstable mass moves forward sliding on a horizontal surface. At $t = 20$s (figure 26a) point 13 on the toe of the slope "feels" the second change in slope of the basal failure surface and it reaches peak conditions. The remaining stages of the motion are shown in figure 26b. A new process of progressive failure damages the intact rock reaching the position of the second "virtual" shearing plane. At the end of the motion only a small volume of rock roughly located at the position of material points 7 and 8 remains intact. Points 1 to 6, located below the residual mobilized friction, have been unloaded by the new configuration of the slide.

4.3. Final remarks

The example analysed, inspired in a well-known case, illustrates the expected behaviour of brittle rock masses subjected to the kinematics of a compound slide. Internal shearing is probably required for the stability in case of reactivation of the landslides, especially if the basal surface exhibits a low frictional resistance (high plasticity clays).

Progressive failure phenomena are controlled by the kinematics of the motion, which, in turn, are dependent on the geometry of the sliding surface. Exhaustion of available peak strength and the sudden transition to residual conditions explain the initial acceleration of the landslide.

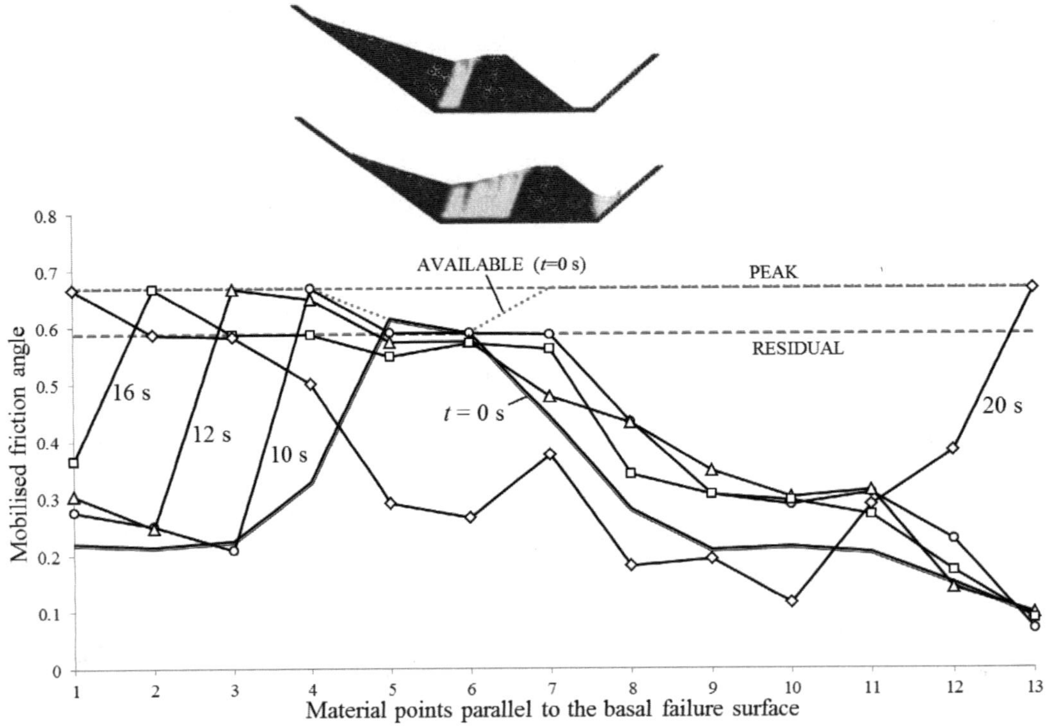

(a)

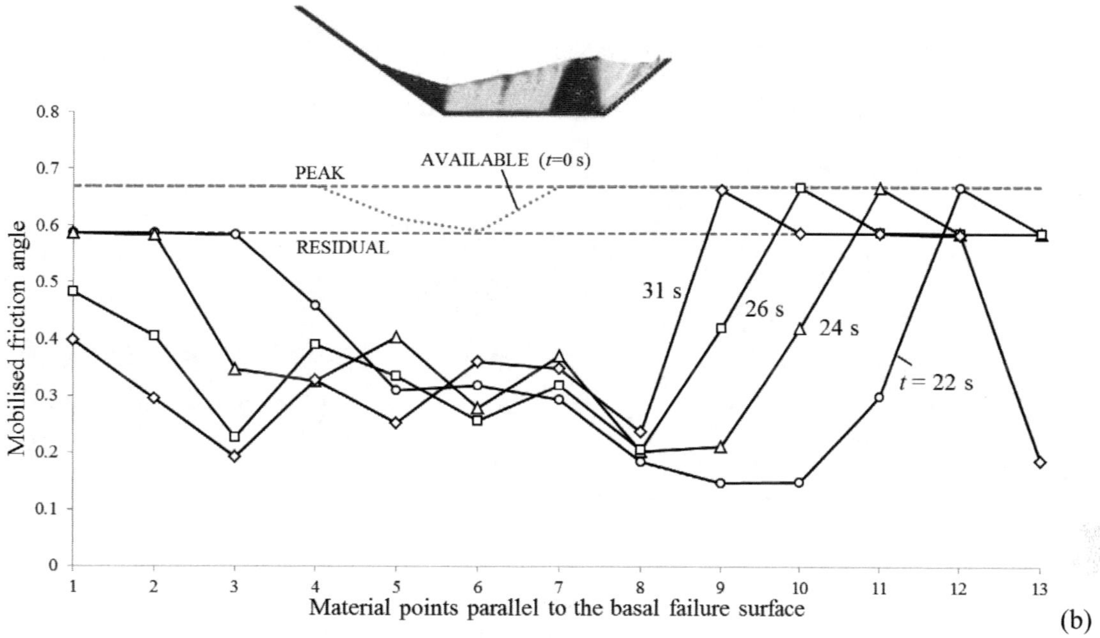

Figure 26. Evolution of mobilized friction angle on selected points. (a) Early stages of deformation; (b) final stages of deformation

In the simple geometrical configuration analysed, the development of an admissible mechanism is internally solved by a single shearing plane, which originates at the sharp transition between the two planes of different inclination. This was a simplified "a priori" choice in the thermomechanical analysis described by Alonso & Pinyol [19], which now receives some support. However, relatively minor changes in the geometry of the sliding surface will most probably lead to different admissible mechanisms. It has also been shown that the very high velocity reached by slides such as Vaiont can hardly be explained only by internal mechanisms of progressive failure. The analysis presented is also a support to phenomena leading to a zero friction basal surface, such as the thermal pressurization of the pore water in the main sliding surface.

5. Rainfall induced landslides

The general formulation for unsaturated soils described in section 2 is applied to simple slope geometry (figure 27). The case is directly based, especially as far as soil properties is concerned, on the analysis of a case history ("Collapse of compacted soil: Girona road embankments, Spain") presented by [26].

The set of material properties is given in table 3. The computational mesh is made of 3,654 tetrahedral elements. 7,593 material points define the soil mass. A small damping factor (α = 5%) was used in calculations. The time step was $\Delta t = 2 \times 10^{-4}$s.

The water retention curve (a one-to-one relationship between suction and degree of saturation) for wetting conditions is shown in figure 28. The points in the plot are actual laboratory determinations of the embankment soils: residual soils from granite. They are described as sandy clays of low to medium plasticity.

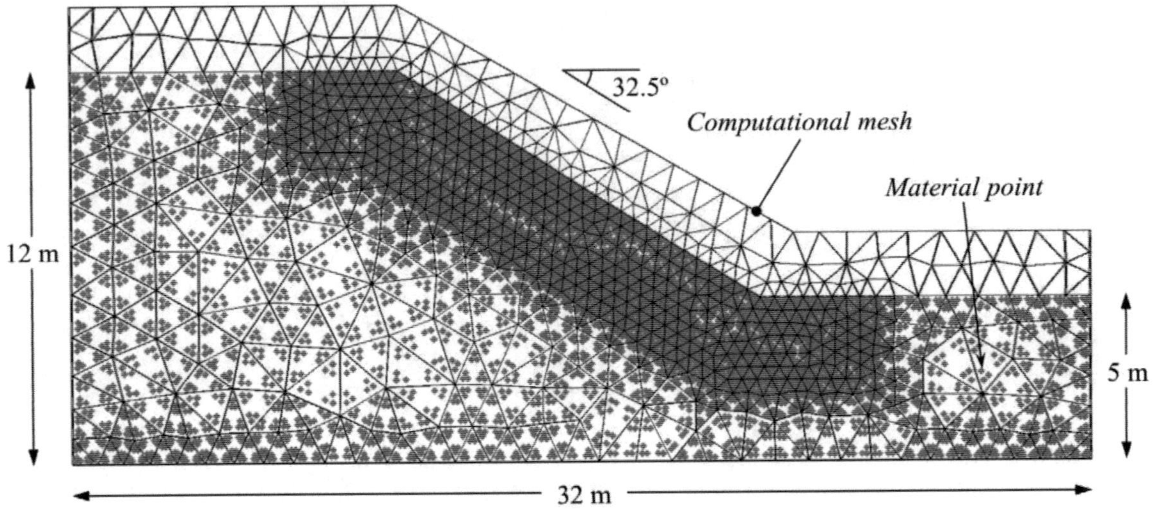

Figure 27. Slope geometry for the rain infiltration case.

Table 3. Mesh details and material properties.

Element type	Tetrahedron
Number of elements	3,654
Number of material points	7,593
Damping factor, α	0.05
Time step	2×10^{-4}s
Solid density, ρ_s	2,700kg/m^3
Porosity, n	0.35
Poisson ratio, ν	0.33
Liquid density, ρ_l	1,000kg/m^3
Liquid density, ρ_g	1kg/m^3
Liquid bulk modulus, K_l	100MPa
Gas bulk modulus, K_g	0.01MPa
Liquid viscosity, μ_l	10^{-3}kg/m·s
Gas viscosity, μ_g	10^{-6}kg/m·s
Intrinsic permeability liquid, k_l	10^{-10}m^2
Intrinsic permeability gas, k_g	10^{-11}m^2
Young modulus, E	10MPa
Effective cohesion, c'	1kPa
Effective friction angle, φ'	20°
Δc_{max}	15kPa
B	0.0007
A	0.0001

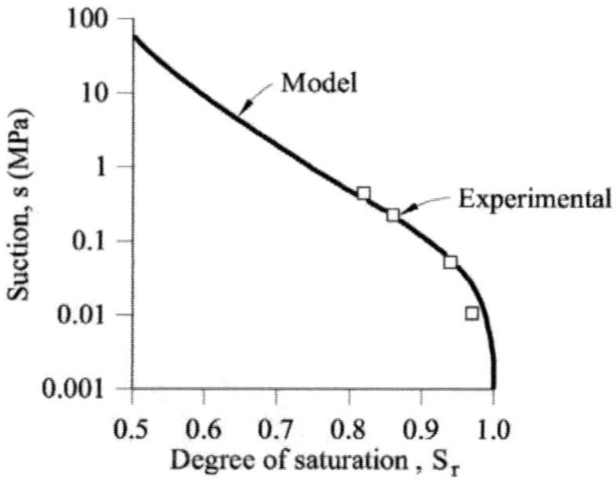

Figure 28. Water retention curve.

The embankment was compacted in summer time and measured water contents in samples recovered indicated that the initial suction was in the range of 700–800 kPa. The initial state of the simulation performed was achieved by imposing equilibrium conditions under boundary suction of 800kPa and a gravity force.

The slopes experienced in the field frequent shallow slides (as well as internal collapse, not analysed here) when they received a series of heavy rains in a relatively short intervals (1–3 days).

The model slope was subjected to a boundary saturation (suction = 0) which was shorter than the actual field conditions. Model permeability was increased to reduce calculation times.

The effect of suction reduction (see section 2) on strength is controlled by the A, B and Δc_{max} parameters in table 3. In other words, a reduction in suction from 800kPa to saturated conditions reduces the cohesion from 67kPa to 1kPa. Friction reduces slightly (less than 1°).

5.1. Model results

The effect of boundary wetting is shown in figure 29. As suction reduces throughout the slope, a shallow shear band develops. The four points selected to show the soil displacements (S_1, D_1 at the toe of the slope and S_2 D_2 at mid height) illustrate the slope behaviour. Maximum displacements parallel to the slope surface develop at mid height in shallow locations. Points at the slope foot remain essentially still: they are covered by soil made unstable at higher elevations. At the end of the deformation, the slope angle decreases a few degrees.

Shallow positions (S_2, S_1) experience a faster reduction in suction and a faster increase in saturation than deeper points. This is shown in figure 30. The first stages of wetting induce fast changes in suction and saturation because of the high pressure gradients involved. The rate of change of suction and saturation decreases continuously, in a non-linear way, during the wetting process.

Information on the dynamics of the motion is one of the key features of MPM. Wetting the slope changes the stress-pore water pressure conditions during a certain time in which sliding displacements are very small. The instability develops suddenly when suction has reduced to low values at shallow depths. This is illustrated in figure 31, which could be compared with figure 30 (a series of common time instants have been indicated in both figures). Dynamic variables depend markedly on the position of the selected control points within the slope. The point at mid slope in a shallow position displaces more and reaches higher maximum velocities and accelerations than other points in the slope. The final part of the motion is characterized by a slow reduction in velocity (slow deceleration). Note that the concept of run-out requires some previous conventions to be defined unambiguously. In general terms, all the points involved in the motion travel different distances. One possible definition of run-out is the distance between the toes of the initial and final slope geometry. In the example solved this distance is 2.5m. However, the maximum (horizontal) displacement of surface points at mid-height is more than 4m.

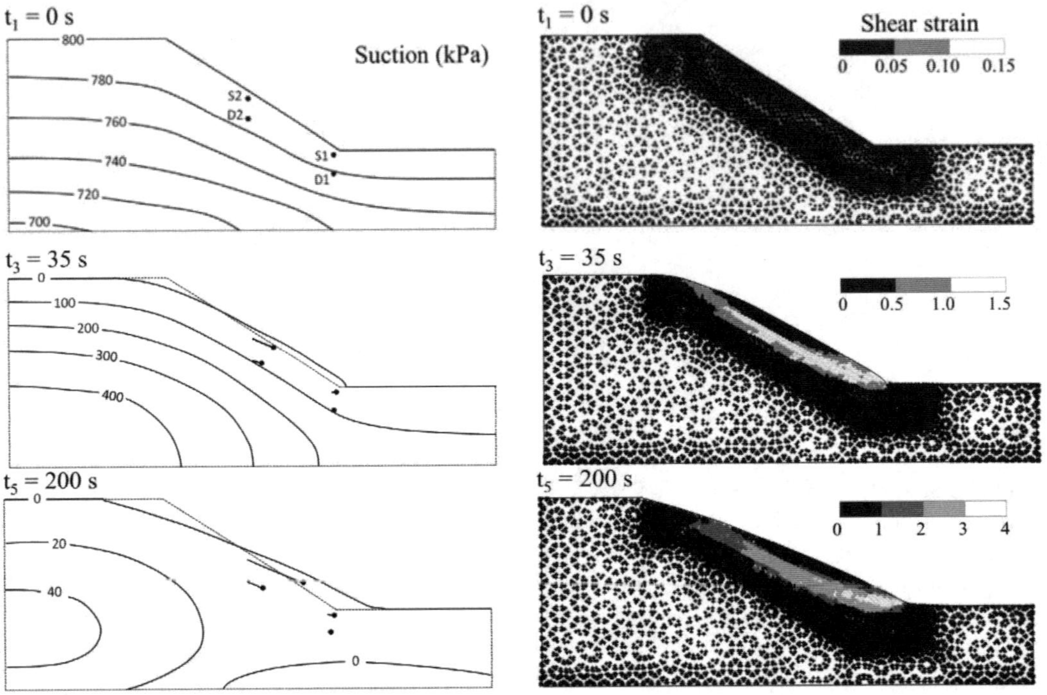

Figure 29. Evolution of suction distribution and deformation of the slope.

The stress paths experienced by the four reference points are also of interest. They are given in a (net mean stress: $\overline{p}$; deviatoric stress: q) plane in figure 32. Also plotted in these two figures are the strength envelopes for different suctions.

Point S1, at the toe of the slope is highly stressed and it is initially ($t = t_1$) at yielding conditions. When wetting starts, the stress state in this point changes fast. Local yielding explains the drop in net mean stress (total stress in this case). In addition q decreases following the evolution of strength envelopes. When point S1 is "buried" by the unstable slope at times $t > t_2$, the mean total stress increases and the point remains at yielding conditions for a saturated state.

Point S2 is initially in an elastic state far from yielding conditions for s = 800kPa. Suction reduction brings the yielding envelope to the proximity of the stress state of S_2, which does not change much during the slope motion because point S2 is above the sliding surface.

Point D1 is in a stable location, below the sliding surface and away from the singularity provided by the toe of the slope. Therefore it is initially under elastic conditions, far from the current yield locus at $t = t_1$. Wetting brings down the yield surface and induces a small increase in (q, p). The subsequent sliding adds soil weight over point D1, which remains in an elastic state. Finally, point D2 is initially under elastic conditions but when suction decreases it reaches yielding conditions, which remain during the subsequent sliding (D2 is at the shear band).

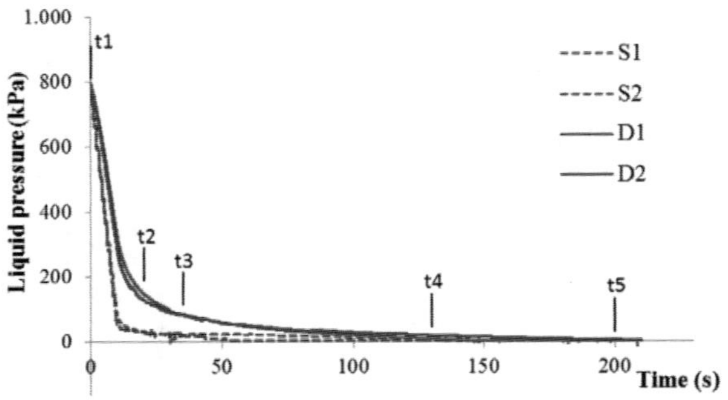

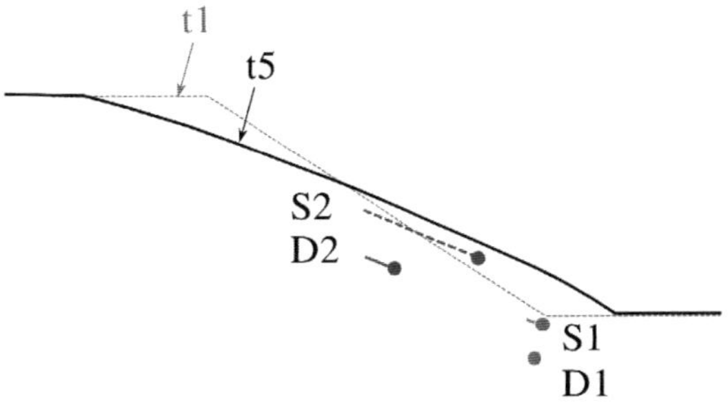

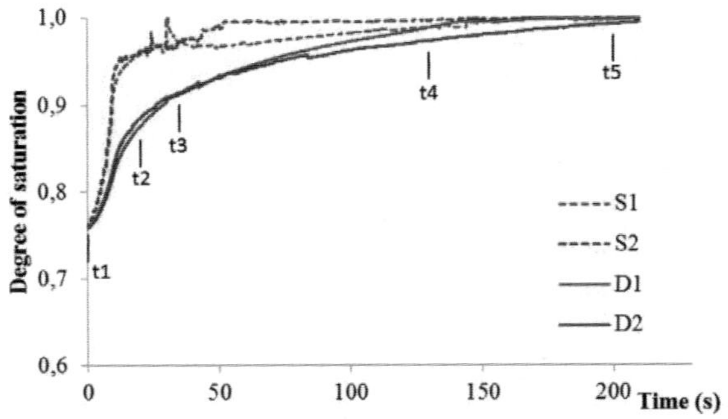

Figure 30. Evolution of suction and degree of saturation for some selected points.

MPM suffers from oscillations of field variables because of the dynamic nature of the formulation. Unbalanced forces, either internal or external, generate waves that cross the domain and may reflect at boundaries. Decrease of stress levels due to softening may be one cause for oscillations. Calculated results show these oscillations. This is shown in figure 33a, which shows the evolution of the deviatoric stress of points D1 and D2. Point D1 shows these oscillations at the later stages of wetting. However, the main reason for this behaviour may be another one. Figure 33b shows evolution of degree of saturation of the two points. Point D1 reaches saturation beyond time t_4. In fact, the initiation of the oscillatory behaviour seems to coincide with the condition $S_r = 1$ (saturated conditions).

It turns out that the bulk modulus of compressibility of an unsaturated soil increases fast when saturation approaches. This is illustrated in figure 34. The curve in the figure is described by the equation (21) [27].

$$K_m = K_s + \frac{1}{\dfrac{nS_l}{K_l} + \dfrac{n(1-S_l)}{K_g}} \tag{21}$$

$$K_{sat} = K_s + \frac{K_l}{n} \tag{22}$$

where K_l, K_s and K_g are the bulk modulus of liquid, solid and gas and S_l the degree of saturation.

The stiffness of the mixture increases abruptly towards the stiffness for saturated conditions (equation (22)), for degrees of saturation in excess of $S_l = 0.99$. This is explained because the bulk modulus of gas is several orders of magnitude smaller than the modulus of liquid water and soil skeleton. Figure 33 shows that the instabilities manifest when the degree of saturation approaches 1.

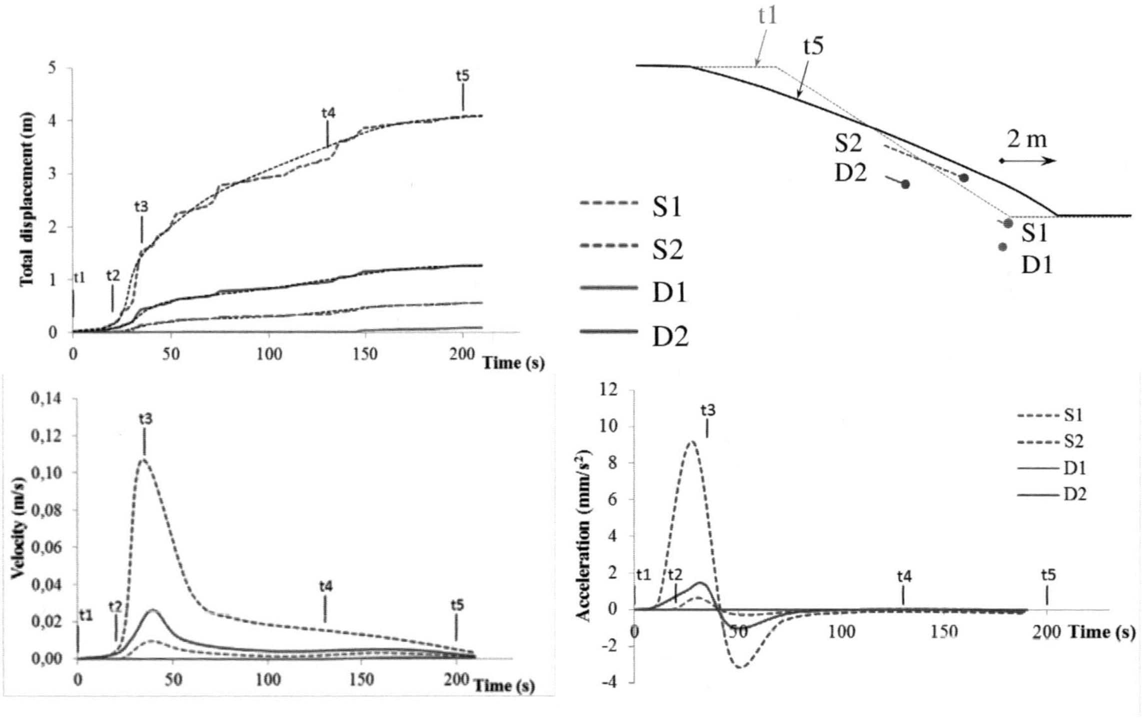

Figure 5.5. Dynamics of the motion.

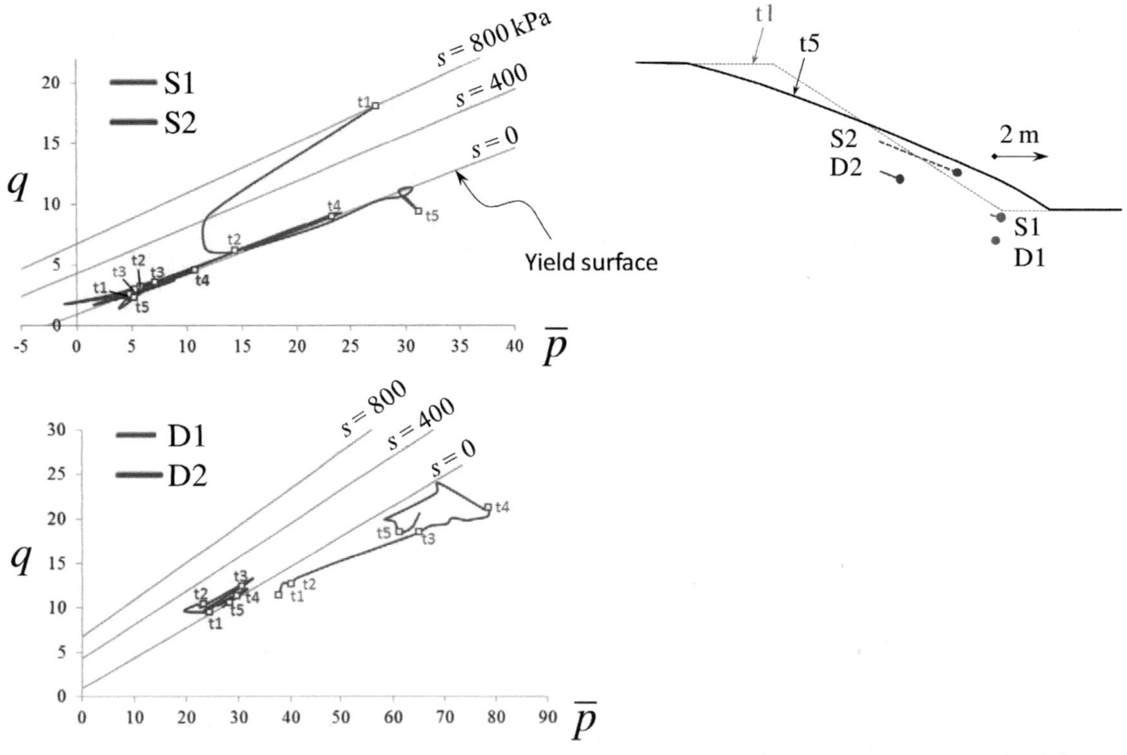

Figure 32. Stress-suction paths of four points within the slope.

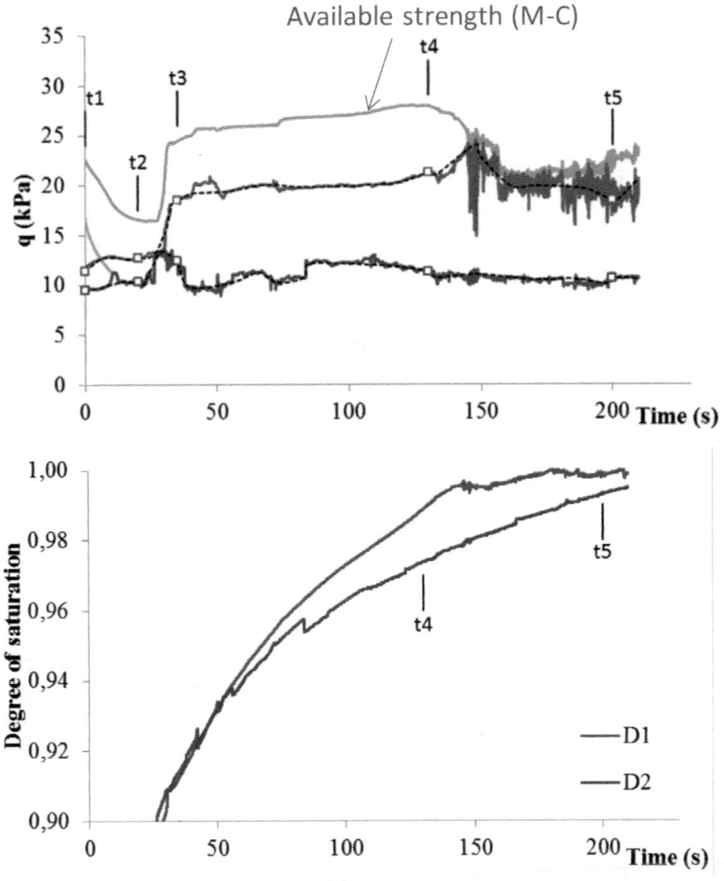

Figure 33. Evolution of deviatoric stress and degree of saturation of points D1 and D2.

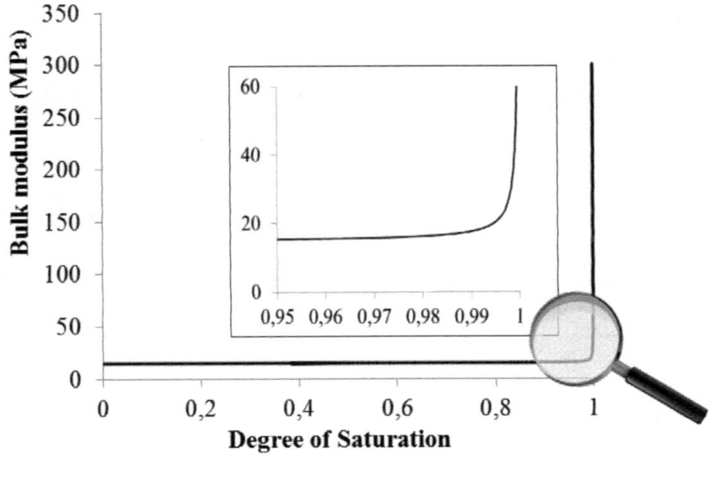

Figure 34. Bulk modulus of unsaturated soil in terms of degree of saturation.

6. Concluding remarks

The MPM and its different formulations for multiphase materials are currently being developed and applied to a variety of geotechnical situations. The MPM concept offers some advantages: being a method for the analysis of continua, it allows the incorporation of the experience gained in finite element analysis; soil behaviour is introduced through well-known constitutive equations (unlike pure particle methods); its dynamic formulation introduces inertia forces and therefore it is well adapted to analyse post-failure motions and, in the case of landslides, to predict run-out and velocity, two variables directly associated with risk.

A comprehensive MPM formulation for three phase materials coexisting in single material points was presented. Dry and saturated soils are particular cases. The soil models introduced are essentially elastoplastic Mohr-Coulomb formulations. Relevant features are strain softening behaviour and suction dependant strength parameters.

The resulting model provides a unified procedure to examine the static conditions for equilibrium, the transition to an accelerated motion and the propagation of the landslide.

Three cases have been analysed, all of them based on real cases:

The Selborne failure experiment in overconsolidated brittle high plasticity clays provided a further insight into a well-known case of progressive failure development. The MPM model, based on published material properties, slope geometry and hydraulic overpressure inducing failure provided remarkably consistent results. Run-out was successfully reproduced.

The second case examines the role of internal shearing in compound landslides. Vaiont landslide was analysed from this perspective. This is also a case of progressive failure and a final accelerated motion. The damage of the rock could be followed during the sliding motion. It was also found that the very high estimated velocity of the slide could only be reproduced by a zero friction at the basal surface, a result that supports the thermal pressurization hypothesis to explain the full drop of shear strength at the sliding surface.

The third case analyses rainfall effects on a slope whose properties correspond to an embankment slope heavily damaged by intense rainfall. Stress paths of significant points within the slope indicate the complex interaction between stress and suction changes. The dynamic post-failure analysis illustrated the evolution of the slope geometry. Some numerical difficulties are thought to be a consequence of the very fast increase in bulk modulus of the soil when degree of saturation approaches unity.

References

[1] Sulsky D, Chen Z and Schreyer H L 1994 A particle method for history-dependent materials. *Comp. Met. App. Mec. Eng.* **118** 179–96

[2] Sulsky D, Zhou S-J and Schreyer H L 1995 Application of a particle-in-cell method to solid mechanics. Comp. Phys. Comm. **87** 236–52

[3] Zhang H W, Wang K P and Chen Z 2009 Material point method for dynamic analysis of saturated porous media under external contact/impact of solid bodies. *Comp. Met. App. Mec. Eng.* **198** 1456–72

[4] Zabala F and Alonso E E 2011 Progressive failure of Aznalcóllar dam using the material point method. *Géotech.* 61 795–808

[5] Jassim I, Stolle D and Vermeer P 2013 Two-phase dynamic analysis by material point method. *Int. J. Num. Anal. Meth. Geomec.* **37** 2502–22

[6] Yerro A, Alonso E and Pinyol N 2015 The material point method for unsaturated soils. *Géotech.* **65** 201–17

[7] Abe K, Soga K and Bandara S 2013 Material point method for coupled hydromechanical problems. *J. Geotech. Geoenv. Eng.* **140** 1–16

[8] Bandara S and Soga K 2015 Coupling of soil deformation and pore fluid flow using material point method. *Comp. Geotech.* **63** 199–214

[9] Fredlund D G, Morgensten N R and Widger R A 1978 The shear strength of unsaturated soils. *Can. Geotech. J.* **15** 313–21

[10] Delage P and Graham J 1996 Mechanical behaviour of unsaturated soils. *Proc. 1st Int. Conf. Unsaturated Soils* (Paris, France, 6–8 Sep 1995) ed E E Alonso and P Delage (Rotterdam, Balkema) pp. 1223–56

[11] Yerro A, Alonso E and Pinyol N 2014 Modelling progressive failure with MPM. *Proc. 8th Eur. Conf. Numerical Methods in Geotechnical Engineering* (Delft, The Netherlands, 18–20 June

2014) ed M A Hicks et al (London, CRC Press) pp 319–23
[12] Alonso E and Gens A (19??)
[13] Cooper M, Bromhead E, Petley D and Grant D 1998 The Selborne cutting stability experiment. *Géotech.* **48** 83–101
[14] Grant D 1996 *Instrumentation systems for and failure mechanisms of an induced slope failure project*. University of Southampton
[15] Cooper M 1996 The progressive development of a failure slip surface in over-consolidated clay at Selborne, UK. *Proc. 7th Int. Symp. Landslides* (Trondheim, Norway, 17–21 June 1996) ed K Senneset (Rotterdam: Balkema) vol 2 pp 683–688
[16] Hungr O, Leroueil S and Picarelli L 2014 The Varnes classification of landslide types, an update. Landslides **11** 167–194
[17] Pinyol N M and Alonso E E 2010a Fast planar slides. A closed-form thermo-hydro-mechanical solution. *Int. J. Num. Anal. Meth. Geomech.* **34** 27–52
[18] Hendron A J and Patton F D 1987 The Vaiont slide. A geotechnical analysis based on new geologic observations of the failure surface. *Engineering Geology* **24** 475–491
[19] Voigt B and Faust C 1982 Frictional heat and strength loss in some rapid landslides. *Géotech.* **32** 43–54
[20] Vardoulakis I 2002 Dynamic thermo-poro-mechanical analysis of catastrofic landslides. *Géotech.* **52** 157–171
[21] Veveakis E, Vardoulakis I and Di Toro G 2007 Thermoporomechanics of creeping landslides: The 1963 Vaiont slide, northern Italy. *J. Geoph. Res* **112** F03026
[22] Pinyol N M and Alonso E E 2010 Criteria for rapid sliding II. *Eng. Geol.* **114** 211–227
[23] Pinyol N M and Alonso E E 2010 Fast planar slides. A closed-form thermo-hydro-mechanical solution. *Int. J. Num. Anal. Meth. Geomech.* **34** 27–52
[24] Cecinato F and Zervos A 2012 Influence of thermomechanics in the catastrophic collapse of planar landslides. *Can. Geotech. J.* **49** 207–225
[25] Alonso E E and Pinyol N M 2010 Criteria for rapid sliding I. A review of Vaiont case. *Eng. Geol.* **114** 198–210
[26] Alonso E E, Pinyol N M and Puzrin A M 2010 *Geomechanics of Failures. Advanced Topics* (The Netherlands, Springer)
[27] Santamarina J C 2001 *Soils and waves* (Berlin, Wiley)

Landslides falling onto a shallow erodible substrate or water layer: an experimental and numerical approach

GB Crosta[1], FV De Blasio[1], M Locatelli[1] S Imposimato[2] and D Roddeman[2]

[1] Department of Earth and Environmental Sciences, University of Milano Bicocca, Italy.

[2] FEAT, Netherlands

E-mail: giovannibattista.crosta@unimib.it

Abstract. Landslides often collapse in areas covered by alluvial deposits forming an erodible layer. This erodible substrate may deform plastically under the intense shear stress of the landslide mass. In other cases, the collapse occurs onto a water basin or tidal flat, creating impulse water waves whilst the landslide may be lubricated by a water layer underneath. In either cases the presence of a medium underneath the landslide will change its dynamics introducing complex processes. While frictional, dry masses and taluses generally hamper the landslide motion. In this work, we present some experiments mimicking the collapse of a landslide onto shallow erodible or water layers. The landslide is simulated with a granular material (sand or gravel) flowing on an incline (35-66°) followed by a horizontal sector covered with a granular bed 1 to 2 cm thick or with a 0.5-1 cm of water. Monitoring evolution in time allows us to describe in detail the process of fluidization of the material at impact, the generation of impact waves, and the erosion process. Concerning impact on a sand layer, the apparent friction coefficient (H/L) is found to increase with the chute slope angle and with the thickness of the erodible layer, and to decrease with the volume. At low slope angles, the material accumulates backwards in a shock wave mode, while at larger slope angles (>45°) it accumulates by prograding forward. A granular avalanche falling from the slope is partially reflected at the sharp slope break where erosion occurs and then propagates initially as a wave partially eroding the superficial material. Folding and thrusting occur within the dense shear flow and the erodible layer. Experiments with a water layer show that the dynamics depends much on the permeability of the granular avalanche. FEM numerical simulations replicate and allow to describe and understand both the spreading and the erosion, and internal deformation recorded in the erodible layer. Experimental findings are compared with real rock avalanches, flowslides and snow avalanches characteristics and morphological features. A medium with low permeability may be lubricated by the presence of water, resulting in a front acceleration and a final double-ringed deposit.

1. Introduction

Field studies and observations on the behavior of a landslide collapsing onto erodible substrates are scarce even if it is known that presence of a medium with very different characteristics may be extremely important in modifying the further flow of the landslide. Because the erodible substrate remains completely sheltered underneath the landslide mass, a closer examination is difficult (Hewitt, 2006; von Poschinger and Kippel, 2009). In other cases, the erodible substrate has been pushed ahead of the landslide and its flow is studied in detail. At Arvel (Switzerland) a large rockfall–avalanche (1922) pushed alluvial sediments forward, folding them (Choffat, 1929; Jaboyedoff 2003; Crosta et al., 2008) in a tectonic style, with arched ridges and grooves up to 7 m high. The average slope angle was 63.3° for a drop height, H, of 240 m and maximum runout, L, of 337m (H/L= 0.71). Accounting for this frontal material as if it belonged to the landslide itself would correspond to a total H/L of 0.35 (Crosta et al., 2009). Other examples of collapse on a flat area are those of the Frank landslide (H=600 m, slope angle = 47°), the Ollague landslide (Crosta et al., 2013, H = 707 m, angle 38.2°, total H/L = 0.25), the dry Las Colinas landslide (Crosta et al., 2005) collapsed along a 30-35° slope (H = 160 m, L = 715 m, H/L = 0.22).

Ancient rock avalanches were affected by a basal erodible layer (McSaveney et al., 2000) like at Flims (Switzerland) where alluvial material of the Bonaduz Gravel Formation is cut by Pavoni tubes suggesting the presence of pressurized fluids in the basal layer (Erismann and Abele, 2001; Von Poschinger, 2006). The wet alluvial sediments are considered a cause for the extraordinary propagation of the Flims landslide. The Marocche (Trentino province, northern Italy) are examples of landslides collapsed on flat alluvial areas in prehistoric or historic times. Some of them dammed locally the drainage system, creating enduring lakes as in Molveno (H = 1100 m, slope angle = 45°).
Similar geometrical constraints have been observed along many Martian slopes of the Valles Marineris gorge (Lucchitta, 1979). We will show that this choice of the geometry leads to interesting predictions for the mobility of the landslides.

The importance of the properties of the basal sliding layer becomes apparent for landslides falling onto glaciers. Numerous case studies, such as Sherman in Alaska (Post, 1967; McSaveney, 1976) and the Thurwieser peak in Italy (Sosio et al., 2009), show that the friction coefficient may decrease by one order of magnitude due to the melting of ice (Schneider et al., 2011; Sosio et al., 2012; De Blasio, 2014). Moreover, the dynamics of such events may be relevant in reconstructing the past environment of Mars, since the presence of a lubrication layer of some sort has been invoked to explain the extraordinary mobility of landslides on the red planet (Lucchitta, 1979; De Blasio, 2011).
A peculiar situation concerns the collapse onto a water basin such as a lake, river, or the sea. We focus on landslides collapsing on very shallow water like a tidal flat, where water provides a thin lubricating medium, rather than the case of large landslides plunging inside deep basins (Crosta et al., 2015). Most events recorded on shallow waters involve chalk cliff collapses onto flat tidal areas in northern Europe (Hutchinson, 2002; Dupperet et al., 2006; Günther and Thiel, 2009), but also large volcanic rock avalanches travelling into lakes (Sosio et al., 2011). Recognizing that vertical failures result in shorter runouts than those occurring along a more gently slope, Hutchinson (2002) suggests that the former involve harder and less porous rocks, perhaps less prone to water remolding. However, the abrupt slope change and the possible role of the water layer in the dynamics are not considered.

The physical processes occurring when a landslide collapses onto an erodible layer or water have not been studied systematically both from a geological viewpoint and concerning the mechanics of flow. Firstly, there is a physically-related problem directly linked to the hazard assessment. The entrainment of the underlying material results in an increase of the overall volume of the moving mass and likely decreases the velocity of the landslide as a consequence of momentum conservation. Thus, it is uncertain whether a landslide entraining an erodible substrate will be more dangerous as consequence of the increased mass or a sudden change in behavior, or less dangerous because of its diminished speed. Moreover, the erodible sediments in front of a potential landslide may act as a protective buffer, decreasing the mobility, or as a lubricant (see next point) or even jetting ahead of the landslide, splashing inhabited zones.

Secondly, the rheology of the erodible layer plays an important role. While some kinds of material (e.g. cohesive, highly frictional, dry masses) may perhaps hamper the movement of the landslide (see Crosta et al., 2008, 2013b), others such as water-rich alluvium, glacial material, cohesive silt and clay, loose saturated fine sands, may act sometimes as a lubricant, transferring the stress from the frictional base to a weaker layer. In terms of final fate of the eroded material, it can be entrained, dragged, sheared, ploughed, bulldozed by the landslide, or even remain unaffected if the base of the landslide is strongly decoupled from the substrate material due to the presence of lubricating material.

Numerous experiments have been performed to reproduce the collapse of landslides and or in general geophysical granular flows (Denlinger and Iverson, 2001; Forterre and Pouliquen, 2008; Lacaze et al., 2008; Lajeunesse et al., 2004, 2005; Mangeney et al., 2007, 2010; Crosta et al., 2008, 2013). These experiments investigated the flow of a granular medium along a flume under controlled conditions while monitoring the flow velocity, change in time of the shape of the travelling mass, and the examination of the final geometry. Most experiments concentrated on the final shape of the deposit or the evolution on smooth surfaces (e.g. Gray et al., 1999; Pudasaini and Hutter, 2006) while fewer have so far been devoted to the flow onto a bed of granular material (Dufresne, 2012; Crosta et al., 2008; 2009; Farin, 2014; Rooley, 2011) and no experiments — to our knowledge — on a thin water layer.

In this work, we simulate a landslide with a series of experiments in which a small-scale granular flow collapsing onto an erodible layer or shallow water. The variety of natural geological processes and topographic situations implies that many different kinds of erodible layers may affect the flow of real landslides. Transferring these situations from the field to the laboratory, we envisage four possible interesting layer types to use in the experiments (Table 1).

1. A **dry granular bed** is the simplest possible system. A granular material endowed of Mohr-Coulomb frictional behavior may affect the friction coefficient at the base of the landslide. In the field, such system may be a reasonable model for landslides falling onto dry rock debris. The particle size, shape, and angularity as well as the soil bulk density can control the consequent behavior.
2. More dramatic changes might be expected if the entrained material has fundamentally different rheological properties. A material of **low shear strength**, especially if **water-rich** may severely affect the behavior of the overlying mass. From the experimental point of view, this situation requires a wet and slightly cohesive erodible layer behaving as a sensitive clay. This may be a good model for rock avalanches falling onto alluvial sediment (Hungr and Evans, 2004; Chen et al., 2006; McDougall, 2006; Crosta et al., 2008, Crosta et al., 2013b). Needless to say, in an attempt to simulate alluvial materials the complexity of the erodible experimental material may easily escalate. For example, layering, porosity and compaction, permeability, coarse versus fine fraction relative content, frictional and cohesive properties, particular rheological behaviors (e.g. liquefiable, dilatant, drained/undrained) will likely control the quantity of material affected by the landslide motion and its response.
3. A **water layer** may be used to simulate the collapse of a landslide onto a shallow water layer, a tidal flat, or a lake.
4. The collapse onto **ice** cannot be easily simulated in the laboratory not only because it would force the experimenter to a low temperature setting (Schneider et al. 2011), but also because the stress developed by a simulated small-scale laboratory flow is insufficient to re-create significant effects at the base of a real landslide travelling on ice and ice melting. Although for completeness, we mention ice as an important natural lubricant for landslides in glaciated environments, experimentation with this material will thus not be conducted in this work.

We present a set of results with controlled experimental tests in which dry sand or gravel reaches a horizontal plate where an erodible layer of sand (point 1) or a thin water layer (point 3). The real material conditions and the relative experimental settings are summarized in Table I, while those with low shear strength material (point 2) and ice (point 4) are deferred to a later study.

Like in other experimental tests, we utilize a bi-linear geometry consisting of an inclined plate followed by a horizontal portion. This is a simple geometry to use in the experiments and data analysis, and as described above it is also a good model for the geometry of many rock avalanche collapses (De Blasio and Crosta, 2015) (Fig. 1).

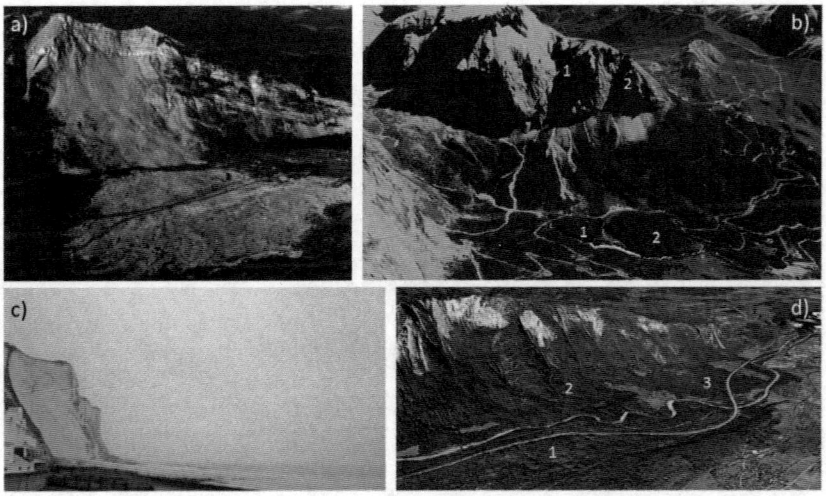

Figure 1 Some field examples of landslides collapsed onto flat area covered with a medium. A: Frank slide; b) Pordoi Pass, Dolomites; Dover, Chalk cliffs; d) Oberschutt, Austria.

We will show that this bi-linear geometry affects significantly the results of the experiments because the impact at high velocity of the granular flow with the basal horizontal layer/board influences the further dynamics more than previously appreciated, with effects that may find an equivalent in real rock avalanches.

Table 1 – Possible erosive layers onto which landslides may collapse and experimental analogs.

Type of erodible layer	***Geological setting***	***Examples***	***Experimental analog***	***This work***
Dry granular	Colluvium	Talus in dry climates *(Crosta et al., 2014)*	Sand	Yes
Wet cohesive	Wet alluvium, tidal flat	Arvel *(Jaboyedoff 2003; Crosta et al., 2008)*	Sand and wet clay	No
Water	Shallow water	Chalk collapses in northern Europe *(Hutchinson, 2002; Dupperet et al., 2006)*	Water	Yes
Ice	Glacier	Landslides on glaciers *(Post, 1967; McSaveney, 1976)* on permafrost, probably on Mars	Ice	No. The basal shear stress of experimental granular flow too low to produce ice melting

2. Experiments

2.1. Experimental setup

The experimental setup consists of a smooth board (1.2 m long) with possibility of setting a desired slope angle between 30° and 65° (Fig. 2). A box containing the granular material is fixed at the upper end of the plate. A mechanical gate controlled by a couple of springs is used to release the material.

The gate opening time, much shorter than the collapse time of the granular material, as checked by multiple testing with high speed video recording, ensures a negligible influence of the mechanism of release. The tilted plate is joined to a smooth horizontal sector where the flowing material come to rest. The plate may be covered by the erodible layer (Fig. 2B). For experiments with water, the plate is made impermeable and filled uniformly with a water layer. Two Plexiglas side boards allow a direct observation of the experiments from the sides, beyond the lateral confinement of the material. Two high speed cameras (60 frames/s and 600 frames/s) and a laser profilometer (profiling frequency:120 Hz) placed in front of the apparatus are used for monitoring the flow characteristics, and the flow evolution in space and time.

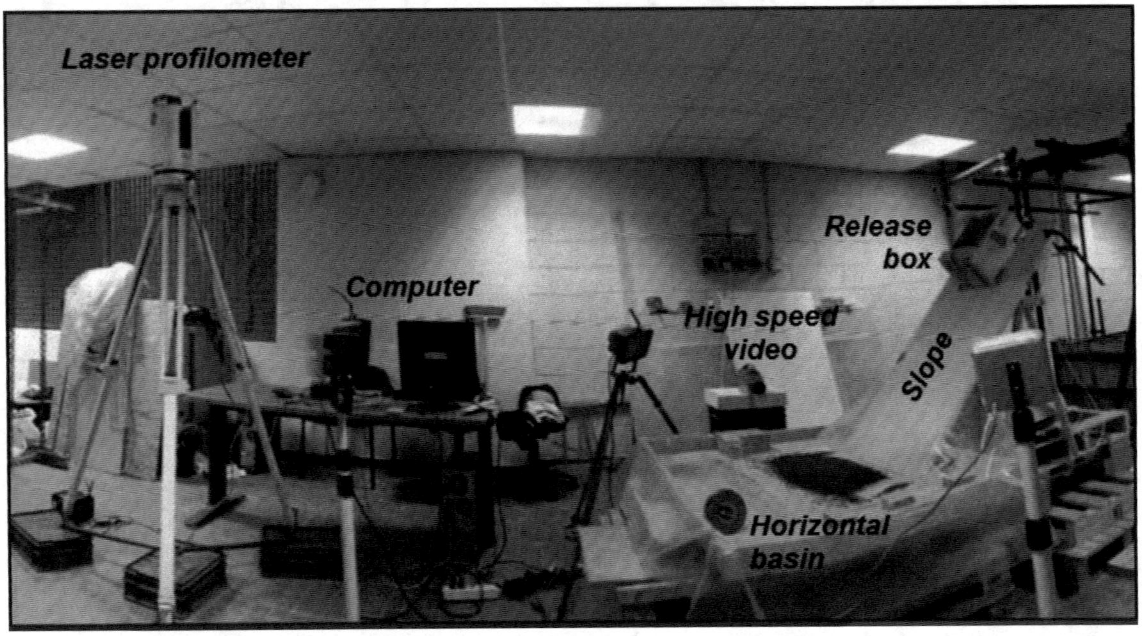

Figure 2. The experimental apparatus consists of a release box where the granular material is placed, a slope along which the flow takes place, and an horizontal basin which may or not be filled by erodible sand or water. The high speed video and laser profilometer are computer-controlled.

The erodible sand layer is positioned by hand pluviation and then leveled avoiding further compaction. The adopted materials consist of uniform Hostun silica sand (C_u = 1.57; D_{50} = 0.32 mm; mean unit weight = 1.42 g/cm^3) or rounded gravel for both the released mass and the erodible layer. In some experiments, we used subangular/subrounded gravel (C_u = 1.5; D_{50} = 2.5 mm; mean unit weight = 1.48 g/cm^3) for the granular flow. Released volume varies between 1.5 and 9.6 liters. The physical and mechanical parameters of interest to these experiments are reported in Table 2.

Table 2 - Physical and mechanical properties of the granular materials used in the experiments. The friction angle has been assessed by: i) depositing sand cones on different surfaces (smooth or sanded wooden board) and surveying their geometry and inclination by a high resolution laser scanner, ii) inclining a board with a homogeneous sand layer till avalanching was observed, and iii) direct shear testing.

Material	Angle of repose (°)		Angle of avalanching (°)		Internal friction angle (°) direct shear test		Mean Bulk Density (g/cm^3)	Use in the experiments
	smooth	*sandy*	*smooth*	*sandy*	*peak*	*residual*		
Hostun	31	33	25.4	31.7	37.7	31	1.42	Erodible layer

sand								±0.1	Granular flow
gravel	32.9	36	22.2	27	-	-	1.48 ±0.1		Granular flow

2.2. *Experimental results*

2.2.1. Bulk experimental deposit

Figure 3 shows the experimental deposit obtained with two different slope angles, θ, of 40° and 60°, without the erodible basal layer. Even though the runout increases with the slope angle, it should be remembered that the fall height increases, too. A better characterization of the mobility in terms of the ratio between fall height and runout will be considered in the next section. The bulk, triangle-shaped thick deposit at the slope break exhibits a series of transversal grooves and furrows. In case of a smooth surface (e.g., Plexiglas rather than wood) and high sloping angle, the tail of the deposit detaches completely from the sloping ground and the slope break, with a very elongated deposit profile. The presence of the basal erodible layer causes a shape change of the deposit, which becomes lunate with steeper lee-ward and up-ward sides. Moreover, the erodible layer decreases the runout compared to the case of smooth slope. A decrease in inclination of the sloping part of the path is associated to deposits progressively propagating uphill in a triangular shaped and stepped deposit.

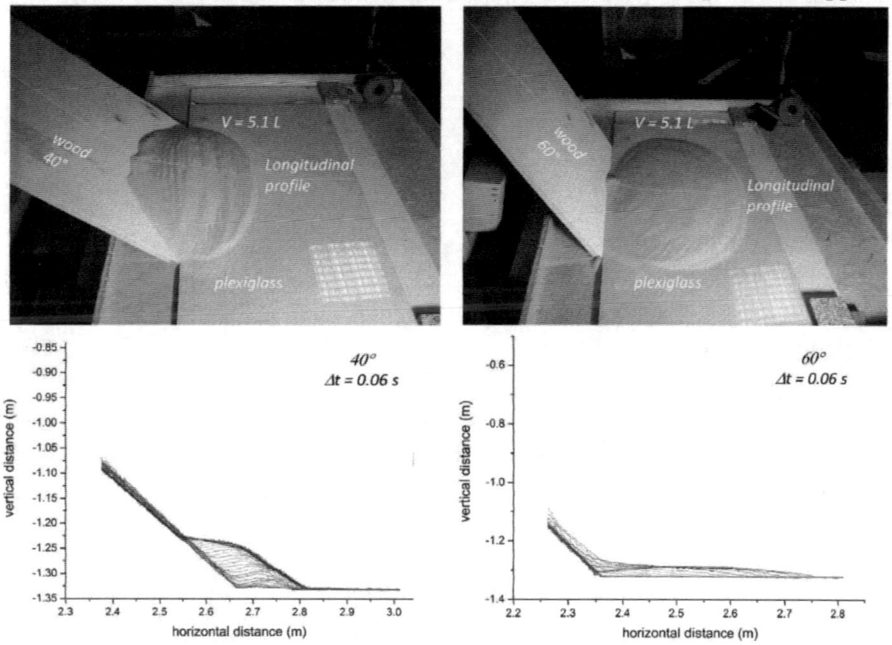

Figure 3. Aspect of the final deposit of some experiments and corresponding laser profiles at different times during the accumulation at two different slope angles: 40° (left) and 60° (right).

The inclination of the frontal part of the deposit decreases progressively (e.g. 22.4° to 14.8° in Fig. 4) with the increase in slope angle (45° to 60°, in Fig. 4), suggesting that a different mechanism reshapes the frontal part of the deposit in the last depositional phase. A much less inclined front is observed in the case of propagation on a smooth horizontal surface (e.g. 9.8°), the upstream facing side of the deposit maintaining approximately the same inclination through the various tests (e.g. 16.9° to 17.4° with the erodible layer and 19.2° on smooth base in Fig. 4). All these values are well below the angle of avalanching as determined for this material by laboratory tests (see table 2). Furthermore, the smoother superficial morphology for the deposits is very well shown by the slope angle along the deposit profile, where for the case of the 45° slope without basal layer the change in slope angle occur smoothly along the entire deposit length. Some interesting observation about the final morphology of the deposit can be done when coarse gravel is released instead of using Hostun sand. Similar to the case for the sand flows, the deposit grows upslope with decreasing slope angle, resulting in a rough conic geometry with the apex aligned with the deposit centerline. The deposit length increases slightly with the slope angle but not the total horizontal distance. While at low slope angles some of the gravel outruns the bulk deposit, an increase of slope produces a wedge of material pushed or upraised well in front of the coarse material (Fig. 4).

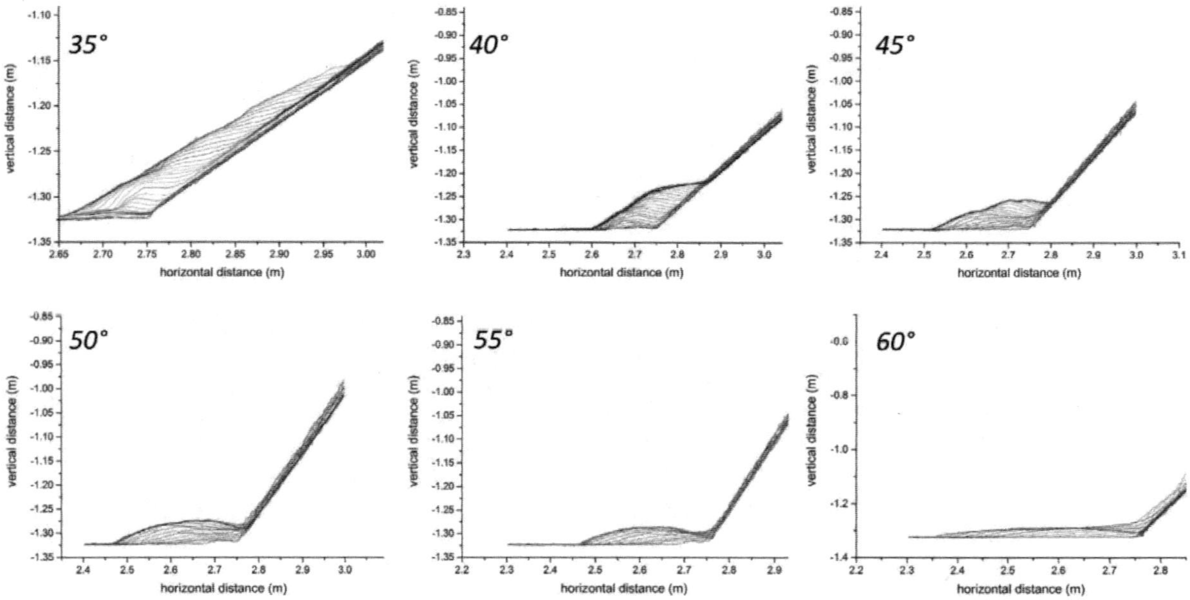

Figure 4. Change in the style of accumulation of the granular flow as the slope angle is progressively increased for experiments with Hostun sand.

Figure 5 shows the difference between an experiment without (a) and with (b) erodible layer. Note that the presence of the layer changes the style of deposition and at the same time decreases the runout.

2.2.2. Interpretation of the bulk experimental deposit

It is customary to introduce the Fahrboschung (FB) as a proxy for the mobility of a landslide or an experimental granular flow. The FB, which is equivalent to the apparent friction coefficient encountered by the granular mass during the flow (Scheidegger, 1973), is defined as the ratio H/L between the vertical drop (H) and horizontal spread (L) of the landslide, calculated adopting as initial point the rear end of the mass prior to failure, and as end point the maximum distance reached by the tip of the failed mass. Results for an experimental mass show that this ratio is close to the friction coefficient of the granular-bed interface (Erismann and Abele, 2001). The value of the FB ratio can be affected by an initial granular mass positioned with very high aspect ratio (Lajeunesse et al., 2004, 2005; Lube et al., 2005; Crosta et al., 2009; Lucas et al., 2014) because part of the available energy derives from the potential energy of the column. However, this effect is minimal in this study because the ratio between the vertical size of the containing box and the vertical fall of the center of mass is very small.

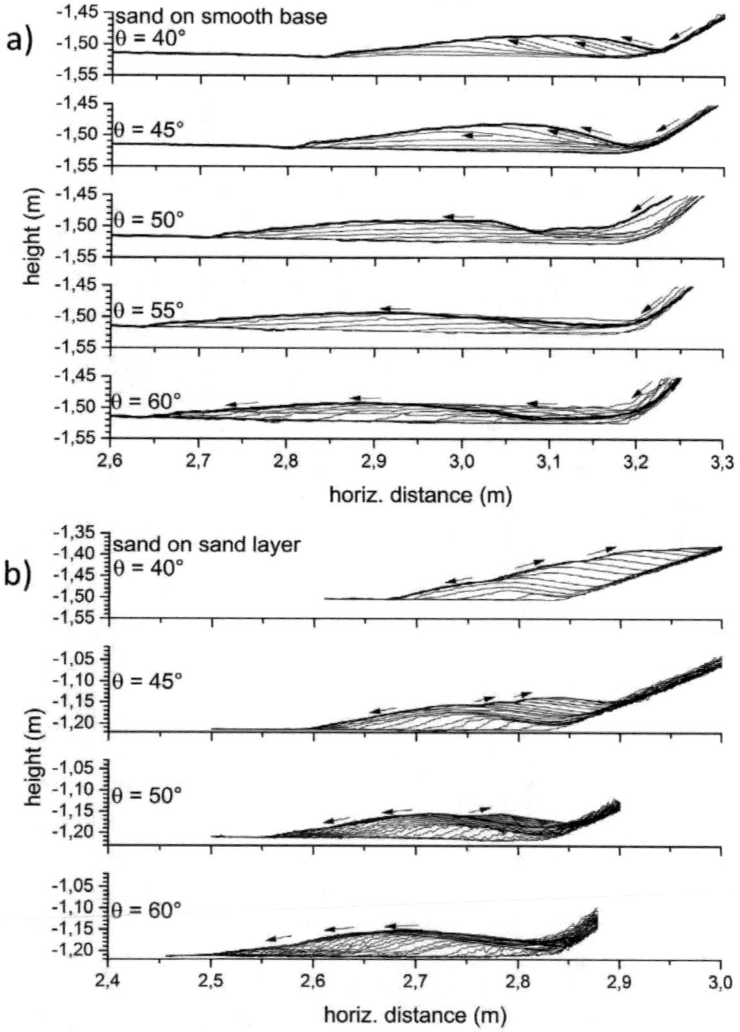

Figure 5. Difference in the style of deposition of a sand granular flow between: a) smooth base, and: b) a sand layer is present. Arrows represent the general trend of evolution of the flow/deposition.

Figure 6 shows the FB of the granular mass as a function of the different slope angles for our experiments with both different granular materials and different basal materials. Note that in all cases the FB increases with the slope angle, indicating lower mobility. We explain this effect noticing that at the slope break most of the vertical component of the momentum is transferred to the earth surface and is lost. Note also that this systematic data confirms that the presence of an erodible layer increases the ratio H/L.

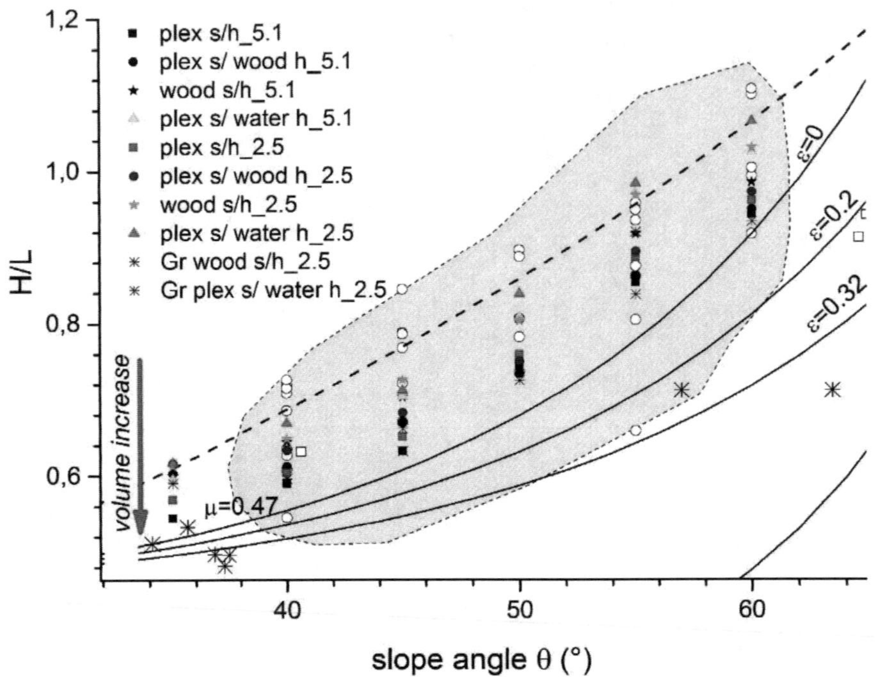

Figure 6. Ratio H/L measured for our experiments under different conditions: type of material (sand where not declared otherwise gravel, Gr); type of material (Plexiglas, plex, or wood) of slope (s) or horizontal portion (h); released volume (2.5 L, 5.1 L). Model results as proposed in eq. (4) are reported for a friction coefficient of 0.47 and different values of the restitution coefficient.

In the absence of an erodible layer, it is assumed that the horizontal component of the granular flow momentum is conserved at the slope break, but the vertical one is absorbed at the break of slope. We also assume that, due to the irregular bouncing on the horizontal surface, part of the vertical momentum is transformed into horizontal momentum. This component sums up as a fraction ε of the vertical component of the velocity, where ε is a coefficient of restitution. The final equation for the velocity of the front immediately after the break of slope is

$$U_x = \sqrt{2gL'(sin\theta - \mu cos\theta) +}\ (cos\theta + \varepsilon\ sin\theta) \qquad (1)$$

where L' is the board length. The run-out on the horizontal board can be calculated as

$$R_H = \frac{L' \sin^3\theta}{\mu}\left[\left(1 - \frac{\mu}{tan\theta}\right)\left(\frac{1}{tan\theta} + \varepsilon\right)^2\right] \qquad (2)$$

and the total length travelled (first on the inclined board, then on the horizontal area) becomes

$$R = L'\ cos\theta + \frac{U_x^2}{2g\mu} \qquad (3).$$

Considering that the vertical drop is $H = L'\ sin\theta$, the relationship for the ratio between the total drop height and the runout becomes

$$\frac{H}{L} = \frac{\tan\theta}{1+[cos\theta+\varepsilon\ sin\theta]^2\left(\frac{\tan\theta}{\mu}-1\right)} \quad (4)$$

Eq. (3) shows that the ratio H/L increases monotonously with the slope angle, implying that a landslide from a steep terrain loses mobility at the slope break compared to one travelling on a smoothly varying terrain.
Figure 6 shows also a comparison of our experimental data and the relation (4). Evidently, the simple model is capable of capturing some of the dynamics at the slope break. We come back to this plot in the section devoted to the comparison with field data.
We now consider the decrease of mobility and thus an increase of friction coefficient resulting from the experiments in the presence of an erodible layer. The acceleration of the center of mass of the granular mass along the ramp is:

$$\frac{du}{dt} = g(sin\theta - \mu cos\theta) \quad (5)$$

where $\mu = \tan\phi$ is the friction coefficient. The flow reaches the lower plateau with a velocity

$$u_0 = \sqrt{2gR(sin\theta - cos\theta\mu)} \quad (6)$$

where R is the inclined runout length. In the absence of the erodible substrate, the velocity as a function of the position x along the horizontal plate would be obtained by integration of

$$\frac{du}{dt} = -\mu g \quad (7)$$

which gives

$$u = \sqrt{u_0^2 cos\theta - 2gx\mu} \quad (8)$$

where $x=0$ corresponds to the position of the slope break. In the presence of the erodible layer, the granular mass entrains part of the layer down to a depth D', accelerating the eroded material to a finite velocity. For momentum conservation, this implies that the total velocity of the mass after entrainment u_{entr} must decrease. After the collapse upon the horizontal plane, we will have:

$$u_{entr}(M + M') = u_0 cos\theta M \quad (9)$$

$$M = SD\rho; \quad M' = SD'\rho \quad (10)$$

where S is the basal surface area of the flow, ρ the bulk density of the granular material, D and D' are the flow thickness before and after entrainment, respectively, and M and M' the respective masses. The velocity as a function of the position is as

$$u = \sqrt{u_{entr}^2 - 2g\mu x} \quad (11)$$

We find that the horizontal runout of the mass in the presence of the erodible layer L' is diminished with respect to the case of no layer L, which gives an apparent friction coefficient increasing like

$$\mu(D') \cong \mu(D' = D)[1 + \frac{u_0^2 cos\theta}{g\mu R}\frac{D'-D}{D}] \cong \mu(D' = D)[1 + 0.2\frac{D'-D}{D}] \quad (12)$$

It is found that 1 cm of erodible material may result in a 10-20% increase of the apparent friction coefficient. Thus, if the friction angle without erodible bed is about 30° (see Table 2), a 1 cm-thick erodible layer increases the apparent friction angle to about 34°, which is close to what observed for example for the curve in Fig. 6.

Note that the apparent friction coefficient appears to saturate as a function of the erodible layer thickness. This indicates a limit depth D_{MAX} of granular material that the material can entrain.

2.2.3. Erosion and thrust structures in the experimental deposit

To understand the geometry of the internal displacement, we have inserted some colored layers within the basal sand and studied how the whole sand deposit was deformed by the impact and deposition of the granular flow (Fig. 7a). Three layers of Hostun sand are prepared in this way (from bottom to top): uncolored, orange, and blue. Even when colored, the sand maintains the same mechanical characteristics. After the experiment performed releasing a granular (uncolored) Hostun sand at 66° slope, the resulting deposit has been wetted to provide the cohesion necessary for cutting sections through it.

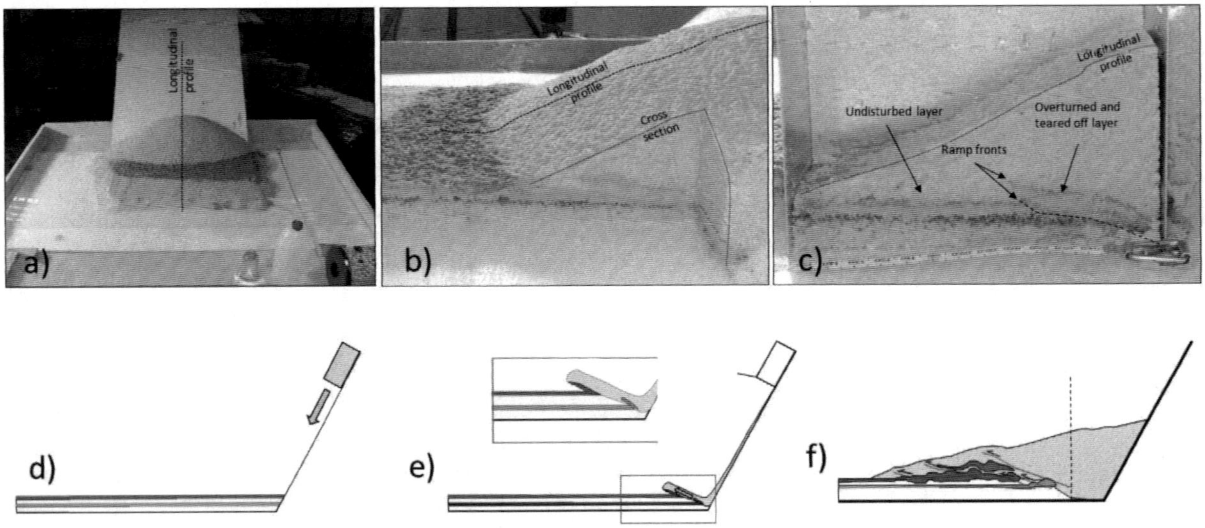

Figure 7. Top: experiment in which one light blue and one orange layer of sand are sandwiched between uncolored sand (initial configuration schematized in d). b) and d) cross sections cut through the deposit to show the internal structure of the deposit. e) and f) scheme of the impact phase and erosion with formation of a sand jet and final geometry of the deposit with internal structures.

The orange layer appears locally doubled in sections both parallel and perpendicular to the direction of flow. The blue layer appears folded and sheared, and locally sandwiched between the two and with a clear increase in thickness and an upper irregular limit (Fig. 7). Figures 7b) and 7c) evidence the lack of both the colored layers in the upstream side of the deposit suggesting the erosive action played by the granular flow descending the slope and hitting the basal layer. The blue layer seems eroded and re-deposited along the basal movement surface. On the other hand, the orange layer at the foot of the slope has been stripped and entrained inside the flow, remaining inside the upper cap of the travelling mass and partially inverting the stratigraphic sequence (orange–blue-orange in Fig. 7e). In some of the longitudinal cross cuts multiple shear planes, associated thrust fold-like features have been observed.

Experiments at low slope angles, in which the granular flow cannot dig deeply enough into the granular bed, do not produce layer doubling.

2.2.4. Numerical models

Such experiments show in a very direct manner the complexity of the final product of erosion, but are not capable to show how erosion occurs. Examining the colored experiments with a high-speed camera is not much useful, since the processes underneath the erosive layer are obviously hidden by the granular flow itself. For this reason, we ran 2D numerical simulations for the same situation using a finite element code (Roddeman, 2011, see Appendix A for a basic introduction). The results (Fig. 8) show clearly the generation of a steep front and a jump of the avalanche flow above the basal layer.

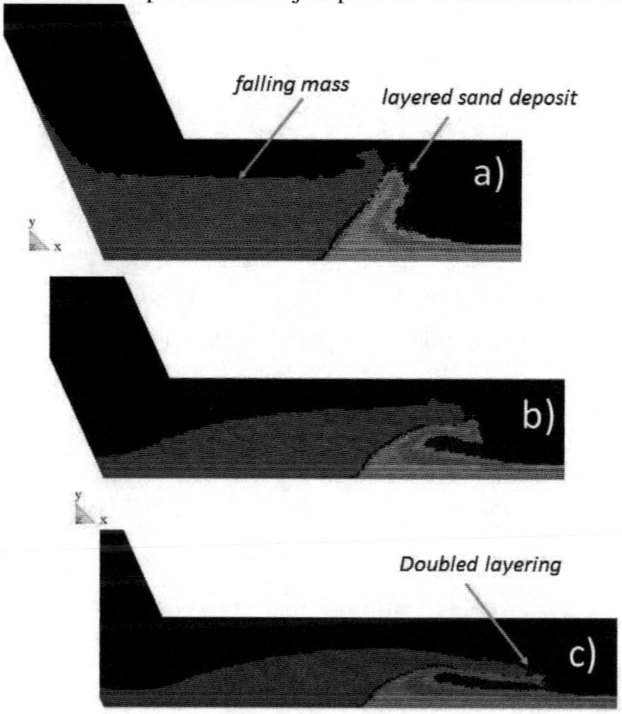

Figure 8. Numerical 2D simulation of the erosion process. Progressive erosion, folding, internal shearing and deposition are shown at three successive time steps.

The basal layer is pushed away at the front of the avalanche, along a steep contact surface, and starts to be eroded from the surface and folded at depth (Fig. 8a). The upper part of the front starts climbing up against the basal sandy material, along a steep and straight shear band, and to overpass the crest generated in the basal layer. In figure 8b) the decreasing velocity of the material climbing at the front and eroding the layer can be observed, as well as the high velocity of the material still rushing down the slope and pushing the front from the back. The maximum velocity at the impact is about 3.2 – 3.5 m s-1 computed by assuming a basal friction angle of 25° (3.57 m s-1) and 31° (3.44 m s-1).

Note how the simulations clearly show that the front of the avalanche material, slowed by the folding basal layer, is completely overturned and the blue layer is doubled (Fig. 8c). The velocity vectors suggest that the evolution of the flow overriding the basal material in a sort of front instability. The final two steps of the simulation are characterized by the elongation of the deposit and of the S-shaped contact band between the avalanche material (light blue) and the overturned basal layer.

2.2.5. Experiments with water

Experiments with water layer have been performed with flows of Hostun sand or rounded gravel falling onto a 0.5 or 1 cm thick water layer. The results in Fig. 9 with sand show the formation of a raised margin upon impact with water. Water shoveled at the front is thrust sideways and propagates in ripples away from the impact region, depriving the travelling sand of some of its energy. A sort of hydroplaning starts, whereby the granular flow floats at the top of the water layer, which prevents friction against the base. However, hydroplaning ceases after a run of few centimeters, after which sand plunges progressively to the bottom of the pool and comes to rest. The three phases can be seen also in the spatiotemporal diagram (Fig. 9d).

The dynamics of impact with water, even in a relatively simplified experimental setting like the present one, may be fairly complex. Hydroplaning in a material with relatively low permeability like fine sand is likely to be initiated by the impact force against the water surface. The pressure at the surface of water is of the order

$$P = \frac{1}{2}\rho_W U^2$$

where ρ_W is the density of water and U is the velocity of the granular flow against the medium. The impact force bouncing the granular medium upwards is proportional to P and to the area A of the surface in contact with water. Because the force of impact F = P A is greater if the impact is oblique rather than head-on (since the area of contact is larger), the granular flow colliding with water at a skewed angle may be thrust upwards by the impact and thus remains at the water surface for a certain time before collapsing to the bottom of the pool.

Figure 10a shows the final deposit obtained after removing the water from the pool. Note the double-ringed appearance where the deposit appears in two ropes with absence of deposition in between. The frontal ring may be the remnant of the frontal hydroplaning portion on the fast frontal wave. Some gravel particles detach from the rest of the granular flow, as is also evident in the filmed sequence. A comparison with real deposits is intriguing. Fig. 10b reproduces in greater scale the same deposit as in Fig. 10a.

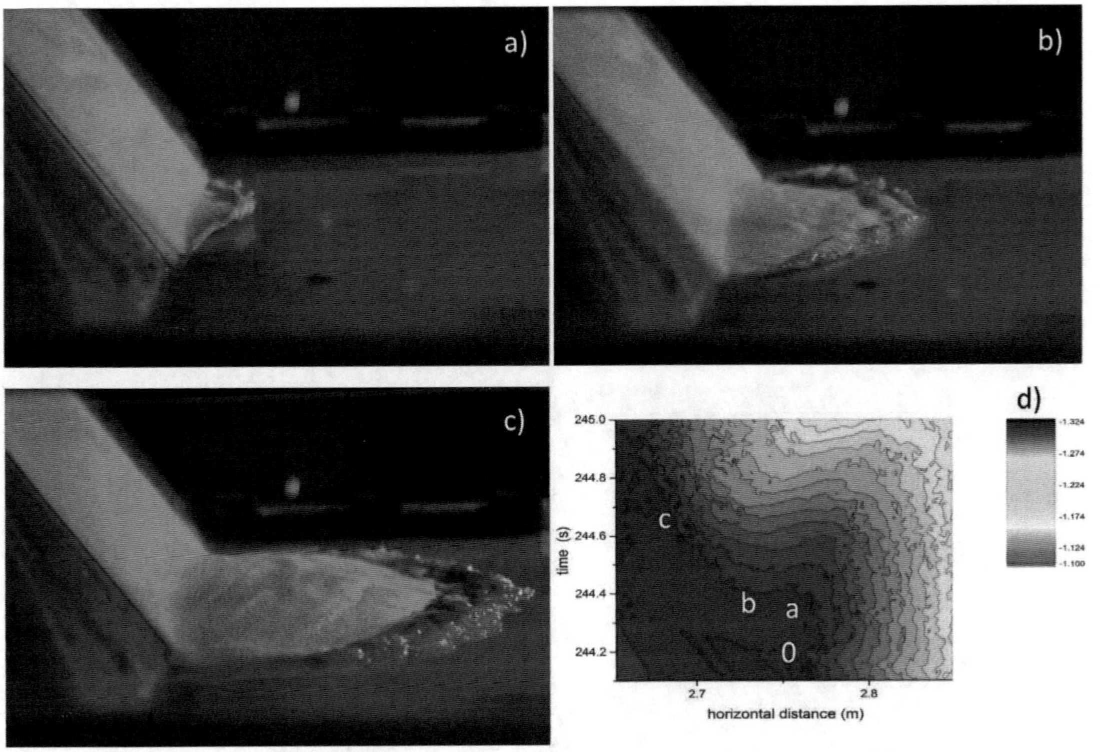

Figure 9. Experiment with a granular flow of Hostun sand onto a shallow water layer at three successive time steps (from a to c). a) post impact phase with frontal part raised above water; b) advancing front with frontal water wave and irregular bumpy front surface; c) well developed frontal wave partially washing back on the deposit and progressive grow of the deposit by successive small waves recognizable as shallow fronts. In this phase water appears through the dry sand at the front. d) Spatiotemporal diagram for the previous experiment showing the first front arrival (0) and the three successive phases a to c.

Note the similarity with the experimental deposit (Fig. 10a), showing that the formation of a double ring may occur also in actual cliff collapses onto a tidal flat. The head detachment in subaqueous landslides and debris flows is well known both from pool experiments and is commonplace in some clay-rich subaqueous debris flows (e.g., Ilstad et al., 2004). However, to our knowledge this is the first time that a similar head detachment is obtained for granular flows collapsing on thin water.

We also simulated the collapse of the sand flow onto a thin water layer using the same FEM-ALE 2D numerical model used for the simulation of sand onto sand experiments. The results in Fig. 11 show the formation of a water layer underneath the granular medium. Note that the water is endowed of pressure field in the simulations, so that the calculations can take into account the stagnation pressure, which is at the origin of hydroplaning (Mohrig et al., 1998).

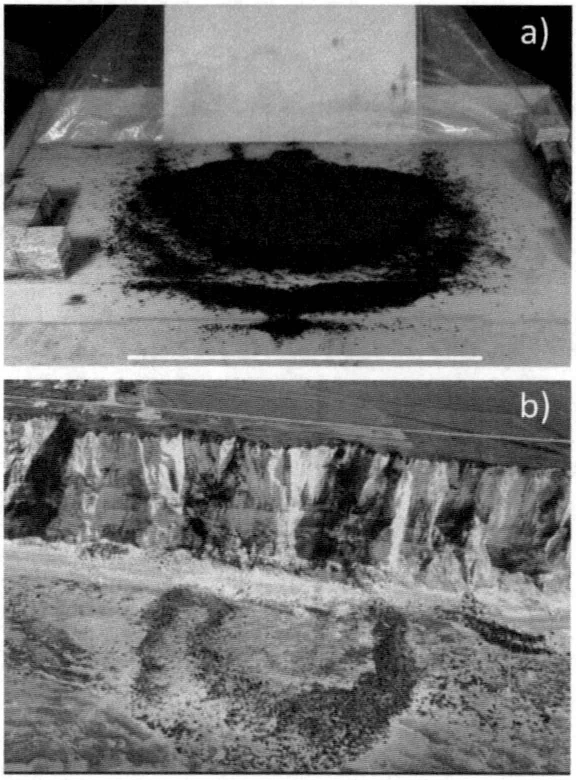

Figure 10. Comparison between the experimental deposit obtained with fine gravel and a layer of 1 cm of water (a) and a chalk flow which collapsed onto shallow water (b).

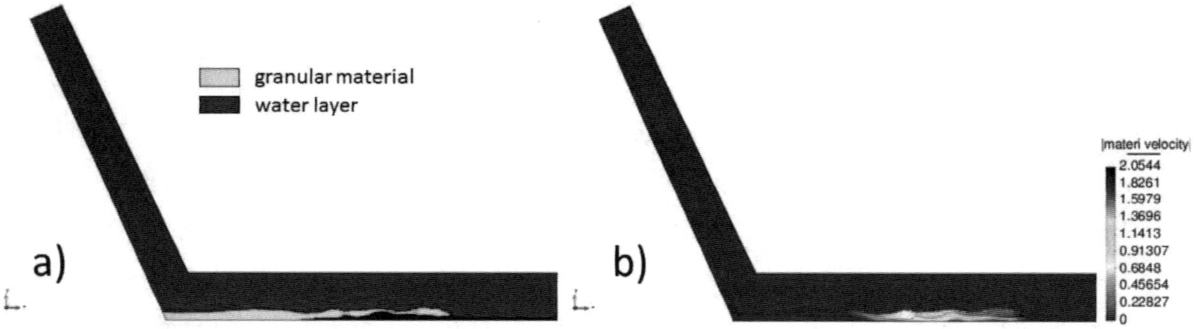

Figure 11. Numerical results for a 2D model for a granular mass hitting a shallow water layer at a late phase of spreading. a) materials at contact (green: granular mass; blue: shallow water layer); b) velocity field at the same time step as a). Hydroplaning of the granular mass is simulated as resulting by 2D conditions and an impermeable flowing material.

3. Comparison with field data

Let us now compare in more detail the results for the ratio H/L as from our experiments with real field data for rock avalanches. Fig. 12 shows the results of the experiments and rock avalanche data on the same plot, together with the results of the simple analytical equation H/L = tan $\alpha = 0.618\ \theta + 9.8$. Note that the real data fall in two regions: rock avalanches occupy the lowermost region of the graph. Some data about the nearly-vertical collapse of large masses in the Yosemite valley are presented in the literature by Wieczorek et al. (1999), and Stock and Uhrhammer (2010), and these plot in the same area of a selection of chalk cliff collapses (Hutchinson, 2002; Duperret et al., 2006). They exhibit a much higher value of H/L. Data of our experiments fall in the region between these two data sets, above the data for real large rock avalanches and to the left of data for vertical cliff collapses. Many rock avalanches of small to medium size show a behavior similar to that of the experiments (e.g., rock avalanches just to the left of the experimental data, Fig. 12). We suggest a continuity of the data regions and, despite the differences in scale, material, and environment, a similarity between these different granular flows. The obvious differences between all these sets of data are summarized here. Our experimental flows are dry, while chalk collapses are wet; rock avalanches are mostly dry, but the fall on the alluvial plain can certainly wets their basal layers. Fragmentation of the granular mass is absent in the experiments given the low impact energy, but might alter the propagation of a real rock avalanche (McSaveney and Davies, 2007; Bowman, 2013), especially in correspondence of the slope break (De Blasio and Crosta, 2015). Considering these differences, the fact that the data exhibits a continuity with experimental flows is remarkable.

Figure 12 also shows the prediction of the simple model introduced by Eq. 1. This model can explain the increase in the value of H/L as a function of the slope, even though different parameters should be adopted for the three different data sets. Real data can be fitted with relatively low friction coefficient and a coefficient of restitution of the order 0.2-0.3. With such values, both data sets for rock avalanches and chalk flows are approximately fitted by one single curve. Remarkably, this implies that the sharp increase in the ratio H/L for nearly vertical collapses is well predicted by this simple analysis without invoking any material property of the material (chalk or granite for the Yosemite collapse).

Experimental data, however, require higher values of the friction coefficient, like the one measured in the laboratory. Following a modified version of the $\alpha\beta$-model by Lied and Bakkehoi (1980) for snow avalanches, we seek a correlation between the angle of slope θ and the angle of Fahrboschung α = tan-1(H/L) in the form $\alpha = K\ \theta + B$, where K and B are fitting constants. This empirical formula, shown in Fig. 12 with a thick dashed line, appears to fit rather well our data. Here the fitting coefficients for the experiments are much different from those found to range between 0.77 and 0.94 for snow avalanches (Mc Clung, 2001). This is not surprising considering the complexity of snow avalanches in terms of wetting, cohesion, and air entrainment. However, the analysis shows that the origin for the increase of H/L with slope angle has a common origin.

Another interesting feature observed in the experiments is the presence of a depression or groove left behind the deposit at the base of the inclined slope.

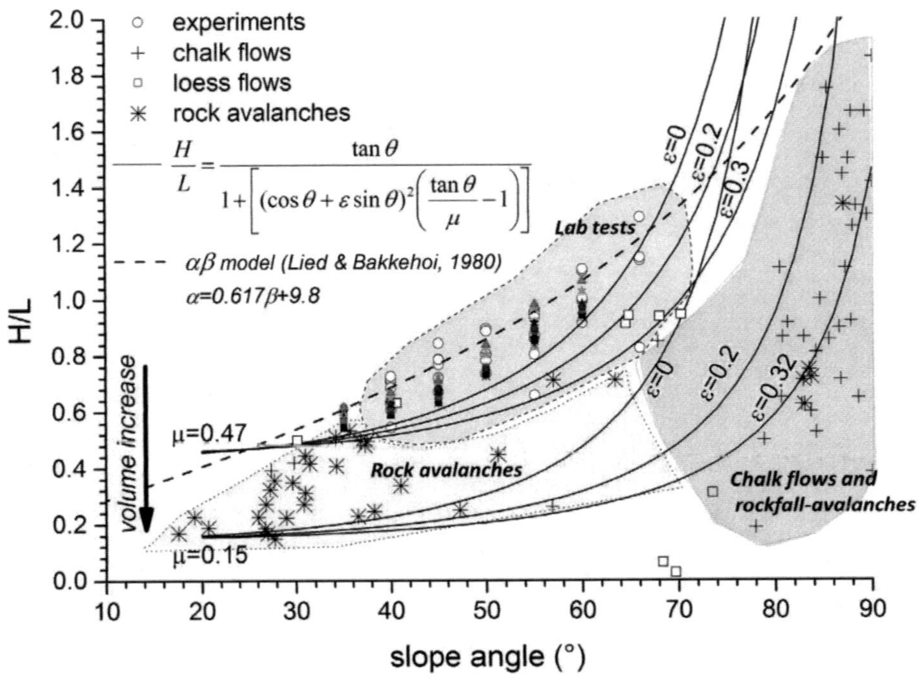

Figure 12. Ratio H/L measured for lab experiments presented in this contribution and for real rock and snow avalanches and chalk flows gathered in the same figure. Curves represent the results of the proposed model considering the transfer of vertical to horizontal momentum for two different coefficients of friction and 3 values of restitution coefficient (0, 0.2, 0.3)

This is sometimes associated to the detachment of the deposit from the inclined slope because of the long runout and is commonly observed in case of low basal friction on the horizontal sector of the path. In presence of a basal erodible layer, a depression in correspondence of the slope toe is clearly associated with the erosion of part of the basal layer by the highly energetic flow and the forward prograding deposition for steep slopes. In the other cases a backward propagating shock wave covers the slope toe. Similar structures have been described by other authors analyzing flows from steep slopes. For example, Hutchinson (2002) describes impact hollows formed by the debris of chalk falls just out from the cliff foot, where fall deposits continue in long tongue like flows, but he does not provide an explanation for the phenomenon. Similar structures have been described (Smith et al., 1994, Fitzharris and Owens, 1984 and Owen et al., 2006) at the toe of steep slopes affected by snow avalanches and Corner (1980) introduced three types of impact-associated depressions. He suggests that with increasing energy, impact tongues may acquire the shape of pits, and then of pools for the most energetic impacts. Corner (1980) also observes the strong erosion in the proximal slope toe sector, the striking similarity between the eroded volume and the frontal curved ridge deposit, and suggests the role of the abrupt change in slope as the main controlling factor for the emergence of such features.

4. Conclusions

This paper presents the results of experiments where a granular mass released along a slope reaches a horizontal plate. In some of the experiments, the plate is devoid of any granular medium. In this case, we find that the presence of a break of slope along the topography affects the run-out of the rock avalanche. The slope break is particularly important in this respect, as it will cause a loss of momentum perpendicular to the basal layer. Numerical and physical experiments show increased run-out when the slope break is smoothened, so confirming the importance of the slope break (Manzella and Labiouse, 2013; Manzella et al., 2013).

The novelty of the present work lays in the study of the interaction of the granular flow with an erodible layer resting on the horizontal plate, or with a thin water layer. In the first case, the formation of a granular wave controlling the erosion of the basal layer and the subsequent flow propagation, modes of accretion, and the final geometry of the deposit were observed in relation to the slope angle and the momentum of the flow. In general, the presence of an erodible dry layer hinders (if not in metastable conditions or close to instability) the further flow of the granular mass, but increases its size. Thus, the experiments provide indications that a granular flow falling onto a loose and dry erodible layer will generally decrease its speed and run-out in a real landslide, at least when an abrupt slope change exists. The physical reason is related to the fact that upon erosion the rock avalanche needs to transfer momentum to the erodible layer, which causes a speed decrease.

Numerical simulations by a FEM-ALE (arbitrary lagrangian-eulerian) code reproduce extremely well the dynamics observed both in time and space, including the erosion of the basal layer and geometry of the final deposit. This suggests that numerical modelling alone could be used to further investigate these events especially when beyond the reach of analogical experiments, for example at larger scale. Experiments with sand flows onto a thin water layer show that the sand may hydroplane, travelling for some distance detached from the base and that the front may detach from the body, all features shown also by numerical simulations and consistent with the field data.

We conjecture that when the basal layer is composed of a wet rather than dry medium, a lubrication effect may occur. Similarly, landslides falling onto glaciers are accompanied by a melting of ice and entrainment of water with a net increase in mobility (Schneider et al., 2011,Sosio et al., 2012; De Blasio, 2014). The gain of mobility due to lubrication may thus compensate or even outdo the effect of the momentum transfer. Similarly, a fall into a cohesive medium with non-Newtonian rheological properties such as alluvial muds may result in a strong lubrication depending on the yield strength of the bed material. It would be important to perform experiment with a wet layer with variable percentage of mud from pure water to dense mud.

Acknowledgements
The research was supported by PRIN Project: Time - Space prediction of high impact Landslides under changing precipitation regimes.

Appendix: Details on the numerical modeling
Following previous experience by the authors (Crosta et al., 2005; 2006, 2008, 2009, 2013a, 2013b, 2015) in simulating both small and large scale granular flows with different characteristics, the same FEM numerical approach (Crosta et al., 2009; Roddeman, 2013) has been applied to the problem presented in this paper. Because of the large displacements and deformations occurring within the material, a Lagrangian finite element method would be subjected to high mesh distorsion and inaccuracy. The adopted approach, presented in Roddeman (2013) and Crosta et al. (2009, 2015) is a combined Eulerian-Lagrangian method which does not distort the FE mesh and demonstrated to guarantee accurate calculation results. We assume an elasto-plastic behavior with a standard Mohr-Coulomb yield law but other constitutive laws are available. The adopted isoparametric finite elements are 3-noded triangles in two dimensional simulations and eight-noded hexahedrals in three dimensions.

The initial equilibrium stress state for the granular mass contained in the releasing box is reached through quasi-static time stepping so that no inertial effect is introduced. Then gravity is incrementally applied in successive time steps. The release is simulated by deleting a containing wall so initiating the mass movement. The dynamic equilibrium equations are then solved for the entire area of the spreading material and without controlling conditions both for the onset and the evolution of the material erosion and deposition.

3. References
[1] Chen, H., Crosta, G. B., Lee C.F. (2006) Erosional effects on runout of fast landslides,

debris flows and avalanches: a numerical investigation. Geotechnique, 56, 5, 305–322.

[2] Choffat, P. (1929), L'ecroulement d'Arvel (Villeneuve) de 1922, Bull. Soc. Vaudoise Sci. Nat., 57(1), 5 – 28.

[3] Corner, G. D., (1980) Avalanche impact landforms in Troms, north Norway. Geografiska Annaler, 62A: 1-10

[4] De Blasio, F.V. (2014) Friction and dynamics of rock avalanches travelling on glaciers. Geomorphology 213, 88-98.

[5] De Blasio, F.V., Crosta G.B., (2015) The effect of slope geometry on rock avalanche fragmentation. submitted

[6] Crosta G.B., S. Imposimato, Roddeman, D.G. (2006) Continuum numerical modelling of flow-like landslides. Landslides from massive rock slope failure, Evans, S.G., Scarascia Mugnozza, G., Strom, A., Hermanns, R., (eds) NATO Science Series, Earth and Environmental Sciences, 49, 211-232

[7] Crosta, G. B., Imposimato, S., Roddeman, D.G., (2009) Numerical modeling of 2-D granular step collapse on erodible and nonerodible surface. J. Geophys. Res., 114, F03020.

[8] Crosta, G. B., S. Imposimato, D. Roddeman, S. Chiesa, Moia, F. (2005), Small fast moving flow-like landslides in volcanic deposits: The 2001 Las Colinas Landslide (El Salvador), Eng. Geol. Amsterdam, 79(3– 4), 185–214, doi:10.1016/j.enggeo.2005.01.014.

[9] Crosta, G.B., Imposimato, S., Roddeman, D.G. (2008) Numerical modelling of entrainment/deposition in rock and debris-avalanches. Eng. Geology, 109, 1-2, 135-145.

[10] Crosta, G.B., Imposimato, S., Roddeman, D. (2013a) Interaction of Landslide Mass and Water Resulting in Impulse Waves. In Landslide Science and Practice, Vol. 5: Complex Environment. Margottini, C., Canuti, P., Sassa, K. (Eds), ISBN 978-3-642-31426-1,28 DOI 10.1007/978-3-642-31427-8, Springer, 49-56

[11] Crosta, G.B., Imposimato, S., Roddeman, D., Frattini, P. (2013b) On Controls of Flow-Like Landslide Evolution by an Erodible Layer. In Landslide Science and Practice, Vol. 3: Spatial Analysis and Modelling. Margottini, C., Canuti, P., Sassa, K. (Eds), ISBN 978-3-642-31426-1,28 DOI 10.1007/978-3-642-31427-8, Springer, 263-270

[12] Crosta, Hermanns, R.L:, Valbuzzi, E., Frattini, P., Valagussa, A. (2014) Large slope instabilities in Northern Chile and Southern Peru. Geophysical Research Abstracts,

EGU2014-16279

[13] Cruden, D.M., Hungr, O. 1986. The debris of Frank slide and theories of rockslide-avalanche mobility. Canadian Journal of Earth Sciences, 23, 425-432.

[14] Denlinger, R. P., Iverson R. M. (2001), Flow of variably fluidized granular masses across three-dimensional terrain: 2. Numerical predictions and experimental tests, J. Geophys. Res., 106(B1), 553 – 566, doi:10.1029/2000JB900330.

[15] Dufresne A., (2012) Granular flow experiments on the interaction with stationary runout path materials and comparison to rock avalanche events. Earth Surf. Processes Landforms, 37, 1527-1541

[16] Duperret, A., Genter, A., Martinez, A., Mortimore, R.N. (2006) Coastal chalk cliff instability in NW France: role of lithology, fracture pattern and rainfall. In Mortimore, R. N. and Duperret, A. (eds), Coastal Chalk Cliff Instability. Geological Society, London, Engineering Geology Special Publications, The Geological Society of London 20, 33-55.

[17] Fitzharris, B. B., Owens, I. F., (1984) Avalanche tarns. Journal of Glaciology, 30(106): 308-312.

[18] Forterre, Y., O. Pouliquen (2008), Flows of dense granular media, Annu. Rev. Fluid Mech., 40, 1– 24.

[19] Gray, J. M. N. T. Wieland, M., Hutter K. (1999) Gravity-driven free surface flow of granular avalanches over complex basal topography. Proc. R. Soc. Lond., A, 455, doi: 10.1098/rspa.1999.0383Günther, A., and Thiel, C. (2009). Combined rock slope stability and shallow landslide susceptibility assessment of the Jasmund cliff area (Rügen Island, Germany). Nat. Hazards Earth Syst. Sci., 9, 687–698, 2009

[20] Hakonardottir, K. M., Hogg, A. J., Johannesson, T., Kern, M., Tiefenbacher, F. (2003a). Large scale avalanche braking mound and catching dam experiments with snow. A study of the airborne jet. Surveys in Geophysics, 24, 543–554.

[21] Hakonardottir, K. M., Hogg, A. J., Johannesson, T., Tomasson, G.G. (2003b) A laboratory study of the retarding effects of braking mounds on snow avalanches. Journal of Glaciology, 49(165), 191–200.

[22] Hewitt, K. (2006) Rock avalanches with complex run out and emplacement, Karakoram Himalaya, Inner Asia. S.G. Evans et al. (eds.), Landslides from Massive Rock Slope Failure, Springer. Printed in the Netherlands 521–550.

[23] Hungr O, Evans SG. (2004) Entrainment of debris in rock avalanches; an analysis of a long run-out mechanism. Geol. Society of America Bulletin 116(9–10): 1240–1252.

[24] Hutchinson, J.N. (2002) Chalk flows from the coastal cliffs of northwest Europe. Geological Society of America, Reviews in Engineering Geology, 2002, 15, 257-302Ilstad, T. De Blasio, F.V., Elverhøi, A., Harbitz, CB, Engvik, L., Longva, O. and Marr, J. (2004). On the frontal dynamics and morphology of submarine debris flows. Marine Geology. 213, 481-497.

[25] Jaboyedoff, M. (2003), The rockslide of Arvel caused by human activity (Villeneuve, Switzerland): Summary, partial reinterpretation and comments of the work of Ph. Choffat (1929): L'ecroulement d'Arvel (Villeneuve) de 1922. Bull. SVSN 57, 5 – 28, Open File Rep. 3, Int. Indep. Cent. of Clim. Change Impact on Nat. Risk Anal. in Mt. Areas, Lausanne, Switzerland. (Available at http://www.quanterra.org/erosion hazard.htm.)

[26] Lacaze, L., J. C. Phillips, Kerswell, R. R. (2008), Planar collapse of a granular column: Experiments and discrete element simulations, Phys. Fluids, 20, 063302.1–063302.12, doi:10.1063/1.2929375.

[27] Lajeunesse, E., A. Mangeney-Castelnau, Vilotte, J.P. (2004), Spreading of a granular mass on a horizontal plane, Phys. Fluids, 16, 2371–2381, doi:10.1063/1.1736611.

[28] Lajeunesse, E., J. B. Monnier, Homsy, G.M. (2005), Granular slumping on a horizontal surface, Phys. Fluids, 17, 103302.1 – 103302.15, doi:10.1063/1.2087687.

[29] Legros, F. (2002) The mobility of long-runout landslides, Engineering Geology, 63, 301-331.

[30] Lied, K., and S. Bakkehoi (1980). Empirical calculations of snow-avalanche run-out distance based on topographic parameters. J. Glac. 26, 165-177.

[31] Liestol, 0., (1974) Avalanche plunge pool effect. Norsk Polarinstitutt Arbok 1972, 179-181.

[32] Lube, G., H. Huppert, S. Sparks, and A. Freundt (2005), Collapses of two dimensional granular columns. Phys. Rev. E, 72, 041301.1–041301.10, doi:10.1103/PhysRevE.72.041301.

[33] Lucchitta, B. K. (1979), Landslides in Valles Marineris, Mars, J. Geophys. Res., 84(B14), 8097–8113, doi:10.1029/JB084iB14p08097.

[34] Ma, A.,Liao, H., Ning, C., Feng, Z. (2014) Stability analysis of a high slope along a loess plateau based on field investigation and numerical analysis. K. Sassa et al., (eds), Landslide Science for a safer Geoenvironment, Springer, vol. 1, 451-458

[35] Mangeney, A., L. S. Tsimring, D. Volfson, I. S. Aranson, and B. Bouchut (2007), Avalanche mobility induced by the presence of an erodible bed and associated

entrainment, Geophys. Res. Lett., 34, L22401,

[36] Manzella, I., Einstein, H. H., Grasselli, G. (2013). DEM and FEM/DEM modelling of granular flows to investigate large debris avalanche propagation. In: C. Margottini et al. (Eds.). Landslide Science and Practice, Vol. 3. Springer Verlag, Berlin-Heidelberg, 247- 253.

[37] Manzella, I., Labiouse, V. (2013). Empirical and analytical analyses of laboratory granular flows to investigate rock avalanche propagation. Landslides (2013) 10:23–36, DOI 10.1007/s10346-011-0313-5

[38] Manzella, I., Labiouse, V. (2008a): Qualitative analysis of rock avalanches propagation by means of physical modelling of not constrained gravel flows. Rock Mechanics and Rock Engineering, 41 (1), 133–151

[39] McClung, D.M. (2001) Extreme avalanche runout: a comparison of empirical models. Can. Geotech. J. 38: 1254–1265, DOI: 10.1139/cgj-38-6-1254

[40] McConnell, R.G. and Brock, R.W., (1904). Report on great landslide at Frank, Alberta, Canada. Can. Dep. Inter., Annu. Rep., 1902—1903, Part 8, 17 pp.

[41] McDougall, S. (2006) A new continuum dynamic model for the analysis of extremely rapid landslide motion across complex 3D terrain. PhD thesis, Univ British Columbia, 253 pp.

[42] McSaveney MJ, Davies TRH, Hodgson KA (2000) A contrast in deposit style and process between large and small rock avalanches. In: Bromhead E, Dixon D, Ibsen M–L (eds) Landslides in research, theory and practice. Thomas Telford Publishing, London, pp 1053–1058

[43] McSaveney, M.J., Davies, T. (2007) Rockslides and their motion. In Sassa K., Fukuoka F., Wang F., Wang G. (eds), Progress in landslide science, Springer, Berlin, 113-133

[44] Mohrig, D., Ellis, C., Parker, G., Whipple, K.X., and Hondzo, M. (1998). Hydroplaning of submarine debris flows. GSA Bulletin 110, 387-394.

[45] Owen, G., Matthews, J. A., Shakesby, R. A., He, X. (2006), Snow-avalanche impact landforms, deposits and effects at Urdvatnet, southern Norway: implications for avalanche style and process. Geografiska Annaler: Series A, Physical Geography, 88: 295–307. doi: 10.1111/j.0435-3676.2006.00302.x

[46] Post, A. (1967) Effects of the March 1964 Alaska Earthquake on Glaciers. USGS Professional Paper 544–D, 54 p.

[47] Pudasaini, S., Kroner, C. (2008). Shock waves in rapid flows of dense granular

materials: Theoretical predictions and experimental results. Phys. Rev. E 78, 041308.

[48] Pudasaini, S.P., Hutter, K. (2006). Avalanche Dynamics: Dynamics of Rapid Flows of Dense Granular Avalanches., ISBN: 3-540-32686-3, Springer-Verlag, 602 pp.

[49] Roddeman, D.G. (2011), TOCHNOG user's manual. FEAT, 255 pp, www.feat.nl/manuals/user/user.html.

[50] Scheidegger, A.E. 1973. On the prediction of the reach and velocity of catastrophic landslides. Rock Mechanics 5, 231-236.

[51] Schneider, D., Huggel, C., Haeberli, W. and Kaitna, R. (2011). Unraveling driving factors for large rock–ice avalanche mobility. Earth Surf. Process. Landforms, 36: 1948–1966. doi: 10.1002/esp.2218

[52] Smith, D.J. , McCarthy, D.P., Luckman, B.H. (1994) Snow-Avalanche Impact Pools in the Canadian Rocky Mountains Arctic and Alpine Research, Vol. 26, No. 2, 1994, pp. 116-127

[53] Sosio, R., Crosta, G., & Hungr, O. (2008). Complete dynamic modeling calibration for the Thurwieser rock avalanche (Italian Central Alps). Engineering Geology, 100(1-2), 11-26.

[54] Sosio, R., Crosta, G.B., Hungr, O. (2011) Numerical modeling of debris avalanche propagation from collapse of volcanic edifices. Landslides, DOI: 10.1007/s10346-011-0302-8.

[55] Sosio, R., Crosta, G.B, Chen, J.H, Hungr, O. (2012) Modelling rock avalanche propagation onto glaciers. Quaternary Science Reviews (Impact Factor: 4.08). 07/2012; 47:23-40. DOI:10.1016/j.quascirev.2012.05.010

[56] Stock, G. M. and Uhrhammer, R. A. (2010), Catastrophic rock avalanche 3600 years BP from El Capitan, Yosemite Valley, California. Earth Surf. Process. Landforms, 35: 941–951. doi: 10.1002/esp.1982

[57] Strom, A. (2006) Morphology and internal structure of rockslides and rock avalanches: grounds and constraints for their modelling. S.G. Evans et al. (eds.), Landslides from Massive Rock Slope Failure, Springer, 305–326.

[58] von Poschinger, A., Kippel, Th (2009) Alluvial deposits liquefied by the Flims rock slide. Engineering Geology, Vol. 103, 1, 50-56.

[59] Wieczorek GF, Morrissey MM, Iovine G, Godt J. 1999. Rock-fall potential in the Yosemite Valley, California. US Geological Survey Open-file Report 99–578.Yugsi Molina, F. X., Hermanns, R. L., Crosta, G. B., Dehls, J., Sosio, R., Sepúlveda, S.A.

(2012) Geomorphological analysis, monitoring and modeling of large rock avalanches in northern Chile (Iquique area) for regional hazard assessment. Geophysical Research Abstracts, Vol. 14, EGU2012-8925, 2012

Research-informed design, management and maintenance of infrastructure slopes: development of a multi-scalar approach

S Glendinning, P R Helm, M Rouainia, R A Stirling[1]; J D Asquith, P N Hughes, D G Toll[2]; D Clarke, W Powrie, J Smethurst[3]; D Hughes, R Harley, R Karim[4]; N Dixon, C Crosby[5]; J Chambers, T Dijkstra, D Gunn[6]; K Briggs, D Muddle[7]

1 School of Civil Engineering and Geosciences, Newcastle University, UK
2 School of Engineering and Computing Sciences, Durham University, UK
3 Faculty of Engineering and Environment, University of Southampton, UK
4 School of Planning, Architecture and Civil Engineering, Queens University Belfast, UK
5 School of Civil and Building Engineering, Loughborough University, UK
6 British Geological Survey, Keyworth, UK
7 Department of Architecture and Civil Engineering, Bath University, UK

E-mail: stephanie.glendinning@ncl.ac.uk

Abstract. The UK's transport infrastructure is one of the most heavily used in the world. The performance of these networks is critically dependent on the performance of cutting and embankment slopes which make up £20B of the £60B asset value of major highway infrastructure alone. The rail network in particular is also one of the oldest in the world: many of these slopes are suffering high incidents of instability (increasing with time). This paper describes the development of a fundamental understanding of earthwork material and system behaviour, through the systematic integration of research across a range of spatial and temporal scales. Spatially these range from microscopic studies of soil fabric, through elemental materials behaviour to whole slope modelling and monitoring and scaling up to transport networks. Temporally, historical and current weather event sequences are being used to understand and model soil deterioration processes, and climate change scenarios to examine their potential effects on slope performance in futures up to and including the 2080s. The outputs of this research are being mapped onto the different spatial and temporal scales of infrastructure slope asset management to inform the design of new slopes through to changing the way in which investment is made into aging assets. The aim ultimately is to help create a more reliable, cost effective, safer and more resilient transport system.

1. Introduction and background to the research

The UK's transport infrastructure is one of the most heavily used in the world. The UK rail network takes 50% more daily traffic than the French network; the M25 between junctions 15 and 14 carries 165,000 vehicles per day; London Underground is Europe's largest at 402 km. The performance of these networks is critically dependent on the performance of cutting and embankment slopes which make up £20B of the estimated £60B asset value of major highway infrastructure alone. However, many of these assets are old; London Underground is the world's oldest underground system and much of the rail network was constructed more than a century ago. Many of these slopes are now suffering from excess deformation and high incidents of instability (which is showing evidence of increasing with time). These problems cause delays and subsequent economic loss to UK industry, and the general public, and pose a safety hazard. The risk of derailment from slope failures is the greatest risk faced by the railway infrastructure. Between August 2003 and December 2009, a total of 429 incidents of failure of Network Rail (NR) earthworks were recorded across the rail network, causing 66 days in delays. It is not only safety but economics that is important: emergency repairs cost 10 times that of planned works. Therefore the impact of reducing delays caused by slope failure will be highly significant in both direct economic and indirect social and economic terms.

At the same time, the UK is poised to start investing billions of pounds in new transport infrastructure, with the proposed high speed line between London and Birmingham (HS2) representing the most high profile of cases. The construction of earthworks for high-speed rail causes additional challenges as they will suffer increased dynamic loading from the higher loads imposed by the trains. Furthermore, inspection and maintenance, currently an integral part of the management of earthworks assets, will need to be undertaken with little or no human access and this will demand new remote techniques. Design of such structures will therefore need to ensure long-term stability whilst also minimizing land-take as the route that the line will take passes through highly valuable estate.

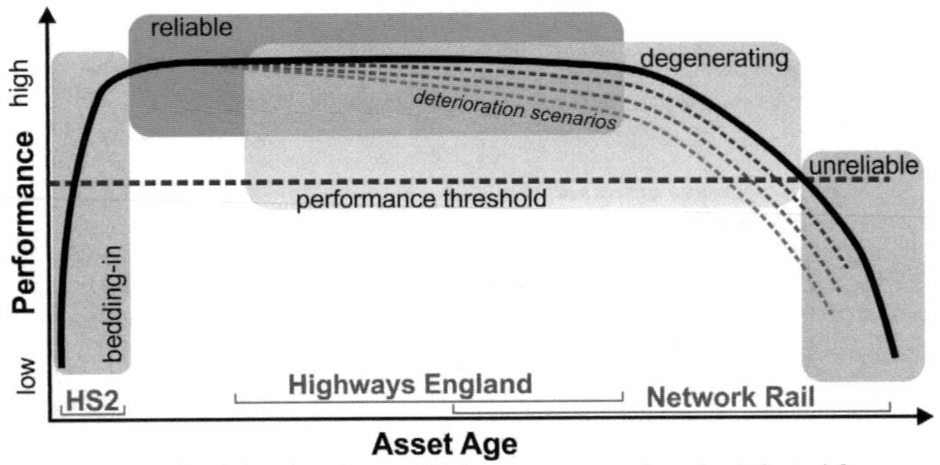

Figure 1. Generalised deterioration model for transport earthworks (adapted from Thurlby, 2013).

Figure 1 illustrates the performance of the asset in relation to its age and summarises the challenge for owners and operators of transport earthworks. This challenge is slightly different for each of the asset owners: Network Rail needs to be able to reliably identify which assets are most unreliable or deteriorating the fastest in order to target resources most effectively at repair; Highways England needs to allocate resources at repair but also to understand how smaller interventions can slow down deterioration and maintain reliability for longer; HS2 need to understand how to design for longer life and to recognise the signs of deterioration through remote monitoring/inspection. However, they all have in common the need to better understand the mechanisms and rate of deterioration, and whether this is likely to increase in the future. They also need to understand possible triggering events which precede catastrophic failure and the most effective mitigation and adaptation/remediation strategies and techniques. Additionally, they all need to make cases for investment, either to justify the design for a new build, as in the case of HS2, or for more pro-active maintenance and repair to improve resilience.

The research described in this paper was developed in collaboration with asset owners and their consultants to ensure that there is a synergistic relationship between the underpinning science of the research and the knowledge required to better design, manage and maintain a safe and resilient transport network. To this end our research has been structured around 3 physical scales, as illustrated in Figure 2.

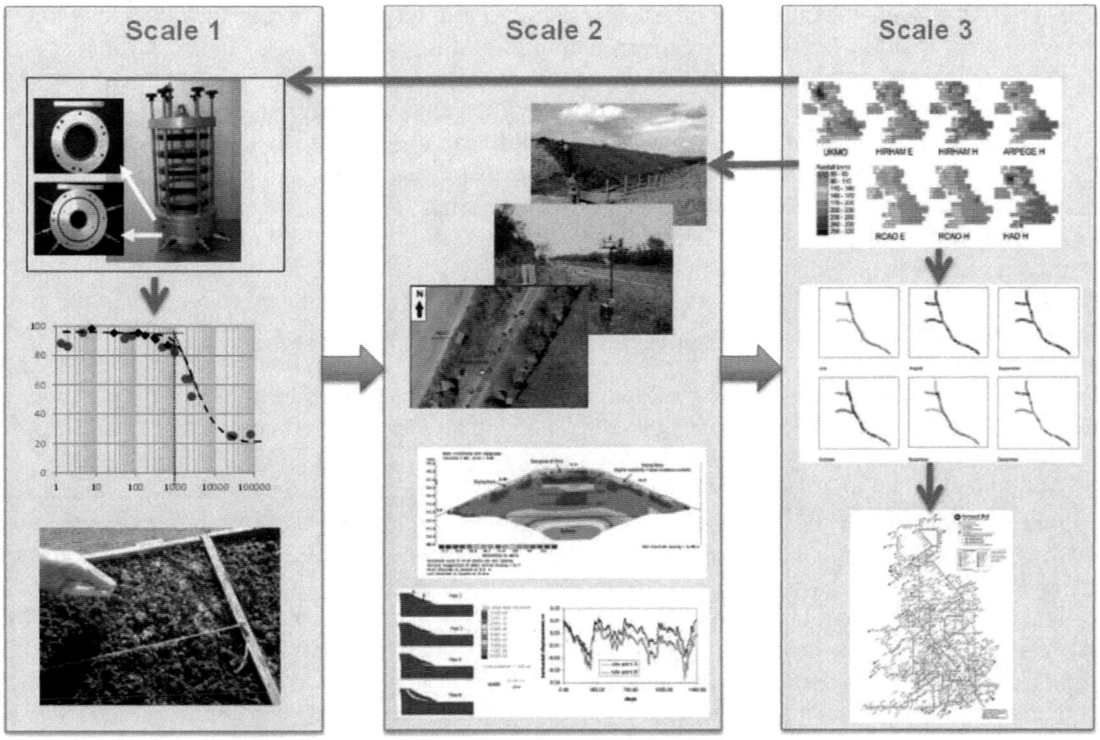

Figure 2. Research structure and integration across scales.

The aims are to understand materials (particularly clay fills) and the soil-water-vegetation system behaviour (scale 1), and how this affects the behaviour of single slopes (scale 2) that make up a transport network (scale 3). We also aim to understand these systems under current and future, generated weather event sequences and to be able to integrate the understanding across these 3 scales. These three scales map effectively onto the knowledge required by asset owners in that: Scale 1 enables advice to be provided on material properties that can be used in design; Scale 2 maps onto operational asset management at the level of a single asset and the design of monitoring and remediation measures; Scale 3 feeds firstly into Tactical Asset Management at the level of groups of assets to set regional performance targets, and then into Strategic Asset Management at the scale of a whole portfolio of assets, across a whole business, to enable setting of targets to achieve strategic goals and to inform investment priorities.

This paper sets out the progress made to date in the research on the iSMART (infrastructure Slopes: Sustainable Management And Resilience Assessment), which is 2 years in to a 3.5 year programme. The paper is structured around the scales illustrated in Figure 2 and will aim to comment on the integration between these scales and the messages we have been able to provide to the project stakeholders.

2. Weather Impacts

It is now widely accepted that climate change is occurring and that this will affect the processes and parameters that determine the stability of slopes. There remains, however, significant uncertainty in forecasting these changes in the medium and long-term. The UK has experienced a sequence of extreme weather events in recent years that have resulted in large increases in the number of both constructed and natural slope failures. Winter storms in December 2013 to February 2014 resulted in above average rainfall that brought widespread impacts and damage to the UK including landslides and flooding. This followed the record rainfall in April to July 2012 breaking previous records, and

November 2012 a sequence of heavy rainfall events resulted in one of the wettest weeks in England in the last 50 years (www.metoffice.gov.uk/climate/uk/interesting/november-2012). As a direct result, four times the number of landslides was recorded in Great Britain in the months of July, November and December 2012, above monthly averages for previous years (Pennington et al. 2014, 2015). On the UK rail network failure rates increased from the average of 50 per year to 150 slope failures in 2012 and 170 failures in 2013 (Brown, 2014). While these periods of extreme weather cannot be directly attributed to climate change, the resulting impacts demonstrate the importance of understanding slope failure processes linked to weather, and they provide examples of events that are forecast to occur more frequently in a changing climate.

Dijkstra and Dixon (2010) discuss the critical design cases for the stability of slopes. In the short term, slope systems can be regarded to be in a steady state with a relatively stable set of processes and parameters determining slope stability. Presently, most design and management of slopes is carried out on the basis of this assumption, sometimes enhanced by using a contingency if the site is particularly sensitive and long-term stability is questionable. These relatively crude assessments provide results that, through experience, can be translated into appropriate designs, and operation and management strategies. This may be acceptable for some design cases, however, for earthwork slopes that affect infrastructure operation, the design life of these assets is some 60 to 120 years and thus covers a period over which climate conditions are forecast to change significantly. Traditional slope models have a very limited capability to incorporate climate information (often only through applying higher water pressures or an additional factor of safety) and are therefore not suited to address long term assessments of slope stability variations in any detail. It is plausible that during the design life of a slope the magnitude of change in climate controls (mainly expressed in terms of changes in precipitation, temperature and sea level rise) will be such that the processes controlling stability will change significantly. Recurrence intervals between key critical events may shorten, recovery times of slopes may not be fully available and the system may enter a stage of severe instability (Dijkstra and Dixon 2010). Changes in antecedent pore pressures and alteration of trigger event magnitudes will lead to a change in the frequency, distribution and mode of earthwork slope failures, as a consequence of combined changes in the triggers (e.g. precipitation events) and conditioning or preparatory factors (e.g. the antecedent groundwater conditions).

The UK Climate Projections 2009 (UKCP09) are the current best available information on forecast climate change in the UK. The projections show that in the 2080s, and under a medium emissions scenario: all areas of the UK warm on average relative to the 1961-90 baseline; summers warm more than winters, particularly in southern England; mean daily maximum and minimum temperatures increase across the UK in both summer and winter; and average annual precipitation changes little across the UK, but winter precipitation increases in western regions while summer precipitation decreases in many, but not all, parts of the UK (Murphy et al. 2009). UKCP09 addresses the uncertainties of earlier projections by producing probabilistic outputs using a multi-model ensemble; running one model over and over with different parameterisations, and including runs from other models (e.g. Collins, 2007). It uses three pre-determined emissions scenarios (Low, Medium and High Emissions) and the resultant probabilistic projections are targeted 30-year "decades" from 2020s to 2080s (the 'baseline' is maintained as the period from 1961-1990). Projections are available at 25km resolution for the whole of UK, averaged over each month, seasonally and annually (Jenkins et al. 2009; Murphy et al. 2009).

To provide further detail, a weather generator has been produced that builds on the EARWIG weather generator developed at the Universities of Newcastle and East Anglia (e.g. Kilsby et al. 2007; Jones et al. 2009). Typical daily precipitation values for 1961-1990 are used as input and users are able to turn these into statistically equivalent daily values for some time in the future, using transformation functions in the form of rainfall characteristics and UKCP09 monthly climate change projections (Wilks and Wilby 1999). This is relevant to a resolution of 5km. In addition, it is also possible to analyse the output by applying user-defined daily weather thresholds. Access to UKCP09 information is through a web interface that allows users to create customised image products. These

IOP Conf. Series: Earth and Environmental Science **26** (2015) 012005 doi:10.1088/1755-1315/26/1/012005

probabilistic forecasts of future climate form a promising development in the quality of the climate data for input into slope stability models (Jenkins et al. 2009; Murphy et al. 2009; see also www.ukcip.org.uk for further up-to-date information). However, the challenge for slope stability modelling communities is to develop their capabilities further to reach a status where the full potential on offer from the climate modellers can be used effectively in the broad assessment (i.e. not sporadic and site specific) of long-term slope stability development (Dijkstra and Dixon 2010).

At present there is insufficient knowledge available to be able to model slope failure processes on a probabilistic basis. Therefore, following the approach developed by Dijkstra et al. (2014) a number of weather event sequences are extracted from the ensemble outputs of the weather generator using the high-emissions scenario outputs from UKCP09 centred on the 2050s. This ensemble output comprises 100 runs of a 30-year period (i.e. 3000 sets of annual weather event sequences), which provides a set of weather years that can be used to interrogate modelled slope performance. The weather event sequences are a primary input into the coupled hydrological- vegetation-soil behaviour modelling (described in Section 5). As a first step to develop and assess the modelling approach we have used the historical weather data for one of our field sites at Newbury, UK. The extended period of detailed monitoring has enabled the data to be categorised statistically in terms of return period and it is fortuitous that the data contain both extremely hot and dry, and extremely wet summer seasons. The approach is summarised in Section 3.

Now that the modelling approach has been validated against field data, a systematic parametric study will be undertaken. In this case, all combinations of annual precipitation and annual temperature for 3000 weather event sequences were plotted, determining for the resultant point cloud the shape of the convex hull, and taking a sample of 16 sequences from this convex hull. These sequences thus represent combinations of average annual precipitation (very wet to very dry) and temperature (very hot to very cold), providing suitable input series to evaluate model responses.

3. Field monitoring and experimentation

3.1 Field sites

Because of the extremely long expected life-times over which earthworks assets exist, a deep understanding of their behaviour can only be gained through long-term monitoring programmes. Long-term data sets from a number of exemplar embankment and cutting sites are being used to calibrate whole-slope numerical models (which are described in Section 5). Data are being collected from a number of sites set up by the project partners prior to the start of the iSMART project. The sites are illustrated in Figure 3 and summarised as follows:

- Newbury, Berkshire: an 8 m high highway cut slope in London Clay, built 1999, with rough grass and shrub cover (Smethurst et al, 2012).
- BIONICS, Newcastle: a purpose built embankment research facility, comprising a 6 m high, 90 m long Glacial Till embankment built in 2006 with extensive in-ground instrumentation and climate control over parts of the slope (Hughes et al, 2009; Glendinning et al, 2014).
- Loughbrickland, Northern Ireland: A 10 year old, 24 m high road cutting in Glacial Till overlying fractured, permeable shale, built and instrumented in 2003 to measure pore water pressure/suction and near surface water content changes (Carse et al, 2009).
- Laverton, Gloucestershire: a railway embankment about 6 m high constructed around 1900 via end tipping of local Charmouth Mudstone.
- Craigmore, Northern Ireland: a 22 m high, 150 year old steep railway cutting in Glacial Till overlying low permeability granite bedrock.
- Hawkwell, Southend: an 8 m high London Clay embankment, built c1880. Mature trees on the slope were felled in March 2007 to give the present rough grass and shrub cover.

Figure 3. Monitored Field sites.

The first four of these have been the focus of field experiments and modelling during the project so far. The sites have a wide geographical distribution within the UK, and are intended to reflect a range of slope and material types, are of differing age and construction quality, and have different vegetation covers. The intention is that they reflect the broad range of slopes typical of UK highway and rail systems.

The following measurements are being made at sites:

- Climate: sites have locally installed weather stations to measure rainfall, and the parameters required to calculate potential evapotranspiration (temperature, relative humidity, wind speed and solar radiation). Climate stations have also been installed to investigate the variations with slope aspect. Cut-off trenches have been installed to measure runoff and interflow through the topsoil, which allows a full water balance to be estimated based on rainfall, evapotranspiration and interflow/runoff.
- Pore-water pressures and suctions are being measured using flushable piezometers, and tensiometers closer to the surface. High suction tensiometers and other devices are being used to measure suctions over 80 kPa, which occur in the clay soils close to the ground surface in a dry summer.
- Soil water content is being measured using a neutron probe, point measurement devices that measure resistivity or permittivity of the soil, and electrical resistivity tomography (ERT) which uses surface arrays to obtain two or three-dimensional distributions of moisture in the ground.
- Lateral displacements are being measured using standard inclinometers and a ShapeAccel Array (SAA; a continuously datalogged in-place device), and vertical displacements using magnet extensometers.

Not all sites contain all of the instruments described above, however under the iSMART project a small number of additional devices have been installed to allow clearer comparisons between sites. Where possible measurements have been recorded at short time intervals using data loggers.

3.2 Utilisation of long-term data sets.

Many of the sites have now been monitored for several years, which has allowed changes in soil water content, pore water pressures/suctions, and displacements to be linked to particular sequences of weather received over the duration of the monitoring period. Smethurst et al (2012) use simple water balance models to show how soil water content and pore water pressures can be linked to the climate. They then use a long-term rainfall record of 42 years to show that over a six-year period of monitoring at the Newbury site it received both fairly dry and wet extremes (for example, 2007 had the wettest summer in the record, while 2003 was a 1 in 10 year dry summer). As clay slopes react to longer periods of climate, this can be used to define the range of soil water content and pore water pressures that result from these more extreme periods of weather. This process is illustrated in Figure 4, and includes profiles of pore water pressure/suction associated with particular return periods for climatic events, based on the data measured at Newbury.

This careful characterisation of the climate in conjunction with long-term monitoring of pore water pressures and displacements has ensured that:

- The worst case (or near worst case) wet winter pore water pressures are captured;
- Cycles of pore water pressures at depth, likely to drive both shallow and deeper-seated progressive failure mechanisms in slopes, are characterised;

The iSMART project is continuing work to monitor and characterise the behaviour of the exemplar sites and to build long-term data-sets for use by the research community. Additionally, the BIONICS site is being used to study the effects of imposed 'designed' rainfall events more closely. Furthermore, the ERT array installed within the BIONICS embankment, which allows us to visualise the water regime continuously through the embankment cross section, is being used to investigate the effects of surface features, such as cracks, on the sub-surface water regime.

While data collected above can be used to ensure that the numerical models correctly characterise the changes in pore water pressures and effectives stress, and incremental increases in strain (or displacement), none of the exemplar slopes is actively failing, and thus there is more uncertainty about the ability of models to correctly determine failure. Wider sets of data are therefore also being provided by the project stakeholder partners, through access to their databases, to provide evidence of the number of, mode, and likely trigger event for failures in clay slopes across the railway and highway networks. These will be used to benchmark the results of the numerical parametric study (Section 5).

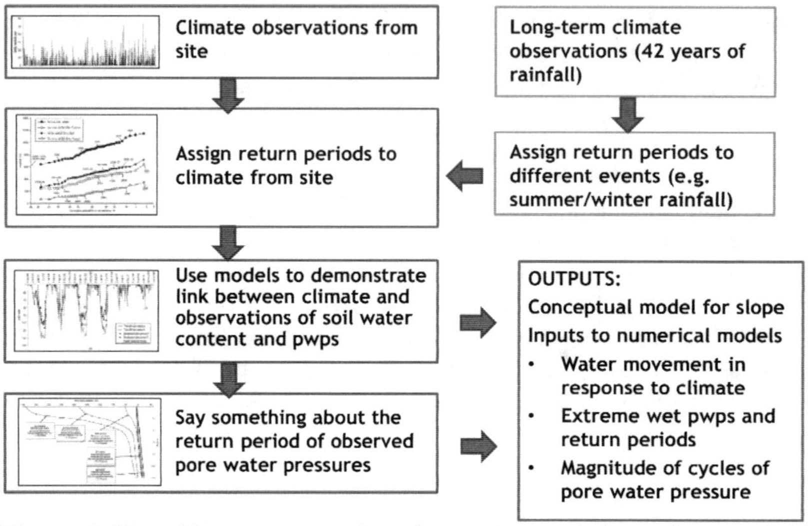

Figure 4. Use of long-term weather observations to inform slope modelling.

4. Materials testing

The central aim of the laboratory testing work is to both provide parameters required for the numerical modelling and to gain a deeper understanding of the effects of weather event sequences on the deterioration of soil properties. Of particular interest is to understand soil water retention behaviour for engineered slopes, soil volume change under suction cycles for our fill materials and changes to the mechanical properties of engineered fills due to the effects of desiccation, vegetation and climate. This advanced and detailed understanding of the mechanical properties of clay fills, including the deformation and deterioration mechanisms produced by cycles of pore water pressures informs numerical models as described in Section 5. Further materials testing is being conducted at field scale, particularly for the determination of permeability, where soil fabric and mass characteristics play an important role.

4.1. Laboratory testing

4.1.1 Soil water retention behaviour. Soil water retention behaviour is important to determine as it governs and explains many observations in behaviour of unsaturated soils. A Soil Water Retention Curve (SWRC) for a soil consists of water content, either gravimetric or volumetric, plotted against suction, as shown in Figure 5.

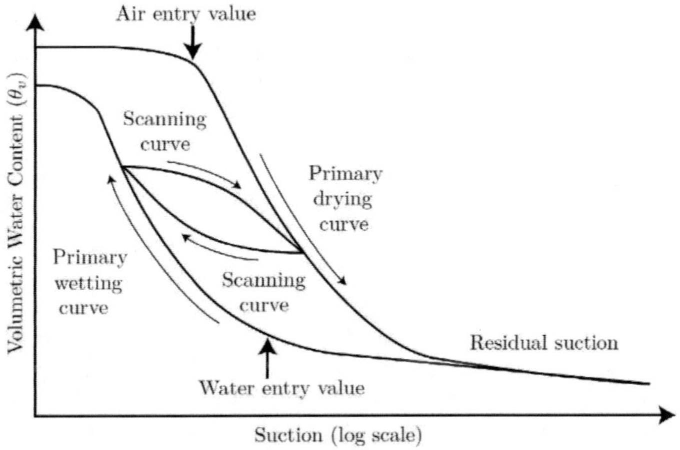

Figure 5. Hysteretic soil water retention behaviour (after Toll, 2012).

The curves consist of a primary wetting and drying path with transitions between these two states occurring via scanning curves, meaning a soil may be in nearly any state between the two primary curves. When a SWRC is determined for a soil, it can be used for estimating permeability functions more accurately and for interpreting shear strengths in unsaturated soils.

Several methods exist for determining a SWRC. Three methods have been employed in the laboratory on material from the iSMART project; these are the point pressure plate apparatus, high capacity tensiometers and the filter paper method. Laboratory results shown in Figure 6 show data recorded using the tensiometer system. As can be seen in the figure a change in behaviour can be observed as samples undergo cycles of wetting and drying, potential mechanisms for which have been investigated using optical techniques described in Section 4.1.3.

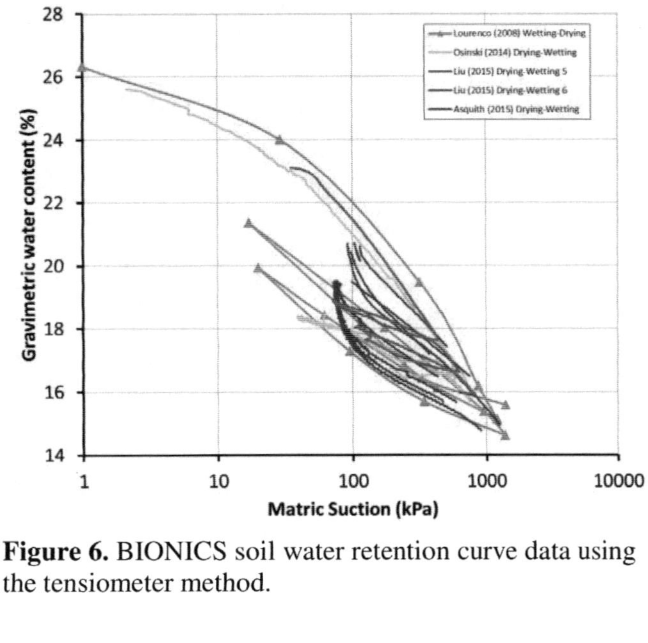

Figure 6. BIONICS soil water retention curve data using the tensiometer method.

*4.1.2 Mechanical behaviour***.** The mechanical behaviour of engineered fills will vary over life span of the earth structure that they form. These variations occur on a number of temporal scales, short term changes due to extreme wetting or drying events, seasonal changes (summer drying / winter wetting) and cumulative aging effects of exposure to many cycles of wetting and drying over decades. Mechanical testing conducted within iSMART aims to identify the mechanisms causing these changes and quantify their effects. This is being achieved through a programme of triaxial and direct tensile testing.

Triaxial Behaviour - influence of the degree of saturation on the strength of fill. Specimens were formed from material taken from the BIONICS site and compacted at differing gravimetric water contents of 15%, 20% and 22%, with nine being tested in drained saturated conditions and a further twenty seven in constant water unsaturated conditions, with suctions measured using high capacity tensiometers. The high water content samples were representative of the in-situ water content of the fill and the lower value was the optimum moisture content of the soil found from compaction testing. The unsaturated samples were tested both compacted at the original water content and also at the two other water contents after undergoing wetting or drying.

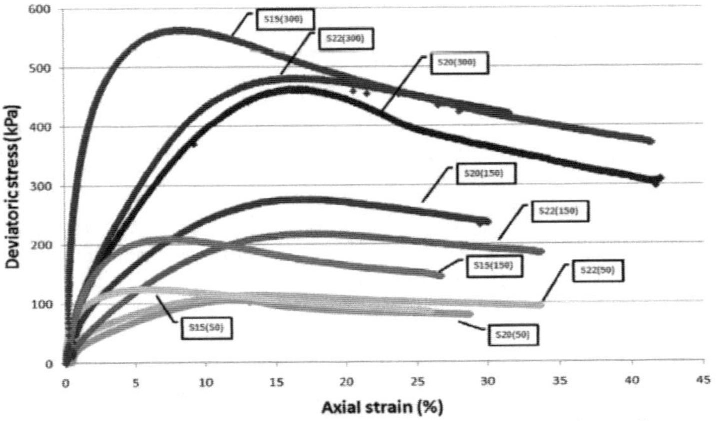

Figure 7. Stress strain plot for saturated triaxial testing of BIONICS material (after Mendes, 2011).

The results for the saturated tests are shown in Figure 7, and the unsaturated test results shown in Figure 8. The unsaturated tests showed that samples with a lower water content had increased strength due to suction, whilst increasing the water content reduced the peak strength. With all samples at 15%, clear strength peaks could be observed with a shearing plane forming in the failed samples whilst samples with a higher water content failed by bulging in a plastic ductile manner. In summary, it was observed that at lower water contents samples behaved with higher strengths, higher stiffnesses and this can be attributed directly to the higher suctions within the samples.

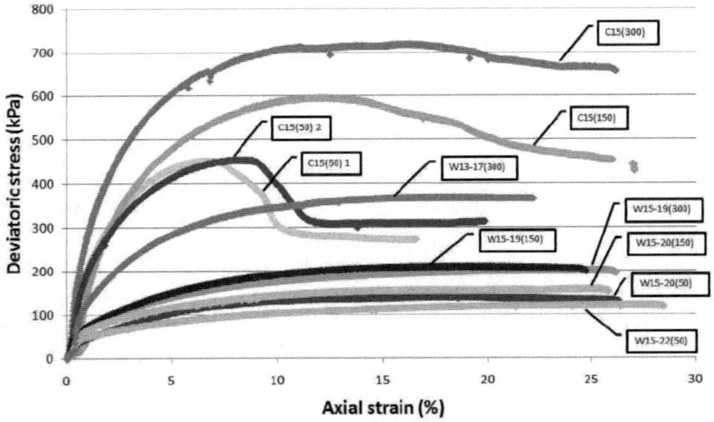

Figure 8. Stress strain plot for unsaturated testing of BIONICS material (after Mendes, 2011).

Drained stress path testing – strength reduction due to cycles of pore water pressure and viscoplastic straining (creep). A series of triaxial tests has shown that a reduction in soil strength occurs when pore water pressures are cycled at effective stresses below the peak failure stress. Cuttings in over-consolidated clay are known to be susceptible to progressive failure, with softening of the toe and the development of a rupture surface into the slope, potentially due to cycles of pore pressure cycles in response to weather events. Laboratory testing on reconstituted samples has been carried out to investigate the time-dependent behaviour of till. The research has shown that under elevated close to constant deviator stresses (70-90% peak strength), the till can accumulate a significant amount of shear strain due to pore water pressure cycling of ±5kPa (Harley et al., 2014). Furthermore, it has been shown that till may also reach its ultimate state simply by creeping when held under constant effective stresses conditions close to failure (Carse, 2014; Harley et al., 2014). Laboratory observations also indicate that the creep rate increases depending on how much a material has softened. Figure 9 shows data where pore water pressures were cycled (±5kPa) at 80% peak strength, and data where pore pressure was held constant at 80% peak strength. Note the acceleration of creep rate due to pore pressure cycling in comparison to a sample left to rest. The sample which was left to rest accumulated a significant amount of shear strain and caused a reduction of ~20kPa in peak strength of the material.

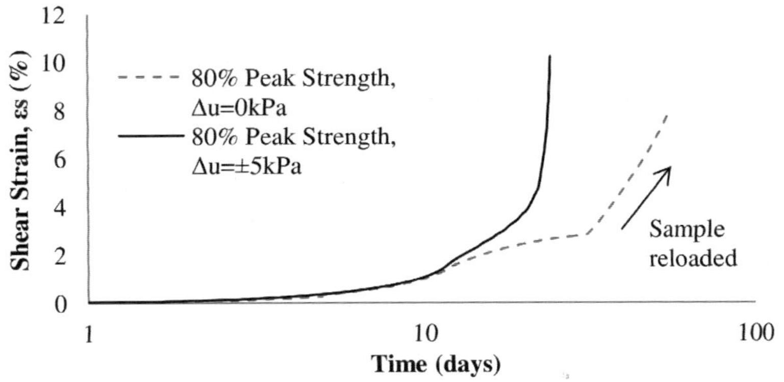

Figure 9. Development of shear strains under constant deviator loading.

Quick undrained triaxial behaviour – strength and stiffness reduction due to cycles of wetting and drying. A series of unconsolidated undrained triaxial tests have been performed on 38mm diameter, 76mm long samples of material from the BIONICS site compacted in the laboratory to densities representing field conditions (Glendinning et al, 2014; Hughes et al, 2009). A detailed description of the methodology may be found in Hen-Jones et al. (2014). The authors acknowledge that quick undrained triaxial tests will not capture the true unsaturated shear strength of the material as no suction measurements are made but the trend of strength changes does provide insight into the long term performance of the material.

Figure 10 shows the results of two cycles of wetting and drying on the compacted fill specimens. Shear strength can be seen to increase with reduction in water content and subsequently decrease with increase in water content. During the wetting phase, lower water contents are achieved than the drying phase for the same water contents and the lower strengths are exhibited for both wetting and drying phases in the second cycle when compared to the first implying a permanent change in material property.

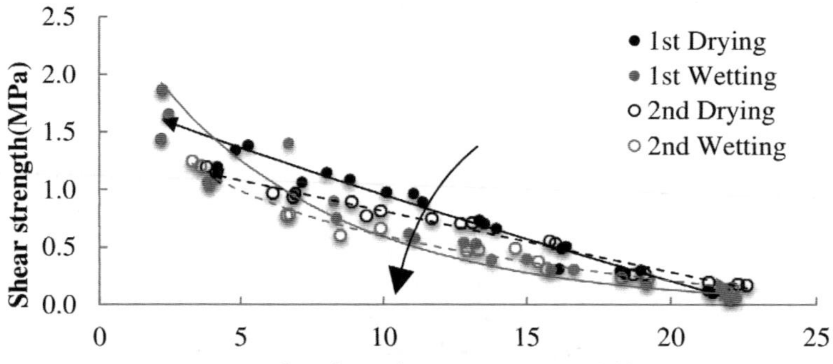

Figure 10. Relationship between undrained shear strength and gravimetric water content during cycles of wetting and drying.

Tensile behaviour. Cracking has an impact on soil mass permeability, strength and stiffness throughout the slope profile and hence slope failure susceptibility. Knowledge of the tensile strength in varyingly saturated conditions is essential to understand the initiation, propagation and ultimate extent of cracking. This has implications for the degradation of near surface material that governs the critical and changing interaction between engineered infrastructure slopes, vegetation and the atmosphere.

Recent investigations into the relationship between tensile strength and soil water content upon repeated drying and wetting cycles in material won from the BIONICS site have been conducted using a direct tensile strength test modification to standard direct shear apparatus. The mechanics and capabilities of this test method are further described by Stirling et al. (2015).

The tensile strength trend with hydraulic cycling is presented in Figure 11, after Stirling et al. (2014) and shows a similar trend of deterioration with increasing cycles of wetting and drying behaviour to the undrained shear strength data presented above.

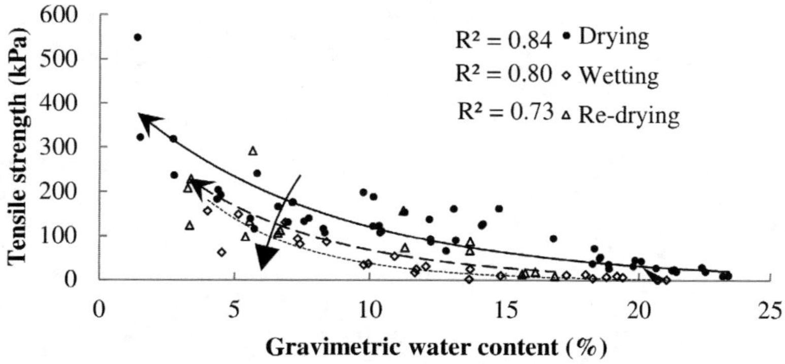

Figure 11. Tensile strength relationship upon initial drying, wetting and re-drying (after Stirling et al, 2014).

4.1.3 Fabric and Micro-structure. To investigate the potential causes of the deterioration in material properties observed in the triaxial and tensile tests, micro-structural changes were investigated on specimens post-testing using an Environmental SEM. This technique enabled the temperature and pressure of the immediate atmosphere around the specimen to be prescribed in order to control humidity and ultimately, influence drying rate. A detailed methodology will be described in Hen-Jones et al. (2015).

Figure 12 displays a clearly visible crack produced as a result of desiccation. Upon closer inspection of this feature (Figure 12b), particles lining the crack wall are found to have aligned during drying-induced shrinkage and have created a distinct coating to the crack surface. After the specimen was removed from the E-SEM and rehydrated by the application of a distilled water droplet and left to homogenise overnight, the specimen surface was again surveyed. Figure 12c exhibits a much more hydrated clay texture and is centred about a relic crack feature. This location is further magnified in Figure 12d and demonstrates both the partial closure and apparent infilling of a previously wider crack aperture. However, such a feature has remained identifiable and is likely to be a product of the permanent realignment of particles at the crack wall and infilling of loose clay particles. It is therefore anticipated that this discontinuity be exploited upon subsequent exposure to drying. The potential of these micro-cracking features to be the cause of the deterioration of clay fills is being further investigated in the laboratory, in conjunction with an investigation of their implications for the long-term mass behaviour of a clay slope through field observations.

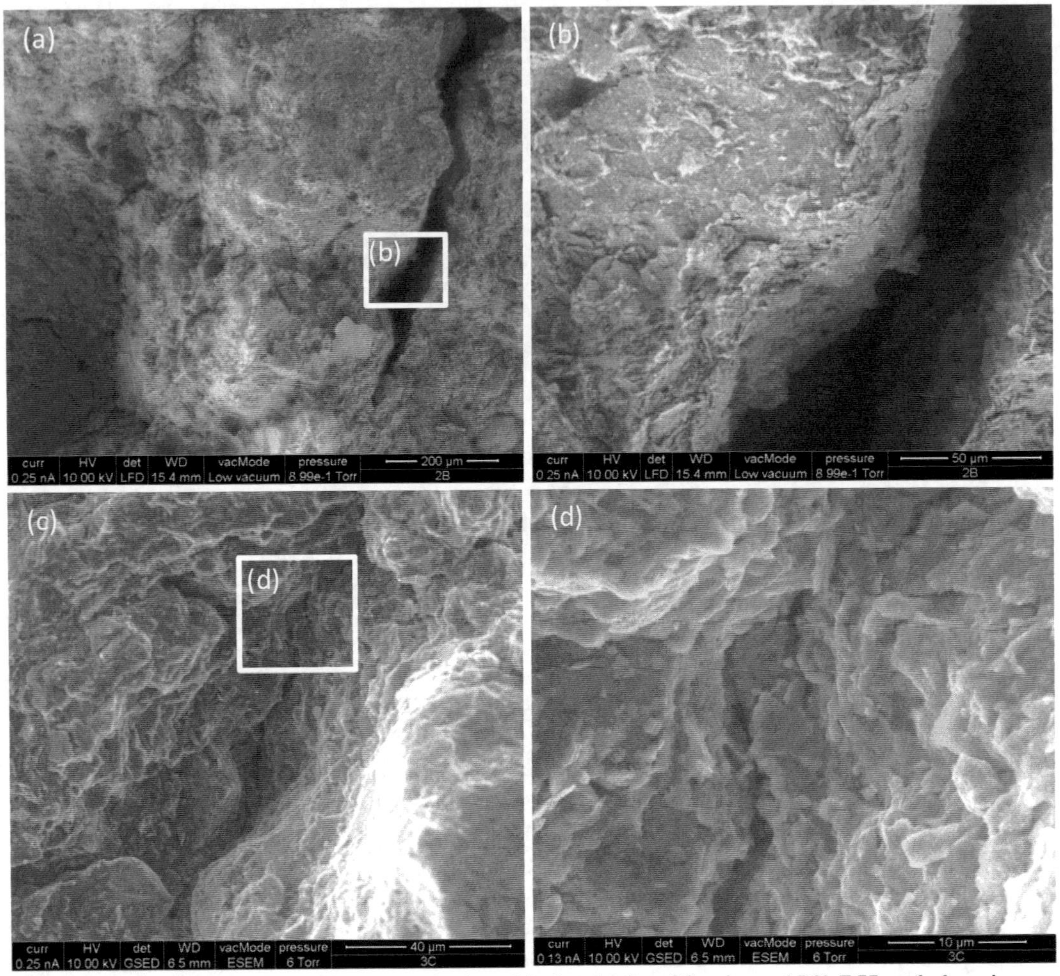

Figure 12. E-SEM images taken at two sites (a) site 1 after 90 min at 10% RH and showing the position of (b) desiccated clay particles and fracture wall under higher magnification, (c) site of reduced crack aperture following re-wetting and the position of (d) hydrated particles under higher magnification (after Hen-Jones et al. 2015).

4.2 Field testing

In addition to the laboratory testing, a series of field experiments is being used to investigate the relationship between laboratory and field-scale material behaviour. In particular, permeability is being tested in the field within the near-surface zone which is likely to have a significant influence on changes in water pressures deeper within the slope. The extent of summer clay cracking is being recorded through observations from carefully excavated trial pits, and at the BIONICS site, changes in the width of large cracks are being continuously measured using strain gauges. Surface permeability is being measured using Guelph, double ring and purposely designed infiltrometers. Initial measurements suggest that bulk permeability increases by around three orders of magnitude close to the ground surface during the summer (e.g. Glendinning et al, 2014). This increase in near-surface permeability significantly influences the ability for infiltration into the slope, and thus potentially high near- surface pore water pressures during wetter weather in autumn.

Field data is also being used to determine the SWRC in the field and to investigate the potential for change over time due to deterioration effects. Further field investigations are also being used to determine the depth of influence of wetting and drying cycles and hence the depth and rate of deterioration in the near surface zone of slopes. This understanding will then be incorporated in to numerical models, which will in turn be able to determine the long-term deterioration in stability.

5. Numerical Modelling

The numerical modelling undertaken as part of iSMART has 3 broad aims:

- To extend and validate whole slope numerical models capable of capturing critical climate-geology combinations;
- To assess the long-term behaviour of slopes and to develop an understanding of climate triggers to slope failure;
- To develop adaptation strategies and help design monitoring regimes.

In order to achieve these aims it is necessary to understand how weather sequences control the spatial and temporal pore water pressure variations in a slope and in turn understand the effect these cycles have on ultimate and serviceability limit states.

The first stage of this research has been to better understand how to model the weather-vegetation-soil-slope hydrological system. This was undertaken using a number of software codes at three differing universities. These include FLAC and SHETRAN, VADOSE/W and SEEP/W at the universities of Newcastle, Bath and Queen's Belfast respectively. Models were selected based on past experience of both researchers and stakeholders. The aim was not to develop new codes or one 'super' model, but rather to use established methods with known capabilities, in combination, to investigate the many different and varied aspects of the problem.

This exercise required a number of numerical modelling validation exercises to be run. The first of these was a simple one-dimensional drainage exercise using the problem geometry, material properties, boundary and initial conditions along with the results were taken from Forsyth et al. (1995). For reasons of brevity, this will not be reported here; the lessons learned will be included in the discussion in Section 5.2.

5.1 Modelling Newbury cutting

A hydrological modelling exercise was firstly performed as an extension of the one-dimensional caisson model, adding surface water fluxes due to weather and surface vegetation along with a two-dimensional slope geometry. The soil hydrological parameters were based on those published within Smethurst et al (2006). The meteorological parameters were derived from an on-site weather station and gaps were in-filled from nearby Met office weather station data. The modelled data produced was compared to site monitored pore water pressure data.

Coupled hydrological-mechanical modelling was undertaken using SHETRAN and FLAC Two-Phase flow only, utilising the methodology developed by Rouainia et al. (2009) in which simulated present and future climate weather time series were applied in a hydrological model of the Newbury cut-slope (using SHETRAN) to derive pore pressure variations over time. These pore pressures were in turn applied to a mechanical model in FLAC to observe the resultant deformation behaviour. A parametric study was then undertaken to investigate the influence of the soil's saturated permeability and pre-failure stiffness on slope stability.

In the modelling work undertaken to date as part of iSMART the majority of the meteorological parameters are derived from on-site weather stations or where this is not possible (or data is missing) taken from nearby Met Office weather stations. Other data (for the coupled hydrological-mechanical modelling) has been derived using a weather generator (Kilsby et al., 2007) based around the UKCIP02 data (Hulme et al., 2002). Further work is to be undertaken using future climate data derived using the updated weather generator based around the UKCP09 scenarios (Kilsby et al., 2009), as described in Section 2.

5.2 Lessons learned so far.

The caisson provided a number of useful lessons regarding the process of undertaking the modelling. In particular, modelling mesh discretisation affects the thickness of a modelled wetting zone during infiltration of water into a soil mass, which in turn has implications for the gradient of effective stress change and hence shear strength where infiltration and gravity driven flow occur within a model. This will affect the shear strength profile within a slope model with a narrower front producing a more sudden change. The initial profile of pore water pressures and degree of saturation used within a model has the capability to affect the rate of progression of a wetting front into the soil. This is a function of both the relative unsaturated permeability and additional volume of water required to saturate pore spaces. Gravitational drainage alone is not enough to generate significant suctions. To generate suctions greater than the soil's field capacity, water losses due to root water uptake and evapo-transpiration are required, which of course are an integral element of the slope modelling.

The hydrological modelling of the Newbury cutting demonstrated that it was possible to capture the broad trends with regard to the magnitude of pore pressure changes (i.e. the maximum and minimum values) and their timing in response to weather events in all three models. However the monitored data appeared to respond more rapidly to discrete events than could be captured by the modelling.

A significant outcome of the exercise was an understanding of the influence of the vegetation parameters used on the modelled pore pater pressure. The choice of vegetation parameters (where explicitly accounted for in the models) can have very significant implications for the derived rates of evapo-transpiration and the resultant surface suctions.

The leaf area index (LAI) in the VADOSE/W model (ratio of leaf area to underlying ground surface area) which in turn controls the proportion of solar energy that acts towards transpiration from the vegetation vs evaporation from the soil surface had a significant effect on the generated suctions.

An initial constant LAI value was used which allocated too much solar radiation to transpiration and in turn caused excess root water uptake and larger suction generations at greater depth than were observed in the field. A more realistic seasonally varying LAI gave a better fit to the monitored data than the original constant value by reducing energy allocated to transpiration and root water uptake for periods where vegetation canopy cover reduces. As such care must be taken when specifying vegetation parameters to account for their seasonally varying nature. Care must be taken when employing vegetation models to ensure that calibration is performed as relaying on suggested or default parameters in modelling is likely to yield poor comparison with field monitored data.

In order to derive a close match between the field pore water pressure data and those modelled using SHETRAN an increase in the permeability of the soil mass forming the cut-slope to a relatively high value of $1e^{-7}$ m/s was required (compared to a measured value of $1e^{-11}$ m/s). It was also found that the modelled evapo-transpiration from the slope was yielding significantly higher suctions than those recorded on site. This latter point was found to be due to the wind speeds used during the calibration exercise, which were in turn taken from a Met office weather station located away from the slope on open ground. The slope itself is sheltered by a stand of trees along the crest and so it was necessary to downscale the wind speed to better approximate site conditions. These two changes significantly improved the match between modelled and monitored suction generation. This exercise (along with the previous hydrological modelling study of the same slope) demonstrated the importance of model calibration against field monitored data and also the importance of site specific data.

In order to look at the potential effects of a future climate on stability, weather time series were generated for both a present and future climate based on UKCP02 climate scenarios (Hulme et al., 2002). These were used as this work extended previous studies by Rouainia et al which were carried out prior to the release of UKCP09. The weather time series were applied to the hydrological model and the resultant pore pressures then input into the mechanical model. The modelling demonstrated that for a future climate, the elevated summer temperatures led to the generation of higher surface suctions and in turn larger shrink-swell cycles causing a higher rate of material softening and progressive failure than observed for a present climate model. This can be seen in Figure 13. The sudden elevated displacements in the future climate model occurring in years three and five (as well as at ultimate failure) were caused by periods of elevated rainfall which in turn caused elevations in the peak pore water pressure at depth beyond normal values.

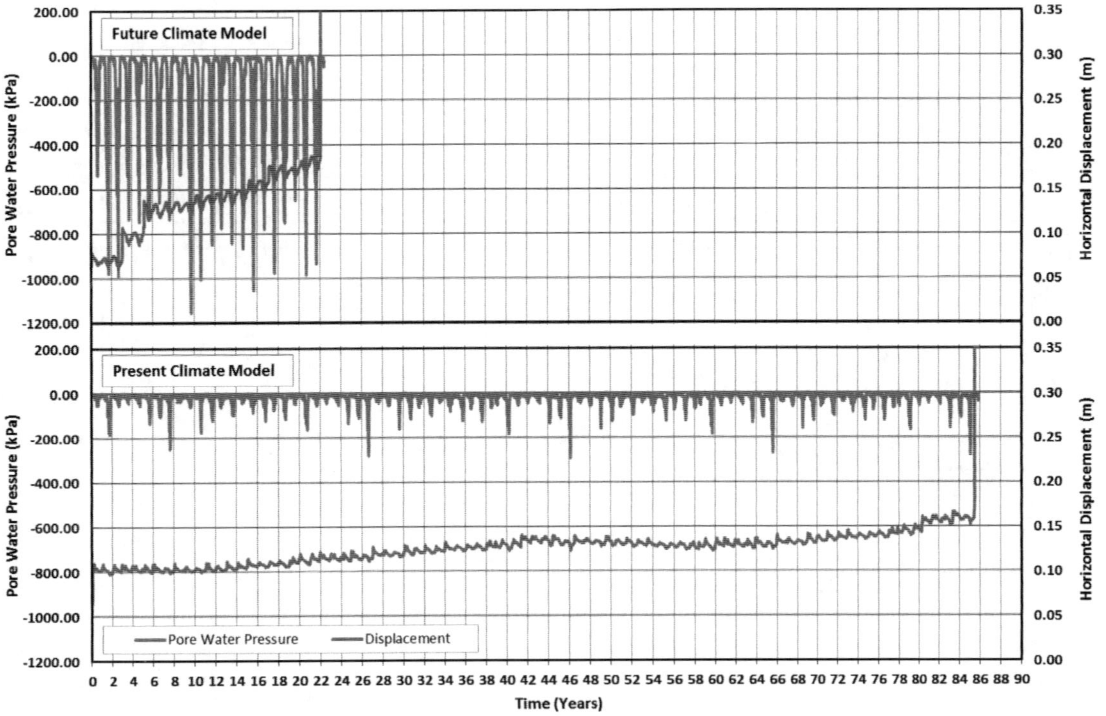

Figure 13. Mid-slope displacement and surface pore water pressure flux for a cut-slope subject to an applied present and future climate (after Rouainia et al. 2015).

In addition to permeability and vegetation, the sensitivity of the model outputs to pre-failure stiffness was investigated. In the modelling undertaken in this work, the Young's modulus (E) of the material was made dependent on the mean effective stress (p'), based on a stiffness multiplier (E_0) as per the work of Potts et al. (1997), see equation (1).

$$E = E_0 \frac{(p' + 100)}{100} \tag{1}$$

The adopted value of this multiplier and hence the pre-failure stiffness was shown to have a significant impact on progressive failure and slope stability. This is due to the fact that increased stiffness leads to an increase in the rate that shear strength is mobilised within the soil mass before significant strain-softening is able to occur in areas where yielding is observed. As such the higher stiffness values should lead to more rapid mobilisation of strength and so cause a reduced rate of progressive failure. The above mechanism was originally described in work undertaken by Ellis and O'Brien (2007), see Figure 14.

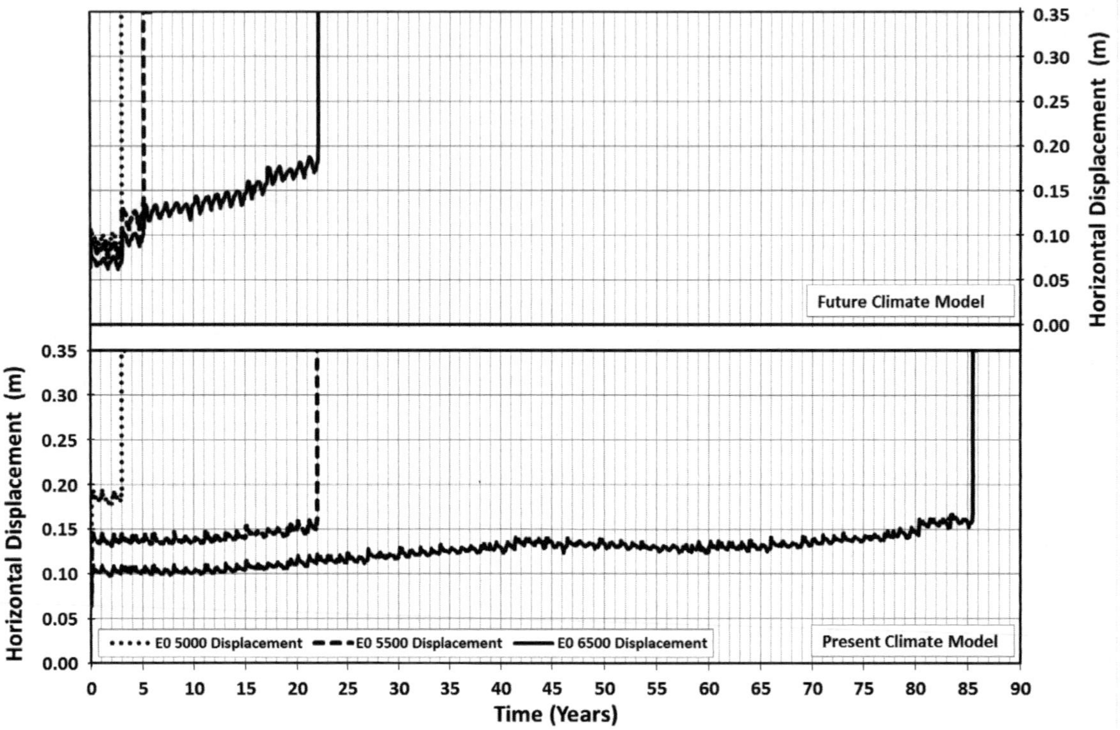

Figure 14. Effect of pre-failure stiffness on cut-slope stability due to imposed present and future climates.

The results of the modelling this far have provided useful insights into the sensitivity of the model outputs to a number of parameters. This has been useful to inform the future field and laboratory testing programmes. Furthermore, it has provided confidence that the models can capture some of the observed behavior in the field and provided an indication that both changes in climate and weather event sequences will change the current performance of engineered slopes. This will necessarily require a change in the way in which these slopes must be managed in the future.

The aim for future work is to use the validated models to investigate systematically how differing soil types, slope geometries and types of vegetation cover effect slope hydrology and related deformations under differing climate and weather scenarios.

The aim will be to identify the relative hazard posed to differing infrastructure slopes by climate change and also to identify patterns of weather which lead to or increase slope instability to allow broad lessons to be learned which can be applied across a range of infrastructure scales from individual slopes to a network scale.

6. Towards network-scale

As discussed in Section 1, iSMART operates across three scales; the soil-water-vegetation system of a single slope (scale 1) and the systems of slopes (scale 2) that make up a transport network (scale 3). The project is now developing methods to integrate the results from scales 1 and 2 into a whole-network model to advance systems-scale understanding of future network responses and vulnerabilities. To date, much of iSMART's activities have focused on achieving a better understanding of what parameters are representative of asset conditions and what process models describe best what is observed at the field sites. In the first phase of the research, rigorous numerical modelling at the site level has been used to bring together field and laboratory-scale data and to examine the consequences for slope performance, responding to actual and forecasted weather event sequences. This process has informed the direction of the laboratory and field work to quantify unknown parameters and understand complex processes. The translation of this information into appropriate serviceability and ultimate limit state of the assets is now being developed in discussion with the stakeholders and this will inform the condition appraisal of systems of slopes. Once this is better understood, it will enable the assessment of the spatial and temporal variability of the condition of engineered earthworks for UK transport networks. In turn, as discussed in Section 1, this knowledge will provide network operators with advice on asset design, management, maintenance and investment strategies (Figure 15).

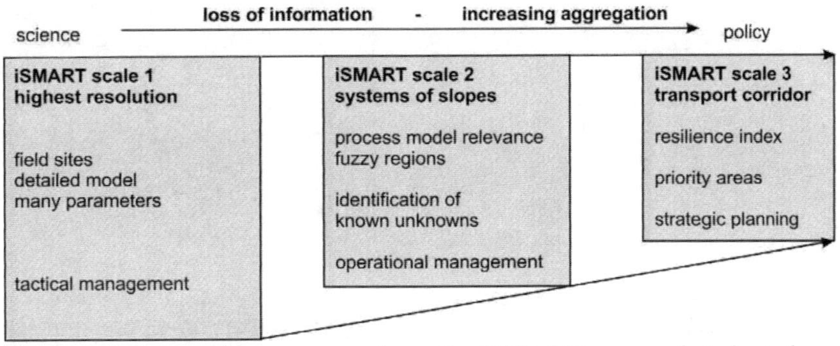

Figure 15. Integration across scales – the iSMART approach (adapted from Dijkstra et al. 2015).

As the pathway of upscaling the iSMART research is based on rigorously tested physical process models, it will be feasible to incorporate (and evaluate the outcomes of) changes in climate and environmental conditions. This is a significant advantage over current industry approaches that have assessed the vulnerability of transport infrastructure slopes on the basis of largely empirical information or expert elicitation (e.g. URS 2010). These models are very important as these use the currently available dataset and knowledge to enable some degree of prioritisation of earthworks asset conditions (slotting into scale 2 of the diagrams depicted in Figures 2 and 15). However, these result in relatively static products where it is difficult to evaluate the changes in conditions over time as climate/environmental change progresses (see Dijkstra and Dixon 2010). iSMART therefore focuses on continuity in the analysis, allowing detailed process models (scale 1) to inform slope system condition assessments (scale 2) and this, in turn to inform transport network performance (scale 3).

There is an increasingly fuzzy spatial relevance associated with the upscaling process. At the highest resolution, it is envisaged that data are available that are highly site specific, feeding into detailed process models. iSMART is delivering critical monitoring/observational datasets against which process models can be tested. However, distillation of site data and process understanding leads to the identification of common parameters and processes that represent systems of slopes of a similar nature (embankment/cutting/age/geology). These have a much larger spatial relevance, but this relevance will fade the further removed the asset conditions are from our field sites. It is important to note that this is not strictly a spatial distance, but a distance expressed as a parameter/process similarity function and this thus leads to complex patterns of relevance along the network (Figure 16). The current research does not claim to offer all answers to the known unknowns, but this exercise will highlight the relevance of the project to the transport network as a whole and indicate where further work on the characterisation of parameters and process is still required. An important step forward is the improved understanding of critical combinations of geology, asset condition, geometry and vegetation impacts.

Further integration of the understanding of the depths of pore water pressure cycling (gained from field monitoring) will enable the understanding of soil deterioration processes (from laboratory testing) be built into the numerical models, which will incorporate future climate scenarios. Hence, asset deterioration will be modelled to enable better estimations of current asset condition based on age, and projection of asset behaviour under future weather and environmental conditions.

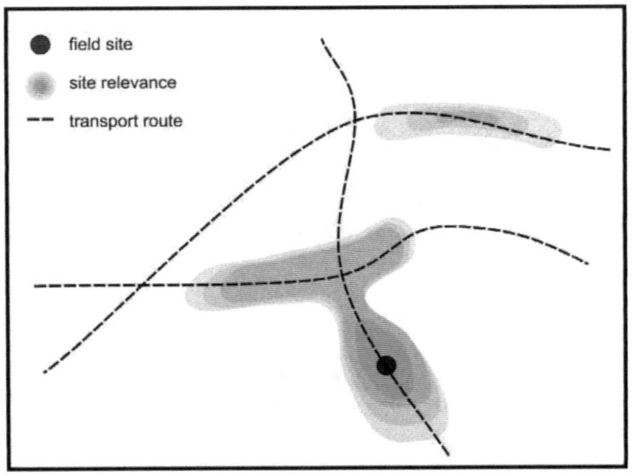

Figure 16. Conceptual diagram of the upscaling process. Field sites will deliver property/parameter information and process relevance that extends beyond the boundary of these sites. However, the relevance of this information will fade with 'distance' (indicated by the increasingly lighter shades of grey).

Validated models that incorporate the greater complexity outlined above will form the basis of the production of a national scale map of the forecasted capacity reduction of transport networks due to slope instability caused by climate effects at several future time slices. In turn, the models will enable testing of the benefits of different stability and maintenance solutions under a changing climate regime and thus enable development of robust adaptation strategies in conjunction with a formal stakeholder dialogue. The long-term vision of the project is to enable a 4D model of transient water movement in infrastructure slopes under a range of current and future environmental scenarios based on a fundamental understanding of earthwork material and system behaviour, which can be used to create a more reliable, cost effective, safer and more sustainable transport system.

7. Acknowledgements

This paper is an output from iSMART, a collaborative research project funded by the UK Engineering and Physical Sciences Research Council (Grant number EP/K027050/1). The team gratefully acknowledges the support received from all stakeholders and academic partners.

8. References

Brown, M. 2014. Personal Communication

Carse, L. (2014) The geomechanical response of cut slopes in glacial till to climatically driven pore-pressure cycling and hydrogeology. PhD Thesis, Queen's University Belfast, United Kingdom.

Carse, L., Mclernon, M., Hughes, D.A.B., Sivakumar, V., Barbour, S.L. 2009. The magnitude and variation of pore pressures in glacial till slopes and their effect on slope stability. 62nd Canadian Geotechnical Conference, CGS

Clarke, B. 2012. Glacial soils. In: Burland, J., Chapman, T., Skinner, H. and Brown, M. ed., *ICE Manual of Geotechnical Engineering: Volume 1 - Geotechnical Engineering Principles, Problematic Soils and Site Investigation*, 1st ed. London: ICE Publishing, pp.363-387.

Collins, M. 2007. Ensembles and probabilities: a new era in the prediction of climate change. Phil. Trans. R.Soc, 365, 1957-1970.

Dijkstra, T.A. & Dixon, N. (2010). Climate change and slope stability: Challenges and approaches. Quarterly Journal of Engineering Geology and Hydrogeology, 43, 4, 371-385.

Dijkstra, T.A., Dixon, N., Crosby, C.J., Frost, M.W., Gunn, D.A., Fleming, P.R., & Wilks, J.H. 2014.

Forecasting infrastructure resilience for climate change. Proceedings Institution of Civil Engineers Transport, 167(5), 269-280.

Ellis, E.A. and O'Brien, A.S. 2007. Effect of height on delayed collapse of cuttings in stiff clay. *Geotechnical Engineering*. **160**(2):73-84.

Forsyth, P.A., Wu, Y.S. & Pruess, K. 1995. Robust numerical methods for saturated-unsaturated flow with dry initial conditions in heterogeneous media. *Advances in Water Resources*. **18**(1):25-38.

Glendinning, S., Hughes, P., Helm, P.R., Chambers, J., Mendes, J., Gunn, D., Wilkinson, P., Uhlemann, S. 2014. Construction, management and maintenance of embankments used for road and rail infrastructure: implications of weather induced pore water pressures. Acta Geotechnica, 9, 5, 799-816.

Harley, R., Sivakumar, V., Hughes, D., Karim, M.R.., Barbour, S.L., (2014). Progressive deformation of glacial till due to viscoplastic straining and pore pressure variation. GeoRegina 2014, 67th Canadian Geotechnical Conference Proceedings.

Hen-Jones RM, Hughes PN, Glendinning S, Gunn DA, Chambers JC, Wilkinson PB, Uhlemann S. Determination of moisture content and soil suction in engineered fills using electrical resistivity. In: Khalili K, Russell A, Koshghalb A, ed. Unsaturated Soils: Research and Applications. London: CRC Press, 2014.

Hen-Jones RM, Hughes PN, Stirling RA, Glendinning S, Gunn DA, Chambers JC. Seasonal effects on geophysical-geotechnical relationships and their implications for Electrical Resistivity Tomography. *Journal of applied geophysics*. (In Preparation)

Hughes, P.N., Glendinning, S., Mendes, J., Parkin, G., Toll, D.G., Gallipoli, D., Miller, P.E. 2009. Full-scale testing to assess climate effects on embankments. Proceedings of the Institute of Civil Engineers: Engineering Sustainability 162. ES2, 67-79.

Hulme, M., Jenkins G.J., Lu, X., Turnpenny, J.R., Mitchell, T.D., Jones, R.G., Lowe, J., Murphy, J.M., Hassell, D., Boorman, P., McDonald, R., Hill, S. 2002. *Climate Change Scenarios for the United Kingdom: The UKCIP02 Scientific Report*. Norwich, Tyndall Centre for Climate Change Research, University of East Anglia.

Jenkins, G.J., Perry, M.C. & Prior M.J.O. 2009. The climate of the United Kingdom and recent trends. Met Office Hadley Centre, Exeter, UK. 121pp.

Jones, P. D., Kilsby, C. G., Harpham, C., Glenis, V., Burton, A. 2009, UK Climate Projections science report: Projections of future daily climate for the UK from the Weather Generator. University of Newcastle, UK, 48p.

Kilsby, C.G., Jones, P.D., Burton, A., Ford, A.C., Fowler, H.J., Harpham, C., James, P., Smith, A. and Wilby, R.L. 2007. A daily weather generator for use in climate change studies. *Environmental Modelling and Software*. 22(12): 1705-1719.

Kilsby, C.G., Burton A., Glenis, V., Jones, P.D. and Harpham, C. 2010. UK Climate Projections science report: Projections of future daily climate for the UK from the Weather Generator. Newcastle University.

Mendes, J. 2011. Assessment of the impact of climate change on an instrumented embankment – an unsaturated soil mechanics approach. PhD thesis, School of Engineering and Computer Sciences, Durham University, UK.

Murphy, J.M., Sexton, D.M.H., Jenkins, G.J., Boorman, P.M., Booth, B.B.B., Brown, C.C., Clark, R.T., Collins, M., Harris, G.R., Kendon, E.J., Betts, R.A., Brown, S.J., Howard, T. P., Humphrey, K. A., McCarthy, M. P., McDonald, R. E., Stephens, A., Wallace, C., Warren, R., Wilby, R., & Wood, R. A. 2009. UK Climate Projections Science Report: Climate change projections. Met Office Hadley Centre, Exeter, 194p.

Perry, J. 1989. *A survey of slope condition on motorway earthworks in England and Wales*. Research Report 199. Transport Research Laboratory.

Perry, J., Pedley, M. and Brady, K. 2003a. *Infrastructure Cuttings – condition appraisal and remedial treatment (C591)*. Construction Industry Research and Information Association, London.

Perry, J., Pedley, M. and Reid, M. 2003b. *Infrastructure embankments – condition appraisal and remedial treatment (C592).* Construction Industry Research and Information Association, London.

Potts, D.M., Kovacevic, K. and Vaughan, P.R. 1997. Delayed collapse of cut slopes in stiff clay. *Proceedings of the Institution of Civil Engineers: Géotechnique*. **47**(5):953-982.

Rouainia, M., Davies, O., O'Brien, A. and Glendinning, S. 2009. Numerical modelling of climate effects on slope stability. *Proceedings of the Institution of Civil Engineers: Engineering Sustainability*. **162**(2):81-89.

Rouainia, M., Helm, P.R., Davies, O. and Glendinning, S. (In preparation). Numerical modelling of meteorologically induced pore water pressure variations and their effect on the stability of an infrastructure cutting in London Clay for both present and future climates. *Geotechnqiue*. (In preparation).

Smethurst, J. A., Clarke, D. & Powrie, W. 2006. Seasonal changes in pore water pressure in a grass covered cut slope in London Clay. *Proceedings of the Institution of Civil Engineers: Géotechnique.* **56**(8):523–537.

Smethurst, J. A., Clarke, D. and Powrie, W. 2012. Factors controlling the seasonal variation in soil water content and pore water pressures within a lightly vegetated clay slope. *Proceedings of the Institution of Civil Engineers: Géotechnique.* **62**(5):429-446

Stirling RA. 2014. Multiphase modelling of desiccation cracking in compacted soils. PhD Thesis, Newcastle University, UK.

Stirling RA, Hughes PN, Davie CT, Glendinning S. Cyclic relationship between saturation and tensile strength in the near-surface zone of infrastructure embankments. In: Khalili K, Russell A, Koshghalb A, ed. Unsaturated Soils: Research and Applications. London: CRC Press, 2014, pp.1501-1505.

Stirling RA, Hughes PN, Davie CT, Glendinning S. 2015. Tensile behaviour of unsaturated compacted clay soils – A direct assessment method. Appl. Clay Sci. DOI: 10.1016/j.clay.2015.04.011

Thurlby, R. Managing the asset time bomb: a system dynamics approach. Proceedings of the ICE - Forensic Engineering, 166 (3), 134 –142

Toll, D.G. 2012. The behaviour of unsaturated soils. Chapter 5 in A Handbook of Tropical Residual Soil Engineering, (eds. Huat, B.B.K., Toll, D.G. & Prasad, A.) London: Taylor and Francis, ISBN: 9780415457316.

Toll, D. G., Mendes, J., Gallipoli, D., Glendinning, S., & Hughes, P. N. 2012. "Investigating the impacts of climate change on slopes: field measurements." Geological Society, London, Engineering Geology Special Publications, 26(1), 151-161.

Trenter, N.A. 1999. *Engineering in Glacial Tills (C504).* Construction Industry Research and Information Association, London.

Wilby, R.L.. 2007. A daily weather generator for use in climate change studies. Environmental modelling and software, 22, 1705-1719.

Wilks, D.S. & Wilby, R.L. 1999 The weather generation game: a review of stochastic weather models. Prog. in Phys. Geog., 23, 329-357.

Reliability-based Assessment of Stability of Slopes

C. Hsein Juang[1*], Jie Zhang[2], Wenping Gong[1]

[1] Glenn Department of Civil Engineering, Clemson University, Clemson, SC 29634, USA

[2] Department of Geotechnical Engineering, Tongji University, Shanghai 200092, China

*Email: C. Hsein Juang, hsein@clemson.edu

Abstract. Multiple sources of uncertainties often exist in the evaluation of slope stability. When assessing stability of slopes in the face of uncertainties, it is desirable, and sometimes necessary, to adopt reliability-based approaches that consider these uncertainties explicitly. This paper focuses on the practical procedures developed recently for the reliability-based assessment of slope stability. The statistical characterization of model uncertainty and parameter uncertainty are first described, followed by an evaluation of the failure probability of a slope corresponding to a single slip surface, and the system failure probability. The availability of site-specific information then makes it possible to update the reliability of the slope through the Bayes' theorem. Furthermore, how to perform reliability-based design when the statistics of random variables cannot be determined accurately is also discussed. Finally, case studies are presented to illustrate the benefit of performing reliability-based design and the procedure for conducting reliability-based robust design when the statistics of the random variables are incomplete.

1. Introduction

The uncertainties in both the solution models and the input parameters make it impossible to capture the true performance (or stability) of a slope with a single factor of safety (FOS), which is the outcome from a deterministic analysis. For example, slopes with the same FOS value may exhibit different risk levels depending on the variability of the solution models and soil parameters (Li and Lumb 1987). Subsequent to the pioneering researches such as Wu and Kraft (1970), Matsuo and Kuroda (1974), Alonso (1976) and Tang et al. (1976), extensive studies have been conducted on the probabilistic assessment of slope stability, including reliability analysis of slope failure along a given slip surface (e.g., Liang et al. 1999; El-Ramly et al. 2002), searching for the most critical slip surface (e.g., Bhattacharya et al. 2003; Xue and Gavin 2007), system reliability analysis (e.g., Oka and Wu 1990; Ching et al. 2009; Zhang et al. 2011&2013), effect of spatial variability of soil properties on slope stability (e.g., Griffiths and Fenton 2004; Cho 2007; Griffiths et al. 2009; Li et al. 2015), effect of environment action on slope reliability (e.g., Babu and Murthy 2005; Zhang et al. 2005; Frattini et al 2009; Santoso and Phoon 2011), and reliability-based design of slopes (Christian et al. 1994; Roh and Hong 2009; Salgado and Kim 2014). The outcome from a reliability analysis is regarded as an important input for the qualitative risk assessment and management of landslide hazards (e.g., Mostyn and Li 1993; Nadim et al. 2005; Cassidy et al. 2008; Zhang et al. 2012; Huang et al. 2013).

The probabilistic framework is generally considered effective for uncertainty quantification, incorporating information from various sources, and expressing the degree of belief about slope stability. Over the past decades, reliability methods have gradually evolved into a useful and practical tool that is complementary to the existing deterministic slope stability analysis. Nevertheless, the development in the field of probabilistic assessment of slope stability is still not well known in the general community of slope engineering. For many engineers, the probabilistic methods are sometimes considered as attractive but difficult to implement, which limits its application in practice. In an effort to bridge such a gap, the goal of this paper is to introduce practical methods and examples for addressing the following important questions in slope reliability analysis:

- How uncertainties in slope stability analysis can be characterized?
- How the failure probability of a slope can be calculated efficiently both for an individual slip surface and for the slope as a system?
- How the performance information can be used to update the reliability of a slope?
- How to implement reliability-based design with incomplete probabilistic information?

The structure of this paper is as follows. First, methods for estimating the uncertainties of solution models and soil parameters are presented. Second, how to evaluate the reliability of a single slip surface and multiple slip surfaces are introduced. Third, the update of the reliability-based slope stability analysis result using the past performance information is outlined. Finally, two application examples are studied to illustrate the use of probabilistic methods in the design of slopes in practice.

2. Uncertainties Characterization in Slope Stability Analysis

2.1. Model uncertainty

In that the solution models adopted in slope engineering are only abstractions of the real world, model uncertainties always exist (Ang and Tang 1984; Cheung and Tang 2005; Zhang et al. 2009). If the model uncertainty is not considered, the prediction from a solution model might be biased. In principle, the model uncertainty (i.e., uncertainty of solution models) can be determined by a systematic comparison between model predictions and observed performances (i.e., field observations or model tests) (Gilbert and Tang 1995; Zhang et al. 2009). Such a comparison is, however, difficult, as in slope engineering the model input parameters can hardly be determined with certainty. Thus, the model uncertainty must be analyzed considering uncertainties in the model input parameters.

Let $g(\boldsymbol{\theta})$ denote a general slope stability model, where $\boldsymbol{\theta}$ denote the uncertain input parameters in this model. Assume the observed performance d can be related to the model prediction $g(\boldsymbol{\theta})$ and model error ε as follows:

$$d = g(\boldsymbol{\theta}) + \varepsilon \tag{1}$$

Assume further that ε is a normal variable with a mean of μ_ε and a standard deviation of σ_ε. The task of model uncertainty characterization is then reduced to the estimation of μ_ε and σ_ε. Let $f(\mu_\varepsilon, \sigma_\varepsilon)$ denote the prior probability density function (PDF) of μ_ε and σ_ε. Based on Bayes' theorem, the posterior PDF of μ_ε and σ_ε can be evaluated as follows (Zhang et al. 2009):

$$f(\mu_\varepsilon, \sigma_\varepsilon \mid \mathbf{d}) \approx k f(\mu_\varepsilon, \sigma_\varepsilon) \prod_{i=1}^{n} \frac{1}{\sqrt{2\pi\left(\sigma_\varepsilon^2 + \sigma_{g(\boldsymbol{\theta}_i)}^2\right)}} \exp\left[-\frac{1}{2}\frac{\left(d_i - \mu_{g(\boldsymbol{\theta}_i)} - \mu_\varepsilon\right)^2}{\sigma_\varepsilon^2 + \sigma_{g(\boldsymbol{\theta}_i)}^2}\right] \tag{2}$$

where k is a normalization constant to make the derived posterior PDF valid; d_i is the ith observation; and $\mu_{g(\boldsymbol{\theta}i)}$ and $\sigma_{g(\boldsymbol{\theta}i)}$ are the mean and standard deviation of $g(\boldsymbol{\theta}_i)$ caused by uncertainties in $\boldsymbol{\theta}_i$.

The above formulation was adopted in Zhang et al (2009) to study the model uncertainty associated with the Morgenstern-Price method (Morgenstern and Price 1965) for slope stability analysis, in which the test results of 20 model slopes examined in centrifuge by Kim (1980) were analyzed. The prior distribution of μ_ε is assumed to be a normal variable with a mean of 0 and a standard deviation of 0.5, while the prior distribution of σ_ε is assumed as a lognormal variable with a mean of 0.5 and a standard deviation of 1.0. After the values of $\mu_{g(\boldsymbol{\theta})}$ and $\sigma_{g(\boldsymbol{\theta})}$ are determined, the posterior statistics of μ_ε and σ_ε are then calculated based on the Markov Chain Monte Carlo (MCMC) simulation, and the results are summarized in Table 1. A comparison of the prior and posterior statistics of these variables shows a significant reduction in the uncertainties in μ_ε and σ_ε. Table 1 shows that the mean of the model error of the Morgenstern-price method is close to zero, indicating this method is almost unbiased. The standard deviation of the model error is about 4.4%. For comparison, Duncan and Wright (1980) determined that any limit equilibrium method satisfying all conditions of equilibrium could predict the FOS within an accuracy of 5% from what is considered the correct answer.

Table 1. Comparison of statistics of model error calibrated based on centrifuge tests

	μ_ε		σ_ε		ε	
	Mean	Std. Dev.	Mean	Std. Dev.	Mean	Std. Dev.
Prior	0	0.5	0.5	1	0	1.225
Posterior	-0.002	0.010	0.042	0.008	-0.002	0.044

The data in Table 1 show the model error associated with the Morgenstern-Price method used to analyze the model slopes tested in Kim (1980). When the Morgenstern-Price method is used to study slopes in the field, the statistics of the model error might be different. Nevertheless, when the observed performance data of similar slopes in a region are available, Equation (2) can also be used to calibrate the model error associated with the slope stability model for analysing the slopes in this region.

2.2. Parameter uncertainty

The uncertainty associated with the parameters of the slope stability model can often be assessed through laboratory and in-situ tests (e.g., Vanmarcke 1977; Einstein and Baecher 1983; Lacasse and Nadim 1996; Duncan 2000). When slope failure is observed, back analysis is also useful in deriving the uncertainties of the model input parameters (e.g., Gilbert et al. 1998; Tang et al. 1999). At the moment of slope failure, the FOS of the slope is equal to unity. Based on the normal assumption of the model error in Equation (1) as mentioned previously, if μ_ε and σ_ε are known, the chance to observe the slope failure event can be written as follows:

$$L(\boldsymbol{\theta} \mid F) = \phi\left(\frac{g(\boldsymbol{\theta}) + \mu_\varepsilon - 1}{\sigma_\varepsilon}\right) \tag{3}$$

where F denotes the event of observing slope failure. Let $f(\boldsymbol{\theta})$ denote a PDF that represents the knowledge about $\boldsymbol{\theta}$ prior to the back analysis. According to Bayes' theorem, the posterior PDF of $\boldsymbol{\theta}$ is

$$f(\boldsymbol{\theta} \mid F) = \frac{\phi\left(\dfrac{g(\boldsymbol{\theta}) + \mu_\varepsilon - 1}{\sigma_\varepsilon}\right) f(\boldsymbol{\theta})}{\int \cdots \iint \phi\left(\dfrac{g(\boldsymbol{\theta}) + \mu_\varepsilon - 1}{\sigma_\varepsilon}\right) f(\boldsymbol{\theta}) d\boldsymbol{\theta}} \tag{4}$$

The posterior PDF given by Equation (4) generally does not have an analytical expression. Therefore, for an arbitrary prior distribution, samples of the posterior distribution can be conveniently obtained using MCMC simulation (Juang et al. 2012a; Wang et al. 2013a). When $\boldsymbol{\theta}$ is multivariate normal, its posterior distribution can be approximated by a multivariate normal distribution based on the system identification theory (Zhang et al. 2010), as described below. Let $\boldsymbol{\mu}_{\boldsymbol{\theta}}$ and $\mathbf{C}_{\boldsymbol{\theta}}$ denote the prior mean and prior covariance of $\boldsymbol{\theta}$, respectively. Let $\boldsymbol{\mu}_{\boldsymbol{\theta}|\mathbf{d}}$ and $\mathbf{C}_{\boldsymbol{\theta}|\mathbf{d}}$ denote the posterior mean and covariance matrix of $\boldsymbol{\theta}$, respectively. In that the slope stability model $g(\boldsymbol{\theta})$ is often quite linear along a given slip surface, $\boldsymbol{\mu}_{\boldsymbol{\theta}|\mathbf{d}}$ and $\mathbf{C}_{\boldsymbol{\theta}|\mathbf{d}}$ can be determined approximately using the system identification (SI) for linear systems as follows (Zhang et al. 2010; Wang et al. 2013a):

$$\boldsymbol{\mu}_{\boldsymbol{\theta}|\mathbf{d}} \approx \boldsymbol{\mu}_{\boldsymbol{\theta}} + \mathbf{C}_{\boldsymbol{\theta}}\mathbf{H}^{\mathrm{T}}\left(\mathbf{H}\mathbf{C}_{\boldsymbol{\theta}}\mathbf{H}^{\mathrm{T}} + \sigma_{\varepsilon}^{2}\right)^{-1}\left(1 - g\left(\boldsymbol{\mu}_{\boldsymbol{\theta}}\right) - \mu_{\varepsilon}\right) \tag{5}$$

$$\mathbf{C}_{\boldsymbol{\theta}|\mathbf{d}} \approx \left(\frac{\mathbf{H}^{\mathrm{T}}\mathbf{H}}{\sigma_{\varepsilon}^{2}} + \mathbf{C}_{\boldsymbol{\theta}}^{-1}\right)^{-1} \tag{6}$$

$$\mathbf{H} = \left.\frac{\partial g(\boldsymbol{\theta})}{\partial \boldsymbol{\theta}}\right|_{\boldsymbol{\theta}=\boldsymbol{\mu}_{\boldsymbol{\theta}}} \tag{7}$$

where $g(\boldsymbol{\mu}_{\boldsymbol{\theta}})$ is the predicted FOS calculated at point $\boldsymbol{\mu}_{\boldsymbol{\theta}}$; and $\mathbf{H}$ is a row vector representing the sensitivity of $g(\boldsymbol{\theta})$ with respect to $\boldsymbol{\theta}$ at point $\boldsymbol{\mu}_{\boldsymbol{\theta}}$.

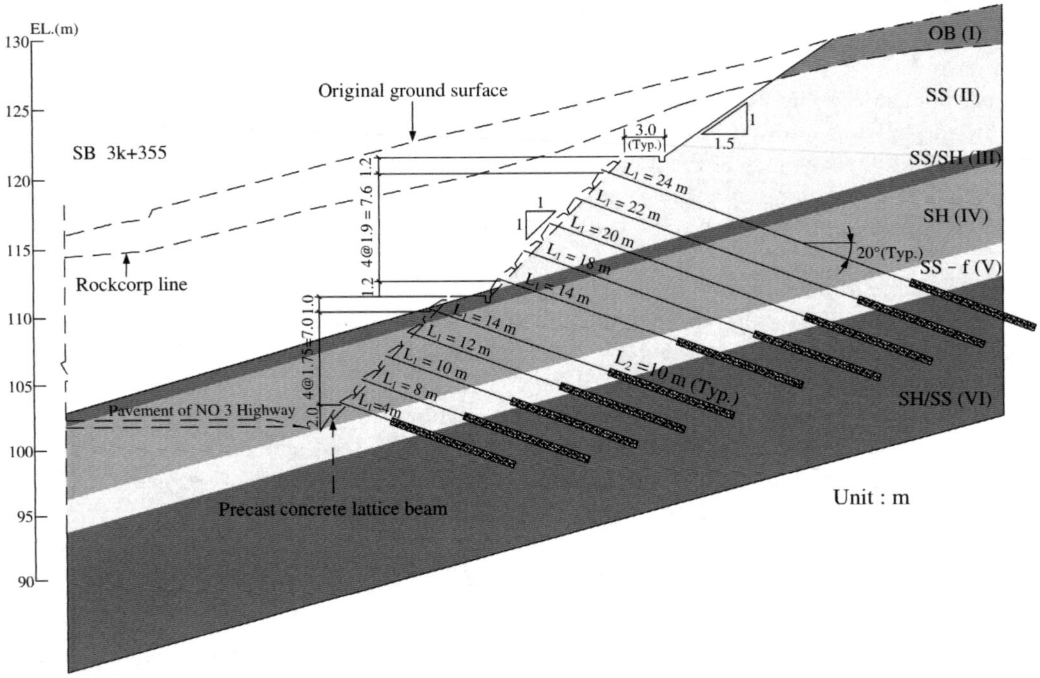

Figure 1. Geometry and geological profile of a cross-section of the slope studied in Wang et al. (2013a) [OB (I): Overburden soil; SS (II): Sandstone with vertical joints; SS/SH(III): Alternations of thin sandstone and shale; SH(IV): Dark gray shale; SS-f (V): Sandstone with trace fossils; and SS/SH(VI): Alternations of thin sandstone and shale]. (Wang et al. 2013a)

A probabilistic back analysis example adapted from Wang et al. (2013a) is illustrated in Figure 1. The cut slope, formed by highway construction, failed suddenly approximately 12 years after completion. The weather was sunny at the time of the landslide and there was no occurrence of tremors, thus excluding both heavy rainfall and earthquake as the trigger. The location of the slip surface was determined from a site investigation, which can be simplified as a single plane failure surface. The FOS of the rock slope can be analyzed using the method developed by Hoek and Bray (1981). For this slope, the friction angle of sliding surface φ (°) and the design forces of all rock anchors T (ton) are the major uncertain variables to be back analyzed, i.e., $\boldsymbol{\theta} = \{\varphi, T\}$. For demonstration purposes, the prior distribution of φ and T is assumed as a bivariate normal distribution with the statistics listed in Table 2. The model error is assumed to follow a normal distribution with mean $\mu_\varepsilon = 0.05$ and standard deviation $\sigma_\varepsilon = 0.07$. This input information is then used to derive the back analysis results using the SI method, as listed in Table 2. Comparing the prior and posterior mean values, both the mean values of φ and T are reduced. The updated outcomes also show a reduction in the uncertainties of φ and T.

Table 2. Prior and posterior statistics of φ and T in the back analysis of slope failure

	φ (°)		T (ton)	
	Mean	Std. Dev.	Mean	Std. Dev.
Prior	21	3.15	60	22.8
Posterior	12.88	1.49	33.05	20.86

3. Reliability Analysis of Slopes

A consideration of the uncertainties involved in the slope stability analysis results in an uncertainty in the FOS of the slope. In such a case, either failure probability or reliability index can be used to measure the stability of the slope. The traditional approach for computing the reliability index of a slope involves minimizing the FOS over a range of possible failure surfaces and then calculating the reliability index related to this surface (Chowdhury et al. 1987; Honjo and Kuroda 1991; Christian et al. 1994; Wolff et al. 1995). However, the slip surface that yields the minimum FOS (termed deterministic most critical slip surface) does not always coincide with the slip surface that yields the minimum reliability index (known as the probabilistic most critical slip surface) (e.g., Hassan and Wolff 1999; El-Ramly et al. 2002; Zhang et al. 2013). Indeed, a soil slope may have many potential slip surfaces, which means that the failure probability of the slope is often larger than that corresponding to any single slip surface (Cornell 1971; Oka and Wu 1990; Chowdhury and Xu 1995). Therefore, system reliability is more appropriate for analyzing the reliability of a slope (Chowdhury and Xu 1995; Ching et al. 2009; Low et al. 2011; Li et al. 2003; Zhang et al. 2013; Zeng and Jimenez 2014).

3.1. Reliability analysis along a single slip surface

Determining the probabilistic most critical slip surface is of significant importance in slope reliability analysis, which is possible through a rigorous constrained optimization algorithm. Although the Hassan and Wolff (1999) procedure is empirical, Zhang et al. (2013) found that it was quite accurate for many applications. Let θ_i denote the ith element of $\boldsymbol{\theta}$ with $\mu_{\theta i}$ and $\sigma_{\theta i}$ respectively being the mean and standard deviation of θ_i. Let n denote the dimension of $\boldsymbol{\theta}$ and let $\boldsymbol{\theta}^{(i)} = \{\mu_{\theta 1}, \mu_{\theta 2}, \ldots, \mu_{\theta i} - \sigma_{\theta i}, \ldots, \mu_{\theta n}\}$. Hassan and Wolff (1999) suggested the following procedure to search for the probabilistic most critical slip surface:

- For $i = 1, 2, \ldots, n$, search the slip surface with minimum FOS using $\boldsymbol{\theta}^{(i)}$ as input parameters. As such, n FOS values as well as n slip surfaces can be obtained;
- Among the n slip surfaces obtained, the slip surface with the minimum FOS is the one with the maximum failure probability, i.e., the probabilistic most critical slip surface. The argument for this is that when the random variable contributing the most to the uncertainty in the computed FOS takes a low value (but of reasonable likelihood) in its possible range, the location of the critical surface is dominated by this random variable.

After the probabilistic most critical slip surface is identified, its reliability index can be evaluated using existing reliability methods such as the first order reliability method (FORM) (e.g., Ang and Tang 1984). However, as deterministic computer programs are widely used for slope stability analysis, it is advantageous to be able to use the existing stand-alone deterministic computer programs for the reliability analysis. To this end, the response surface method is adopted. It is noted that along a given slip surface, the relationship between FOS and shear strength parameters is often quite linear, which can be approximated by a second order polynomial as follows (Zhang et al. 2013):

$$g'(\boldsymbol{\theta}) = b_0 + \sum_{i=1}^{n} b_i\theta_i + \sum_{i=1}^{n} b_{n+i}\theta_i^2 \tag{8}$$

where b_i (i = 0, 1, …, $2n$) is the unknown coefficient. There are ($2n$ + 1) unknown coefficients in Equation (8). To calibrate these coefficients, the FOS of a slip surface is first evaluated at the following ($2n$ + 1) points: $\{\mu_{\theta1}, \mu_{\theta2}, \ldots, \mu_{\theta n}\}$, $\{\mu_{\theta1} \pm m\sigma_{\theta1}, \mu_{\theta2}, \ldots, \mu_{\theta n}\}$, $\{\mu_{\theta1}, \mu_{\theta2} \pm m\sigma_{\theta2}, \ldots, \mu_{\theta n}\}$, …, and $\{\mu_{\theta1}, \mu_{\theta2}, \ldots, \mu_{\theta n} \pm m\sigma_{\theta n}\}$, where m is a parameter selected by the user. By equating the FOSs calculated with Equation (8) to those calculated using the numerical slope stability model at the selected points, the unknown coefficients can be solved. The calibrated second order polynomial can then be used as the performance function for the subsequent slope reliability analysis using the existing reliability methods.

3.2. System reliability

The number of potential slip surfaces of a slope can be large. The failure probability of the most critical slip surface only provides a lower bound value for the system failure probability of the slope. Nevertheless, due to the high correlation among FOS of various slip surfaces, the system reliability of a soil slope is often governed by a few "representative" slip surfaces. Several methods have been suggested for identifying the representative slip surfaces (e.g., Ji and Low 2012; Zhang et al., 2013). In the Hassan and Wolff method, each uncertain variable is decreased from its mean value to locate a slip surface where the location of the slip surface is governed by that random variable, resulting in n slip surfaces. In the recent study by Zhang et al. (2013), it has been shown that these n slip surfaces can indeed be used as representative slip surfaces for the system reliability analysis. Thus, the system reliability of a slope can then be analyzed using the following procedure:

- Identify the representative slip surfaces using the Hassan and Wolff procedure;
- Approximate the relationship between FOS and uncertain variables using Equation (8) for each representative slip surface;
- Calculate the failure probability based on the fitted second-order polynomials using Monte Carlo simulation (MCS).

As an example, a slope adapted from Feng and Fredlund (2011) as shown in Figure 2 is studied. Through analysing 9,261 potential slip surfaces, the failure probability of the most critical slip surface is 0.120 (COV = 2.7%) and the system failure probability of the slope is 0.187 (COV = 2.1%) based on MCS with 1 × 10^4 samples. Using the Hassan and Wolff procedure, the identified representative slip surfaces are shown in Figure 2. As slip surfaces 1 and 2 are the same, there are two representative slip surfaces in this example. Slip surface 3 is the same as the most critical slip surface. Based on the two identified representative slip surfaces as shown in Figure 2, the failure probability of the slope is 0.184 with a COV of 2.1% using Monte Carlo simulation with 1 × 10^4 samples, consistent with the system failure probability calculated based on a 9261 slip surfaces.

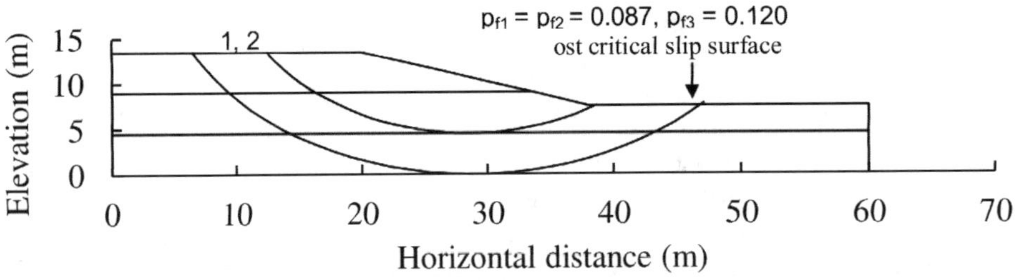

Figure 2. Representative slip surfaces identified by the Hassan and Wolff method (Zhang et al. 2013)

4. Updating Slope Stability with Observed Performance

Common information inherent in many slope reliability analyses includes knowledge learned from geotechnical explorations and tests, engineering judgment, published data, and regional experiences. In addition to the above knowledge, another important source of information is the past performance of the slope of concern: the slope survived or failed at a certain state previously. Such performance information is the outcome of a full-scale test directly performed on the slope under investigation, and could provide valuable information for the safety assessment, upgrading analysis, and remedial design of the slope. This section intends to illustrate how the site-specific past performance information can be used to update the reliability of slope.

4.1. Consideration of survival information

Let FOS_1 denote the FOS of the slope with known performance at a previous moment. If the slope did not fail, it means $FOS_1 > 1$. Let $p_{f2|FOS1>1}$ denote the conditional failure probability of the slope at a future moment considering its past survival information. Mathematically, it can be written as:

$$p_{f2|y_1>1} = P\left(\mathrm{FOS}_2 < 1 \mid \mathrm{FOS}_1 > 1\right) \tag{9}$$

Using the rule of conditional probability, $p_{f2|FOS1>1}$ becomes:

$$p_{f2|y_1>1} = \frac{p_{f2} - p_{f12}}{1 - p_{f1}} \tag{10}$$

In Equation (10), p_{f1} and p_{f2} are failure probabilities at the previous moment and at a future moment without considering the past performance information, which can be calculated with the conventional reliability methods such as first order reliability method (FORM) (e.g., Low et al. 2011); and p_{f12} is the probability indicating a simultaneous violation of the limit-state functions of $FOS_1 - 1 = 0$ and $FOS_2 - 1 = 0$, which can be computed as follows (Zhang et al. 2011):

$$p_{f12} \approx (a+b)\left(\frac{\pi - v}{\pi}\right) \tag{11}$$

$$a = \Phi(-\beta_1)\Phi\left(-\frac{\beta_2 - \rho_{12}\beta_1}{\sqrt{1-\rho_{12}^2}}\right) \tag{12}$$

$$b = \Phi(-\beta_2)\Phi\left(-\frac{\beta_1 - \rho_{12}\beta_2}{\sqrt{1-\rho_{12}^2}}\right) \tag{13}$$

$$v = \cos^{-1}\rho_{12} \tag{14}$$

$$\rho_{12} = \frac{\mathbf{Y}_1^* \mathbf{R}_\mathbf{Y}^{-1} \left(\mathbf{Y}_2^*\right)^\mathrm{T}}{\beta_1 \beta_2} \tag{15}$$

where β_1 and β_2 are the FORM reliability indexes of the slope at the previous moment and at the future moment, respectively; $\mathbf{Y}$ = reduced variables of $\boldsymbol{\theta}$; $\mathbf{R}_\mathbf{Y}$ = correlation matrix of $\mathbf{Y}$; and $\mathbf{Y}_1$* and $\mathbf{Y}_2$* = the design points of the slope at the state with known performance and at the future state, found with FORM, respectively.

To illustrate the effect of survival information on reliability updating, an infinite slope example is adapted from Zhang et al. (2011), as shown in Figure 3. The FOS of the slope can be calculated using the infinite slope model (Wu and Abdel-Latif 2000). Let $m = h_w/h$ denotes the normalized water table. Let m_1 denote the level of groundwater at which the slope has survived and let m_2 denote the maximum groundwater table within a certain design period. Due to the presence of observation error, statistical uncertainty, and modelling uncertainty, the values of m_1 and m_2 may not be exactly known. Consider a general case where both m_1 and m_2 are uncertain. Suppose that m_1 is normally distributed with a mean of μ_{m1} and a standard deviation of 0.05 and that m_2 is normally distributed with a mean of 0.5 and a standard deviation of 0.05.

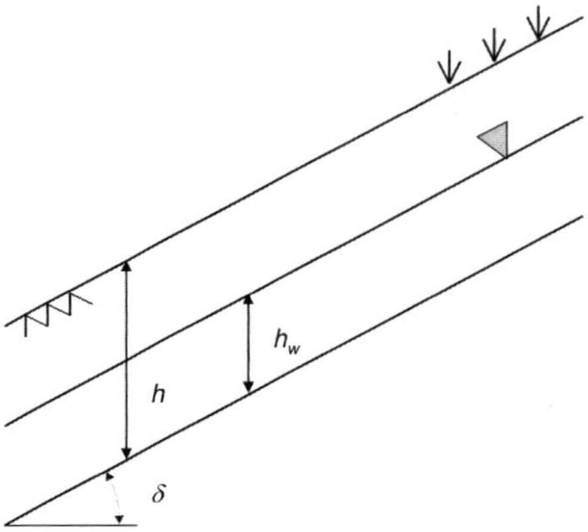

Figure 3. An existing slope for reliability assessment (Zhang et al. 2011)

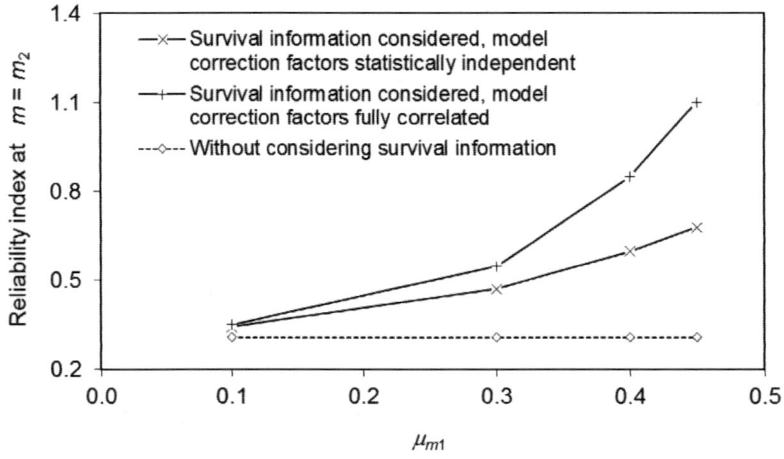

Figure 4. Reliability index of the infinite slope at $m = m_2$ when μ_{m1} takes various values (Zhang et al. 2011)

Let ε_1 and ε_2 denote the model correction factors in computing FOS_1 and FOS_2, respectively. Consider first the case where ε_1 and ε_2 are statistically independent. Figure 4 shows the reliability indexes of the slope at $m = m_2$ considering the survival information of $FOS_1 > 1$. After considering the survival information, the reliability index of the slope at $m = m_2$ increases, indicating that we are more confident in the safety of the slope. Figure 4 also shows that the amount of improvement in reliability index increases with μ_{m1}, implying that the updating effects will be more obvious when the slope survives a more critical state. Hence, ignoring the survival information would underestimate the reliability index of a slope and may result in an uneconomical design. Assuming the model correction factors (i.e., ε_1 and ε_2) are fully correlated, the reliability index of the slope at $m = m_2$ is re-evaluated and the results are also plotted in Figure 4. With this latter assumption, the assessed reliability index is larger than that obtained by assuming ε_1 and ε_2 being statistically independent given the same μ_{m1}.

This is because the assumption of model correction factors being fully correlated implies a stronger association between the performances of the slope when the groundwater is at various levels; hence, the past survival information will be more useful in updating the reliability of the slope at $m = m_2$.

4.2. Consideration of failure information

Suppose it is known that a slope has failed at a previous state, i.e., $\text{FOS}_1 = 1$. Let $p_{f2|\text{FOS1}=1}$ denote the failure probability of the slope at a future moment considering its past failure information. Based on the conditional probability theorem, $p_{f2|\text{FOS1}=1}$ can be written as:

$$p_{f2|y_1=1} = \frac{P(\text{FOS}_2 < 1 \cap \text{FOS}_1 = 1)}{P(\text{FOS}_1 = 1)} \tag{16}$$

Let β_1 and β_2 denote the reliability indexes of the slope at the previous moment and at the future moment without considering the information of past performance. Let ρ_{12} denote the correlation coefficient between the FOSs of the slope at these two moments. The conditional failure probability shown in Equation (16) can be estimated based on β_1, β_2, and ρ_{12}, which has been given in Madsen et al. (1986), as follows:

$$p_{f2|\text{FOS}_1=1} = \Phi\left(-\frac{\beta_2 - \rho_{12}\beta_1}{\sqrt{1-\rho_{12}^2}}\right) \tag{17}$$

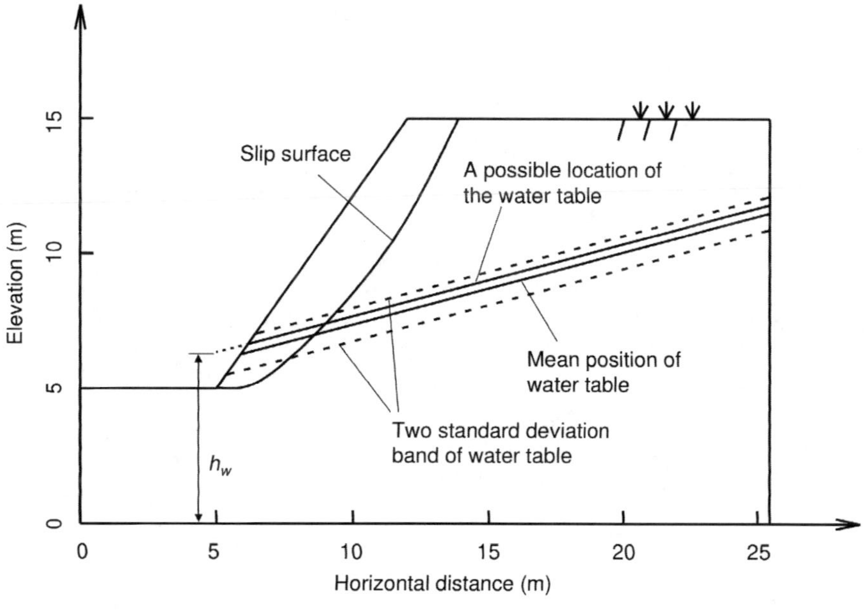

Figure 5. A failed cut slope for repair design (Zhang et al. 2011)

To show the impact of past failure information on reliability evaluation, Figure 5 shows the geometry of a failed cut slope that has a slope angle of 55°. The exact position of groundwater table at the moment of slope failure is unknown, and is assumed as normally distributed around its mean position with a standard deviation of 0.3 m, as shown in Figure 5. The location of the groundwater table is characterized by the elevation of the projection line of the groundwater table at the slope toe, which is denoted as h_w (Figure 5). Based on the previous assumption about the groundwater table, h_w is normally distributed with a mean of 6 m and a standard deviation of 0.3 m. Suppose the lowering of the slope angle is used to repair the slope. Based on Equation (15), the correlation coefficients between the FOSs of the slope at the angle of 55° and at a new slope angle are all around 0.74 when the angle of the new slope varies within the range of 35° to 50°. Based on Equation (17), the reliability indices of the slope at various slope angles, considering the past failure information, are calculated and plotted in Figure 6.

The reliability index of the new slope at various slope angles with and without considering the past failure information is shown in Figure 6. Here the past failure information can significantly affect the reliability indexes of the slope at other slope angles. When the slope angle is 55°, the reliability index of the slope, considering the failure information, is lower than that obtained without considering the previous failure information, which is more consistent with failure of the slope at this particular angle. As the slope angle decreases, the reliability index of the slope considering the failure information increases more rapidly than without considering the failure information. This inclusion rather than omission of failure information thus reduces the uncertainty in the slope stability analysis, making the reliability of the slope more sensitive to the change in slope angle. To reach a target reliability index of 3.0, we only need to flatten the slope to approximately 40° when the previous failure information is considered. For comparison, if the previous failure information is not considered, the slope would need to be flattened to about 35° to meet the same target reliability index.

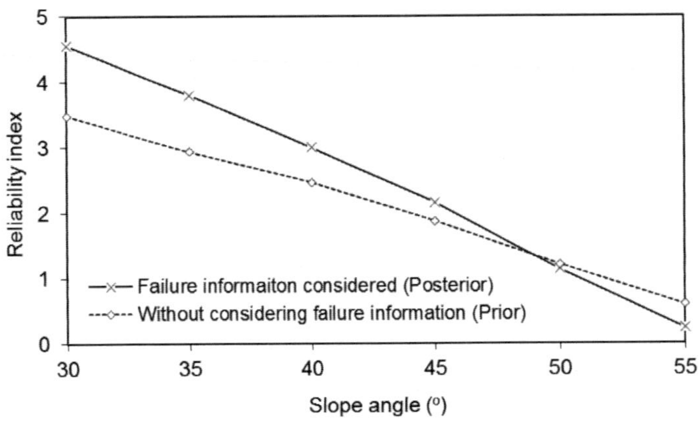

Figure 6. Reliability index of the slope at various slope angles with and without considering past failure information (Zhang et al. 2011)

5. Reliability-Based Design of Slopes with Incomplete Probabilistic Information

Most probabilistic methods require detailed statistics of uncertain variables, which are sometimes difficult to obtain due to budgetary constraints for site investigation and field and laboratory tests. The results from a probabilistic analysis, however, may depend on the input statistics of the uncertain variables. As such, while an initially acceptable design may not satisfy the target reliability index or failure probability if the variation of the geotechnical properties is underestimated; an initially cost-efficient design may not be cost-efficient if the variation of the geotechnical properties is overestimated (Juang et al. 2012b&2013). To address such a dilemma, the robust design concept initiated by Taguchi (1986) for improving product quality and reliability in industrial engineering is recently explored for design of slopes (Juang et al. 2012b&2103; Wang et al. 2013b; Gong et al. 2015), in which the easy-to-control parameters can be systematically adjusted such that the performance of a slope becomes insensitive to the statistics of variables that are hard to determine accurately. Naturally, higher design robustness is achieved at a higher cost. Thus, a multi-objective optimization considering both cost and robustness is needed to select the optimal designs among those in the acceptable design space. The robust reliability-based design methodology will be illustrated later in this paper with a slope in Hong Kong.

6. Application Example 1: Reliability-Based Design of Slopes

6.1. Engineering background

The #6 landslide of Shang-Shan Highway lies in Zhejiang Province, China. From May to June in 2000, the slope deformation accelerated due to continuous rainfall, with an obvious overall landslide occurring on June 10 of that year. The typical geological cross section A2-A2′ of #6 landslide, shown in Figure 7, is adopted for the following reliability analysis. The post-failure investigation indicates that the slip surface of the landslide occurred along the surface between the overlying soil and the bedrock as marked in Figure 7. The landslide mass is composed of gravel and gravelly clay. After the failure event, a series of engineering measures including drainage holes and reinforced piles were adopted to stabilize the slope in December, 2001. This landslide is reanalyzed here to illustrate the benefit of preforming reliability-based design.

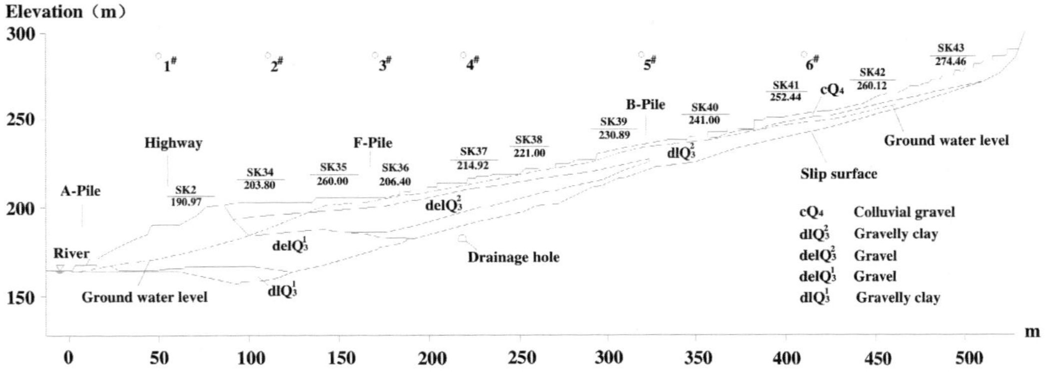

Figure 7. Cross section A2-A2′ of #6 landslide

6.2. Quantification of uncertainties

An analysis of the slope stability requires determining the variances in cohesion (c), the friction angle (φ) of the slip surface materials as well as the pore pressure ratio (r_u) at the moment of the slope failure. Extensive laboratory tests determined mean values of c and φ were 21.6 kPa and 7.3°, respectively, and standard deviations of c and φ were 9.0 kPa and 3.0°, respectively. The determination of the pore pressure ratio r_u at the moment of slope failure seems difficult because the ground water level was not measured then. On the basis of the monitoring data of the ground water level after the slope failure, the mean value of r_u was estimated as 0.35, and the standard deviation of r_u was assumed as 0.1. Let $\boldsymbol{\theta} = \{r_u, c, \varphi\}$. Based on the above analysis, the mean and standard deviation of $\boldsymbol{\theta}$ are $\boldsymbol{\mu}_{\theta} = \{0.35, 21.6, 7.3\}$ and $\boldsymbol{\sigma}_{\theta} = \{0.1, 9, 3\}$, respectively. Assuming that the elements of $\boldsymbol{\theta}$ are statistically independent, the Morgenstern-Price method is then used for the slope stability analysis. For this particular slope, the mean and standard deviation of the model error is assumed as 0.02 and 0.07, respectively.

6.3. Back analysis of slope failure

The slope failure can provide valuable information about the parameters of the slope. Assuming that $\boldsymbol{\theta}$ follows the multivariate normal distribution, the probabilistic back analysis method derived by Equations (5)-(7) is applicable. Figure 8 shows the spreadsheet template for implementing the above probabilistic back analysis method. The **H** vector is obtained based on sensitivity analysis of the slope stability model around the mean point. Based on Figure 8, the posterior values of mean and standard deviation of $\boldsymbol{\theta}$ are $\boldsymbol{\mu}_{\theta|d} = \{0.323, 26.126, 9.356\}$ and $\boldsymbol{\sigma}_{\theta|d} = \{0.096, 7.574, 2.03\}$, respectively. Although the elements of $\boldsymbol{\theta}$ are uncorrelated in the prior distribution, they are correlated in the posterior distribution, which is a reflection of the constraint that the FOS of the slope at the moment of failure should be equal to unity.

	A	B	C	D	E	F	G	H	I	G	K
1		μ_θ	C_θ					r_u	c	ϕ	
2	r_u	0.35	0.01	0	0		H	-0.712	0.0147	0.06	
3	c	21.6	0	81	0		$g(\mu_\theta)$	0.752			
4	ϕ	7.3	0	0	9		d_{obs}	1			
5											
6		μ_ε	σ_ε				$HC_\theta H^T$				
7		0.02	0.07				0.055084				
8											
9		H^TH					$\mu_{\theta\|d}$	$C_{\theta\|d}$			
10		0.5072	-0.01	-0.043			0.322929	0.0092	0.1414	0.064	
11		-0.01	0.0002	9E-04			26.12589	0.1414	57.364	-10.74	
12		-0.043	0.0009	0.004			9.35598	0.0642	-10.74	4.122	
13											

Figure 8. Spreadsheet of probabilistic back-analysis of slope failure

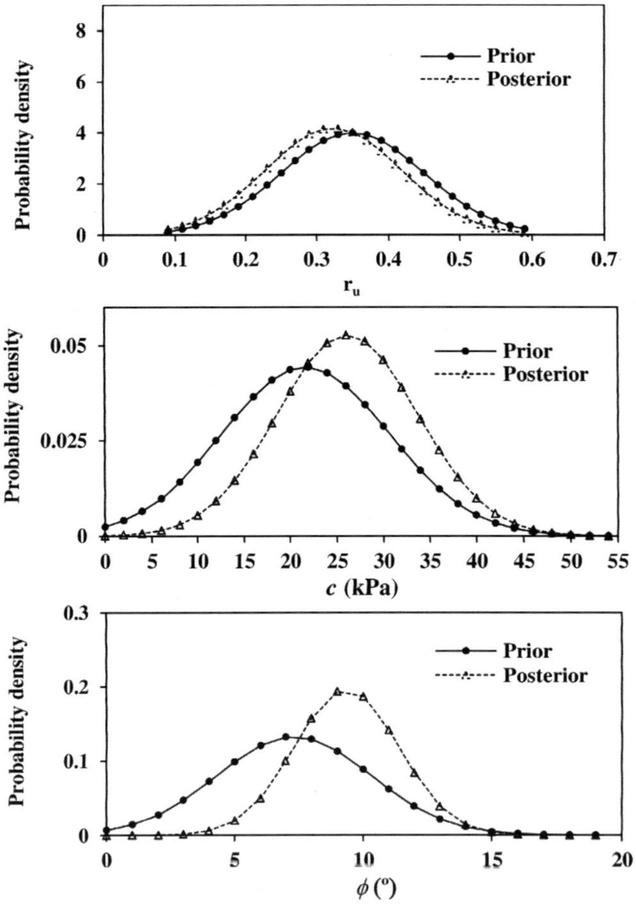

Figure 9. Prior and posterior probability density of r_u, c and ϕ

As $\boldsymbol{\theta}$ is assumed as multivariate normal, the posterior distribution of $\boldsymbol{\theta}$ is also the multivariate normal. Figures 9(a), 9(b), and 9(c) compare the prior and posterior PDF of for r_u, c and ϕ, respectively. Note that after a back analysis, the obvious reduction in the spread of the posterior distributions indicates a corresponding reduction in the uncertainties in these input parameters. The degree of updating for different variables, however, differs. The probabilistic back analysis algorithm can automatically allocate the observed information to different variables considering their relative importance and contribution to the uncertainty of the overall performance of the slope.

6.4. Design of reinforcing piles

To restore the stability of the landslide, we provide as an example the use of reinforcing piles for slope stabilization. First designed using the traditional deterministic analysis, the adopted values of c, ϕ, and r_u are 26 kPa, 9.1°, and 0.25, respectively. Based on Chinese Specifications for Design of Highway Subgrades (JTGD30-2004), the target FOS is 1.3. This deterministic analysis in turn determined that a lateral resisting force of 14,334 kN is required to meet the target FOS.

Reliability methods are then used for design of the reinforced piles. To show the effect of back analysis, both the prior distribution of $\boldsymbol{\theta}$ and the posterior distribution of $\boldsymbol{\theta}$ are used to calculate the relationship between resisting force of the piles and the reliability index of the landslide, as shown in Figure 10. As the total resisting force increases from 5,000 kN to 15,000 kN, the reliability index changes from about 0.4 to 1.7 based on the prior distribution, and changes from 1.2 to about 3.8 based on the posterior distribution. If a resisting force of 14,334 is applied, the reliability index of the landslide will be only 1.65 based on the prior distribution of $\boldsymbol{\theta}$, which may be considered as inadequate. However, if the posterior distribution of $\boldsymbol{\theta}$ is employed, the reliability index of the slope is approximately 3.75 when the resisting force is 14,343, which can often be considered as adequate. The reliability method can not only be used to assess the reliability of a design, but can also be used to determine the required resistance to achieve a certain target reliability level. For instance, to meet a target reliability index of 3.0, the required resisting force is 12,500 kN.

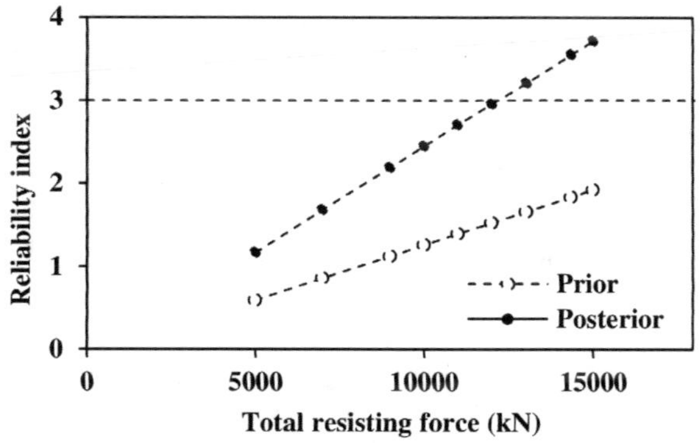

Figure 10. Reliability index at various lateral resisting forces

7. Application Example 2: Reliability-Based Robust Design of Slopes

7.1. Engineering background

The second case history in this article details the use of the robust design approach on the Sau Mau Ping rock slope in Hong Kong. The rock mass of the Sau Mau Ping slope is un-weathered granite with sheet joints, which is formed by the exfoliation processes during the cooling of granite. An initial and elemental analysis by Hoek (2006) determined that the Sau Mau Ping slope was composed of a single unstable block with a water-filled tension crack involving only a single failure mode. The geometry of the Sau Mau Ping slope is illustrated in Figure 11. Following Hoek (2006), the slope before remediation has a height H of 60 m and a slope angle Θ of 50°. The potential failure plane is inclined at 35° (ψ = 35°) and the unit weight of rock (γ) is 2.6 ton/m^3, which are assumed to be fixed values. A deterministic model with a single failure mode developed by Hoek and Bray (1981) is adopted herein for the case study of the Sau Mau Ping slope, the detailed formulation for which can be found in Hoek (2006) and Wang et al. (2013b&2015).

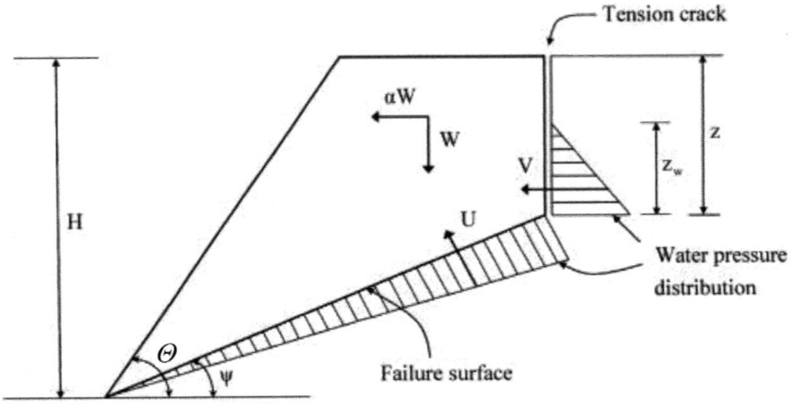

Figure 11. Illustration of geometry for plane sliding of Sau Mau Ping rock slope (Wang et al. 2013b)

There are two approaches for the remedial design of a rock slope: (i) geometry design, in which the slope height and slope face angle are reduced, and (ii) reinforcing of the rock slope using rock bolts and cables. Stabilizing rock slopes by means of geometry design is generally more cost-efficient than the use of cables and bolts (Hoek 2006). For ease of illustration, we will focus on the geometrical design of the Sau Mau Ping slope, with slope height H and slope angle Θ as design parameters. Robust reliability-based design of rock bolts for the Sau Mau Ping slope can be found in Wang et al. (2015).

7.2. Quantification of uncertainties

As summarized by Hoek (2006), five uncertain variables are considered in the reliability analysis of the Sau Mau Ping slope: the cohesion of rock discontinuities (c), friction angle of rock discontinuities (ϕ), tension crack depth (z), percentage of the tension crack depth filled with water (i_w), and gravitational acceleration coefficient (α). Following the previous studies in Hoek (2006) and Low (2007), c, ϕ and z are assumed to follow the normal distribution, and i_w and α are assumed to follow the truncated exponential distributions with mean and upper and lower bounds listed in Table 3. The mean values of these uncertain parameters can usually be adequately estimated even with a small sample size (Wu et al. 1989); however, the coefficient of variation (COV) and the correlation of coefficient (ρ) of these uncertain parameters are difficult to ascertain (Gong et al. 2014). In past studies (Hoek 2006; Low 2007), COV and ρ of these uncertain parameters were primarily estimated by engineering judgment or based on published literature.

According to a survey of literature by Lee et al. (2012), the COV of cohesion c, denoted as COV[c], typically ranges from 10% to 30%; the COV of friction angle ϕ, denoted as COV[ϕ], typically varies in the range from 10% to 20%. Furthermore, the coefficient of correlation for c and ϕ, denoted as $\rho_{c,\phi}$, is likely to be negative varying from -0.2 to -0.8; and the coefficient of correlation for z and i_w , denoted as $\rho_{z,iw}$, may vary from -0.2 to -0.8. Then, the mean and the COV values of these statistics are estimated. The mean of COV[c], denoted as $\mu_{COV[c]}$, is 0.20 and the COV of COV[c], denoted as $\delta_{COV[c]}$ is 17%; the mean of COV[ϕ], denoted as $\mu_{COV[\phi]}$, is 0.14 and the COV of COV[ϕ], denoted as $\delta_{COV[\phi]}$, is 12%; the mean of $\rho_{c,\phi}$, denoted as $\mu_{\rho c,\phi}$, is -0.50 and the COV of $\rho_{c,\phi}$, denoted as

$\delta_{\rho c,\phi}$, is 25%; and, the mean of $\rho_{z,iw}$, denoted as $\mu_{\rho z,iw}$, is -0.50 and the COV of $\rho_{z,iw}$, denoted as $\delta_{\rho z,iw}$, is 25%.

Table 3. Statistics of uncertain variables for the Sau Mau Ping rock slope (Wang et al. 2013b)

Random variables	Probability distribution	Mean	Std. dev.
c	Normal	10 ton/m^2	–
ϕ	Normal	35°	–
z	Normal	14 m	3 m
i_w	Exponential with mean 0.5, truncated to [0, 1]		
α	Exponential with mean 0.08, truncated to [0, 0.16]		

7.3. Robust design of slope geometry for slope stabilization

For the two design parameters of the Sau Mau Ping slope, the slope height H may typically range from 50 m to 60 m, and slope angle Θ may typically range from 44° to 50°. For convenience of construction, slope height H may be rounded to the nearest 0.2 m and slope angle Θ may be rounded to the nearest 0.2°. Thus, a discrete design space is adopted herein: H may assume 51 discrete values and Θ may assume 31 discrete values in their possible ranges, which yields a total of 1,581 candidate designs.

When the statistics of the random variables are uncertain, the calculated failure probability of the slope is also uncertain. Within the framework of the robust reliability-based design, a target failure probability (i.e., p_T = 0.0062) can be set as the safety constraint and used to screen the unsatisfactory designs based on the mean value of the failure probability (μ_p). The standard deviation of the failure probability (σ_p) can be used to gauge the robustness, with a smaller standard deviation of the failure probability signaling higher design robustness. In this example, the cost of a rock slope design is approximated as the volume of rock mass that is to be excavated (Duzgun et al. 1995). The design is considered as most competitive when it is optimal to both cost efficiency and design robustness while satisfying the safety constraint. Mathematically, the robust reliability-based design can be formulated as a multi-objective optimization problem as follows

Find $\mathbf{d} = [H, \Theta]$

Subject to: $H \in \{50\text{m}, 50.2\text{m}, 50.4\text{m}, \ldots, 60\text{m}\}$ and $\Theta \in \{44°, 44.2°, 44.6°, \ldots, 50°\}$

$\mu_p < p_T = 0.0062$

Objectives: Minimizing the standard deviation failure probability (σ_p),

Minimizing the cost for rock slope design

NSGA-II (Deb et al. 2002) can be used to solve the above multi-objective optimization, in which the designs optimal to both objectives (cost and robustness) are searched iteratively in the discrete design space. With this algorithm, 89 non-dominated optimal designs are selected into a Pareto Front, as shown in Figure 12. As can be seen from this figure, an apparent trade-off relationship exists between design robustness and cost efficiency, and greater design robustness can only be attained at the expense of a higher cost. The Pareto Front is useful as a design aid for the final decision making especially when an acceptable cost or robustness level is specified. For instance, when the maximum acceptable cost is set at 200 units, 38 designs on the Pareto Front are deemed acceptable (Figure 12). The design with the lowest σ_p value in this cost range is the best design based on the robustness criterion, and in this scenario, it is a design with H =50 m and Θ= 50°. The optimal design of H =50 m and Θ= 50° has a cost of 195.1 units and a standard deviation of failure probability of 1.25×10^{-3}.

Furthermore, the feasibility robustness index β_β, defined as follows, may be computed and used as a decision making aid.

$$P[p_f - p_T \leq 0] = \Phi(\beta_\beta) \geq P_0 \tag{18}$$

The feasibility robustness index β_β is readily computed for each of the 89 non-dominated optimal designs on the Pareto Front. As expected, the design with higher feasibility robustness costs more. When a target feasibility robustness level (β_β^T) is selected, the least-cost design that satisfies the feasibility robustness level can easily be identified, which renders a final design using the proposed robust design method. As an example, Table 4 shows the final designs identified using the robust design method for different levels of target feasibility robustness.

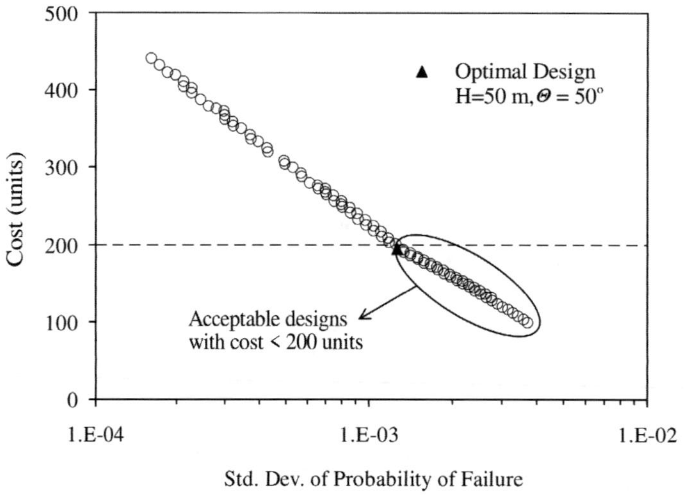

Figure 12. Pareto Front obtained by NSGA-II and selection of final optimal design based on the acceptable cost level (Wang et al. 2013b)

Table 4. Selected final designs at various target feasibility robustness levels (Wang et al. 2013b)

β_β^T	P_0	H (m)	Θ (°)	Cost (units)
0.5	69.15%	54.0	50.0	107.0
1.5	93.32%	51.4	49.8	170.6
2.5	99.38%	50.0	45.4	378.9

8. Concluding Remarks

The characterization and assessment of the effect of uncertainties on slope stability evaluation are important in the analysis and design of slopes. In recent years, the probabilistic methods have gradually evolved into a practical tool for stability analysis of slopes considering uncertainties, and simple yet sound methods are available for addressing challenging problems that are hard to answer within a deterministic framework. In this paper, we have introduced and illustrated practical methods developed recently to address important problems related to reliability-based assessment of slope stability. The following conclusions can be drawn from this paper.

- Probabilistic methods are now available for characterizing the uncertainty associated with the slope stability model considering the presence of parameter uncertainties. The model error associated with a slope stability model can be determined if the field performances of similar slopes in a region are available. For a specific slope, simple probabilistic back analysis methods can be used to update the probabilistic distribution of uncertain variables based on the observed failure. These methods provide practicing engineers tools for better uncertainty characterization as the basis for slope reliability analysis.

- The most critical slip surface of a slope can be found with the empirical procedure suggested by Hassan and Wolff quite accurately. After the most critical slip surface is found, the failure probability can be calculated based on a second-order polynomial-based response surface method. The Hassan Wolff procedure can also be used to locate representative slip surfaces that govern the system failure probability of a slope. Combined with response surface method, the Hassan and Wolff method can be extended into a practical tool for evaluating the system reliability of a slope efficiently.
- The past performance information of a slope can be viewed as the outcome of a full-scale test directly performed on the slope under investigation, and can be directly used for reliability updating. When the past performance information is considered, the uncertainty in the slope stability analysis can be reduced, and the reliability of the slope can be evaluated more realistically.
- When the probabilistic distribution of random variables cannot be determined accurately, the reliability-based design approach can be implemented as a multi-objective optimization problem by considering design robustness as one additional design objective. In such a case, a set of non-dominated optimal designs that satisfy the safety (reliability) reθuirement can be found, which show the trade-off relationship between design robustness and cost efficiency. The robust design method can be a useful design aid in the environment of high uncertainties. It complements the reliability-based methods for the analysis and design of a slope.

Acknowledgements
The second author wishes to acknowledge the support from the Natural Science Foundation of China (41372275) and the Shanghai Rising-star Program (15QA1403800).

References

[1] Alonso E E 1976 Risk analysis of slopes and its application to slopes in Canadian sensitive clays *Geotechnique* **26** 453-72
[2] Ang A H S and Tang W H 1984 *Probability Concepts in Engineering Planning and Design: Decision, Risk and Reliability* Vol. 2 (New York: Wiley)
[3] Babu G L S and Murthy D S N 2005 Reliability analysis of unsaturated soil slopes *J. Geotech. Geoenviron. Eng.* **131(11)** 1423-28
[4] Bhattacharya G, Jana D, Ojha S and Chakraborty S 2003 Direct search for minimum reliability index of earth slopes *Comput. Geotech.* **30(6)** 455-62
[5] Cassidy M J, Uzielli M and Lacasse S 2008 Probability risk assessment of landslides: A case study at Finneidfjord *Can. Geotech. J.* 45(9) 1250-67
[6] Cheung R W M and Tang W H 2005 Realistic assessment of slope reliability for effective landslide hazard management *Geotechnique* **55(1)** 85-94
[7] Ching J, Phoon K K and Hu Y G 2009 Efficient evaluation of reliability for slopes with circular slip surfaces using importance sampling *J. Geotech. Geoenviron. Eng.* **135(6)** 768-77
[8] Cho S E 2007 Effects of spatial variability of soil properties on slope stability *Eng. Geol.* **92(3)** 97-109
[9] Chowdhury R N, Tang W H and Sidi I 1987 Reliability model of progressive slope failure *Geotechnique* **37(4)** 467-81
[10] Chowdhury R N and Xu D W 1995 Geotechnical system reliability of slopes *Reliab. Eng. Syst. Saf.* **47(3)** 141-51
[11] Christian J T, Ladd C C and Baecher G B 1994. Reliability applied to slope stability analysis *J. Geotech. Eng.* **120(12)** 2180-207
[12] Cornell C A 1971 First-order uncertainty analysis of soils deformation and stability *Proc. 1st Int. Conf. on Application of Statistics and Probability to Soil and Structural Engineering* (Hong Kong: Hong Kong University) pp 130-43
[13] Deb K, Pratap A, Agarwal S and Meyarivan T 2002 A fast and elitist multi-objective genetic

algorithm: NSGA-II *Trans. Evol. Comput.* **6(2)** 182-97
[14] Duncan J M and Wright S G 1980 The accuracy of equilibrium methods of slope stability analysis *Eng. Geol.* **16(1)** 5-17
[15] Duncan J M 2000. Factors of Safety and Reliability in Geotechnical Engineering *J. Geotech. Geoenviron. Eng.* **126(4)** 307-16
[16] Duzgun H S B, Pasamehmetoglu A G and Yucemen M S 1995 Plane failure analysis of rock slopes: a reliability approach *Int. J. Surf. Min. Reclam. Env.* **9(1)** 1-6
[17] Einstein H H and Baecher G B 1983. Probabilistic and statistical methods in engineering geology, Part I. Explorations *Rock Mech. Rock Eng.* **16(1)** 39-72
[18] El-Ramly H, Morgenstern N R and Cruden D M 2002 Probabilistic slope stability analysis for practice *Can. Geotech. J.* **39** 665-83
[19] Feng T and Fredlund M 2011 *SVSLOPE: Verification Manual* (Saskatoon: SoilVision Systems Ltd)
[20] Frattini P, Crosta G and Sosio R 2009 Approaches for defining thresholds and return periods for rainfall-triggered shallow landslides *Hydrol. Process* **23(10)** 1444-60
[21] Gilbert R B, Wright S G and Liedtke E 1998 Uncertainty in back-analysis of slopes: Kettleman Hills case history *J. Geotech. Geoenviron. Eng.* **124(12)** 1167-76
[22] Gilbert R B and Tang W H 1995 Model uncertainty in offshore geotechnical reliability *Proc., 27th Offshore Technology Conf.* (Houston: Society of Petroleum Engineers) pp 557-67
[23] Roh G and Hong H P 2009 Calibration of information-sensitive partial factors for assessing earth slopes *J. GeoEng.* **4(3)** 93-102
[24] Gong W, Luo Z, Juang C H, Huang H, Zhang J and Wang L 2014 Optimization of site exploration program for improved prediction of tunneling-induced ground settlement in clays *Comput. Geotech.* **56** 69-79
[25] Gong W, Wang L, Khoshnevisan S, Juang C H, Huang H and Zhang J 2015 Robust geotechnical design of earth slopes using fuzzy sets *J. Geotech. Geoenviron. Eng.* **141(1)** 04014084
[26] Griffiths D V and Fenton G A 2004 Probabilistic slope stability analysis by finite elements *J. Geotech. Geoenviron. Eng.* **130(5)** 507-18
[27] Griffiths D V, Huang J and Fenton G A 2009 Influence of spatial variability on slope reliability using 2-D random fields *J. Geotech. Geoenviron. Eng.* **135(10)** 1367-78
[28] Hassan A M and Wolff T F 1999 Search algorithm for minimum reliability index of earth slopes *J. Geotech. Geoenviron. Eng.* **125(4)** 301-8
[29] Hoek E and Bray J 1981 *Rock Slope Engineering* (3rd ed) (London: Institution of Mining and Metallurgy)
[30] Hoek E 2006 *Practical Rock Engineering* Chapter 7: A slope stability problem in Hong Kong and Chapter 8: Factor or safety and probability of failure. http://www.rocscience.com/hoek/PracticalRockEngineering.asp
[31] Honjo Y and Kuroda K 1991 A new look at fluctuating geotechnical data for reliability design *Soils Found.* **31(1)** 110-20
[32] Huang J, Lyamin A V, Griffiths D V, Krabbenhoft K and Sloan S W 2013 Quantitative risk assessment of landslide by limit analysis and random fields *Comput. Geotech.* **53** 60-7
[33] Ji J and Low B K 2012 Stratified response surfaces for system probabilistic evaluation of slopes *J. Geotech. Geoenviron. Eng.* **138(11)** 1398-406
[34] Juang C H, Luo Z, Atamturktur S and Huang H 2012a Bayesian updating of soil parameters for braced excavations using field observations *J. Geotech. Geoenviron. Eng.* **139(3)** 395-406
[35] Juang C H, Wang L, Atamturktur S and Luo Z 2012b Reliability-based robust and optimal design of shallow foundations in cohesionless soil in the face of uncertainty *J. GeoEng.* **7(3)** 43-55
[36] Juang C H, Wang L, Liu Z, Ravichandran N, Huang H and Zhang J 2013 Robust geotechnical design of drilled shafts in sand: New design perspective *J. Geotech. Geoenviron. Eng.*

139(12) 2007-19

[37] Kim M M 1980 *Centrifuge Model Testing of Soil Slopes* (Boulder: Univ. of Colorado)

[38] Lacasse S and Nadim F 1996. Uncertainties in characterizing soil properties. *Uncertainty in the Geologic Environment: From Theory to Practice. ASCE Geotechnical Special Publication* vol 58, ed C D Shackleford, P P Nelson and M J S Roth (Reston: ASCE) pp 49-75

[39] Lee Y F, Chi Y Y, Juang C H and Lee D H 2012 Reliability analysis of rock wedge stability–a knowledge-based Clustered Partitioning (KCP) approach *J. Geotech. Geoenviron. Eng.* **138(6)** 700-8

[40] Li D Q, Jiang S H, Cao Z J, Zhou W, Zhou C B and Zhang L M 2015 A multiple response-surface method for slope reliability analysis considering spatial variability of soil properties *Eng. Geol.* **187** 60-72

[41] Li K S and Lumb P 1987 Probabilistic design of slopes *Can. Geotech. J.* **24(4)** 520-31

[42] Li L, Wang Y, Cao Z and Chu X 2013 Risk de-aggregation and system reliability analysis of slope stability using representative slip surfaces *Comput. Geotech.* **53** 95-105

[43] Liang R Y, Nusier O K and Malkawi A H 1999 A reliability based approach for evaluating the slope stability of embankment dams *Eng. Geol.* **54** 271-85

[44] Low B K 2007 Reliability analysis of rock slopes involving correlated nonnormals *Int. J. Rock Mech. Min. Sci.* **44(6)** 922-35

[45] Low B K 2008 Efficient probabilistic algorithm illustrated for a rock slope *Rock Mech. Rock Eng.* **41(5)** 715-34

[46] Low B K, Zhang J and Tang W H 2011 Efficient system reliability analysis illustrated for a retaining wall and a soil slope *Comput. Geotech.* **38** 196-204

[47] Madsen H O, Krenk S and Lind N C 1986 *Methods of Structural Safety* (Englewood Cliffs: Prentice-Hall)

[48] Matsuo M and Kuroda K 1974 Probabilistic approach to the design of embankments *Soils Found* **14** 1-17

[49] Ministry of Transport P.R. China 2004 *Specifications for Design of Highway Subgrades. JTGD30-2004* (Beijing: Ministry of Transport P.R. China)

[50] Mostyn G R and Li K S 1993 Probabilistic slope stability-state of play *Proc. Conf. on Probabilistic Methods in Geotechnical Engineering (Canberra, Australia, 10-12 Feburary 1993)* ed K S Li and S-C R Lo (Rotterdam: Balkema) pp 89–110

[51] Morgenstern N R and Price V E 1965 The analysis of the stability of general slip surfaces *Geotechnique* **15(1)** 79-93

[52] Nadim F, Kvalstad T J and Guttormsen T. 2005. Quantification of risks associated with seabed instability at Ormen Lange *Mar. Petrol. Geol.* **22(1–2)** 311-8

[53] Oka Y and Wu T H 1990 System reliability of slope stability *J. Geotech. Eng.* **116(8)** 1185-9

[54] Phoon K K and Kulhawy F H 1999 Characterization of geotechnical variability *Can. Geotech. J.* **36** 612-24

[55] Salgado R and Kim D 2014 Reliability analysis of load and resistance factor design of slopes *J. Geotech. Geoenviron. Eng.* **140(1)** 57-73

[56] Santoso A M, Phoon K K and Quek S T 2011 Effects of soil spatial variability on rainfall-induced landslides *Comput. Struct.* **89(11-12)** 893-900

[57] Taguchi G 1986 *Introduction to Quality Engineering: Designing Quality into Products and Processes* (New York)

[58] Tang W H, Yucemen M S and Ang A S 1976 Probability-based short term design of soil slopes *Can. Geotech. J.* **13(3)** 201-15

[59] Tang W H, Stark T D and Angulo M 1999 Reliability in back analysis of slope failures *Soils Found.* **39(5)** 73-80

[60] Vanmarcke E H 1977 Probabilistic modeling of soil profiles *J. Geotech. Eng.* **103** 1227-46

[61] Wang L, Hwang J H, Luo Z, Juang C H and Xiao J 2013a Probabilistic back analysis of slope failure–A case study in Taiwan *Comput. Geotech.* **51** 12-23

[62] Wang L, Hwang J H, Juang C H and Atamturktur S 2013b Reliability-based design of rock slopes–A new perspective on design robustness *Eng. Geol.* **154** 56-63

[63] Wang L, Gong W, Luo Z and Juang C H 2015 Reliability-based robust geotechnical design of rock bolting system for slope stabilization *IFCEE 2015 © ASCE 2015* pp 1926-35

[64] Wolff T F, Hassan A, Khan R, Ur-Rasul I and Miller M 1995 Geotechnical reliability of dam and levee embankments *Technical report prepared for U.S. Army Engineer Waterways Experiment Station–Geotechnical Laboratory* (Vicksburg)

[65] Wu T H, Tang W H, Sangrey D A and Baecher G B 1989 Reliability of offshore foundations-state-of-the-art *J. Geotech. Eng.* **115(2)** 157

[66] Wu T H and Abdel-Latif M A 2000 Prediction and mapping of landslide hazard *Can. Geotech. J.* **37(4)** 781-95

[67] Wu T H and Kraft L M 1970 Safety analysis of slopes *J. Soil Mech. Found. Div.* **96(2)** 609-30

[68] Xue J F and Gavin K 2007 Simultaneous determination of critical slip surface and reliability index for slopes *J. Geotech. Geoenviron. Eng.* **137(7)** 878-86

[69] Zhang J, Zhang L M and Tang W H 2009 Bayesian framework for characterizing geotechnical model uncertainty *J. Geotech. Geoenviron. Eng.* **135(7)** 932-40

[70] Zhang J, Tang W H and Zhang L M 2010 Efficient probabilistic back-analysis of slope stability model parameters *J. Geotech. Geoenviron. Eng.* **136(1)** 99-109

[71] Zhang J, Zhang L M and Tang W H 2011 Slope reliability analysis considering site-specific performance information *J. Geotech. Geoenviron. Eng.***137(3)** 227-238

[72] Zhang J, Huang H W, Juang C H and Li D Q 2013 Extension of Hassan and Wolff method for system reliability analysis of soil slopes *Eng. Geol.* **160** 81-8

[73] Zhang L L, Zhang L M and Tang W H 2005 Rainfall-induced slope failure considering variability of soil properties *Geotechnique* **55(2)** 183-8

[74] Zhang S, Zhang L M, Peng M, Zhang L L; Zhao H F and Chen H X 2012 Assessment of risks of loose landslide deposits formed by the 2008 Wenchuan earthquake *Nat. Hazards Earth Syst. Sci.* **12(5)** 1381-92

[75] Zeng P and Jimenez R 2014. An approximation to the reliability of series geotechnical systems using a linearization approach *Comput. Geotech.* **62** 304-9

IOP Conf. Series: Earth and Environmental Science **26** (2015) 012007 doi:10.1088/1755-1315/26/1/012007

Preliminary experimental study on three-dimensional contact behavior of bonded granules

M J Jiang[1,2,3,5], S L Jin[1,2,3], Z F Shen[1,2,3], W Liu[1,2,3] and M R Coop[4]
[1] State Key Laboratory for Disaster Reduction in Civil Engineering, Tongji University, Shanghai, China, 200092
[2] Key Laboratory of Geotechnical and Underground Enigineering of Ministry of Education, Tongji University, Shanghai, China, 200092
[3] Department of Geotechnical Engineering, College of Civil Engineering, Tongji University, Shanghai, China, 200092
[4] City University of Hong Kong, Hong Kong, China

E-mail: mingjing.jiang@tongji.edu.cn

Abstract. In order to explore the microscopic contact behaviour of structured sands, devices were developed for specimen preparation and for carrying out tests on mechanical contact behaviour of three-dimensional (3D) bonded spheres. The specimen preparation device can cement two separate aluminium hemispheres by epoxy adhesive with accurate size control. The auxiliary loading devices can carry out compression, tension, shear, bending and torsion tests and any of their combinations. The experimental results show that the peak shear force, bending moment and torque of the bonded hemispheres are normal force dependent; that is, they first increase with the normal force and then decrease after the normal force exceeds a critical value.

1. Introduction

Slope stability has attracted wide range of research using equilibrium methods [1], in-situ monitor methods [2], physical models [3] and numerical simulations [4-5]. The discrete element method (DEM), first proposed by Cundall and Strack [6], has been increasingly applied as well in recent years. Compared with the complex macroscopic constitutive models and parameters in finite element method (FEM) simulations, only a simple and reasonable microscopic contact model and relatively few parameters are required in DEM. More importantly, DEM can naturally simulate large deformations and failure of slopes. Some slope failures can be attributed to the degradation of cementitious geomaterials. Utili and Nova [7] studied the evolution of natural cliffs subject to weathering with DEM. Jiang and Murakami [8] carried out a DEM simulation of the full-process slope failure with focus on very-rapid and extremely-rapid landslide process due to the loss of inter-particle bonding. In these simulations, a simple and reliable bonding contact model is one of the key issues.

A few researchers have investigated the mechanical behaviour of cemented granules through experiments and accordingly some bonding contact models have been proposed for DEM simulations [9, 10]. Delenne et al. [11] firstly presented an experimental verification of a DEM simulation of cemented granules. Kirsch et al. [12] presented a novel experimental setup invented to measure the mechanical behaviour of solid bridges between double-particle systems in the millimetre range. Jiang

[5] To whom any correspondence should be addressed.

et al. [13, 14] performed an experimental study on the contact behaviour of idealized granules (cemented aluminium rods) with different bond materials (epoxy adhesive and calcium aluminate cement) and different bond geometries (thin and thick bond modes). The mechanical behaviour of cemented rods was tested in simple and complex loading tests (tension, compression, pure shearing, pure rolling, and shear-rolling tests). The test results were used to examine the microscopic contact model which was employed in two-dimensional DEM simulations of structured sands [15].

Following the work in [13, 14], the main objective of this study is to carry out a preliminary experimental investigation of the mechanical behaviour of three-dimensional cemented granules, idealized as a pair of hemispheres glued together by epoxy adhesive. For this purpose, the devices to prepare and test the cemented granules are introduced first. The contact behaviour of the cemented hemispheres was tested in different loading modes, i.e., tension, compression, shear, bending and torsion tests. The effects of normal force on shear, bending and torsion strengths were examined.

2. Sample preparation

The aluminium hemispheres were machined from aluminium alloy blocks with an elastic modulus of 70 GPa and a Poisson's ratio of 0.3, which can match the properties of quartz sand particles. The epoxy adhesive has an average tensile strength of 16 MPa [13]. The epoxy adhesive was chosen because of its stable mechanical behaviour and high workability.

Figure 1 presents the sample preparation device which can be used to prepare high-quality reproducible cemented hemisphere pairs. The preparation device is composed of a bearing platform, three moulding modules and three fixing plates, which can be assembled by bolts.

A systematic preparation process was established after a large numbers of trials. First, the three moulding modules are assembled and a thin layer of petroleum jelly was evenly spread on the hemispherical concavities to make it easy to remove the prepared cemented granules from the moulding modules. Second, the hemispheres are cleaned with acetone and are placed into the lower six concavities (facing upward at this stage). Then, the bearing platform is fixed on the moulding modules with four bolts. The assembled device is then turned upside down so that the upper six concavities are exposed. Third, a certain amount of epoxy adhesive is filled into the cylindrical grooves. Then, another six hemispheres are put into the upper concavities and the three fixing plates are fixed on the moulding modules with nine bolts. Finally, the preparation device is kept in an electric oven at 100℃ for 1.5 h to shorten the solidification time of epoxy adhesive from 24 h to 1.5 h. After cooling, the six prepared pairs of cemented hemispheres are carefully removed from the device and placed in a sealed box under a constant temperature of 20℃ and humidity of 80% to cure for 20 days.

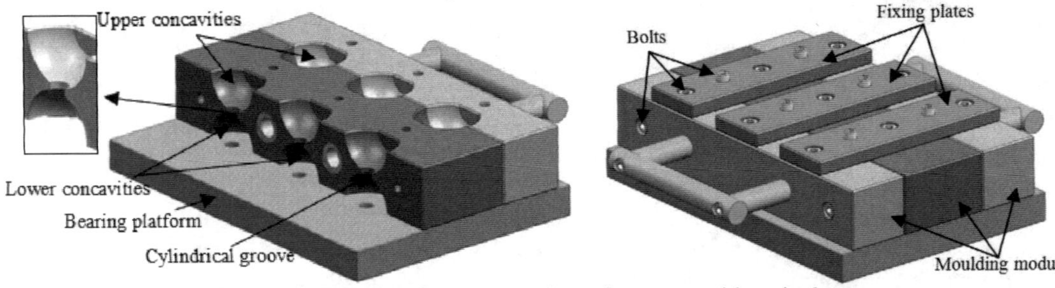

Figure 1. Devices for preparation of cemented hemispheres

3. Loading tests

3.1 Compression tests

The compression tests were carried out using a compression-auxiliary device, as shown in Figure 2(a). The auxiliary device is a rigid block with a cylindrical hollow. Four "windows" are opened to observe the deformation of the cement material. The cemented hemisphere pair was inserted into this hollow

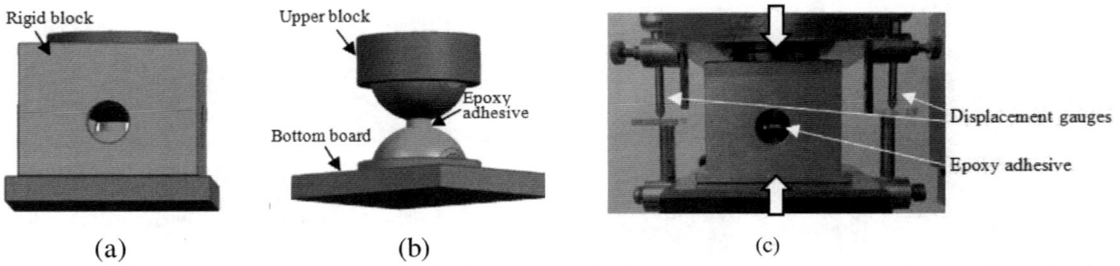

Figure 2. Compression test: (a) and (b) 3D schematic diagrams of the compression-auxiliary device; (c) physical test

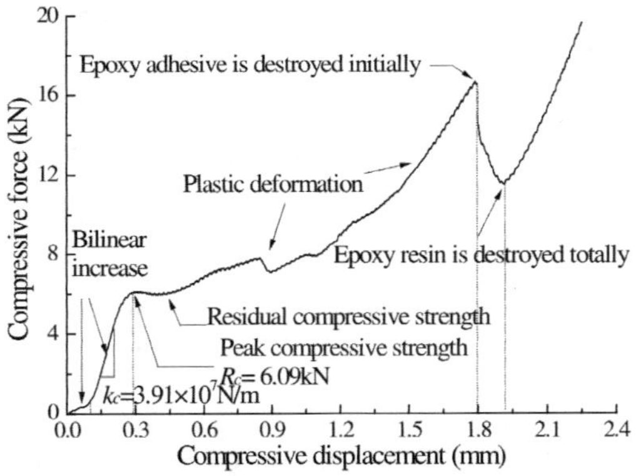

Figure 3. Compressive force-displacement relationship of the cemented hemispheres

so that any lateral displacement was constrained. The cemented hemisphere pair was then seated on the bottom board and capped with the upper block. The assembled compression-auxiliary device was then seated on the pedestal of the biaxial compression system and centered for the test. Petroleum jelly was used to reduce friction at interfaces. During the compression test, the pedestal was continuously moved upward at a constant rate of 0.1 mm/min. Two displacement gauges were attached on both sides to measure compression displacement. The vertical load was measured by a load cell installed inside the biaxial compression system.

Twenty-five cemented hemisphere pairs have been tested in compression. Figure 3 presents a typical relationship between the compressive force and displacement obtained in one test. Figure 3 shows that the compressive force initially increases bi-linearly up to the first peak, R_c, when the compressive displacement is 0.29 mm. The bi-linearity is possibly due to initial contact defects between the cemented hemispheres and the compression-auxiliary device. The compressive force then decreases slowly to a residual strength, followed by the second-stage nonlinear increase up to another peak when the compressive displacement reaches 1.8 mm. The epoxy adhesive experiences a clear plastic drum-like deformation mode in this stage. Then, the cementation is abruptly damaged and the two hemispheres start to contact.

In the compression test, the measured compressive stiffness k_c is 3.91×10^7 N/m and the peak compressive strength R_c is 6.09 kN. By normalizing this compressive strength with the cross-sectional area A of the cylindrical bond between the two hemispheres, the compressive strength expressed in stress is σ_c = 77.5 MPa ($\sigma_c = R_c/A$, $A = \pi(d/2)^2$, d = 10.0 mm). The compressive Young's Modulus E is 0.54 GPa ($E = k_c \cdot H_{center}/A$, H_{center} = 2.0 mm).

3.2 Tension tests

The tension tests were carried out with a tension-auxiliary device, as illustrated in Figure 4. The device includes two rigid blocks with grooves where the cemented hemispheres can be fixed. Four fixing bolts are used to fix hemispheres in the other direction. Four vertical rods are arranged in such a way as shown in Figure 4(b) that the compressive forces from the biaxial compression system act to push the two blocks apart, leading to a tensile force in the epoxy adhesive. The four rods can move freely in the pilot holes.

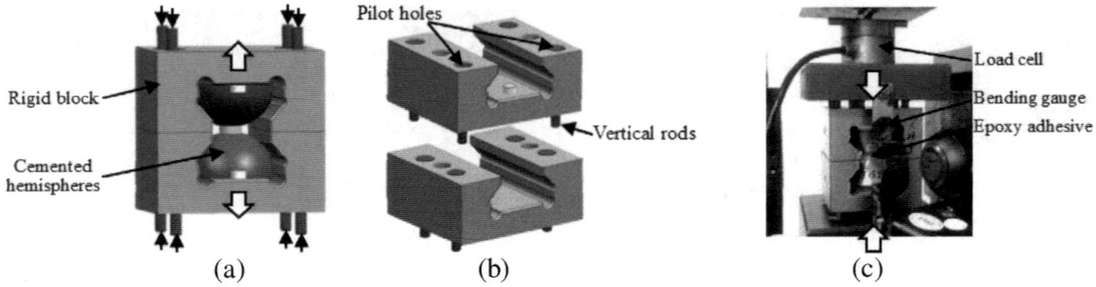

Figure 4. Tension test: (a) and (b) 3D schematic diagrams of the tension-auxiliary device; (c) physical test

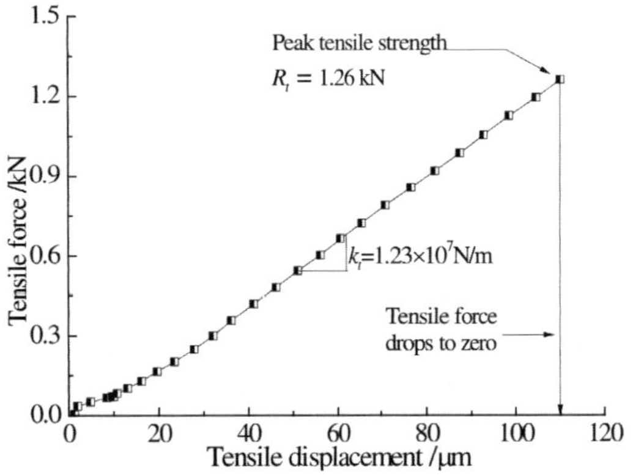

Figure 5. Tensile force-displacement relationship of the cemented hemispheres

During the tension tests, the vertical load, measured by a load cell, was applied at a constant tension rate of displacement of 0.1 mm/min. Two bending gauges (with higher accuracy than a displacement gauge [20, 21]) were attached on both sides to record the tensile displacements. Figure 5 presents the tensile force-displacement relationship of the cemented hemispheres in one tension test. The tensile force increases linearly until the tensile displacement reaches 0.11 mm, and then drops to zero suddenly as the bond split apart in the middle.

In the tests, the measured tensile stiffness k_t is 1.23×10^7 N/m, and the peak tensile strength R_t is 1.26 kN. The tensile strength expressed as a stress is σ_t = 16.06 MPa ($\sigma_t = R_t/A$, $A = \pi(d/2)^2$, d= 10.0 mm). The tensile Young's Modulus E ($E=k_t \cdot H_{center}/A$, H_{center}= 2.0 mm) is 0.29 GPa, smaller than the value obtained from the compression test.

3.3 Shear tests

The shear tests were carried out with a shear-auxiliary device, as illustrated in Figure 6. Similar to the tension-auxiliary device, this device includes two rigid blocks with grooves to fix the hemispheres. A vertical insert that is fixed in a rectangular groove in the upper block is used to transfer normal force from the biaxial compression system to the cemented hemispheres. Two horizontal inserts are fixed on the upper and lower blocks respectively so that the horizontal forces applied by the biaxial compression system can shear the cemented hemispheres. The two horizontal rods are aligned along the same axis that passes the center of the epoxy adhesive. Before shearing, a vertical (normal) force smaller than R_c in Figure 3 was first applied. This vertical force was maintained constant throughout the test by a servo-control mechanism. Then, the horizontal (shear) force was applied at a constant shearing rate of 0.1 mm/min. The shear force was measured by two load cells, attached on the two horizontal pedestals of the biaxial compression system. Two displacement gauges were used to record the shear displacement.

Figure 7 presents the shear force-displacement relationship of the cemented hemispheres under different normal forces (i.e., F_n = 0.2, 1, 2, 3, 4 kN). The shear force initially increases linearly to a

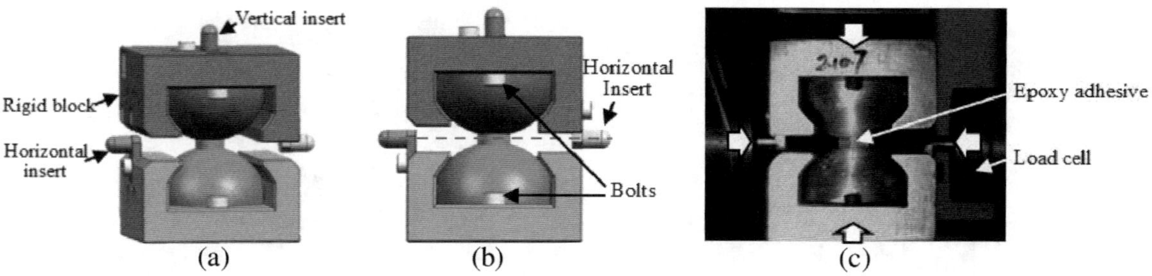

Figure 6. Shear test: (a) and (b) 3D schematic diagrams of the shear-auxiliary device; (c) physical test

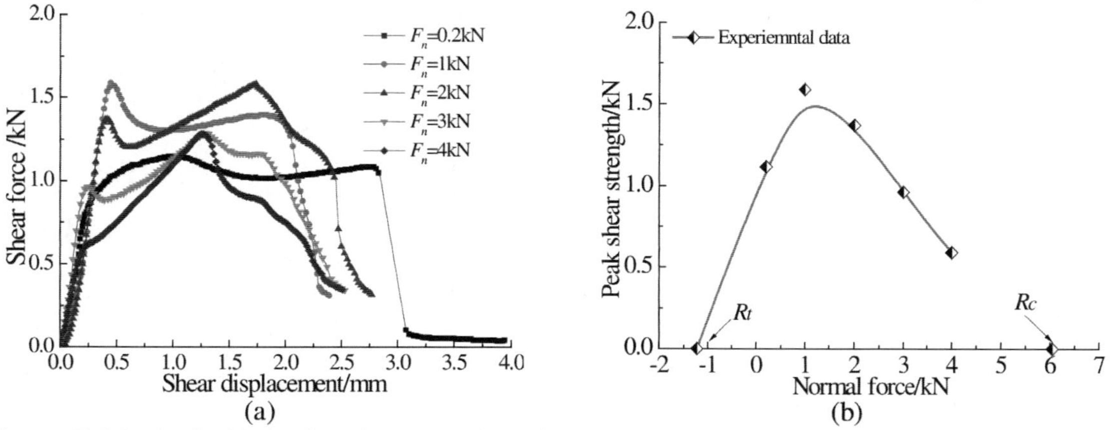

Figure 7. Mechanical behavior of cemented hemispheres in shear tests: (a) shear force-displacement relationship; (b) variation of the peak shear strength versus the normal force

peak value with the shear displacement, and then decreases to a residual strength with a little further shear displacement. Possibly due to friction, the shear force continues to increase with the shear displacement until the bond becomes detached from sphere surfaces. Figure 7(b) shows that the normal force has a significant effect on the shear strength. The peak shear strength first increases with the normal force and then decreases after the normal force exceeds a critical value of 1.2 kN. In shear tests conducted on sand particles artificially bonded with Portland cement, Nardelli & Coop [16] observed qualitatively similar shear force-displacement curves, but the effect of the normal force was only to increase the shear force monotonically, perhaps because the normal forces were not large enough to be close to the peak compressive strength.

3.4 Bending tests

The bending tests were carried out using a bending-auxiliary device, as shown in Figure 6. The bending-auxiliary device is similar to the shear-auxiliary device except that both the two horizontal inserts are fixed on the upper block with a vertical interval (forming a bending arm) so that the horizontal forces applied by the biaxial compression system can bend the cemented hemispheres. Two load cells were used to measure the horizontal forces which were then multiplied by the bending arm to obtain the applied bending moment. Two displacement gauges were used to measure the rotational displacements of the cemented hemispheres, which can be used to obtain the rotational angle. Similarly to the shear test, a vertical (normal) force smaller than R_c in Figure 3 was applied before bending and maintained constant throughout the following test.

Figure 9(a) presents the moment-rotational angle relationship of cemented hemispheres obtained with different normal forces (i.e., F_n = 0.2, 1, 2, 3 kN). The bending moment increases linearly up to a peak value until the rotational angle reaches nearly 1.5°, and then decreases with the rotational angle. Note that the rotational stiffness under different normal forces is almost the same, whereas the peak

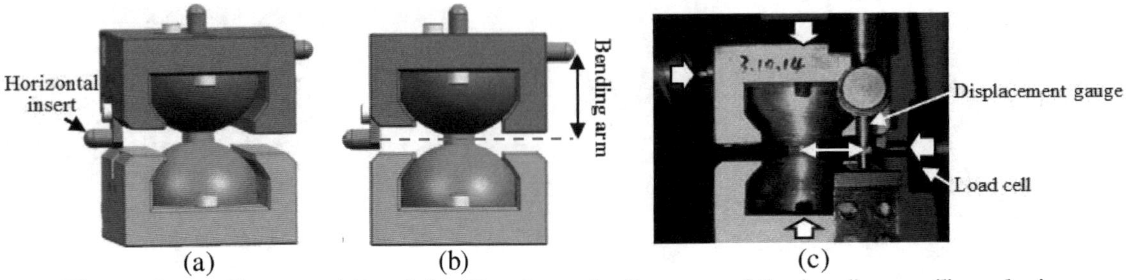

Figure 8. Bending test: (a) and (b) 3D schematic diagrams of the bending-auxiliary device; (c) physical test

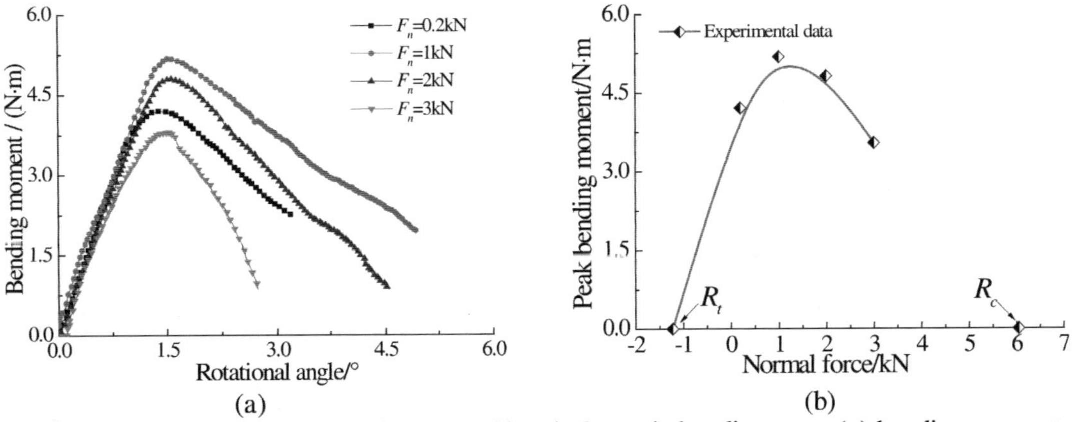

Figure 9. Mechanical behavior of cemented hemispheres in bending tests: (a) bending moment-rotational angle relationship; (b) variation of the peak bending moment versus the normal force

bending moments vary. Figure 9(b) shows that the peak bending moment increases up to a peak value until the normal force approaches 1.2 kN, and then decreases with the normal force.

3.5 Torsion tests

The torsion tests were carried out using a torsion-auxiliary device, as shown in Figure 10. The torsion-auxiliary device is similar to the shear-auxiliary device except that both the two horizontal inserts are fixed on the upper block with a horizontal interval (forming a torsion arm) so that the horizontal forces applied by the biaxial compression system can twist the epoxy adhesive. Two load cells were used to measure the horizontal forces which were then multiplied by the torsion arm to obtain the applied torque. Two displacement gauges were used to measure the horizontal displacements of the two horizontal rods, which were then used to obtain the torsional angle. Similar to the shear test, a vertical (normal) force smaller than R_c in Figure 3 was first applied and maintained constant throughout the torsion test. Figure 11(a) presents the torque-torsional angle relationship of the cemented hemispheres

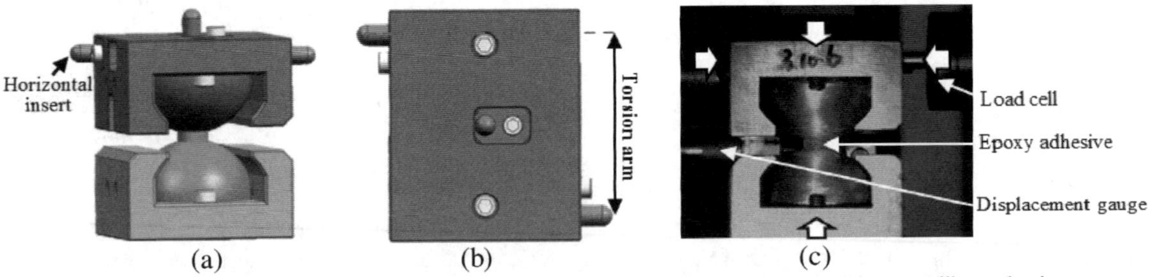

Figure 10. Torsion test: (a) and (b) 3D schematic diagrams of the torsion-auxiliary device; (c) physical test

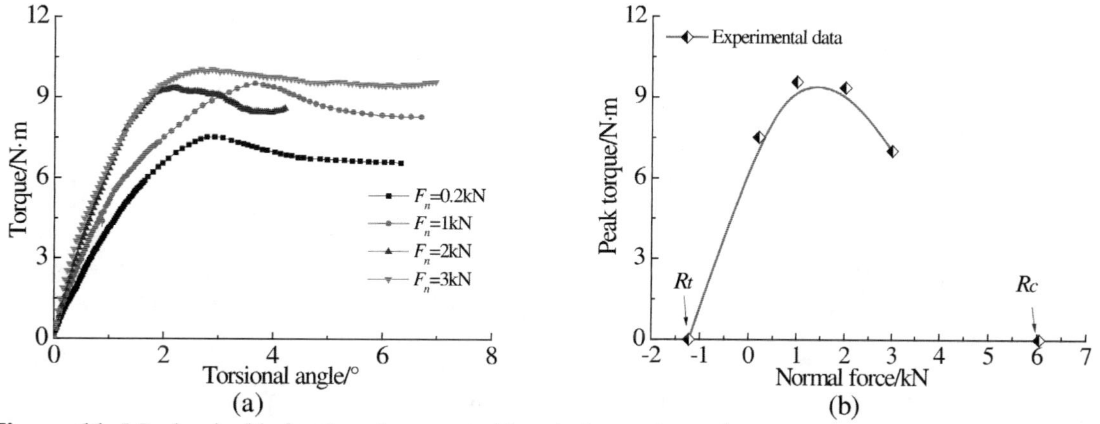

Figure 11. Mechanical behavior of cemented hemispheres in torsion tests: (a) torque-torsional angle relationship; (b) variation of the peak torque versus the normal force

under different normal forces (i.e., F_n = 0.2, 1, 2, 3 kN). Figure 11(a) shows that the torque increases up to a peak strength and then drops to a residual strength slowly. Figure 11(b) presents the relationship between the peak torque and the normal force. The peak torque increases firstly until the normal force reaches a critical value F_{ncr}= 1.2 kN, and then decrease with the increase of normal force.

4. Conclusion

An experimental study on the three-dimensional mechanical behavior of cemented granules, idealized as a pair of aluminum hemispheres glued by epoxy adhesive, was carried out in different loading modes (compression, tension, shear, bending and torsion tests). Several conclusions can be reached. (a) The preparation device is effective to prepare cemented hemispheres of high quality. The auxiliary devices in different loading modes are useful to transfer loads from a biaxial compression system to the cemented hemisphere pair in the desired way. (b) In the compression test, the compressive force increases with compression bi-linearly to the first peak strength, and after a slight decrease, continues increasing to the second peak value until the bond is broken. In the tension test, the tensile force increases linearly up to its peak with tension and then drops abruptly to zero. (c) The normal force has a significant influence on the peak strengths in shear, bending and torsion tests. That is, the peak shear force, bending moment and torque first increase with the normal force and then decrease after the normal force exceeds a critical value.

Acknowledgement

This study is supported by the State Key Lab. of Disaster Reduction in Civil Engineering (No. SLDRCE14-A-04), the China National Funds for Distinguished Young Scientists (No. 51025932), the Major Project of Chinese National Programs for Fundamental Research and Development (973 Program) (No. 2014CB046901) which are all appreciated.

References

[1] Duncan J M, Wright S G 1980 The accuracy of equilibrium methods of slope stability analysis *Eng. Geol.* **16** 5-17
[2] Xu X, Shang M, Deng Y H 2014 Analysis about the classification of Landslide Monitoring Method *Appl. Mech. Mater.* **638-640** 360-364
[3] Milledge D G, Bellugi D, McKean J A, et al. 2014 A multidimensional stability model for predicting shallow landslide size and shape across landscapes *J. Geophys. Res. Earth Surf.* **119** 2481-2504
[4] Yashima A, Moriguchi S, Uzuoka R, et al. 2013 Large Deformation Analysis for Costal Geo-

Disasters Using Continuum and Discrete Modeling *Springer Geol.* **62** 13-29
[5] Fuchs M, Torizin J, Kühn F 2014 The effect of DEM resolution on the computation of the factor of safety using an infinite slope model *Geomorphology* **224** 16-26
[6] Cundall P A, Strack O D L 1979 Discrete numerical-model for granular assemblies *Géotechnique* **29** 47-65
[7] Utili S, Nova R 2008 DEM analysis of bonded granular geomaterials *Int. J. Numer. Anal. Methods Geomech.* **32** 1997-2031
[8] Jiang M J, Murakami A 2012 Distinct element method analyses of idealized bonded-granulate cut slope *Granular Matter* **14** 393-410
[9] Jiang M J, Yu H S, Leroueil S 2007 A simple and efficient approach to capturing bonding effect in naturally microstructured sands by discrete element method *Int. J. Numer. Methods Eng.* **69** 1158-1193
[10] Jiang M J, Yu H S, Harris D 2006 Bond rolling resistance and its effect on yielding of bonded granulates by DEM analyses *Int. J. Numer. Anal. Methods Geomech.* **30** 723-761
[11] Delenne J, El Youssoufi M S, Cherblanc F, et al. 2004 Mechanical behaviour and failure of cohesive granular materials *Int. J. Numer. Anal. Methods Geomech.* **28** 1577-1594
[12] Kirsch R, Bröckel U, Brendel L, et al. 2011 Measuring tensile, shear and torsional strength of solid bridges between particles in the millimeter regime *Granular Matter* **13** 517-523
[13] Jiang M J, Sun Y G, Xiao Y 2012 An experimental investigation on the mechanical behavior between cemented granules *Geotech. Test. J.* **35** 678-690
[14] Jiang M J, Sun Y G, Li L Q, et al. 2012 Contact behavior of idealized granules bonded in two different interparticle distances: An experimental investigation *Mech. Mater.* **55** 1-15
[15] Jiang M J, Yan H B, Zhu H H and Utili S 2011 Modeling Shear Behavior and Strain Localization in Cemented Sands by Two-Dimensional Distinct Element Method Analyses *Comput. Geotech.* **38** 14-29
[16] Nardelli V, Coop M R 2015 An experimental investigation of the micromechanical behaviour of cemented sand particles *Proc. TC105 ISSMGE Int. Symp. Geomech. from Micro to Macro* (*IS-Cambridge September* 2014) **2** 1013-1017

IOP Conf. Series: Earth and Environmental Science **26** (2015) 012008 doi:10.1088/1755-1315/26/1/012008

Understanding the effects of inter-particle contact friction on the elastic moduli of granular materials

K. Taghizadeh, N. Kumar, V. Magnanimo & S. Luding

MultiScale Mechanics, CTW, MESA+, University of Twente, Enschede, Netherlands

E-mail: `k.taghizadehbajgirani@utwente.nl`

Abstract. Understanding the mechanical stiffness of closely packed, dense granular systems is of interest in many fields, such as soil mechanics, material science and physics. The main difficulty arises due to discreteness and disorder in granular materials at the microscopic scale which requires a multi-scale approach. The Discrete Element Method (DEM) is a powerful tool to inspect the influence of the microscopic contact properties of its individual constituents on the bulk behavior of granular assemblies. In this study, the isotropic deformation mode of polydisperse packings of frictionless and frictional spheres are modeled by using DEM, to investigate the effective stiffness of the granular assembly. At various volume fractions, for every sample, we determine the stress and fabric incremental response that result from the application of strain-probes. As we are interested first in the reversible, elastic response, the amplitude of the applied perturbations has to be small enough to avoid opening and closing of too many contacts, which would lead to irreversible rearrangements in the sample. Counterintuitively, with increasing inter-particle contact friction, the bulk modulus decreases systematically with the coefficient of friction for samples with the same volume fraction. We explain this by the difference in microstructure (isotropic fabric) the samples get when compressed to the same density.

1. INTRODUCTION

Granular materials behave differently from usual solids or fluids and show peculiar mechanical properties like dilatancy, history dependence, ratcheting and anisotropy, see [21] and references therein. The behavior of these materials is highly interesting, non-linear and involves irreversibility (plasticity), possibly already for very small strains, due to rearrangements of the elementary particles [7, 17, 2, 11]. The concept of an initial purely elastic regime (at very small strain) for granular assemblies is an issue still under debate in the mechanical and geotechnical communities [4]. Recent works [5, 22, 13] show that along with the macroscopic properties (stress and volume fraction) also the structure, quantified by the fabric tensor [10, 11] plays a crucial role, as it characterizes, on average, the geometric arrangement of contacts, i.e. the microstructure of the particle packing. In particular, when the material is sheared, anisotropy in the contact network develops, as related to the opening and closing of contacts, restructuring, and the creation and destruction of force-chains. The anisotropic state is at the origin of interesting observations on wave propagation in sheared granular media. However, for isotropic deformations, shear is not active so that we can focus on the interplay between contact friction and the isotropic fabric.

In order to investigate the evolution of the incremental response as a function of loading (in different directions) we performed so-called strain probing tests along an axisymmetric deformation path [12], resembling what happens in physical experiments [16]. The directional-dependent data can be collected in stress and fabric response envelopes as defined in [8] in the case of both small (elastic) and large (elasto-plastic) probes. In the case of a finite assembly of frictionless particles, in simulations, a finite elastic regime can always be detected and the elastic stiffnesses can be measured by means of an actual, very small strain perturbation [15].

In this short paper, we focus on the isotropic deformation mode only, but try to address the question on what the effects of friction are before, in ongoing research, we also address the different deformation directions axial-symmetric and plane-strain. By isolating elasticity we aim to identify the kinematics at the microscale that leads to either macroscopic elasticity or plasticity. The final goal is to improve the understanding of inter-particle contact friction on the microstructure and thus on the elastic bulk modulus in dense frictional particle systems and to guide further developments for new constitutive models based on microstructure information, as presented by [11] for frictionless samples, complementing the uni-axial compression tests by [10].

2. NUMERICAL SIMULATION

The Discrete Element Method (DEM) [14, 11] can help to understand the deformation behavior of particle systems. At the basis of DEM are force laws that relate the interaction force to the overlap and tangential displacement of two particle contact surfaces. If all forces $\mathbf{f}_i$ acting on particle i, from all sources, are known, the problem is reduced to the integration of Newton's equations of motion for the translational and rotational degrees of freedom, see for example [19] for more details. $\frac{\mathrm{d}}{\mathrm{d}t}(m_i \mathbf{v}_i) = \mathbf{f}_i + m_i \mathbf{g}$,

2.1. Contact model

For the sake of simplicity, the linear visco-elastic contact model for the normal component of force is used. The simplest normal contact force model is given by $f^n = k\delta + \gamma\dot{\delta}$, where k is the spring stiffness, γ is the contact viscosity parameter, $\delta = (d_i + d_j)/2 - (\mathbf{r}_i - \mathbf{r}_j) . \hat{\mathbf{n}}$ is the overlap between two interacting species i and j with diameters d_i and d_j, with contact normal vector $\hat{\mathbf{n}} = (\mathbf{r}_i - \mathbf{r}_j) / |(\mathbf{r}_i - \mathbf{r}_j)|$ and $\dot{\delta}$ is the relative velocity in the normal direction. In order to reduce dynamical effects and shorten relaxation times, an artificial viscous background dissipation force $\mathbf{f}_b = -\gamma_b \mathbf{v}_i$ proportional to the moving velocity $\mathbf{v}_i$ of particle i is added, resembling the damping due to a background medium, as e.g. a fluid. The tangential force model introduced by [14] is used and thus will not be detailed here.

The standard simulation parameters are, $N = 4096 (= 16^3)$ particles with average radius $\langle r \rangle = 1$ [mm], density $\rho = 2000$ [kg/m^3], elastic stiffness $k = 10^8$ [kg/s^2], particle damping coefficient $\gamma = 1$ [kg/s], background dissipation $\gamma_b = 0.1$ [kg/s]. The tangential stiffness and viscosity are set as $k_t/k = 0.2$ and $\gamma_t/\gamma_n = 0.2$, while the coefficient of friction is varied. Note that the polydispersity of the system is quantified by the width ($w = r_{\max}/r_{\min} = 3$) of a uniform size distribution, where $r_{\max}$ and $r_{\min}$ are the radii of the biggest and smallest particles respectively. For details about other time scales present in the system, see [9, 11].

2.2. Macroscopic (tensorial) quantities

Here, we define averaged tensorial macroscopic quantities – including strain-, stress- and fabric (structure) tensors – that provide information about the state of the packing and reveal interesting bulk features.

By speaking about the strain-rate tensor $\dot{\mathbf{E}}$, we refer to the external strain that we apply to the sample. The isotropic part of the infinitesimal strain tensor ε_v is defined as: $\varepsilon_\mathrm{v} = \dot{\varepsilon}_\mathrm{v}\mathrm{d}t = -(\varepsilon_{xx} + \varepsilon_{yy} + \varepsilon_{zz})/3 = \mathrm{tr}(-\mathbf{E})/3 = \mathrm{tr}(-\dot{\mathbf{E}}\mathrm{d}t)/3$, where $\varepsilon_{\alpha\alpha} = \dot{\varepsilon}_{\alpha\alpha}\mathrm{d}t$ with $\alpha\alpha = xx, yy$ and zz as the diagonal components of the tensor in the Cartesian $x - y - z$ reference system. The trace integral of $3\dot{\varepsilon}_v$ is denoted as ε_v, the true or logarithmic volumetric strain, i.e., the volume change of the system, relative to the initial reference volume, V_0.

On the other hand, from DEM simulations, one can measure the 'static' stress in the system as

$$\boldsymbol{\sigma} = (1/V) \sum_{c \in V} \mathbf{l}^c \otimes \mathbf{f}^c, \tag{1}$$

averaged over all the contacts in the volume V of the dyadic products between the contact force $\mathbf{f}^c$ and the branch vector $\mathbf{l}^c$, where the contribution of the kinetic fluctuation energy has been neglected [9, 10]. The isotropic component of the stress is the pressure $P = \mathrm{tr}(\boldsymbol{\sigma})/3$.

In order to characterize the geometry/structure of the static aggregate at microscopic level, we measure the fabric tensor, defined as

$$\mathbf{F} = \frac{1}{V} \sum_{\mathcal{P} \in V} V^{\mathcal{P}} \sum_{c \in \mathcal{P}} \mathbf{n}^c \otimes \mathbf{n}^c, \tag{2}$$

weighted according to $V^{\mathcal{P}}$, the particle volume for particle $\mathcal{P}$, which lies inside the averaging volume V, and $\mathbf{n}^c$ is the normal unit branch-vector pointing from center of particle $\mathcal{P}$ to contact c [11]. Note that in this work we use $k^* = k/(2\langle r \rangle)$ to non-dimensionalize the stress, i.e. $\boldsymbol{\sigma}^* = \boldsymbol{\sigma}/k^*$.

2.3. Sample preparation and test procedure

In this subsection, we first describe the preparation procedure and then the details of the numerical isotropic test. The preparation procedure is an essential step in any physical/numerical experiment to obtain reproducible and reliable results, especially when friction is involved. The initial configuration is such that spherical particles are randomly generated, with low volume fraction and rather large random velocities in a periodic 3D box, such that they have sufficient space and time to exchange places and to randomize themselves. The initial configuration is obtained by, first compressing a granular gas up to a volume fraction below the jamming fraction. The system is then relaxed to allow the particles to dissipate kinetic energy and achieve a zero-pressure static configuration [9, 11]. This is followed by a compression-decompression isotropic cycle to a desired maximum volume fraction $\nu_{\max} = 0.82$ [9, 11]. This procedure is applied for several packings of polydisperse spheres with different coefficients of friction while the randomization procedure was the same for all the samples. Note that the preparation is carried out with strain-control, where at every time-step the particles are first moved according to the momentary strain-rate and then allowed to relax according to the momentarily acting forces (for one time step) before the next step.

3. BULK MODULUS

We now study the evolution of the incremental behavior during isotropic compression as function of the micro property contact-friction. Various configurations, at different volume fraction above jamming along the unloading branch, the first one being near jamming point and last one highly dense packed, are chosen for all the packings and sufficient relaxation is applied to allow the particles to achieve a static configuration in mechanical equilibrium.

Further on we probe the samples, by applying small strain perturbations and measuring the incremental (stress and fabric) responses [15, 9, 10, 11] to investigate the effects of inter-particle contact friction on the elastic bulk modulus. For each state we apply an identical strain-rate with amplitude $\dot{\varepsilon}_v = 10.0$ [μs^{-1}] applied. After probing the configurations, the bulk modulus is obtained for the granular assembly as the ratio between the measured increment in pressure and the applied isotropic strain

$$B = \delta P^* / 3\delta\varepsilon_v \ . \tag{3}$$

3.1. Evolution of the bulk modulus

As we are interested in the elastic response, we first have to identify the elastic regime, the marginal regime and the plastic regime [11]. For the results presented below, the applied infinitesimal strain step is kept small enough to avoid strong, irreversible particle rearrangements, while plasticity develops in the sample as soon as those rearrangements happen. The response of the bulk modulus against the amplitude of the applied isotropic $3\delta\varepsilon_v$ strain is depicted for a chosen configuration during probing by using the so-called incremental stress response in Fig. 1 to test the rearrangements argument. It is clear that, B stays practically constant for small amplitudes ($3\delta\varepsilon_v < 10^{-4}$) and the regime can be considered to be elastic. By increasing the amplitudes of the small perturbation $3\delta\varepsilon_v$, B starts to increase non-linearly as soon as the first change in the number of contacts happens. It is interesting to notice that for B, the elastic regime expands when the friction coefficient increases (data not shown).

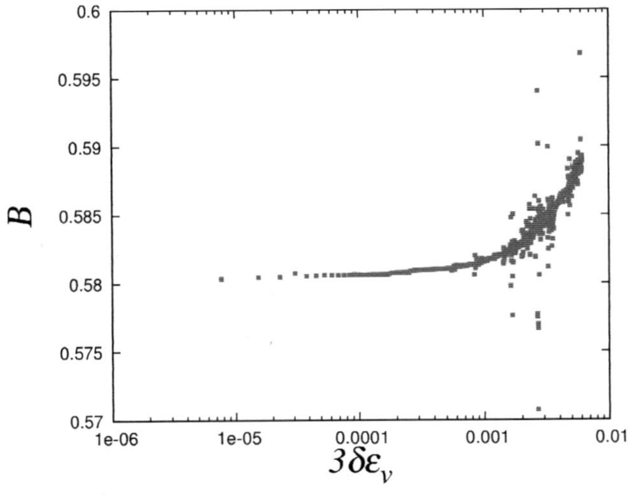

Figure 1. Evolution of the bulk modulus B with applied isotropic amplitudes $3\delta\varepsilon_v$ for a system with $\mu = 1$ and volume fraction $\nu = 0.813$

3.2. Effect of inter-particle contact friction on the bulk modulus

In this subsection, the effect of inter-particle contact friction on the mechanical response of packings will be analyzed, following the strain-probes methodology presented in section 2.3. In Fig. 2(a), we plot the variation of the bulk modulus B, with volume fraction for packings with different coefficients of friction μ. For higher densities, the bulk modulus increases by decreasing the friction coefficient for samples with the same volume fraction. This is related to a higher average number of contacts for samples prepared with low friction (given the rather low tangential stiffness k_t used here). For low densities, the behavior is opposite, since the higher friction at the contact leads to a lower jamming volume fraction. Thus, frictional and frictionless curves show qualitatively different slopes, associated with the activation of tangential forces.

In general, one can observe three regimes, frictionless, low and high friction. However, when the data are plotted not against volume fraction, but against the isotropic fabric, see Fig.2(b), all data have about the same slope. The low friction regime behaves similar to the frictionless one at the loose states (close to the jamming volume fraction) and it follows the behavior of high friction systems when far from the jamming point.

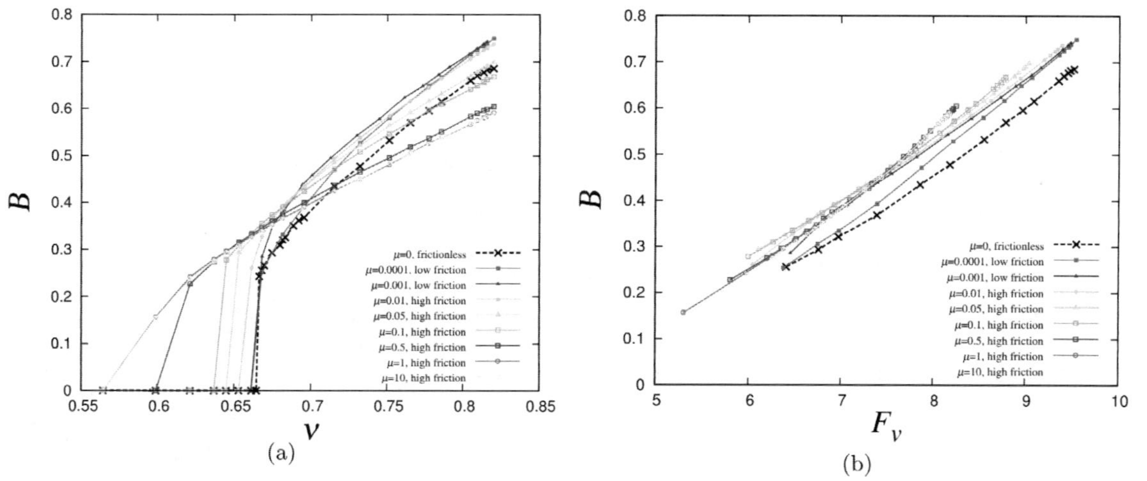

Figure 2. a) Evolution of the bulk modulus B with volume fraction ν f or different coefficient of friction, μ, as shown in the legend. b) Evolution of the bulk modulus B as scaled with the isotropic fabric F_v f or various volume fractions and different μ, as shown in the legend.

4. CONCLUSIONS

In a triaxial box, the bulk modulus used to describe the incremental, elastic constitutive behavior of granular material can be measured by applying small strain perturbations to relaxed states. Those states can have experienced different deformation history, or the particles can have different properties, like the coefficient of inter-particle contact friction. The goal of this paper was to understand the effect of friction on the macroscopic bulk modulus and its relation to the micro-stucture of the packing.

Concerning the probing, the amplitude of the applied perturbations has to be small enough to avoid particle rearrangements and thus to get the elastic response, whereas large amplitudes develop plasticity in the sample due to contact and structure rearrangements between particles. A relation between the macroscopic elastic response and the micro-structure was investigated, by considering both the bulk modulus B and volumetric fabric F_v, respectively.

Surprisingly, frictionless packing do not follow the same trend as frictional packings. The tangential force plays a crucial role in establishing the contact network and thus for the mechanical behavior of granular materials. This is a big step towards realistic materials; since frictionless particles are an unrealistic limit-case material, the tangential force has to be taken into account, even though it carries relatively small magnitude of force, in the case of low friction coefficients.

Extension of the work to investigate the influence of inter-particle contact friction on the shear modulus is in progress, using the same small perturbation probing and allowing for many different modes of deformation. Additional future work will focus on establishing a constitutive model for frictional particles, involving the elastic moduli measured and the relation between effective moduli and the microstructure tensor (fabric).

5. ACKNOWLEDGEMENTS

The financial support of the European-Union Marie Curie Initial Training Network, T-MAPPP, funded by FP7 (ITN 607453), is appreciated, see http://www.t-mappp.eu/ for more information.

References

[1] F. Alonso-Marroquín, S. Luding, H. J. Herrmann, and I. Vardoulakis (2005). Role of anisotropy in the elastoplastic response of a polygonal packing. *Phys. Rev. E.*, 71(5), 051304(1-18).

[2] J.P. Bardet (1994). Numerical simulations of the incremental responses of idealized granular materials. *Int. J. of Plasticity*, 10(8), 879-904.

[3] F. Calvetti, G. Viggiani, and C. Tamagnini (2003). A numerical investigation of the incremental behavior of granular soils. *Rivista Italiana di Geotecnica* 37(3), 11-29.

[4] I. Einav (2012). The unification of hypo-plastic and elasto-plastic theories. *International Journal of Solids and Structures*, 49(11), 1305-1315.

[5] A. Ezaoui and H. Di Benedetto (2009). Experimental measurements of the global anisotropic elastic behaviour of dry Hostun sand during triaxial tests, and effect of sample preparation. *Geotechnique*, 59(7), 621-635.

[6] F. Froiio and J. N. Roux (2011). Incremental response of a model granular material by stress probing with DEM simulations. In *Proc. Particles 2011* pp. 183-197.

[7] J. D. Goddard (1990). Nonlinear Elasticity and Pressure-Dependent Wave Speeds in Granular Media. *Proc. R. Soc. Lond. A*, 430(1878), 105-131.

[8] G. Gudheus (1979). A Comparison of Some Constitutive Laws for Soils Under Radially Symmetric Loading and Unloading. In *Proc. 3rd Conf. Numerical Methods in Geomechanics*, pp. 1309-1323.

[9] O. I. Imole, N. Kumar, V. Magnanimo, and S. Luding (2013). Hydrostatic and Shear Behavior of Frictionless Granular Assemblies Under Different Deformation Conditions. *KONA*, 30, 84-108.

[10] O. I. Imole, V. Magnanimo, and S. Luding (2014). Micro-Macro Correlations and Anisotropy in Granular Assemblies under Uniaxial Loading and Unloading, Phys. Rev. E 89, 042210.

[11] N. Kumar, S. Luding & V. Magnanimo (2014). Macroscopic model with anisotropy based on micro-macro informations. *Acta Mechanica*, 225(8), 2319-2343.

[12] N. Kumar, S. Luding, and V. Magnanimo (2014). Incremental stress and microstructural response of granular soils under undrained axisymmetric compression, in: "Geomechanics from Micro to Macro", Kenichi Soga, Krishna Kumar, Giovanna Biscontin, Matthew Kuo (Eds.), CRC Press/Balkema, UK, ISBN: 978-1-138-02707-7, pp. 115-121.

[13] L. La Ragione and V. Magnanimo (2012). Contact anisotropy and coordination number for a granular assembly: A comparison of distinct-element-method simulations and theory. *Phys. Rev. E.*, 85(3).

[14] S. Luding (2008). Cohesive frictional powders: Contact models for tension Granular Matter 10(4), 235-246.

[15] V. Magnanimo, L. La Ragione, J.T Jenkins, P. Wang & H.A. Makse (2008). Characterizing the shear and bulk moduli of an idealized granular material. *Europhys. Lett.*, 81, 34006.

[16] D. Muir Wood, T. Sadek, L. Dihoru, M.L. Lings, H. Javaheri (2006). Deviatoric stress response envelopes from multiaxial tests on sand. In *Soil Stress Behavior: Measurements, Modeling and Analysis*, Springer, pp. 253-262.

[17] L. Sibille, F. Nicot, F.-V. Donzoe, and F. Darve (2009). Analysis of failure occurrence from direct simulations. *Eur. J. Env. Civ. Eng.*, 13(2), 187-201.

[18] C. Thornton and L. Zhang (2010). On the evolution of stress and microstructure during general 3D deviatoric straining of granular media. *Géotechnique*, 60(5), 333-341.

[19] T. Weinhart, R. Hartkamp, A.R. Thornton and S. Luding (2013). Coarse-grained local and objective continuum description of three-dimensional granular flows down an inclined surface. *Phys. Fluids*, 25, 070605.

[20] W. Wu and D. Kolymbas. Hypoplasticity then and now (2000). In *Constitutive Modelling of Granular Materials*, Springer-Verlag, pp. 57-101.

[21] A. Yu, K. Dong, R. Yang, and S. Luding, Powders and Grains 2013, American Institute of Physics, 2013, *AIP Conference Proceedings 1542*, ISBN: 978-0-7354-1166-1.

[22] J. Zhao and N. Guo (2013). Unique critical state characteristics in granular media considering fabric anisotropy. *Géotechnique*, 63(8), 695-704.

IOP Conf. Series: Earth and Environmental Science **26** (2015) 012035 doi:10.1088/1755-1315/26/1/012035

Centrifuge Tests of Dyke Collapses on Soft Subsoil

J Fern[1], D de Lange[2], C Zwanenburg[2] and J A M Teunissen[2]

[1] University of Cambridge, Department of Engineering, Trumpington Street, Cambridge CB2 1PZ, United Kingdom

[2]Deltares, Geo-Unit, Boussinesqweg 1, 2629 HV Delft , the Netherlands

E-mail: [1] `jf497@cam.ac.uk`

E-mail: [2] `dirk.delange@deltares.nl`

Abstract. This paper presents the results of an experimental investigation on the failure or collapse of dykes with a strong contrast of stiffness between the dyke itself and its foundation layers. In deltaic regions, dykes are commonly built out of stiff sandy or clayey materials and rest on soft foundations layers (*e.g.* soft clay or peat). Their interaction is largely unexplored and the failure mechanism unknown. The tests presented in this paper aim to highlight the difference in failure mechanisms. The tests consisted mainly of stiff dykes on soft subsoils but a few soft dykes on stiff subsoil were also investigated. The dykes were made out of *Speswhite* clay or *Baskarp* sand whereas the subsoil was made out of silicon or clay. The silicon was used to replicate a soft elastic subsoil for which it was possible to control the stiffness. The model was then subjected to increasing gravity up to 100 G-level. The results show that slope failures take place in the dyke when the foundations layers are stiff. It is believed to be the consequence of a build up of pore pressure due to the volumetric contraction of the dyke. On the other hand, soft foundation layers underwent large settlements which in turn deformed the stiff dyke without any slope failure taking place. However, diffused sheared zones were observed in the core of the dyke.

1. Introduction

The Netherlands relies on a vast network of dykes to protect its low-lands from flooding. The rapid urban expansion and the rise of sea levels are putting ever increasing economic and social pressures on these dykes. Many of them were built centuries ago with locally available materials and rest on the natural subsoils which in some areas are very soft [1, 2]. Little is known about the failure mechanism of dykes involving two different materials with a strong contrast in stiffness. This is typically the case when dykes are built on peat. However, some research has been carried out on the behaviour of embankments resting on soft soil [3, 4] but focussed on the failure of the subsoil rather the failure of the embankment itself. In order to understand the influence of the stiffness of the subsoil on the failure mechanism of dykes, a series of centrifuge tests were carried out in which dykes were made out of clay or sand and placed on stiff or soft subsoil. Centrifuge tests offer a cost and time effective way of investigating such behaviours. The design of the centrifuge tests was based on a first attempt to model the collapse of dykes in a centrifuge [7]. Full scale experiments have already been carried out in previous studies such as the IJkdijk project [5] or the field test near Bergambacht [6]. However, these tests mobilised a lot of human, technical and financial resources and are therefore limited to a single case. Centrifuge testing is an alternative with which it is possible to carry out numerous tests within

Figure 1. The beam Geo-Centrifuge at Deltares, the Netherlands

Figure 2. Model A with a stiff sandy dyke and a soft silicon subsoil

a short period of time. In the present study, nine centrifuge tests were carried out in which a different combination of materials, respectively for the dyke and the subsoil, were investigated. The models were subjected to increasing gravity up to 100 G-force. The results show single and multiple failure surfaces resulting in a slope failure or in very large deformation with diffuse shear zones.

2. Geo-centrifuge, model description and test programme

The nine centrifuge tests were carried out at Deltares, the Netherlands, with their Geo-Centrifuge (Figure 1). It is a 300 G-force beam centrifuge with an arm length of 5.5 m and a model weight capacity of 2 tonnes. It allows to test large models in various conditions. In this present study, the tests consisted in modelling the failure of dykes on soft subsoils. The material of the dyke and its subsoil were changed successively in order to investigate the different cases in which a contrast of stiffness would occur. Six tests were carried out on a clayey dyke made out of *Speswhite* clay and resting on a silicon block. The silicon block was used to replicate soft soils in which the deformations are large. Natural soft soils such as peat are difficult to model in centrifuge tests. Three different silicon blocks were used with respectively different stiffnesses - a soft block, a medium-stiff block and a stiff block. Three additional tests were carried on a sandy dyke made out of *Baskarp* sand and resting on a soft silicon block or on a remoulded natural clay called *Oostvaardersplassen* clay [8].

2.1. Model preparation

All models were built in a rectangular strong box (H 450 mm x H 870 mm x W 200 mm) with a glass face. They were built in a plane strain configuration. The dimensions of the dyke and the thickness of the subsoil changed between tests with a silicon subsoil (Model A) and those with the *Oostvaardersplassen* clay (Model B). The latter was smaller in order to reduce the preconsolidation time due to temporal constraints of the project. Preconsolidation is a necessary step in centrifuge testing in order to initialise the correct effective stress state. Figure 3 illustrates the dimensions of the models and summarises the different tests.

The clayey dykes were made out of *Speswhite* clay which is a fined grained kaolin with a low permeability ($K_{sat} = 10^{-9}$ m/s). It was prepared in a stiff tub and consolidated with a vertical pressure of 50, 100 or 150 kPa. It was then extracted from the tub with a vacuum crane, cut into a block and placed in the strong box on the silicon block (Fig. 4A-B). The faces of the strong box were smeared with Vaseline in order to reduce the friction between the model and the strong box. The contact between the dyke and the subsoil was carefully prepared in order

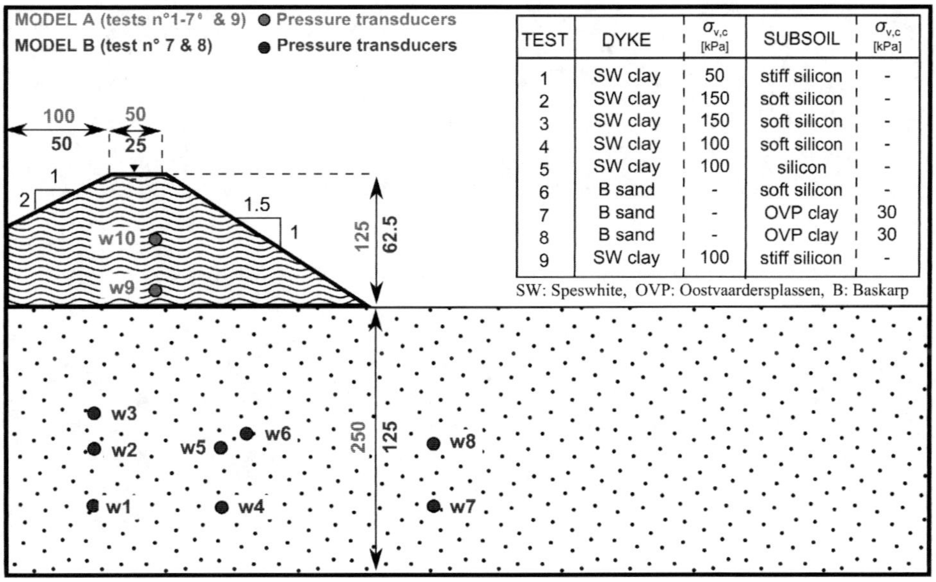

TEST	DYKE	$\sigma_{v,c}$ [kPa]	SUBSOIL	$\sigma_{v,c}$ [kPa]
1	SW clay	50	stiff silicon	-
2	SW clay	150	soft silicon	-
3	SW clay	150	soft silicon	-
4	SW clay	100	soft silicon	-
5	SW clay	100	silicon	-
6	B sand	-	soft silicon	-
7	B sand	-	OVP clay	30
8	B sand	-	OVP clay	30
9	SW clay	100	stiff silicon	-

SW: Speswhite, OVP: Oostvaardersplassen, B: Baskarp

Figure 3. Centrifuge model with description of the material used in each tests

to obtain a perfect contact between them. Once the block was in place, the *Speswhite* clay was trimmed into a dyke (Fig. 4C-D). Speckles were then blown on the face of the dyke in order to facilitate Particle Image Velocimetry (PIV) analysis. The front and rear panel of the strong box were mounted and the container was sealed until the test was performed the next day.

The sandy dykes were made out of *Baskarp* sand which is uniformly graded silica sand with a dominant grain size of 0.1 mm. The sandy dyke was directly built in the strong box. The silicon block was first placed in the strong box and the sides of the strong box mounted. Then, the sand was placed using the drizzle method [9] in which the sand was pluviated in de-aired water (Figure 5A). This method allowed to control the density and homogeneity of the sand as well as providing repeatability of the tests. The dykes were built to achieve a relative density of 90% $\pm$ 5%. The different layers of sand were coloured in order to facilitate the PIV analysis (Figure 5B). The sand was then left to drain by gravity. The unsaturated conditions allowed some suction tension to appear. The sand was then trimmed to the desired dimensions (Figure 5C-D). The strong box was then covered and the test took place the next day.

The silicon subsoil was prepared in the form of blocks and to achieve a specific stiffness. Then, a block was chosen according to its stiffness and placed in the strong box. The dyke was then built on it. The mechanical behaviour of silicon is complex (*e.g.* viscosity). However, the influence of the time dependent behaviour of silicon was limited due to the short testing time (100 minutes). It was supposed that the silicon was incompressible with a Poisson ratio (ν) of 0.5 and has an elastic behaviour. The secant elastic modulus (E) was determined by means of element testing and were respectively for the soft, medium-stiff and stiff blocks 114 kPa, 565 kPa and 1,377 kPa.

Two tests with sandy dykes were carried out with a subsoil made out of a natural clay called *Oostvaardersplassen* clay. It is a soft silty clay which was remoulded from a natural clay in a tub and subjected to a preconsolidation pressure of 30 kPa (Figure 6A). It was then trimmed to the desired dimension (Figure 6B) and placed in the strong box with a vacuum crane (Figure 6C). The block was then trimmed into a dyke. Speckles were then blown on the subsoil to facilitate the PIV analysis (Figure 6D). The dyke was then built following the aforementioned procedure.

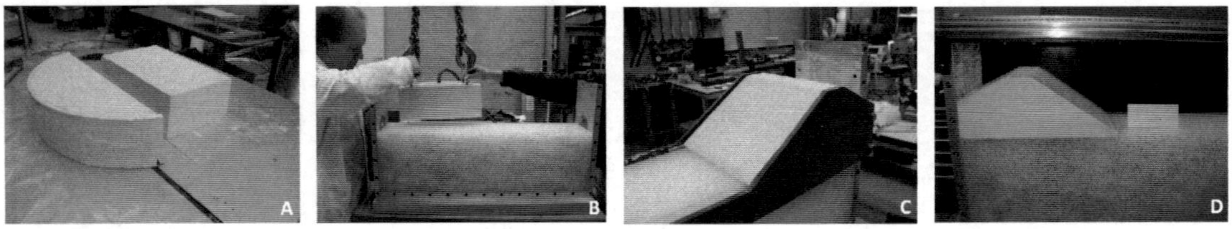

Figure 4. Preparation of the *Speswhite* clay dykes

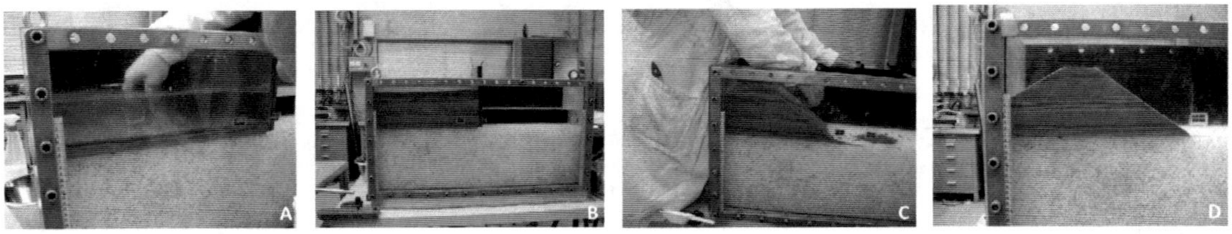

Figure 5. Preparation of the *Baskarp* sand dykes

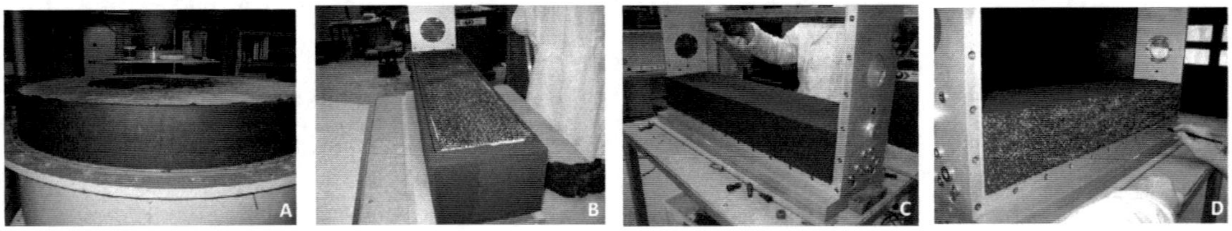

Figure 6. Preparation of the *Oostvaardersplassen* clay subsoil

The strong box was sealed and stored until the next day when the tests took place.

2.2. Test programme

Figure 3 shows a table with the different tests. The first series of tests with Model A focussed on the behaviour of dykes made out of *Speswhite* clay (Tests 1-5 and 9) or Baskarp sand (Test 6) and resting on silicon subsoil. The second series of tests with Model B focussed on dykes made out of Baskarp sand and a subsoil made out of a natural clay called *Oostvaardersplassen* clay (Test 7 and 8).

The model was subjected to a linear increase in gravity acceleration from 1 to 100 G-force in a 100 minutes. Following the scaling laws [10], an increase the stresses which in effect results in a difference in the size of the prototype being tested. With other words, the tests covered a wide range of prototypes and allowed the identification of a critical prototype size. Additionally and for operational reasons, a vacuum is created in the centrifuge chamber at 20 G-force. No other loads were applied to the model.

2.3. Instrumentation

The models were equipped with different arrays of instruments. The deformation of the dyke and the subsoil was tracked with PIV. Four cameras were mounted on the container - 3 on the side view and 1 on the plan view. In order to use PIV, the models were sprinkled with speckles or coloured in layers. Additionally, a LVDT with a range of 50 mm and accuracy of ± 0.3% was mounted to measured the displacement at the crown of the dyke. Pore pressure transducers with a maximum range of 300 kPa ± 0.1% were also placed in the soil in order to track both positive and negative pore pressures. However, some difficulties arose during testing. Some sensors lost contacts with the soil and others seem to have been affected by the vacuum in the centrifuge chamber. The vacuum was necessary to decrease the friction between the centrifuge and the ambient air and to allow it to spin up to a 100 G-force.

3. Results of centrifuge tests

3.1. Test 1: Speswhite clay ($\sigma'_{v,0} = 50$ kPa) dyke on - stiff silicon

The first tests consisted of a soft *Speswhite* clay dyke ($\sigma'_{v,0} = 50$ kPa) on a stiff silicon subsoil (E = 1377 kPa). The initial water content (w) of the clay was 63% and was higher than expected. The undrained strength (s_u) was measured with a pocket penetrometer and had a value of 3 kPa. The dyke was very soft. Two pore pressure transducers were placed in the dyke and showed negative initial pore pressures of -4 kPa and -8 kPa. It implies that some suction strength enhancement can be expected.

During the test, the dyke exhibited large deformations. At only 4 G-force, it experienced a brittle slope failure as shown in Figure 7.①. The mobilised mass flowed on the silicon block. The large deformation of the dyke out ranged the LVDT measurements and no readings were possible. As the acceleration carried on, the dyke continued its large deformation and a second slide took place. This mechanism carried on until the end of the tests by which point the entire dyke was flattened. The pore pressure measurements showed a rapid development of positive pore pressures which suggests that the failures took place in undrained conditions (conservation of the pore water mass within the soil). The build up of these pressures was caused by the volumetric compaction of the dyke.

3.2. Test 2: Speswhite clay ($\sigma'_{v,0} = 150$ kPa) - stiff silicon

The second tests consisted of an overconsolidated *Speswhite* clay dyke ($\sigma_{v,0} = 150$ kPa) and a stiff silicon subsoil (E = 1377 kPa). The initial water content (w) was 53% and its undrained strength (s_u) was 12 kPa which was significantly higher than for Test 1. The initial pore pressures measured by the two transducers were -46 kPa and -57 kPa.

As the gravity was increased in the centrifuge, the weight of the dyke was increased and the stiff silicon subsoil settled which caused a differential displacement of the dyke. Additionally, some plastic volumetric deformation took place within the dyke itself, albeit limited. The PIV analysis showed a displacement of 17 mm at the crown of the dyke at 99 G-force. At a 100 G-force, the dyke underwent large deformation including a circular failure surface which resulted in a very brittle slope failure. Both the PIV analysis and the LVDT readings showed that the failure was fast developing. Figure 7.② shows the dyke after the failure and in which the failure surface is clearly visible. It is believed that the increased strength and stiffness, in comparison to Test 1, limited the volumetric deformation and additional loading was required to develop sufficient positive pore pressure to cause the failure. Furthermore, the initial pore pressures were significantly more negative. The overconsolidation of the *Speswhite* clay also caused additional brittleness to the material.

3.3. Test 3: Speswhite clay ($\sigma'_{v,0} = 150$ kPa) - soft silicon

The third test consisted in the same overconsolidated *Speswhite clay* ($\sigma_{v,0}$ = 150 kPa) as in Test 2 but with a soft silicon base (E = 114 kPa). The initial water content (w) was 52% and the undrained shear strength (s_u) was 10 kPa. The initial pore pressures measured by the two transducers were +3 kPa and -37 kPa.

Figure 7. Results of the centrifuge tests

Figure 8. Test 4: Shear strains from PIV analysis at 100 G-force

Figure 9. Test 5: Shear strains from PIV analysis at 48 G-force

As gravity was increased in the centrifuge, the mass of the dyke increased and the soft silicon base deformed substantially and formed a dip in which the dyke sat without any slope failure taking place. The dyke exhibited some shearing towards the end of the simulation. It is believed to be caused by excessive deformation of the subsoil and the influence of the stiff boundary conditions (strong box). Figure 7.③ shows the final deformation of the dyke and its subsoil in which the shear bands are marked in red doted lines. Despite the absence of a slope failure, the dyke has undergone such a large deformation that, at prototype scale, flooding of the low-lands is to be expected.

3.4. Test 4: Speswhite clay ($\sigma'_{v,0} = 100$ kPa) - soft silicon

The fourth test was the same as Test 3 but in which the *Speswhite* clay was consolidated at a lower load ($\sigma_{v,0} = 100$ kPa). The silicon subsoil was soft ($E = 114$ kPa) as for Test 3. The initial water content (w) was 55% and the undrained strength (s_u) 7 kPa. The initial pore pressures measured by the two transducers were -21 kPa and -24 kPa. The results show a similar behaviour as for Test 3. The dyke and its subsoil underwent large deformations without any slope failure taking place. Figure 8 shows the contour plot of the shear strains obtained from the PIV analysis. It clearly shows that, despite the absence of a slope failure, the dyke did develop multiple shear surfaces. Figure 7.④ shows the deformation at a 100 G-force. The deformation of the subsoil formed a dip in which the dyke sat. The final configuration of the dyke is similar to the one observed in Test 3.

3.5. Test 5: Speswhite clay ($\sigma'_{v,0} = 100$ kPa) - medium-stiff silicon

The fifth test consisted of the same dyke as for Test 4 (*Speswhite* clay, $\sigma_{v,0} = 100$ kPa) but with a medium-stiff silicon base ($E = 565$ kPa). The initial water content (w) was 55% and the undrained shear strength (s_u) was 8 kPa. The initial readings of the two pore pressure transducers were -28 kPa and -30 kPa. The results show that the silicon base was deformed by the weight of the dyke. However, the dyke itself underwent larger deformation. At 48 G, two circular failure surfaces developed simultaneously in the dyke and slide in a rotational manner. Figure 9 shows a contour plot of the shear strains and in which the two failure surfaces are clearly visible. Unlike Test 4, the shear deformation was concentrated in these two failure surfaces. Figure 7.⑤ shows the final configuration of the model in which the two failure surfaces are clearly visible.

3.6. Test 6: Baskarp sand ($D_r = 90\%$) - soft silicon

The sixth test consisted of a sandy dyke made out of *Baskarp* sand and a soft silicon base ($E = 114$ kPa). The sand was compacted to achieve a relative density D_r of 90% which gave it a stiff and brittle mechanical behaviour.

The results show that during the increase in gravity, the de-aired water present in the dyke was drained out by gravity. This was due to the increase in weight of the water and facilitated by the increase in permeability. Suction tension was hence increased and an enhancement of the strength and brittleness of the sand is to be expected. Figure 7.⑥ shows the final deformation of the dyke in which the groundwater table is visible. During the tests, the weight of the sand increased which increased the load on the soft silicon base. The settlement was in the range of the one with a clayey dyke. This settlement obliged the dyke to deform with it. The dyke developed a series circular failure surfaces allowing it to rotate like a knee-joint and can be seen in Figure 7.⑥. Unlike clay, dense sand does not compact significantly upon isotropic compression but dilates when sheared. Consequently, it was difficult to saturate the pores and develop positive pore pressure. No slope failure was noticed.

3.7. Test 7 and 8: Baskarp sand ($D_r = 90\%$) - Oostvaardersplassen clay

The seventh and eighth tests were duplicates. They consisted of the same sandy dyke as for Test 6 but resting on a remoulded natural soft clay called *Oostvaardersplassen* clay. The subsoil had an initial water content (w) of 149% and an undrained shear strength (s_u) of 7 kPa. Two pore pressure transducers were placed in the clay and the initial pore pressures were zero. During the test, the water in the dyke drained out as for Test 6. However, the behaviour of the clayey subsoil differed from the silicon subsoil. The clayey subsoil underwent a continuous plastic deformation with an overall contraction. The entire subsoil is believed to have yielding possibly due to liquefaction (zero effective stress). However, in both Test 7 and 8, shear surfaces developed in the clayey subsoil and propagated to the dyke. The large deformation of the subsoil obliged the sandy dyke to deform and develop shear surface to allow it to deform. In Test 6, a straight shear band developed to the surface whereas in Test 8 circular shear bands developed as in Test 6. These shear bands of Test 7 and 8 can be seen respectively in Figure 7.⑦ and 7.⑧.

3.8. Test 9: Speswhite clay ($\sigma'_{v,0} = 100$ kPa) - stiff silicon

The ninth test was the continuation of Test 4 and 5 in which the dyke is made out of an lightly overconsolidated *Speswhite* clay ($\sigma_{v,0} = 100$ kPa) and rested respectively on a soft and medium-stiff silicon base. In Test 9, it rested on a stiff silicon subsoil ($E = 1377$ kPa). The initial water content (w) was 56% and the undrained shear strength (s_u) was 6 kPa. The readings of the two pore pressure transducers showed pressures of -26 and -28 kPa.

The results show very little deformation until 40 G-force after which some large deformation took place. A curved failure surface then propagated from the base to the crown of the dyke and part of the dyke slid as a block. At 80 G-force, a second failure took place but with a circular failure surface. The deformation of the stiff silicon subsoil was smaller than for Test 2. This may be due to the hardening of the silicon in between the tests.

4. Discussion of results and conclusion

The results show that the stiffness of the subsoil plays an important role in the collapse mechanism of the dyke. Stiff subsoils allowed the dyke to develop a failure surface which rose to a slide. Soft subsoils deformed excessively which formed a dip in which the dyke sat. The imposed deformation of the subsoil to the dyke led to diffused sheared zones or 'knee-joint' shear bands giving the dike more freedom in movement. Figure 10 shows the LVDT measurement of the displacements at the crown. It clearly shows that the soft sub-soil leads to a progressive and continuous deformation of the subsoil and hence the dyke sitting on it. On the other hand, the results show a stiffer response with the stiff silicon subsoil. The brittle failure of the dyke can be seen with the appearance of a plateau. It is interesting to note that the preconsolidation pressure of the *Speswhite* clay influenced its undrained strength and hence when the failure

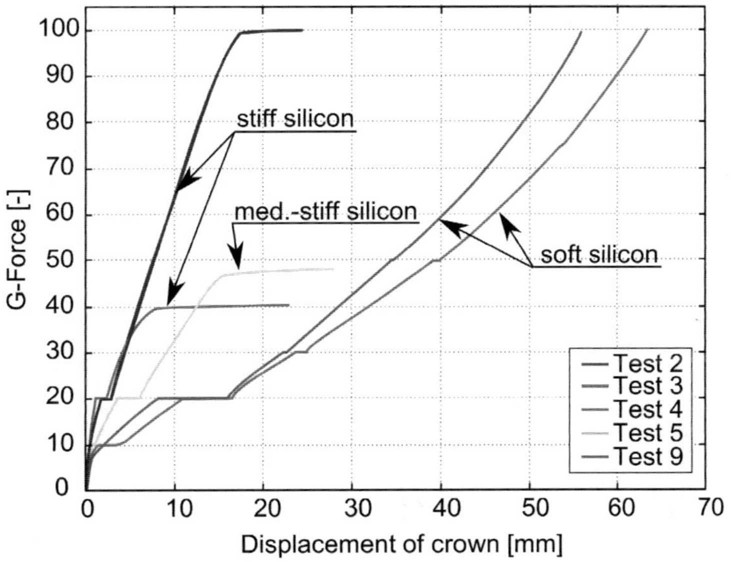

Figure 10. Displacement of the crown during loading for the clayey dykes

will took place. Test 5 with a medium-stiff silicon subsoil and an intermediate preconsolidation pressure of the *Speswhite* clay confirms this thought. The displacement rate controlled by the subsoil is in between the stiff and the soft case and the failure takes place at similar G-force as Test 9 which has a stiffer subsoil.

It is unclear if the failures took place in undrained (without any groundwater flows) or drained (with groundwater flows)conditions and if the pores were saturated or unsaturated. It is believed that the clayey dykes were largely undrained due to low permeability. However, the pores were initially unsaturated but, as the gravity increased, the pore space decreased and positive pore pressure was able to develop. The readings of the pore pressures confirmed this behaviour. Therefore, it is believed that the failures in the clayey dykes were undrained failures. On the other hand, it is believed that the sandy dykes remained unsaturated throughout the entire tests. Its high permeability was increased as the gravity was increased. Therefore, it is believed that the failures in the sandy dykes are drained failures. For all tests, the initial pore pressure were negative implying that the dyke is unsaturated. The presence of menisci, an inter-particle bonding force, provides an enhancement of the hydro-mechanical properties of the material.

The stiffness of the subsoil is demonstrated with the silicon blocks. The stiff block led to brittle failures as the dyke was free to move on the subsoil. In contrast, soft subsoil settled forming a dip in which the dyke sat. The PIV analyses showed the emergence of diffused sheared zones in the core of the dyke. The dip in which the dyke sat prevented any mass from moving. The two cases with a the *Oostvaardersplassen* clay subsoil showed that the development of shear surface in the subsoil propagate to the dyke.

The current study highlighted the influence of the subsoil, partial saturation and material properties in the collapse mechanism of dyke. The next phase of the study will cover cases of loose sandy dykes on stiff subsoils. It will be followed by a detailed investigation of the unsaturated conditions and the effect of a drop of air pressure in the centrifuge chamber. A three-phase numerical investigation with the Material Point Method is currently under process.

Acknowledgment
The authors acknowledge the Dutch Ministry of Public Works (Rijkswaterstaat-WVL) and the HHNK Water Authority for commissioning and financing the centrifuge tests. They would also like to thank the European Unions Seventh Framework Programme for research, technological development and demonstration under grant agreement no PIAP-GA-2012-324522 and the Swiss National Science Foundation under grant agreement P1SKP2 158621 for additional funding for the analysis of the data. The authors would also like to thank the reviewers for their fruitful comments.

References

[1] van Geel B, Hallewas DP and Pals J P 1983 A late holocene deposit under the Westfriese Zeedijk near Enkhuizen (Prov. of Noord-Holland, The Netherlands): Palaeoecological and archaeological aspects. *Review of Palaeobotany and Palynology.* **38** 269-335.
[2] Pleijster E J, van der Veeken C and Robbert J 2015 *Dutch Dikes* (Rotterdam: nai010)
[3] Indraratna B, Balasubramaniam AS, Balachandran S 1992 Performance of Test Embankment Constructed to Failure on Soft Marine Clay. J Geotech Eng **118**12-33.
[4] Rowe RK, Gnanendran CT, Valsangkar AJ, Landva AO 2001 Performance of a test embankment constructed on an organic clayey silt deposit. Can Geotech J **38**1283-96.
[5] Zwanenburg C, Den Haan E J, Kruse G A M and Koelewijn A R 2012 Failure of a trial embankment on peat in Booneschans, the Netherlands *Géotechnique* **62**(6) p 479-90.
[6] Koelewijn A R. and Van M A. 2002 Stability analysis for embankments prone to uplift induced failure *8th International Symposium on Numerical Models in Geomechanics* eds G N Pande & Pietruszcak (London: Taylor & Francis)
[7] Zwanenburg C 2012 Dike Failure by large slip planes *12th Baltic Sea Geotechnical Conference*, Rostock
[8] Hjortneas-Pedersen A G I and Broers H 1993 Material properties of Oostvaardersplassen Clay *Delft Geotechnics* SE-704441/1
[9] Rietdijk J, Schenkeveld F M , Schaminée P E L and Bezuijen A 2010 The drizzle method for sand sample preparation *Physical Modelling in Geotechnics* (London: Taylor & Francis) p. 267-72
[10] Taylor RN 2005 *Geotechnical Centrifuge Technology* (London: Chapman & Hall)

Effect of Groundwater Table Rising and Slurry Reduction during Diaphragm Wall Trenching on Stability of Adjacent Piles

A.Mohamed[1]
Geotechnical institute – TU Bergakadmie Freiberg – Germany

E-mail: ahmed.mohamed01@feng.bu.edu.eg

Abstract. The process of diaphragm wall trenching normally affects the surrounding environment. The piles near a diaphragm wall trench are commonly affected by the trenching process as well. During trenching the slurry level and the natural groundwater level are assumed to be constant. However flooding may cause damage to the slurry trenches due to the increase of the groundwater table. Another possible scenario for the trench failure is due to reduction of the slurry level. The reduction of the slurry level could be due to the presence of cavities or a very coarse soil layer. Piles located near a trench could be affected greatly if the trench is subjected to reduction of slurry or increase of the groundwater level. This research focuses on studying numerically the stability of piles adjacent to the diaphragm wall during the trenching process, especially in cause of slurry reduction or increase of groundwater level. The slurry reduction were simulated numerically with the finite different analysis and compared with previous laboratory work. The increase of groundwater level is simulated for a case study in Giza, Egypt. Groundwater level was assumed to increase in the area. Piles are generally affected by the trenching process. The behaviour of the pile is related to its position from the slurry trench. The stability of the pile may not be affected greatly by a normal and successful trenching process. However slurry reduction or increase in the groundwater level may cause a great effect on the stability of the nearby piles. Trenching in general causes an increase in pile settlement, horizontal displacement and bending moment. The pile skin friction and end bearing are affected as well. The percentage of the change in slurry or groundwater levels affects piles deflection and bending moment.
Key words: slurry trenches; pile; finite difference; finite element; skin friction; end bearing; diaphragm wall.

1. Introduction

The use of diaphragm walls inside cities is now very common. Such walls could be located near a deep foundation, especially in the crowded modern cities. During trenching, the soil near the trench is expected to be deformed. The settlement due to trenching was previously discussed and summarised by [1] . The same study was made by [2] for several metro stations in Cairo. The settlement equation due to trenching was then derived. [3],[4], [5] and [6] studied the deformation due to diaphragm wall trenching including the trenching effect on lateral earth pressure coefficient. From such findings, the soil deformation due to diaphragm

[1] PhD student – Geotechnical institute – TU Bergakadmie Freiberg – Germany

wall trenching is in most cases relatively small and its effect on the surrounding structures is limited. However, a collapse during trenching of a diaphragm wall due to flooding and water table rising was observed and discussed by [7]. [8] Investigated the collapse of two slurry trench panels in underground metro station due to heavy rainfall that caused groundwater rising.

Drop in slurry level is also possible if the slurry penetrated into the soil. The study conducted by [9] on the filter cake showed that if the soil is very coarse, the penetration of slurry into such soil is expected and could be a reason for reducing the slurry level. Additionally the slurry level should be 1.5m higher than the groundwater level [10]. [11] Performed a field test to study the effect of trenching on adjacent piles in mixed soil layers. It was found that piles were deflected due to trenching; however the settlement of piles and soil was the same so that the bearing capacity of piles didn't show noticeable change. [12] and [13] performed a parametric study using centrifuge tests to investigate the effect of slurry trenching panels on adjacent single pile. The slurry reduction effect was the purpose of such models. A case study for slurry trenching and deep excavation near a deep foundation was investigated by [14] and [15]; the ground surface settlement due to trenching was the main investigation issues.

The collapse shape of the trench is expected to be the key rule influencing the nearby piles. If the failure of the slurry wall is in the upper part it may cause a horizontal upper movement on the pile equivalent to an additional horizontal upper load on pile. On other hand, if the trench collapses in the lower level it will probably causes maximum horizontal movement of the pile but not at its top. Piles are expected to moved together with the surrounding soil [16].

Studying the effect of trench collapse due to slurry reduction or rise of groundwater table on the piled foundation is the purpose of this paper. A comparison was conducted between the numerical simulation and the centrifuge model test results presented by [12]. On the light of such comparison a wider numerical parametric study was conducted to investigate the effect of slurry reduction and groundwater rising on the adjacent pile foundation. In addition the effect of reduced slurry pressure at some levels was also included in such study.

2. Centrifuge Model Tests and Remarks

Centrifuge model tests were conducted by [12] to investigate the effect of trenching on single pile in dry sand (no groundwater table). Reducing the slurry in a trench was investigated instead of investigating the effect of high groundwater table on the shallow trench instability. There were eight successful centrifuge tests made in such work.

2.1. Test discretion

The sand density and friction angle were 32° and 15.55 kN/m^3. It was placed in a strong box of dimension 700mm x 400mm x 470mm to depth of 400mm equivalent to 30m. The pile in the tests was simulated using aluminium pipe with outer diameter of 12mm (0.9 m in real scale). The pile had to penetrate about 10mm in the soil to reach a final depth of about 250 mm equivalent to 18.5m. The trench depth and thickness were 350mm (26.25m) and 16mm (1.2m) in all tests; however the length was 40mm (3m) and 80mm (6m). The guide wall was also simulated. The average slurry density was 10.97kN/m^3. The pile was chosen to be at offset distance of 3.5D, 5.6D and 7.7D where D is the pile diameter.

2.2. Test procedures

After placing the sand in the strong box, the trench was filled with water. Latex membrane was used to prevent penetration of water to the sand. Pile penetrated then into the sand and the load used for penetration was released keeping a normal load on the pile. The water was replaced with liquid equivalent to the slurry. Finally reduction for the slurry was made.

2.3. Remarks about the test

The absence of groundwater table is not common in most of the soil conditions. Using the water inside the trench in the initial condition affected the real initial state. The water inside a trench provides horizontal pressure of $\gamma_w z$ (9.8z), while the soil pressure should equal to $k_o \gamma_s z$ (7.3z). In this case horizontal displacement had occurred in the soil direction.
Two dimensions stability analysis suggested by [7] could be used to find out the minimum slurry depth that keeps the trench stable. The factor of safety could be calculated from equation 1.

$$FS = P_s / P_f. \tag{1}$$

The slurry force Ps calculated as

$$Ps = \gamma_s Z_s^2/2 \tag{2}$$

γ_s and Z_s are the slurry density and depth, respectively. The horizontal acting force is calculated from equation 3.

$$P_f = [\gamma Z^2/2(\sin\theta - \cos\theta \tan\phi) + U \tan \phi]/ (\cos\theta + \sin \theta \tan \phi) \tag{3}$$

γ is the soil density and Z is the trench depth. The value of θ (the inclination of the plane of failure) was used as $45^o + \phi/2$. As the water pressure U is equal to zero and friction angle ϕ equal 32° the horizontal acting force P_f equal 62.7kN/m`. In this case the slurry level could be only 3.5m (87% reduction in slurry) to achieve a factor of safety equal unity. However, this solution considered conservative one as it ignores the arching effect around trench ends. The comparison made by [17]showed that [7] two dimensional method provides the lowest factor of safety value. Accordingly, three dimension analytical solution by [18] for such case provides a factor of safety equal 1.35 for the 6m length trench and 2.35 for the 3m length trench.
From the above the trench could sustain global failure if the slurry reduced up to about 90%. The horizontal movement of the trench face could in this case govern the stability of the trench. However, the above equations didn't deal with the local failure. In realty filter cake is formed during trenching but indeed it is not strong enough to form a structural membrane; it only prevents grains' penetration inside the trench [19]. In contrast, [9] showed that filter cake is important for the stability in the case of sand, as the filter cake permits a full slurry pressure and it could also prevent local failure.
The slurry in the tested trench was reduced up to about 60% for the long trench and 80% for the shorter one. The used latex membrane should have played the role of the filter cake but indeed it could provide some additional strength for the trench wall surface. It is possible that the pattern of soil deformation was affected by the presence of such membrane. The detected failure in the experiment was due to large deformations that could damage the model pile. However, the tests provide a very good overview of the effect of trenching on piles. Therefore, numerical analyses were used to simulate such tests with the proper approximation in order to point out the reliability of the numerical models.

3. Numerical Simulation for the Tests

Table 1 shows a small description of each test. The numerical simulation of the tests was made using commercial software FLAC3D® which based on Finite difference analysis. The numerical simulation of tests CKC3, CKC6 and CKC16 was similar. Accordingly, a total of 6 numerical models were made for such simulation and comparison.

Table 1. Summary of successful centrifuge tests.

Test	*Panel length (m)*	Pile offset distance (m)	Number of Concrete panels	*Slurry density (kN/m³)*	*segmental pile resonse*
CKC1	6.0	6.93	-	10.97	N/A
CKC3	6.0	3.15	-	10.95	N/A
CKC6	6.0	3.15	-	10.89	N/A
CKC7	6.0	5.04	-	10.92	N/A
CKC13	3.0	3.15	1	10.92	A
CKC14	3.0	3.15	2	10.91	A
CKC15	3.0	3.15	-	10.92	A
CKC16	6.0	3.15	-	10.93	A

Note :The dimension is in the prototype scale

3.1. Modeling of soil

The soil was modelled using strain hardening softening soil model. The used friction angle was 32°; the sand density was 15.55kN/m^3. The strain hardening softening soil model used in FLAC required relation between mobilized friction angle and plastic strain which can be calculated according to (Byrne, 2003) as:

$$\varepsilon_p = \frac{P_{ref}}{\beta G^e_{ref}} \times \frac{sin\emptyset}{R_f}\left((1 - \frac{sin\emptyset_m}{sin\emptyset} R_f)^{-1} - 1\right) \& G^e_{ref} = \frac{E^{ref}_{ur}}{2(1+\vartheta_{ur})} \quad (4)$$

Where ϕ_m is the mobilized friction angle, ϕ is the ultimate friction angle, ε_p is plastic strain, ε_f is the strain needed to mobilize the limit friction angle, P_{ref} is the reference pressure, R_f is the failure ratio, β is calibration factor, G^e_{ref} elastic tangent shear modulus and ϑ_{ur} is the undrained poisons ratio.

3.2. Modeling of guide wall, pile and latex membrane

The guide wall was modelled as an elastic soil element in order to simulate the1.5mm (11.5cm – prototype scale) thick aluminium steel plate. The chosen properties for such element were equivalent to the used aluminium material.

The implemented Pile element in FLAC was used to model the pile. Material and coupling spring properties were used to define such model. The pile was solved with stiffness matrix which divided it into finite elements. Between the pile and soil grid frictional and normal interaction accrued. The shear and normal coupling stiffness were defined according to equation 2 provided by FLAC manual [21]. The friction angle for the spring was chosen according to the values of the mobilized interface friction angle [12].

The latex membrane was modelled using the shell element. The thickness of the shell element was chosen to be 10cm with a small value of deformation modulus. The used shell element was mainly to prevent the local failure as the latex membrane was probably functioned.

3.3. Numerical Construction Stages

The numerical stages of construction were chosen to simulate each laboratory test. All the tests were used to simulate a single panel except two tests. These two tests were used to simulate multiple panels including concrete panels. The general simulation process is placing the sand in the strong box. Theoretically, the pile should be placed before the diaphragm wall. However, the water was placed first inside the latex membrane to keep the position of the diaphragm wall. The pile installation took place after water filled the latex membrane. A load was applied to make the pile penetrates 10mm inside the sand and then the load was removed. The water inside the trench was replaced with liquid which represent the slurry. Finally, the slurry is to be reduced in stages until noticeable failure occurred. Slurry is to be replaced by concrete if a multiple panel is to be simulated. More details about the test could be found in [12] and [13].

The numerical simulation phases were chosen to simulate the centrifuge tests. The guide wall was initially simulated by replacing soil properties with concrete properties at its location. Soil elements in the trench position were removed in stages of 5m approximately. Water pressure is simultaneously applied on the trench wall. Pile penetrated after that in the soil to approximately 75cm. The water pressure is then replaced by the slurry pressure of 10.97kN/m^2. The pressure is then removed gradually to simulate the slurry reduction process. The concrete panels were simulated by reactivating the soil in such location and replace its properties by concrete properties. General geometry, construction stages and typical mesh model is shown in Figure1.

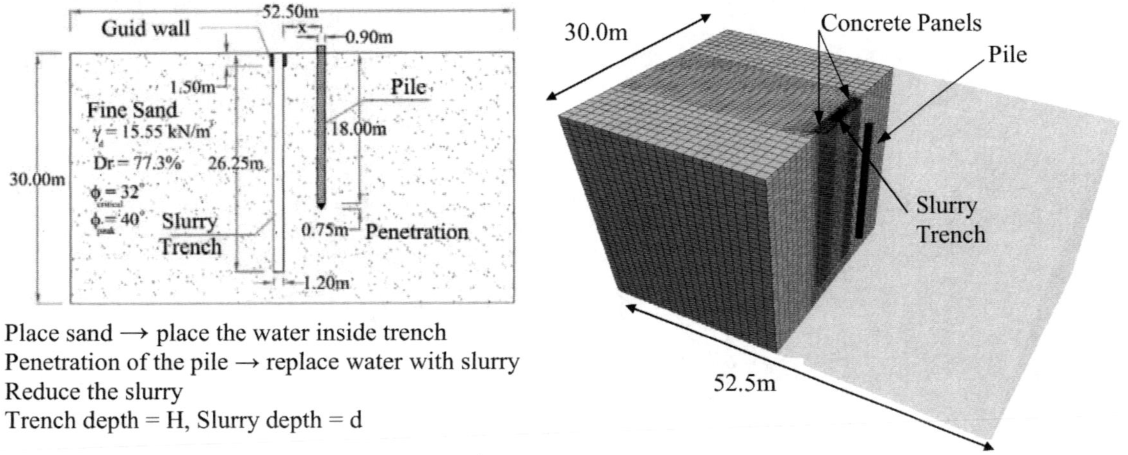

Figure 1. General geometry, construction stages and typical mesh used for simulation.

4. Results and comparison

This section discussed and compared results from finite different analysis and centrifuge model tests. Soil settlement and pile behaviour in different construction stages are the focus of the comparison.

4.1. Results relating to soil settlement

The comparison between centrifuge tests and Finite difference results regarding settlement of soil due to pile installation and slurry reduction are presented in this section. The settlement due to pile installation is shown in Figure 2. The settlement from the numerical analysis and from the test was in a good match. Figures 3, 4 and 5 show the settlement due to slurry reduction. Generally, the settlement due to slurry reduction from numerical analysis was quit in a good contrast with laboratory work. However, the settlement from the numerical analysis was decreased with distance shorter than that from laboratory tests. The shape of settlement from numerical analysis was in better contrast with [1]. The settlement for the 3.0m length trench was much lower than the trench with 6.0m length; this indicates that the length of the trench have a great effect on soil settlement.

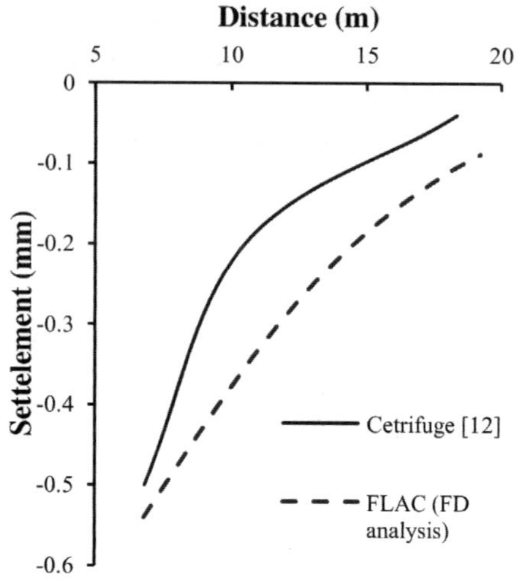

Figure 2. Typical Settlement due to pile installation.

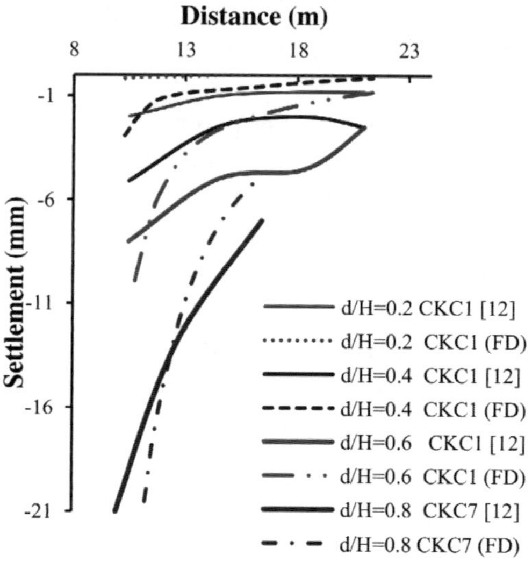

Figure 3. Settlement due to Slurry reduction (CKC1 and 7).

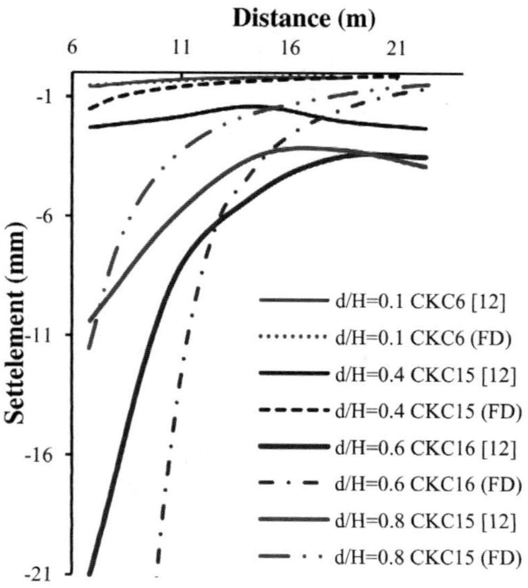

Figure 4. Settlement due to Slurry reduction (CKC 6, 15 & 16).

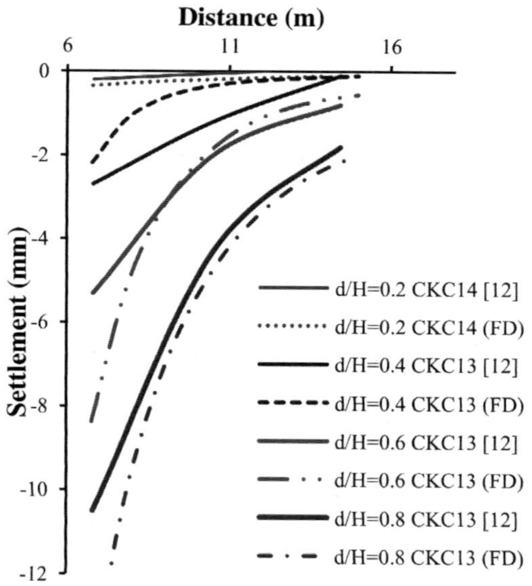

Figure 5. Settlement due to Slurry reduction (CKC 13 & 14).

4.2. Results of pile horizontal movement and bending moment

The horizontal displacement and bending moment of piles due to slurry reduction from numerical and laboratory tests are presented and compared in this section. It can be obvious from Figures 6 and 7 that horizontal displacement and bending moment from FLAC analysis and laboratory experiments have almost the same shape; however, the values are different. The values of the laboratory tests were probably affected by eccentric loading which have probably caused a disturbance in the horizontal displacement and bending moment values of the pile. The direction of horizontal displacement in CKC 15 for the last stage was illogical as the displacement was moving against the wall. Generally, the values of bending moment at failure were relatively high compared to the numerical analysis.

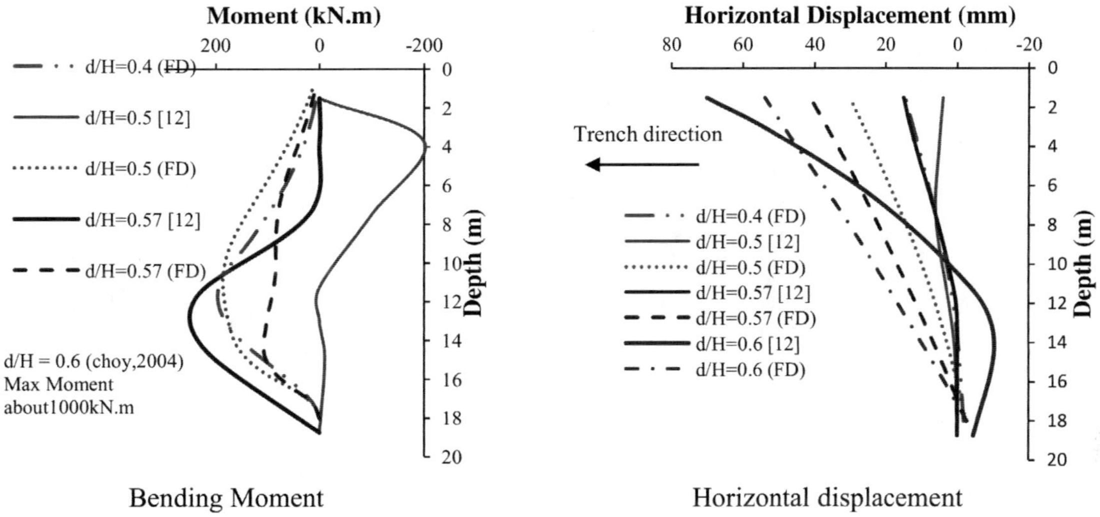

Figure 6. Pile horizontal displacement and Bending Moment due to slurry reduction (CKC16).

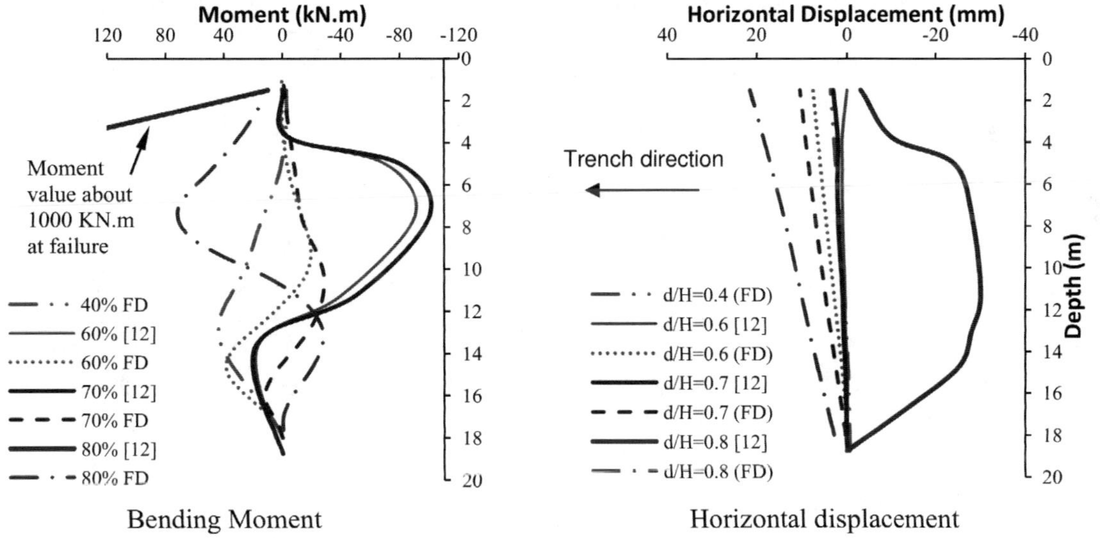

Figure 7. Pile horizontal displacement and Bending Moment due to slurry reduction (CKC15).

4.3. Results of pile shear force and base load

The shaft frictions during slurry trenching are presented in Figures 8 and 9.The values from the laboratory tests and the numerical analysis show that shaft friction increases with slurry reduction. The last stage of construction shows a higher difference between the laboratory results and the numerical analysis. The collapse that happened in the laboratory experiment and didn't happen in the numerical analysis is the reason behind that noticeable difference. However, the trend of force is almost identical. The bearing capacity of pile decreased with slurry reduction. The percentage of decreasing predicted from the laboratory and numerical analysis was almost identical. The values of bearing capacity doesn't show noticeable change until the slurry reduced to about 40% (10.5m). According to [12], the skin friction increased due to the decrease of bearing capacity.

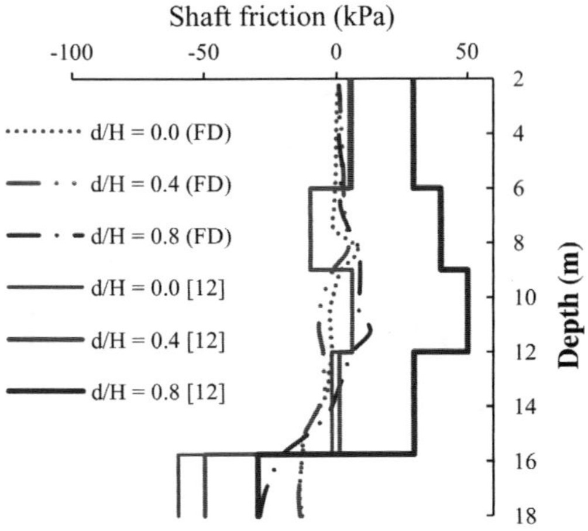

Figure 8. Pile shaft friction during slurry reduction (CKC 15).

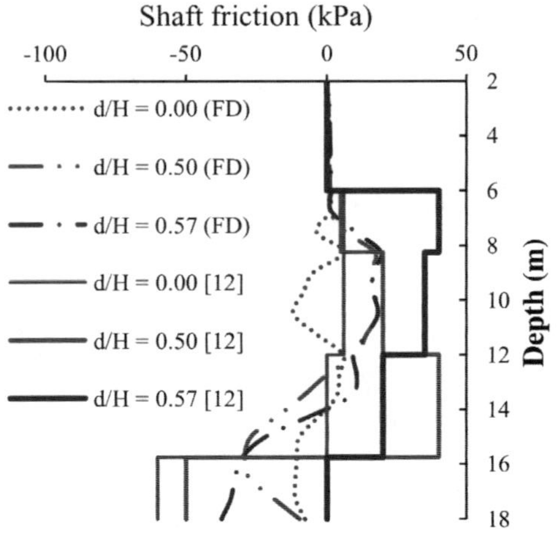

Figure 9. Pile shaft friction during slurry reduction (CKC 16).

5. Assumption of rising water table in a case study

A case study in Giza, Egypt is discussed in this section. The case study is described in details by Abdel-Rahman& El-Sayed (2002, 2009). It was a deep excavation constructed near piled foundation. The settlement due to diaphragm wall trenching was measured and simulated in the normal case. In this research, a model is made to simulate the same case with raising the water table. The effect of such rise on the ground settlement and pile behaviour is to be discussed and compared with the normal case. The soil was mainly sand with density ranging between 17.0 and 20.0kN/m^3. The friction angle varies between 28° and 36°. The upper soil layers were Fill and silty Sand while the lower soil layers were medium dense and dense Sand.

5.1. Modeling the case history

A 3D finite element Analysis (PLAXIS 3D®) was used in simulating the case history. The water level in normal condition was 2.0m below the ground surface and it was simulated that it rises up to the ground level as in the flood case. During trenching process of the last panel, the water table is assumed to increase and reach the ground surface. The other panels in this case were already filled with concrete. Figure 10 discussed the used mesh model.

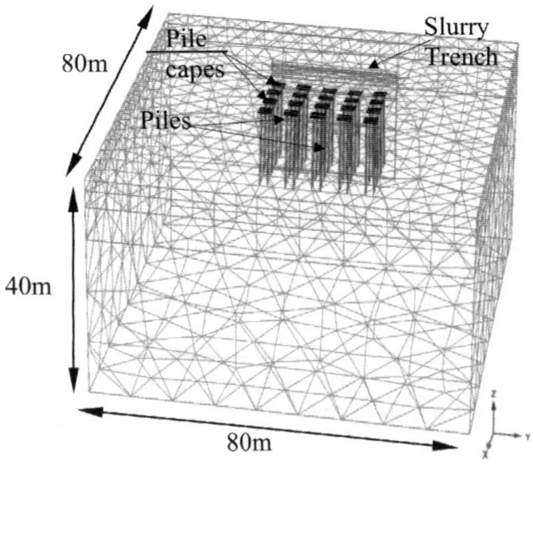

Figure 10. Mesh used in modelling the soil layers is 15-node triangular element with 15 Gaussian points. Hardening soil model used to model the soil layers while Elastic soil model was used to simulate the concrete panels. The piles were simulated using embedded pile element. Plate elements were used to simulate pile caps. The building load was distributed on the pile caps.

5.2. Settlement comparison and pile numerical results

The plane showing sections and panels is presented in Figure 11. All panels were assumed to be filled with concrete except panel P-20. Settlement of soil due to trenching and water rising from finite element and field data are presented in Figure 12. Soil generally heaved due to water rising; on the other hand piles 1, 2 nearest to the excavated panel P-20 showed a noticeable horizontal movement during water rising as shown in Figure 13. The shape of bending moment was also affected with water rising. The bearing capacity was almost constant; however, skin friction decreased for pile 1 and pile 2 to values about 45% and 25%, respectively. This reduction was significant and may cause failure to the pile. The reduction in skin friction was due to the decrease of the earth pressure coefficient. The skin friction didn't increase as in the previous section because the end bearing was not changed.

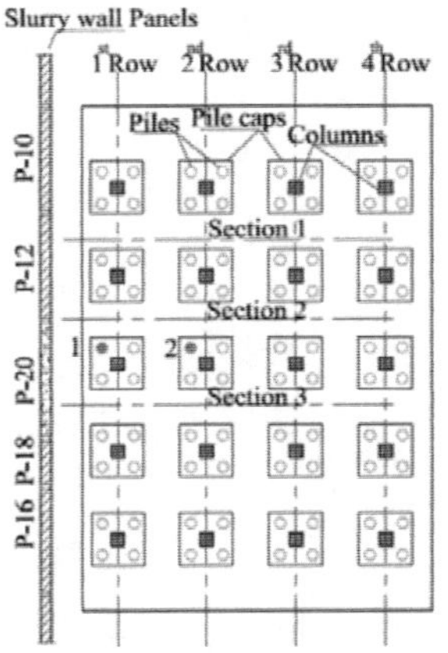

Figure 11. Piles and slurry trench

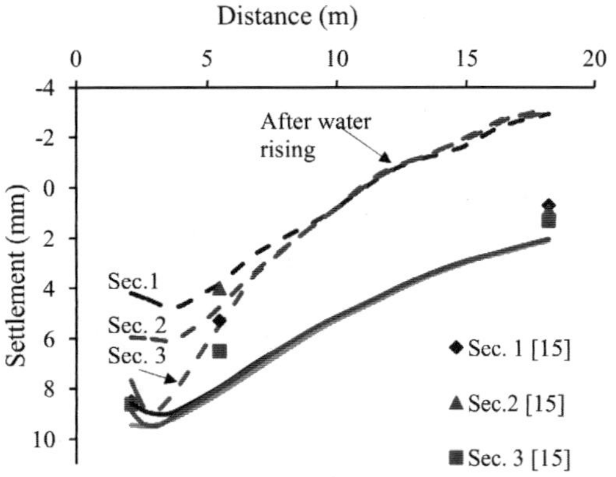

Figure 12. Soil settlement due to trenching and water rising.

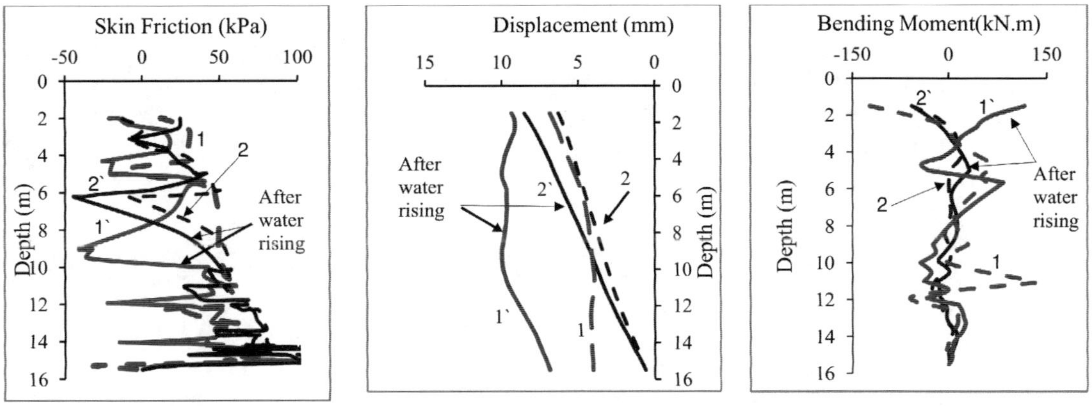

Skin friction Horizontal displacement Bending moment

Figure 13. Piles skin friction, Horizontal displacement and bending Moment before and after water rising

6. Numerical parametric study

A simple parametric study based on numerical analysis was presented in this section. Finite difference analysis was used for such study. The main purpose of the parametric study is to find out the critical situation that could cause damage or high movement to the pile element near the trench. The possibility of slurry reduction or changes in groundwater levels was simulated for different pile positions from the trench. Additional simulation was made for slurry reduction at some locations which represent the case of presence of very course soil layer or cavities at these locations.

6.1. Main parameters and models
A total of 48 model analyses were made in this parametric study. The general variables are described in Figure 14. The soil was modelled using strain hardening softening soil model, while the pile element was used in modelling the pile. The modelling phases started by placing the pile in its position. Then the soil element at trench location was removed and replaced with slurry pressure.

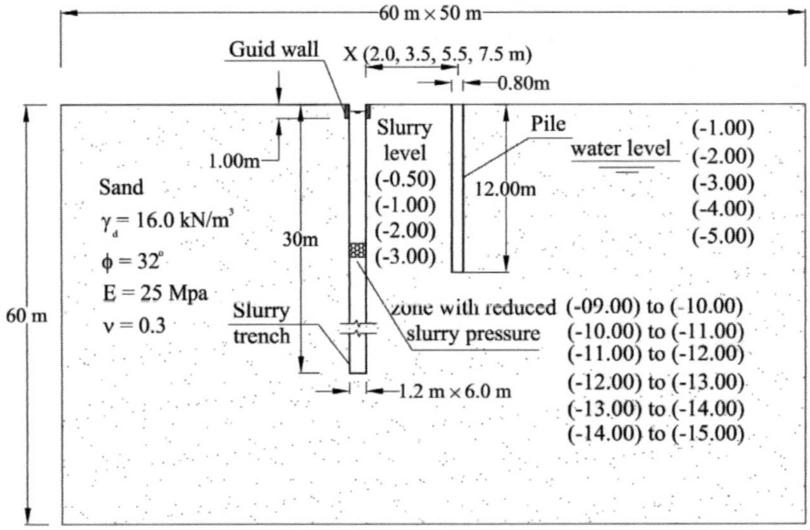

Figure 14. Parametric study main variables.

6.2. Parametric study results
The effect of different parameters on pile horizontal displacement and bending moment are presented and discussed in this section. The effects of slurry and groundwater level are presented in Figure 15 and 16. The groundwater level effect was studied for a pile at distance of 3.5m from the trench with a slurry level of 0.5m. The horizontal deformation increases by the increase of water level. The shape of the horizontal displacement varies according to the different water levels. Generally the pile tip is moving more than its top in all groundwater levels except for levels (-4.00) and (-5.00). The change in groundwater table causes the bending moment shape to change; without a noticeable change in its values. The groundwater table was fixed to be 2.0m below ground surface in order to find out the effect of slurry level. The bending moment and horizontal displacement are almost not effected by changing slurry level up to 2m below ground surface. The pile top shows a relatively significant movement toward the trench as the slurry level was 3.0m below the ground surface. The bending moment also increases in the positive direction when the slurry level was 3.0m below ground surface. The reason for the big changes in pile behaviour at slurry level of 3.0m is due to the failure of the trench at such level.
Effects of reduced slurry pressure at some levels are presented in Figure 17 and 18. The reduction in slurry pressure is presented for a pile distance of 3.5m, water level of 2.0m and slurry level of 1.0m. The maximum displacement and bending moment recorded for reduced pressure at levels between 12.0m and 13.0m (pile tip location).
The distance of pile from the trench is considered to be a major factor that affects the pile behaviour. The horizontal displacement and bending moment of the pile are reduced with distance from the trench. The horizontal displacement for piles at 3.5m distance from the trench is about 25%less than

piles at distance 2.0m from the trench. While piles at distance 7.5m from the trench is about 62% less than the nearest piles to the trench. The shape of bending moment is greatly affected by pile location. Additionally, the values of bending moment were higher for piles near the trench than piles at distance from the trench.

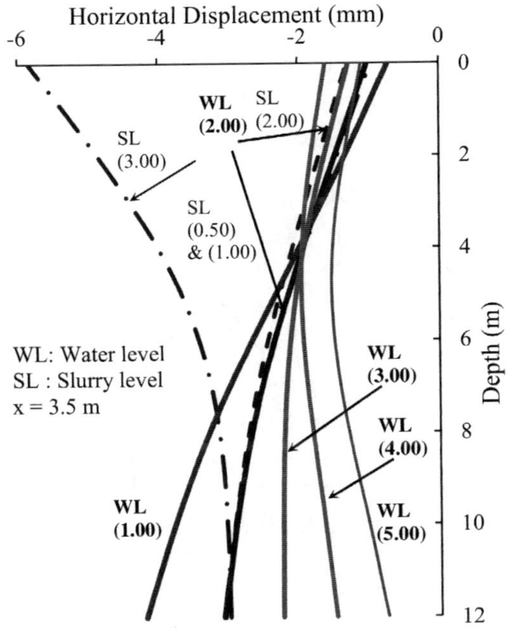

Figure 15. Effect of water and slurry level on pile Horizontal displacement

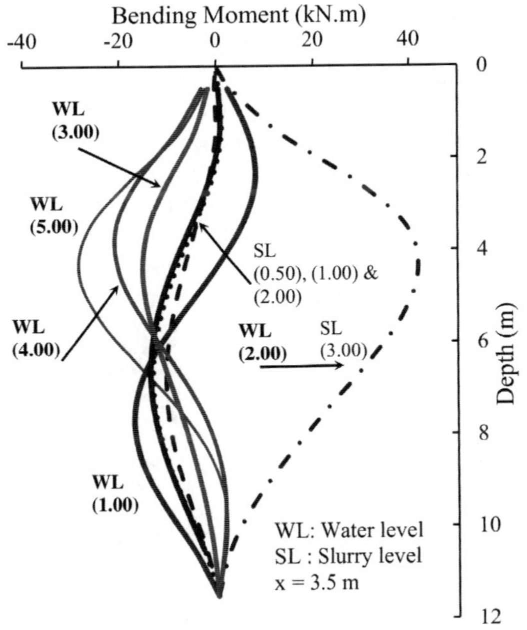

Figure 16. Effect of water and slurry level on pile Bending moment

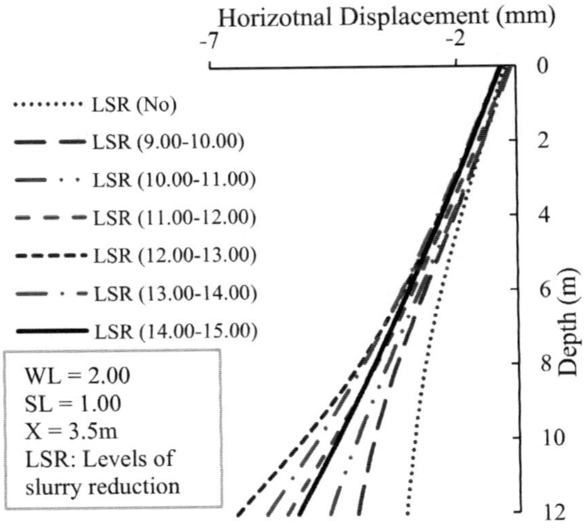

Figure 17. Effect of slurry level reduction at some levels on horizontal displacement.

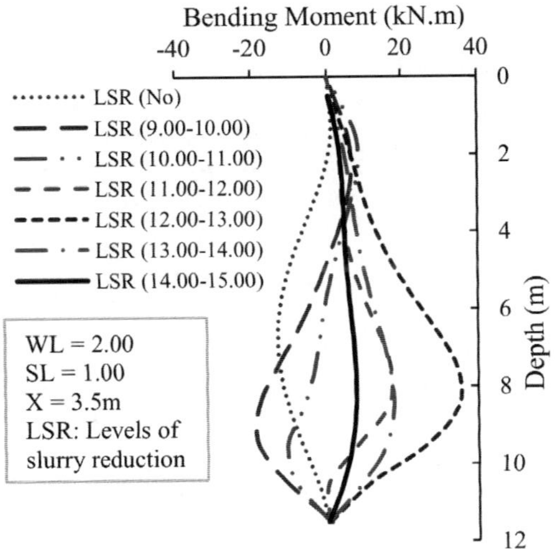

Figure 18. Effect of slurry level reduction at some levels on Bending moment.

7. Conclusions and recommendation

The results of the finite difference and centrifuge model tests regarding soil settlement was in better agreement than results regarding pile horizontal displacement and bending moment. The pile results from centrifuge model tests were not reliable at some cases, as the pile could be subjected to eccentric load during the centrifuge model test. However, the results from the centrifuge and finite difference analysis provided the same general trend of pile behaviour. The horizontal displacement and bending moment of the pile increased with slurry reduction. The skin friction of the pile increased due to the decrease in end bearing. The pile end bearing decreased when the slurry reduced to about 40%. The trench length has a significant effect on the pile horizontal movement and bending moment.

Finite element results and field results were in a good agreement regarding soil settlement during trenching. The soil heaved due to water rising. However, piles near the excavated panel moved horizontally. The shape and values of bending moment for such piles were changed. The skin friction values show reduction contradicted with centrifuge model tests and finite difference results because the pile end bearing in this case was almost constant.

The changes of the groundwater level and slurry level were studied in the numerical parametric study. The slurry level at level 1.0m below the groundwater table caused the piles within distance of 0.2H from the trench to deflect significantly. Such deflection could cause instability to the pile and the whole structural system. Groundwater rising causes an increase in piles deflection. The shape of pile deflection was affected by the groundwater level. The reduction of slurry pressure at some levels causes additional deformation and bending moment for the pile. The reduction of slurry pressure at pile tip level has the highest effect on the pile horizontal movement and bending moment.

The increase of water table due to flood or other reason could cause damage to the piles near the slurry trench. The slurry level is the greatest parameter that could cause collapse in trench and damage to the nearby piles. The failure of the trench may not always cause a failure on the nearby pile; the failure shape is a main factor affecting the behaviour of the nearby piles. Precautions should be made regarding slurry level during trenching near deep foundations include keeping the slurry trench in its level. If the groundwater rises, an increase of the slurry level is highly recommended. A high quality soil investigation is required to observe any lenses or cavities of course soil that could cause a leakage of slurry into the soil and reduces its pressure in such locations.

It is recommended to widen the parametric study to include the effect on pile length and diameter. The pile group should be also taken into consideration. Different soil properties should be included in such study. The effect of trenching on pile skin friction and end bearing should be better clarified.

8. References

[1] Clough, G. W. and O'Rourke, T. D. (1990): Induced Movements of Insitu Walls. In *Design and Performance of Earth Retaining Structures, ASCE Geotechnical Special Publications* 25, pp. 439–470.

[2] Hamza, M.M., Atta, A. and Roussin, A. (1999): Ground Movements Due to the Construction of Cut-and-Cover Structures and Slurry Shield Tunnel of the Cairo Metro. In *Tunnelling and Underground Space Technology* 14 (3), pp. 281–289.

[3] Farmer, I. W. and Attewell, P. B. (1973): Ground movements caused by a bentonite-supported excavation in London Clay. In *Géotechnique* 23 (4), pp. 577–581.

[4] Ng, C. W.W.; Yan, R. W.M. (1999): Three-dimensional modeling of a diaphragm wall construction sequence. In *Géotechnique* 49 (6), pp. 825–834.

[5] Poh, T. and Wong, I. (1998): Effects of Construction of Diaphragm Wall Panels on Adjacent Ground: Field Trial. In *Journal of Geotechnical and Geoenvironmental Engineering* 19 (2), pp. 285–300.

[6] Tsai, J.-S., Jou, L.-D., and Hsieh, H.-S. (2000): A full-scale stability experiment on a diaphragm wall trench. In *Canadian Geotechnical Journal* 37 (2), pp. 379–392.

[7] Morgenstern, N., and Amir-Tahmasseb, I. (1965): The stability of a slurry trench in cohesionless soils. In *Géotechnique* 15 (4), pp. 387–395.

[8] Khare, M. G., Ramanathan, M. (2014): Investigation of Collapse of Slurry Trench in Underground Metro Construction. In. The fifth International Conference on Foresic Engineering. London, April, 2014.

[9] Filz, G. M., Adams, T. & Davidson, R. R. (2004): Stability of long trenches in sand supported by bentonite-water slurry. In *Journal of Geotechnical and Geoenvironmental Engineering ASCE* 130 (9), pp. 915–921.

[10] Puller, M. (2003): Deep excavations: a practical manual. 2nd edn. London: Thomas Telford.

[11] Wit, de J. C. W. M. & Lengkeek, H. J. (2002): Full scale test on environmental impact diaphragm wall trench installation in Amsterdam – the final results. In. Proceedings of the 3rd international Symposium on Geotechnical Aspects of Underground Construction in Soft Ground. Toulouse, France.

[12] Choy, C. K. (2004): Installation effects of diaphragm walls on adjacent piled foundations. PhD dissertation. University of Cambridge, UK.

[13] Choy, C. K., Standing, J. R. and Mair, R. J. (2007): Stability of a loaded pile adjacent to a slurry-supported trench. In *Géotechnique* 57 (10), pp. 807–819.

[14] Abdel-Rahman, A. H.; El-Sayed, S. M. (2002): Settlement Trough Associated with Diaphragm Wall Construction in Greater Cairo. In *the Journal of the Egyptian Geotechnical Society.*

[15] Abdel-Rahman, A. H.; El-Sayed, S. M. (2009): Foundation subsidence due to trenching of diaphragm walls and deep braced excavations in alluvium soils. In. proceeding of the 17th international geothechnical engineering conference. Alexandria, Egypt (3), pp. 1935–1938.

[16] Cole, K. W. (1980): The south abutment at Kessock Bridge, Scotland. In. Proc. IABSE Conf,. Vienna, pp. 1099–1105.

[17] Tsai, J. and Chang, J. (1996): Three-dimensional stability analysis for slurry-filled trench wall in cohesionless soil. In *Canadian Geotechnical Journal* 33 (5), pp. 798–808.

[18] Fox, P. J. (2004): Analytical solutions for stability of slurry trench. In *Journal of Geotechnical and Geoenvironmental Engineering ASCE* 130 (7), pp. 749–758.

[19] Filz, G. M., Boyer, R. D. and Davidson, R. R. (1997): Bentonite-Water Slurry Rheology and Cutoff Wall Trench Stability. In. In Situ Remediation of the Geoenvironment. Minneapolis, Minnesota, United States, pp. 139–153.

[20] Byrne, P. M., S. S. Park and M. Beaty. (2003): Seismic Liquefaction: Centrifuge and Numerical Modeling. in FLACand Numerical Modeling in Geomechanics. In *Proceedings of the 3rd International FLAC Symposium, Sudbury, Ontario, Canada*, pp. 321–331.

[21] Itasca (2012): Fast Lagrangian Analysis of Continua in 3 Dimensions (FLAC 3D) version 5 User's Guide. 5th ed. Minneapolis, Minnesota, USA: Itasca Consulting Group Inc.

Methods for the physical measurement of collisional particle flows

Devis Gollin[1], Elisabeth Bowman[1] and Paul Shepley[1]

[1]Department of Civil and Structural Engineering, University of Sheffield, Mappin Street, Sheffield, S1 3JD, United Kingdom.

E-mail: devis.gollin@sheffield.ac.uk

Abstract.
Particle image velocimetry (PIV) and particle tracking velocimetry (PTV) are used in this paper to test their ability in measuring kinematic properties of granular flows, such velocity fluctuations and granular temperature. A small inclined chute geometry was used here to reproduce flows encompassing different flow regimes. An experimental investigation of a dry free-surface flow composed of almost spherical monodisperse ceramic beads is presented. The two image velocimetry techniques are directly applied to images captured in a region of the flow where an apparent uniform steady regime was observed. Our results shows that PTV is capable of measuring the movement of individual particles resulting in estimations of granular temperature that can be compared with other studies. In contrast, PIV tends to damp the magnitude of the random component of the velocities, which in turn produces lower values of granular temperature.

1. Introduction

In geophysical contexts, granular flows are encountered in various forms. Examples are granular chute flows, rock avalanches, landslides and debris flows. In general, a granular flow involves one or more flow regimes: a slowly deforming or quasi-static flow governed primarily by inter-particle friction, a dense flow regime where particles interact by long-lived contacts as well as inter-particle collisions, and a strongly agitated regime dominated by particle collisions. This reflects the inherent complexity and the multitude of heterogeneities present in bulk flow that makes these systems very difficult to study as a whole.

Over the last half-century, granular flows have been investigated by means of diverse analytical, experimental and numerical studies. Efforts have been made to formulate mathematical models that try to capture the complex behaviour of granular flows (e.g. [1]). However, a single unified model that includes all the aforementioned regimes is yet not available. The research and development of new models is continuous and there is a strong need for experimental data in order to verify such models. The final goal is to produce theories and models that incorporate notions of the behaviour produced by the relative configuration of the particles which results from complex frictional and also collisional particle interactions.

It is common to treat granular flows, such as debris flows, according to their frictional or viscous behaviour but little is known about the effect induced by collisional interactions to the flow dynamics. Macroscale models that use depth averaging methods are the principle approach for the dynamic modeling of debris flows [2-3]. However, difficulties arise when accounting for the microscopic behaviour of debris flow materials in terms of transportation kinematics and collision interaction. Consequently, depth averaging methods may result in a unclear picture of the debris flow dynamics. Ultimately, careful consideration to the microscale mechanics should also be considered to achieve a more comprehensive understanding of these granular flows.

From an experimental point of view, non-intrusive techniques such as particle image velocimetry (PIV) and particle tracking velocimetry (PTV) are now widely used for measuring behaviour of granular flows in laboratory experiments. Their development over the past decades has made them practical tools to estimate particle displacement and single particle movements. Such measurements have become useful for checking and refining mathematical theories, especially those based on stresses originated by the collisions of particles such as the kinetic theory of granular flows. This theory relates the random components that generate particle stresses through collisions to the concept of granular temperature. The latter can be used to characterize the flow regimes and infer the ability of a granular medium to flow. However, estimations of granular temperature are rarely found in the literature and in any case are difficult to interpret.

In this paper PIV and PTV are applied to dry free-surface flows of almost spherical monodisperse ceramic beads down an inclined chute geometry. We use the two image velocimetry techniques and compare their ability to estimate kinematic properties of the granular flows, e.g., velocity fluctuations and granular temperature. The flows are relatively less complex than natural debris flows, although the primary motivations for this work are to test PIV and PTV and to provide new experimental evidence that may help in the formulation of unified models that encompass the whole range of flow regimes. After a brief introduction on the two image analysis techniques, apparatus, and experimental procedure we present our results and discuss their significance.

2. PIV and PTV for granular flows

Two main methods can be distinguished for the measurement of instantaneous velocity fields from a set of recorded images, namely, particle image velocimetry (PIV) and particle tracking velocimetry (PTV). Each technique presents strengths and weaknesses; which method one decides to use strongly depends on the experimental set-up and flow conditions [4].

PIV is a well-established techniques that uses statistical cross-correlation to determine the spatial shift of particles from subimages. PIV provides exhaustive analyses although some limitations exist. One of the problem encountered is gradient biasing caused by the development of velocity gradients in the flow. Faster moving particles leave the subimage region and the remaining slower particles act to bias the displacement estimate [5]. A further limitation regards the size of the interrogation region. It plays an important role in determining the achievable spatial resolution of the flow. There is a limit on how large or small this interrogation region can be for adequate measurements. Subimages smaller than the particle diameter do not produce sufficient textural information and increment the level of noise. Conversely, larger interrogation area implies that the measure will refer to an ensemble of particles.

The other technique (PTV) uses matching algorithms to recognize the trajectory of particles from a prior identification of their centroid coordinates. The first step of the PTV analysis is the detection of the particle centres. It has a strong influence on the quality of the recovered information and can result in a clear limitation of this method. Indeed, PTV contains individual errors which are mostly related to the estimate of the particle centroids which may induce physically unrealistic disturbances to the velocity vectors. PTV has been historically classified as suitable for images of low particle density. However, the combination of current algorithms for particle detection [6-8] with algorithms used to solve the temporal matching problem [9], have also provided good performance for images with a moderately high particle density. In any case, intrinsic centroiding errors may occur and a trade-off between particle concentration and imaging delay has to be found. An appropriate choice of the temporal resolution (i.e., the frame speeds) is important. Long term steps between successive frames damps high-frequency fluctuations, whereas short time steps lead to large displacement errors [10].

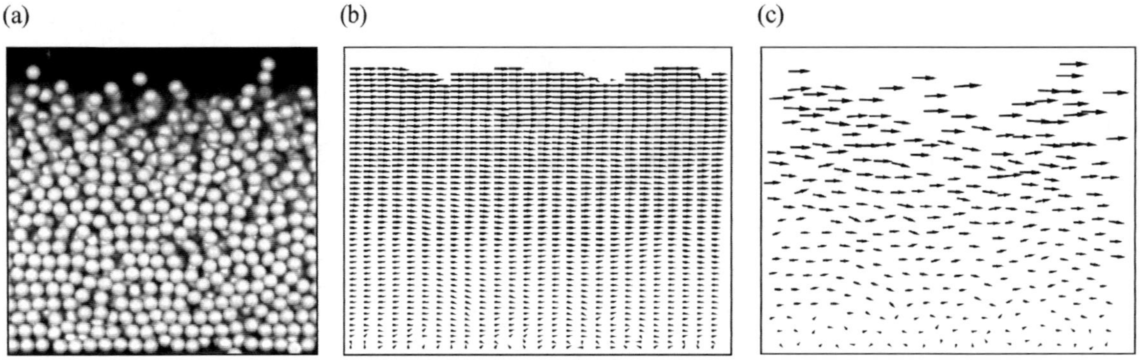

Figure 1. Pattern-based particle velocimetry methods: (a) Example of image extracted from the bead flows; (b) PIV velocity vector output produce by the cross-correlation routine between to frame images; (c) PTV displacement vectors (up-scaled) obtain by resolving the matching algorithm.

For the present work we selected two imaging algorithms that we deemed appropriate for the study of granular flows. A modified version of the PIV algorithm [11] used by [12] was employed directly to the captured images and was found sufficient to determine the velocity fields. This PIV method is adequate to determine the vector field via the cross-correlation analysis between interrogation areas in the first and second images (Fig. 1(b)). The software allows a static mesh to be kept in the same location while the granular mass is flowing. Within the mesh, columns of patches are drawn. User-defined slope-normal and slope-parallel spacing of the patches determine the amount of overlap and the number of patches within the mesh. The number of columns is dependent on the size of the patches chosen and the user-defined width of the mesh. The instantaneous velocity between two frames is finally obtained as the spatial average of the patches belonging to the same row.

For the PTV analysis we used a cross-correlation-based algorithm for PTV, coupled with a relaxation method proposed by Brevis et al. [9]. This integrated system has the ability to ease the matching problem by applying a relaxation algorithm where the cross-correlation method shows low reliability. An example of processed velocity vectors is shown in Fig. 1(c). To detect the particle centres, the particle-positioning approach [8] was found to yield good results. More specifically, the accuracy of this method was tested in a series of bench-scale experiments consisting of a container sliding unidirectionally along a track in a supporting base plate. Known increment translations were applied by a manual micrometer. By collating the displacement vectors and calculating the standard deviation, the error magnitude in the direction parallel to the plane of sliding can be seen. The estimated standard deviation error was found of the order of 0.05 pixel. Particle detection was also implemented with a Circular Hough Transform (CHT) algorithm (e.g. [13]) in order to remove the background particles from the analysis. Use of the Circular Hough Transform (CHT) is a robust technique mainly used to detect spherical/elliptical-shaped particles from noisy images. However, this implementation resulted in sporadic detection of background particles although these occasional outliers were removed by the matching algorithm. Finally, binning was used to extract velocity profiles. The flow was subdivided into non-overlapping horizontal slices, typically of thickness one particle diameter. The mean velocity was then constructed by taking the ensemble-average streamwise and transverse velocities in each bin.

Within PIV and PTV analyses a transformation from camera to real world coordinates is required. Before each test a calibration grid was located at the wall surface exactly where the images were captured. With this information an approach based on Brevis and Garcia-Villalba [14] was used. Using a binarized grid image, the calibration marks were detected with subpixel accuracy and the image coordinates were subsequently associated with its real x and y coordinates by means of an interpolation technique that uses a Delaunay triangulation approach.

3. Apparatus and experimental procedure

A small inclined chute has been developed in order to study dry granular flows (Fig.2). The flume is a sloping rectangular 10 cm wide and 150 cm long channel that can tilt from horizontal up to 45°. At one end, the material is held inside a hopper which can contain approximately 10 kg of material. A double-slider gate mechanism enables control of the releasing flow height which in turns determines the thickness of the descending mass. The thickness is well controlled although it is unlikely to be equal to the opening. The basal roughness is made by covering the flume bottom with the same material used to reproduce the granular flows. The walls are made of transparent material allowing observation to be taken throughout the entire length of the chute.

The material used is almost spherical ceramic beads. It was selected because of its spherical shape which is typically assumed in the majority of the granular flow theories. Ceramic beads are nominally $1.4 - 1.6$ mm in diameter with a bulk density of $2.43\ \mathrm{g\,cm^{-3}}$. The static friction angle, determined from a tilt test, was 24°.

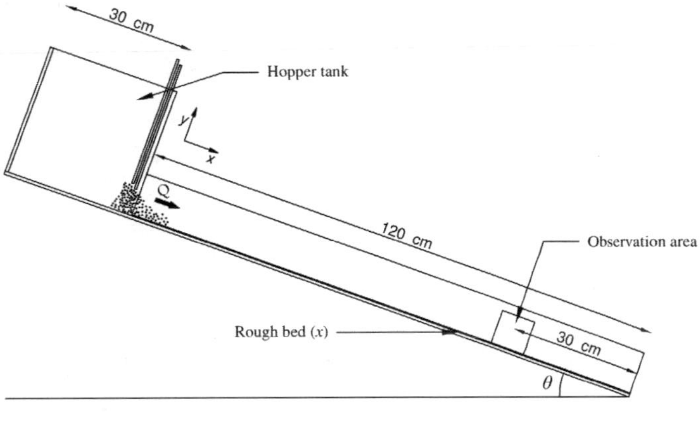

Figure 2. Sketch of the inclined chute geometry used to study dry granular flows (side view).

A high-speed camera (Phantom Miro 310) supplemented with two light sources was positioned 30 cm before the outlet. In this region the flows are more likely to reach an apparent steady regime. A steady state is needed as the analyses performed here are fully valid only if the flows are uniform in the direction parallel to the main plain of sliding. The uniformity was checked by visual inspection. We selected temporal regions where flow exhibited no detectable variations in the depth profile. Typical nominal flow depths are approximately 20 particle diameters. Video were recorded at a frame rate of 4000 fps with a camera resolution of 448 x 448 pixels. The frame rate was selected in order to minimize the gradient biasing in PIV and at the same time to avoid losing low-frequency fluctuations in PTV. However, further verifications and testing are required to identify which temporal resolution is the most suitable for the image velocimetry analyses described here.

From the image information, the instantaneous velocities were computed by the PIV and PTV algorithms and the following granular flow field properties calculated: mean velocities, velocity fluctuations and granular temperature. Using PIV, Reynolds et al. [15] showed that granular temperature may be measured by statistical analysis. The standard deviation of the velocities can be used as a good indicator of the velocity fluctuations. Moreover, if the mean velocity of all particles is constant, the variance of the velocity may characterise the particle granular temperature. Notably, a patch size of the order of the particles should generate the most accurate values of granular temperature [15]. As the material is almost uniform in size, we deemed accurate to use a patch size that covers the entire pixel surface of a particle. Specifically, a patch 32 x 32 pixel patch was chosen to analyse the flows of ceramic beads. Similarly, for the PTV analysis, granular temperature is given by the variance of the fluctuation velocity calculated by subtracting the mean velocity from the velocities between two image frames.

4. Results

In this section our intention is to demonstrate the results that can be obtained from these types of dry granular flows. For clarity we present only one flow which was released down the inclined at an inclination of 32°. It serves well to represent the flow properties attainable by using PIV and PTV techniques. Furthermore, the results reported here correspond to measurements and ensemble-averages based on a sampling period of 0.5 s.

Distributions of the mean and velocity fluctuations for PIV and PTV are given in Fig. 3. In the figure the transverse (U) (normal to the plain of sliding) and streamwise (V) velocity components are shown separately. The mean velocities reach their maximum at the free-surface (approximately 28 mm) while decaying almost linearly to 0 at the base. A small dissimilitude arises at the free-surface of Fig. 3(b). This is explainable by the fact that PTV analysis discards the particles in the background resulting in slower velocities than PIV. Fluctuation

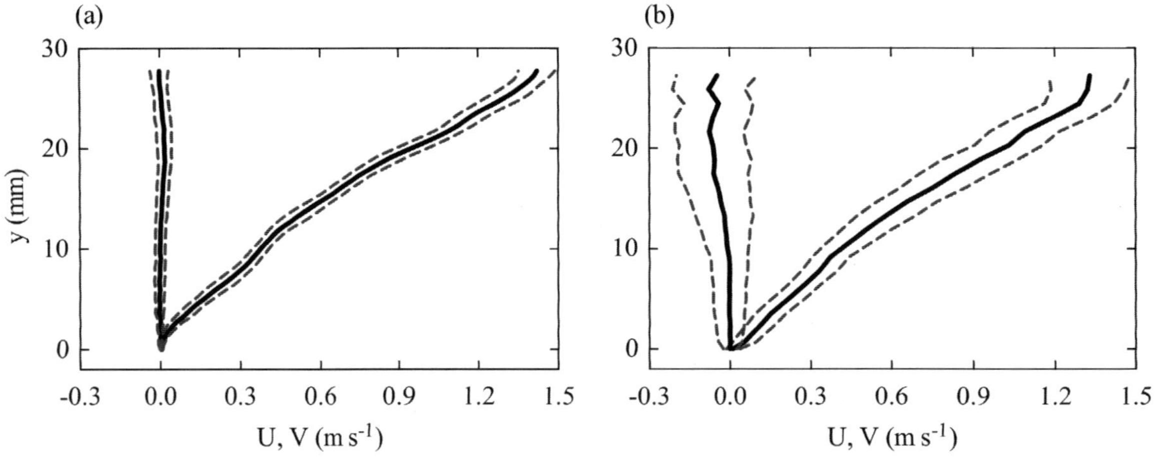

Figure 3. Ensemble-averaged raw velocity data over a sampling period of 0.5 s for (a) PIV and (b) PTV. U and V are the transverse and streamwise velocity components, respectively. Mean velocities (solid line) and fluctuation velocities (dashed line) plotted on both sides of the corresponding means.

IOP Conf. Series: Earth and Environmental Science **26** (2015) 012017 doi:10.1088/1755-1315/26/1/012017

velocities progressively increase towards the free-surface where the flow becomes more diluted. Interestingly, PTV gives rise to a much larger velocity fluctuation magnitude whereas PIV tends to damp the fluctuations in both directions. It seems that the average procedure followed by the PIV algorithm used here is unable to capture the random component of the velocities producing values that are very close to the mean profiles.

Granular temperature was extracted as the second-order statistics, i.e. the square of the velocity fluctuation; temperature was then taken as the non-stationary component of the local mean velocity. The fluctuation velocity gradient perpendicular to the flow direction (U) showed non-negligible effects. Thus, in the calculation we have taken both U and V components into account. For the PIV and PTV analyses, profiles of granular temperature are shown in Fig. 4. Temperature is zero at the flume bottom while reaching its maximum at the free-surface indicating that flow changes from a slowly deforming flow, mainly dominated by friction, to a more diluted collisional regime.

The most striking result concerns the magnitude of the granular temperature. When compared, values of temperature measured by the two methodologies differ almost by a factor of 9. PTV exhibits a granular temperature that is comparable to values found in the literature. For instance, Hanes and Walton [16] estimated granular temperature at the sidewall of up to 0.01. In saturated granular flows for material sliding over a dense layer of particles Armanini et al. [17] measured granular temperature of 0.05 at the surface. The PIV analysis damps the velocity fluctuation which in turn produces lower values of granular temperature. On the other hand, by measuring the movement of single particle, PTV appears to be the method of choice able to capture the collisional nature of the flow.

Differences can also be seen in the shape of the temperature profiles. While PIV produces different uncorrelated peaks with no clear trend (Fig. 4(a)), PTV shows a linear profile up to nearly 20 mm (Fig. 4(b)). Towards the free-surface, peaks arise but they are most probably related to the generation of granular temperature (e.g. the streaming mechanism [18]) and the removal of information due to the routine followed by the PTV algorithm. By removing the background information, fewer particles are found and if a longer sampling period is analysed, a smoother more linear profile might be generated in this part of the flow.

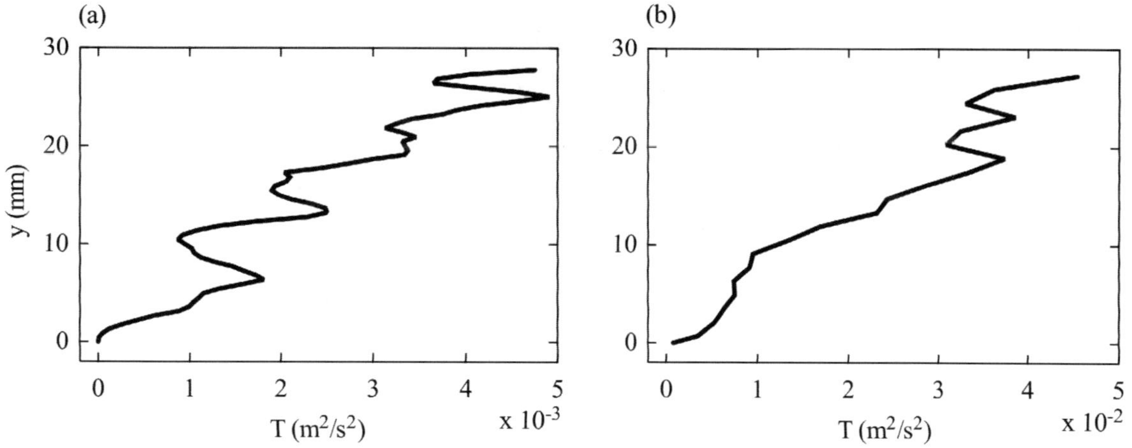

Figure 4. Granular temperature profiles. Comparison between experimental measurements obtained by (a) PIV and (b) PTV.

5. Concluding remarks

Experimental dry granular flows were reproduced in a small inclined chute geometry. Two non-invasive techniques, namely particle image velocimetry (PIV) and particle tracking velocimetry (PTV) were tested to verify their validity in estimating kinematic property of granular flows, such as velocity fluctuation and granular temperature. We found that PTV analysis is a more reliable method to extract such measurements. Indeed, by detecting the movement of individual particles, PTV is able to provide higher-accuracy measurements as shown by our results. Conversely, PIV yields averaged vectors resulting in underestimations of the random component of velocity and granular temperature. However, further investigations have to be done. We need to determine the adequate temporal resolution useful to minimize centroiding errors. If we are able to ascertain the accuracy of the PTV method, it might be possible, by comparison, to identify the appropriate configuration (e.g., in terms of spatial resolution of the interrogation region) to achieve the same results using PIV. Additionally, flows were recorded a the boundary conditions where the sidewalls are known to damp the generation of fluctuating motion. It would be interesting to study the motion at different cross-sections by using a non-intrusive technique that allows the visualization of the three-dimensional movement of particles. Ultimately, the measurements provided here may be processed to obtain results that could be useful for checking and refining mathematical models of granular flow.

Acknowledgments

The authors wish to thank Wernher Brevis for sharing the PTV algorithm and for his valuable advice. We are also grateful for the assistance of David R. Callaghan and Alex Cargill of the Department of Civil and Structural Engineering at the University of Sheffield for construction of the flume apparatus.

References

[1] Armanini A, M Larcher E N and Dumbser M 2014 *Adv. Water Resour.* **63** 1
[2] Savage S B and Hutter K 1989 *J. Fluid Mech.* **199** 177
[3] Iverson R M and Denlinger R P 2001 *J. Geophys. Res-Sol. Ea.* **106** 537
[4] Pereira F, Stuer H, Graff E C and Gharib M 2006 *Meas. Sci. Technol.* **7** 1680
[5] McKenna S P and McGillis W R 2002 *Exp. Fluids* **32** 106
[6] Takehara K and Etoh T 1989 *J. Vis.* **1** 603
[7] Ohmi K and Li H 2000 *Meas. Sci. Technol.* **11** 313
[8] Capart H, Young D L and Zech Y 2002 *Exp. Fluids* **32** 121
[9] Brevis W, Nino Y and Jirka G H 2011 *Exp. Fluids* **50** 135
[10] Jesuthasan N, Baliga B R and Savage S B 2006 *KONA* **24** 15
[11] White D J, Take W A and Bolton M D 2003 *Geotechnique* **53** 619
[12] Sanvitale N and Bowman E T 2014 *Geomech. from Micro to Macro Conf. Proc.* **1** 1553
[13] Tang J and Ren J 2013 *J. Comput. Inform. Syst.* **9** 187
[14] Brevis W and Garcia-Villalba M 2011 *J. Hydraul. Res.* **49** 586
[15] Reynolds G K, Nilpawar A M, Salman A D and Hounslow M 2008 *J Powder Technol.* **182** 211
[16] Hanes D M and Walton O R 2000 *Powder Technol.* **109** 133
[17] Armanini A, Capart H, Fraccarollo L and Larcher M 2005 *J. Fluid Mech.* **532** 269
[18] Campbell C S 2006 *Powder Technol.* **162** 208

Centrifuge model study of thresholds for rainfall-induced landslides in sandy slopes

V. Matziaris, A.M. Marshall, C.M. Heron and H.-S.Yu

Nottingham Centre for Geomechanics, University of Nottingham, University Park, NG7 2RD, Nottingham, UK

E-mail: Vasileios.Matziaris@nottingham.ac.uk

Abstract. Rainfall-induced landslides are very common natural disasters which cause damage to properties and infrastructure and may result in the loss of human life. These phenomena often take place in unsaturated soil slopes and are triggered by the saturation of the soil profile due to rain infiltration which leads to the decrease of effective stresses and loss of shear strength. The aim of this study is to determine rainfall thresholds for the initiation of landslides under different initial conditions. Model tests of rainfall-induced landslides were conducted on the Nottingham Centre for Geomechanics geotechnical centrifuge. Initially unsaturated plane-strain slope models made with fine silica sand were prepared at varying densities at 1g and accommodated within a centrifuge container with rainfall simulator. During the centrifuge flight at 60g, rainfall events of varying intensity and duration, as well as variation of groundwater conditions, were applied to the slope models with the aim of initiating slope failure. This paper presents a discussion on the impact of soil state properties, rainfall characteristics, and groundwater conditions on slope behaviour and the initiation of slope instability.

1. Introduction

Rainfall-induced landslides are very common natural disasters which cause damage to properties and infrastructure and may result in the loss of human life. These phenomena often take place in unsaturated soil slopes and are triggered by the saturation of the soil profile due to rain infiltration which leads to the decrease of effective stresses and loss of shear strength. Often, rainfall-induced landslides develop into debris flows which can experience high velocities and result in catastrophic outcomes. The study of the initiation conditions for rainfall-induced landslides is very important in order to predict these phenomena and be able to develop reliable real-time warning systems in the future.

Physical modelling plays a fundamental role in the development of the understanding of the triggering mechanisms of landslides and is performed in order to validate theoretical and empirical hypotheses [1]. Many studies have been performed in the past regarding the centrifuge modelling of slope behaviour under rainfall conditions ([2], [3], [4], [5], [6]).

A research project being conducted at the University of Nottingham aims to use geotechnical centrifuge modelling to better understand how the initiation of rainfall-induced landslides is affected by rainfall intensity and the initial state of the soil. Small-scale slope models constructed with fine-grained sand at an angle of 50° are tested within a climatic chamber at a centrifugal acceleration of $60g$. Under the increased gravitational field, controlled rainfall conditions are applied to the slopes and deformations are recorded. Rainfall parameters (intensity and duration), void ratio, and initial boundary and groundwater conditions are all controlled or measured during the tests. Deformations are determined by taking digital images of the slopes through a transparent wall which forms one side of the centrifuge container and processing the images using image analysis techniques.

2. Apparatus and instrumentation

Model tests were conducted using the Nottingham Centre for Geomechanics 2 m radius, 50g-Tonne beam type geotechnical centrifuge facility (Figure 1). The centrifuge can operate at 100-g (at 1.7 m radius) with a payload of 500 kg and up to 150-g at a reduced payload of 330 kg [7]. A data acquisition system enables the transfer of sensor readings from the centrifuge model to the control room through a fibre optic rotary joint. De-aired water, used in rainfall simulations, is supplied to the model through a hydraulic rotary union. The model container is placed on the swinging platform and the centrifuge is rotated at an appropriate speed in order to achieve the desired g level. A key issue in centrifuge modelling is the use of appropriate scaling factors to relate model-scale parameters to their prototype-scale equivalents. Table 1 presents the scaling factors applicable to the current study [8].

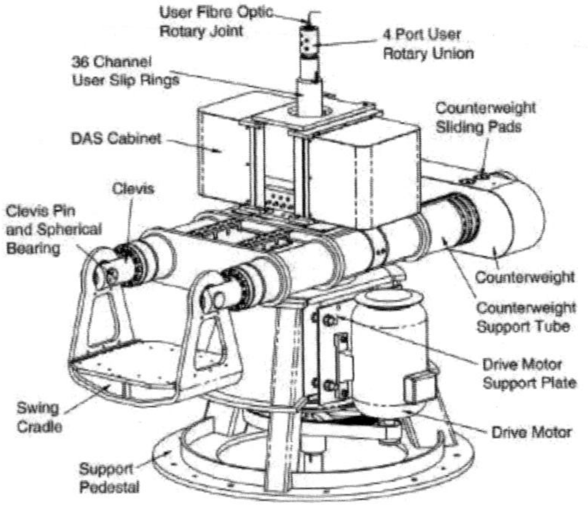

Figure 1. Main components of the University of Nottingham geotechnical centrifuge [7].

A plain strain strong box with inside dimensions of 600×400×200 mm (*length* × *height* × *width*) is used for the model slope tests. It includes a transparent Perspex front wall that allows for viewing of the soil model. Pore pressures and suctions were measured using miniature pressure transducers, however results obtained from these instruments were inconclusive and are therefore not presented in this paper. The centrifuge model includes the following systems, which are also illustrated in Figure 2:

- A rainfall system, consisting of a series of spraying/misting nozzles attached to aluminium rods (extensions) and located below a sealed lid. The length of the extensions was set so that nozzles are located approximately 70 − 100mm above the soil surface. This distance was found to be optimal for minimizing the impact of Coriolis effects and preventing soil erosion [4]. Nozzles are interconnected using a network of pipes. A supply water tank, located outside the centrifuge, is connected to the nozzles via the hydraulic rotary joint. A flow measurement device and an on/off solenoid valve are located between the water tank and the nozzles.
- A drainage system consisting of two drainage boxes located on both sides of the slope model. The boxes are perforated and allow water to move through them but prevent soil from being

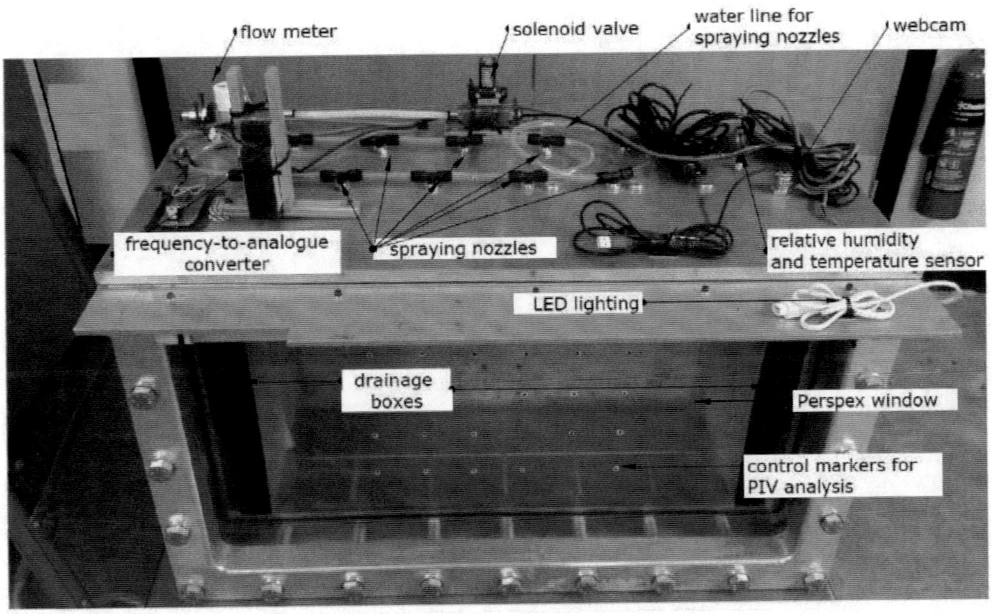

Figure 2. Main components of centrifuge container for modelling of rainfall-induced landslides.

washed away. The drainage boxes are connected to two standpipes located at the back side of the box and allow the water to be removed from the chamber.

- A groundwater table system which consists of two standpipes located outside the climatic chamber and connected through valves with the drainage boxes. Overflow at different heights of the standpipes offer the ability to control the water level within the climatic chamber and, therefore, control the water table height in the slope model.
- A recording system, including two digital cameras capturing images of the soil slope model through the transparent Perspex window. These images are processed using geoPIV [9] in order to determine the soil deformations that occur during the tests. Also, a web-cam is placed within the climatic chamber, under the lid, capturing video of the slope face during the test and allowing confirmation of the plain strain behaviour of the model.

3. Experimental program

Soil slope models are tested at 60g gravitational acceleration and consist of Leighton Buzzard (LB) sand (fraction E). This is a fine, uniform silica sand with $D_{10} = 0.095$ mm, $D_{50} = 0.12$ mm and uniformity coefficient ($U = D_{60}/D_{10}$) equal to 1.39 [10]. A Scanning Electron Microscopy (SEM) image and grain size distribution are shown in Figures 3 and 4, respectively. The critical state friction angle of LB sand was found to be 30° ([11]). Finally, the coefficient of saturated permeability was determined using the falling head permeability test and was approximately 5×10^{-5} m/sec.

Slope models are prepared within the centrifuge container to a height of 150 mm. The model is prepared by dry sand pouring, followed by saturation of the soil, and subsequent desaturation. Subsequently, soil mass remains in a partially saturated state, with the water content ranging between 8% and 11%, allowing for the slope to be cut to an inclination of 50° using a sharp cutting tool.

Table 1. Scaling factors for the parameters involved in the current study. L stands for units of length and T for units of time [8].

Parameter	Dimension	Model	Prototype
Length (macroscopic)	L	1	N
Seepage velocity (soil permeability)	L/T	1	N^{-1}
Seepage time (macroscopic)	T	1	N^2
Total rain	L	1	N
Rainfall duration	T	1	N^2
Rainfall intensity	L/T	1	N^{-1}
Seepage time (microscopic)	T	1	N
Hydraulic gradient (macroscopic)	L/L	1	N^{-1}
Hydraulic gradient (microscopic)	L/L	1	N^{-1}

Figure 3. SEM photo of the Leighton Buzzard Sand, fraction E.

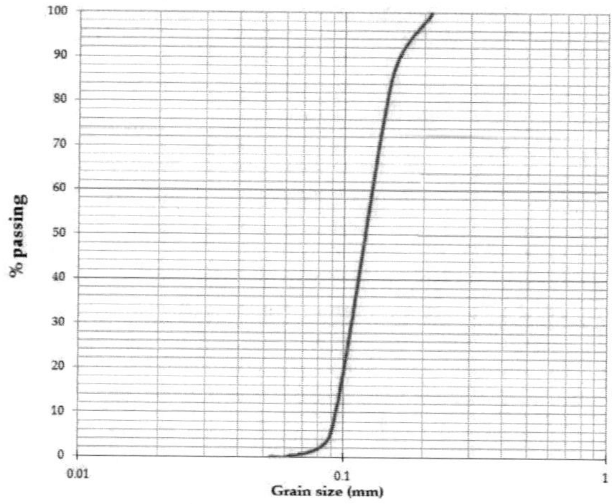

Figure 4. Grain size distribution

Three different sets of centrifuge tests are currently being performed, involving different rainfall and groundwater scenarios, as described in Table 2. Relative intensity, I_r, is the ratio between the prototype rainfall intensity and the saturated soil permeability. I_r gives a measure of the rate of rainfall provided to the slope in relation to the maximum rate at which water can infiltrate the soil. The test scenarios indicated in Table 2 correspond to the descriptions provided in Figure 5.

Digital images were captured throughout the centrifuge tests and geoPIV analyses were performed in order to determine soil displacements.

4. Results and discussion

For discussion purposes, the times and displacements mentioned in this section all refer to scaled prototype values unless otherwise indicated.

4.1. Scenario 1

In Scenario 1 tests (Figure 5a), rainfall is applied to the slopes while both drainage valves are kept open, allowing drainage to take place at both sides of the model. Therefore, a horizontal water table is formed and maintained just below the slope toe.

Table 2. Centrifuge test program.

Test	Test scenario	Relative density [%]	Void ratio [-]	Prototype I [mm h^{-1}]	I_r [-]
1.1	1	52 (*loose*)	0.81	7.4	0.04
1.2	1	52 (*loose*)	0.81	27.7	0.15
1.3	1	52 (*loose*)	0.81	87.7	0.50
2.1	1	100 (*dense*)	0.65	7.4	0.04
2.2	1	100 (*dense*)	0.65	27.7	0.15
2.3	1	100 (*dense*)	0.65	87.7	0.50
1.4	2	52 (*loose*)	0.81	87.7	0.50
2.4	3	100 (*dense*)	0.65	0	0

I is prototype rainfall intensity.
Relative intensity, $I_r = Ik^{-1}$, where k is saturated permeability.
$e_{min} = 0.65$, $e_{max} = 0.99$.

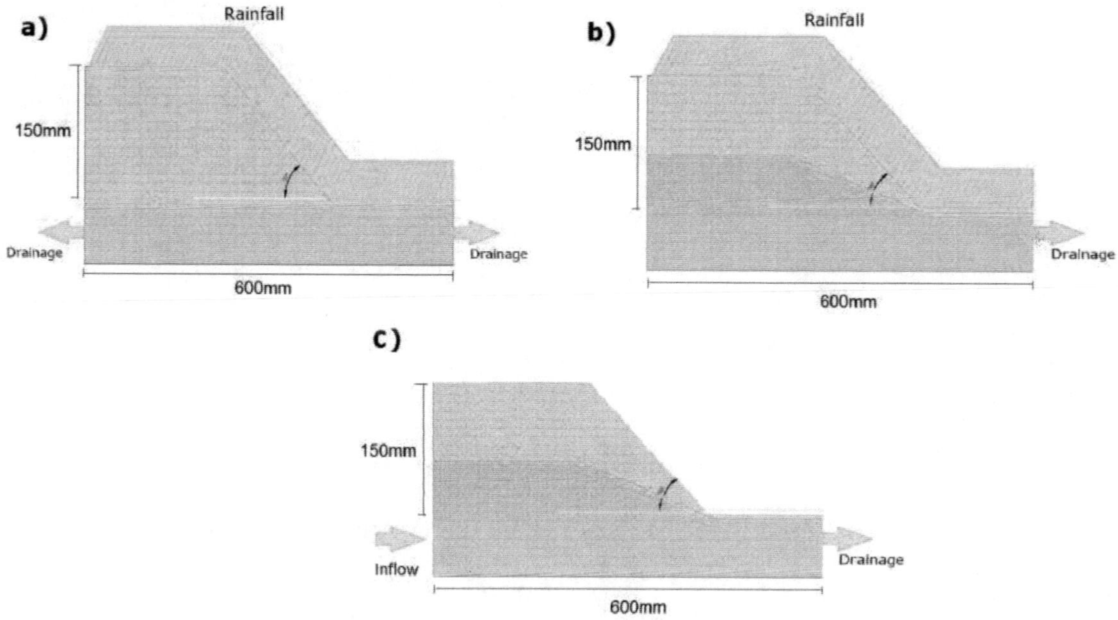

Figure 5. Test scenarios. a) Scenario 1: rainfall applied with drainage on both sides; b) Scenario 2: rainfall applied with drainage on down-slope side only; c) Scenario 3: no rainfall, water inflow from up-slope, outflow from down-slope.

Slope models, under loose and dense states, were subjected to light, moderate and heavy rainfall conditions, i.e. $7.4mm/hr$, $27.7mm/hr$ and $87.7mm/hr$, respectively. Before rainfall was applied to the models, the soil mass experienced excessive desaturation due to the enhanced acceleration field. The water content was determined through separate test to be close to 3%, corresponding to a degree of saturation (S_r) of 10%.

For the cases of the light and moderate rainfall events, slopes showed only minor deformation after 70 days of rainfall and no global failures occurred. This indicates that the low intensity rainfall events on sandy slopes have minor effects even after prolonged periods of rainfall. On the other hand, during heavy rainfall events ($I = 87.7mm/hr$) larger displacements occurred to the slopes which progressively increased with time, as shown in Figure 6. In the case of the dense slope, a local failure occurred near the slope toe after 1800 hours of constant rainfall (Figure 7). However, no global failure occurred in either case.

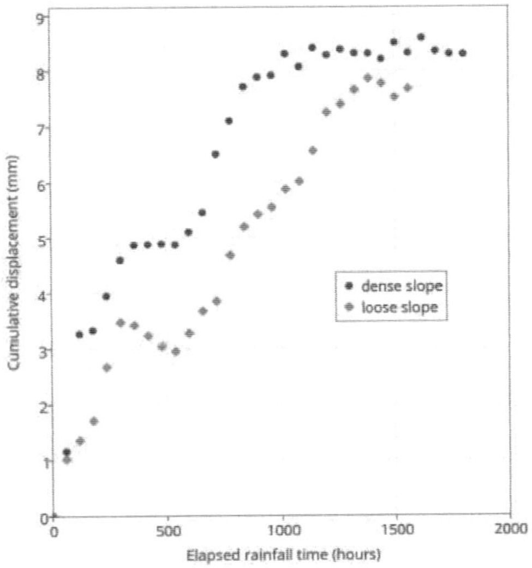

Figure 6. Cumulative displacement at slope crest during heavy rainfall event for loose and dense slopes (tests 2.3 and 1.3, respectively). Rainfall duration and settlement appear in prototype scale.

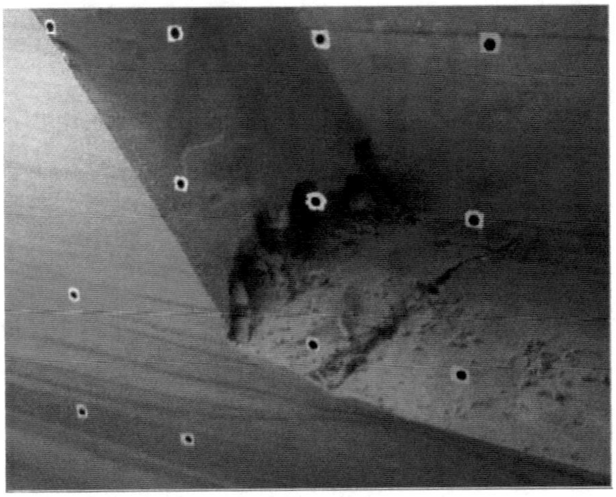

Figure 7. Local failure at the slope toe after 1800 hours of rainfall ($I = 87.7mm/hr$).

The displacements shown in Figure 6 are very small compared to the size of the slope. The exact cause of these displacements is not yet known, therefore further work is being conducted in order to determine if these deformations are caused due to the rainfall events or due to other phenomena, such as creep. The results reported here support the statement that initiation of rainfall-induced landslides, especially under unsaturated conditions, is very difficult to replicate through centrifuge modelling ([12]). However, alternative mechanisms for landslide initiation have been proposed, such as groundwater level rise ([13]) and adverse layering ([2], [14]).

4.2. Scenario 2

In the Scenario 2 test, rainfall with the higher intensity was applied to the loose slope, however the drainage from the upslope side was closed and an inclined water table formed (Figure 5b). The soil mass above the groundwater table was again in a residual water content state before rainfall. In this case, a progressive failure occurred, starting from the slope toe, as shown in Figure 8.

In Figure 8a, a horizontal water table formed just before rainfall was applied. After 68 hours of rainfall the water table was inclined, crossing and passing close to the slope toe (Figure 8b). After an additional 24 hours, the first failure occurred near the slope toe (Figure 8c), while a number of subsequent instabilities took place soon afterwards (Figures 8d and 8e). The final

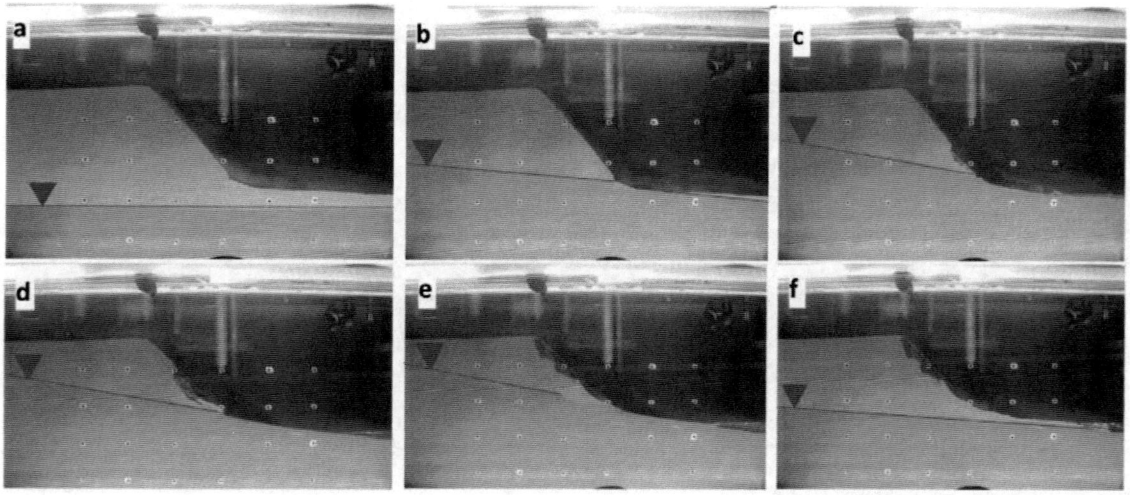

Figure 8. Progressive slope failure for test 1.4 (Scenario 2).

slope failure is shown in Figure 8f and took place 144 hours after the initiation of rainfall. These images illustrate that the base of the rupture surfaces passed near to the location of where the water table met the soil surface. Since the soil mass above the groundwater table is in a partially saturated state, it can be concluded that rapture surfaces formed just above the boundary of zero pore water pressure or where soil suctions have been reduced. The groundwater flow will also have contributed to the initiation of slope instability.

4.3. Scenario 3

In the Scenario 3 test, no rainfall was applied to the slope, which was initially in a dense state. A water inflow line was connected to the up-slope side of the container providing the model with the necessary amount of water to form an inclined water table (Figure 5c). Test results can be seen in Figure 9.

As was the case for the Scenario 2, progressive failure took place during the rise of the groundwater table. However, in this case the soil mass above the groundwater table during the whole test procedure is in a residual water content state due to the absence of rainfall. In both Scenarios 2 and 3, failure surface had a similar geometry.

5. Summary

A series of centrifuge tests on soil slope models have been performed on the Nottingham Centre for Geomechanics centrifuge in order to evaluate the impact of rainfall intensity and initial soil and groundwater conditions on slope stability. Soil slopes, consisting of fine silica sand, were subjected to varied rainfall and initial conditions at 60g centrifugal acceleration. Three Scenarios have been simulated in order to evaluate the impact of rainfall on landslide initiation. Initially, three different rainfall intensities were applied to loose and dense slopes while the water table remained at a constant level. Results showed that more intense rainfall events cause greater soil displacement, but did not always result in full slope collapse. Failure occurred only in models in which the groundwater table rose above the slope toe. Further testing is ongoing in order to examine additional configurations and to obtain a better understanding of the initial conditions of the partially saturated soil at elevated gravity (i.e. saturation and pore suctions).

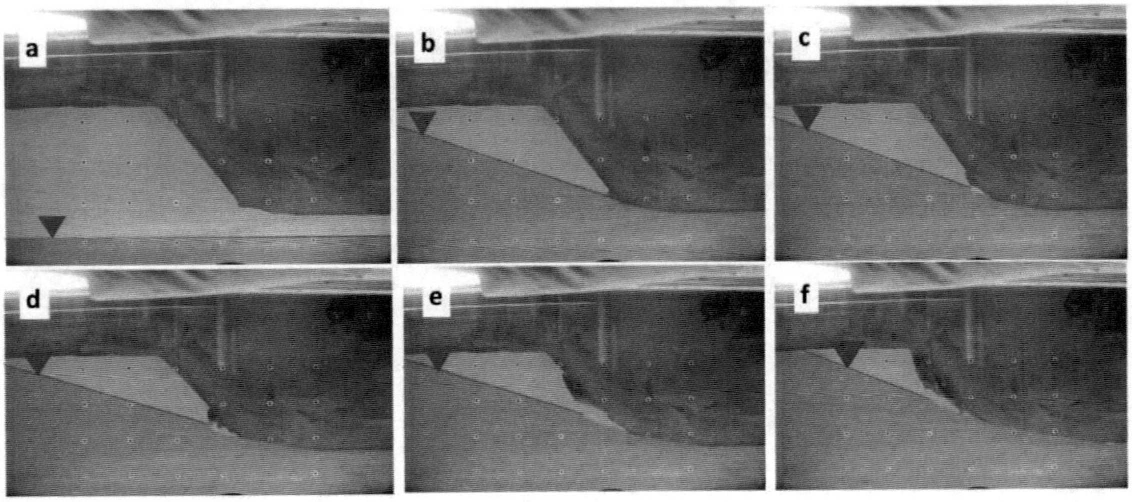

Figure 9. Progressive slope failure test 2.4 (Scenario 3).

Acknowledgments

Research leading to these results has received funding from the People Programme (Marie Curie Actions) of the European Union's Seventh Framework Programme FP7/2007-2013/ under REA grant agreement n° 289911.

References

[1] Wood D M 2004 *Geotechnical Modelling* (London: Taylor & Francis Group).
[2] Take W A, Bolton M D, Wong P C P and Yeung F J 2004 Evaluation of landslide triggering mechanisms in model fill slopes *Landslides* **1**(3) 173-84
[3] Take W A and Bolton M D 2011 Seasonal ratcheting and softening in clay slopes, leading to first-time failure *Géotechnique* **61**(9) 757-69
[4] Askarinejad A, Laue J, Zweidler A, Iten M, Bleiker E, Buschor H and Springman S M 2012 Physical modelling of rainfall induced landslides under controlled climatic conditions *Eurofuge 2012* Delft, The Netherlands
[5] Milne F D, Brown M J, Knappett J A and Davies M C R 2012 Centrifuge modelling of hillslope debris flow initiation *CATENA* **92** 162-71
[6] Tamate S, Suemasa N and Katada T 2012 Simulation of precipitation on centrifuge models of slopes *Int. J. of Physical Modelling in Geotechnics* **12**(3) 89-101
[7] Ellis E A, Cox C, Yu H S, Ainsworth A and Baker N 2006 A new geotechnical centrifuge at the University of Nottingham, UK 6th *Int. Conf. Physical modelling in Geotechnics* ed C W W Ng et al (London: Taylor & Francis Group) pp 129-33
[8] Askarinejad A, Beck A and Springman S M 2015 Scaling law of static liquefaction mechanism in geocentrifuge and corresponding hydromechanical characterization of an unsaturated silty sand having a viscous pore fluid *Canadian Geotechnical Journal* **52**(6) 708-20
[9] White D J, and Take W A and Bolton M D 2003 Soil deformation measurement using particle image velocimetry (PIV) and photogrammetry *Géotechnique* **53**(12) 619-31
[10] Matziaris V, Marshall A M and Yu H S 2014 Centrifuge model tests of rainfall-induced landslides *Recent Advances in Modelling Landslides and Debris Flows* (Springer Series in Geomechanics and Geoengineering) ed W Wu (Switzerland: Springer International Publishing) pp 73-83
[11] Cai Z Y 2001 A comprenhensive study of the state-dependent dilatancy and its application in shear band formation analysis *PhD thesis* (Hong Kong:University of Science and Technology)
[12] Take W A 2014 Physical modelling of instability and flow in loose granular slopes 8th *Int. Conf. Physical Modelling in Geotechnics* ed C Gaudin and D J White (London:Taylor & Francis Group) pp 113-24
[13] Zhang M and Ng C W W 2003 Interim Factual Testing Report I - SG30 & SR30 (Hong Kong:UST)
[14] Lee Y S, Cheuk C Y and Bolton M D 2008 Instability caused by a seepage impediment in layered fill slopes *Canadian Geotechnical Journal* **45**(10) pp 1410-25

An experimental model for slopes subject to weathering

C Voulgari
PhD Student, *University of Warwick*,
School of Engineering, Library Rd, CV4 7AL, Coventry

E-mail: C.Voulgari@warwick.ac.uk

Abstract. In this paper, an experimental prototype model to study the influence of cracks on the morphologic evolution of natural cliffs subject to progressive retreat induced by weathering is presented. A set of small scale laboratory tests is designed to investigate weathering induced successive landslides. Weathering is applied to the slope model by wetting the slope crest through a rainfall simulator device. The moisture content and the suction of the soil during the tests are monitored by soil moisture sensors and tensiometers that are buried inside the slope model. High resolution cameras record the behaviour of the slope model and GeoPIV is used to analyse the frames and obtain the deformations of the slope model during the tests. After a short time of rainfall, vertical cracks appear in the slope model with significant vertical deformations developing. Experimental results indicate that there is a strong connection between moisture content - thus degree of weathering - and the occurrence of a landslide. A prediction model of slope failures can be introduced based on the observed moisture content response of the slope models.

1. Introduction

An important cause of landslides is weathering, which often leads to the progressive retreat of exposed natural cliffs. Modelling weathering induced cliff retreat has recently received considerable attention by the engineering community due to increasing coastal erosive processes caused by climate change and increased environmental awareness [1-3].

Weathering tend to turn hard rocks into soft rocks characterised by higher void ratios and reduced bond strengths; then soft rocks are transformed into granular residual soils typically by destroying the bonds keeping the rock grains together [4]. Note that also other natural processes, e.g. methane hydrate dissociation in methane hydrate bearing sands [5], can be conceptualized as weathering. Therefore, a common feature of weathering processes is the degradation of the mechanical properties of the slope's geomaterials (rocks and soils), which in turn leads to the occurrence of successive landslides. This paper aims at investigating and developing an engineering model of the morphological evolution of natural cliffs with limited tensile strength subject to weathering induced landslides.

There are many studies investigating weathering induced landslides. With regard to numerical methods, stability analyses performed by finite element with strength reduction technique [6, 7] are traditionally employed. However, recent developments in contact detection algorithms for 3D convex blocks [8, 9] have made it possible to use also the discrete element method for the analysis of 3D slopes subject to strength reduction [10]. With regard to experimentation, most laboratory experiments require large slope models [11-14] to realistically simulate weathering induced slope failures. Centrifuge modelling [15-17] is also widely used to investigate the failure of slopes under different conditions. A vibrating sand box in the study of Katz and Aharonov [18] was used to induce cracks in soils and investigate the instability of heterogeneous slopes. Model test is also an important approach, in order to investigate the behavior of a slope under certain conditions. In the case of rainfall induced slope failures, model test provides the opportunity to observe the processes of infiltration, movement of the weathering front, deformation and failure [19, 20]. In each case, different techniques were applied to degrade the properties of the soil, while a number of different monitoring devices were used in order to record soil characteristics and the behavior of the slope during the tests. Sensors were vastly used to record the changes in the characteristics of the soil (moisture content, pore pressure, temperature etc.) and digital cameras to record the displacements and the changes in the slope's front. The results of most of these studies suggest that measurements of the changes of moisture content can be a factor to predict the slope's movement, while the initiation of the slopes failure is mainly caused with a decrease of the cohesion of the soil, thus making the cohesion the most critical property of the soil to measure, when it comes to slope stability. Not many examples of field experiment studies are found in the literature as they demand continuous monitoring of the characteristics of the slope and soil and are time and cost intensive.

This paper aims at investigating the morphological evolution of natural cliffs subject to progressive retreat induced by weathering, through a set of small scale experiments, with the ultimate objective of developing an engineering model of the phenomenon. To this end, slope models are built in a transparent flume and the weathering is applied to the slope model through light rainfall. The aim of this experiment is to develop a physical model to replicate retrogressive slope failures due to weathering. A set of experiments is carried out to soil with limited tensile strength; after a short period of rainfall and due to degradation of the soil strength; vertical cracks appear in the slope model and significant vertical deformations start to occur around the crack, until failure is reached.

Table 1. Geotechnical properties of the materials used in the experiments

Geotechnical Properties of calcarenite [21]

γ_d (kN/m^3)	γ_{sat} (kN/m^3)	Gs (-)	N (-)	E (-)
13.92	18.82	2.73	0.49	0.96

Geotechnical Properties of glass beads

γ_d (kN/m^3)	Gs (-)	Micron range
14.715	2.55	840-590

2. Test material

The aim is to develop a model that describes the relation between weathering process and the development landslides, for this reason small scale slope models are employed. Weathering was induced by applying light rainfall on the slope until failure was observed. First a suitable material had to be chosen. To this end, preliminary tests to select the appropriate material were carried out. Small cylindrical soil specimens (13mm diameter and 28mm height) of several different soil mixtures, containing calcarenite powder, gypsum, silica sand, glass beads and water were created and tested under certain load until they collapsed. The small size of the specimens, allowed more than 20 alternative mixtures to be tested without spending large amounts of material. Since the main interest was the response of the material under the influence of water, the specimens were tested under half of their dry collapse load, while at the same time water was added gradually at the base of the specimen, starting from 2 mm for the first two minutes. The material that fitted the desired characteristics was selected to carry out the main experiments for this research.

The mixture for the slope model consists of 43.5% w/w calcarenite, 43.5% w/w glass beads and 13% w/w water. The geotechnical properties of the calcarenite and the glass beads used for the formation of the slope models are illustrated in Table 1.

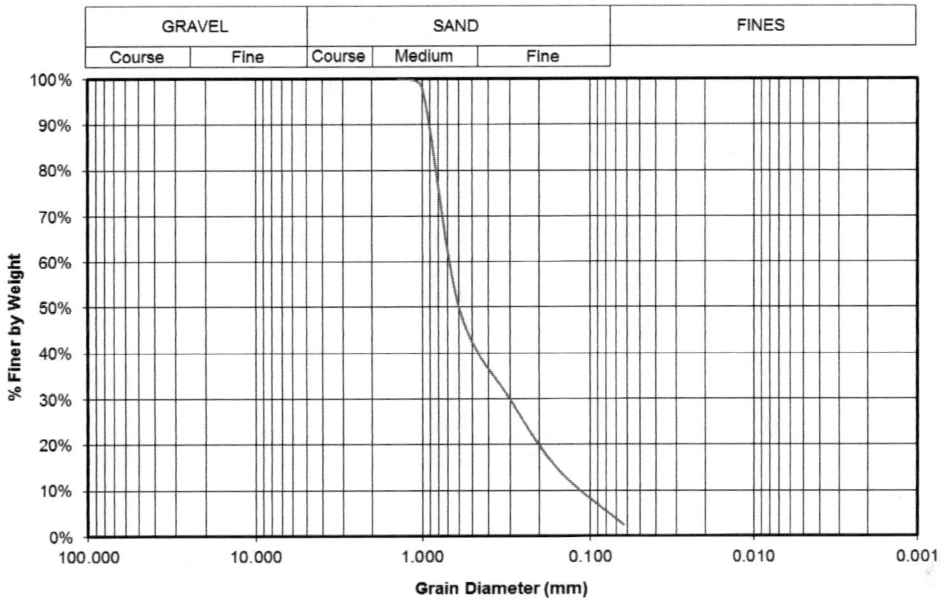

Figure 1. Grain Size Distribution Curve

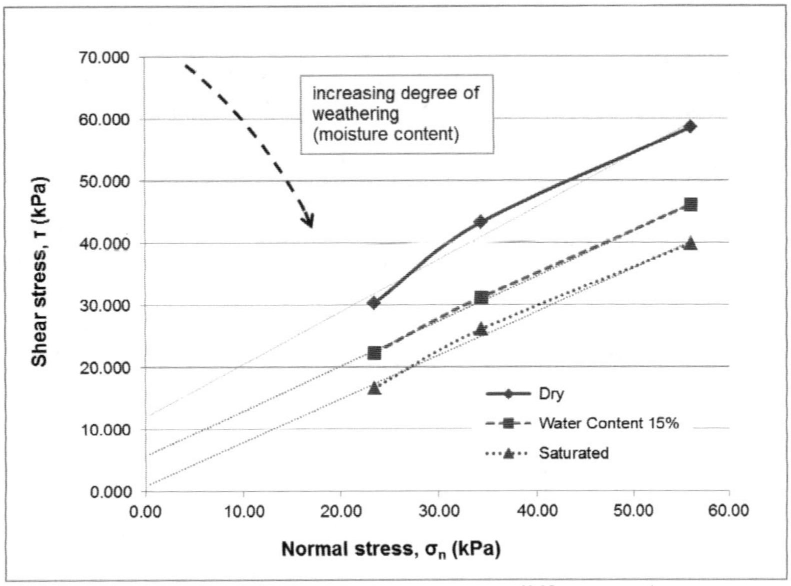

Figure 2. Weathering of calcarenite and glass beads mixture – different moisture contents.

The Grain Size Distribution performed with sieve analysis is presented in Figure 1. The effective particle size D_{10} of the mixture is 0.12mm, the average particle size D_{50} is 0.6mm and the uniformity coefficient $Cu=D_{60}/D_{10}$ is 5.83. The coefficient of permeability can be estimated using Hazen's approximation ($k=100D_{10}^2$) and is $k=1.44x10^{-2}$cm/s.

Direct shear box test were performed in the laboratory to identify the relation between the degree of saturation and the strength characteristics of the material. As stated in [22], soil strength can be expressed by two parameters, cohesion c and internal friction angle φ, according to the Mohr-Coulomb yield criterion and weathering causes mainly a decrease in the cohesion and to a much lesser extent a decrease in the friction angle of the soil. The direct shear box tests justify how weathering can be applied in the mixture through the addition of water. Specimens of the exact same characteristics (ingredients and density) as those used in the main experiments and with different moisture contents, therefore various degrees of weathering were tested and the failure envelopes were obtained and are illustrated in Figure 2. As shown in Figure 2, an increase in water content results in a decrease of the strength characteristics of the material. The decrease of the cohesion is greater than that of the friction angle.

3. Dimensional analysis

Since a model test is employed to study the progressive failure mechanism of a slope, a number of physical parameters that are used have to be related between the model and the real slope. The parameters include soil and geometric properties, rainfall intensity, velocity of the weathering front and characteristic times. To ensure complete similarity between the model test and the real case, the model and prototype must be geometrically similar, and all the independent parameters must be identical between the model and the real slope [23].

In Table 2, the relationship between the physical quantities of the model and the real slope, obtained by dimensional analysis is illustrated. The scaling factor (λ) between the length of the real slope (L_r) and the slope model (L_m) is $\lambda = L_r/L_m$ and the materials used in the model slope have the same unit weight with the real slope.

Weathering processes can have different velocities depending on the type of weathering. The changes and the evolution of natural slopes range from decades to thousands years depending on the type of soil and the weathering processes. To manage to model the weathering of a slope in the laboratory, a scaling factor should be chosen that will allow the progressive failure of the slope model to occur in laboratory time.

Table 2. Dimensional analysis of the physical quantities in model test

Physical Quantity	Relationship
Length	$L_m = (1/\lambda)\ L_r$
Slope angle	$\theta_m = \theta_r$
Unit weight	$\gamma_m = (1/\rho)\ \gamma_r = \gamma_r$
Friction angle	$\varphi_m = \varphi_r$
Pore water pressure	$u_m = (1/\lambda)\ u_r$
Strain	$\varepsilon_m = \varepsilon_r$
Time	$t_m = (1/\lambda)\ i_r$
Rainfall intensity	$i_m = i_r$
Hydraulic conductivity	$k_m = (1/\lambda)\ k_r$

Note that λ, and ρ are the scaling factors for length and unit weight respectively. In this study the value of chosen to be λ was equal to 100 and ρ was equal to 1.

4. Experimental Setup

4.1. Main test apparatus

The experimental apparatus is illustrated in Figure 2. The main apparatus consists of a soil container made of 10mm thick Plexiglas walls and 20mm Plexiglass underlying layer supported by a steel frame, a rainfall simulation device, soil moisture sensors, tensiometers and two high speed cameras to record the movement of the slope model during the tests. The container, with transparent sides, is 120cm long, 12 cm wide and 50 cm high (inner dimensions).

Starting at longitudinal distance x=25.00 cm from one end of the flume and at latitudinal distance y=38.00 cm from the top of the flume, a mesh of small holes is drilled. The holes are located 3.00 cm far from one another (vertically and horizontally) and are used to place the soil moisture sensors and the tensiometers. The rainfall simulator is placed above the container to provide steady and uniform rainfall on the model. The high resolution cameras are also used to monitor the test process from both sides of the flume.

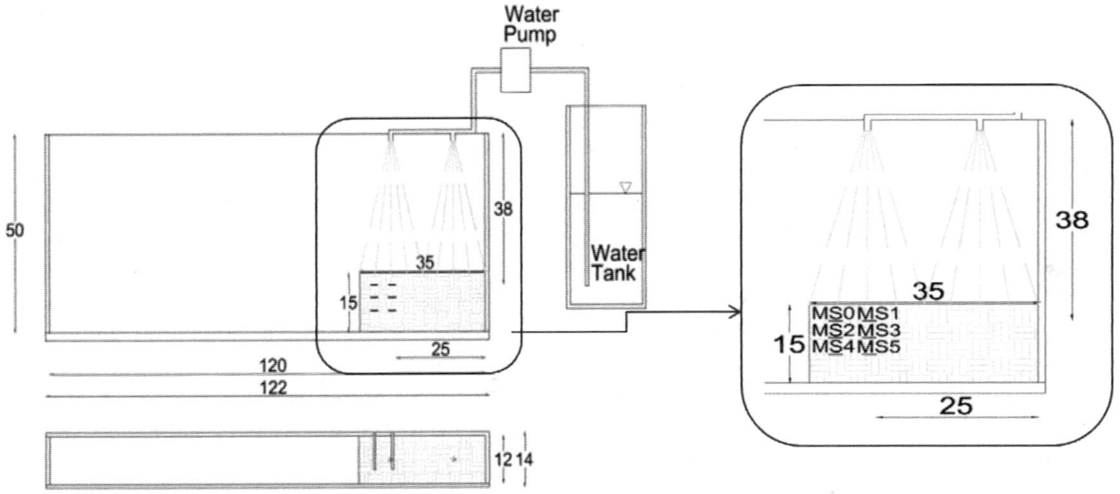

Figure 3. Arrangement of the experimental apparatus and soil moisture sensors' location.

4.2. Rainfall simulation device

Two different systems are used to generate the artificial rainfall, depending on the desired rainfall intensity. One system consists of a 1x12v diaphragm pump, a control set including a pressure gauge and a pressure regulating valve, plastic pipes and two M1 - Mini nozzles and the other system consists of two GSC Air atomizing nozzles connected to a water tank with liquid flow controlled by suction height. Both systems are placed directly above the container and guarantee uniformity and accuracy. The uniformity of the nozzles for both systems was tested using small containers and measuring the flow rate at different pressures / suction heights.

4.3. Slope model

For the experiments, homogeneous slope models made of the material described above are constructed. Although non planar slope profiles are more common in nature and they may be more stable than planar ones [24], all the experiments were performed with planar slopes for sake of simplicity. The dimensions of the slope models are: height: varying between 10 and 25cm, width: 12cm and length: 35cm. For the construction of homogeneous slope models, the materials are manually mixed and layered inside the box in 15 layers of certain weight. Each layer is compacted into 1.00cm to achieve uniformity of the slope model, obtaining a wet density of 1.70g/cm^3. The soil moisture sensors and the tensiometers are buried inside the slope model during the construction phase, so as to cause the least possible disturbance to the soil.

Transparent silicon oil is applied on the sides of the container to avoid friction between the slope model and the container. When the model is prepared, rainfall is induced to the model by applying low intensity rainfall to the slope's surface through the rainfall simulation device set directly above the slope model.

Three different slope angles will be tested. For the preliminary results only vertical slopes with different slope heights are examined to verify the repeatability of the procedure and that it is not scale dependent.

4.4. Measurements

4.4.1. GeoPIV analysis
During the event, two high resolution cameras are aimed at both sides of the container, to record the process of the test and the movement of the soil. The cameras have a video resolution of 1920 x 1080 pixels and a maximum frame rate of 25 fps. The main reason for using two cameras is verify that the wetting front is moving homogeneously through the model at both sides; the displacement vectors are obtained by analysing the images of the video from one side via Particle Image Velocimetry (PIV). Frames are extracted every second from the video and are analysed in order to measure the displacement vectors of the soil and the velocity of the wetting front, without causing any disturbance to the slope model.

The displacement vectors are obtained using GeoPIV, a Matlab module that implements Particle Image Velocimetry (PIV) in a manner suited to geotechnical testing [25, 26]. The analysis technique used is based on the principles of PIV, which is a digital image-based surface displacement measurement method that compares a reference image to a series of deformed images. It measures whole velocity fields by taking two images in successive time instants and calculating the distance individual particles travelled within these instants. Since the accuracy of displacement measurements, using particle image analysis is strongly dependent on the surface contrast of the soil, some of the glass beads were painted blue and white to be used as markers, in order to have unique and easy to track patches.

4.4.2. Matlab Image Processing
Red coloured water is used for the rainfall and its movement can be tracked through Matlab image processing, giving the velocity and the position of the water front (Figure 4) at any time. This analysis is necessary to confirm that the water front is moving homogeneously inside the slope model by comparing the height of the 'red region' at a certain time at different locations along the slope. These results are also compared to the readings from the soil moisture sensors to validate their accuracy.

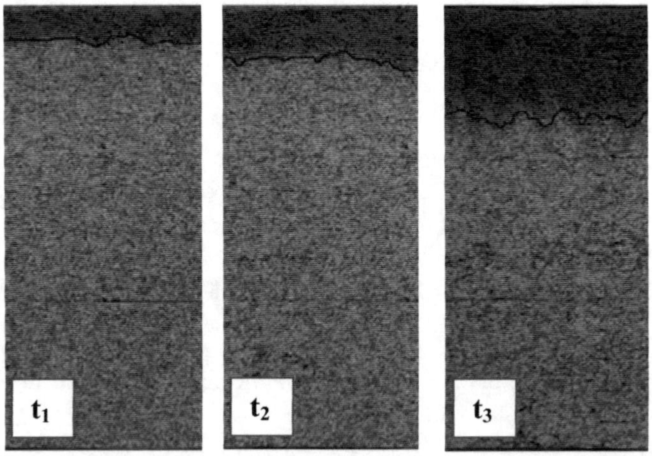

Figure 4. Water front movement through matlab image processing during the test.

4.4.3. Soil moisture sensors
VH400 Soil moisture sensor probes are buried horizontally in one side of the slope model at the locations shown in Figure 3. The accuracy of these sensors is 2% at 25°C and they measure volumetric

soil moisture content, using transmission line techniques to detect the dielectric constant of the soil. The size of these sensors is very small, 9.4cm long and 0.7cm thick, thus causing minimal disturbance to the surrounding soil. Since reliable measurements are very important while performing experiments, two SM300 soil moisture and temperature sensors are also buried inside the slope model for independent measurements. These sensors are even smaller than the VH400 sensors causing least disturbance to the soil, their accuracy is 2.5% for the moisture content and 0.5°C for the temperature and they give continuous reading. The SM300 sensors are used in combination with the others to ensure correct readings. The water content of the soil is recorded at frequency 1/s.

Soil moisture content tests took place in the laboratory to calibrate the output voltages of the sensors prior to the main tests. Moreover, some preliminary tests for the response time of these sensors were conducted, using small samples of the same density as that of the slope model and with known water contents which showed that the sensors responded accurately in less than 1s of contact with the soil.

4.4.4. Tensiometers
The suction of the soil can have a considerable influence on the slope stability [27]; therefore it is essential to determine the suction of the slope during the test to evaluate the likelihood of a slope failure. In order to capture the changes in the soil suction during the tests, it is crucial to use reliable miniature sensors that will cause the least disturbance to the slope models. Two SWT-5x tensiometers, that are specially designed for point measurements monitoring both positive and negative pore pressures, are employed; the accuracy of these tensiometers is ± 0.5 kPa, over the range +100 to -160 kPa, while the small size of the shaft (5mm diameter and 5mm length) causes very little disturbance to the soil.

5. Preliminary tests
Fourteen experiments have been carried out to test the designed apparatus and the experimental procedure, and given the procedure described in this paper, up to two progressive failures are observed. Results suggest that the procedure is repeatable and not scale dependent. Tests were performed for slopes with different heights, under different rainfall intensities, with and without sensors or soil particles glued on the bottom of the tank to observe the effect of all these parameters on the behaviour of the soil slope. Successive failures are observed for some combinations of H and I_r (where H is the height of the slope model and I_r the rainfall intensity) in laboratory times. When rainfall intensity is too low, especially for short heights, only one failure is observed.

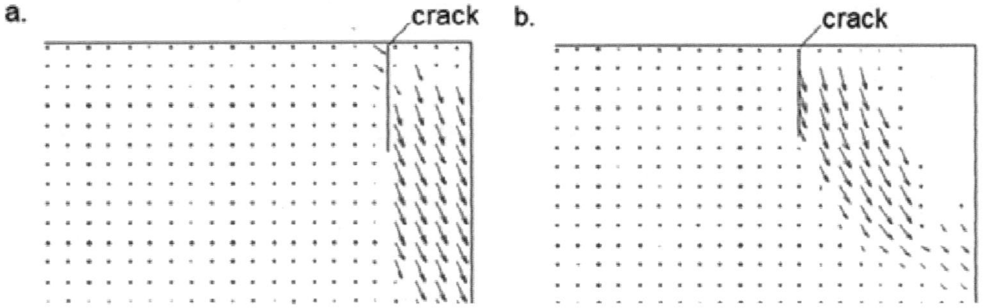

Figure 5. Displacement vectors at the initiation of a. the first and b. the second failure event for a vertical slope by GeoPIV analysis

An example of the displacement vectors of the failure events according to the GeoPIV analysis during a test on a vertical slope with H=15cm and I_r=18mm/h is plotted in Figure 5. The formation of two cracks is observed. Cracks can significantly reduce slope stability [28]. Through the analysis the location of each crack and failure mechanism can be tracked, the deformation of the slope and the velocity of the sliding mass can be calculated. From all the analyses it turns out that vertical deformations at the initiation of the failure are greater that horizontal ones.

Combining the time histories of the volumetric moisture content of the slope from the soil moisture sensors with the displacements from the GeoPIV analysis useful information can be extracted for the relation between the moisture content (thus the degree of weathering) and the deformation of the slope.

6. Conclusions

A series of small scale tests for the investigation of the successive failures of soil with limited tensile strength has been designed and tested under different conditions to verify that the procedure is repeatable and not scale dependent. A soil container to accommodate the slope models has been constructed as well as a rainfall simulator device to provide homogeneous wetting of the slope surface. Soil moisture sensors and tensiometers have been carefully selected and tested to ensure reliable measurements, while GeoPiv and Matlab image analyses were employed to carefully track the movement of the water and investigate the deformation of the slope during the tests.

The described setting is satisfying to study the retrogressive retreat of the slope, as up to two successive failures are observed on the slope model during the tests. The behaviour of the slope and the effect of weathering on the retrogressive retreat of soil slope are modelled. Crack formation is observed just before the failure occurred. Given the decrease on the soil strength with the increasing moisture content a prediction model of slope failure due to weathering can be introduced based on the observed moisture content.

7. References

[1] Utili S. and Crosta G., "Modeling the evolution of natural cliffs subject to weathering: 1. Limit analysis approach," *Journal of Geophysical Research: Earth Surface (2003–2012),* vol. **116**, p. F01016, 2011.

[2] Bray M. J. and Hooke J. M., "Prediction of soft-cliff retreat with accelerating sea-level rise," *Journal of Coastal Research,* vol. **13**, pp. 453-467, 1997.

[3] Utili S. and Crosta G., "Modeling the evolution of natural cliffs subject to weathering: 2. Discrete element approach," *Journal of Geophysical Research-Part F-Earth Surface,* vol. **116**, p. F01017, 2011.

[4] Nova R., Castellanza R., and Tamagnini C., "A constitutive model for bonded geomaterials subject to mechanical and/or chemical degradation," *International Journal for Numerical and Analytical Methods in Geomechanics,* vol. **27**, pp. 705-732, 2003.

[5] Jiang M., Zhu F., Liu F., and Utili S., "A bond contact model for methane hydrate-bearing sediments with interparticle cementation," *International Journal for Numerical and Analytical Methods in Geomechanics,* vol. **38**, pp. 1823-1854, 2014.

[6] Huang M. and Jia C. Q., "Strength reduction FEM in stability analysis of soil slopes subjected to transient unsaturated seepage," *Computers and Geotechnics,* vol. **36**, pp. 93-101, 2009.

[7] Zheng H., Liu D., and Li C., "Slope stability analysis based on elasto-plastic finite element method," *International Journal for Numerical Methods in Engineering,* vol. **64**, pp. 1871-1888, 2005.

[8] Boon C., Houlsby G., and Utili S., "A new algorithm for contact detection between convex polygonal and polyhedral particles in the discrete element method," *Computers and Geotechnics,* vol. **44**, pp. 73-82, 2012.

[9] Boon C., Houlsby G., and Utili S., "A new contact detection algorithm for three-dimensional non-spherical particles," *Powder Technology,* vol. **248**, pp. 94-102, 2013.

[10] Boon C., Houlsby G., and Utili S., "New insights into the 1963 Vajont slide using 2D and 3D distinct-element method analyses," *Geotechnique,* vol. **64**, pp. 800-816, 2014.

[11] Bachmann D., Bouissou S., and Chemenda A., "Influence of weathering and pre-existing large scale fractures on gravitational slope failure: insights from 3-D physical modelling," *Natural Hazards and Earth System Science,* vol. **4**, pp. 711-717, 2004.
[12] Jia G., Zhan T. L., Chen Y., and Fredlund D., "Performance of a large-scale slope model subjected to rising and lowering water levels," *Engineering Geology,* vol. **106**, pp. 92-103, 2009.
[13] Tohari A., Nishigaki M., and Komatsu M., "Laboratory rainfall-induced slope failure with moisture content measurement," *Journal of Geotechnical and Geoenvironmental Engineering,* vol. **133**, pp. 575-587, 2007.
[14] Huang C. C., Lo C. L., Jang J. S., and Hwu L. K., "Internal soil moisture response to rainfall-induced slope failures and debris discharge," *Engineering Geology,* vol. **101**, pp. 134-145, 2008.
[15] Ling H. I., Wu M. H., Leshchinsky D., and Leshchinsky B., "Centrifuge modeling of slope instability," *Journal of Geotechnical and Geoenvironmental Engineering,* vol. **135**, pp. 758-767, 2009.
[16] Wang R., Zhang G., and Zhang J. M., "Centrifuge modelling of clay slope with montmorillonite weak layer under rainfall conditions," *Applied Clay Science,* vol. **50**, pp. 386-394, 2010.
[17] Zhang G., Qian J., Wang R., and Zhang J. M., "Centrifuge model test study of rainfall-induced deformation of cohesive soil slopes," *Soils and foundations,* vol. **51**, pp. 297-305, 2011.
[18] Katz O. and Aharonov E., "Landslides in vibrating sand box: What controls types of slope failure and frequency magnitude relations?," *Earth and Planetary Science Letters,* vol. **247**, pp. 280-294, 2006.
[19] Wang G. and Sassa K., "Factors affecting rainfall-induced flowslides in laboratory flume tests," *Geotechnique,* vol. **51**, pp. 587-599, 2001.
[20] Chen R. H., Kuo K. J., and Chien W. N., "Failure Mechanism of Granular Soil Slopes under High Intensity Rainfalls," *Journal of GeoEngineering,* vol. **7**, pp. 021-031, 2012.
[21] Castellanza R., Parma M., Pescatore V., and Silvestro G., "Model footing load tests on soft rocks," *ASTM geotechnical testing journal,* vol. **32**, pp. 262-272, 2009.
[22] Kimmance G., "Computer aided risk analysis of open pit mine slopes in kaolin mined deposits," Ph. D thesis, Imperial College London (University of London), 1988.
[23] Cengel Y. A., *Fluid mechanics*: Tata McGraw-Hill Education, 2010.
[24] S. Utili and R. Nova, "On the optimal profile of a slope," *Soils and foundations,* vol. **47**, pp. 717-729, 2007.
[25] White D. and Take W., "GeoPIV: Particle Image Velocimetry (PIV) software for use in geotechnical testing," 2002.
[26] White D., Take W., and Bolton M., "Soil deformation measurement using particle image velocimetry (PIV) and photogrammetry," *Geotechnique,* vol. **53**, pp. 619-631, 2003.
[27] Fredlund D., Morgenstern N., and Widger R., "The shear strength of unsaturated soils," *Canadian Geotechnical Journal,* vol. **15**, pp. 313-321, 1978.
[28] Utili S., "Investigation by limit analysis on the stability of slopes with cracks," *Geotechnique,* vol. **63**, pp. 140-154, 2013.

Evaluating the capability of a critical state constitutive model to predict the collapse potential of loose sand

A. Azizi[1] G. Bella[2] and I. Farshchi[3]
[1]Polytechnic University of Milan, Milan, Italy
[2]Polytechnic University of Turin, Turin, Italy
[3]Linton University College, Negeri Sembilan, Malaysia

Many catastrophic flow failures in granular soil slopes are believed to be caused by a rise in pore water pressure associated with substantial loss of soil shear strength. This failure mechanism is known as prefailure instability or static liquefaction. Constant shear (CS) and consolidated undrained (CU) triaxial tests can reproduce stress paths, in which such instability may occur before reaching the failure. In the present study, a previously proposed critical state constitutive model was first used to simulate the behavior of loose saturated sand in CU tests. It was then employed to predict the instability of loose sand subjected to the CS loading. Under such loading, loose dry sand initially experience small volume increase, and then start to contract substantially. In saturated sand, such contractions can lead to the generation of pore water pressure and sudden decrease of shear strength. The capability of the model to predict the onset of the volume contraction and collapse potential of loose dry sand was examined by comparing the model predictions with experimental results of CS tests. The comparison showed that the effect of initial void ratio, consolidation and deviatoric stresses on behavior of loose dry sand can be well predicted by the model.

1. Introduction

Static liquefaction is known to be responsible for many disasters related to failures of geotechnical structures, such as flow failures of the submarine Nerlerk berm (Lade 1993) or north dike of Wachusett dam (Olsen et al. 2000). This failure mechanism is defined as the prefailure instability of the soil associated with a sudden loss of soil shear strength, resulting from an increase in pore pressure within the soil mass and a subsequent reduction in its effective stress. Consequently, flow failures initiated by static liquefaction leads to catastrophic damages as they rapidly develop with no noticeable warning.

The prediction of soil bahavior due to changes in stress state is a fundamental step in any geotechnical engineering project. Regarding to the widespread application of computational tools and important implication of instability in geotechnical engineering, the need for constitutive models capable of reproducing the soil behavior in stress paths, which leads to instability, is evident.

The present study provides a brief review on previous works concerning the instability of granular material, followed by introducing a previously proposed critical state constitutive model. The results of triaxial tests and relevant constitutive modeling are then represented. The model was first used to simulate the behavior of loose saturated sand in isotropic consolidated undrained (ICU) and constant

[1] Corresponding Author: Arash Azizi
Ph.D student, Polytechnic University of Milan, Milan, Italy
Piazza Leonardo da Vinci, 32, 20133 Milano
email: arash.azizi@polimi.it

shear (CS) tests. The capability of the model to predict the collapse potential of loose dry sand was then examined by comparing model predictions with experimental results of CS tests.

1.1 Instability of loose sand in experimental tests
Static liquefaction was first characterized as a sudden increase of pore water pressure and decrease of shear stress of loose saturated sand, induced by monotonic compression in undrained triaxial tests (e.g., Castro 1969, Sladen et al. 1985, Lade 1992, Yamamuro and Lade 1997, Leong et al. 2000, Wanatowski and Chu 2007). These studies illustrated that the deviatoric stress of a loose sand sample increases with little associated deformation, until the deviatoric stress is reached to the peak point before reaching the failure. Beyond the peak point, strain softening occurs as the pore pressure continues to increase. The Sample continues to deform until it reaches to the failure as shown in figure 1 (CU). Lade (1992) introduced the instability line (IL) passing through the peak points of undrained effective stress paths, in which the collapse of the granular material may occur above the stress ratio corresponding to the instability line. Figure 1 shows also the typical behavior of loose sand when the drainage is permitted. In consolidated drained (CD) compression triaxial test, the shear strength of the loose sample increases monotonically until it reaches the critical state strength, represented by the critical state line (CSL). It implies that the instability does not occur under monotonic compression loading as the drainage is permitted.

The prefilure instability can be triggered not only by an increase in external load, but also by a reduction in the effective mean stresses. An increase in pore water pressure resulting from infiltration of rainfall or snowmelt leads to a decrease in the mean effective normal stress, while the vertical load due to the weight of overlying soil does not change. It was observed that such a stress path can lead to the failure even under drained condition (e.g., Olsen et al. 2000). The CS triaxial test under drained condition, in which a constant deviatoric stress is applied to the top of a sample while the confining stress is decreased, reproduces a similar stress path. Numerous studies have been carried out to investigate the instability of both saturated and dry loose sand under the CS loading (e.g., Sasitharan et al. 1993, Skopek et al. 1994, Anderson and Riemer 1995, di Prisco and Imposimato 1997, Gajo et al., 2000, Chu et al. 2003, 2012, Daouadji et al. 2010). Previous studies have shown that cohesive, and medium to dense granular soils generally dilate when subjected to CS loading, while loose sand initially experience small or no volume change, and then start to contract substantially as failure is approached. If sand is saturated and drainage cannot occur fast enough, such contractions can lead to the generation of substantial pore water pressures and loss of strength. The same can happen in the field under drained condition, and the loss of soil strength may lead to the sudden, catastrophic failure of slopes. As shown in figure 1b, such instability takes place well above the instability line.

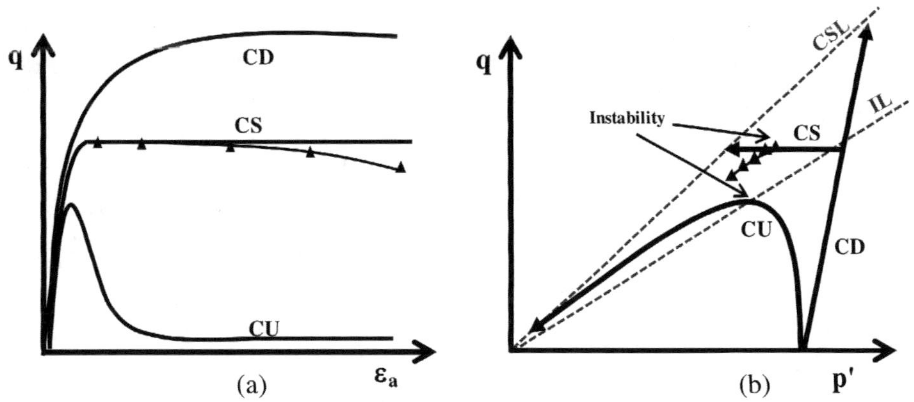

Figure 1. Typical behaviors of loose sand in triaxial consolidated drained (CD), consolidated undrained (CU) and constant shear (CS) tests: (a) the deviatoric stress-axial strain behaviors (b) the stress paths.

Beyond the instability, the saturated sample is unable to follow the prescribed stress path, experiencing loss of controllability (Nova 1994).

1.2 Critical state constitutive model (Imam et al. 2005)

The critical state constitutive model proposed by Imam et al. (2005) was adopted in this study. This model was developed with emphasis on taking into account important aspects of loose sand behavior. The constitutive relationships of the model are presented in the Appendix. This model consists of five elements: a capped yielding surface YS (equation [1]); a Mohr Coulomb failure criteria (equation [8]) which is a function of the friction angle and the current state parameter (Been and Jefferies, 1986); a flow rule based on Nova and Rowe's (1962) stress-dilatancy relationships combined with a modified form of equations used by Manzari and Dafalias (equation [9]); a hardening law (equation [10]); isotropic elasticity.

Figure 2a shows the capped yield surface (YS) of the model for different types of triaxial tests, namely, isotropic consolidated compression (IC), anisotropic consolidated extension (ACE) and anisotropic consolidated compression (ACC). The yield stresses and stress ratios are normalized to the mean normal stress at consolidation p_c. The Imam model is in fact a kinematic model, in which he yield surface changes its shape depending on the stress path.

Figure 2b shows the YS of isotropically consolidated sand with stress ratios at critical state (M_{CS}), at failure (M_f) and at peak deviatoric stress (M_P). The stress ratios M_P mentioned above control the shape of the yield surface, taking into account the effects of void ratio and mean normal stress, while the size of the yield surface is controlled by p_c. Figure 2b shows that in loose sand the stress ratio M_P is smaller than M_f. The peak stress ratio decreases with void ratio and confining pressure p_c and it takes values much smaller than M_{CS}.

The anisotropy is represented by another parameter, namely, α, in which the tangent of the yield surface is parallel to the q-axis. The parameter α is equal to zero in isotropic consolidation (IC), non-zero for anisotropic consolidation (AC). This parameter depends on the stress ratio at consolidation η_C, but it is assumed to be equal for compression and extension loading.

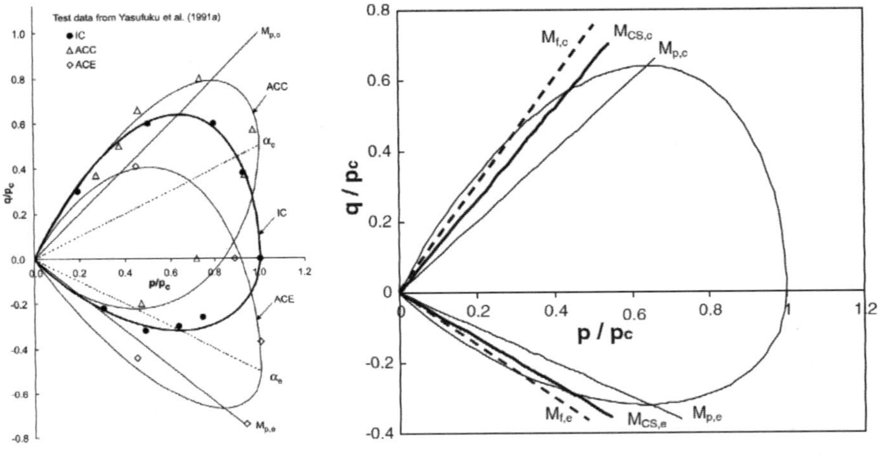

(a) (b)

Figure 2. (a) Yield surfaces for Yasufuku sand in isotropic consolidation (IC), anisotropic consolidation in compression (ACC) and anisotropic consolidation in extension (ACE) (b) Yield surface of isotropically consolidated sand with stress ratios corresponding to peak deviatoric stress, critical state and failure (Imam et al., 2005).

Table 1. Parameters of the model

Type	Parameter	Value
Peak state	φ_μ	24
	κ_p	1.2
Stress-dilatancy	φ_{CS}	33
	κ_{PT}	0.6
Plastic hardening	h	1
	c	0.024
Elasticity	G_r	5100
	K_r	17200
CSL	Γ	1.0121
	λ	0.0364

The model employs 10 parameters to simulate the sand behavior in triaxial compression loading over a wide range of void ratios and stresses. The parameters of the model used in this study are presented in table 1. These parameters were calibrated using the behavior of two loose samples in CU tests.

2. Material and sample preparation

All tests were carried out on a local uniform, subangular to angular, quartzic sand called the Firoozkooh No. 161 sand. The properties of the sand are listed in table 2. Samples were prepared using the moist-tamped technique, in which the samples with relatively 2 percent water content were gently tamped at 8 layers in order to obtain homogenous samples. The moist-tamped technique allows preparing samples with high void ratios. As reported by Ishihara (1993), the surface tension between soil grains generates a particular structure in soil samples prepared by the moist-tamped technique, so that upon undrained shearing there is a significant volumetric contraction. This is relevant to the aim of the present study.

Most CS tests were carried out on dry samples (or low moist samples as the water content of the samples was about 2 percent). However, a few CS and also all CU tests were carried out on saturated samples. For saturation, CO_2 was passed through the samples and followed by de-aired water percolation. Finally, the samples were assumed to be saturated if the B-Skempton value obtained at subsequent steps, in which the cell pressure and back pressure were increased simultaneously, was greater than 98 percent. For details, see Azizi et al. (2009).

Table 2. Properties of sand tested

Property	Firoozkooh No. 161
G_s	2.65
e_{max}	0.964
e_{min}	0.548
D_{50}	0.26
C_c	1.13
C_u	1.9
Angularity	Subangular to angular

3. Instability in Isotropic Consolidated Undrained Compression Test

Following saturation, the samples were consolidated isotropically to the effective stress desired for testing. Undrained tests were conducted under strain-controlled mode at a uniform strain rate of 1% per min, and the samples were sheared to large strain, mostly greater than 25%. As an undrained condition, samples were not allowed to drain during the shearing phase.

The results of ICU tests were used to obtain the critical state of the Firoozkooh sand. It is believed that for undrained behavior of any given sand, the shear strength at the critical state is only the function of void ratio. The slope of the CSL (M_{CS}) achieved by undrained tests was estimated to be about 1.28, which corresponds to the mobilized friction angle (ϕ_{CS}) of 33 degrees.

Figures 3a and 3c illustrate behavior of three samples in undrained monotonic compression loading. All samples were prepared at very low densities to ensure contractive responses and consolidated to the same confining pressures (250 kPa). The void ratios obtained at the end of consolidation (e_c) in tests CU87, CU96 and CU89 were 0.967, 0.922 and 0.903 respectively. Figure 3c shows that the ratio of q/p′ at the peak point of the effective stress path (M_P) decreases with an increase of void ratio. The experimental results showed that the stress ratio M_P changed not only due to void ratio variation, but also with changes of consolidation stress. M_P decreased as the state parameter at consolidation (ψ_c) increased. It implied that the slope of instability line of loose sand samples obtained from ICU tests decreased with an increase of void ratio and consolidation pressure.

Figures 3b and 3d show the model predictions for the behavior of the same samples. As illustrated the experimental results and the model predictions were consistent. The model was able to capture the effect of void ratio on the position of instability line in undrained isotropic compression tests. In fact, the model computes the stress ratio M_p which is a function of the state parameter (ψ) (equations 4 and 5), and employs M_p to control the shape of the yield surface.

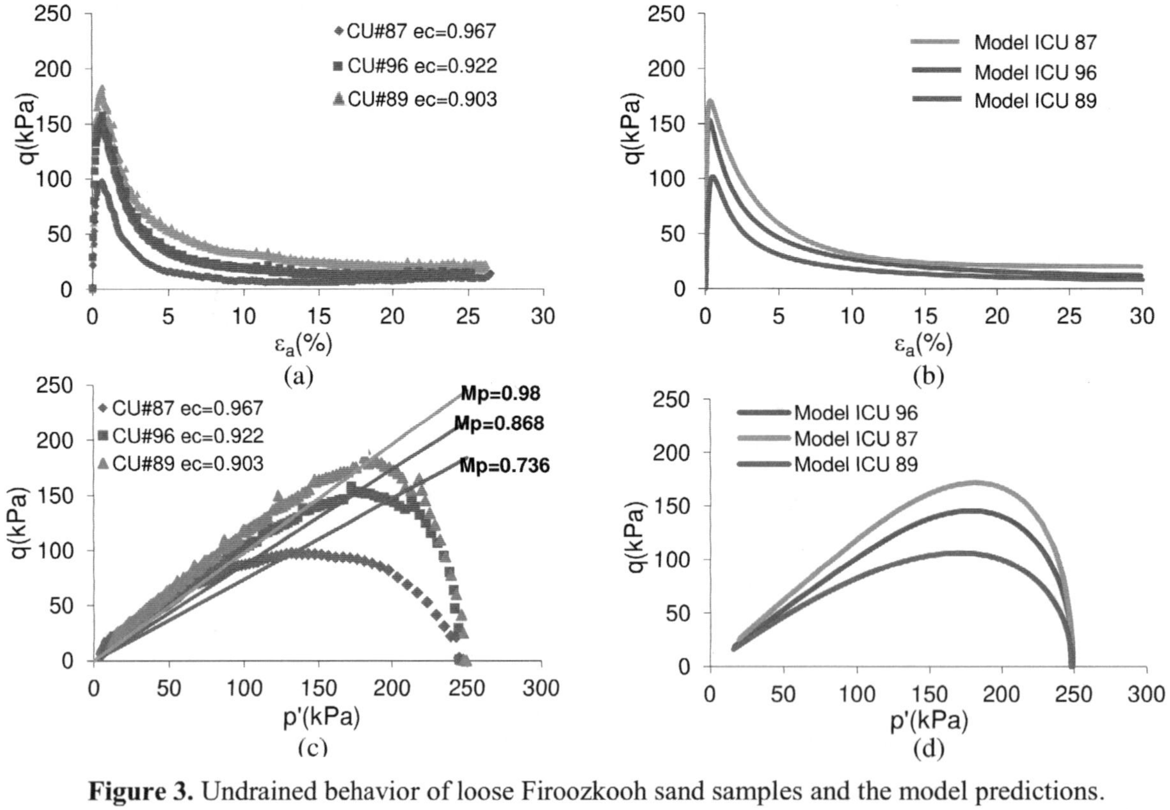

Figure 3. Undrained behavior of loose Firoozkooh sand samples and the model predictions.

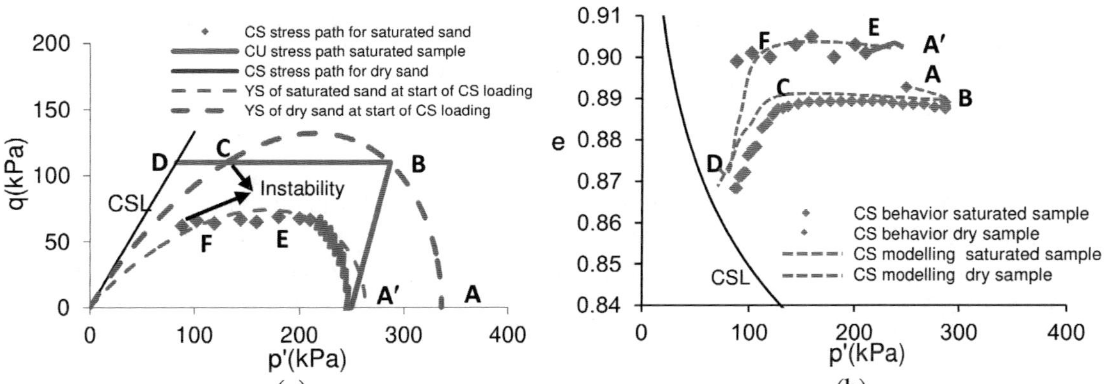

Figure 4. Behavior of dry and saturated loose sand in CDS loading and the model predictions (a) Stress paths (b) Void ratio changes.

4. Instability in Constant Shear Test

The stress path ABD in figure 4a represents CS loading, in which the dry sample was first isotropically consolidated to 250 kPa. The axial load was then increased (A-B) until the deviatoric stress reached to 110 kPa at point B. The deviatoric stress was then kept constant while the mean normal stress was reduced. The reduction in mean normal stress was continued until the soil state reaches the critical state in D. Figure 4b represents the variation of the void ratio for the same dry sample. Following the volume reduction caused by applying the axial load (A-B), the sample was slightly expanded while the cell pressure was decreasing (B-C). However, the sample started to contract substantially for further reduction in the cell pressure (point C). The same microstructural collapse of dry sand was observed by Skopek (1994).

Figure 4a shows also the yield surface of the dry sample established after applying the deviatoric stress and at the beginning of the CS loading. As pointed out by Imam et al. (2007), this yield surface passes through points B and C. In fact, the sample experiences unloading associated with a volume expansion while the CS loading is applied between B and C. As soon as the stress path reaches to the yield surface at point C, the instability occurs associated with volume contractions due to hardening of the yield surface, and it lasts until it reaches to CSL.

The CS test was also performed on saturated sample (stress path A′EF), in which the soil was first subjected to a conventional strain-controlled undrained triaxial compression loading until the desired deviatoric stress (63 kPa) was reached while mean effective stress was 217 kPa. Shearing was then stopped and the deviatoric load applied to the top of the sample was kept constant at the current state. The back pressure equal to the pore pressure at the end of undrained phase were applied to ensure no volume change of the sample would occur, and then second phase of the test began by gradually increasing the back pressure while drainage valve was opened and the cell pressure and deviatoric stress were kept constant. Therefore, the void ratio of the sample remained approximately unchanged from that achieved during isotropic consolidation (figure 4b). This resulted in the application of an almost constant deviatoric stress to the top of the sample, while the mean effective normal stress was being reduced under drained conditions. Approaching to the mean effective stress of 86 kPa at point F, saturated sample suddenly collapsed. The instability happened in such a rapid speed that it was impossible to record any data.

The mobilized friction angles at stress ratio where the collapse starts were similar for both saturated and dry samples. This mobilized friction angles were well below that of failure at the critical state (33 degrees). The friction angles reflect the stress states on YS that was established at the start of CS loading (point B for dry and point E for saturated sample), and this YS depends on both the current void ratio and the current stress state. Although the initial void ratio of the saturated sample (e=0.901) was greater compared to that of dry sample (e=0.882), but the hardening of the YS and decrease of the stresses in undrained compression stress path resulted in state parameter and M_p similar to those of dry sample at the start of the CS loading. As a result, the saturated sample also experienced the elastic unloading within the stress path between E and F inside this YS until it approached to other side of the YS where the collapse occurred at point F.

5. Validating the capability of the model to predict the collapse of dry sand in CS tests

Although the sudden failure of saturated samples did not allow recording any data but it was possible to study the collapse behavior of dry samples. Thus, some CS tests carried on dry samples in order to examine the performance of the model considering the effect of different parameters on behavior of dry samples subjected to the CS loading.

5.1 Void ratio

Fig. 5 shows the results of CS tests on samples of Firoozkooh sand consolidated to 250 kPa and subjected to a constant deviatoric stress of 110 kPa. The void ratios obtained after consolidation (e_c) were different because the samples were prepared using different initial dry densities. The behaviors of these samples indicated that the samples experienced less volume contraction under CS loading as their void ratios obtained after consolidation (e_c) decreased. It may be noticed that the collapse potential for a dense material (e_c = 0.84) vanished and it experienced a continuous volume expansion. The onset of the volume contraction also occurred at smaller mean effective stresses and higher mobilized friction angles for samples with lower void ratios.

The effect of void ratios on the behavior of dry sand in the CS loading can be interpreted by the geometry of the YS established at the end of applying deviatoric stresse, i.e. at start of CS loading. The sample with higher void ratio had a higher state parameter, resulting in the yield surface with lower stress ratio M_p. As shown in figure 5a, the stress path approached to the YS at higher mean effective stress upon applying the CS loading for samples with higher void ratios. As a result, the distance between the onset of the collapse and the CSL, in which the volume contraction occurred, was greater. As the void ratio and state parameter of the samples decreased, M_p increased. Therefore, it established the YS with smaller distance between the onset of the collapse and the CSL leading to experience very small or no contraction during CS loading (samples with e_c=0.839).

The comparison between model predictions and experimental results showed that the model can predict the onset and severity of the collapse of dry samples in CS loading by taking into account the effect of the state parameter on the shape of the yield surface.

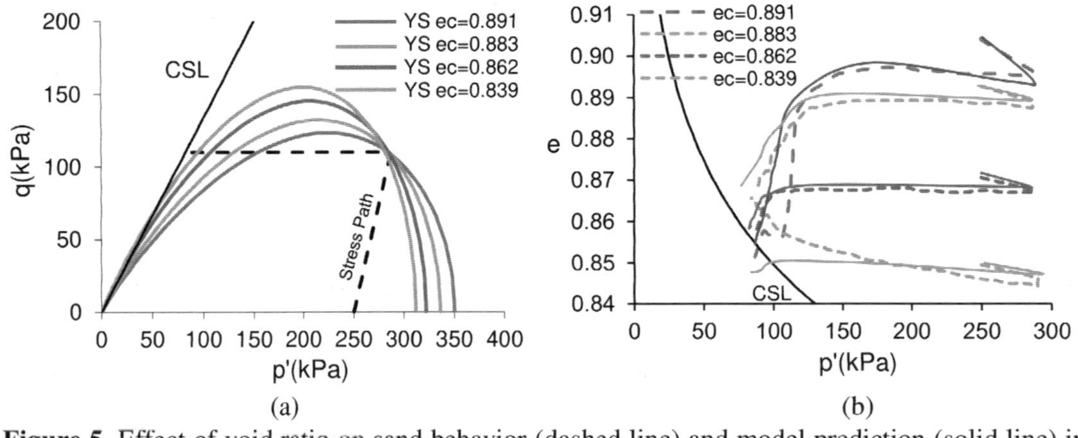

Figure 5. Effect of void ratio on sand behavior (dashed line) and model prediction (solid line) in the CS loading: (a) Stress paths (b) Void ratio changes.

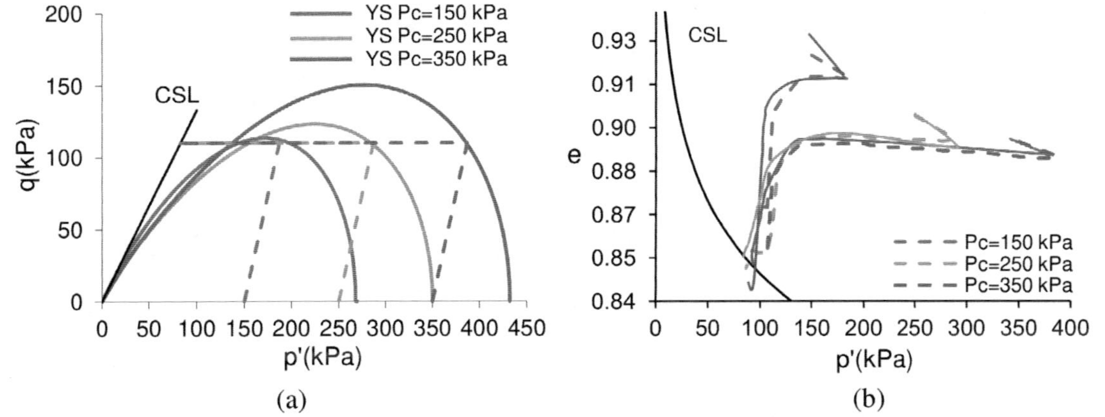

Figure 6. Effect of consolidation stress on sand experimental behavior (dashed line) and model prediction (solid line) in the CS loading: (a) Stress paths (b) Void ratio changes.

5.2 Confining pressure

The Samples were prepared using the same dry density and then consolidated to the mean normal stresses of 150, 250, and 350 kPa. Due to the different consolidation pressures, void ratios of the samples obtained after consolidation were not the same. Following the consolidation, all samples were subjected to the same deviatoric stress of 110 kPa.

The test results shown in figure 6 indicate that these samples had similar collapse potential as the volume contractions initiated at about the same mean normal stress and the amount of these volume contractions were approximately the same.

The state parameter controls the shape of the YS of this model. The samples with the higher confining pressures exhibited lower void ratio at the end of consolidation, and the higher confining pressure appeared to compensate for the decrease in void ratio. It implies that these samples had similar state parameters at the beginning of CS loading regardless of their confining pressures. It can be noticed from the YS of these samples (figure 6a) that they established yield surfaces with similar shapes but different sizes. The size of the elastic region developed at higher consolidation stress was larger, resulting in more unloading during the initial part of the CS loading. However, they reached to the similar mean normal stress where they met again their yield surfaces at the end of unloading. Consequently, they experienced the same amount volume contractions to reach to the CSL.

5.3 Deviatoric stress

The Samples were prepared at similar void ratio and consolidated to the same confining pressure of 250 kPa. They were then subjected to deviatoric stresses of 94, 110, 153, and 200 kPa, followed by the CS loading. The results of these tests shown in figure 7 indicated that in general, for samples subjected to higher deviatoric stress, as the confining pressure was decreased, the volume contractions were greater, and they started earlier.

Fig. 7a shows yield surfaces of the samples at the beginning of the CS loading. The state parameter, M_p and the shape of yield surfaces were the same for these samples. However, they established bigger yield surfaces as the applied deviatoric stress increased. It can be comprehended from the stress paths and the corresponding yield surfaces that the unloading distance during CS loading decreased for the samples with greater deviatoric stresses. The volume contractions commenced earlier and the collapse potential enlarged as the deviatoric stress applied before CS loading increased.

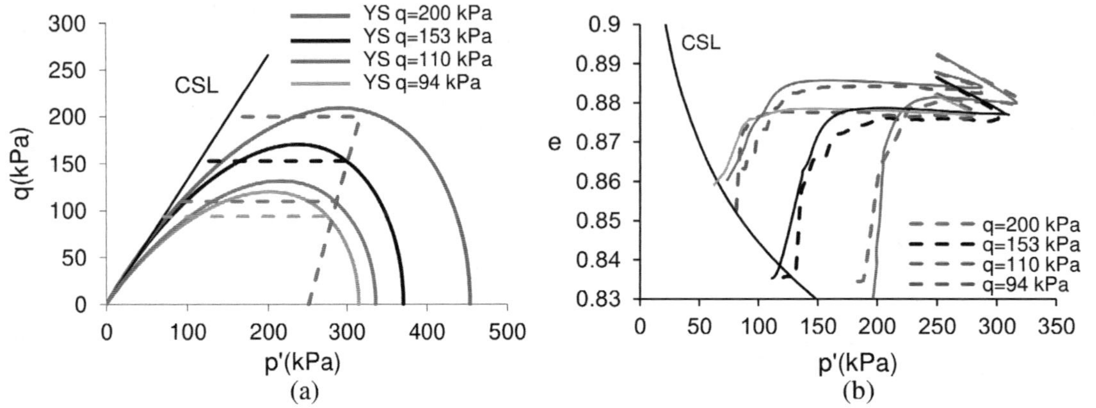

Figure 7. Effect of deviatoric stress on sand experimental behavior (dashed line) and model prediction (solid line) in the CS loading: (a) Stress paths (b) Void ratio changes.

6. Conclusion

The instability of loose saturated sand samples in undrained compression and constant shear tests was predicted by using the critical state constitutive model (Imam et al., 2005). The model employs the cap yield surface in which the shape of the yield surface is controlled by the instability line (stress ratio M_p), considering the effect of void ratio, mean normal stress and anisotropy. The instability and the behavior of loose sand in isotropic compression undrained tests were well predicted by the model.

The saturated loose sample subjected to the CS stress path exhibited a run-away failure due to the sudden increase of pore water pressure. The sample followed elastic unloading within the yield surface, established at the end of the undrained shearing, until it met the other side of the yield surface. The collapse occurred so fast that it was impossible to record any data.

The sudden increase of the pore water pressure in saturated sample was the consequence of the structural collapse and the volume contraction of dry samples in CS tests. In fact, the samples experienced unloading while the confining pressure was decreasing resulted in predicting elastic swelling by the model. It continued until the stress path reached the yield surface where it continued to show volume contraction as the yield surfaced hardened. The dry samples showed different collapse potential in tests performed at different initial void ratios, consolidation and deviatoric stresses. The onset of the collapse and the amount of volume contractions predicted by model were consistent with experimental results.

7. Appendix

The constitutive relationships of the model used in this study are summarized here. In general terms, the capped yield surface YS is defined as follow:

$$f = (\eta - \alpha)^2 - M_\alpha^2 \left[1 - \left(\frac{p}{p_c}\right)^{\frac{1}{2}}\right] = 0 \quad [1]$$

where:

$$M_\alpha^2 = (5M_p - \alpha)(M_p - \alpha) = 0 \quad [2]$$

p_c is the mean normal stress.
If the sand is isotropically consolidated (IC), α=0 and [1] becomes:

$$\eta^2 - 5M_p^2 \left[1 - \left(\frac{p}{p_c}\right)^{\frac{1}{2}}\right] = 0 \quad [3]$$

For triaxial compression TC, considering $\varphi_{p,c}$ as the friction angle at peak of undrained effective stress path in TC, the stress ratio $M_{p,c}$ is given by:

$$M_{p,c} = \frac{6sin\varphi_{p,c}}{3-sin\varphi_{p,c}} \quad [4]$$

whereφ_{μ} is the friction angle corresponding to zero state parameter ($\psi = 0$) in triaxial compression. The friction angles $\varphi_{p,c}$ is given by:

$$sin\varphi_{p,c} = sin\varphi_{\mu} - k_p\psi \quad [5]$$

where k_p is the model parameters, while ψ is the state parameter.
Considering φ_{cs} as the friction angle at the critical state and k_f as a constant parameter for sand in TC (k_f =0.75), the Mohr Coluomb failure criteria is defined using the following equation:

$$sin\varphi_f = sin\varphi_{cs} - k_f\psi \quad [6]$$

The failure friction angle φ_f obtained from [6] can be easily used to compute the maximum stress ratio $M_{f,c}$ according to [4] at the current soil state. Stress ratios can be different from the current stress ratio η, only at critical state $M_f = M_p = \eta$. The flow rule is the stress-dilatancy relationship proposed by Nova (1982) because of its simplicity for sands:

$$d = \frac{d\varepsilon_p^p}{d\varepsilon_p^q} = A(M_{cs} - \eta) \quad [7]$$

The rate of dilatancy is d, the volumetric plastic and deviatoric strains increments are $d\varepsilon_p^p$ ($d\varepsilon_p^p = d\varepsilon_1 + 2d\varepsilon_3$) and $d\varepsilon_p^q$ ($d\varepsilon_p^q = 2(d\varepsilon_1 - d\varepsilon_3)/3$) depending from the major ($d\varepsilon_1$) and minor ($d\varepsilon_3$) plastic strain increments. The term A is a material parameter (typically$A_c = 0.9$) for TC

$$A_c = \frac{9}{(9-2M_{pt,c}\eta+3M_{pt,c})} \quad [8]$$

It is worth to note that if A=1 eq. [7] became the same of Modified Cam Clay (Rocsoe et al. 1963), while stress ratios and $M_{pt,c}$ is given by:

$$sin\varphi_{pt,c} = sin\varphi_{cs} + k_{pt}\psi \text{ for TC} \quad [9]$$

The hardening law is expressed as follow:

$$\frac{\delta p_c}{d\varepsilon_q^p} = \frac{hG}{(p_f-p_c)_{ini}}(p_f - p_c) \quad [10]$$

where the mean normal stress at failure p_f is given by:

$$p_f = \frac{p}{\left[1-\frac{(M_f-\alpha)^2}{M_\alpha^2}\right]^2} \quad [11]$$

The term h is a non-dimensional material parameter depending to the sand stiffness during the shearing phase, G is the elastic shear modulus and $(p_f - p_c)_{ini}$ is the difference of $(p_f - p_c)$ after consolidation and before the shearing phase. Using G_r and K_r as reference values, elastic shear moduli G and elastic bulk moduli K are given respectively by:

$$G = G_r \frac{(2.973-e)^2}{1+e}\left(\frac{p}{p_a}\right)^{1/2} \quad [12]$$

$$K = K_r \frac{(2.973-e)^2}{1+e}\left(\frac{p}{p_a}\right)^{1/2} \quad [13]$$

Reference values are obtained from elastic moduli corresponding to the atmospheric pressure p_a.

NOTATION:

a_p, a_{PT}: Difference between sinϕat peak point of the yield surface and at PT at ψ=0 in TC and TE, respectively

d: Soil dilatancy

e, e_{cs}: Current and critical state void ratios, respectively

f: Yield function

G, G_a: Elastic shear modulus at current and atmospheric mean normal stresses, respectively

h: Material parameter related to plastic shear stiffness

K, K_r: Elastic and reference bulk moduli, respectively

k_p, k_f, k_{PT}: Slope of variation of $\sin\phi_p$, $\sin\phi_f$ and $\sin\phi_{pt}$ with state parameter, respectively

M_{cs}, $M_{cs,c}$: Stress ratios q/p at critical state, and its values in TC, respectively

$M_p, M_{p,c}$: Stress ratio at the peak point of the yield surface, and its values in TC, respectively

M_μ, M_f: Stress ratio q/p corresponding to interparticle friction and failure, respectively

p, p_a, p_c, p_p, p_f: Efective mean normal stress, and its values at atmospheric pressure and at consolidation, respectively

q: Deviatoric stress

α: Stress ratio q/p at which tangent to yield surface is perpendicular to the p-axis

$\varepsilon_1, \varepsilon_3$: Major and minor principal strains respectively

$\varepsilon_p, \varepsilon_q, \varepsilon_p^p, \varepsilon_q^p$: Volumetric and shear strains and their plastic components, respectively

ϕ_f: Mobilized friction angle at failure

$\phi_{PT}, \phi_{PT,c}$: Friction angle at PT, and its values in TC, respectively

$\phi_p, \phi_{p,c}$: Friction angle at peak point of the yield surface, and its values inTC, respectively

ϕ_μ, ϕ_{cv}: Inter-particle and constant volume friction angles, respectively

η, η_c: Stress ratio q/p at current stress state and at consolidation, respectively

ψ: State parameter ($\psi = e - e_{cs}$)

8. References

[1] Anderson, S. A. and Riemer, M. F. 1995.Collapse of Saturated Soil Due to Reduction in Confinement. Journal of Geotechnical Engineering, ASCE, 121(2): 216–219.

[2] A. Azizi, R. Imam, A. Soroush, R. Zandian “Behavior of Sands in Constant Deviatoric Stress Loading”, International Symposium on Prediction and Simulation Methods for Geohazard Mitigation (2009), Kyoto, Japan.

[3] Azizi, A. 2009. Experimental study and modelling of of sand behavior in constant deviatoric stress loading. MSc thesis, Amirkabir University of Technology.

[4] Been K. and Jefferies M.G., (1985). A state parameter for sands.Gèotechnique, Vol. 35, No. 2, pp. 99-112.

[5] Been K., Jefferies M.G., Hachley J.E., (1991). The critical state of sands.Gèotechnique, Vol. 41, No. 3, pp. 365-381.

[6] Castro G., (1969). Liquefaction of Sands.PhD thesis, Harvard Soil Mechanics Series 81, Harvard University, Cambridge, Massachusetts.

[7] Castro, G., Seed, R. B., Keller, T. O. and Seed, H. B. 1992. "Steady State strength analysis of the Lower San Fernando Dam Slide. Proc. of the 53rd Canadian Geotechnical Conference, Montreal, pp. 169–176.

[8] Chu, J., Lo, S-C. R., and Lee, I. K. (1993). "Instability of granular soils under strain path testing." J. Geotech. Eng., 119(5), 874- 892.

[9] Chu, J., andWanatowski, D. (2009). "Effect of loading mode on strain softening and instability behavior of sand in plane-strain tests." J. Geotech. Geoenviron. Eng., 135(1), 108-120.

[10] Dafalias, Y., and Hermann, L.R. 1982. Bounding surface formulation of soil plasticity. In Soil mechanics-transient and cyclic loads. Edited by G.N. Pande and O.C. Zienkiewics. John Wiley & Sons, Chichester. Pp. 253-282.

[11] Daouadji, A., AlGali, H., Darve, F., and Zeghloul, A. (2010). "Instability of granular materials: Experimental evidence of diffuse mode of failure for loose sands." J. Eng. Mech., 136(5), 575-588.

[12] di Prisco, C., and Imposimato, S. (1997). "Experimental analysis and theoretical interpretation of triaxial load controlled loose sand specimen collapses." Mech. Cohes.-Frict. Mater, 2(2), 93-120.

[13] Eckersley, J. D. (1990). "Instrumented laboratory flowslides." Géotechnique, 40(3), 489-502.

[14] Gajo, A., Piffer, L., and de Polo, F. (2000). "Analysis of certain factors affecting the unstable behaviour of saturated loose sand." Mech. Cohes.-Frict. Mater, 5(3), 215-237.

[15] Imam, S. M. R., Morgenstern, N. R., Robertson, P. K., and Chan, D. H. 2005.A Critical State Constitutive Model for Liquefiable Sand. Canadian Geotechnical Journal, 42: 830–855.

[16] Imam, S. M. R., Morgenstern, N. R., Robertson, P. K., and Chan, D. H. 2002.Yielding and flow liquefaction of loose sand. Soils and Foundations, 42(3): 19-31.

[17] Imam, S. M. R., Morgenstern, N. R., 2007.Modeling sand Behavior in constant deviatoric stress loading. Soft Soil Engineering, pp. 389–395.

[18] Ishihara, K. 1993. Liquefaction and flow failure during earth-quakes. Geotechnique, 43(3): 351-415.

[19] Jefferies M.G., (1993). Nor-sand: a simple critical model for sand. Gèotechnique, Vol. 43, No. 1, pp. 91-103.

[20] Lade P.V., (1992). Static instability and liquefaction of loose fine sandy slopes. Journal of Geotechnical Engineering, Vol. 118, No. 1, pp. 51-71.

[21] Lade, P.V.: Initiation of static instability in the submarine Nerlerk berm, Canadian Geotechnical Journal, 30, 6, (1993) 895-904.

[22] Leong, W. K., Chu, J., and Teh, C. I. (2000). "Liquefaction and instability of a granular fill material." Geotech. Test. J., 23(2), 178- 192.

[23] Olson, S.M., Stark, T.D., Walton, W.H., Castro, G.: 1907 Static liquefaction flow failure of the North Dike of Wachusett Dam, Journal of Geotechnical and Geoenvironmental Engineering, 126, 12, (2000) 1184-1193.

[24] Nova, R., 1982. Static A constitutive model for soil under monotonic and cyclic loading. In Soil mechanics-transient and cyclic loads. Edited by G.N. Pande and O.C. Zienkiewics. John Wiley & Sons, Chichester. Pp. 243-373.

[25] Nova, R. (1994). "Controllability of the incremental response of soil specimens subjected to arbitrary loading programs." J. Mech. Behav. Mater., 5(2), 193-201.

[26] Riemer, M. F. and Seed, R. B. Factors Affecting Apparent Position of Steady State Line. Journal of Geotechnical and Geoenvironmental Engineering,ASCE, 123(3): 281–287.

[27] Rowe, P.W. 1962.The stress-dilatancy relation for static equilibrium of an assembly of particles in contact. Pro.of the Roy.Soc.A269: 500–527.

[28] Sasitharan, S., Robertson, P. K., Sego, D. C. and Morgens-tern, N. R. 1994, Collapse behavior of sand, Canadian Geotechnical Journal, 30(4), pp.569-577.
[29] Skopek, P., Morgenstern, N. R., Robertson, P. K., and Sego, D. C. (1994). "Collapse of dry sand." Can. Geotech. J., 31(6), 1003-1008.
[30] Wanatowski, D., and Chu, J. (2007). "Static liquefaction of sand in planestrain." Can. Geotech. J., 44(3), 299- 313.
[31] Yamamuro J.A. and Lade P.L., (1997). Static liquefaction of very loose sands.Canadian Geotechnical Journal, Vol. 34, pp. 905-917.

An Experimental study of the initial volumetric strain rate effect on the creep behaviour of reconstituted clays

Meghdad Bagheri[1], **Mohammad Rezania**[1,*], **Mohaddeseh Mousavi Nezhad**[2]
[1] Nottingham Centre for Geomechanics, University of Nottingham, NG7 2RD, UK
[2] Civil Engineering Research Group, School of Engineering, University of Warwick, CV4 7AL, UK

*Email: Mohammad.Rezania@nottingham.ac.uk

Abstract. Clayey soils tend to undergo continuous compression with time, even after excess pore pressures have substantially dissipated. The effect of time on deformation and mechanical response of these soft soils has been the subject of numerous studies. Based on these studies, the observed time-dependent behaviour of clays is mainly related to the evolution of soil volume and strength characteristics with time, which are classified as creep and/or relaxation properties of the soil. Apart from many empirical relationships that have been proposed in the literature to capture the rheological behaviour of clays, a number of viscid constitutive relationships have also been developed which have more attractive theoretical attributes. A particular feature of these viscid models is that their creep parameters often have clear physical meaning (e.g. coefficient of secondary compression,C_α). Sometimes with these models, a parameter referred to as initial/reference volumetric strain rate, $\dot{v}_0$ has also been alluded as a model parameter. However, unlike C_α, the determination of $\dot{v}_0$ and its variations with stress level is not properly documented in the literature. In an attempt to better understand $\dot{v}_0$, this paper presents an experimental investigation of the reference volumetric strain rate in reconstituted clay specimens. A long-term triaxial creep test, at different shear stress levels and different strain rates, was performed on clay specimen whereby the volumetric strain rate was measured. The obtained results indicated the stress-level dependency and non-linear variation of $\dot{v}_0$ with time.

1. Introduction

In recent decades, the population growth in urban areas and the subsequent increase in the demand for accommodation and new infrastructure (e.g. buildings, roads, railways, and tunnels) have resulted in the infrastructure construction on soils of marginal quality such as soft to very soft soils which are known to behave in such a way that are highly influenced by the time effects. Soft soils such as clays undergo continuous compression with time even after excess pore water pressure is dissipated. These deformations (creep), which mainly occur over a long duration of time, cause difficulties for the serviceability limit state design of infrastructures and if this is not controlled, it may lead to potential geohazards and infrastructure instabilities. A significant number of studies have been conducted over the past few decades to develop empirical correlations and constitutive models capable of capturing the rheological behavior of clays. This has led to the introduction of various empirical correlations (e.g. [1]) and the emergence of diverse conceptual theoretical frameworks such as the concept of isotaches [2], the overstress theory [3], and the concept of non-stationary flow surface [4], from which advanced constitutive time- and rate-dependent soil models were developed. Among these, there are viscid constitutive relationships (e.g. [5]) which their creep parameters, mainly coefficient of secondary compression, C_α, and initial/reference volumetric strain rate, $\dot{v}_0$ possess clear physical meanings. A thorough understanding of the variations of these key parameters with time and stress level can further support the applicability of viscid models for accurate prediction of the rheological behavior of natural clays in field conditions. C_α, has received significant attention in the past 50 years and a substantial amount of information is available in the literature to describe its variation with various influencing factors such as time, consolidation pressure, soil structure, and sustained loading [6, 7, 8]. Unlike C_α, volumetric strain rate, $\dot{v}$, and its reference value, $\dot{v}_0$, have been the subject of a very few studies in the past. In one of the first studies of the creep behavior of normally consolidated clays under drained and undrained triaxial conditions, Singh and Mitchell [9] reported that the volumetric creep strain rate increases with the applied deviatoric stress. Tavenas et al. [10] studied the effect of stress state (p', q) during the triaxial creep tests on lightly overconsolidated undisturbed Saint-Alban clay. It was observed that there was a slight increase of volumetric strain rate with effective mean stress (p'), as well as an increase in axial strain rate (ε_d) with increase in deviatoric stress (q). These reports along with some other related information in the literature approve the stress-dependency of $\dot{v}$. Furthermore, there is general agreement that during the secondary compression, the volumetric strain rate, $\dot{v}$, varies linearly with the rate of change of time logarithm. However, a few studies can be found in the literature which can suitably demonstrate stress dependency and evolution with time of $\dot{v}_0$ during the secondary compression. In this paper we endeavor to experimentally investigate this parameter under triaxial conditions.

2. Triaxial creep test

2.1. Apparatus

A fully automated triaxial testing system was utilized to study the evolution of volumetric strain rate during the creep test. The triaxial system is based on the classic Bishop-Wesley type stress path cell which enables control of stresses directly on the sample. The fundamental system hardware elements are shown in Figure 1. The system is composed of a cell pressure/volume controller, a back pressure/volume controller, a lower chamber pressure/volume controller as a hydraulic axial loading driver, high accurate transducers and a data acquisition system. The apparatus can meet the needs of the creep test under constant stress and the stress relaxation test under constant strain, respectively. Accurate monitoring of specimen volume change, stresses and strains are obtained from the feedback of the automatic controllers and transducers.

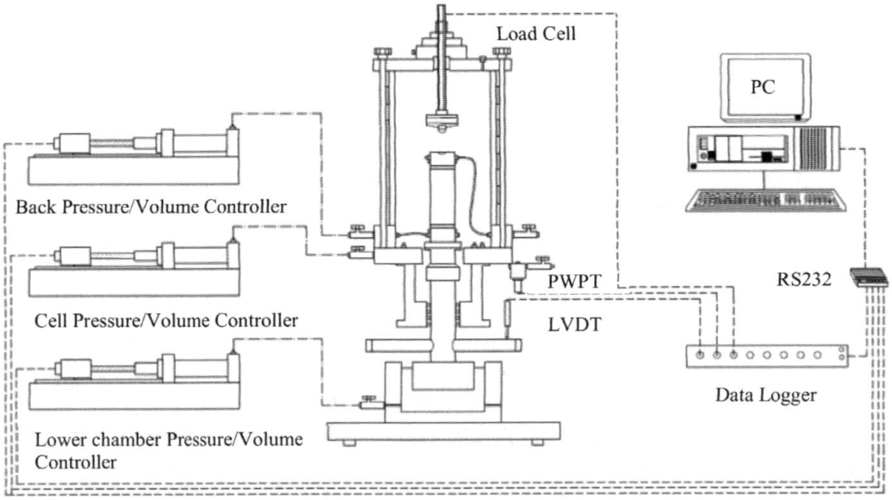

Figure 1. Automatic stress-path triaxial testing system

2.2. *Testing material and sample preparation*

The material investigated in this study was London Clay extracted from an excavation site in The Isle of Sheppey, UK. The Properties of the tested material are given in Table 1. A triaxial specimen of 50mm in diameter and 99mm in height was prepared by the moist tamping method (dynamic compaction). The soil specimen was compacted at the initial water content of 20 % to a specific volume of 1.95. The soil specimen was compacted dynamically in 9 layers on the base pedestal of the triaxial cell. In order to achieve better uniformity, the compaction method proposed by Ladd [11] was adopted. In this method, the uniformity of the soil specimen can be achieved by compacting the top, middle, and bottom thirds proportions of the specimen at 4% denser than the desired dry density, equal to the desired dry density, and 4% looser than the desired dry density, respectively.

Table1. Index properties of the testing material (London Clay)

Liquid Limit (%)	Plastic Limit (%)	Plasticity Index (%)	Specific Gravity	In-situ Water Content (%)
78	24	54	2.67	30-34

2.3. *Testing program and procedures*

To study the stress-level dependency and the time evolution of the volumetric strain rate, a multistage triaxial drained creep test was carried out on the re-compacted specimen. The test procedure involved the following three stages; (1) Saturation: the specimen was saturated gently by using low rates of saturation ramp, i.e. ramping cell pressure and back pressure at the same time and at the same rate so the apparent state of effective stress (here 20 kPa) was maintained. The Saturation stage was completed after approximately 10 days under maximum cell and back pressures of 700 kPa and 680 kPa respectively. (2) Consolidation: the specimen was consolidated isotropically under two levels of mean total stress p_1=100 kPa, and p_2=200 kPa. (3) The Incremental drained creep test: with the value of p being constant the specimen was loaded at a constant rate of strain to certain stress states (p, q). Appropriate loading rates were considered for the loading stages so the development of excess pore pressure was kept at a very low range. Each effective stress increment was sustained for a fixed period of 120 hours and the resultant creep deformations and volume changes were monitored for every hour. De-stilled de-aired water was used for this experiment in order to minimize the effects of algae growth and air dissolution on the volume change measurements during the long-term test.

3. Results

The overall test duration including saturation, consolidation, incremental loading, and 4 creep stages took about 35 days. Therefore, the obtained results were corrected for the effects of water absorption into the cell body as well as the creep expansion of the cell's body. Figure 2 presents the incremental loading and creep stages of the test on the graph of deviator stress, q, versus axial strain, ε_d.

Figures 3 and 4 present the volumetric strain – time and the volumetric strain rate – time responses at different applied deviatoric stresses, respectively. The Stress-level dependency and time-dependency of both v and $\dot{v}$ are clearly observed on these graphs. The magnitude and rate of volumetric strain tends to increase with the deviatoric stress. In terms of time-dependent response, unlike the volumetric strain which shows a linear increasing trend with time, the volumetric strain rate is decreased non-linearly with elapsed time. These findings are in good agreement with other works performed on various clayey soils and cited in the literature. The slight deviation of the plots of v and $\dot{v}$ for q=10.2 kPa in Figures 3 and 4 can be interpreted as a possible fundamental modification in the soil structure and the development of reverse resistance as a result of aggregates structuration.

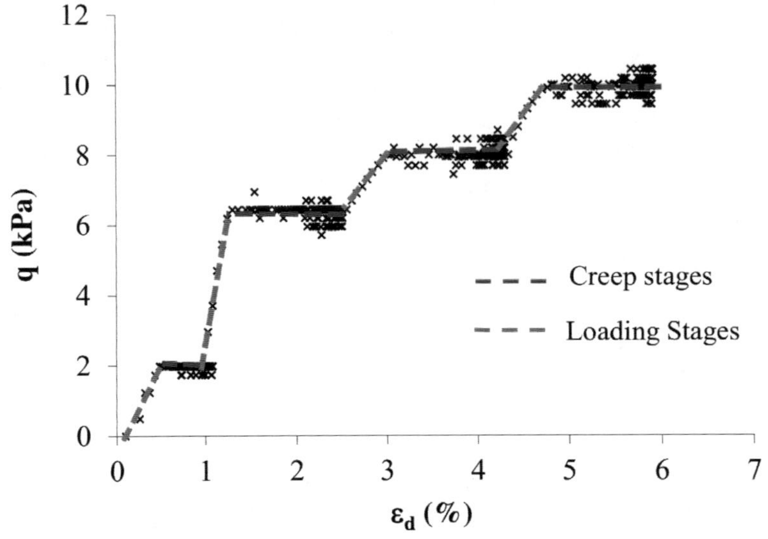

Figure 2. Stress – strain relationship during incremental loading and creep stages.

Figures 5 and 6 show the variation of $\dot{v}_0$ with time and stress-level respectively. Here, the $\dot{v}_0$ is referred to as the volumetric strain rate immediately before the change in loading. Note that the $\dot{v}_0$ values are based on the average volumetric strain rates in the last two hours of each creep stage. Interestingly, it can be observed in Figure 5 that the $\dot{v}_0$ values tend to increase with time but with a decreasing rate. In addition to the time-dependency, the stress-dependency of $\dot{v}_0$ is also approved as shown in Figure 6. It is worth to mention here that, although the $\dot{v}_0$ values tend to increase with the increase in deviatoric stress at the initial stages of creep at constant stress, but without obtaining more data points from the final test stages, it is hard to comment on the overall effect of the deviatoric stress on the$\dot{v}_0$, and whether the trend exhibits a decreasing or an increasing rate. However, these observations are sufficient to contradict the general assumption of $\dot{v}_0$ as a material constant of the viscid theory proposed initially by Sekiguchi [12] and modified by Sekiguchi and Ohta [5].

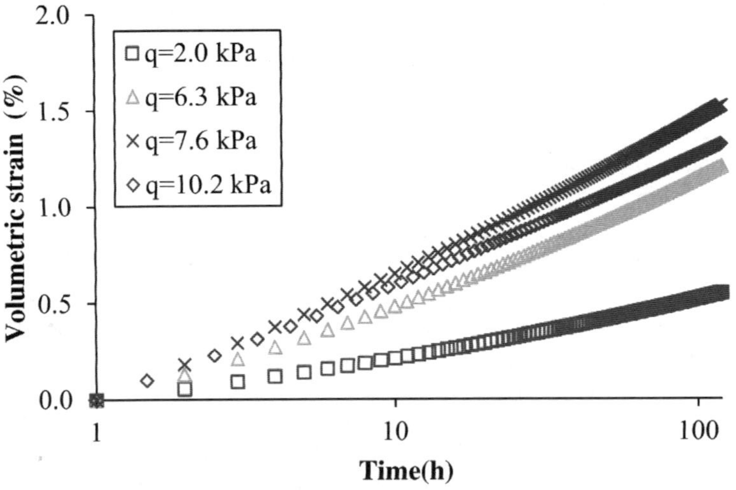

Figure 3. Volumetric strain – time response at different deviator stresses.

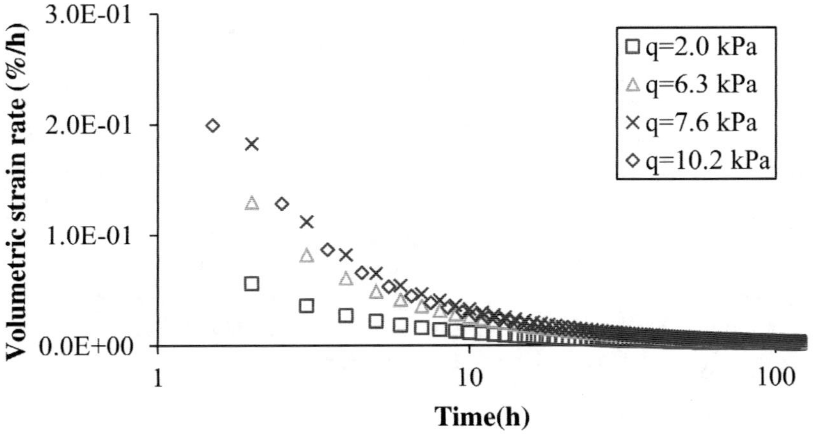

Figure 4. Volumetric strain rate – time response at different deviator stresses.

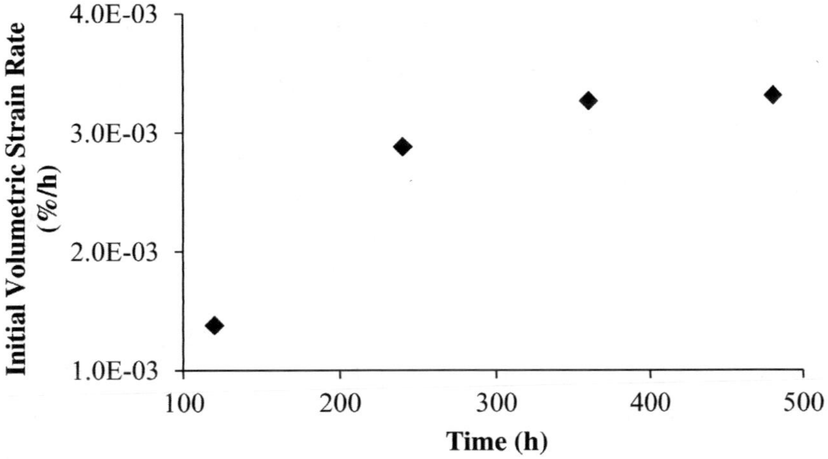

Figure 5. Evolution of initial volumetric strain rate with time.

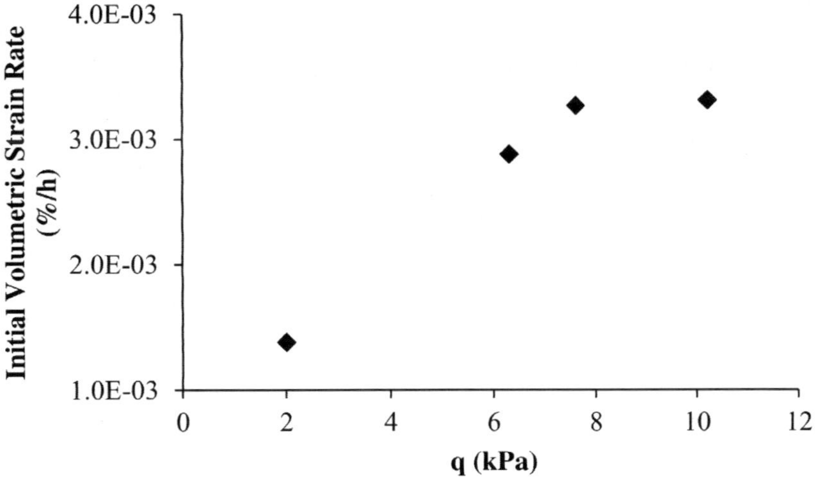

Figure 6. Stress-level dependency of the initial volumetric strain rate.

4. Conclusion

A multi-stage triaxial creep test was performed on the re-compacted London Clay specimen in order to explore the variations of $\dot{v}_0$ with stress level and time. Although the volumetric strain – time and volumetric strain rate – time responses were found to be in good agreement with the literature, the observed time- and stress-level-dependency of the $\dot{v}_0$ were found to be in contradiction with the general assumption of $\dot{v}_0$ as a material constant. Nevertheless, the findings of this work may not be generalised at this stage and could be limited to the studied material and mode of sample preparation. Given the fact that complicated mechanics of the long-term behaviour of soft natural soils is almost certainly behind the progressive (creep) failure of geo-structures found on these soils, further studies are required to investigate the variations of $\dot{v}_0$ over wider ranges of stress states (p, q), strains, and timescales. Research on further classification of this parameter based on additional experimental investigations is currently underway in The Nottingham Centre for Geomechanics at the University of Nottingham.

References

[1] Singh A and Mitchell J K 1969 Creep potential and rupture of soils *Proc. 7th Int. Conf. Soil. Mech. Found. Eng.* (Mexico) pp 379-384

[2] Suklje L 1957 The analysis of the consolidation process by the iso-taches method *Proc. 4th Int. Conf. Soil. Mech. Found. Eng.* (London) vol 1 pp 200-206

[3] Perzyna P 1963 The constitutive equations for work-hardening and rate sensitive plastic materials *Proc. Vibr. Pro.* **4** pp 281-290

[4] Olszak W and Perzyna P 1970 Stationary and nonstationary viscoplasticity Inelastic behavior solids *Battelle Institute Materials of Science Colloquia* (Ohio: Columbus and Atwood Lake)

[5] Sekiguchi H and Ohta H 1977 Induced anisotropy and time dependency in clays *Proc. 9th Int. Conf. Soil. Mech. Found. Eng.* (Tokyo) vol 1 pp 229-238

[6] Mesri G 1973 Coefficient of secondary compression *J. Soil. Mech. Div. ASCE* **99** pp 123-137

[7] Mesri G and Godlewski P M 1977 Time and stress-compressibility interrelationship *J. Geotec. Eng. Div. ASCE* **103** pp 417-430

[8] Fodil A, Aloulou W and Hicher P Y 1997 Viscoplastic behaviour of soft clay. *Geotechnique,* **47** pp 581-591

[9] Singh A and Mitchell J K 1968 General stress-strain-time function for soils *J. Soil. Mech. Div. ASCE* **94** pp 21-46

[10] Tavenas M, Leroueil F, La Rochelle P and Roy M 1978 Creep behaviour of an undisturbed lightly overconsolidated clay *Can. Geotech. J.* **15** pp 402-423Ladd R S 1978 Preparing test specimens using undercompaction *Geotech. Test. J.* **1** pp 16-23Sekiguchi H 1977 Rheological characteristics of clays *Proc. 9th Int. Conf. Soil. Mech. Found. Eng.* (Tokyo) vol 1 pp 289-292

Response of shallow geothermal energy pile from laboratory model tests

A Marto[1, 3] **and A Amaludin** [2]

[1] Professor, Faculty of Civil Engineering, Universiti Teknologi Malaysia, 81310 UTM Skudai, Johor, Malaysia

[2] Tutor, Civil Engineering Programme, Faculty of Engineering, Universiti Malaysia Sabah, Jalan UMS, 88400 Kota Kinabalu, Sabah, Malaysia.

[3] E-mail: aminaton@utm.my

Abstract. In shallow geothermal energy pile systems, the thermal loads from the pile, transferred and stored in the soil will cause thermally induced settlement. This factor must be considered in the geotechnical design process to avoid unexpected hazards. Series of laboratory model tests were carried out to study the behaviour of energy piles installed in kaolin soil, subjected to thermal loads and a combination of axial and thermal loads (henceforth known as thermo-axial loads). Six tests which included two thermal load tests (35°C and 40°C) and four thermo-axial load tests (100 N and 200 N, combined with 35°C and 40°C thermal loads) were conducted. To simulate the behaviour of geothermal energy piles during its operation, the thermo-axial tests were carried out by applying an axial load to the model pile head, and a subsequent application of thermal load. The model soil was compacted at 90% maximum dry density and had an undrained shear strength of 37 kPa, thus classified as having a firm soil consistency. The behaviour of model pile, having the ultimate load capacity of 460 N, was monitored using a linear variable displacement transducer, load cell and wire thermocouple, to measure the pile head settlement, applied axial load and model pile temperature. The acquired data from this study was used to define the thermo-axial response characteristics of the energy pile model. In this study, the limiting settlement was defined as 10% of the model pile diameter. For thermal load tests, higher thermal loads induced higher values of thermal settlement. At 40°C thermal load an irreversible settlement was observed after the heating and cooling cycle was applied to the model pile. Meanwhile, the pile response to thermo-axial loads were attributed to soil consistency and the magnitude of both the axial and thermal loads applied to the pile. The higher the thermo-axial loads, the higher the settlements occurred. A slight hazard on the model pile was detected, since the settlement occurred was greater than the limiting value when the pile was loaded with thermo-axial loads of 40°C and 200 N. It is therefore recommended that the global factor of safety to be applied for energy pile installed in firm soil should be more than 2.3 to prevent any hazard to occur in the future, should the pile also be subjected to thermal load of 40°C or greater.

1. Introduction

The shallow geothermal energy pile system is a type of sustainable geostructure technology that is designed to achieve energy efficient space heating and cooling, for residential and commercial buildings; while satisfying load bearing requirements of the underlying foundation [1]. The mechanism of a geothermal energy pile system uses its surrounding soil as an intermittent energy storage medium. A normal pile foundation may be converted into a geothermal energy pile by incorporating a heat exchanger pipes containing heat exchanger fluids. As such, the heat exchanger piles are able to facilitate the transportation of ground thermal energy from buildings to the soil (and vice versa) via these heat exchanger pipes embedded in the piles [2][3].

[3] To whom any correspondence should be addressed.

Nowadays, the geothermal energy pile system has been implemented all around the world, for example in Austria [4], Switzerland [5], United Kingdom [6], Japan [7] and China [1]. Even though the use of this sustainable geostructure has gained acceptance in the said countries, the engineering community has reservations with regards to the effect of thermal loads on the performance of this system during its operation. This concerns stems from the limited knowledge of the performance for this system, and also the absence of an established geotechnical design method designed specifically for this novel technology.

In recent times, the behavior of the geothermal energy piles under thermal loads, and coupled axial and thermal loads (thermo-axial loads) has been the subject of research of European, North American and Asian researchers alike. Some physical model research endeavors include full-scale testing of energy piles conducted by [4], [5] and [6], centrifuge tests conducted by [8] and [9], and single gravity laboratory model testing conducted by [10] and [11]. Meanwhile, the numerical models of energy pile systems, particularly on soil-structure interaction were developed by [12], [13] and [14] in order to obtain a deeper understanding of the geothermal energy pile behaviour under operational conditions. For the purpose of this study, the laboratory tests on model of shallow geothermal energy piles were able to give an insight on the performance of energy piles in a tropical weather condition found in the Malaysian environment. This paper presented the effect of thermal loads on the behavior of energy piles based on laboratory model tests on firm kaolin soil. The study had focused on the thermally-induced settlements in order to quantify the effects of additional thermal loads imposed on axially loaded energy piles.

2. Materials and Methodology

This section presents the soil material and research methodology used in this study, by describing the laboratory work carried out to achieve the aim and objectives.

2.1. Material

In order to investigate the behaviour of energy piles using laboratory model, commercially available kaolin soil was used. Prior to the start of the main testing stage, relevant standard laboratory tests were conducted to obtain the basic properties and undrained shear strength of the compacted cohesive soil. The laboratory test results are shown in Table 1.

Table 1. Physical Indices of Kaolin soil

Parameter	Soil Type	Liquid Limit (%)	Plastic Limit (%)	Plasticity Index (%)	Specific Gravity (G_S)	USCS Classification
Description/ Value	Cohesive Soil	38	27	11	2.64	ML

The kaolin soil used for this study was similar to that used by [8] and [9], since both researchers had used the soil that was classified also as ML. Furthermore, the S300 kaolin soil was chosen for this study because it possesses a high fines content to enable the observation of thermal consolidation and low plasticity to prevent changes in soil-pore water interactions due to temperature changes, as stated by [9]. Meanwhile, the standard Proctor compaction test had shown that the soil used in this study had a maximum dry density (MDD) value of 1.63 Mg/m^3, with an optimum moisture content (OMC) value of 17%.

2.2. Experimental Setup

In general, model pile tests are commonly used to study soil-structure interaction mechanism by allowing researchers to assess the behaviour of model piles in a controlled environment [10]. The size of the energy pile laboratory model developed for this study was quite similar to that developed by [9] and [15], but slightly smaller than the model developed by [11]. Moreover, the laboratory models developed by the said researchers were meant for tests conducted with cohesionless soils, while this study used cohesive soil instead.

As such, this laboratory model pile test in single gravity conditions was developed to determine the behavior of a model energy pile subjected to thermal and thermo-axial loads in cohesive soil. The laboratory model energy pile test system comprises of four (4) components, namely the axial load control system, thermal load control system, soil container and the model pile.

In particular, the axial load control system comprised of a stress-controlled system provided using a pneumatic cylinder, while the load applied to the model pile head was monitored using a load cell. The thermal load system consisted of a metallic U-tube that was installed within the model pile length to emulate a uniform temperature distribution, which was then connected to a temperature bath to enable the thermal load to be controlled.

Meanwhile, the model of soil and the model energy pile assembly was designed by scaling down an energy pile found in practice. The typical value of energy pile diameter ranges from 0.6 to 1.5 m, while the pile length ranges from 10 to 25 m [15]. For the purpose of this study, a representative energy pile prototype with a diameter of 1.35 m and length of 13.5 m was chosen. Therefore, the model foundation chosen for this study was made of stainless steel with a diameter of 15 mm and 150 mm embedded length. Figure 1 shows the prototype and model dimensions while the soil chamber used in this study is shown in Figure 2.

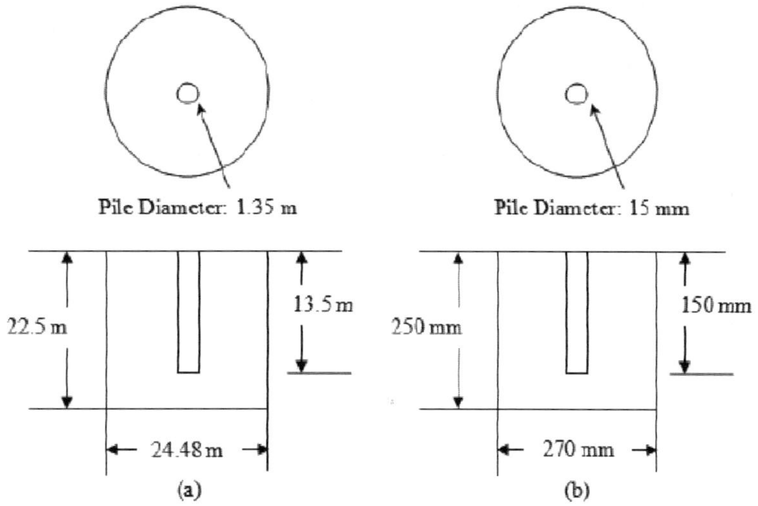

Figure 1. Soil and energy pile dimensions: (a) Prototype and (b) Model

Figure 2. Soil chamber

2.3. Experimental Method

To prepare the soil model, the S300 kaolin powder was oven dried at 105°C for 24 hours to prevent any biological activity within the sample, as explained by [16]. The model pile was fixed in the centre of a 270 mm inner diameter cylindrical acrylic plexiglass, while the model soil was compacted around the pile by using a steel tamper. The chosen target dry density was 1.39 Mg/m^3 (corresponding to a relative density of 90%). In this type of pile installation, the method chosen to install the model pile was similar to that of the installation of non-displacement piles [4].

Through the loading test conducted on the model pile, it yielded an ultimate bearing load of 460 N. The ultimate load was used to calculate the actual factors of safety (FOS), based on the chosen values of axial loads used in this study, which were 100 N and 200 N. The global factor of safety was then calculated by dividing the ultimate load with the axial loads imposed on the pile head, as shown in Table 2.

Table 2. Factor of safety for imposed axial loads

Ultimate Load	Axial Load	
	100 N	**200 N**
460 N	4.6	2.3

Next, the undrained shear strength of the model soil were obtained by carrying out the unconfined compression tests (UCT) on the prepared model soil. Standard UCT were conducted on samples with 38 mm diameter and 76 mm height. The average undrained shear strength obtained was about 37 kPa and the average water content of the soil was about 17%.

Upon the completion of the model pile and soil preparations, the experimental equipment (temperature bath, load cell and LVDT) were setup, followed by the application of the axial load and finally the application of the thermal loads (one cycle of heating and cooling). All observed data were recorded on a data logger. To measure the vertical displacement of the model pile, an LVDT was set on a steel plate fixed to the pile head. A total of two thermal load tests and four thermo-axial load tests were performed.

For thermal loads series (chosen temperature values were 35°C and 40°C), no axial loads were applied on the model pile. The reason behind choosing the 35°C and 40°C for the thermal load values stems from the typical conditions where energy piles are subjected to thermal loading with temperature varying in the range between 0-50°C [4].

In the thermo-axial load tests, the chosen axial loads (100 N and 200 N, representing working loads with the application of global FOS of 4.6 and 2.3 respectively to the ultimate load) were applied first. The thermal load was applied immediately after the settlement of the soil due to axial loading has remained constant. The thermal cycle consists of increasing the pile temperature to 35°C and 40°C from the room temperature (approximately 27°C), and subsequently reduced back to 27°C. Only one thermal cycle was applied for each test.

3. Experimental Results and Discussion

This section features the results and analysis of tests conducted. As previously mentioned, laboratory tests were carried out to determine the physical and the mechanical properties of the S300 kaolin material, followed by the thermal load and thermo-axial load tests on the model piles. Results of the tests on the energy pile models were presented and analyzed, based on the varying thermo-axial loads.

3.1. Thermal Load Test

Heat transfer through soils is through soil grains, water, and pore air. Yet, the transfer mainly occur through the soil particles, owing to the much higher thermal conductivity of the solids relative to the water and air. In general, increase in temperature will result in thermal expansion of soil grains as well as pore fluid [17]. Hence, the soil will heave as the temperature increases and settle as the temperature decreases. However, the results from this study shown in Figure 3 reveals a different trend. The experimental data obtained from the thermal load tests showed a downward trend of the pile head settlement. It is also observed that the pile response appeared to be thermo-elastic [18] in which the settlement increased as the temperature increased but decreased back to almost the same value as the temperature was reduced during the cooling period.

The results of the said thermal load tests seemed to contradict the findings of which where the pile was not axially loaded, the results show pile head heave during heating, and settlement during cooling [10], [12] and [19]. This is because when a heating cycle is applied to a pile, it expands, and any axial deformation will be opposed by shaft restraint at the soil-pile interface. Furthermore, the expansive deformations generate a "negative" shaft friction, whereby the pile expands up relative to the surrounding soil over the upper half of the pile [19]. Consequently, it is in agreement with the outcome of the thermal load test conducted by [10]. They concluded that the thermal strains generated consistently negative (expansive) strains on the pile head during heating, since it is permitted to freely expand upward under the constant applied thermal load.

For the thermal loading test series in this study, there should not be any axial load imposed on the model pile. Upon checking, it was detected that the pile was packed with the load application system before subjected to thermal load. Residual axial loads contributed from the model pile cap and load plunger, both placed directly on the model pile head may have inadvertently contributed to the initial axial loads. As stated by [5] and [6], the dilation of the model pile during heating causes additional stresses to be developed at the pile toe, which eventually induces the thermal settlement. This is in agreement with the results obtained from this study.

The designated limiting settlement was defined as 10% of the model pile diameter, which amounts to 1.5 mm. Results showed that the maximum settlement for the model pile heated to 40°C only reached a value of 0.14 mm, while the maximum settlement for the model pile heated to 35°C reached a value of 0.1 mm. Hence, both thermal settlement values were observed to be relatively small, which were less than 1% of the pile diameter. An hysteresis phenomenon [11], that was a distinct heating and cooling paths, was also observed in the cycle. Consequently, the irreversible thermal settlement phenomenon that was observed during the cooling cycle of the 40°C thermal load was in line with the findings of [5] and [6]. However the value was only 0.01 mm (less than 0.1% of the pile diameter), which could be considered as negligible.

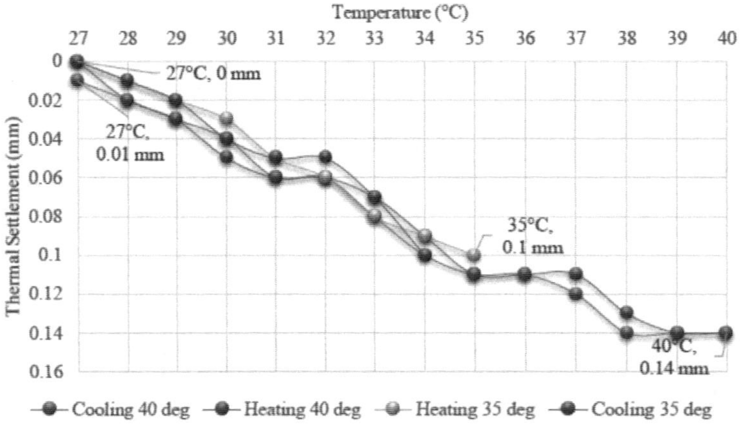

Figure 3. Model pile head settlement due to thermal load at 35°C and 40°C

3.2. Thermo-Axial Load Test

The thermo-axial settlements due to thermal load of several combinations of both axial and thermal loads were shown in Figure 4. The thermo-axial settlement at 35°C thermal load, where combined with an axial load of 100 N yielded 0.67 mm, while for axial load of 200 N the settlement produced was 0.87 mm. Similarly, for the thermo-axial settlement at 40ºC thermal load, when combined with an axial load of 100 N the resultant settlement was 1.34 mm, and for the 200 N axial load the settlement amounted to 1.54 mm. It was found that only one thermo-axial settlement exceeded the limiting settlement (1.5 mm), which was the 200 N axial load combined with 40ºC thermal load.

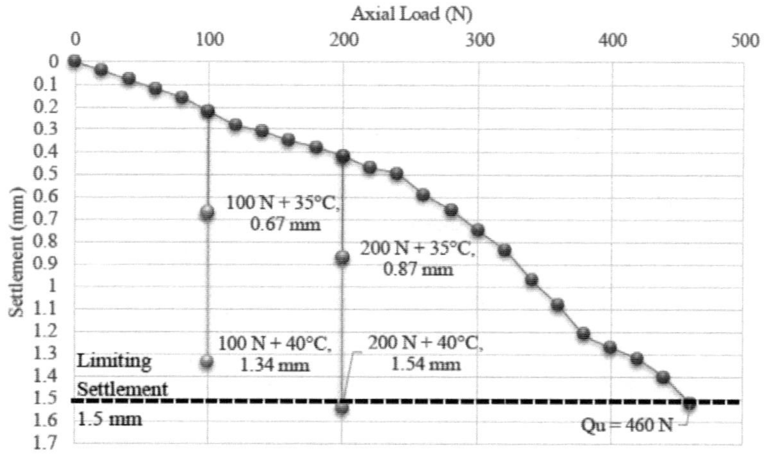

Figure 4. Thermo-axial settlements with respect to settlement at ultimate load

Results showed in Figure 4 indicated that the global FOS of 2.3 applied to the ultimate load was found to be deemed not adequate if the pile was also to withstand thermal load of 40ºC or more. Therefore, it could be recommended that for high thermal loads (40˚C or more), the working load for geothermal pile should be reduced, hence applying higher FOS. This is to accommodate the effects of thermally induced settlements, which were observed to give a greater damaging effect when combined with axial loads, hence the failure hazard could be prevented. The global FOS to be applied on energy pile installed in firm kaolin soil should be more than 2.3 to ensure that the thermo-axial settlement does not exceed the limiting value.

Based on the plotted graphs in Figure 4, generally the thermo-axial settlement curve was somewhat similar to the thermal settlement curve that indicated a steady downward trend as the heat was gradually transferred from the pile to the soil. As stated by [19], the pile response to thermal loads and thermo-axial loads are attributed mainly to ground conditions, end restraint conditions and the magnitude of thermal load applied to the pile. The initial axial load contributes additional stresses at the pile toe, which was reflected in the larger values of thermo-axial settlement, compared to the thermal settlement alone.

Reflecting back on the thermal loading results, there was the needs for addressing the limitation of the current experimental setup. As such, on-going research works will focus on eliminating possible sources of shock loads and initial loads during the experimental setup.

4. Conclusion

In this study, the behaviour of geothermal energy pile model embedded in firm cohesive soil, subjected to varying thermal load and thermo-axial loads were determined using a single gravity laboratory model tests. Although some limitation occurred in the study but the results could provide an insight on the behavior of energy piles, since the current state of knowledge regarding this novel technology is limited, particularly in Malaysia. A total of six tests were performed on kaolin clay compacted to 37 kPa undrained shear strength, to evaluate the behavior of the energy pile model under different thermo-axial loads. The thermal loads of up to 40˚C induced very small values of thermal settlement, whereby it was less than 1% of the pile diameter. The irreversible thermal settlement phenomenon was also observed during the cooling cycle of the 40°C thermal load. The model pile's response subjected one cycle of heating and cooling was also appeared to be thermo-elastic.

Meanwhile, the pile response to thermo-axial loads were attributed to soil consistency and the magnitude of both the axial and thermal loads applied to the pile. The higher the thermo-axial loads, the higher the settlements occurred. It can be concluded that the global factor of safety (FOS) of 2.3 applied to the ultimate bearing capacity was found to be deemed not adequate if the pile was also to withstand 40°C thermal load or more. Hence to avoid hazard due to large settlement, a greater FOS is necessary.

In view of the limitation of this study, future research considerations include the addition of thermocouple strain gauges to monitor the strain changes along the model pile, and investigating the effects of several thermal cycles on the behavior of energy piles models to represent a better simulation of the actual operational conditions of a geothermal energy pile system.

5. Acknowledgement

The authors would like to extend their appreciation to the Ministry of Science, Technology and Innovation (MOSTI), Malaysia for funding the study, which is part of the ScienceFund Project No.: 03-01-06-SF1185, entitled "Shallow Geothermal Energy Pile Bearing Capacity in Clay". Acknowledgements are also due to the Universiti Teknologi Malaysia for providing the necessary facilities to undertake the experimental works carried out in this research.

References

[1] Gao J, Zhang X, Liu J, Li K and Yang J 2008 Numerical and experimental assessment of thermal performance of vertical energy piles: An application. *Applied Energy* **85(10)** : 901-10.

[2] Brandl H 2006 Energy Foundations and other Thermo-Active Ground Structures. *Géotechnique* **56(2)** : 81-122.

[3] Man Y, Yang H, Spitler J D and Fang Z 2011 Feasibility study on novel hybrid ground coupled heat pump system with nocturnal cooling radiator for cooling load dominated buildings. *Applied Energy* **88(11)** : 4160-4171.

[4] Brandl H 2009 "Energy Piles Concepts", *Deep Foundations on Bored and Auger Piles* - Van Impe and Van Impe, Taylor & Francis Group, London, ISBN 978-0-415-47556-3, pg. 77 – 95.

[5] Laloui L, Nuth M and Vulliet L 2006 Experimental and numerical investigations of the behaviour of a heat exchanger pile. *International Journal for Numerical and Analytical Methods in Geomechanics* **30 (8)** : 763–81.

[6] Bourne-Webb P J, Amatya B, Soga K, Amis T, Davidson C and Payne P 2009 Energy Pile Test at Lambeth College, London: Geotechnical and Thermodynamic Aspects of Pile Response to Heat Cycles. *Géotechnique* **59(3)** : 237–48.

[7] Laloui L and Di Donna A 2013 *Energy Geostructures: Innovation in Underground Engineering*, ISTE-Wiley Publishing, United Kingdom.

[8] Nagano K 2007 Energy pile system in new building of Sapporo City University in thermal energy storage for sustainable energy consumption. *NATO Sciences Series II., Mathematics, Physics and Chemistry, Vol. 234: Thermal Energy Storage for Sustainable Energy Consumption: Fundamentals, Case Studies and Design*, pg. 245-53. Dordrecht: Springer Netherlands.

[9] Rosenberg J E 2010 *Centrifuge Modeling of Soil-Structure Interaction in Thermo-Active Foundations*. M Sc Thesis, University of Colorado Boulder.

[10] Stewart M A and McCartney J S 2012 Strain Distribution in Centrifuge Model Energy Foundations. *GeoCongress 2012*. Oakland, CA. pg. 4376-85.

[11] Wang B, Bouazza A and Haberfield C 2011 Preliminary Observations from Laboratory Scale Model Geothermal Pile Subjected to Thermo-Mechanical Loading. *Geo-Frontiers 2011*. Dallas, TX. pg. 430-39.

[12] Kalantidou A, Tang A M, Pereira J-M and Hassen G 2012 Preliminary Study on the Mechanical Behavior of Heat Exchanger Pile in Physical Model (Technical Note). *Géotechnique* **62(11)** : 1047 –51.

[13] Knellwolf C, Peron H and Laloui L 2011. Geotechnical analysis of heat exchanger piles. *Journal of Geotechnical and Geoenvironmental Engineering* **137 (10)** : 890-902.

[14] Ouyang Y, Soga K and Leung Y F 2011 Numerical Back-analysis of Energy Pile Test at Lambeth College, London. *Geo-Frontiers 2011*. Dallas, TX. pg. 440-49.

[15] Shin H, Jeon J and Lee S 2014 Numerical Study on Ground Heat Exchange System *Advances in Civil, Environmental, and Materials Research (ACEM 14)* Busan,South Korea.

[16] Wang B, Bouazza A, Barry-Macaulay D, Singh M R, Webster M, Haberfield C and Chapman G. 2012 Field and Laboratory Investigation of Heat Exchanger Pile. *GeoCongress 2012*. Oakland, CA. pg. 4396-4405.

[17] Mitchell J K and Soga K 2005 *Fundamentals of Soil Behavior*, 3rd Edition, John Wiley and Sons, Inc., United Kingdom

[18] Effendi R 2007 *Modelling of the settlement interaction of neighbouring buildings on soft ground.* PhD Thesis, University of Sheffield.

[19] Amatya B L, Soga K, Bourne-Webb P J, Amis T and Laloui L 2012 Thermomechanical Behaviour of Energy Piles. *Géotechnique* **62(6)**: 503 –19.

Engineering properties of cement/lime-stabilized Egyptian soft clay

M A Mansour[1], A M Samieh[2] and H E Matter

Civil Engineering Dept, Faculty of Engineering, Helwan University, Cairo, Egypt

monamansor83@gmail.com

Abstract. Soft clay formations are extensively located in many coastal areas around the world. The significant high compressibility and low shear strength of these formations impose challenging engineering problems. The deep cement/lime-mix-in-place method is one of the ground improvement techniques exhibiting successful use in stabilizing soft clay. Analysis and design of the deep mixing systems necessitate the identification of the additive content, the proportions of the lime to cement and the characteristics of the stabilized clay. This paper investigates experimentally the influence of adding lime and cement or cement alone, as stabilizing additives, on the engineering behavior of an Egyptian soft clay extracted from the north delta region. A series of laboratory tests were carried out considering, different additive contents of 8, 10, 12, and 14% of the dry weight, with different proportions of lime to cement of 50:50, 25:75 and 0:100. A series of unconfined compression strength tests were performed after different periods; one week, four weeks and 8 weeks, to assess the effect of curing period on the stabilized clay response. In addition, one dimensional consolidation tests were carried out to evaluate the compressibility properties of the stabilized clay. This study declared that the use of an additive content in the range of 12% and more is recommended to improve the characteristics of the considered Egyptian clay. It was pointed out that addition of lime and cement to soft clay significantly increases the strength characteristics and significantly reduces the compressibility characteristics of such clay.

1. Introduction

Inevitable coastal urbanization in many countries has compelled engineers to construct infrastructure, including major highways, over soft marine and alluvial deposits. Construction on soft clay formations encounters the problems of excessive total and differential settlements, large lateral displacement and inadequate bearing capacity. It is therefore crucial to treat soft clay deposits prior to construction activities in order to avoid the undesirable behaviour and the subsequent potential damages to structures. Nowadays, different techniques of soft clay stabilization are successfully employed. The injection method using chemical grouting in the ground has been widely applied to improve soft soil formations. Lime and cement admixture has been extensively used in both shallow and deep stabilization to improve inherent properties of soil such as strength and deformation behaviour. This process was developed simultaneously in Sweden and Japan in the 1970's [1], [2] and [3]. The cement/lime columns developed using dry jet mixing (DJM) that pneumatically delivers cement or

[1] Associate Professor of Geotechnical Engineering
[2] Professor of Geotechnical Engineering

lime powder into ground and mixes it with soil to form soil-cement/soil-lime column, were preliminary reported to be successfully executed in practice in 1980 and 1982 [4], [5], [6], [7] and [8]. This technique is recommended when time is of major concern [9].

Analysis of structures resting on deep mixed soil columns requires identification of such columns characteristics. This need triggered the research and laboratory studies all-over the world to identify the engineering characteristics of local clays when mixed with lime and/or cement. Laboratory test results indicated that, cement is more effective than lime in stabilization of organic soils [6]. It was found that, increasing plasticity index of the clay decreases its improvement readiness, [10] and [8]. Generally, as the amount of the added stabilizer increases, the strength of the treated soil increases with time [11] and [12].

The current research investigates, experimentally, the impact of lime and cement or cement alone, as stabilizing additives, on the shear strength and compression of an Egyptian soft clay extracted from the north delta region. It is aimed to identify engineering characteristics of soft clay when mixed with different contents of lime and cement, considering different proportions of the lime and cement or cement alone. A series of unconfined compression strength tests were performed after different periods to assess the effect of curing period on the stabilized clay response. In addition, one dimensional consolidation tests were carried out to evaluate the compressibility characteristics of the stabilized clay.

2. Experimental work description

2.1. Natural soil characteristics

The soil considered in the current research is an Egyptian soft clay extracted from the north delta region, Damanhur city, at a depth of around 2m below ground surface. It was classified as silty clay with some sand. The liquid limit of the soil ranged from 65 % to 67 % and the plastic limit value ranged from 27% to 34% which indicates clay with high plasticity. Grain size distribution of the natural soil sample is shown in figure 1.

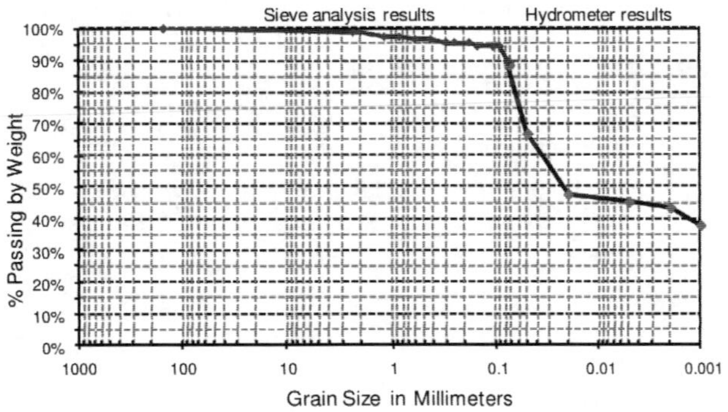

Figure 1. Sieve analysis and hydrometer analysis results

2.2. Method of preparing the lime–cement–soil specimens

A bulk of the soil was dried and then crushed to a powder form. The dry soil that passes from the 4.75mm sieve was mixed with the required percentage of additives (lime and/or cement). The water was then added to the mix with an amount corresponding to 75% of the dry weight of soil. The used water content is within the range recommended by [13] for laboratory testing of chemically stabilized clay, to be between the liquid limit and twice the liquid limit of the clay. This water content value was found to give a high workability mix for the clay considered in the current study. However, such high water content makes preparing samples with no additives for testing, not visible.

2.3. Experimental program

The additives content values examined in the current study were 8%, 10%, 12% and 14%, expressed as a percent of the dry weight of soil. Three proportions of lime to cement were considered; 50:50, 25:75 and 0:100. Table 1 illustrates the considered combinations of the additive contents and lime/cement proportions.

Table 1. Combinations of additive contents and the proportions of lime and cement considered in the study.

Additive contents (%)	Lime content (%)	Cement content (%)
	0	100
8	25	75
	50	50
	0	100
10	25	75
	50	50
	0	100
12	25	75
	50	50
	0	100
14	25	75
	50	50

3. Analysis of experimental results

The unconfined compressive strength tests were performed at curing periods of 7, 28, and 56 days. Three samples for each curing time were prepared. Whereas, oedometer tests were carried out after curing time of 28 days only.

3.1. Results of unconfined compression strength tests

The unconfined compression tests were carried out on 12 different mixtures at three curing periods of 1, 4 and 8 weeks. A summary of the peak strength results obtained for each specimen is presented in Table 2.

Table 2. Summary of the unconfined test results.

Lime content (%)	Cement content (%)	Additives content, (%) dry weight	Curing time (weeks)		
			1	4	8
			Peak strength (kPa)	Peak strength (kPa)	Peak strength (kPa)
50	**50**	8	17.36	39.05	59.3
		10	30.37	45.56	85.33
		12	98.35	147.52	182.23
		14	138.12	251.66	323.97
25	**75**	8	195.25	274.8	390.50
		10	254.55	334.1	435.34
		12	371.70	512	585.75
		14	543.81	604.56	759.31
0	**100**	8	325.42	448.36	513.44
		10	361.58	529.35	571.3
		12	533.69	943	992.17
		14	549.6	989.27	1157.05

Figure 2 shows the soil strength gain with time due to adding various additives contents having different lime to cement ratios. The summarizing table 2 and figure 2 reveal that, with the increase of the additives content, the strength of the stabilized clay increases, firstly with a low rate till 10% content. Then, more increase in additives content, results in increasing the clay strength with a higher rate.

The figure shows also that, for the same additives content, the increase of the cement proportion significantly increases the improved clay strength. The soil strengths corresponding to lime/cement ratio of 25:75 are about 2.3 to 6.6 times those corresponding to lime/cement ratio of 50:50. Moreover, soil samples treated with cement only (L:C = 0:100) has gained strengths equal 3.6 to 8.7 times the strengths obtained using 50:50 lime/cement ratio.

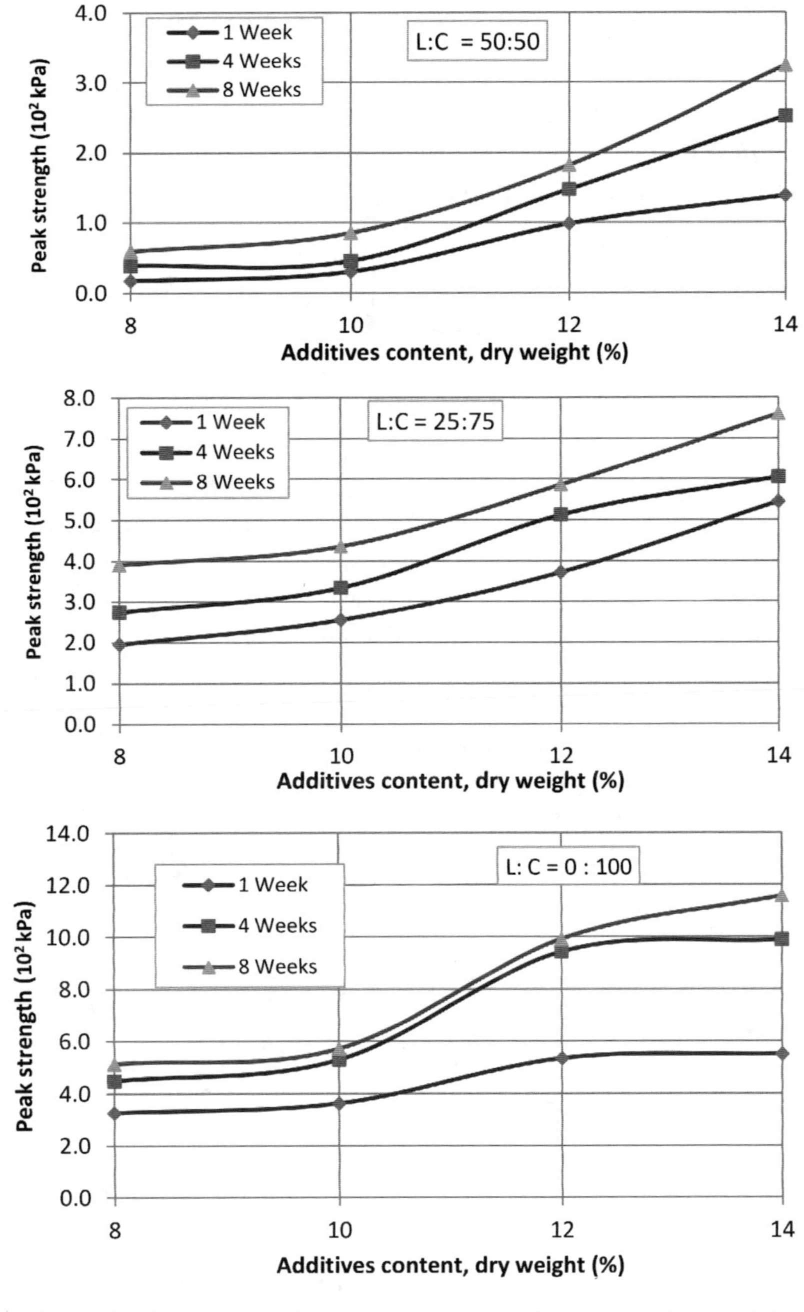

Figure 2. Relationship between additives content and peak strength of the stabilized soil at different curing periods for different lime to cement ratios

For the examined clay, the obtained results exhibits that the use of additives content in the range of 12% or more with higher cement proportion, is recommended to obtain a significant increase in the soil strength. In addition, cement proportion is more effectual in increasing the improved soil strength.

3.2. Deformation / secant modulus

The secant modulus of elasticity (E_{50}) calculated at stress equal to half the unconfined strength, was determined for all the tested specimens. For the specimens tested at 28 days, the variation of the deformation modulus with the additives contents and considering different ratios of the lime to cement, is presented in figure 3. The figure exhibits that, the deformation modulus values were found to increase with increasing the additives content and with increasing the cement proportion, which has a considerable impact.

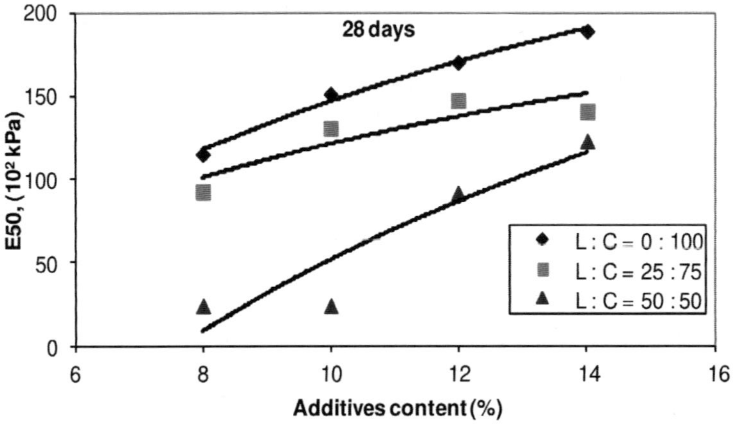

Figure 3. Variation of deformation modulus with the additive content, considering different ratios of the lime to cement after 28 days.

3.3. Consolidation characteristics of the stabilized clay

A series of consolidation tests for the considered stabilized Egyptian clay specimens were carried out for all the considered contents of the additives and all lime to cement ratios. The specimens were subjected to axial stress up to 1600 kPa. It worth mentioning that this stress level is much higher than the working stress level on the deep mixing columns under earth embankments, that ranges between 500 to 600 kPa [14].

3.3.1. Void ratio (e).

Figure 4 shows the variation of void ratio (e) with the stress applied to the stabilized soil specimens. The figure delineates that, with the increase of the additives content, the soil compressibility slightly decreases. Whereas, the use of high proportions of cement in the additive, efficiently decreases the compressibility of the stabilized clay.

3.3.2. Volume compressibility (m_v).

The coefficient of volume compressibility (m_v) of the stabilized clay samples was estimated from the consolidation test results. Figure 5 presents the variation of the m_v-values with the applied stress level for the considered additive contents and for each of the lime to cement ratios. For a given stress level, the figure delineates that the m_v–values decreased with the increase of the cement proportion in the additive content. The figure illustrates, also, that with the increase of the additive content the m_v–values are decreased. In other words, the constrained modulus (E_c) of the stabilized clay ($E_c = 1/ m_v$) is increased with the increase of the cement content and increase of the additive content.

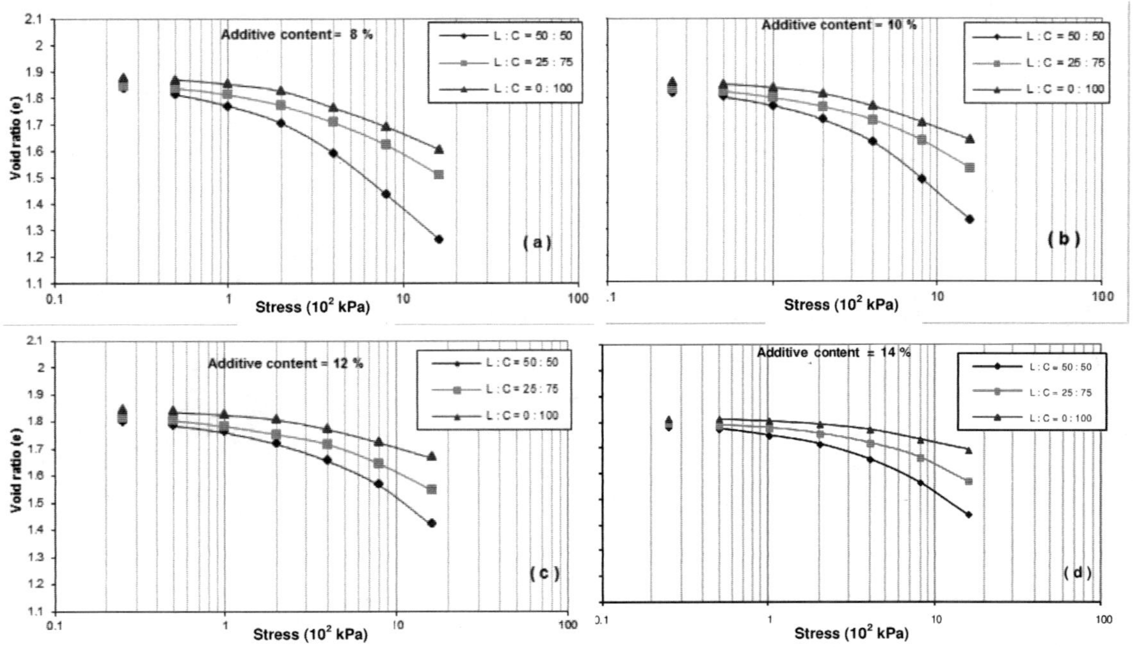

Figure 4. Variation of the void ratio with the applied consolidation stress at different additives content and different lime to cement ratios

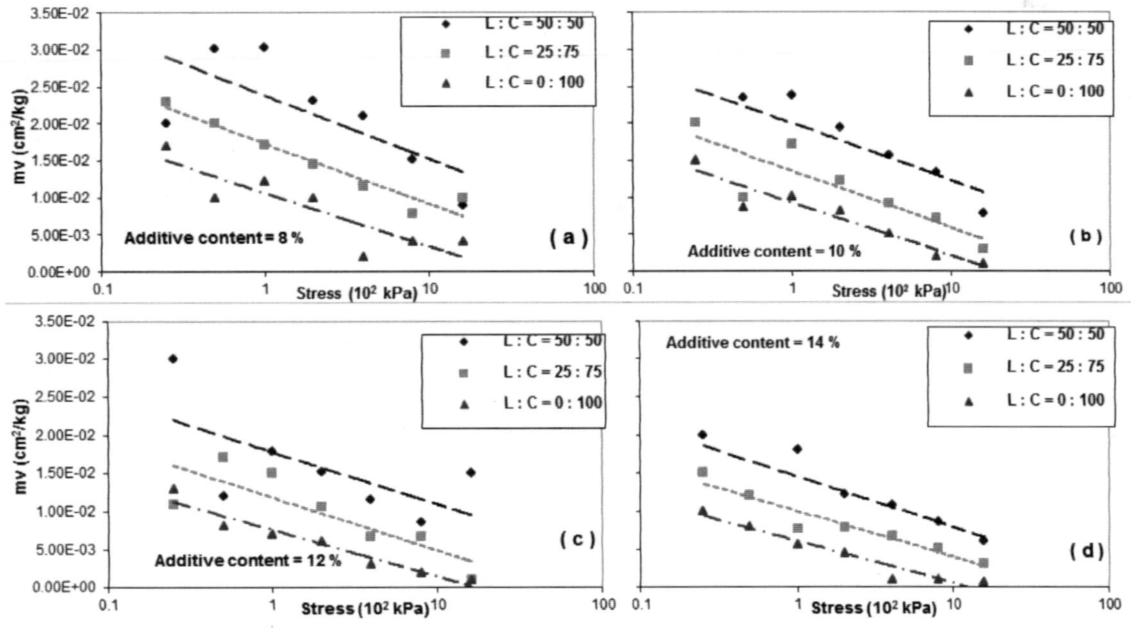

Figure 5. Variation of the volume compressibility coefficient with the applied stress at different additives content and different lime to cement ratios

3.3.3. Consolidation coefficient (c_v).

The results of the consolidation tests were analyzed using the square root of time method (Taylor's method) to assess the C_v–values of the tested specimens. In the current study, the loading progress was adopted to reach 1600 kPa. Figure 6 shows the variation of the C_v–values with the applied stress level for different additive contents and at each lime to cement ratios, after 4 weeks of curing. The figure shows that the C_v–values are generally increased with increasing the additive content. Whereas, the impact of the consolidation stress level on decreasing the C_v–values is insignificant. Therefore, table 3 proposes average values of the consolidation coefficient for the considered clay specimens.

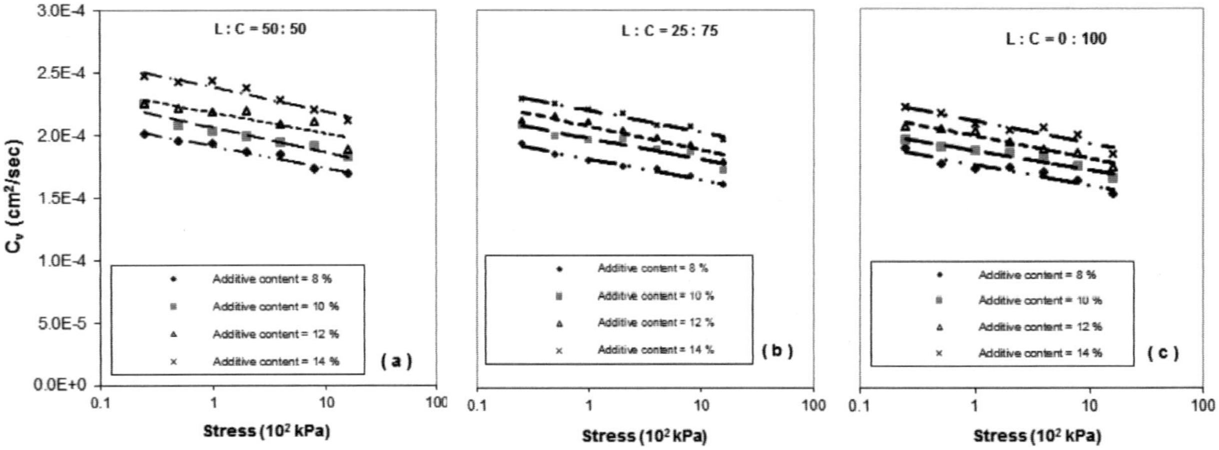

Figure 6. Variation of the consolidation coefficient with the applied level for different additive contents

Table 3. C_v–average values for a stress range of 25 kPa to 1600 kPa, after 4 weeks curing time.

Additive contents, dry weight (%)	Lime content (%)	Cement content (%)	Average value of the coefficient of consolidation (Cv) cm²/sec
8	50	50	$1.83*10^{-3}$
10	50	50	$1.97*10^{-3}$
12	50	50	$2.10*10^{-3}$
14	50	50	$2.30*10^{-3}$
8	25	75	$1.74*10^{-3}$
10	25	75	$1.90*10^{-3}$
12	25	75	$2.00*10^{-3}$
14	25	75	$2.13*10^{-3}$
8	0	100	$1.70*10^{-3}$
10	0	100	$1.81*10^{-3}$
12	0	100	$1.94*10^{-3}$
14	0	100	$2.06*10^{-3}$

3.3.4. Coefficient of permeability (k).

Direct measurement of the permeability coefficient is not possible, it is calculated from the consolidation coefficient (C_v) and the volume of compressibility coefficient (m_v) obtained from the oedometer tests using the following relation:

$$k = C_v . m_v . \gamma_w \quad (1)$$

The permeability coefficient (k) of lime-cement stabilized samples were found for the considered progress of loading range. Figure 7 presents the variation of the k–value with the applied stress level for different additive contents and for each of the lime to cement ratios, after 4 weeks of curing. The figure delineates that k–value is generally decreases with increasing the applied stress. Also, the permeability coefficient of the samples stabilized with cement only is the least. In other words, addition of lime makes to increase the material permeability and this increase is proportional to the lime content.

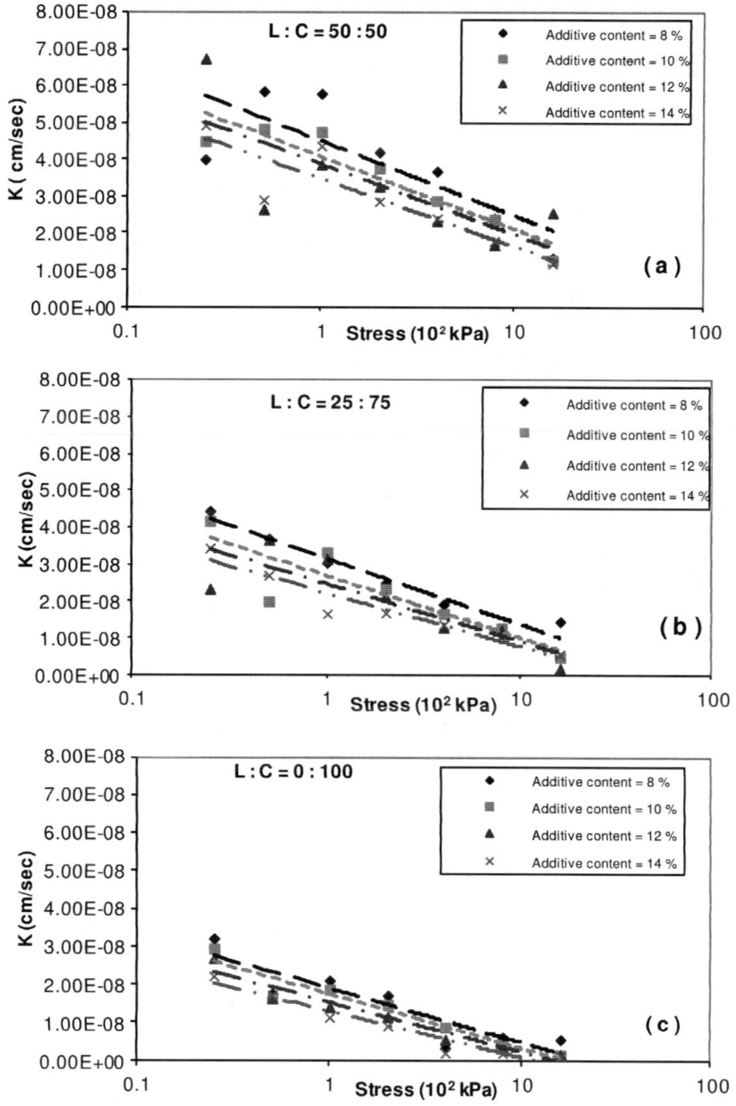

Figure 7. Variation of permeability coefficient with the applied stress level for different additive contents and different lime to cement ratios

4. Summary and conclusions

The work described in this paper is an attempt to identify the unconfined compressive strength as well as the consolidation characteristics of an Egyptian soft clay when mixing different contents of lime and cement at different proportions. From the results of the experimental work, the following conclusions can be obtained:

- The unconfined compressive strength and the stiffness of the treated samples increased with the increase of the additive contents.
- The addition of the cement significantly increased the unconfined compressive strength and the stiffness while the unconfined compressive strength and the stiffness are decreased when part of the cement is replaced by lime.
- The strength gain and the stiffness of the stabilized clay specimens increase with time.
- For the considered clay, additives contents of 12% and more are recommended to reach a significant increase in strength of the stabilized clay.
- With the increase of the additive content, the coefficient of volume compressibility (m_v) was found to decrease while the coefficient of consolidation (C_v) was found to increase.
- For the clay under consideration, it was found that as the lime proportion increased (in comparison to the cement proportion), the coefficient of volume compressibility, the coefficient of permeability (k) and the coefficient of consolidation (C_v) were increased.
- For the clay under consideration, it was found that with the increase of the consolidation pressure, the void ratio, volume compressibility coefficient, permeability coefficient, and the consolidation coefficient, were decreasing.

In general, the trend of the current research results compared well with the results introduced by previous studies. However, researches showed that, the type of the treated clay was of major impact on the values of the improved characteristics [11], [15]. Therefore, the additive content and lime/cement ratio should be optimized for each individual clay formation to attain the desired strength. It was recorded by [16] that, the optimized cement content value required to improve compressibility features of an investigated clay, was 15%, which is near the percentage of additive content got in this research (12%). As concluded in the current work, lime stabilized clay possessed higher increase of permeability coefficient. Whereas cement mixed specimens exhibited slightly lower permeability [11].

5. References

[1] Terashi M,Tanaka H and Okumura T (1979) "Engineering properties of lime- treated marine soils and deep mixing method" *The 6th Asian Regional conf. on Soil Mech. and Found Eng.*, Vol **1**, pp 191-194.

[2] Kawasaki T, Niina A, Saitoh S, Suzuki Y and Honjo Y (1981) "Deep mixing method using cement hardening agent" *Proc. of the 10th Int. Conf. on Soil Mech. Found. Eng.*, Stockholm, pp 721-724.

[3] Suzuki Y (1982) "Deep chemical mixing method using cement as hardening agent" *Proc. Symp. on Soil and Rock Improvement Techniques Including Geotextiles, Reinforced Earth and Modern Piling Method,* AIT, Bangkok, Thailand.

[4] DJM Research Group (1984) "The manual for the dry jet mixing method", pp 20-31.

[5] Chida S (1982) "Dry jet mixing method" State of the Art of Improvement Methods for Soft Ground, *Japanese Society of Soil Mech. and Found. Eng.*, pp 69-76.

[6] Miura N, Koge Y and Nishida K (1986) "Application of a deep mixing method with quicklime for the Ariake clay ground" *Journal of Japanese Society of Soil Mechanics and Foundation Engineering*, Vol **34**, No 4, pp 5-11.

[7] Cox J B (1981) "The settlement of a 55 Km long highway on soft Bangkok clay" *Proc The 10th Int. Conf. of Soil Mech. and Found. Eng.*, Stockholm, pp 101-104.
[8] Broms B B (1986) "Stabilization of soft clay with lime and cement columns in Southeast Asia" Applied Research Project RP10/83, Nanyang Technical Institute, Singapore.
[9] Holm G (1999) "Keynote lecture: Application of dry mix methods for deep soil stabilization" *Proc. Int. Conf. on Dry mix Methods for Deep Soil Stabilization*, Stockholm, Sweden, pp 3-11.
[10] Woo M (1971) "Cement and lime stabilization of selected Lateritic soils" Thesis No 409, Asian Institute of Technology, Bangkok, Thailand.
[11] Ahnberg H, Holm G, Holmqvist L, and Ljungcrantz C (1994) "The use of different additives in deep stabilization of soft soils" *Int. Conf. on Soil Mech. and Found. Eng.,* pp 1191-1194.
[12] Bergado D T, Anderson L R, Miura N and Balasubramaniam A S (1996) "Soft ground improvement in Lowland and other environments" USA: ASCE Press, pp 235-243.
[13] Lorenzo G A and Bergado D T (2004). "Fundamental parameters of cement-admixed clay-new approach". *Journal of Geotechnical and Geoenvironmental Eng.*, ASCE, Vol **130**, No. 10, pp 1042-1050.
[14] Lin K Q and Wong I H (1999) "Use of deep cement mixing to reduce settlement at bridges approaches". ASCE, *Journal of Geotechnical and Geoenvironmental Eng.,* **125**(4)., pp 309-320.
[15] Kivelo M. (1998) "Stabilization of embankments on soft soils with lime/cement columns" Doctoral Thesis, Royal Institute of technology, Sweden.
[16] Uddin K., Balasuramaniam A.S. and Bergado D.T. (1997). "Engineering behavior of cement-treated Bangkok soft clay". *Geotechnical Eng.* Vol **28**, No 1, pp 89-119.

IOP Conf. Series: Earth and Environmental Science **26** (2015) 012054 doi:10.1088/1755-1315/26/1/012054

Correlation between index properties and electrical resistivity of hydrocarbon contaminated periodic marine clays

Preeti Tiwari[1] and M.V Shah[2]
[1]P.G Student,Applied Mechanics, L.D College of Engineering, Ahmedabad-380015,India
[2]Assistant Professor, Applied Mechanics, L.D College of Engineering, Ahmedabad-380015, India
E-mail: mvs2212@yahoo.co.in

Abstract. Hydrocarbon contamination is a measure issue of concern as it adversely affects the soil inherent properties viz. index properties and strength properties.The main objective of this research work is to determine Electrical resistivity to study and correlate with soil index properties and engineering propertiescontaminated with hydrocarbon at the rate of 3%, 6% and 9% for the period of 15, 30 45 and 60 days and compare it with the results obtained for non-contaminated marine clay. Electrical resistivity of virgin marine clay (bentonite which is expansive in nature) and hydrocarbon contaminated clay for each percent of contamination is obtained in the laboratory for each period and its co-relation with index properties and engineering properties is proposed. CEC, EDAX tests were performed to evaluate the effect of ions of montmorillonite clays and their penetrability into hydrocarbon- clay matrix. The correlations at the end of each period for each percentage of contamination thus enabled to integrate index properties of non-contaminated and hydrocarbon contaminated marine clays with Electrical resistivity.

1. Introduction

The past few years have witnessed tremendous growth in petroleum sector all over the world. This has led to hydrocarbon contamination of soil situated in the vicinity of the petroleum spillage areas which is a serious problem for the environment as it adversely affects the soil inherent properties viz. index properties and strength properties. Hydrocarbons contaminated clay resulting in large amounts from the treatment of crude-oil and oil Refineries represent a great geo-environmental pollution threat. Hydrocarbon soil contamination is very common in marine clays. Petrochemicals, hydrocarbons (oil spills), are extensively liberated onto the surrounding soil and get penetrated into the deep ground soil further affecting its physico-chemical properties. With this process of large-scale exploitation of coastal and offshore oil, coastal soils have encountered oil pollution. And this legacy of industrialization which has resulted in a prevalence of hydrocarbons

IOP Conf. Series: Earth and Environmental Science **26** (2015) 012054 doi:10.1088/1755-1315/26/1/012054

contamination of surrounding soils, necessitates the evaluation and analysis of the hazardous effects of these contaminants on soils as well. Electrical resistivity is one of the perturbed properties in the presence of hydrocarbons in the soil. Some authors have correlated Electrical resistivity with varied soil parameters: 1. Moisture content 2.Specific gravity 3. Atterberg'slimit 4. C and ϕ (G. Kibria and M. S. Hossain (2012), Mohd Hazreek Bin Zainal Abidin et.al (2013) and Z Abu-Hassanein et.al (1996). In the present research work, efforts have been made to obtain a correlation between Electrical resistivity of non-contaminated and periodic (15,30,45 and 60 days) hydrocarbon contaminated marine clays at the rate of 3%,6% and 9% with index properties of non-contaminated and hydrocarbon contaminated bentonite clays so as to evaluate whether electrical resistivity can be used as a tool to study the geotechnical properties of areas subjected to hydrocarbon contamination.

2. Material and experimental setup

2.1 soil resistivity box

Soil resistivity is obtained in laboratory by fabricating soil resistivity box (10 cm × 5cm × 5cm) in accordance with ASTM G187. For laboratory test an individual soil sample is compacted in the resistivity box in a predetermined unit weight and moisture content. LCR meter, D.C voltage and current source, can be utilized for the resistivity measurements. To obtain the correlations, soil resistivity tests are conducted using different geotechnical considerations. Using Ohm's law,
Electrical resistance, $R = \Delta V / I$
Electrical resistivity, $\varrho = AR/L = \Delta V A / IL$
Thus electrical resistivity of the soil sample can be obtained using the above equation.

Fig 2.1 Typical Two-Electrode Soil Box (Empty and Full)

2.2 Clay sample and hydrocarbon sample

Source of the Bentonite clay was identified and procured commercially from Baroda, Gujarat. The disturbed soil samples were put in plastic bags, and then transported to the soil laboratory. The natural soil was classified visually and experimentally as clay of high plasticity (CH) according to unified classification soil system (IS classification). For the present study the hydrocarbon contaminated soil was procured from one of the major petroleum unit. The sample was in saturated form mixed with various hydrocarbons.

3. LABORATORY TESTING

3.1 GC-MS (Gas chromatography &Spectroscopy)

To identify the type of hydrocarbon, GC-MS test was performed at Department of Chemistry, Gujarat University and the hydrocarbon was identified as Dodecane ($C_{12}H_{26}$).

3.2 Index properties

The index properties of non-contaminated bentonite clay (with montmorillonite as basic mineral) and periodic (15, 30, 45 and 60 days) hydrocarbon contaminated clay was obtained in accordance with IS code for each percentage of contamination i.e., 3%, 6% and 9%.

Non-contaminated clay			Hydrocarbon contaminated clay											
			3%				6%				9%			
			DAYS OF CONTAMINATION											
Parameters	Symbol	B_{NC}	15	30	45	60	15	30	45	60	15	30	45	60
Liquid limit	LL	335	290	300	325	338	250	285	290	310	275	292	312	327
Plastic limit	PL	113	120	118	115	113	128	125	123	120	122	120	117	115
Shrinkage limit	SL	57	17	19	21	22	38	41	45	47	21	22	25	29
Specific gravity	G	2.05	2.4	2.7	2.8	2.9	2.1	2.3	2.4	2.5	2.2	2.4	2.6	2.7
Free swell index	FSI	60.25	54	51	49	48	52	49	45	44	50	47	42	41

Table1: Index properties of contaminated and non-contaminated bentonite clay

3.3Electrical Resistivity

The electrical resistivity is obtained in accordance with ASTM G187 and the values are obtained for non-contaminated and hydrocarbon contaminated marine clay at the rate of 3%, 6% and 9% using constant DC source having voltage of 35V (provides most desirable and comparable results) for the purpose of setting correlations and evaluation of various geotechnical properties.

Table 2: Electrical resistivity of contaminated and non-contaminated bentonite clay

Non-contaminated clay			Hydrocarbon contaminated clay											
			3%				6%				9%			
			DAYS OF CONTAMINATION											
Parameter	Symbol	B_{NC}	15	30	45	60	15	30	45	60	15	30	45	60
Electrical resistivity	ρ	2.7	3.75	3.45	3.27	3.12	4.28	3.83	3.52	3.31	4.03	3.58	3.31	3.21

Fig 3.1 Measurement of electrical resistivity

4. RESULTS AND DISCUSSION

4.1 Electrical resistivity v/s period of contamination and rate of hydrocarbon contamination

The Electrical resistivity increases for hydrocarbon contaminated bentonite clay by 28%, 40%, and 33% for 3%, 6% and 9% of hydrocarbon contamination respectively (Fig 4.1) at the period of 15 days of contamination. As the period of contamination increases to 60 days, Electrical resistivity decreases by 17%, 23% and 20% for 3%, 6% and 9%.Hydrocarbons are non-polar in nature floats over the water layer around the clay particles and hence they do not participate in ion exchange mechanism. Current is flow of charges and since hydrocarbon is hindering with the flow of charges, therefore results in increased resistivity.

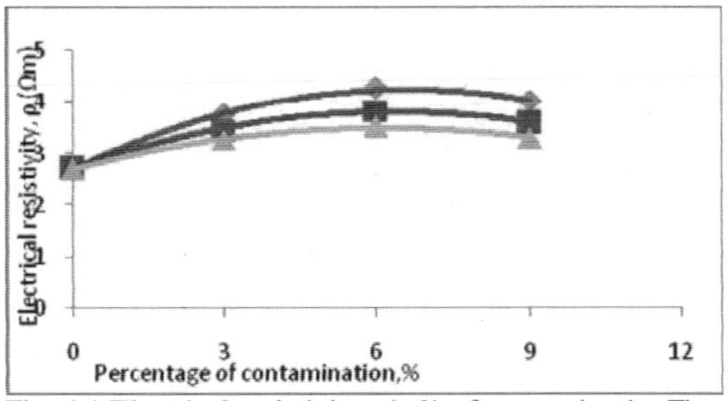

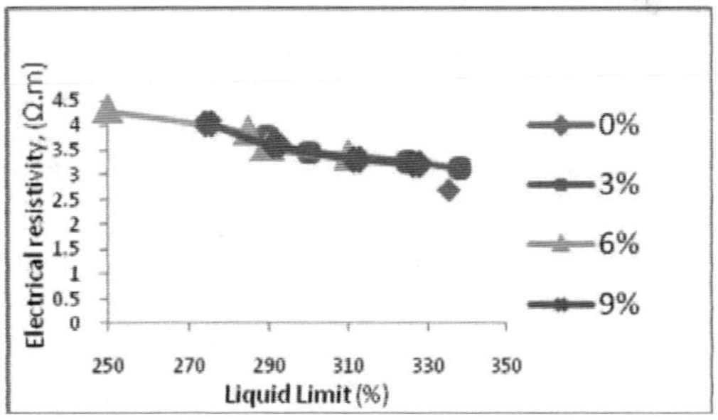

Fig. 4.1 Electrical resistivity v/s % of contaminationFig. 4.3 Electrical resistivity v/s liquid limit

4.2Comparison between Electrical resistivity and liquid limit

Electrical resistivity decreases with increase in liquid limit (fig. 4.2) and LL increases as the period of hydrocarbon contamination increases by 14.2%, 19%, and 16% for 3%, 6% and 9% respectively. LL of non-contaminated clay is higher than hydrocarbon contaminated clay.
Electrical resistivity decreases as the liquid limit increases as higher liquid limit states for higher cation exchange capacity and greater the cation exchange capacity, lower is the resistivity.

4.3Comparison between Electrical resistivity and Plastic limit

Electrical resistivity increases with increase in plastic limit (fig. 4.3) and PL decreases as theperiod of contamination increases. PL of non-contamination clay is lower than that of hydrocarbon contaminated bentonite clay. Presence of hydrocarbon leads to increase in plastic limit. The interaction between bentonite clay and hydrocarbon which has C-H layer over it prevents the fast shrinkage of clay, causing increase in plastic limit.

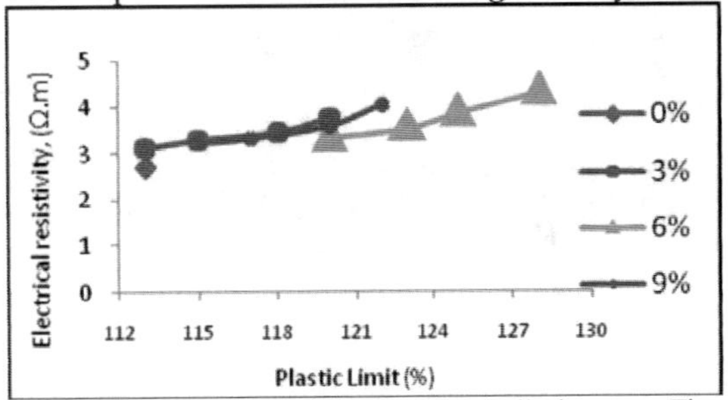

Fig. 4.3 Electrical resistivity V/s Plastic limit

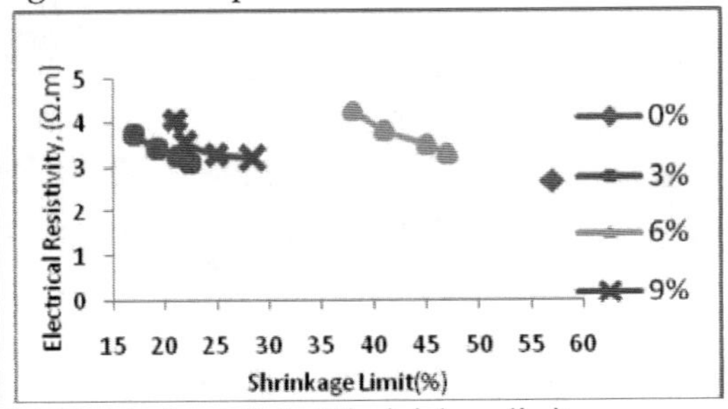

Fig. 4.4 Electrical resistivity V/s shrinkage limit

4.4Comparison between Electrical resistivity and Shrinkage limit

Electrical resistivity increases with increase in shrinkage limit (fig. 4.4) and SL increases as the period of contamination increases. SL of non-contamination clay is higher than that of hydrocarbon contaminated bentonite clay. Electrical resistivity

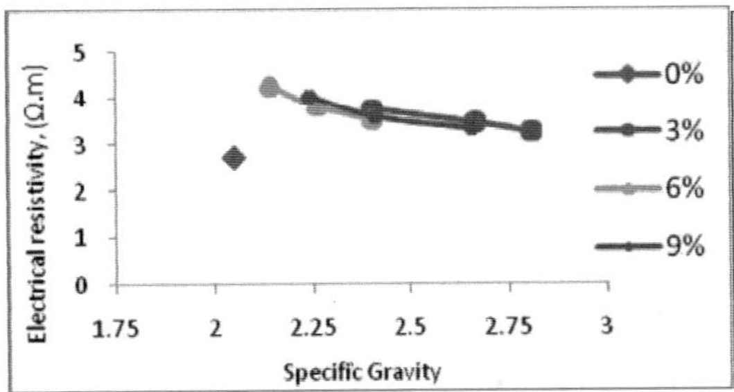

Fig. 4.5 Electrical resistivityV/s specific gravity

% of contamination	CEC (meq./100 gm)_0 days	CEC (meqv/100 gm)_15 (days)	CEC (meqv/100 gm)_60 (days)
3	65.4	30.2	40.5
6	65.4	25.5	45.4
9	65.4	28.4	42.5

Table 3: CEC v/s Percentage of contamination

4.5Comparison between Electrical resistivity and specific gravity

Electrical resistivity decreases with increase in specific gravity (fig. 4.5) and specific gravity increases as the period of contamination increases. Specific gravity of non-contamination clay is lower than that of hydrocarbon contaminated bentonite clay. Presence of hydrocarbon leads to increase in specific gravity.

4.6Comparison between CEC (Cation Exchange capacity) and period of contamination and Electrical resistivity

CEC values for hydrocarbon contaminated clays decreases by 53.8%, 61%, & 56.57% for 3%, 6% and 9% respectively in comparison with non- contaminated clays after 15 days. CEC values decreases by 38%, 30% & 35% for 3%, 6% and 9% respectively after 60 days of contamination. Presence of hydrocarbon thus leads to decrease in CEC values in comparison with non-contaminated clays and as the period of contamination increases, CEC values further increase. Electrical resistivity can thus be correlated with the fact that higher the

CEC values, lower is the value of Electrical resistivity as the ion exchange mechanism reduces the resistivity values.

4.7Comparison between Electrical resistivity and free swelling index

Electrical resistivity increases with increase in free swelling index (fig. 4.6) and free swellingindex increases as the period of contamination increases. Free swelling index of non-contamination clay is higher than that of hydrocarbon contaminated bentonite clay. Presence of hydrocarbon leads to decrease in swelling capacity of expansive clays.The hydrocarbons used are volatile in nature due to

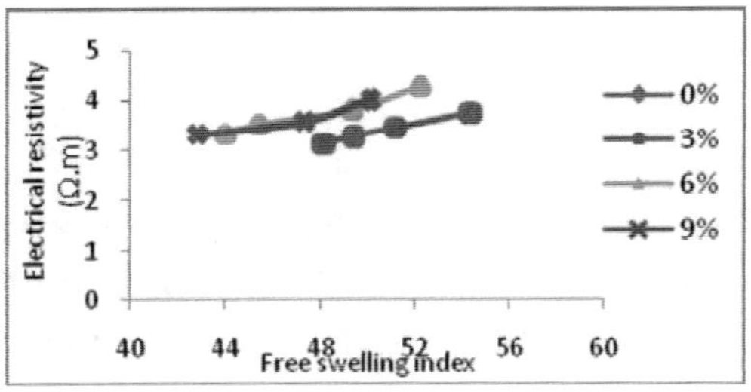

Fig 4.6 Electrical resistivity v/s FSI

which under the influence of these volatile substances, free swelling index reducesandelectrical resistivity increases.

CONCLUSIONS

The above study aimed at correlating index properties of non-contaminated marine clay and hydrocarbon contaminated marine clay with electrical resistivity. The electrical resistivity is found to be directly correlated to liquid limit, plastic limit, shrinkage limit and specific gravity. Also the effects of hydrocarbon contamination on the index properties were evaluated. The hydrocarbon contamination leads to decrease in liquid limit by 2.9%, 13.43%, & 6.8 %, increase in plastic limit by 6%, 11% &7.3% and decrease in shrinkage limit by 70%, 33% and 63% for 3%, 6% and 9% of hydrocarbon contamination respectively. There is an increase in specific gravity by 28%, 18% and 24% for 3%, 6% and 9% of hydrocarbon contamination respectively. There exists a definite correlating pattern of various index properties and Electrical resistivity. Therefore in can be concluded that the electrical resistivity can be used to correlate various index properties of non-contaminated and hydrocarbon contaminated marine clay being a non-destructive tool which can be economical as well as time saving. The above study reveals that there is a major influence of hydrocarbon contamination on engineering behavior of clays. This leads us advance assessment of influence of both percentage contamination and periodic contamination on strength characteristics of such marine clays. India's coastal belt is highly influenced by oil and gas spillages and other volatile toxic and non-toxic intrusion of chemicals demands geotechnical engineer to study the engineering behavior of clays by adapting various types of geophysical methods simulating true field conditions.The actual field application of the present study is that the hydrocarbon contaminated soil samples collected through boreholes through soil samplers at varied depths can be brought to the laboratory and instead of extensive, conventional and destructive soil investigation methods, Electrical resistivity can be used as a tool, values of which can give an idea about soil index and engineering properties. There are numerous types of hydrocarbons, some of which are very complex in nature, whose assessment should be done thoroughly so as to know its long term influence on engineering behavior of clays. Mostly, statistical analysis of such relevant data may not be useful every time unless the soil characterization is done through method like electrical resistivity which replicates soil-water-hydrocarbon interaction. Thus in the present research an attempt is made to know the influence of periodic hydrocarbon contamination particularly on expansive clays using not only conventional analysis but also using physico-chemical analysis (CEC, EDAX) which gives clear insight on various parameters responsible for compressibility and shear strength characteristics which are easy to determine using soil-electrical resistivity inter-relationships developed for various % of contamination.

ACKNOWLEDGEMENTS

The authors are highly thankful to Prof. (Dr) G.P. Vadodariya, Dean, L.D. College of Engineering and Prof. (Dr) S.P. Dave, Head, Applied Mechanics Department, LDCE, Ahmedabad, India for extending their research facilities for this project.

References

[1] W.J Mc Carter,*the electrical resistivity characteristics of compacted clays*,J. Geotech. Engg. Div., Am. Soc. Civil Engg.(1984) 263-267

[2] D.N Singh, Sneha J. Kuriyan,K., ChakravathyMathena, *Relationship between thermal resistivity and electrical resistivity of soil*,Experimental thermal and fluid science, 25(2001) 175-181

[3] Zeyad S. Abu-Hassanein,t Craig H. Benson,2 and Lisa R.Blotz,*Electrical resistivity of compacted clays* J. geotech. engg, **vol. 122**, no.5, may, 1996.122(5)397-406

[4] Tae-Min oh ; Gye-Chun Cho, and Changho Lee,*Effect of Soil Mineralogy and Pore-Water Chemistry on the Electrical Resistivity of Saturated Soils*, J. Geotech. andGeoenviron. Engg, ASCE, ISSN 1090-0241/ 06014012(5)

[5] Abu-Hassanein, Z. (1994). "*Using electrical-resistivity measurements as a quality-control tool for compacted clay liners*," MSc thesis, Univ. ofWisconsin-Madison, Madison, Wis.

[6] Kelly, W. E. "*Electrical Resistivity for Estimating Permeability*," Journal of Geotechnical Engineering Division, ASCE, **Vol. 103**, 1977, pp. 1165–1169.

[7] ASTM–G57–95, "*Standard Test Method for Field Measurement of Soil Resistivity Using the Wenner Four-Electrode Method*," Annual Book of ASTM Standards, **vol. 03.02**, ASTM International, West Conshohocken, PA, 2000, pp. 221–225.

[8] Chaker, V., "*Measuring Soil Resistivity*," ASTM Standardization News, April, 1996, pp. 30–33.

[9] McCarter,W. J., and Desmazes, P., "*Soil Characterization Using Electrical Measurements*," Geotechnique, **vol. 47**, No. 1, 1997, pp. 179–183.

[10] AASHTO. (2005). "*Standard specifications for transportation materials and methods of sampling and testing.*" AASHTO T288-91, Washington, DC.

[11] Okoye, C. N., Cotton, R. R., and O'Meara, D. (1995). "*Application of resistivity cone penetration testing for qualitative delineation of creosote contamination in saturated soils.*" *Geoenvironment 2000*, ASCE, New York, N.Y., 151-160.

[12] Rhoades, J. Raats, P. A. C., and Prather, R. J., "*Effects of Liquid-phase Electrical Conductivity, Water Content and Surface Conductivity on Bulk Soil Electrical Conductivity*," Journal of Soil Science Society of America**, vol. 40**, 1976, pp. 651– 655.

[13] R. J. Kalinski and W. E. Kelly, *"Estimating water content of soils from electrical resistivity,"* *Geotechnical Testing Journal,* **vol. 16**, pp. 323-329, 1993.

[14] S. Sreedeep, A. C. Reshma, and D. N. Singh, *"Generalized relationship for determining soil electrical resistivity from its thermal resistivity,"* *Experimental Thermal and Fluid Science,* **vol. 29**, pp. 217-226, 2005.

[15] Akintorinwa, O. J., &Adesoji, J. I. (2009). "*Application of geophysical and geotechnical investigations in engineering site evaluation. Pacific Journal of Science and Technology, USA, 10*(2), 680-692.

[16] Alhassan H.M. Fagge. S.A., (2013), " *Effects of Crude Oil, Low PointPour Fuel Oil and Vacuum Gas Oil Contamination on the GeotechnicalProperties Sand, Clay and Laterite Soils*", International Journal ofEngineering Research and Applications, **Vol. 3**, January –February 2013, pp .1947-1954.

Comparing kinematically detachable rock masses and rockfall scar volumes

O. Mavrouli and J. Corominas

Department of Geotechnical Engineering and Geosciences, Technical University of Catalonia, BarcelonaTech.

Email: olga-christina.mavrouli@upc.edu, jordi.corominas@upc.edu

Abstract. In rockfall prone areas the evaluation of the risk due to worst case scenarios requires the establishment of maximum thresholds for the expected rockfall volumes. The magnitude of such instabilities is often related to the properties of the jointed rock mass, with the characteristics of the existing unfavorably dipping joint sets playing a major role. The study-site here is the chute of Forat Negre in Andorra. The size distribution of the missing volumes from the scars was calculated using terrestrial laser scanner point cloud data and reaches up to few thousands of m^3. On the other hand, the application of Markland criteria on a Digital Elevation Model of the zone indicated the kinematically detachable rock masses to be up to tens of thousands of m^3. As the size of the scar areas does not indicate the occurrence of such events in the past, the effect of the joint persistence as assumed for the two analyses is discussed here. The areas of the exposed joint surfaces belonging to each discontinuity set are obtained and their use as a measure of the relative persistence of each set is proposed. The average and median length of the sets F3 and F5 (sliding planes) are found to be similar to the average and median spacing of the intersecting set F7 (tension crack), suggesting that the F7 set exerts a control over the persistence of the former ones.

1. Introduction

The rockfall hazard assessment for worst-case scenarios requires the establishment of maximum thresholds for the expected rockfall volumes in an area. The magnitude of such instabilities is related to the properties of the jointed rock mass, with the characteristics of the existing unfavorably dipping joint sets often playing a major role as planes of detachment of the rock mass from the slope face, although not always failures follow the predominant geological structure.

We have been working at a study-site next to the urban area of Santa Coloma, in Andorra, in order to determine the possibility of large rockfall events. Its special interest lies in the intense rockfall activity, posing a threat to buildings, infrastructure and persons at the adjacent built area. Important funds have been invested by the Government of Andorra for the installation of highly dissipative steel fences to retain and decelerate falling blocks. However, a residual risk still remains [1] which is attributed to rock blocks that are not retained by the protective fences, due to excessive energy or bouncing height. The residual risk should additionally include some potential worst case scenarios for rockfalls with a volume of higher order of magnitude than the one commonly observed.

To evaluate the potential of such scenarios and the respective magnitude, we focus on the slope face and the structural characteristics of the jointed rock mass. To this purpose the results from two types of analysis are discussed.

The first analysis [2] aimed at measuring the dimensions of the scars produced by past rockfalls, in order to obtain information on the volume distribution of previous events. The measuring was realized by means of a point cloud obtained with a Terrestrial Laser Scanner, TLS.

The second analysis [3] was performed to detect kinematically detachable rock masses on a Digital Elevation model, DEM, applying the Markland criteria [4] at each cell. The joining of adjacent cells on the DEM, where the same unfavourable discontinuities outcrop, represents unstable zones. The distribution of the potential rockfall volumes in these zones was calculated.

The results from these analyses are compared. Although with the second analysis potential rockfalls of up to some tens of thousands of m^3 are assessed, there are no indications of such events in the slope face. Actually, maximum scar volumes from the first analysis only reach a few thousands of m^3. This suggests the existence of a local factor limiting the mass that can be detached from the slope face.

The characteristics of the discontinuity sets are focused on. The considered sets are those that, intersecting each other, they bound the potentially unstable rock masses and/or form the scar walls. The studied characteristics are the discontinuity persistence and spacing of the different joint sets and the relations between them. Bringing this data into evidence, we discuss the use of the dimensions of the scars and of the kinetically detachable rock masses on the slope surface calculated at [2] and [3] respectively, as an input for the indirect measurement of the joint persistence.

This work forms part of an on-going investigation for the detection of potential large rock masses at the study-site.

Figure 1. The slope above the town of Santa Coloma and the chute of Forat Negre.

2. Study site

The chute of Forat Negre (figure 1) is one of the several active rockfall source areas situated at the slope next to the town of Santa Coloma in Andorra, belonging to the Eastern Pyrenees. It is a steep V shaped couloir formed in the Southern slope of the Enclar peak (2383m), Santa Coloma wall, and it is composed of graniodiorite rock. The couloir extends from 990 m to about 1300 m a.s.l. The rock mass is highly fractured.

The rockfall activity at the Santa Coloma wall is high with an average frequency of 1 event every 2 years. In the last decades (since the 1960^{s}) the maximum recorded rockfall events reached the volume of 150 m^3 (April, 2008) in the chute of Forat Negre and of 1000 m^3 in its neighborhood (Tartera de la Pica, April 1969). The average annual rainfall precipitation is of 1071.9 mm. A direct relation with the rainfall precipitations that might trigger a rockfall has not been established so far. Nevertheless the freeze-thaw weathering process might also play a major role for the rock mass detachment from the slope face.

3. Rockfall scar volumes

The work of [2] aimed at the calculation of the size distribution of past rockfalls, using TLS data, as an alternative to the historical records and reconstructed series. The calculated size distribution represents the volumes of the rockfall masses that have been detached from the slope face in the past leaving a scar bounded by discontinuities. The procedure is based on the assumption that stepped failures sliding over parallel discontinuity surfaces spaced more than 0.2 m did not occur and that each scar on the slope face corresponds at least to an event. This would give an initial estimation of the maximum detached rockfall volume.

A prerequisite for the application of the methodology is the definition of the principal discontinuity sets in the rock mass. Here, the discontinuity sets have been detected following a semi-automatic procedure in which planes are fitted at selected points of the point cloud that fulfil given co-planarity conditions, to form point classes (30 point classes were indentified). The resultant point classes were then merged into 7 main discontinuity sets, based on the scores of a matrix that indicates the percentage of co-existence of these classes on identified discontinuity surfaces both on the point cloud and photos.

Rockfall scars mostly have prismatic shapes, which are formed by the intersection of 4 joint sets. The sets F3 and F5 alternate in the formation of sliding (basal) planes. They are intersected by F1 and F7 which play the role of tension crack (figure 2) although the former may also form rock wedges but less frequently. Table 1 shows the dip direction and dip angle of these sets.

Table 1. Dip direction and dip angle of the discontinuity sets that contribute to the formation of scars.

	Dip direction (°)	Dip angle (°)	Role
F1	54	59	Lateral plane/tension crack
F3	157	56	Basal sliding plane
F5	182	47	Basal sliding plane
F7	141	89	Tension crack

To measure the size distribution of the missing volume from the scars, the points of the point cloud belonging to each one of the four sets were isolated and planes were adjusted to them. These planes correspond to the exposed surfaces of the slope face. Afterwards, their areas were measured as well as their maximum width (along the strike) and length (along the dip direction). The areas of F3 and F5 (basal planes of the scars) were well fitted by a Log-Pearson III distribution. The scar heights are the intersections of the tension cracks F1 and F7. To identify them the distribution of the maximum widths and lengths of F1 and F7 were compared, looking for similarities. The lengths of the F1 and F7 were found to have the same distribution and must correspond to the scar heights. The heights were well fitted by the General Extreme Value distribution. Eventually, the size distribution of the scars was calculated past a Monte Carlo simulation by the multiplication of the scar areas with the scar heights.

The results are shown in figure 3.The maximum scar volumes that have been identified using this method are of the order of some thousands of m^3.

Figure 2. The intersection of the discontinuity sets F3 and F5 (sliding planes) with F5 and F7.

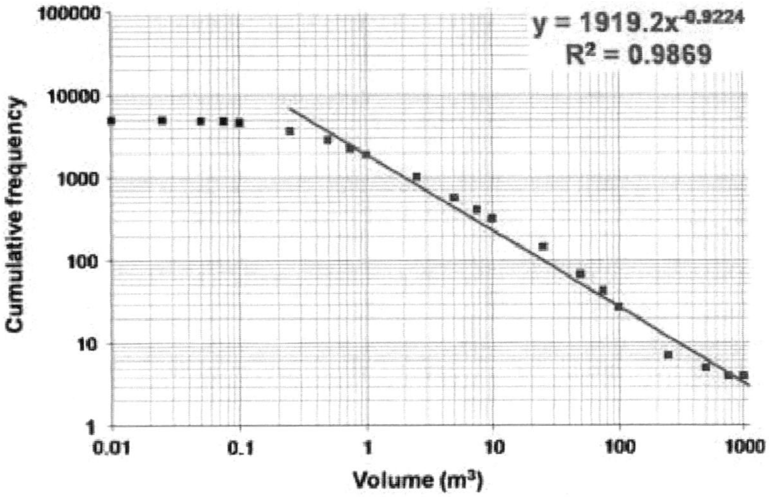

Figure 3. Size distribution of the missing volume from the scars [2].

For the evaluation of the results the fundamental assumption of this procedure should be taken into consideration which is that every scar corresponds, at least, to a detachment event from the rock mass. Hence these results may be interpreted as giving a first estimation of the maximum rockfall volumes. However, this excludes the possibility of larger volumes involving several (stepped) basal surfaces, which cannot be fully disregarded with the available information.

4. Kinematically detachable rock masses

Alternatively, for the evaluation of the size distribution of the kinematically detachable rock masses a Digital Elevation Model DEM derived from a topographical map at 1:5,000 scale was used [3]. The DEM at its raster format has a 1m x1 m cell and covers the entire V zone of the chute of Forat Negre.

The detection of potentially unstable volumes consists in identifying on the DEM large kinematically detachable rock masses, by checking compliance with the Markland criteria [4] at every cell of it. In particular, the discontinuity sets F3 and F5, matching the basal planes of the previous paragraph, were checked against two criteria indicating instability: (i) slope angle>joint dip and (i) slope orientation=dip direction±20°. The cells that fulfilled these conditions were marked on the DEM and adjacent cells were joined to form wider areas A, indicating the zones where a rockfall is possible due to the presence of these unfavourably dipping discontinuities (figure 4).

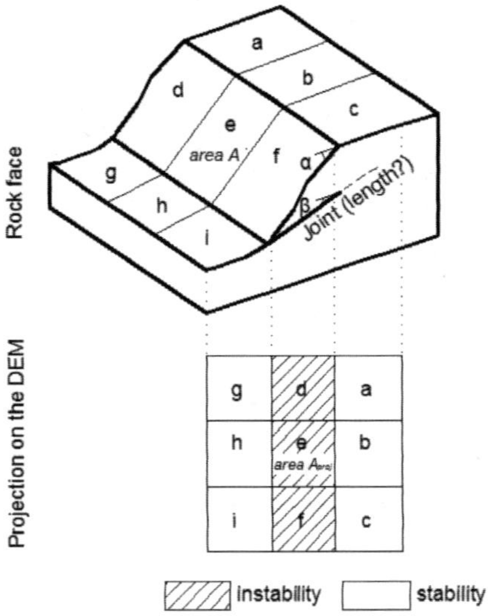

Figure 4. Check of DEM raster cells against compliance with Markland instability criteria and joining of adjacent cells to form wider unstable zones [3]

These zones were afterwards superposed on orthophotos to visually verify the zones of the kinematically detachable rock masses and, also, to delineate smaller masses inside the big ones corresponding to smaller detachments. The most extensive zones can be seen in figure 5. The calculation of the volume of the detachable masses was then simplified, assuming either cubic or prismatic shape of the detachable rock masses, with one side equal to the surface of the area A. For the equivalent prismatic volumes the length L of the joint is equal to half of the height of the cubic volumes. The volumes for cubic and prismatic shape were calculated as $V=A^{3/2}$ and $V=0.5*A^{3/2}$, accordingly. The procedure was repeated separately for F3 and F5. Figure 6 shows the size distribution of the volumes summing up the results for both sets.

For the application of this procedure the presence of the discontinuities F3 or F5 at every cell of the DEM is a prerequisite. Infinite persistence of the joints is implied.

The maximum volumes here are of the order of 50,000 and 25,000 m3 for cubic and prismatic volumes accordingly. The largest basal area was estimated at 1,361 m^2.

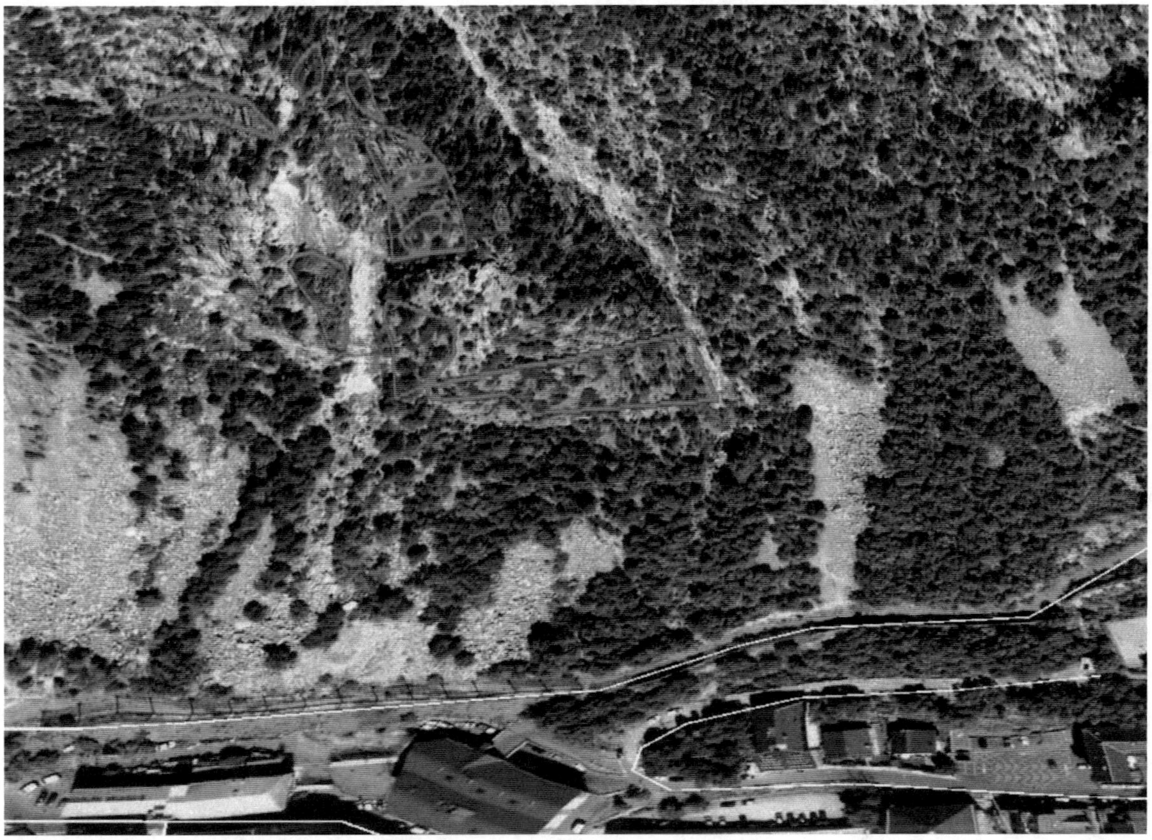

Figure 5. Indicative zones where a rockfall is possible given the presence of continuous unfavourably dipping discontinuities superposed on the orthophoto of the study-site.

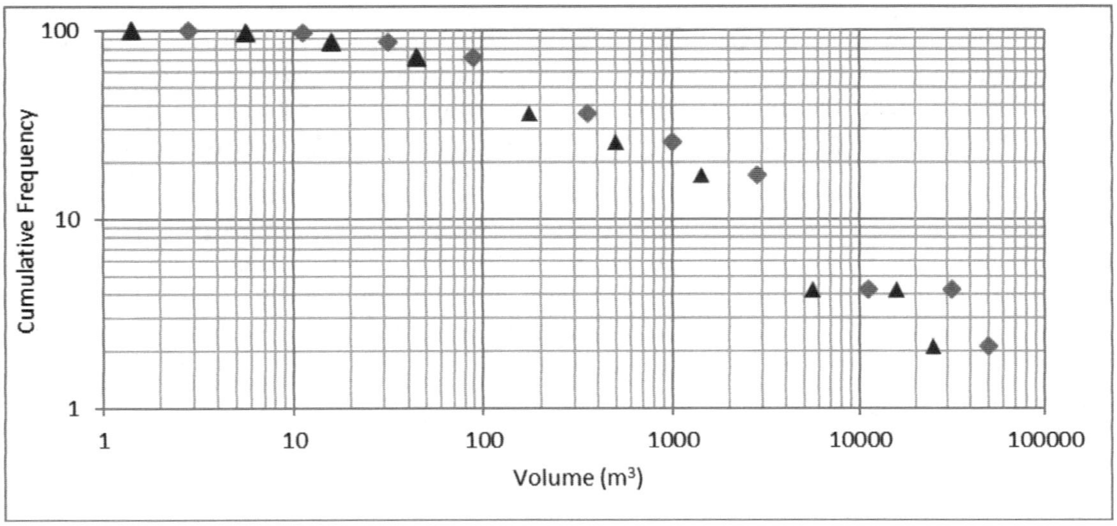

Figure 6. Size distribution of the kinematically detachable rock masses, for cubic (rhombus) and prismatics (triangles) volumes.

5. Comparison of scar volumes with kinematically detachable rock masses

5.1. Conceptual and methodological differences of the two analyses

The calculated maximum volumes based on one hand on the missing scar volumes (of some hundreds up to few thousands of m^3) and on the other on the kinematically detachable rock masses (of some tens thousands of m^3) present differences of one order of magnitude, being smaller in the first case. This is mainly attributed to the differences of the basic assumptions of the two methodologies.

The calculation of the scar volumes assumes the part of the discontinuities that after a rockfall event remained exposed on the face of the slope. Thus the extent of the discontinuity surface that is implicated in the detachment of the rock mass from the slope face is determined by the scar edges and is limited. In contrast, for the detection of the kinematically detachable rock masses, almost infinite persistence of the discontinuities is assumed. Therefore, the extent of the discontinuity surface that is involved in the detachment due to simultaneous or gradual detachment of blocks is larger (figure 7). Hence the formation of large sliding planes up to 1361 m^2 is permitted wherever the instability criteria are fulfilled. Nevertheless such large-extent sliding planes are not observed in situ.

In the light of this evidence the persistence and spacing of the discontinuities are investigated, in order to interpret the limitation in the maximum volume observed in field.

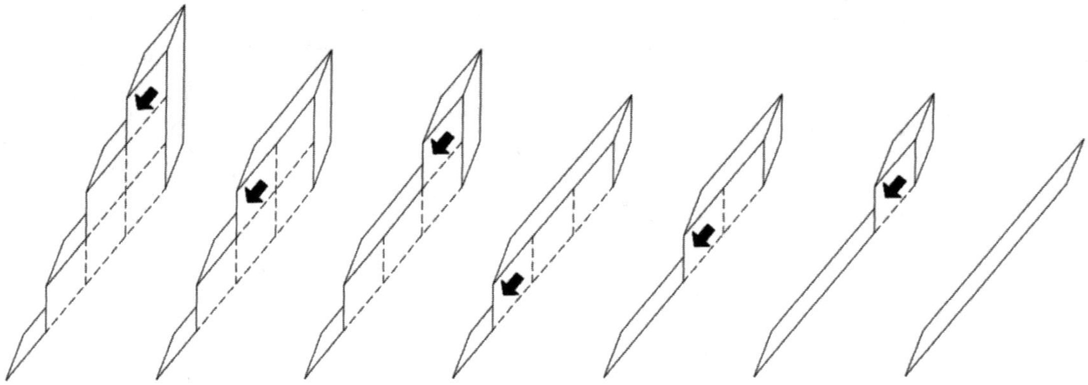

Figure 7. Formation of large sliding planes through the gradual detachment of blocks along the existing discontinuities (the discontinuities inside the rock mass are drawn with dashed line). The simultaneous detachment of all blocks (large unique failure) would give the same result as well.

5.2. Effect of persistence and spacing on exposed surface areas and rockfall volumes

To measure the spacing of the discontinuity sets, the planes defined at [2] as the exposed discontinuity surfaces on the slope face, were used. The spacing distributions were calculated manually measuring the perpendicular distances between successive planes using the software Rhinoceros (254, 34, 104 and 162 measurements for F1, F3, F5 and F7, respectively). These distributions are seen in figure 8. It can be observed that the joint sets are highly present, with spacings of few meters. It is worth noting that in the detachment of the rock masses, F7 that mostly plays the role of the tension crack, has larger spacings than F3 and F5 (basal planes). The obtained average and median spacings are shown in table 2.

Following the same procedure as described at section 3, the length (along the dip) of the scar edges was also calculated for each set. It was calculated automatically as the maximum edge distance along the dip of a plane. Table 2 summarizes the results for the maximum areas, average and median lengths.

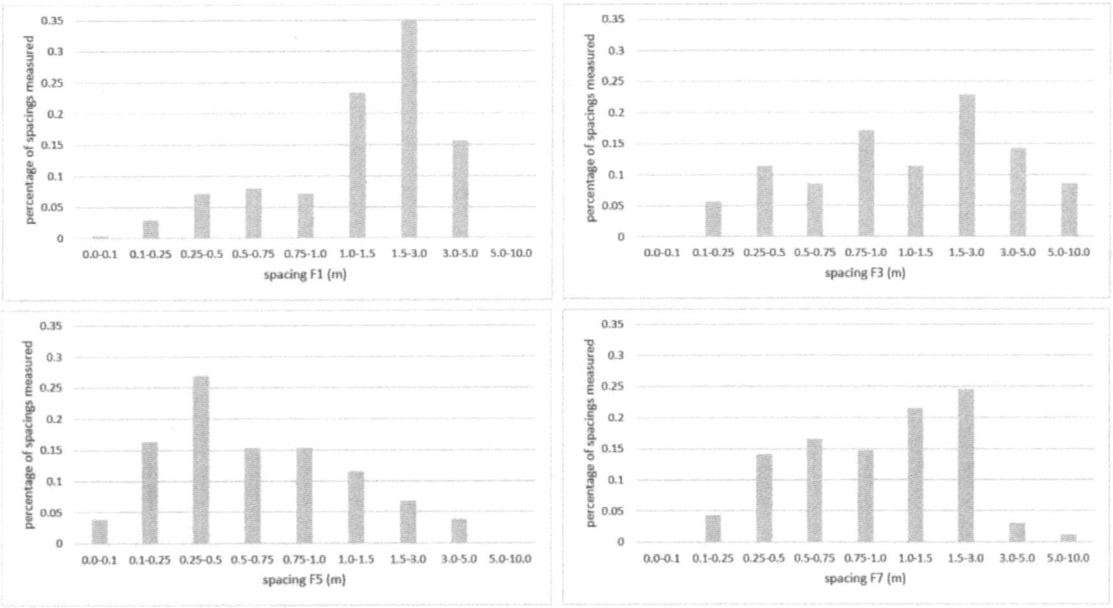

Figure 8. Spacing distributions for sets F1, F3, F5 and F7.

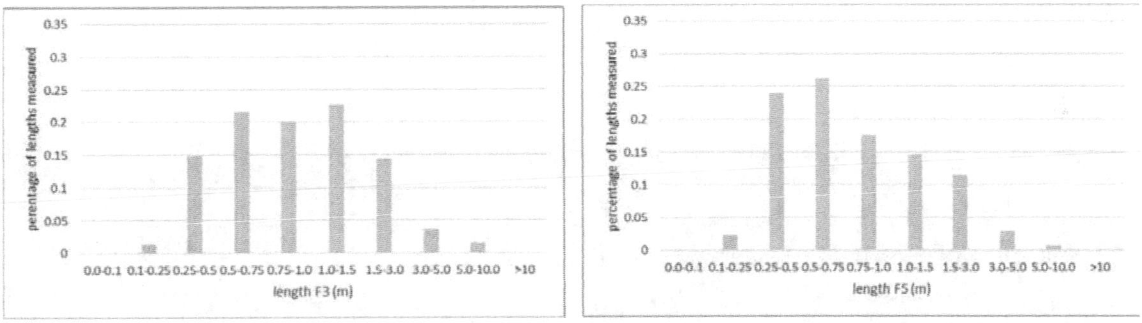

Figure 9. Length distributions for sets F3 and F5.

Table 2. Measured areas, lengths and spacings of the discontinuity sets.

	F1	F3	F5	F7
Max spacing (m)	8.09	6.35	4.11	5.60
Average spacing (m)	2.11	1.84	0.76	1.22
Median spacing (m)	1.63	1.22	0.53	1.00
Max area (m^2)	121	213	144	236
Max length (m)	19.85	19.14	14.65	27.08
Average length (m)	0.96	1.19	0.99	1.45
Median length (m)	0.76	0.89	0.73	1.04

The maximum areas of the exposed surfaces on the scars are those of F7 and F3, of 236 and 213 m^2 respectively. They are followed by the areas of F5 and F1 with 144 and 121 m^2 accordingly. This sorting is consistent with the observations in situ on the exposed surfaces per discontinuity set.

If the sets F3 or F5 had very high to infinite persistence some large basal planes as those delineated in figure 7 would exist. Nevertheless the size of the observed basal planes of the scars is limited. This subsequently implies a limitation on the persistence of discontinuities F3 and F5 due to their intersection with F7 that restricts the detachment of large rockfalls. Additionally, the average and median lengths of F3 and F5 (table 2 and figure 9) are of the same order of magnitude as the spacing of F7. These observations suggest that F7 exerts a control over the length of F3 and F5 and in consequence over the expected rockfall volumes, probably disrupting them as depicted at figure 10.

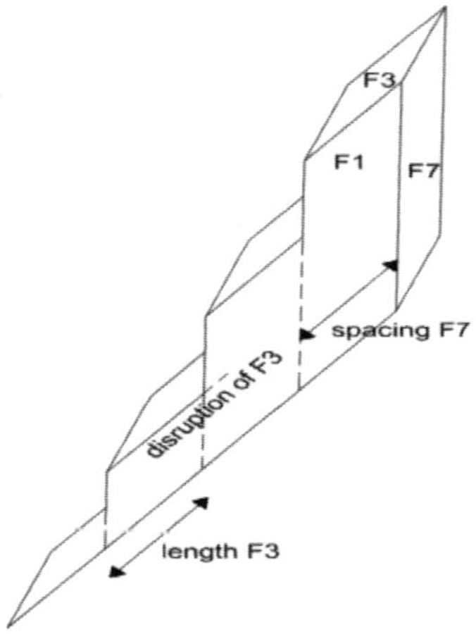

Figure 10. Similar spacing of F3 and F7 indicate that the former is disrupted by the latter.

The maximum measured length of F3 and F5 (19.14 m and 14.65 m respectively) is one order of magnitude higher than the spacing of F7. To interpret the formation of basal planes with that length, the breakage of small rock bridges of intact rock connecting disrupted discontinuities is assumed.

Figure 11. In the yellow circle large exposed surfaces of the discontinuity set F7.

To check the afore-mentioned interpretations, rockfall failures have been observed in situ. The areas belonging to the F7 set are predominant (figure 11). These areas of up to few thousands of m^2, correspond to the largest calculated scar areas, and likely have been formed through the mechanism described in Figure 7. Figure 12 additionally shows a scar where the length of F3 is longer than the spacing of F7. In accordance with the interpretation made here, this scar has been produced as a stepped-path failure with the rock mass sliding over parallel discontinuity surfaces with a maximum distance of 20 cm between them. As verified on the photo this failure involves locally the breakage of small bridges.

Figure 12. A scar with a large basal plane (F3) which is marked by red colour, might be produced as a stepped-path failure after the breakage of small rock bridges, marked with blue colour. The local breakage of the bridges is presented schematically in the lower right hand corner.

6. Conclusions

The results of two analyses for the determination of rockfall volumes are discussed. The first analysis [2] provided the missing volumes corresponding to the rockfall scars and the second one [3] the kinematically detachable rockfall masses according to Markland instability criteria. The obtained results indicate a difference of one order of magnitude between the two maximum calculated volumes, which are greater in the second case. The difference is attributed to the basic assumptions of the two procedures: oppositely to the first analysis, the second one assumes joints of very large to almost infinite persistence.

Based on the geometry of the rockfall scars, the calculation of the area of the exposed surfaces has been possible for each discontinuity set at the study site, the chute of Forat Negre in

Andorra. The discontinuity set F7, acting as tension crack, forms the maximum areas on the slope face (236 m^2), followed by the maximum areas of F3 (213 m^2) that is the principal sliding basal plane.

As in situ observations do not indicate rockfall failures of the order of the calculated kinematically detachable rockfall masses, the persistence of the discontinuity sets was investigated and indicated to have an effect on it. The smaller size of the observed basal planes was interpreted as a limitation in the persistence of discontinuities F3 and F5, when they intersect with F7. The similar average and median lengths of F3 and F5 with the spacings of F7 (1 m approximately) suggest that F7 exerts a control over the length of F3 and F5 and in consequence over the expected rockfall volumes, probably displacing joints belonging to F3 and F5.

In the scars where the maximum length of F3 and F5 (19.14 m and 14.65 m respectively) is higher than the spacing of F7, the formation of the basal planes can be explained by the breakage of small intact rock bridges.

Acknowledgments
This work has been realized with the support of the fellowship "Ayudas Fundación BBVA a Investigadores, Innovadores y Creadores Culturales" by the Foundation BBVA to the first author and within the framework of the research project RockRisk financed by the Spanish Ministry of Economy and Competitiveness (BIA2013-42582-P) and by the Government of Andorra (Edicte de 10/04/2013, BOPA n°18 17/04/2014).

References

[1] Corominas J, Copons R, Moya J, Vilaplana JM, Altimir J and Amigó J 2005 Quantitative assessment of the residual risk in a rockfall protected area *Landslides* 2 (4) pp 43-357

[2] Santana D, Corominas J, Mavrouli O and Garcia-Selles D 2012 Magnitude-Frequency relation for rockfalls using a Terrestrial Laser Scanner *Engineering Geology* 145–146 pp 50-64

[3] Mavrouli O, Corominas J, Jaboyedoff M (2014) Size distribution for potentially unstable rock masses and in-situ rock blocks using LIDAR generated Digital Elevation Models *Rock Mechanics and Rock Engineering* DOI 10.1007/s00603-014-0647-0

[4] Markland JT 1972 A useful technique for estimating the stability of rock slopes when the rigid wedge slide type of failure is expected *Imperial College Rock Mechanics Research Reprints* 19 pp 10

Application of geotechnical and geophysical field measurements in an active alpine environment

D R Lucas[1], K Fankhauser[1] and S M Springman[1]

[1]ETH Zürich, Institute for Geotechnical Engineering, Wolfgang-Pauli-Str. 15, 8093, Zürich, Switzerland

daisy.lucas@igt.baug.ethz.ch, kerstin.fankhauser@igt.baug.ethz.ch, sarah.springman@igt.baug.ethz.ch

Abstract. Rainfall can trigger landslides, rockfalls and debris flow events. When rainfall infiltrates into the soil, the suction (if there is any) is reduced, until positive water pressure can be developed, decreasing the effective stresses and leading to a potential failure. A challenging site for the study of mass movement is the Meretschibach catchment, a location in the Swiss Alps in the vicinity of Agarn, Canton of Valais. To study the effect of rainfall on slope stabilities, the soil characterization provides valuable insight on soil properties, necessary to establish a realistic ground model. This model, together with an effective long term-field monitoring, deliver the essential information and boundary conditions for predicting and validating rainfall-induced slope instabilities using numerical and physical modelling. Geotechnical monitoring, including soil temperature and volumetric water content measurements, has been performed on the study site together with geophysical measurements (ERT) to study the effect of rainfall on the (potential) triggering of landslides on a scree slope composed of a surficial layer of gravelly soil. These techniques were combined to provide information on the soil characteristics and depth to the bedrock. Seasonal changes of precipitation and temperature were reflected in corresponding trends in all measurements. A comparison of volumetric water content records was obtained from decagons, time domain reflectometry (TDR) and electrical resistivity tomography (ERT) conducted throughout the spring and summer months of 2014, yielding a reasonable agreement.

1. Introduction

The Meretschibach catchment (figure 1), situated in the Swiss Alps, within the Canton of Valais, provides an environment in which different kinds of mass movements occur more or less frequently where the slope of inclination is steeper than the angle of repose. Furthermore, this complex field environment consists of a type of soil that is challenging for monitoring and characterisation. Shallow landslides can be triggered within this catchment by a decrease of suction due to rainfall, followed by an increase in pore water pressure with a consequent reduction in effective stress within the gravelly soil. Triggering of a mass movement event can generate debris flows, which have the potential to reach the village of Agarn and cause significant damage to infrastructure, as well as loss of life [1]. Some of these events have been documented [2], but little has been published in the case of slope instabilities triggered in gravelly soils. The project at Meretschibach involves the soil characterisation, including the determination of soil properties at different depths, thickness of soil (depth of bedrock) and the establishing of a realistic ground model. This characterisation forms the base of an effective field monitoring to observe and measure the response of the scree slope to rain infiltration and saturation processes. A long-term monitoring campaign, integrating geotechnical and geophysical methods together with laboratory analyses, allows a progressively more detailed calibration and validation of the ground model. Ultimately a collection of all acquired data is used for numerical and physical modelling to study the mechanism of failure due to rainfall [3], hence enabling a hazard assessment to be made and the implementation of suitable safety measures, along with early warning systems.

Instrumentation is placed at shallow depths within three geotechnical trenches (IT1-3 at elevations of 1868-1914 m.a.s.l., figure 1) in one of the steep scree slopes in the catchment, to perform real-time monitoring of the volumetric water content (VWC), temperature and suction [4, 5]. Electrical resistivity tomography (ERT) measurements were conducted repeatedly throughout the summer months of 2014, on a profile located closely to IT1, to complement the geotechnical data [6], allowing the VWC recordings to be validated in specific locations. Furthermore, the precipitation in the area is measured at two weather stations located within 1 km, at an elevation of 2220 m.a.s.l., and of 1370 m.a.s.l., respectively.

The data acquired in two of the instrumented trenches (IT1 and IT3) are analysed separately and compared to the corresponding data in the other trench. Trends in temperature, as well as VWC in the scree slope's gravelly soil, correspond well to expected seasonal changes. Peaks in VWC can be related to intense rainfall events. Unanticipated variations and deviations can be explained by isolated rockfall events and freezing/thawing cycles. A good correlation between the peaks in various measurements is observed for the times when debris flows occurred in the Meretschibach in July 2014.

A comparison between the VWC measurements performed by time domain reflectometry (TDR) and decagon sensors within each trench is made, and the degree of saturation is calculated. Furthermore, the VWC obtained from the two geotechnical instruments is compared to the VWC estimates obtained from the ERT acquisitions using Archie's law [7, 8]. All trends and values agree well with each other, proving that the three different methods are consistent. The VWC and the suction measurements, provide valuable information to be able to assess changes in soil resistance, particularly in a partially saturated state.

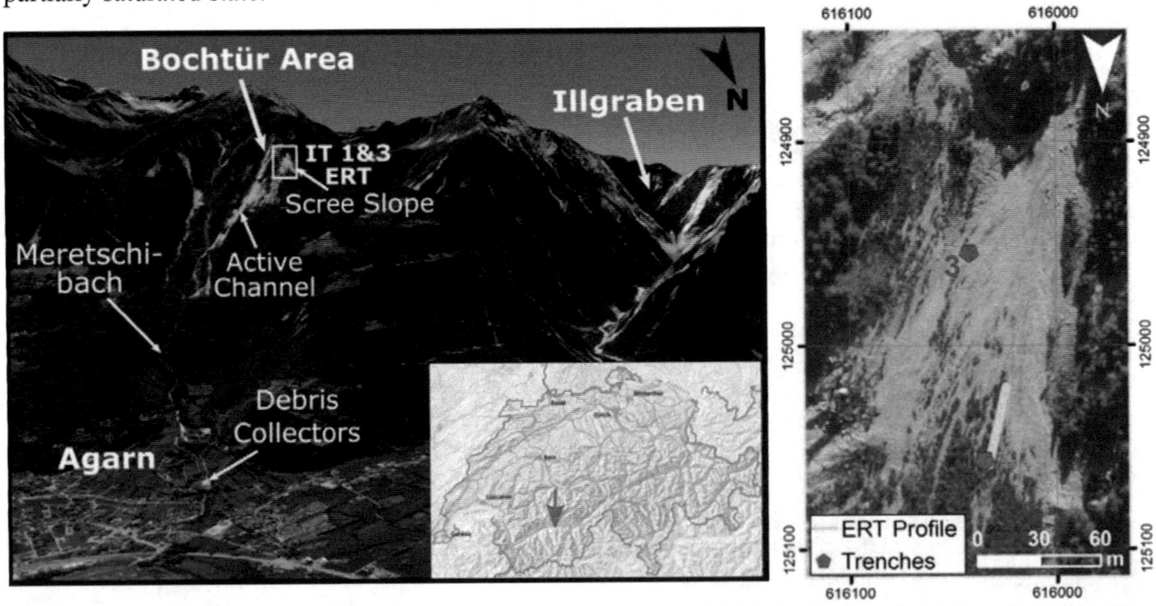

Figure 1. Overview of the field area, located in the canton of Valais, Switzerland (small map). The left image looking to the south shows a view of Agarn, situated on the Rhone valley floor and the Meretschibach catchment on the mountain slopes behind. The most active area within the catchment, the Bochtür area, can be divided into an active channel and a scree slope. The right image shows an enlargement of the scree slope (yellow rectangle) and the locations of the instrumented trenches (IT1 and IT3) are indicated along with the ERT profile on which the monitoring was performed.

2. Soil and ground characteristics

2.1. Soil classification

Figure 1 shows the location of the research site and specifically the location of instrumented trenches 1 and 3 (IT1 – IT3). Samples were taken from three test pits TP1-3, near IT1 to perform grain size analysis. The work was carried out during dry summer weather conditions. The following characteristics were noted: the ground surface was considered to be 'depth zero' for each test pit, at least 4 kg of soil were extracted for each disturbed sample, some cobbles larger than 70 mm and any boulders were left at the site, and roots were found in almost all of the samples.

The soil is classified (figures 2, 3 and table 1) according to USCS and Swiss standard classification SN 670 004-2NA, as gravel with silt and sand (GM). Figure 2, shows the soil profiles at the trenches to a depth of 38 cm. IT1 on the left, consists of a layer of approximately 30 cm thickness (A and B, above the orange dashed line) of gravel with more fines, overlying coarser gravel. This upper layer can further be divided into an upper (A) and a lower (B) part with more and less roots respectively. At IT3, the top layer consists of finer gravel (D) and a transition to slightly coarser gravel (E) can be inferred.

Figure 2. Soil at instrumented trenches IT1 (left) and IT3 (right). The following description is based on field observations. IT1: A (~0-12 cm), gravel with silt, sand and roots; B (~12-25 cm), gravel with silt and sand; C (below ~25-30 cm), coarser gravelly soil with less fines. IT3: D (~0-25 cm), gravelly soil with little content of sand and silt; E (below ~25 cm) change to a coarser gravel.

Table 1. Soil classification from test pits.

Test Pit	USCS Classification	Moisture content %	Percent of fines %	C_u [-]	C_c [-]	G_s [-]
TP1	GM	2.9-3.7	5.7-8.2	73.6-139.3	8.3-14.4	2.68
TP2	GM	3.0-5.4	5.5-10.8	33.3-105.1	7.9-11.2	
TP3	GM	3.7-4.3	6.7-16.4	72.8	9.1	

C_u: Coefficient of uniformity, C_c: Coefficient of curvature; G_s: Specific gravity

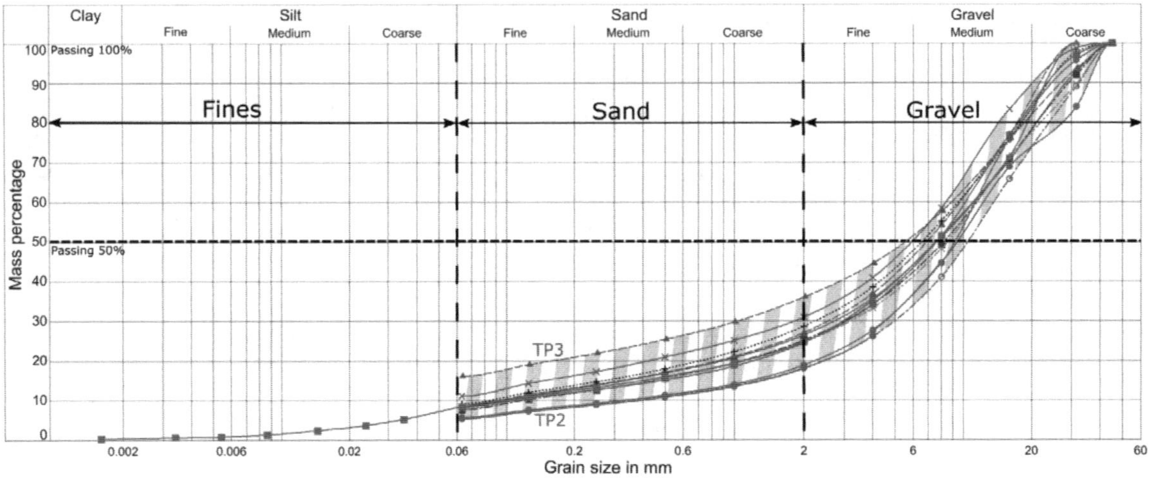

Figure 3. Grain size distribution: samples from Meretschibach scree slope test pits 1-3. A range is marked in grey to illustrate the characteristics and content of fines for the entire set of samples. A sample in green from test pit 3 (TP3) at a depth of 30-40 cm and a sample in red from test pit 2 (TP2) at a depth of 20-40 cm set the upper and lower limit respectively.

3. Monitoring methodology

3.1. Instrumented trench

Selected spots on the slope(s) (i.e. landslide source areas) are monitored *in situ* to measure water infiltration-suction characteristics in the ground, to provide time-series data about infiltration quantities and rates, and the effect on the inter-particle suction, which directly affects the *in situ* effective stress. Suction is measured with tensiometers, providing complementary data to calculate effective stress and hence to assess changing shear stress conditions.

Two vertical monitoring profiles were installed (IT1-IT3), consisting of sets of co-located tensiometers, time domain reflectometry (TDR) soil moisture, decagons and temperature sensors. Freezing/thawing conditions, particularly during the crucial spring snowmelt period, can be identified through measurements of temperature, while the TDRs are calibrated to measure the volumetric water content. An SR50A Campbell weather station was installed and monitored within the study area (precipitation, ambient air temperature, humidity) by the Swiss Federal Institute for Forests, Snow and Landscape research, (WSL), and a Pluvio OTT[2] (solid and liquid precipitation) was installed by ETH to measure total precipitation.

3.1.2. Sensors & calibration

The sensors used in this monitoring set up include: tensiometers, decagons, time domain reflectometry (TDR) and temperature devices. These have been described and used in previous landslide monitoring programmes [3]; technical details are given in table 2.TDR and decagon sensors measure the volumetric water content of the soil in the field and have been used in the past [3, 9], to measure the VWC in unsaturated silty sand, and alpine moraine with good results. The performance of both sensors will be evaluated and compared in the field for application in gravelly soils. The decagons and TDRs were calibrated using the gravelly soil from the field, over the range of volumetric water contents estimated for unsaturated conditions in gravel. However, when comparing measurements between these two types of sensors the cylinder of influence as well as sensitivity to temperature changes should be taken into account.

Table 2. Technical information about the sensors installed.

Sensor	Name	Brand	Length	Measuring range
Decagon	EC-5	Decagon devices	0.089 m	Apparent dielectric permittivity ξ_a 1-50.
TDR	TDR100	Campbell Scientific	0.15 m	Apparent dielectric constant $Ka^{0.5}$.
Tensiometers		Keller AG Company	0.15, 0.30, 0.45 m	80-85 centibar.
Temperature sensors	RTD PT100	Heraeus		-50 to 100 °C.

3.2. ERT

A 2D electrical resistivity survey was carried out on a fixed profile line next to the instrumented trench 1 (figure 1) to complement the TDR and decagon point measurements, achieving a more widespread characterisation of the subsurface. The basic principle of geoelectrical resistivity measurements involves the injection of current into the ground through a pair of electrodes and measuring the resulting potential in the form of a voltage difference at another pair of electrodes along the subsurface current flow. 48 electrodes were planted along the profile line and the Syscal Pro system (IRIS instruments) was used to conduct six ERT measurements (three of which are presented here) in a monitoring phase from May to July, 2014. A Wenner configuration was used for all acquisitions due to the highly resistive environment. The inversion was performed using the BERT (Boundless Electrical Resistivity Tomography) software package [6, 10, 11, 12, 13].

3.3. Archie's law

ERT measurements are highly susceptible to changes in the subsurface water content/saturation as the current injected during a geoelectrical survey flows mostly through the pore water. An empirical relationship between the resistivity of a porous medium to the amount of pores, their connectivity and saturation, as well as the resistivity of the pore filling fluids, was found by Archie [7]:

$$\rho = a\rho_w \Phi^{-m} S^{-n} \tag{1}$$

where ρ is the resistivity of the porous soil, ρ_w is the resistivity of the pore filling fluid, Φ is the soil's porosity and S its saturation which can also be expressed as the ratio between the water content θ and the porosity (eq. 2). Furthermore, a is the tortuosity factor set to 1 for granular soils, m is the cementation factor and n is the coefficient of saturation. The latter two factors are taken from literature (e.g. [7, 14, 15]) and are assumed to be constant: m =1.3 and n = 2.

$$S = \frac{\theta}{\Phi} \tag{2}$$

Archie's law is considered to be valid in this case, as the soil consists mainly of gravel and sand and contains almost no fine material, especially clay (figures 2 and 3 and table 1) and has enabled effective comparisons between geotechnical soil moisture measurements and ERT acquisitions (e.g. [16, 17, 18]). The temporal variation of soil resistivities along the repeatedly measured profile can therefore be related to the temporal variation of soil saturation or water content (under the assumption that the porosity and the pore water resistivity remains constant) as:

$$\frac{\rho_t^{-\frac{1}{n}} - \rho_{t_0}^{-\frac{1}{n}}}{\rho_{t_0}^{-\frac{1}{n}}} = \left[\frac{\rho_t}{\rho_{t_0}}\right]^{-\frac{1}{n}} - 1 = \frac{S_t - S_{t_0}}{S_{t_0}} = \frac{\Delta S}{S_{t_0}} = \frac{\Delta \theta}{\theta_{t_0}} \qquad (3)$$

The profile measured on May 15, 2014 was chosen as the baseline model (at time t_0) and variations in saturation in later acquisitions are relative to this profile. The advantage of this approach is the fact, that none of the constants (except for n) nor the porosity or the resistivity of the pore water need to be estimated to see a seasonal variation in soil saturation. The resulting trend can relatively be compared to the readings obtained from TDR and decagon measurements.

To achieve a more detailed comparison however, the parameters in Archie's law have to be estimated from soil and water samples extracted from the field and analysed through lab and field experiments. The values of the key parameters (Φ and ρ_w) have been estimated at this stage of the project on the basis of preliminary tests and the literature. These can be refined to improve the calculated soil saturation (and VWC) for comparison with field measurements in the trenches.

4. Results and discussion

4.1. Instrumented trench IT1

The data recorded from the end of October 2013 to April 2015 are shown in figure 4, representing variations from four seasons with data from two winter's. The change of volumetric water content at 25 cm depth shows a general agreement between the measurements from the TDRs and decagons for the first eight months. The decagon readings differ from the TDR recordings after the rockfall event (beginning of August), but it is not possible to affirm that the impact of the rockfall affected the results at this depth. Values decrease to a minimum VWC during winter time (end of December to end of March), probably due to the accumulation of snow, leading to a low or non-existent rate of rain infiltration. This situation changes from April 2014 (spring season) through snow melt and rainfall, and continues to fluctuate due to varying seasonal precipitation to the end of 2014. Changes in VWC occur following to rain peaks. There is a gap from October 12 to November 10, 2014, with no precipitation information available at this time. However, it is still possible to see the variations in VWC, most probably due to precipitation.

A comparison of the temperatures in the soil at 15 cm depth and in the data logger, which was considered to be the local exterior temperature on the soil's surface, was also plotted in figure 4. Temperatures are around 0°C, with minimum variation, for both sensors during the winter season, and this trend continues until April 2014. The temperatures increase again in May (spring season), and stay mainly between 10 – 15°C until August 2014 (summer season). They decrease again during November 2014 (autumn) and remain near zero degrees, illustrating an annual temperature curve with a range from approximately -8 to +26 °C.

A rockfall happened at the beginning of August 2014 (figure 4) during an extended period of rain, with a debris flow observed (dashed brown line) at the end of July. At least one large boulder hit the trench breaking the stake and exposing the soil temperature sensor, causing some instability in the readings. An offset between the decagon and TDR readings was also observed. The boulder was removed and the sensor placed in the trench again on August 28, 2014, after which the temperature sensors (but not the decagons and TDRs) were recording similarly again. The VWC increased during January 2015, as water melted in the voids, when temperatures changed from a period of negative temperatures to above 0 °C.

Figure 5 shows the variation of pore water pressure with precipitation (SR50A weather station), recorded by the tensiometers, which were installed in June 2014 at depths of 15, 30 and 45 cm approximately three meters uphill of IT1. The correlation between the loss in suction after rainfall events can clearly be observed, in particular after all tensiometers show a significant increase in pore water pressure during the period of higher frequency in precipitation (July 2014). In addition, it should be noted, that the shallower tensiometer reacts faster the deeper tensiometers after a rainfall event, according to the direction of infiltration with depth.

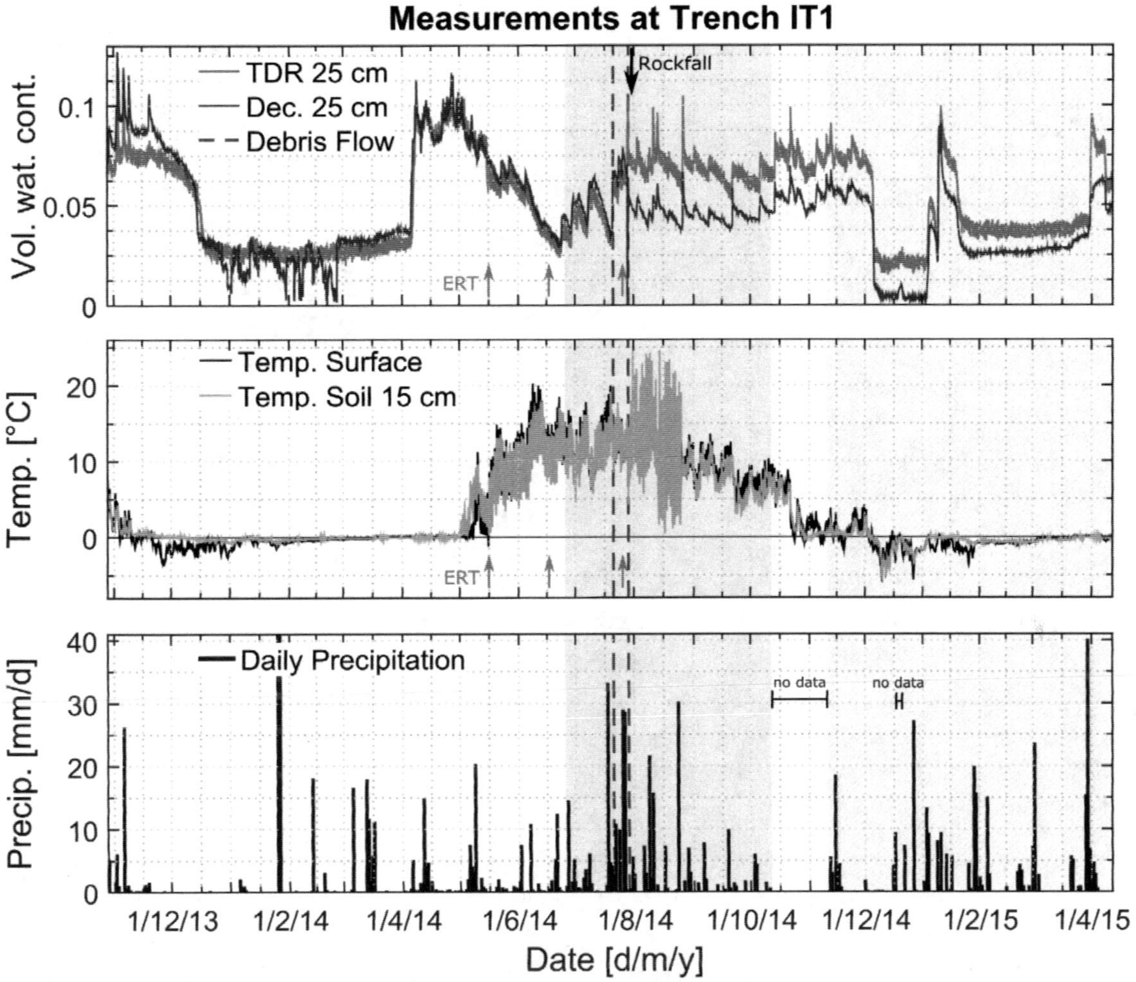

Figure 4. Measurements at IT1. From top to bottom: volumetric water content at 25 cm depth measured with TDR and decagons, temperature at 15 cm depth and in the data logger at the surface and daily total precipitation (28/10/2013-11/04/2015). Precipitation data were not recorded or not available during two periods, indicated in the bottom figure. A rockfall happened at the beginning of August (labelled black arrow), right after the second debris flow event (dashed brown lines indicate the two largest such events). Also indicated are the three ERT data acquisitions (magenta arrows), from which results are presented later (figure 8). The colour change in the background reflects time windows during which tensiometer measurements (orange, figure 5) or readings at instrumented trench IT3 (blue, figure 6) are available. Precipitation data were supplied by WSL during the period of 2013 to October 2014.

4.2. Instrumented trench IT3

IT3 was installed at the end of October of 2014, and mainly reflects the seasonal changes in volumetric water content and temperature during the most recent autumn and winter seasons, with a short period corresponding to the beginning of the spring. Data from two sensors, a TDR and a decagon, located at depths of 15 and 20 cm respectively, are plotted in figure 6 to show the change in VWC with temperature and precipitation. A comparison between the VWC values indicates similar trends in the readings, whether they originated from TDR or decagon, however some discrepancies are noticeable in autumn and winter (December and January) and spring time (April). These changes could be due to a higher sensitivity of the decagons to a period of negative temperatures (from December to mid of January) and the differently sized cylinders of influence for the two types of sensors. TDRs have larger cylinders of influence and could therefore be considered as more representative measurement. Furthermore, a difference in readings might arise from the different installation depths of the two sensors.

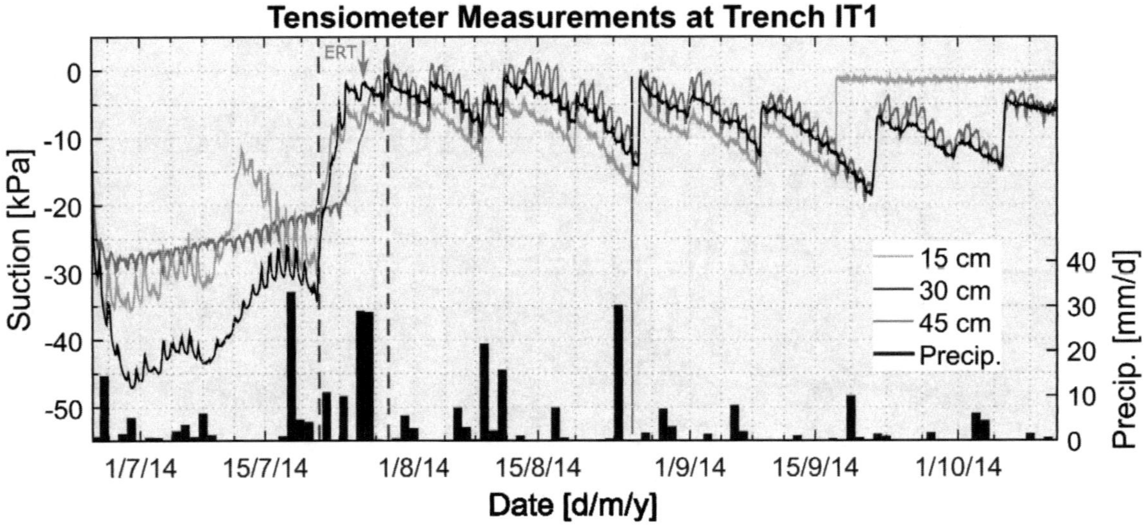

Figure 5. Tensiometer measurements at IT1 and precipitation data (25/6/2014-11/10/2014). Two debris flow events in the Meretschibach are indicated (brown dashed lines) and the ERT acquisition (figure 8) on July 25, 2014 (magenta arrow). Precipitation data were supplied by WSL from the meteostation (SR50A) located at Meretschihorn 2220 m.a.s.l.

Temperatures remained near or below zero during the winter season (end of December to end of March), and the VWC reached minimum values due to minimal precipitation in form of rain (figure 6). An increase in VWC is observed during the beginning to mid-January in the form of two peaks. The occurrence can be explained due to ice or snow melting after a transition from a negative temperature period to 0 °C, or slightly above freezing temperatures. The increase in temperature manifests itself in two corresponding peaks within the surface temperature and a continuous period of higher values in the soil temperature near 0 °C. The peaks in VWC during winter do not correlate with peaks in precipitation, as no such peaks are observable in the remaining winter period (February-March 2015) even though peaks in precipitation (mostly snowfall) can be noted. Peaks of VWC at IT3 show up in a very similar manner in IT1 (figure 4 and 7, discussion in section 4.3).

Changes in VWC are also observed in April 2015 (figure 6), when spring starts, but temperatures still remain close to 0 °C. A correlation to (a) rainfall event(s) is obvious; however thawing cycles could play a role in the increase of VWC in the soil, as the snow at the higher elevation start to melt.

In IT3, the decagon appears to be more sensitive to variation of precipitation and temperature, compared to the TDR. The relatively higher sensitivity of the decagon device is not as clearly detectable in IT1. During winter, at the end of 2014 (figure 6), the decagon records more variation whereas the TDR device remains relatively steady at low VWC values over the entire season.

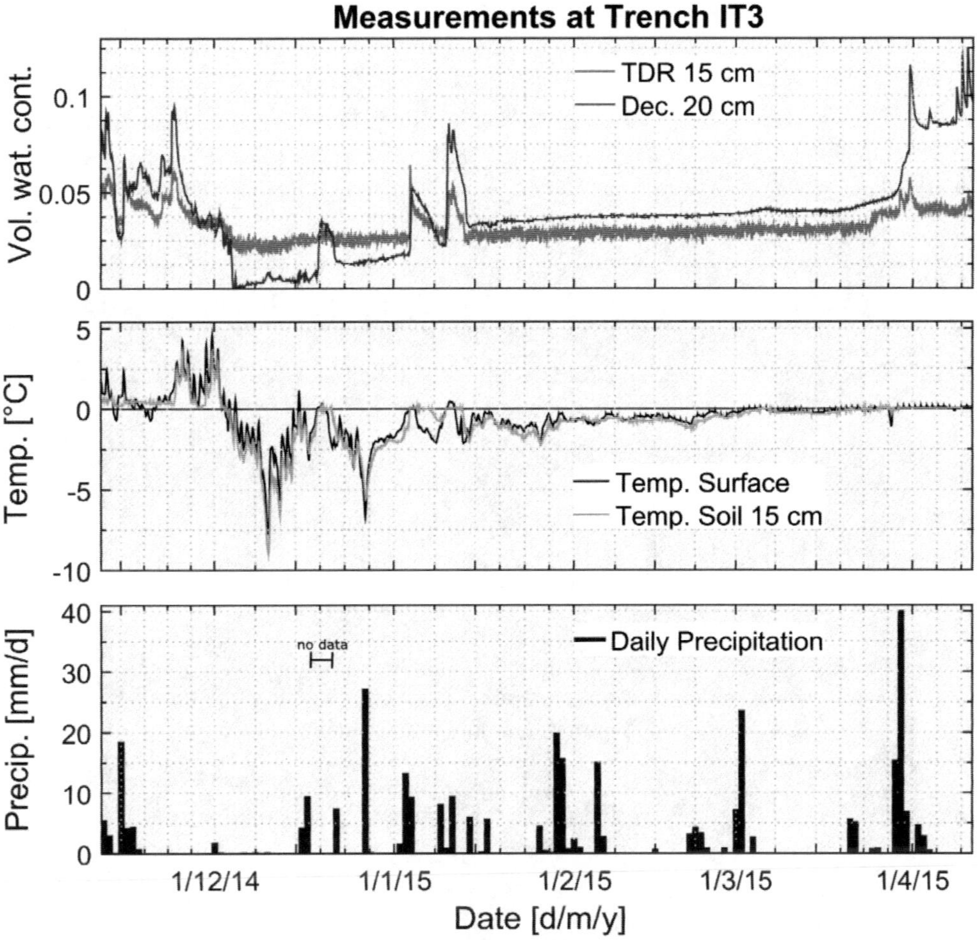

Figure 6. Measurements at IT3. From top to bottom: volumetric water content at 15 and 20 cm depth recorded with TDR and decagon sensors, temperature at 15 cm depth and in the data logger at the surface and daily total precipitation measured with Pluvio OTT^2 at 1370 m.a.s.l. (11/11/2014-11/04/2015). It should be noted, that the meteostation failed to record data during a period of four days (December 18-21, 2014) for unknown reasons.

4.3. Comparison between instrumented trenches

Figure 7 shows the comparison of the volumetric water content at IT1-IT3, located at elevations of 1868 m.a.s.l. and 1914 m.a.s.l. respectively. TDR and decagons were plotted separately, one from each trench over the same period, (as shown in figure 4 and 7, marked with a blue background).

In addition, the precipitation and temperature are plotted in the same figure to facilitate a comparison in terms of correlation. The records of VWC agree within the seasonal changes: the changes in autumn (November- December 2014) are due to precipitation (rain/snow), and during winter when it reaches a minimum, the changes (e.g. the two peaks in beginning to mid-January 2015 marked in grey background) reflect rather temperature variations (thawing - freeze cycles) than actual increases in precipitation. Variations at the beginning of the spring season (April 2015) due to precipitation in the form of rainfall are again observed.

Most interestingly, the two large peaks in VWC data in January 2015 (figure 7, grey background) were recorded in a very similar manner in both trenches. This not only means that both types of instruments expressed a comparable sensitivity most likely to a temperature variation, resulting in thawing (as discussed in section 4.2), but also that this specific event influenced the entire scree slope instead of isolated locations.

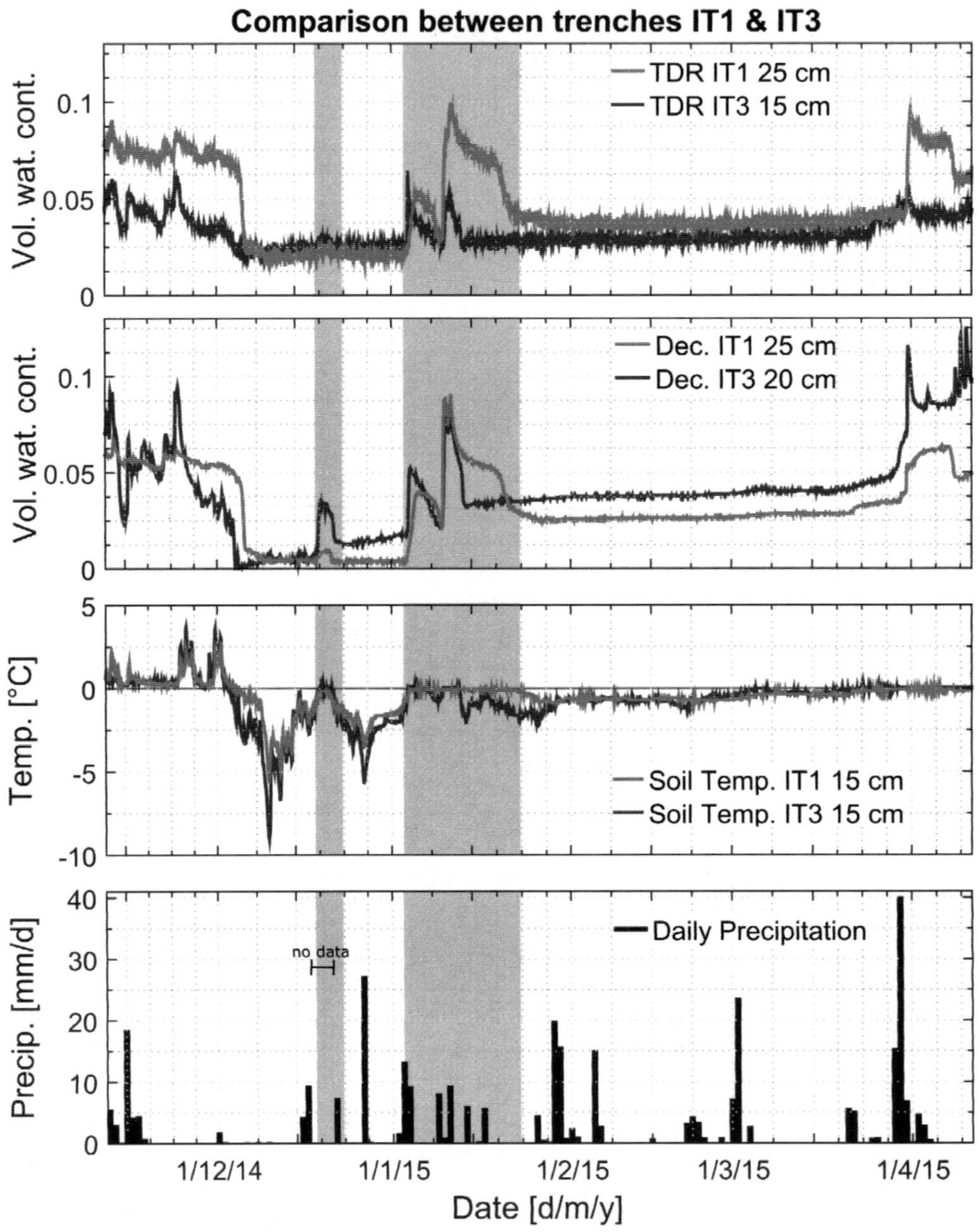

Figure 7. Comparison between TDR and decagon measurements at IT1 and IT3 during the period of 11/11/2014-11/04/2015 (same length of time as in figure 6).

The possibility of isolated and sudden events such as avalanches on the scree slope is also considered for an explanation, although this scenario does not necessarily imply an increase of the VWC in the point of view of the authors. Instead, an explanation for this abrupt increase can be attributed to the difference in the dielectric constant of water and ice (approx., 80 and 3, respectively [6]). According to the Topp equation used for TDR calibration [3], the VWC increases with the dielectric constant. Since the total dielectric constant of a soil can be estimated from its components (solid, water and air), freezing/thawing cycles have an impact on the magnitude of the calculated VWC. A better understanding of this widespread type of event will be possible when photographs from permanently installed cameras at the scree slope near IT3 are retrieved.

Furthermore, a third peak occurring at the end of December 2014 (grey background), was only detected by two decagons, but the peak remained smaller in IT1 compared to IT3. An increase in temperature was measured in both trenches at the same time. This again suggests a greater sensitivity of decagons (as an instrument type) to changes in temperature, in particular after longer periods of subzero conditions. Variations in VWC between trenches, as shown in figure 7, can be due to a difference in the gravel characteristics as described in figures 2 and 3.

4.4. Comparison between ERT and TDR

The resulting geoelectrical tomograms in figure 8, show the subsurface resistivities obtained from inversion of the ERT data acquired on May 15, June 16 and July 25, 2014. All models depict a two-layered subsurface: an upper, relatively low resistive layer, which can be attributed to the gravelly soil and a highly resistive layer underneath, which is assumed to represent the quartzite bedrock.

The precipitation data before each of the acquisitions (magenta arrows in figure 8) show that the conditions were dry and undergoing a drainage phase for the first two campaigns, whereas the data acquired in July was recorded within a period of heavy rainfalls. These seasonal changes are mostly reflected in resistivity changes within the soil layer, in particular as a significant reduction of resistivities in the July model. The saturation relative to the subsurface model obtained in May clearly reflects these conditions; there is a slight near-surface decrease in saturation in June, whereas an increase in saturation is observed within the soil layer in July. This trend in near-surface soil saturation corresponds very well to the volumetric water content measurements (which are directly proportional to saturation) from TDR and decagons installed at shallow depths within trench 1 (lowermost graph in figure 8).

5. Conclusions

Combining geotechnical and geophysical techniques, has lead to greater insight about the characterisation of the ground forming the scree slope and its response to rainfall and snow in terms of saturation and desaturation. A monitoring campaign was designed to investigate this, using different methods consisting of two types of sensors installed in instrumented trenches to determine the VWC. ERT was used to obtain resistivities which can be converted to VWC through Archie's law.

The monitoring of two instrumented trenches (IT1-IT3) gives valuable information regarding the response of the volumetric water content and temperature in the soil due to seasonal weather changes. In this aspect, the VWC increases in spring due to thawing cycles and precipitation in the form of rain, and reaches the lowest values during winter when temperatures, that stay near 0 °C or below, could lead to soil freezing under the protective insulation of a snow cover. The occurrence of isolated peaks in VWC during winter can be explained by thawing after relatively extended periods of subzero temperatures.

There is a good agreement in the data trend between decagons and TDR, but some differences can be observed, either due to rockfall events (IT1) or a greater sensitivity of one of the sensors, e.g. decagons to temperature changes during thawing periods (IT3). Additional sensor specific conditions, such as the cylinder of influence, can account for variations in the readings. The data collected from VWC in both trenches show that the response to seasonal changes are similar for the shallower layers at two different elevations on the scree slope.

The analyses of the volumetric water content time series obtained from instrumentation in the trenches not only showed very plausible seasonal trends and a clear correlation to increases in precipitation, but also compared reasonably well to saturation trends calculated from ERT measurements, conducted on different dates from May-July 2014. Thus, these results illustrate how well geotechnical and geophysical measurements can complement each other. Furthermore, models obtained from different ERT acquisitions consistently showed a subsurface separable into a highly resistive, underlying bedrock at shallow depths and a less resistive gravelly soil layer at the surface.

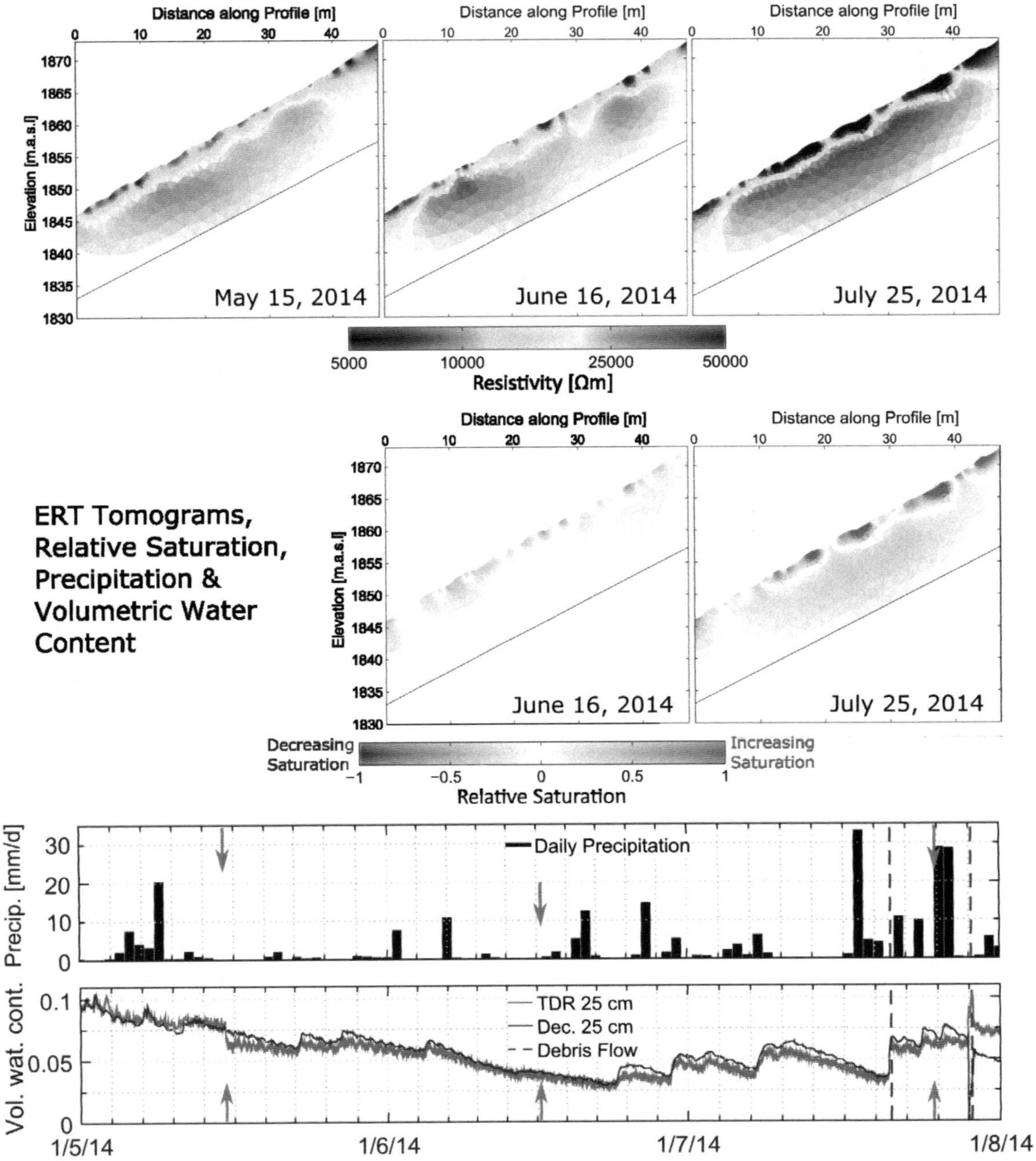

Figure 8. The uppermost row shows the resulting geoelectrical tomograms from May 15, June 16 and July 25, 2014. The resistivity scale beneath the tomograms is the same for every acquisition. The row below shows the increasing and decreasing saturation, relative to the subsurface model obtained on May 15. Finally, the three acquisition dates are indicated (purple arrows) within the precipitation and the volumetric water content measurements from TDR and decagon instrumentation in trench 1, shown in the lower most graphs from 01/05/2014-01/08/2014. Also indicated in both graphs are the debris flow events in July (see figure 4 for description).

6. References

[1] Oggier N 2011 Simulierung von Murgängen mit RAMMS am Beispiel des Meretschibachs *Master Thesis* ETH Zürich & Forschungsanstalt für Wald Schnee und Landschaft (WSL)

[2] Rickenmann D and Zimmermann M 1993 The 1987 debris flows in Switzerland documentation and analysis *Geomorphology* **8(2)** 175-89

[3] Askarinejad A 2013 Failure mechanisms in unsaturated silty sand slopes triggered by rainfall *ETH Dissertation* Nr. 21423

[4] Lucas D R, Askarinejad A A, Herzog R, Bleiker E and Springman S M 2015 Volumetric water content determination by TDR sensors and decagons in gravelly soils *submitted to the XVI European Conference on Soil Mechanics & Geotechnical Engineering (ECSMGE)* Edinburgh

[5] Springman S M, Lucas D R, Oggier N C, Kos A, Fankhauser K and McArdell B, 2015 Study of the seasonal response of a scree slope and debris flow catchment in the Swiss alps *submitted to the XVI European Conference on Soil Mechanics & Geotechnical Engineering (ECSMGE)* Edinburgh

[6] Fankhauser K 2014 Geophysical slope characterization using GPR and ERT in an active debris flow catchment *Master Thesis* EEG ETH Zurich

[7] Archie G E 1942 The electrical resistivity log as an aid in determining some reservoir characteristics *AIME* **146** 54-62

[8] Friedel S, Thielen A, Springman S M 2006 Investigation of a slope endangered by rainfall-induced landslides using 3D resistivity tomography and geotechnical testing *J. of Appl. Geophys.* **60** 100-14

[9] Teysseire P 2005 Geotechnische Eigenschaften von Möranen *ETH Dissertation* Nr. 16322

[10] Günther T, Rücker C and Spitzer K 2006 Three-dimensional modelling and inversion of DC resistivity data incorporating topography - II. Inversion *Geophys. J. Int.* **166** 506-17

[11] Günther T and Rücker C 2013 Boundless electrical resistivity tomography - the user tutorial available at: www.resistivity.net

[12] Rücker C, Günther T and Spitzer K 2006 Three-dimensional modelling of DC resistivity data incorporating topography - I. Modelling *Geophys. J. Int.* **166** 495-505

[13] Günther T 2011 Timelapse ERT inversion approaches and their applications *Int. Workshop on Geoelectric Monitoring Within the Frame of the FWF Project TEMPEL (TRP 175-n21) and the 7th FP European Project SafeLand: Berichte der geologischen Bundesanstalt Nr.93 (Vienna, 30 November - 2 December 2011)* pp 91-7

[14] Lowrie W 2007 *Fundamentals of Geophysics* (New York: Cambridge University Press)

[15] Schön J 1983 *Petrophysik Physikalische Eigenschaften von Gesteinen und Mineralen* (Berlin: Akademie-Verlag)

[16] Brunet P, Clément R and Bouvier C 2010 Monitoring soil water content and deficit using Electrical Resistivity Tomography (ERT)–a case study in the Cevennes area France *J. of Hydro.* **380(1)** 146-53

[17] Lehmann P, Gambazzi F, Suski B, Baron L, Askarinejad A, Springman S M and Or D 2013 Evolution of soil wetting patterns preceding a hydrologically induced landslide inferred from electrical resistivity survey and point measurements of volumetric water content and pore water pressure *Water Resources Research* **49(12)** 7992-8004

[18] Springman S M, Thielen A, Kienzler P and Friedel S 2013 A long-term field study for the investigation of rainfall-induced landslides *Geotechnique* **63(14)** 1177-93

Acknowledgments

The authors are most grateful for funding from the SNF Project No. 200021_144326/1 and Canton Valais, together with the Councils of Agarn and Leuk. The project also contributes to the TRAMM2 programme of the Competence Centre of Environmental Sustainability.

In particular, we wish to acknowledge the support from Pascal Stoebener, WSL, as well as to Professor Dr. Hansruedi Maurer (Exploration and Environmental Geophysics, EGG), Dipl. Ing. Ralf Herzog, Ernst Bleiker and to Nicole Oggier, for giving access to part of the rainfall data used in the analysis.

Strategies for rock slope failure early warning using acoustic emission monitoring

D Codeglia[1], N Dixon[1], GJ Fowmes[1] and G Marcato[2]
[1] School of Civil and Building Engineering, Loughborough University, Loughborough, Leicestershire, LE11 3TU, UK
[2] Italian National Research Council - Research Institute for Geo-Hydrological Protection (CNR-IRPI), C.so Stati Uniti 4, 35127 Padova, Italy

E-mail: d.codeglia@lboro.ac.uk

Abstract. Research over the last two decades has led to development of a system for soil slopes monitoring based on the concept of measuring Acoustic Emission (AE). A feature of the system is the use of waveguides installed within unstable soil slopes. It has been demonstrated that the AE measured through this technique are proportional to soil displacement rate. Attention has now been focused on the prospect of using the system within rock materials. The different nature of the slope material to be monitored and its setting means that different acoustic trends are measured, and development of new approaches for their interpretation are required. A total of six sensors have been installed in two pilot sites, firstly in Italy, for monitoring of a stratified limestone slope which can threaten a nationally important road, and secondly in Austria, for monitoring of a conglomerate slope that can endanger a section of the local railway. In this paper an outline of the two trial sites is given and AE data collected are compared with other physical measurements (i.e. rainfall and temperature) and traditional geotechnical instrumentation, to give an overview of recurring AE trends. These include clear AE signatures generated by stress changes linked to increased ground water levels and high energy events generated by freeze-thaw of the rock mass.

1. Introduction

Landslides cause thousands of fatalities every year, and there is an expectation for this number to grow in the coming years [1]. Global incidence of landslides is rising dramatically as a response to increased climate variability and frequency of extreme weather conditions. At the same time, expanding urbanization, uncontrolled land-use and environmental degradation are increasing the size of vulnerable areas. All these factors result in greater exposure of population, infrastructures and economic activities to landslide risk. It is therefore clear that monitoring techniques able to foresee imminent landslides and provide an early warning are essential in order to undertake actions that lead to risk reduction and disaster prevention.

Research over the last two decades [2] has led to development of a system for early warning of soil slopes instability based on the concept of measuring Acoustic Emission (AE). The AE is measured by a piezoelectric transducer sensor installed on a waveguide. The waveguide is a pipe placed into a borehole drilled within an unstable soil slope and has the specific function to transmit AE waves generated by the mass movement to the sensor. Laboratory tests and field trials have demonstrated [3–5] that detected AE trends are proportional to the rate of deformation (velocity) of the unstable slope. As the system is capable of sending alarms when AE rates exceed a certain level, it is suitable to be used as an early warning of soil slopes instabilities. The monitoring technique developed is called Slope ALARMS (Assessment of Landslides Using Acoustic Real-time Monitoring Systems).

Recently, attention has been focused on the prospect of using the system within rock materials. The different nature of rock, its setting and the different failure modes mean that different acoustic trends are measured. Their interpretation requires development of new strategies comprehending how a specific rock mass system acoustically responds to various external and internal excitations (e.g. temperature, groundwater level changes, seismicity, deformation) and identifying possible sources of noise (e.g. water seepage) in order to discern trends in the AE information that can provide an early warning of collapse. This paper provides a first overview of recurring AE trend patterns detected, considers these in relation to measurements from traditional geotechnical instrumentation installed at the sites and discusses possible associated causes.

2. Acoustic Emission (AE) monitoring of rock slopes

Materials undergoing deformation generate transient sub-audible stress waves due to rapid release of energy, which radiates from the source and propagate through the material. These waves have typically frequencies higher than 20 kHz and are called Acoustic Emission (AE). Monitoring the evolution of AE in a material gives information on the degradation of the material itself. In rocks, AE is generated at different scales by propagation of new fractures and/or displacement along existing discontinuities [6]. Therefore, detection of AE is an indicator of deformation within a rock mass. Increasing AE rates cannot be directly related to an increasing rate of deformation, they could rather be related to accelerating damage events at the micro-scale as forerunners of a macroscopic brittle failure. Evidence of accelerating patterns prior to collapse has been widely found at the laboratory sample scale (e.g. [7]) and precursory event trends are known also at the slope scale [6,8].

Acoustic Emission is measured by means of a system (Figure 1) that consists of three segments: detection, analysis and communication. The detection segment comprises a steel pipe, also called waveguide, and a piezoelectric transducer. The primary function of the waveguide is to increase the zone of influence of the transducer to high frequency waves (i.e. high frequency waves travelling in steel attenuate much less than in a discontinuous rock medium). The waveguide is installed in a borehole, which ideally should reach the stable stratum below any shear surfaces or potential shear surfaces that may form within a slope. In rock slopes the waveguide is grouted into the borehole in order to provide continuity for stress waves to propagate from the rock mass to the steel tube. This is considered to be a passive system, as the grout is not expected to be the primary source of generated AE in a deformation event. In soil slopes a noisy granular backfill is used to generate AE (i.e. an active system). AE waves generated across the waveguide, or in its vicinity, are conveyed by the steel bar to the piezoelectric transducer mounted at the free end of the bar (Figure 1), which converts mechanical signal to electronic signal. The use of a high-pass transducer (i.e. very low sensitivity to low frequencies) ensures that low frequencies, which are generally generated by wind, traffic, anthropic activities and therefore considered background noise, are not recorded.

The electronic signal is subsequently analysed by a sensor (analysis segment of the system), which is actually a computing device with limited functions. The sensor amplifies the signal and applies the band-pass filter that removes frequencies lower than 20 kHz and higher than 30 kHz. Ring Down Counts (RDC) rates are subsequently determined counting the number of times the signal exceeds a pre-determined voltage threshold (Figure 2) within a pre-set period of time, typically set at 15 minutes. For each monitoring period, the sensor compares RDC rates with 4 pre-set alarm thresholds of increasing RDC orders of magnitude.

As soon as one of the thresholds is exceeded, the sensor sends an alert SMS to an assigned person with the corresponding warning status stamped on it (communication segment). The four warning status available are Very slow rate, Slow rate, Moderate rate and Rapid rate, which were selected for describing soil slope movement rates. Alarm thresholds have still not been determined for rock slopes. It is part of the ongoing research to derive relationships between AE rates and level of damage within the rock in order to be able to identify the appropriate criteria for setting threshold levels for early warning purposes. The system works continuously and in real-time providing high temporal resolution information. Power consumption has been optimised in order to work on batteries without maintenance for more than one year, which makes the system suitable to be installed at remote sites.

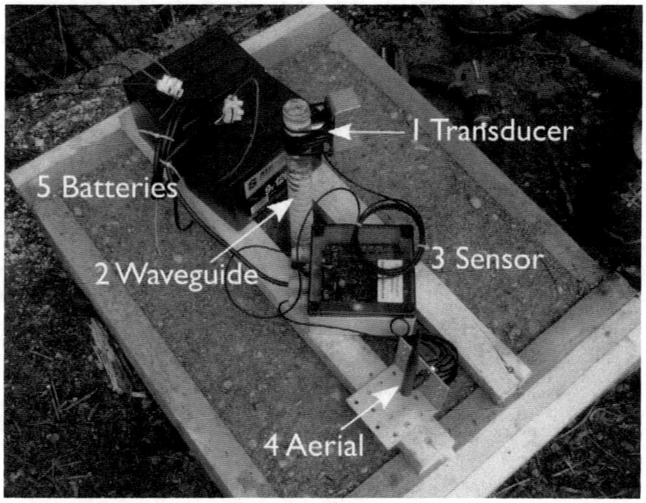

Figure 1. AE monitoring system installation within a slope. AE is measured by a piezoelectric transducer (1) placed at the free end of the waveguide (2) and processed by the sensor (3). In case an alarm is triggered, the information is sent through an aerial (4). The system is battery operated (5). All the equipment is protected with a weatherproof cover.

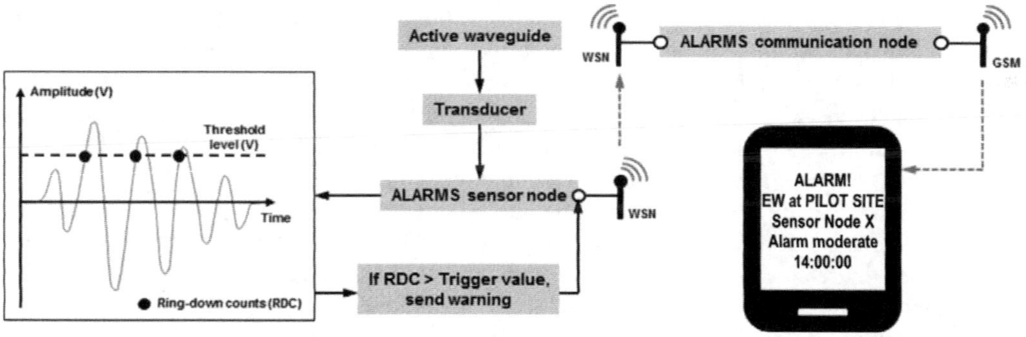

Figure 2. Schematic of operation of the AE monitoring for soil slopes including sensor node and communication system, after [9,10]. In rock slopes the active waveguide is replaced by a grouted passive waveguide.

The distinction between stable/unstable strata to install the waveguide through is not always clear. For example, in intensely fractured rock slopes the rupture surface could develop along one or many pre-existing discontinuities, which are hard to identify. Moreover, some types of rocks experience very small deformation prior to collapse, which makes it very challenging to identify a distinct rupture surface prior to failure without first installing monitoring instrumentation. Boreholes are often drilled across sets of discontinuities for which movements are likely to be expected but without penetrating a definite stable stratum. Also in the case of unstructured rock it is difficult to locate the stable zone, and it can be challenging to distinguish discontinuities that might lead to failure. In this case waveguides are located in areas with potential for instability or near areas that have already experienced failures.

At this point it is very important to highlight that detected AE are not only generated from deformation of discontinuities crossed by the waveguide (i.e. installed directly through them) but also by deformation mechanisms within a volume of rock surrounding the waveguide. Thus, the waveguide location precision is not essential; what is important is to assess the size of the monitored volume of rock surrounding the waveguide. The volume depends on the specific rock mass properties, i.e. attenuation of AE in the rock material [11], structural characteristics and particular grade of fracturing. All these elements affect the attenuation of the signal, which is a measure of the energy loss of wave propagation in a rock mass: the higher the attenuation, the shorter the distance that waves can travel and, thus, smaller the monitored volume surrounding the waveguide; and vice versa. Attenuation within the rock mass is not the only source of loss of energy. Some energy is also dispersed (reflected) as the AE propagates through interfaces between rock, grout and steel as every material has its specific acoustic characteristics.

In the interpretation of AE rates it is important to consider that recorded RDC are a function of a number of factors. Along with attenuation, due to various aspects mentioned above, the characteristics of the AE source (i.e. the magnitude of AE event generated by deformation of the rock mass) and settings of the monitoring equipment (i.e. sensitivity of the transducer, voltage threshold set in the sensor node) all affect the recorded RDC rates. Therefore, AE rates, trends of AE behaviour in time and comparisons between waveguides are relative measures of the deformation phenomena and can be used to provide information on deformation mechanisms rather than absolute values. In the following sections a description is provided of two monitoring sites where the Slope ALARMS system is installed.

3. Trial site 1: Passo della Morte (Northern Italy)

Passo della Morte trial site is situated in North-Eastern Italy, about 3 km east from Forni di Sotto [Latitude 46.398325, Longitude 12.701861] on the left flank of Tagliamento River valley. The AE monitoring system at Passo della Morte was set up in various stages since summer 2010. The site consists of an unstable rock mass in stratified limestone (Calcari scuri stratificati – lower Carnian), steeply lying on massive dolomite (Dolomia dello Schlern – upper Ladinian), which forms the stable underlying bedrock (Figure 3). The limestone outcrop is about 130 m wide and 250 m high, occupying elevations between 650 and 900 m above sea level (a.s.l.). It is relevant to highlight that the Passo della Morte tunnel crosses the limestone rock mass for its entire width, at a constant altitude of 720 m a.s.l. with only shallow cover (0-15 m) on the side towards the slope. The rock mass is throughout its volume divided into blocks with a range of sizes that are mainly isolated from the bedding and three other discontinuity sets, along with several random fractures and small faults.

Two possible slope evolution scenarios have been anticipated [12]: progressive detachment of blocks in the order of 1to 10 m^3 or failure controlled by the bedding involving a part (less than 500,000 m^3) or the entire slope (up to 650,000 m^3). In the latter case it has been demonstrated [13] that the falling material could reach the base of the slope forming a temporary valley obstruction about 20 m high and 200 m long. The instability endangers operation of the National Road 52 but it represents also a considerable threat for settlements and strategic elements, such as bridges, located downstream along the valley if valley damming and sudden release of the accumulated water occurred. Other relevant research studies regarding this site include [14,15].

3.1. Monitoring system

At this site three horizontal waveguides were inserted in boreholes drilled through the steep limestone layers from within the road tunnel (Figure 3). The three 146 mm diameter boreholes were designed with specific functions in mind: AEWG1penetrates for 50 m into the rock away from the slope, reaching the stable stratum of dolomite in the last 12 m; AEWG2 (30 m) and AEWG3 (10 m) penetrate the limestone slabs between the tunnel and the slope surface, to monitor activity of wide openings filled with marl that can be observed daylighting on the slope face. Waveguides inserted and grouted in the boreholes are 50 mm diameter steel pipes, in singular lengths of 3 meters screwed together with connectors to reach the desired total length. The waveguides were equipped with three sensors at different times: AEWG1 has been in place since 16/12/2010, AEWG2 since 27/09/2011 and AEWG3 since 12/10/2012.

IOP Conf. Series: Earth and Environmental Science **26** (2015) 012028 doi:10.1088/1755-1315/26/1/012028

Other than the three AE sensors, several other monitoring instruments are installed at the site (Figure 3). Five Time Domain Reflectometry (TDR) cables of various diameters (22 or 41 mm), a three point rod extensometer, an inclinometer, piezometric sensor, two MEMS accelerometers, a down-hole accelerometer and a seismometer have been installed to monitor displacements of strategic sections and other physical quantities (groundwater level, seismic motion) within the rock mass. Three crackmeters and three GPS benchmarks monitor displacement of key points on the surface. Data recorded since April 2011 has been made available by CNR-IRPI for comparison with RDC trends. Rainfall data are available since December 2010.

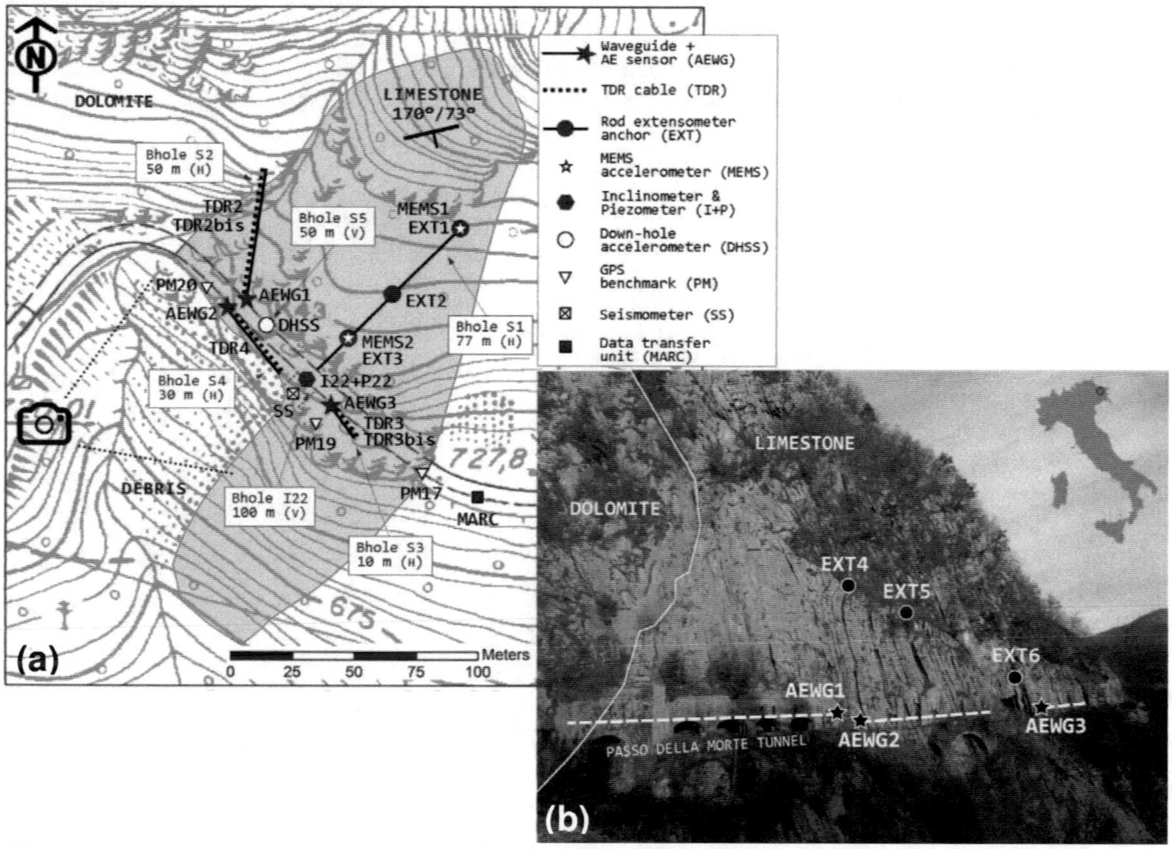

Figure 3. Passo della Morte (Italy): (a) schematic of the unstable limestone rock mass and position of waveguides (AEWG1, 2, 3), along with TDR cables, extensometers, MEMS accelerometers, an inclinometer, piezometer, a down-hole accelerometer and a seismograph station; (b) west rock mass face with projection of the waveguide positions and locations of crackmeters (EXT4, 5, 6) [16].

3.2. AE trends

AE events at Passo della Morte site can be subdivided in three categories based on different AE rate levels: low (<1,000 RDC/hour), medium (1,000-10,000 RDC/hour) and high (10,000-100,000 RDC/hour) rate events. Low AE rate events are rather common throughout the data series. They occur both during rainfall events and during dry periods and their duration is from hours to a few days, with occasional spikes of 400-800 counts per hour that are several times higher than the average 50-100 counts per hour. These events are often recorded by all the three sensors (AEWG1, AEWG2 and AEWG3) although they show different numbers of counts (RDC). Medium AE rate events usually last for a few days and are recorded in particular from one sensor while the other two are silent or show lower activity levels. They don't seem to be associated with particular rainfall events and occur when the piezometric level does not change. These types of events can show a sharp increase at the beginning, or they can gently rise to a peak AE rate, but in both cases the rates usually decrease gradually. High AE rate events are discussed in the following paragraph, as a hypothesis on their genesis can be made.

3.2.1. High AE rate events. Very high energy events are observed in the data series. They are extremely high counts which last generally for one monitoring period (i.e. 1 hour in the example), and certainly for not more than three consecutive monitoring periods. These spiky events can be isolated or grouped in clusters within a few days. Another distinctive trait of such events is that they are recorded simultaneously by all the three sensors, although AEWG3 consistently records acoustic rates higher by about one order of magnitude compared with AEWG1 and AEWG2. The example provided in Figure 4 shows five clusters of spiky events. AEWG3 reaches almost 100,000 counts per hour while the other two sensors barely exceed 10,000 counts per hour at the same time (thus they are hidden in Figure 4 by the AEWG3 data).

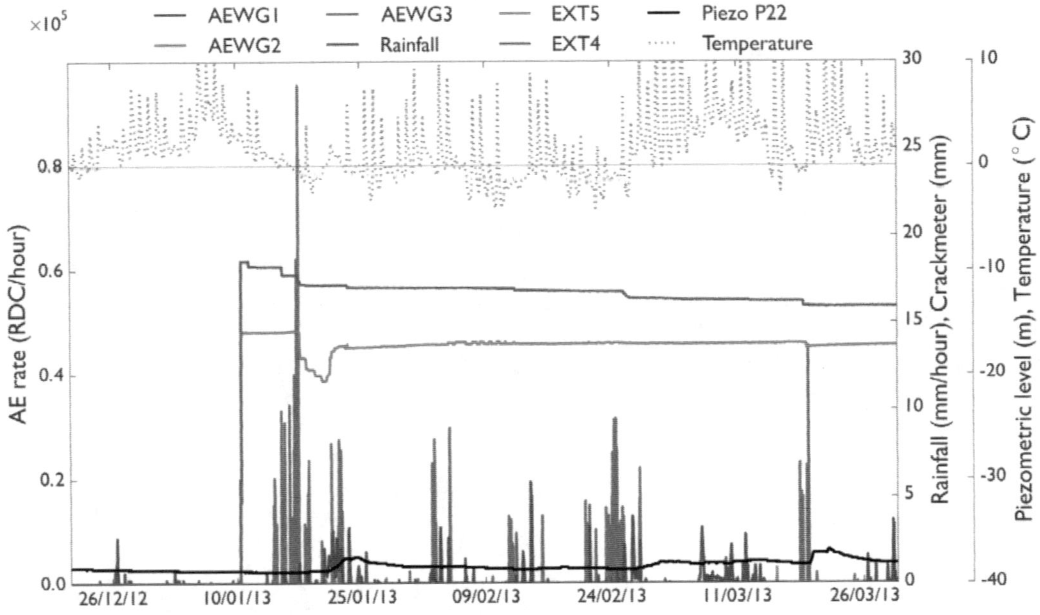

Figure 4. High AE rate events. EXT4, EXT5 were not operating until 10/01/2013. Piezometric level is referred to the well-head placed in the tunnel at an altitude of 720 m a.s.l..

Rainfall and changes in ground water level do not seem to be critical for the appearance of such high RDC trends: in Figure 4 rainfall rarely reaches 5 mm/hour and the piezometric level change is not remarkable (about 2 m). Considering the temperature, there seems to be a relation between RDC and temperatures dropping below 0°C and, apparently, these spiky type of events are less frequent later in the year when the temperatures rise, and they reappear in Jan-Mar 2014 and Jan-Feb 2015. Therefore, the trends could be generated by water freezing into discontinuities surrounding the waveguide. This doesn't explain though why trends are recorded by all the three sensors, although with different orders of magnitude, in the same monitoring period (1 hour) considering that the temperatures stay low for several days.

Other interesting observations regard the crackmeters EXT4 and EXT5, which monitor two of the key discontinuities where they daylight on the slope (see Figure 3 for their location). They experienced operating problems in late 2012 but started to work correctly again on 10/01/2013. The crackmeters work in compression because the limestone slabs tend to tilt backwards (i.e. towards the stable dolomite) in that part of the slope. Figure 4 shows a possible good correlation between displacements recorded by the crackmeters and RDC rates, although this is not always consistent. Some clusters are in fact not linked to any measured displacement, but they could be generated by fractures that are not monitored, and a few displacement events do not seem to match any RDC clusters. Causes for these spiky types of events are still unknown but it appears clear that effects of low temperatures and displacements need to be further investigated, possibly as a combined and/or related effect.

3.2.2. Trends due to changes in groundwater level. A category of events is attributed to changes in water pressure within the rock mass due to variations of piezometric level (Figure 5). It can be observed that a delay of several hours occurs between the start of a rainfall event and the rise of water level, but AE recorded is simultaneous with the water level change. AE rates tend to accelerate when the water table rises and to decelerate, more or less sharply, when it drops. Hence the AE trend follows the stress changes within the rock structure. This behaviour is particularly evident during very intense rainfall events that are typical of the area in the autumn season. They can last for days and induce large variations in the piezometric level (i.e. tens of meters).

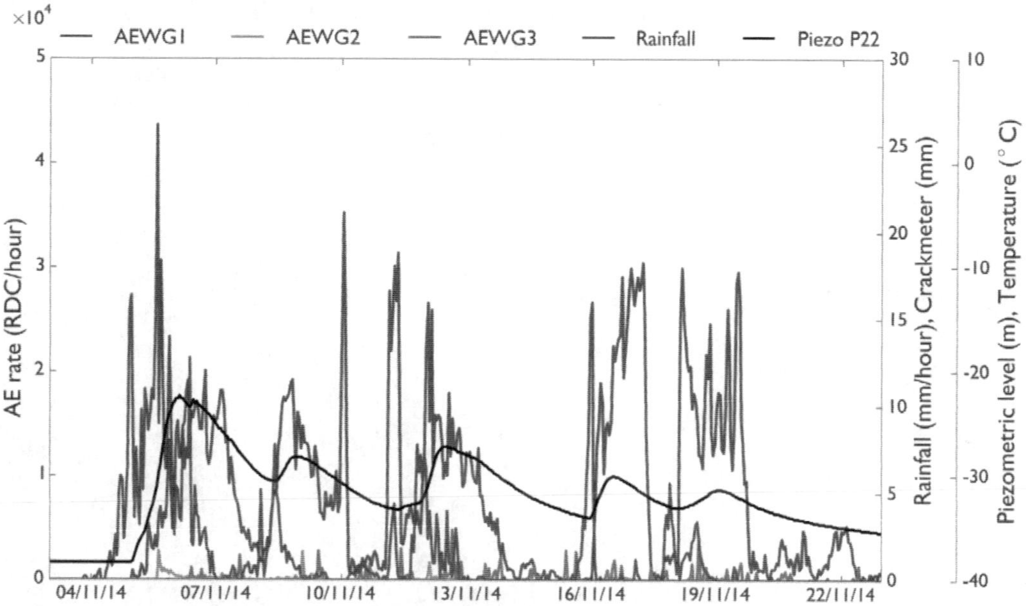

Figure 5. AE rate (RDC/hour) in response to changes in water pressure within the rock mass. Rainfall (mm/h) and piezometric level (m) are also indicated.

For the example shown in Figure 5, the first rainfall event starts at 00:00 on 04/11/2014 after a quiet and dry period; the piezometric level is initially -39 m from the tunnel level, 720 m a.s.l.. As the ground water table response begins at 00:00 on 05/11/2014, the RDC accelerating trends are recorded by waveguide AEWG1. During the previous 24 hours no AE activity is recorded by any of the sensors, although it was raining intensively. This latter observation is of great importance because it excludes the possibility that AE recorded is generated by rainfall seepage within discontinuities or by water flowing onto the waveguide. If this was the case AE would have been recorded earlier. The same behaviour can be observed for the consecutive rainfall events in Figure 5, although RDC doesn't start from zero as there is an overlapping effect of previous events. The groundwater table reaches its maximum (-22 m) at 03:00 on 06/11/2014, which is 17 m higher than the beginning, and after a few hours it starts to decline. AE in this phase remains relatively high because of the continuous water pressure variation (firstly increasing and then decreasing). Finally, AE rates decrease, but do not reach zero because a new rainfall event takes place.

Looking at the whole time series in Figure 5 it is noticeable that AE response follows the change in water level but is not proportional to it and shows occasional spikes. The reason for this is that AE is not generated by the water itself but as a secondary effect. In other words, the rearrangement of stress within the rock mass, which is caused by varied conditions in the water pressure, results in micro-deformations in the rock mass, thus the acoustic stress release. This deformation is a non-linear process and develops in consequent steps of instant energy release and the release intensity depends on the energy previously accumulated.

4. Trial site 2: Grossreifling (Austria)

Grossreifling trial site is situated in Styria, Austria about 1.5 km north of Grossreifling [Latitude 47.673932, Longitude 14.709952] on the left bank of Enns River, which is one of the largest Austrian tributaries of the Danube River. The site consists of a steep conglomerate slope that threatens a section of the railway line St. Valentin-Tarvisio at Km 91,400 (Figure 6). The Grossreifling trial site was set up in April 2014 as a complementary component of the Sentinel for Alpine Railway Traffic (SART) project. SART is a pilot project that aims to increase the safety of alpine railways through reducing the risk of damage to tracks and trains due to rock falls, and providing a cost saving alternative to expensive dynamic rock fall barriers. The system takes advantage of a dual approach: early warning of imminent rockfalls, given by acoustic emission generated within the rock constituting the slope, and detection of rock fall occurrence, provided by a light static catch fence instrumented with movement sensors that give information about the debris that detaches from the slope and impacts the fence (Figure 6). The two subsystems share a common control centre, which issues warnings and alarms to the rail traffic operator, providing enough information to take action, specifically slow down or stop the railway traffic (although this control function is not implemented in the pilot phase).

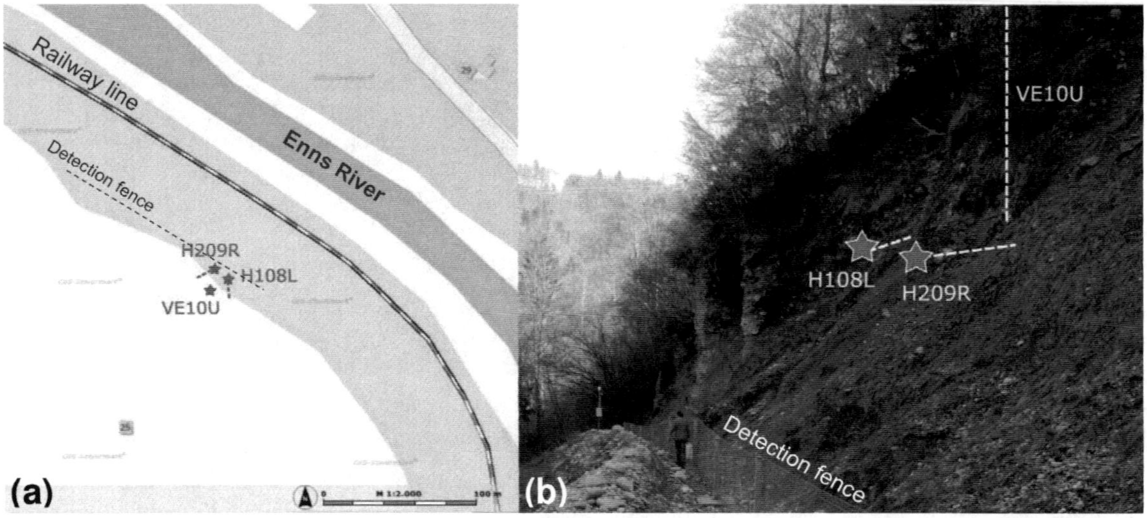

Figure 6. Grossreifling (Austria): (a) site layout map with plan view of the three waveguides (map from gis.steiermark.at); (b) image of the conglomerate slope with projection of the three waveguides and location of the detection fence.

The study slope is about 70 m high, from 435 to 505 m a.s.l.. The conglomerate outcrops only in the upper 15 m of the slope and is almost vertical, with local overhangs. Below this altitude the rock is hidden by debris falling from the top of the slope and accumulating with an angle of about 40°. Considering the geomorphology of the area it can be thought that the conglomerate forms the entire slope, from the top to the bottom. The top of the slope is also covered by 80-100 cm of vegetative soil. Clasts composing the rock are polygenic, with a dominant carbonate fraction, and high degree of roundness. It is possible, though difficult, to recognise a structure organised in flat layers of about 1-1.5 m, not always distinctly separated, which sometimes show a gradation inside them. The granular size range is in general very wide, from sand to boulders, but the percentage combination of grain sizes is variable from layer to layer. The grade of cementation is also irregular, but it does not necessarily follow the layer structure mentioned above. Some highly cemented zones are interposed with very poorly cemented ones, which can also include loose material. Size and distribution of the grains as well as the cementation degree affects the permeability; water seepage and formation of ice stalactites have been observed in winter coming from permeable zones. This suggests that water circulation has preferential routes inside the rock mass. Permeable and poorly cemented zones are probably more weathered due to the washout operated by the flowing water.

4.1. Monitoring system

At Grossreifling two horizontal waveguides (H108L and H209R) and one vertical waveguide (VE10U) were installed (Figure 6). 32 mm threaded self-drilling rods were used as waveguides, which differs from the 50 mm smooth pipes usually installed at other sites. The self-drilling type of rods is quite common in slope surface stabilisation applications but their use as waveguides is innovative. As there is no need to pre-drill a borehole of bigger diameter, the time for installation is greatly reduced as well as the overall cost. The annulus between the rod and borehole wall is filled by pumping grout through the hollow stem to the drill bit thus backfilling towards the slope surface. The downside of this installation approach is that in rock it is not possible to reach a great depth, as bars are quite thin and after about 10 m they struggle to transmit the power needed to the drill bit for progression into the rock mass. From a preliminary study conducted in the laboratory on 3 m rod lengths, there is little difference (for slope monitoring purposes) in AE propagation within the waveguide between threaded and smooth bars. However, attention should be paid when mounting the piezoelectric transducer. Experiments showed that the best coupling between transducer and waveguide is given when the transducer is mounted on a flattened thread (i.e. produced by filing), increasing the area of contact.

Waveguides H108L and H209R are installed horizontally in the conglomerate at altitudes of about 487 m a.s.l. and 486 m a.s.l., respectively, and penetrate into the rock mass for 3 m. They are installed about 5 m apart from one another diverging of an angle of about 45° as they enter into the rock mass. It is important to note that H209R is installed into loose debris for about 0.8 to 1 m. Waveguide VE10U is composed of 4 bars of 3 m each screwed together for a total length of 12 m and penetrates the conglomerate from the top of the slope (500 m a.s.l.). Its bottom is therefore about 1 to 2 m higher in terms of altitude than the two horizontal waveguides and, from a plan perspective, waveguide VE10U is located in between them.

To allow comparison between the three waveguides, piezoelectric transducers and voltage threshold settings are the same for all of them. Other data available at this site for comparison are rainfall, temperature and the record (date and time) of events that hit the fence, along with photos from the cameras triggered by movement sensors installed on the fence. Unfortunately, the rainfall gauge and temperature sensor placed at the top of the slope experienced many power failures losing long periods of data and the measurement frequency has not been constant and regular throughout the year (ranging from about 15 to 70 min), which is not ideal for comparison with AE data collected every 15 minutes. Therefore, rainfall and temperature data have recently been requested from a nearby weather station, but they are still in the process of being obtained.

4.2. AE trends

AE trends recorded by H209R and VE10U sensors are, throughout the data series, consistently higher, by about one order of magnitude, than AE trends recorded by H108L. RDC for the first two sensors are generally in the order of 30,000 to 40,000 counts per hour, with a few occasional spikes that can broadly exceed 100,000 counts per hour. RDC recorded by sensor H108L are normally about 1,000 to 2,000 counts per hour, with few spikes that reach 5,000 counts per hour.

Looking at the whole data series it appears that H209R and VE10U have a high response to rainfall events, and looking closely also at H108L, AE seems to have a relationship with the rainfall, although with a different order of magnitude (Figure 7). The problems with the site rainfall records mentioned above do not make it possible, at this stage, to determine whether there is a delay or not between rainfall and generated AE. This is particularly important because three scenarios are possible: (1) no delay, meaning that AE is generated by direct infiltration of water within the near surface permeable stratum, which is debris for H209R and vegetative soil for VE10U, and if a proportionality is found, this effect can be removed from the data series (this behaviour is not expected for waveguide H108L as it is entirely grouted in rock and also protected from direct rainfall by overhanging conglomerate); (2) delay in the order of 1 hour, AE could be generated by water flowing within the rock mass discontinuities/porosity and the delay is due to the time for infiltration; (3) delay of several hours, due to the time for the water to elevate groundwater pressures and hence generate stress changes within the rock mass. During January 2015 some low level AE trends (400-500 RDC/hour) are recorded by one or all waveguides around 10am-3pm. It is likely that these trends are generated by the slope surface defrosting when temperatures rise during the hours of maximum solar gain; unfortunately rainfall events cannot be excluded as there are no rainfall data available for the period. At this stage further interpretation is not possible until more reliable rainfall and temperature are available.

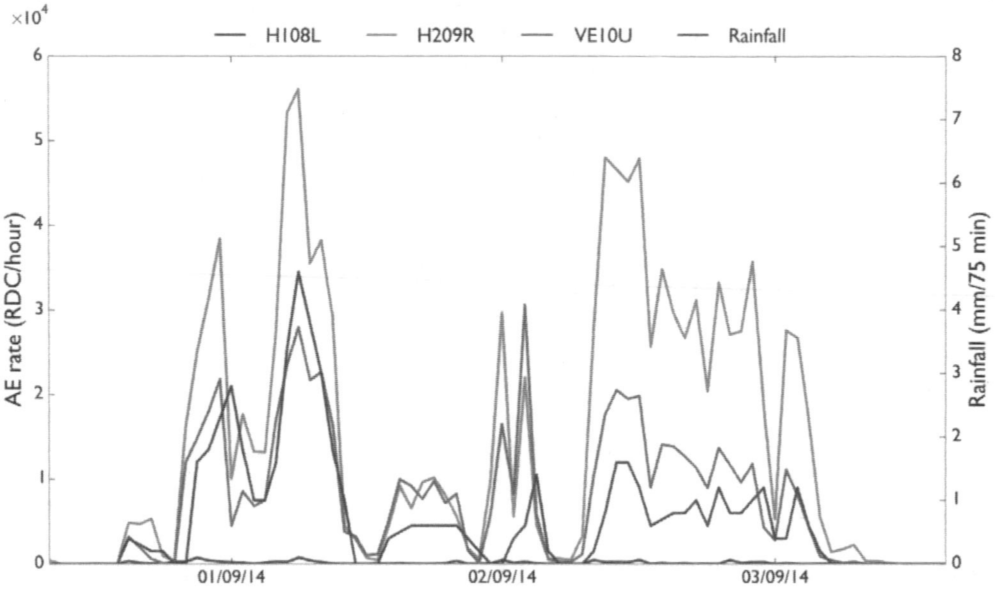

Figure 7. RDC rates in response to rainfall events. RDC values are given every 60 minutes while rainfall monitoring frequency is about 75 min. This is responsible for the plotted rainfall being shifted and hence appearing to arrive after the AE, which is not the case. It is clear that sensors H209R and VE10U with sections in loose material have a much greater response to rainfall than H108L.

5. Summary

A system for monitoring rock slope stability through Acoustic Emission (AE) has been described in this paper using grouted passive wave guides, and details of installations at two trial sites is presented along with an overview of recurring trends associated with deformation mechanisms. Interpretation of two recurring types of events at Passo della Morte (Italy) have been presented: a very high RDC rate event and AE RDC events that follow the evolution of the groundwater level. For the first event type a possible correlation with the movement of discontinuities monitored by EXT4 and EXT5 have been observed, but also temperatures below 0°C could have an influence. In the second type there is a clear response to changes in stress and resulting deformations due to the variation of the water level within the rock mass.

At the Grossreifling site in Austria, interpretation is limited by the lack of reliable rainfall data. However, it is possible to qualitatively identify dependence between rainfall events and AE trends, and low AE trends are possibly generated by thawing of the frozen slope surface during daylight hours in winter.

Field measurements at the two sites and analysis of data series will continue and, in combination with field and laboratory experiments, they will be used for further interpretation of AE generated by deformation mechanisms and consequently to derive relationships between AE and deformation. The ultimate purpose is to identify initial stages of collapse and to develop criteria for setting thresholds that can provide an early warning of the imminent collapse to allow useful action to be taken.

Acknowledgement
The field studies reported in this paper have been funded by a combination of the Engineering and Physical Sciences Research Council (EP/H007261/1), the Research Institute for Geo-Hydrological Protection of the Italian National Research Council (Passo della Morte, Italy), and the VIF 2011 program of the Austrian Research Promotion Agency and The Austrian Railway OeBB (Grossreifling, Austria). Thanks are also due to SART project partners Reinhard Hendricks and Thomas Meisel, INGLAS GmbH, and PhD student Alister Smith who has supported field work at both field sites.

References
[1] Petley D 2012 Global patterns of loss of life from landslides *Geology* **40** 927–30
[2] Dixon N, Kavanagh J and Hill R 1996 Monitoring landslide activity and hazard by acoustic emission *J. - Geol. Soc. China* **39** 437–64
[3] Dixon N, Spriggs M P, Smith A, Meldrum P and Haslam E 2014 Quantification of reactivated landslide behaviour using acoustic emission monitoring *Landslides*
[4] Smith A, Dixon N, Meldrum P, Haslam E and Chambers J 2014 Acoustic emission monitoring of a soil slope: Comparisons with continuous deformation measurements *Géotechnique Lett.* **4** 255–61
[5] Smith A and Dixon N 2014 Quantification of landslide velocity from active waveguide-generated acoustic emission *Can. Geotech. J.*
[6] Hardy H R J 2003 *Acoustic Emission/Microseismic Activity: Volume 1: Principles, Techniques and Geotechnical Applications* ed Taylor and Francis (A. A. Balkema)
[7] Lockner D 1993 The role of acoustic emission in the study of rock fracture *Int. J. Rock Mech. Min. Sci. Geomech. Abstr.* **30** 883–99
[8] Amitrano D, Grasso J and Senfaute G 2005 Seismic precursory patterns before a cliff collapse and critical point phenomena *Geophys. Res. Lett.* **32** L08314
[9] Smith A, Dixon N, Meldrum P and Haslam E 2014 Inclinometer casings retrofitted with acoustic real-time monitoring systems *Gr. Eng.* 24–9
[10] Dixon N, Moore R, Spriggs M, Smith A, Meldrum P and Siddle R 2015 Performance of an Acoustic Emission Monitoring System to Detect Subsurface Ground Movement at Flat Cliffs, North Yorkshire, UK *Engineering Geology for Society and Territory* vol 2 pp 117–20
[11] Koerner R M, McCabe W M and Lord A E 1981 Overview of acoustic emission monitoring of rock structures *Rock Mech. Felsmechanik Mec. des Roches* **14** 27–35
[12] Codeglia D 2011 *Analisi geomeccanica e predisposizione di un sistema di monitoraggio lungo la galleria del Passo della Morte. Undergraduate dissertation.* (University of Trieste, Italy)
[13] Codeglia D 2013 *Analisi numerica tridimensionale tramite modello SPH della frana di Passo della Morte (UD). Master's Thesis.* (University of Bologna, Italy)
[14] Dixon N, Spriggs M, Marcato G and Pasuto A 2012 Landslide hazard evaluation by means of several monitoring techniques, including an acoustic emission sensor *Landslides and Engineered Slopes: Protecting Society through Improved Understanding* ed E Eberhardt, C Froese, A Turner and C Leroueil (CRC Press) pp 1405–11
[15] Teza G, Marcato G, Pasuto A and Galgaro A 2014 Integration of laser scanning and thermal imaging in monitoring optimization and assessment of rockfall hazard: a case history in the Carnic Alps (Northeastern Italy) *Nat. Hazards* **76** 1535–49
[16] Dixon N, Codeglia D, Smith A, Fowmes G and Meldrum P 2015 An acoustic emission slope displacement rate sensor: Case studies *Proceedings of the 9th International Symposium on Field Measurements in Geomechanics, 9–11 September 2015, Sydney, Australia*

Shear wave velocity analysis of a deep seated gravel landslide structure using the microtremor survey method

SU Lijun[1, 2,*], XU Xingqian[1, 3], LIAO Hongjian[4] and GENG Xue-yu[5, 6]

[1] Key Laboratory of Mountain Hazards and Earth Surface Process, Chinese Academy of Sciences Institute of Mountain Hazards and Environment, Chinese Academy of Sciences, Chengdu, Sichuan 610041, China

[2] CAS Center for Excellence in Tibetan Plateau Earth Sciences, Beijing 100101, China

[3] University of Chinese Academy of Sciences, Beijing 100049, China

[4] Department of Civil Engineering, Xi'an Jiaotong University, 28 western Xianning Road, Xi'an Shan Xi 710049, China

[5] School of Civil Engineering, Qingdao Technological University, Qingdao, Shandong Province 266033, China

[6] School of Engineering, University of Warwick, Coventry CV4 7AL, UK

E-mail:sulijun1976@163.com

Abstract. The depth and geometry of potential failure surface is the fundamental for evaluating the mechanisms of a landslide. Traditional techniques to acquire information on potential sliding surface are mainly drilling, pitting, and trenching, but these techniques are time consuming and expensive. In this study, microtremor signals and the dispersion curves of surface wave are extracted from the vertical component of microtremor records using the spatial autocorrelation (SPAC) method to estimate shear wave velocity structure. The results suggest that the buried depth of phyllite bedrock is approximately 47.4m, and the thickness of weathered bedrock layer is about 9.9m at about 57.3m deep, which could be interpreted as the potential sliding surface of this landslide, in accordance with borehole data. The microtremor survey method (MSM) is flexible, non-invasive, relatively quick and deployable on the landslide. It clearly demonstrat that it is an effective tool to improve the drilling success rate, and hence allow a large scale and high density investigation of structure characteristics of a deep seated landslide.

1. Introduction

The conventional ways to acquire underground structure information are mainly artificial excavation and mechanical drilling. Physical and mechanical parameters of rock and soil are measured by indoor mechanical tests with drilling cores to provide some reference for the stability analysis of landslides. These traditional techniques could obtain relatively accurate information for landslide studies. However, excavation or drilling on the unstable slopes are difficult and costly.

In the past two decades, the Microtremor Survey Method (MSM) was increasingly applied in the subsurface investigation as an effective probing tool, such as civil engineering, collapsed columns mapping, boulders and geothermal faults mapping[1,2]. The method assums that microtremor is a stable random process, and we could estimate the shear wave velocity structure using the inversion of Rayleigh Wave dispersion curves extracted from microtremor signals [3]. MSM is featured by high resolution, no artificial sources, cheap and convenient data acquisition, without topographical restriction and particularly for density populated and vibration disturbance area.

Microtremor produced by natural phenomenon and mankind's activity is common on the earth's surface, and frequency is about 1Hz. It is composed mainly of body wave and surface wave (Rayleigh Wave, Love Wave, etc.), and Surface Wave accounts for about seventy percent of total signal power [4]. To provide available geological model for landslide hazards prevention and controls, this study attempted to detect the internal structure, especially potential sliding surface by combining MSM with drilling data.

2. Method

MSM has two fundamental hypotheses: firstly, it is a stationary random process in space and time, and secondly microtremor is composed of different frequency waves where the fundamental mode Rayleigh Wave energy is predominate [5]. There are two methods to exact phase velocity dispersion curves from the vertical components of microtremor: the Frequency-wavenumber (F-K) method and the Spatial Autocorrelation Method (SPAC). The former is constrained by station quantity, coverage area and recording requirement, etc. In turn,The latter can be always used to effectively estimate the shear wave velocity [3].

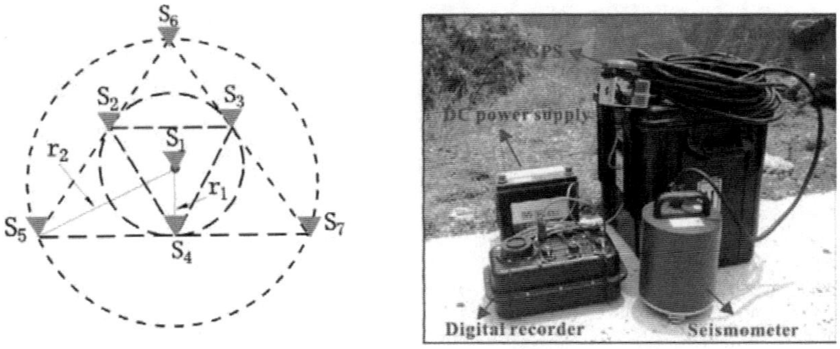

(a) (b)

Figure 1. The microtremor observation layout.
(a) Observation array (b) Mirotremor device

The survey point S_1 in figure 1(a) is the center of observation array, and the other six points are observation points (S_2~S_7) equally spaced on the circumference of two circles with radius r1 and r2, respectively. The seven microtremor survey instruments were used to acquire digital signal simultaneously, and the synchronization was controlled by inner GPS clocks in collector. Each microtremor device is composed of DC power supply, GPS, digital recorder and seismometer sensor in figure 1(b).

The probing depth of microtremor array is 3 to 5 times than observation radius[3,6]. Overburden thickness of this landslide is about 50m according to preliminary engineering geological survey. So we selected inner radius r1 of 5m and outer radius r2 of 10m to meet the depth requirement in this investigation.

3. A case study

3.1. Study site

The survey site was on the Xishan village landslide seated in Tonghua Town, Li County, Aba Zang & Qiang Autonomic State, Sichuan province, China. According to engineering geolocial data in figure 2 , the stratum structure of landslide from top to bottom was Quaternary diluvium, gravel soil, strong weathered phyllite, weathered phyllite and phyllite bedrock ,respectively. The black arrow indicates the direction of this landslide, and the black bold dashed line reports the potential sliding surface speculated from boreholes in figure 2.

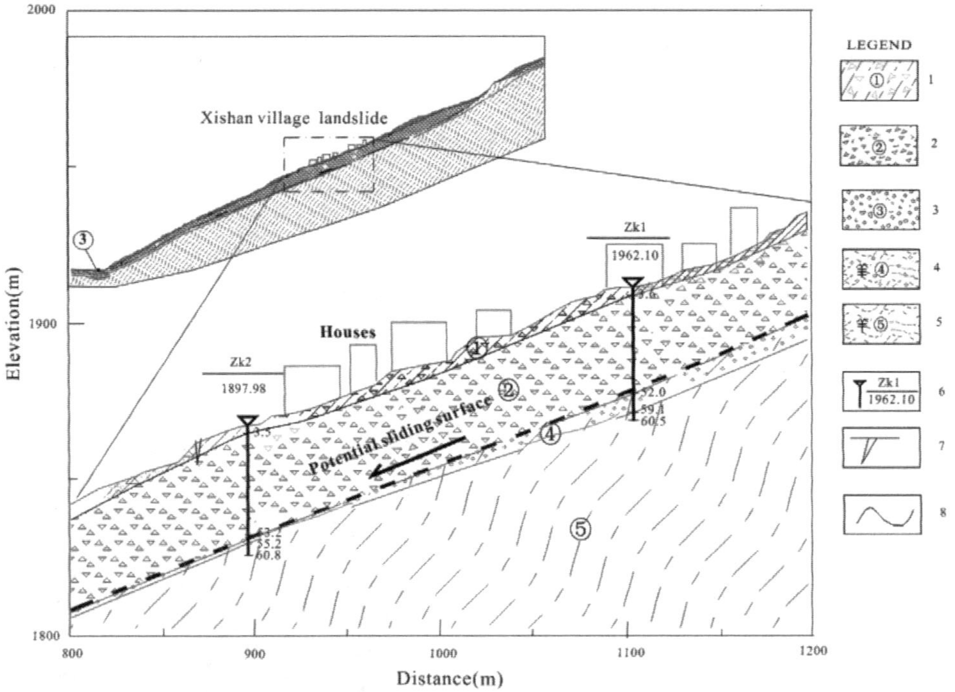

Figure 2. Engineering geological corss-section of Xishan village landslide.
(1) Diluvium layer. (2)Gravel soil layer. (3)Flood alluvial boulder. (4)Weathered phyllite. (5)Phyllite bedrock.(6)Borehole. (7)Crack. (8)Boundary of underlie bedrock.

To verify the surveying effect of MSM for a deep seated gravel landslide, this study selected borehole Zk1 as the research object. On the basis of the previous investigation, we conducted the microtremor array obsevation on the position of Zk1 using the layout and devices in figure 1. The test is avaiable if the probing results estimated from MSM investigation is in accordance with the borehole data on the same positon Zk1.

3.2. Data acquisition

The position relative to the survey point S_1 was measured on the site based on the above selected observation radius. To collect high quality records, we needed to weed out the soil to install broadband digital seismometer sensors CMG-3ESPC and data acquisition device REFTEK72-08A from Chinese Seismic Probe Array Center (CSPAC). Then, the sensors were tested and a sample frequency of 100Hz was selected.

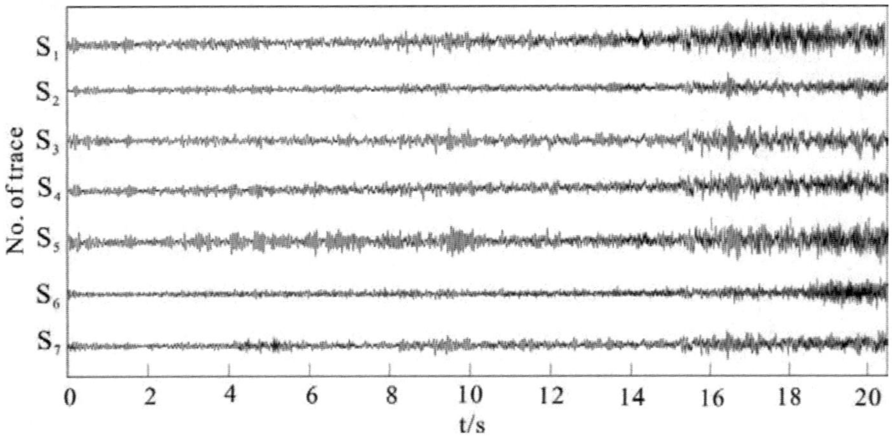

Figure 3. Examples of microtremor records. Each trace report the vertical component of micro vibration on each sensor's position.

The instrumental stabilization time was not lower than 3 hours during data acquisition and available signals were collected at least 2 hours after instrument stabilization. We tried to avoid walking near the stations as far as possible when sensors were recording. A low pass filter with a high cut-off frequency of 10Hz was adopted. The preprocessing of microtremor mainly includes instrument response correction, tilt component removing, remean, resample and band-pass filter, etc.

3.3. Spatial Autocorrelation Coefficient

The spatial autocorrelation coefficients can be calculated using narrowband filter to separate harmonics in time domain[7]. But this method is time-consuming and poorly accurate. Okada (1994)[8] improved the computed speed and accuracy using Fourier transforms (FT) to extract spatial autocorrelation coefficients in frequency domain. The following equation is given by Asten et al. (2006) [9]to calculate spatial autocorrelation coefficients.

$$\rho(\omega, r) = \frac{1}{2\pi}\int_0^{2\pi} \mathrm{Re}\,al\left\{\frac{S_A(r,\omega)S_o^*(\omega)}{\sqrt{\left[S_A(r,\omega)S_A^*(r,\omega)S_o(\omega)S_o^*(\omega)\right]}}\right\}d\theta \qquad (1)$$

Where $S_o(\omega)$ and $S_A(r,\omega)$ are Fourier transform of center point O and point A on the circle, $S_o^*(\omega)$ and $S_A^*(r,\omega)$ are corresponding complex conjugate, respectively.

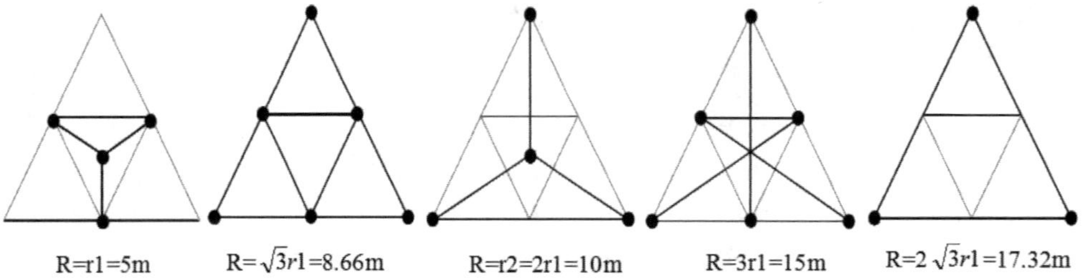

Figure 4. Layout schematic diagram of array combination pattern.

The distance between any two stations was different due to the both observation radius. Five type combinations are listed as follows: r1, $\sqrt{3}$ r1, 2r1, 3r1 and $2\sqrt{3}$ r1 shown in figure 4. The spatial autocorrelation coefficients of these combination patterns were calculated one by one with equation (1). More reliable results could be obtained by the azimuthal average of these coefficients.

3.4. Phase velocity of Rayleigh Wave

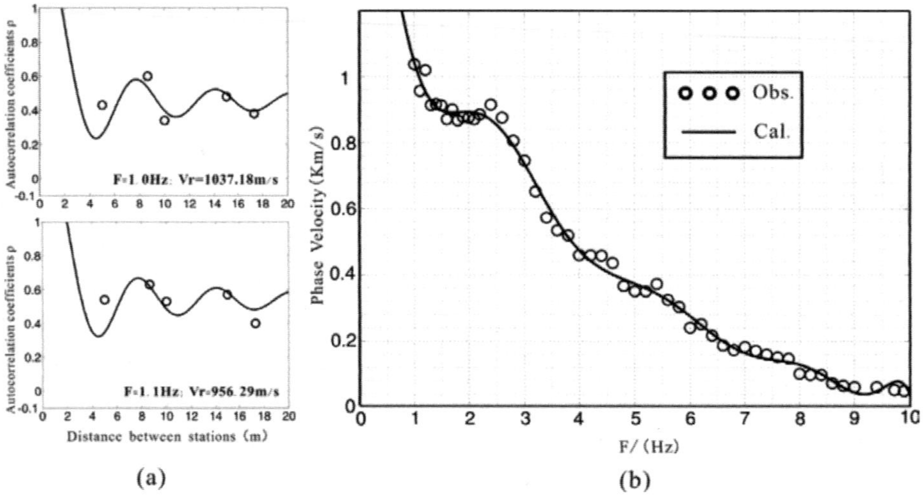

Figure 5. Phase velocity of Rayleigh Wave.

(a) Phase velocity fitted by Bessel function (b) Fitted dispersion curve of Rayleigh Wave.

Based on the above calculation, we fitted spatial autocorrelation coefficients with the Bessel function of first kind and order zero to obtain phase velocity of Rayleigh Wave for different frequency when f was fixed and r was random. The circles in figure 5(a) represent the spatial autocorrelation coefficients corresponding to the distance of 5m, 8.66m, 10m, 15m, 17.32m.

Phase velocity in the range of 10Hz was calculated with the fitting using Bessel function of first kind and order zero as circles shown in figure 5(b). Phase velocity of Rayleigh Wave v_r decreased with increasing frequency corresponding to the shallower probing depth.

3.5. Shear wave velocity structure

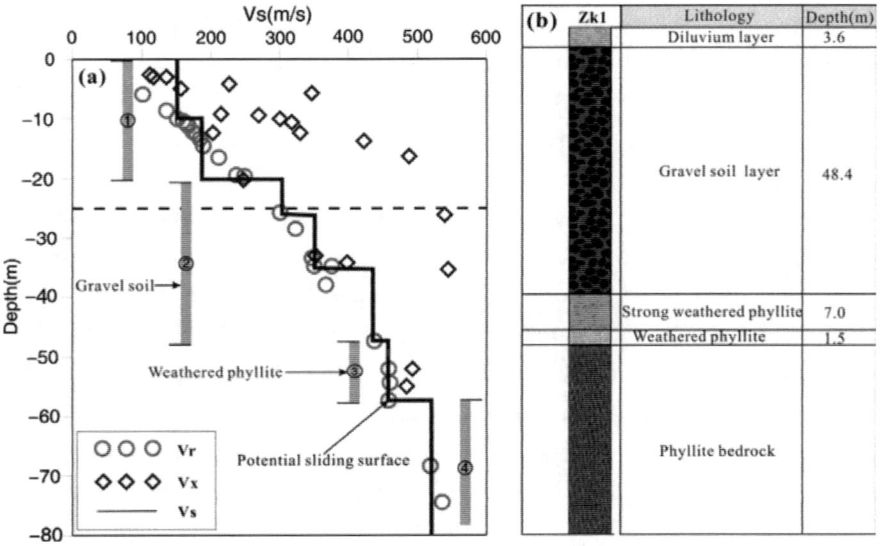

Figure 6. Shear wave velocity estimated by MSM on the position of Zk1. v_r is phase velocity of surface wave, v_x is apparent shear wave velocity, and v_s is shear wave velocity.

Shear wave velocity structure was obtained with the inversion of phase velocity of Rayleigh Wave, and the most popular method is Forking Genetic Algorithm (FGA). This method could search optimum solution of shear wave velocity mainly due to its advantage of fast calculation and strong global searching ability[10]. But it mainly depends on the initial model, and requries the layer number, each layer thickness and the velocity of the initial model.

For landslide investigation, variation of shear wave velocity along depth is interpreted as the basis of stratum structure, so it was not necessary to calculate the absolute value of shear wave velocity,but the apparent shear wave velocity is appropriate [11].

$$v_{x,i} = \left(\frac{t_i . v_{r,i}^4 - t_{i-1} . v_{r,i-1}^4}{t_i - t_{i-1}} \right)^{1/4} \quad (2)$$

Where v_r is phase velocity of Rayleigh Wave, t_i is cycle, v_r-H is converted into v_x-H using above equation. Finally, we transversely interpolated and fitted v_x to acquire the shear wave velocity structure on the position of borehole Zk1.

4. Discussion and Conclusion

Shear wave velocity structure was evaluated with phase velocity dispersion curves exacted from the vertical component of microtremor. In figure 6(a), v_s, v_r and v_x represent the shear wave velocity, phase velocity and apparent shear wave velocity, respectively. MSM has poor resolution for shallow landslide structure (10~20m), but it is more effective for the deeper buried interfaces.

Overburden thickness on the bedrock of Xishan village landslide was about 52m according to borehole data in figure 6(b). Here two obvious interfaces were shown at 47.4m and 57.3m depth on the Vs curve in figure 6(a), and they could be interpreted overburden thickness(180~410m/s) and bedrock interface(>515m/s), respectively. Shear wave velocity of weathered bedrock is about 457.6m/s in the range of 47.4 to 57.5m at depth. However, our survey results could not obtain more information about weathering degree of underlying bedrock due to low coverage density investigation.

Compared with borehole Zk1 in figure 2, the shallow stratum of landslide from top to bottom is Quaternary diluvium(3.6m), gravel soil(3.6~52m), strong weathered phyllite(52~59m), weathered phyllite(59~60.5m)and phyllite bedrock(>60.5m). Potential sliding surface is considered as strong weathered phyllite layer about 52m at depth. If we averaged the thickness in the range of 47.4~57.5m, potential sliding surface estimated from MSM is 52.35m at depth. This indicates that this method has high resolution for detecting deep seated landslide structure,especially for accumulative landslide on the bedrock.

Based on the above analysis and discussion , the results of this study are : (1) MSM is available for deep seated underlie bedrock landslide with high resolution without topographical restriction. (2) MSM is more effective than drilling in terms of probing depth, survey scale and cost. Hence, MSM seems to be more suitable for large scale and high density survey on a gravel landslide.

Acknowledgments
Our study has been supported financially by the National Basic Research program of China (973 program, Grant No .2013CB733201). we would like to thank our friends, colleagues for their invaluable advice, assistances and supports during field experiments and data acquisition.

References

[1] Feng S K 2003 Array microtremor survey and its application to civil Engineering[J]. *Chinese Journal of Rock Mechanics and Engineering*, **6**: 028.

[2] Xu P, Ling S, Li C, et al 2012 Mapping deeply-buried geothermal faults using microtremor array analysis[J]. *Geophysical Journal International*, **188**(1): 115-122.

[3] Okada, H 2003 The Microtremor Survey Method (translated by Koya Suto): *Geophysical Monograph Series*, No.**12**, Society of Exploration Geophysicists.

[4] Toksoz M N, Lacoss R T 1968 Microtremors — mode structure and sources [J].*Science*, **159**:872~873.

[5] Aki, Keiiti 1957 Space and Time Spectra of Stationary Stochastic Waves, with Special Reference to Microtremors [J]. *Bull .Earthq. Res. inst*, **35**(3): 415-456.

[6] Okada H 2006 Theory of efficient array observations of microtremors with special reference to the SPAC method [J]. *Exploration Geophysics*, **37**(1): 73-85.

[7] Chavez-Garcia F J, Rodriguez M, Stephenson W R 2006 Subsoil structure using SPAC measurements along a line. *Bull. seism. Soc. AM.*, **96**(2):729-736.

[8] Okada H, Ling S Q 1994 About a recent study on the surveying geologic structures by using the Microtremor Survey Method. *Report of Hokkaido University* (in Japanese),.

[9] Asten M W 2006 On bias and noise in passive seismic data from finite circular array data processed using SPAC methods. *Geophysics*, **71**(6):V153-V162.

[10] Cho I, Nakanishi I, Ling S Q, et al 1999 Application of Forking Genetic Algorithm fGA to an exploration method using microtremors . *Butsuri-Tansa* (in Japanese), **52**:227-246.

[11] Ling S Q,Miwa S 2006 The evaluation of soil structures by Surface Wave Prospecting Method and Microtremor Survey Method—2004 Mid Niigata Prefecture Earthquake .In: Liu Y Z e d. *A New Technique On Engineering Geophysical Method.* Beijing: Geological Publishing House.80~85.

Using several monitoring techniques to measure the rock mass deformation in the Montserrat Massif

M Janeras[1], JA Jara[1], F López[1], J Marturià[1], MJ Royán[2], JM Vilaplana[2], A Aguasca[3], X Fàbregas[3], F Cabranes[4], JA Gili[4]

[1] Institut Cartogràfic i Geològic de Catalunya (ICGC), Parc de Montjuïc, 08038-Barcelona, Spain
[2] RISKNAT Group, GEOMODELS. Department of Geodynamics and Geophysics, Faculty of Geology, University of Barcelona (UB), 08028-Barcelona, Spain
[3] RSLAB - Research Group on Remote Sensing. Signal Theory and Communications Department, Technical University of Catalonia (UPC), 08034-Barcelona, Spain
[4] Department of Geotechnical Engineering and Geosciences, Technical University of Catalonia (UPC), 08034-Barcelona, Spain

E-mail: marc.janeras@icgc.cat

Abstract. Montserrat Mountain is located near Barcelona in Catalonia, at the north-east corner of Spain, and its massif is formed by conglomerate interleaved by siltstone/sandstone with steep slopes very prone to rock falls. The increasing visitor's number in the monastery area, reaching 2.4 million per year, has pointed out the risk derived from rock falls for this building area and also for the terrestrial accesses, both roads and rack railway. A risk mitigation plan is currently been applied for 2014-2016 that contains monitoring testing and implementation as a key point. The preliminary results of the pilot tests carried out during 2014 are presented, also profiting from previous sparse experiences and data, and combining 4 monitoring techniques under different conditions of continuity in space and time domains, which are: displacement monitoring with Ground-based Synthetic Aperture Radar and characterization at slope scale, with an extremely non uniform atmospheric phase screen because of the stepped topography and atmosphere stratification; Terrestrial Laser Scanner surveys quantifying frequency for unnoticed activity of small rock falls, and monitoring rock block displacements over 1cm; monitoring of rock joints with a wireless net of sensors; and tentative surveying for singular rocky needles with Total Station.

1. Introduction

1.1. Situation and background

Montserrat Mountain is located near Barcelona in Catalonia, at the north-east corner of Spain (figure 1). This isolated massif formed by thick layers of conglomerate interleaved by siltstone/sandstone coming from a Late-Eocene fan-delta, emerges over the Llobregat River with an overall height of 1000 m (from 200 to 1200 m.a.s.l.). This configuration leads to staggered slopes where vertical cliffs of conglomerate alternate with steep slopes (figure 2). The slight tectonic suffered by this massif preserved the horizontality of the stratigraphic layers, but also determined two dominant joint sets which are planar, vertical, orthogonal, very persistent and with spacing usually ranging from

1 – 10 m [1]. A main consequence of this geological structure is that the massif is very prone to rock falls.

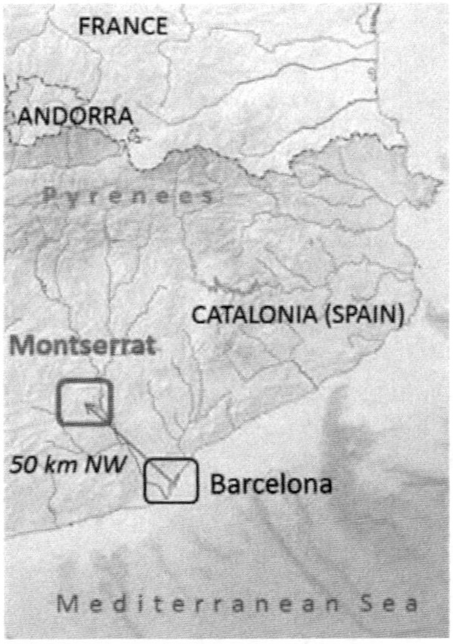

Figure 1. Situation map of Montserrat in Catalonia, NE corner of Spain.

Figure 2. Overview of Montserrat Monastery area and the access by road and railway (SE edge of Montserrat Mountain).

Montserrat massif constitutes a Natural Park (about 3500 ha) and hosts a sanctuary and monastery with a millenarian history and great tradition in Catalonia. The monastery and some touristic premises are placed at the SE edge of the mountain at 700 m.a.s.l. Combining the local and foreign tourism or pilgrimage, the visitor's number in the monastery area is increasing yearly, reaching 2.4 million per year in 2014. Leaving aside this focus, nowadays it is estimated that an additional 0.8 million person per year are circulating for hiking or climbing overall Natural Park.

These hazardous conditions, faced to the concentrated exposition, point out the risk derived from rock falls for this building area and also for the terrestrial accesses, both roads and rack railway ([2], [3] and figure 2).

1.2. Rock fall risk

In this particular massif, rock mass instabilities range 6 orders of magnitude in volume, starting with pebbles disaggregation in the conglomerate, following with slabs and plates, and ending with monolithic rock masses (1000 m^3 or more) delimited by spaced joints. Several recent episodes have highlighted that the frequency is higher than social perception tends to recognize. After [4] and [5], a frequency of about 10 rockfall per year is estimated for magnitude over few tens cubic meters in the overall mountain massif.

In the period 2001-2014, several active and passive protection works have been carried out, mainly along railway and, after several major rock falls, also for the road access. In 2014 a new 3-year plan for global risk mitigation has been started. It is founded by the Catalan Government, directed by Institut Cartogràfic i Geològic de Catalunya (ICGC, the public service for cartography and geology), and supervised by the Patronat de la Muntanya de Montserrat (PMM, the public entity that manages this natural and cultural heritage, under protection) [6].

Figure 3. 2008/12/28 and 890 m^3 rockfall event at Degotalls cliff (right). In the left-hand pictures, the road and the railway cut by blocks can be appreciated.

For the larger events, stabilization measures are only possible in some specific cases; also passive defences, like rock fall fences, offer only partial protection for low magnitude rock falls. Therefore, monitoring is explored as a new line of work in Montserrat, while not feasible as an early warning system (EWS) at present state of experience in site, at least to improve the knowledge on the failure triggering factors, which is a main goal adopted for the current risk mitigation plan at short term. The aim of this hazard control strategy is to get an aid for setting priorities for stabilization works, that might achieve a fairly and sustainable risk management.

In this sense, successive rock falls between 2007 and 2008 at the same place (a 170m high cliff called Degotalls, figure 3) remarked the importance to detect the progression of fractures in the rock mass, and therefore the need of monitoring rock mass deformation and a better comprehension of the instability mechanisms.

2. Monitoring techniques

Landslides are commonly monitored and several early warning systems are implemented successfully, thanks to the displacements of the terrain that often are developed along a certain period before collapse. References about premonitory deformations of rockfall are ever more usual as [7] and [8]. In the Montserrat case, considering the properties of the rock mass, it can be especially difficult to implement such type of approach because the mechanical behaviour seems to be very stiff and/or brittle; the collapse can be reached at low strain. For this reason, high precision measurements will be mandatory. The goal of monitoring action during the current 3-year plan is to improve the knowledge on rock mass behaviour and the comprehension of the instability propagation and the failure preparation, like in [9] and [10]. Thus, in this first current phase different pilot tests are being performed to check the applicability of several techniques depending on each case and site.

To carry out this project, ICGC has established scientific-technical collaboration with specialized research teams of UPC (Technical University of Catalonia) and UB (University of Barcelona). After the analysis and comparison of these first results, new monitoring tests are under study and development for 2015-16. For the first current phase, 4 monitoring techniques were selected with different spatial resolution (punctual or scattered) and temporal acquisition (continuous or discontinuous) as seen in table 1. These 4 techniques and each test performed in Montserrat during

2014 are introduced in next subsections 2.1 to 2.4, but their preliminary results will be summarized in common in section 3.

Table 1. Overview of the 4 monitoring techniques used in the pilot tests.

		Temporal domain	
		Continuous	Discontinuous
Spatial domain	Scattered	Ground Based Synthetic Aperture Radar	Terrestrial Laser Scanner
	Punctual	Rock joint instrumentation	Total Station

2.1. Surveying Total Station

The monitoring with Total Station (figure 4) is the simplest technique presented in this contribution. It was envisaged as a very cheap (low cost) approach to monitor a small number of points in a block. The initial approach is also punctual in the time domain as the measurements are carried out by campaigns through periodic visits.

The Cadireta block (figure 5) was used as a demonstration place to adapt the method and assess its performance. It is a block about 8000 m^3 that is overhanging over the hillslope. Its potential risk only threatens a number of trails and the mountain-climbers visiting the area, and the local road that surrounds the Mountain [11]. Here, our main goal has been to test the monitoring method.

Figure 4. Leica TM30 Total Station used in Montserrat, and lateral view of the Cadireta and the base station (in the road).

A convenient place for setting up the Total Station was the small road bellow the block (figure 4). From there, the measurements were made over a set of prisms distributed over the block and around it (attached to the rear wall, used as references, see figure 6).

Figure 5. The Cadireta block in the Montserrat north face, in the "Agulles" (needles) zone.

As expected, the big slope-distances between the Total Station and the block (520-540 m) prevented to reach a high precision in the coordinates of the prisms (XYZ). The angular precision (about 1 arc second, i.e. 0.3 mgon) produced a significant error in the coordinates (3 to 5 mm). On the other hand, the measured distance is more robust (in the mm-level).

So, the distance difference between points ($D_{ij} = D_i - D_j$) is being used as monitoring observable (figure 4 and 6). D_i is the distance to a control prism, whereas D_j is the distance to the closer reference prism. Several tests have been carried out at the Cadireta site to confirm the actual precision of the method. The results show that the Standard Deviation of ΔD_{ij} is close to 0.8 mm, whereas the Standard Deviation of the 3D coordinates is roughly 5 mm.

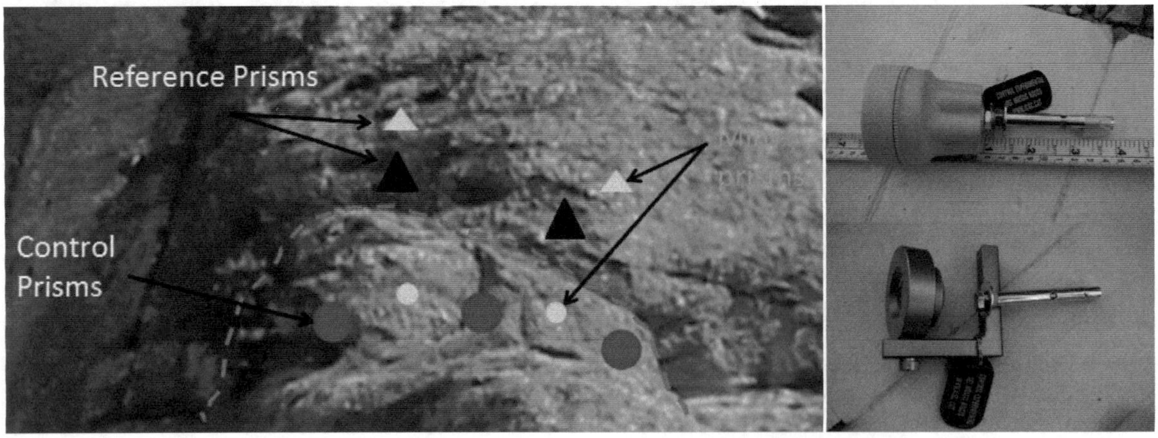

Figure 6. Close view of the top part of the Cadireta block (contoured by dashed red line). The prisms are attached to the block and also to the rear wall (used as reference). Right: standard circular prism or reflector (top) and mini-prism (bottom).

Moreover, this differential procedure has two additional advantages. The first one is related with the correction of the raw distances to take into account the ambient factors. In the Montserrat

Mountain it is very difficult to measure the actual temperature and humidity of the air due to its huge variability in time and in elevation, like exposed in section 2.4 as a critical factor for GBSAR. Through the difference of similar measurements, the observable is quite insensitive to small inaccuracies in the measurement /estimation of the mean temperature and humidity along the laser beam. Secondly, our procedure is also robust in front of small "changes" in the base point; these spurious "movement" might be produced by an eccentricity in the setup of the Total Station or by a real settlement of the ground in the vicinity of the base benchmark.

When writing this contribution we have just acquired the firsts campaigns at the Cadireta site; we plan to carry out a seasonal monitoring during some years in order to characterize the natural behaviour of this block. On the other hand, we plan to spread this simple methodology to other points around the Mountain in order to crosscheck the other techniques.

2.2. Monitoring of rock joints

Monitoring of rock mass movement by installing sensors that directly measure the relative motion of discrete points of individualized blocks is a method of monitoring; when combined with an automatic data acquisition system, the method can provide high precision data and high temporal resolution in real time.

In Montserrat, an experience with this auscultation technique was started on September 2010 to monitor the movement of a huge rock mass (A3-6 block). To that end, 3 displacement extensometric type sensors together with an air temperature sensor were installed to measure movement of this mass relative to the massif. These sensors were wire-connected to a CR800 Campbell Scientific (CS) datalogger provided with a photovoltaic power system and a 3G mobile communications system for remote access. Data are acquired at one sample per minute and stored locally every 30 minutes. This system provides continuous data with ±0.01 mm of resolution and ±0.02 mm of precision.

Currently, after four and a half years of operation, system continues providing continuous data demonstrating its feasibility for long term operation. Some results will be given in section 3.2. Data show elastic displacements, with a full range of about 2.5 mm, clearly related with thermal effects on mass and sensors. However, a major disadvantage of such auscultation technique is the effort, and the cost, of cabling the sensors until the datalogger, especially at Montserrat massif where this task must be carried out by personnel qualified for rope access works.

A project to extend auscultation in Montserrat massif to others potentially unstable blocks was started in 2014, following the priorities previously analysed in [6] and [12]. Taking advantage of experience with A3-6 block, new sensors were installed along the year to monitor the movement of a cluster of blocks called "devil's ceiling" (Diable), where a huge rockfall occurred in XVIth century, as from historical documents (figure 7).

Three different types of displacement sensors have been installed: 13 crackmeters to measure distances between close anchor points (figure 8a), 5 wire crackmeters to measure distances between distant anchor points and 4 tiltmeters, based on MEMS accelerometers, mounted on aluminium beams to measure vertical displacement between anchor points of the beam (figure 8b).

Figure 7. Situation, front and side views of the cluster of blocks called "devil's ceiling" at Montserrat massif. Green spots illustrate the locations of the installed sensors. Each spot represents one or more sensors.

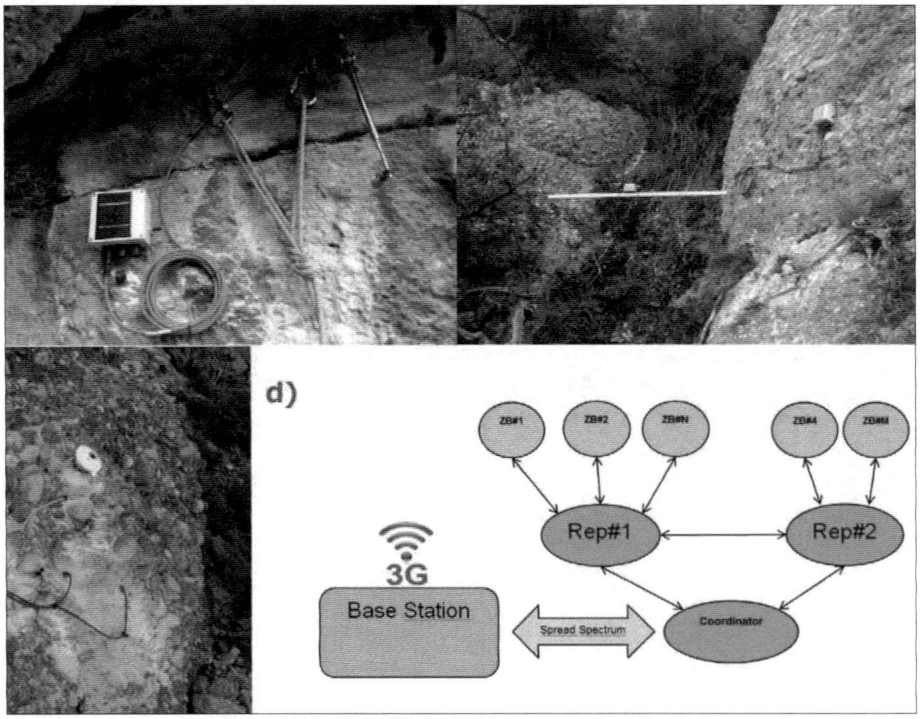

Figure 8. a) Crackmeter installed to measure distance between anchor points located at both sides of a crack, connected to a ZBLogger; b) Tiltmeter installed on a beam and connected to a ZBLogger; c) Soil temperature sensors and pyranometer installed on a face of the block cluster; d) Schema of the ZBLoggers network.

For a deep knowledge of the effect of temperature on the measure, besides a new air temperature sensor, two soil temperature sensors have been installed inside the rock at 20 cm and 40 cm depth near a pyranometer to measure solar (300 to 2800 nm) irradiance over the cluster of blocks (figure 8c). Also, a temperature sensor is installed at each of 4 beams.

To reduce the cost and effort of cabling sensors to one or several CS dataloggers, and the visual impact of it, a new concept of distributed wireless data acquisition network has been implemented, based on ZBLoggers. A ZBLogger is a low power, and low cost, system of ICGC own design with 3 single ended (or 1 differential plus 1 single ended) input channels and wireless connection, based on ZigBee protocols (IEEE 802.15.4 standard), which can communicate with others ZBLoggers and with CS datalogger as well. Eight LR6 1.5V rechargeable batteries and a solar cell of 0.96 W provides the system with enough energy for long term operation (5-10 years, depending of batteries service life in terms of charging cycles).

ZBLoggers digitize signal of input channels, at 1 sample each 10 minutes, with a 16 bits sigma-delta A/D converter. Data from a ZBLogger can be sent directly to a datalogger across the wireless network (800 m maximum in case of direct line of sight), or via others ZBLoggers acting as repeaters (figure 8d). Once data are stored inside the datalogger, they are sent to Data Centre via Internet, and archived into a database. Finally, "NetMon", a proprietary web based tool developed by ICGC, allows available data to be harnessed effectively. Currently, after 6 months without any important issue, the system remains operational and ready to be extended to others blocks during 2015.

2.3. Terrestrial Laser Scanner

The monitoring of several rock cliffs of Montserrat Mountain was carried out through an Ilris-3D (Optech) Terrestrial Laser Scanner (TLS) instrument also called Terrestrial LiDAR, which consists of a laser transmitter and receiver and a scanning device (figure 9). The laser beam is reflected directly onto the rock slope surface, which means that no reflector is necessary. The instrument computes the distance to an object, also called the range, by using the time-of-flight of the laser pulse [13]. According to specifications, ILRIS-3D can reach a maximum range of 700 m for natural slopes and decreases as a function of the material reflectivity and incidence angle to the object. The datasets acquired by this device can provide an accuracy of 7 mm at a distance of 100 m, according to the manufacturer's specifications. The device can acquire a huge number of points, also called point clouds, in a very short time (2,500 points/s), which provides a high density of information (around a few thousand points per square meter).

Figure 9. Data acquisition with the TLS equipment in Montserrat Monastery cliff.

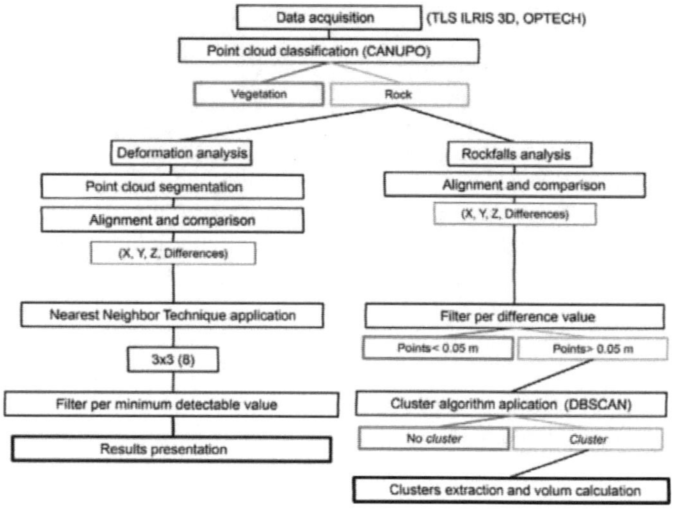

Figure 10. Workflow of the processes applied to detect deformation and rockfalls.

The time series of TLS data in Montserrat Mountain is the longest of all the techniques exposed in this work, so it has special interest in order to detect both rockfalls and deformation from years ago. In Degotalls area the first data acquisition was performed in May 2007 and 9 different fieldwork campaigns were conducted up to July 2014 (table 2). So, in this case the monitoring period span 2630 days. Degotalls study area consists of two different rock cliffs, so for the analysis this study area was separated into Degotalls North (N) and Degotalls East (E). Moreover, the first data acquisition of the Monsterrat Monastery area was performed in February 2011 and 7 data acquisitions were carried out up to July 2014 (table 2), spanning a monitoring period of 1254 days. Firstly, a classification of each point cloud using CANUPO application [14] was performed in order to filter vegetation areas. Then, each dataset was aligned with the previous datasets to detect rockfalls between periods, or with the first dataset to detect accumulated deformations. This procedure was carried out through a preliminary identification of homologous points and a subsequent minimization of the distance between point clouds using the Iterative Closest Points (ICP) algorithm [15]. A refinement of the alignment was obtained through automatic outlier selection: those points with a distance greater than 3 cm were not included in the scan alignment process. The comparison of the different TLS datasets was based on the quantification of the distances between each pair of datasets and was performed using a conventional "point-to-surface distance" methodology implemented in IMInspect software, InnovMetrics PolyWorks v.10.0.

Table 2. Date of TLS acquisition and monitoring periods.

Monastery		Degotalls North		Degotalls East	
Date of acquisition	Cumulative days	Date of acquisition	Cumulative days	Date of acquisition	Cumulative days
		2007.05.11	0	2007.05.11	0
		2009.12.18	952	2009.12.18	952
2011.02.15	0	2010.12.02	1301	2010.12.02	1301
2011.05.12	86	2011.05.12	1462	2011.05.12	1462
2011.12.12	300	2011.12.12	1676	2011.12.12	1676
2012.03.28	407	2012.03.28	1783	2012.03.28	1783
2012.06.22	493	2012.06.22	1869	2012.06.22	1869
2012.11.06	630	2012.11.06	2006	2012.11.06	2006
2013.12.11	1030	2013.12.11	2406	2013.12.11	2406
2014.07.23	1254	2014.10.07	2706	2014.07.23	2630

In order to detect rockfalls or deformation, two different processes were applied (figure 10). In the case of deformation, a Nearest Neighbour (NN) filtering technique [7] was applied in order to improve the accuracy of the scans comparison. Briefly, this technique involves the extraction of the differences between successive scans and the median computation of the NN (NN = 8 in this study). As a result of this process the minimum detectable deformation value was 1 cm. On the other hand, to detect rockfalls a procedure to find clusters was applied [16]. Mainly this process consists of three steps: a) filtering of differences not corresponding to rockfalls; b) application of DBSCAN algorithm (Density-Based Spatial Clustering of Applications with Noise [17]) to remove noise and find different clusters; and c) quantification and volume computation of the single clusters detected corresponding to single rockfalls. Given that the minimum value of used differences was 5 cm, the minimum number of required points of a cluster was defined as 10 and the average point spacing of the scans was between 7 cm and 9 cm, the minimum detectable rockfall volume was defined as 0.001 m^3 (equivalent to 2.5 kg of conglomerate).

2.4. GBSAR

In 2014-2015 a first extensive measurement campaign using a Ground Based Synthetic Aperture Radar (GBSAR) has been performed in Montserrat. Data collected by this kind of SAR sensors allow obtaining terrain reflectivity images, which can be processed by means of differential SAR interferometry (DInSAR) algorithms for the monitoring of deformation episodes, with millimetric accuracy [18] and [19]. DInSAR techniques are based on the use of the phase-differences between multitemporal pairs of complex SAR images of the same area of study, to obtain the terrain displacement undergone in the line-of-sight (LOS) direction.

The aim of the experiment was to take advantage of the wide area coverage (2 km long by 1 km wide) and its high sensitivity to displacement detection (on the order of millimeter). GBSAR sensors are used for the precise monitoring of small-scale phenomena. These kind of instruments have been used for several applications as landslide monitoring, subsidence hazards in urban areas, volcanoes monitoring, etc, [20] and [21], when a classical Orbital DInSAR systems cannot be used because of the bad geometric orientation of the scenario or the revisit time of the Spaceborne sensor is not short enough to monitor the deformation. The main characteristics of the UPC RiskSAR (GBSAR) are summarized in table 3.

Table 3. Main characteristics of RiskSAR equipment (GBSAR system of UPC).

Parameters	**Values**
Carrier Frequency	9,65 GHz
Modulation	triangular CW-FM
Signal bandwidth	100MHz
Polarization	Full polarimetric combinations
range resolution	1,5 m
azimuth resolution	10 mRad
range coverage	1,5Km
azimuth coverage	±30°

The instrument was placed in el Pont de Guilleumes (figure 11), at 380m altitude, its orientation was south and 25° upslope, pointing to two different main rock cliffs, 400 and 800 meters far from the radar. The system operated autonomously but it could be remotely controlled via wireless link. A sequence of 1 measurement per hour was programmed along the 5 months experiment (October 2014 - February 2015).

Figure 13 shows the reflectivity image covered by the sensor projected on topography. From the image, it can be observed the strong reflectivity response corresponding to the main rock cliffs. Different radar calibrators were deployed along the slope to facilitate the image geocoding process. A test with a Polarimetric Active Radar Calibrator (PARC) over a millimetric positioner was performed to validate system capability to detect small displacements (figure 12). Small difference between real and retrieved position can be seen, which are due to thermal stabilization of the internal electronics of the PARC.

Figure 11. RiskSAR system placed at Pont de Guilleumes site with weatherproof cover for temporary use at short term.

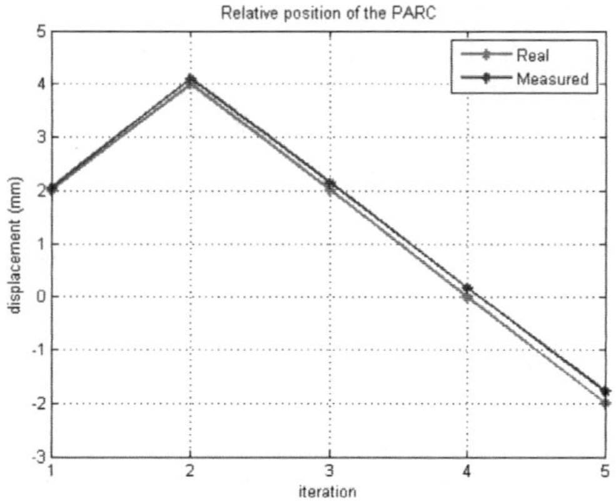

Figure 12. PARC displacement position retrieved after DInSAR processing.

Figure 13. Radar reflectivity image on Google Earth map of the scenario.

For this type of sensors, the atmospheric phase screen (APS) is the most relevant artefact of distortion on the interferometric phase. Hence, in order to apply any DInSAR technique to obtain reliable deformation maps, these atmospheric artefacts must be correctly estimated and consistently compensated for [22] and [23]. Unfortunately it implies that an important set of measurements have to be discarded because the accentuated atmospheric anomalies in mountainous areas as fog, rain, stratification and temperature inversion.

3. Preliminary results and analysis

As corresponding to pilot tests phase, the results obtained until now are only considered as preliminary, pending on reviewing, adjustment or improvement. The interpretation of an eventual rock mass movement is expected to be a combination of: temporary and retrievable displacements as the elastic response to thermal effects and variable actions under failure threshold; and permanent

IOP Conf. Series: Earth and Environmental Science **26** (2015) 012030 doi:10.1088/1755-1315/26/1/012030

cumulative displacements as the plastic signs of progress of failure. This distinction corresponds to the next subsections, preceded by another particular result obtained by the monitoring works.

3.1. Rock fall activity assessment

TLS allows the detection of rockfalls that were not recorded by classical observational tasks, and also to calculate their volume. On the contrary, GBSAR is not able to clearly detect these rapid changes on scenery, because the comparison in terms of the signal's phase. All these unobserved events were characterized by small magnitude, ranging volumes between 0.001 and 0.732 m^3, and associated with pebbles disaggregation and small slabs failure. The rockfalls detected by period could be observed in table 4. In the case of Monastery cliff the total number of rockfalls detected was 44; ranging volumes between 0.001 and 0.45 m^3 (90% of rockfalls minor than 0.05 m^3). In this area, a remark has to be done in the results of periods between June 2012 and December 2013, because some of the detections were derived from the stability tasks carried out in a certain rock block.

Table 4. Rockfall activity characterized with Terrestrial TLS expressed in number of rockfalls detected per period of analysis in each study area.

Monastery		Degotalls North		Degotalls East	
Period	Rockfalls	Period	Rockfalls	Period	Rockfalls
		2007.05–2009.12	11	2007.05–2009.12	18
		2009.12–2010.12	22	2009.12–2010.12	33
2011.02–2011.05	0	2010.12–2011.05	8	2010.12–2011.05	8
2011.05–2011.12	3	2011.05–2011.12	16	2011.05–2011.12	8
2011.12–2012.03	2	2011.12–2012.03	8	2011.12–2012.03	9
2012.03–2012.06	1	2012.03–2012.06	1	2012.03–2012.06	0
2012.06–2012.11	13	2012.06–2012.11	9	2012.06–2012.11	4
2012.11–2013.12	21	2012.11–2013.12	8	2012.11–2013.12	7
2013.12–2014.07	4	2013.12–2014.10	6	2013.12–2014.07	4
Total	44	Total	89	Total	91

On the other hand, the specific results of Degotalls N and E could be also observed in table 4. In order to detect just the unobserved rockfalls, in Degotalls N a filtering of the events caused by the big rockfall occurred in December 2008 had to be carried out. Thus the final results showed a total number of rockfalls of 78; ranging volumes between 0.001 and 0.73 m^3 (90% of rockfalls minor than 0.1 m^3). In the case of Degotalls E the total number of rockfalls recorded was 91; ranging volumes between 0.001 and 0.28 m^3 (95% of rockfalls minor than 0.1 m^3). As an example of results, the rockfalls detected in Degotalls E rockface are presented in figure 14a, as well as their evolution over time (figure 14b) and their volume histogram (figure 14c).

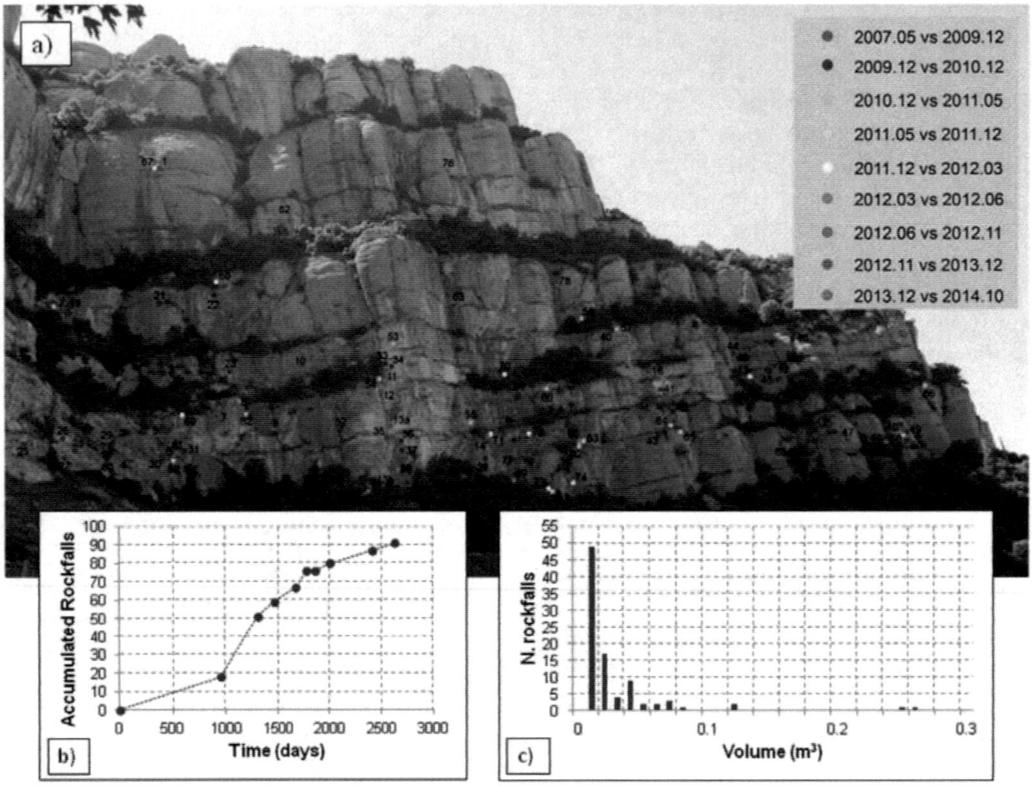

Figure 14. a) Rockfalls in Degotalls E rockface; b) Rockfalls evolution over time; c) Rockfall volume histogram.

Rate of activity seems to be uniform during time but the low time resolution of TLS data does not seem to be the most appropriate. So, for 2015 it is planned to increase the sampling frequency to seasonal control, to check time variability.

3.2. Rock mass elastic/temporal deformation

The register of the monitoring of rock joints in block A3-6 during 4 years (but with some gaps due to technical problems with wildlife interaction) clearly show the annual cycle and also daily oscillation (figure 15), as it can be clearly identified with a FFT analysis. This behaviour can be linked with thermal deformation of rock mass superficial part, including the analysed block, but also thermal effect on sensor must be considered. The mean amplitude of annual oscillation is 1.7, 0.9 and 1.3 mm for sensors 1, 2 and 3 respectively, and 34.6ºC for temperature (figure 15). Sensor 2 and 3 show a completely elastic behaviour recovering the displacement after annual cycle, but for the sensor 1 (a wire extensometer measuring the aperture of the rear joint at the upper part) a slight tendency to accumulate displacement at a rate of 0.33 mm/year can be seen. This fact must be confirmed with longer time-series to determine if it could be a creeping process conducting to a long-term toppling failure.

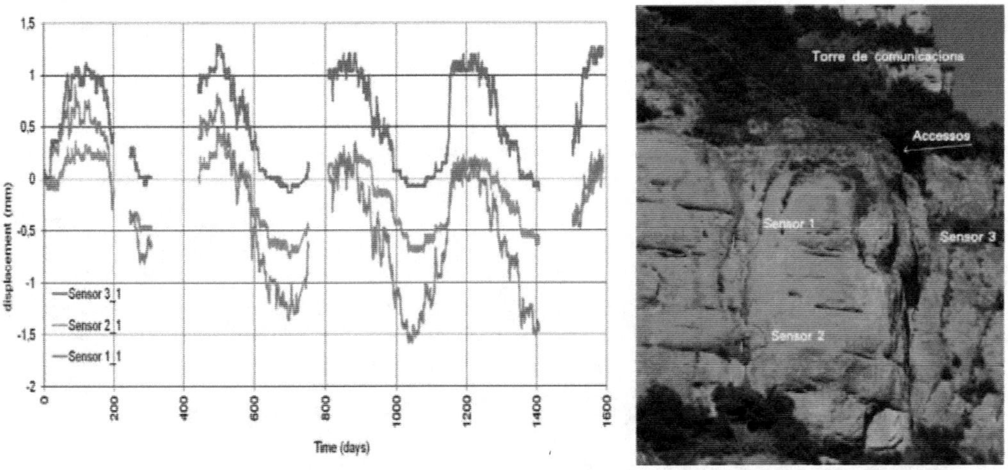

Figure 15. 4-years long register of monitoring of rock joints in block A3-6, where sensors 1 and 2 are wire extensometers orthogonal to joint, and sensor 3 is a bar extensometer parallel to joint.

Thermal effect on sensors is expected to be mainly related with air temperature with rapid heat transfer, whereas the rock mass response should be related to the temperature of the ground. To improve the knowledge in this issue, in the new instrumented block (Diable) thermometers for the air and inside the rock (at depths of 20 and 40 cm) have been installed, as well as a pyranometer (because the sun radiation influences the rock temperature). After the first 3 months of monitoring, the heat flux inside the rock mass is clearly shown in the figure 16, but the data are just starting to be analysed.

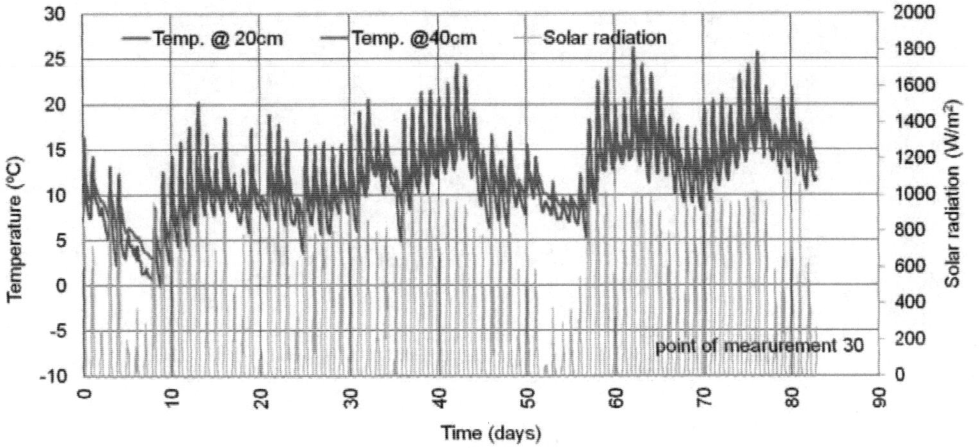

Figure 16. Register for the 3 first months in Diable block: temperature inside the rock mass at 20 and 40 cm, and solar net radiation.

The other monitoring techniques, at present, are not giving a clear result for this level of rock mass deformation. Probably this range of movement (mm level) is under the error for the equipment and site implementation. Under the working conditions, at present we are reaching precisions about 10 mm for TLS, 3 mm for GBSAR and 1 mm for Total Station. Depending on the magnitude of displacement of each rock block related with its structural disposal, the movement will be detected or not with a given technique. For instance, at the Cadireta we plan to complete an annual cycle with seasonal measurements with Total Station priori to drive any conclusion. To improve the precision of both scanning techniques, for 2015 a new test with GBSAR at closer distances and higher frequency in the

Monastery cliff is planned; on the other hand, 2 points for positioning TLS will be added closer to both sectors under deformation analysis.

3.3. Rock mass plastic/permanent deformation

On the detection of precursory deformation, TLS monitoring was able to detect movement in two different blocks located in Degotalls N study area (figure 17). The Block A instability was detected in the period 2007.05vs2009.12, after the big event occurred on December 2008 and next to the fallen block. This movement was characterized by an initial mean displacement of 2 cm and a subsequent stop in the following periods. This behaviour suggests that the movements were caused by the great rockfall occurred on December 2008; in the following periods the displacement stopped because of the appropriate stabilization works carried out on the same block.

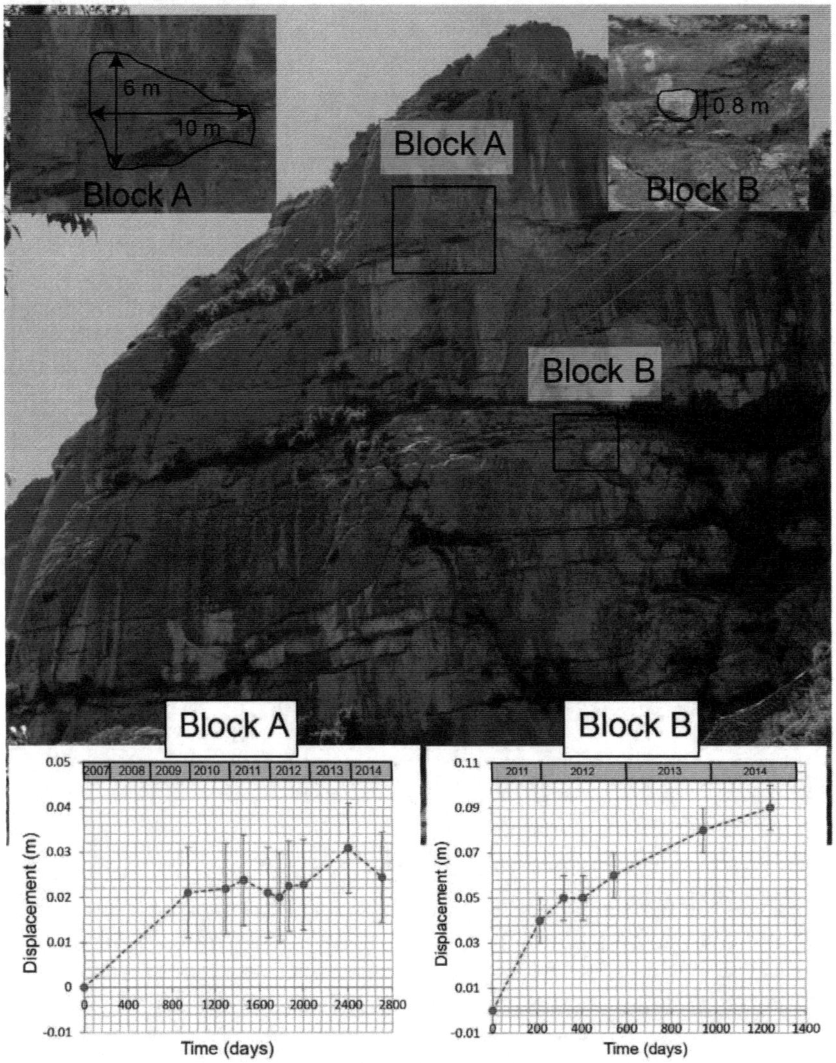

Figure 17. Location of the two blocks where displacement has been detected and their evolution over time, as measured with TLS.

The second instability (Block B) was detected in the period 2011.05vs2011.12 in a block close to a small rockfall occurred in the same period. An initial mean movement of 4 cm was detected and in this case any stabilization tasks were performed so an increment of the displacement has been recorded (figure 17). Until July 2014 the accumulated displacement in this block was 9 cm showing the need to continue with the monitoring surveys, although the TLS seasonal frequency makes difficult to use it as a prevention tool. Fortunately, there is a rockfall fence protecting the road. The total amount of displacement that the block will be able to accumulate before falling will depend on the particular stability conditions of the block; for hard rock massif and isolated rockfall detachment, it is expected to be small [24].

For 2015 it is planned to add several surveying prisms for the block A to contrast the TLS measures with Total Station technique (with slightly higher precision), ultimately to confirm the proper response of the stabilization works. In parallel the feasibility of continuous instrumentation is under analysis.

Finally, for the pilot test of GBSAR at Guilleumes, the displacement map retrieval of the rock cliffs can be carried out with the zero baseline adaptation of the CPT technique [20]. Figure 18 shows a preliminary result obtained using a daily dataset of 34 SAR images. Non significant displacements are observed on the surface of the rock cliffs, showing a stable behaviour during the measurement campaign.

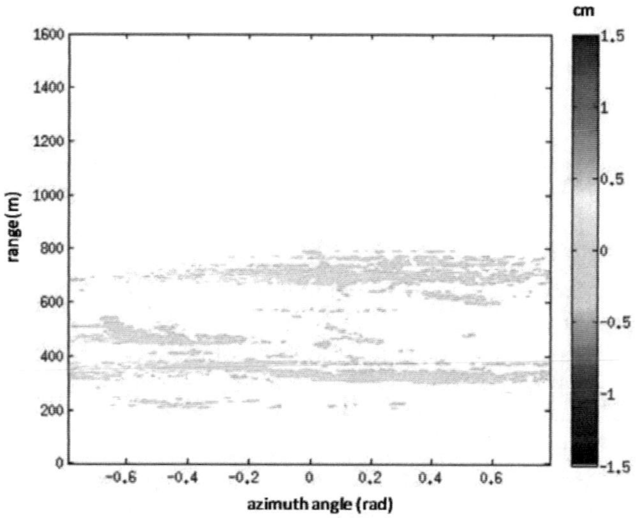

Figure 18. Displacement map (in polar coordinates) retrieved after DInSAR processing. Non significant displacements are observed by the moment.

4. Conclusions

We are on the way to improve our knowledge on the behaviour of the rock mass in Montserrat Mountain and the comprehension of the geomechanical processes leading to rockfall failures. One of the main difficulties to reach this goal is its stiff/brittle behaviour, so it is necessary to achieve the best precision in measurement. However, the applied monitoring techniques are giving first interesting results in this sense. Preliminary observations show two kinds of movements: centimetric displacements in blocks with apparent high instability without detachment, that confirms our hope to be able to analyse premonitory signs of rupture; and millimetric displacements in joints as a daily/seasonal cyclic response, for which it is necessary to determine the thermo-mechanical coupling in the next future.

At the same time, these techniques let us to accurately quantify the rockfall activity, being the TLS a valuable tool to detect the unnoticed rockfall. These data are very useful to assess the magnitude-frequency relationship for a proper hazard assessment.

The different monitoring techniques are providing coherent results, according to each time-series length and increasing experience on its application in Montserrat. Taking into account the best capabilities of each technique, the results show an excellent complementarity of the four techniques that should be explored in deep. After these separate tests, it is planned to establish Monastery cliff (and secondly Degotalls) as pilot master area for the integration of the different techniques for a cross check vision.

Acknowledgments

All these works are supported by the Catalan Government, thanks to the current intervention plan promoted by Patronat de la Muntanya de Montserrat and managed by ICGC. We also acknowledge the logistic support and predisposition from the security service of the monastery and the benedictine community.

Authors and their corresponding research groups, according to each part of the project, acknowledge the support of several projects and institutions as follows: the "ROCKRISK" research project (Reference BIA2013-42582-P, Spanish Ministerio de Economía y Competitividad); the equipment lent by Infraestructures.cat (public company for civil works promotion by the Catalan Government); the CHARMA project (Reference CGL2013-40828-R, Spanish Goverment); the support for GBSAR operation by the Spanish Ministry of Science and Innovation (MICINN) under project TEC2011-28201-C02-01.

References

[1] Alsaker E, Gabrielsen RH and Roca E 1996 The significance of the fracture patterns of Late-Eocene Montserrat fan-delta, Catalan Coastal Ranges (NE Spain) *Tectonophysics* 266 465–91

[2] Fontquerni S, Vilaplana JM, Guinau M, Royán MJ, Barberà M, Janeras M 2012 Los elementos vulnerables expuestos a los desprendimientos de rocas en Montserrat *VIII Congreso Geológico de España* (Oviedo)

[3] Palau J, Janeras M, Prat E, Pons J, Ripoll J, Martínez P and Comellas J 2011 Preliminary assessment of rockfall risk mitigation in access infrastructures to Montserrat *Second World Landslide Forum* (Rome)

[4] Royán MJ and Vilaplana JM 2012 Distribución espacio-temporal de los desprendimientos de rocas en la montaña de Montserrat Cuaternario y Geomorfología 26 **1–2** 151–70

[5] Janeras M, Palau J, Prat E, Pons J, Rodríguez H, Martínez P and Comellas J 2013 Valoración de 10 años de mitigación del riesgo de caída de rocas en el Cremallera de Montserrat *VIII Simposio Nacional sobre Taludes y Laderas Inestables* (Palma de Mallorca: ed E Alonso, J Corominas and M Hürlimann)

[6] Janeras M, Palau J, Prat E and Ripoll J 2011 Montserrat: on a long way to rock fall risk mitigation – First experiences, some lessons and future perspectives *Interdisciplinary Rockfall Workshop* (Innsbruck – Igls: RocExs)

[7] Abellán A, Jaboyedoff M, Oppikofer T and Vilaplana JM 2009 Detection of millimetric deformation using a terrestrial laser scanner: experiment and application to a rockfall event *Nat. Hazards Earth Syst. Sci.* **9** 365–72

[8] Royán MJ, Abellán A, Vilaplana JM 2015 Progressive failure leading to the 3 December 2013 rockfall at Puigcercós scarp (Catalonia, Spain) *Landslides* DOI 10.1007/s10346-015-0573-6

[9] Arosio D, Longoni L, Papini M, Scaioni M, Zanzi L and Alba M 2009 Towards rockfall forecasting through observing deformations and listening to microseismic emissions *Nat. Hazards Earth Syst. Sci.* **9** 1119–31

[10] Royán MJ, Abellán A, Jaboyedoff M, Vilaplana JM and Calvet J 2014 Spatio-temporal analysis of rockfall pre-failure deformation using Terrestrial LiDAR *Landslides* **11** 697-709

[11] Cabranes, F 2015 *Estudio geomecánico de la estabilidad de la Cadireta d'Agulles (Montserrat)* (*Unpublished* B.Sc.Degree Thesis, ETSECCPB, Technical University of Catalonia) p 156

[12] Gallach X 2012 Estudi de susceptibilitat de caiguda de roques a la paret del Monestir de Montserrat a partir de la inspecció d'indicadors d'inestabilitat i d'anàlisi SIG (*Unpublished* M.Sc.Degree Thesis, UB, University of Barcelona) p 32

[13] Petrie G and Toth CK 2008 Introduction to laser ranging, profiling and scanning *Topographic Laser Ranging and Scanning: Principles and Processing* ed J Shan and CK Toth (London: CRC Press / Taylor & Francis) pp 1–28

[14] Brodu N and Lague D 2012 3D terrestrial LiDAR data classification of complex natural scenes using a multi-scale dimensionality criterion: Applications in geomorphology *ISPRS J. Photogrammetry and Remote Sensing* **68** 121–34

[15] Chen Y and Medioni G 1992 Object modelling by registration of multiple range images *Image Vis. Comput.* **10** 145–55

[16] Tonini M and Abellán A 2014 Rockfall detection from LiDAR point clouds: a clustering approach using R *J. Spatial Information Sci.* **8** 95–110

[17] Ester M, Kriegel H, Sander J and Xiaowei Xu X 1996 A density-based algorithm for discovering clusters in large spatial databases with noise *Proc. 2nd Int. Conf. on Knowledge Discovery and Data Mining*

[18] Gabriel AK, Goldstein RM and Zebker HA 1989 Mapping small elevation changes over large areas: Differential radar interferometry *J. Geophys. Res.* vol. 94 **B7** 9183

[19] Massonnet D and Feigl KL 1998 Radar interferometry and its application to changes in the Earth's surface *Rev. Geophys.* vol. 36 **4** 441

[20] Iglesias R, Aguasca A, Fabregas X, Mallorqui JJ, Monells D, López-Martínez C and Pipia L 2015 Ground-Based Polarimetric SAR Interferometry for the Monitoring of Terrain Displacement Phenomena–Part I: Theoretical Description *IEEE Journal of Selected Topics in Applied Earth Observations and Remote Sensing* **8** 980–93

[21] Iglesias R, Aguasca A, Fabregas X, Mallorqui JJ, Monells D, López-Martínez C and Pipia L 2045 Ground-Based Polarimetric SAR Interferometry for the Monitoring of Terrain Displacement Phenomena–Part II: Applications *IEEE Journal of Selected Topics in Applied Earth Observations and Remote Sensing* **8** 994–1007

[22] Pipia L, Fabregas X, Aguasca A and Lopez-Martinez C 2008 Atmospheric artifact compensation in ground-based DInSAR applications *IEEE Geosci. Remote Sens. Lett.* **5** 88–92

[23] Iglesias R, Fabregas X, Aguasca A, Mallorqui JJ, López-Martínez C, Gili JA and Corominas J 2014 Atmospheric phase screen compensation in ground-based SAR with a multiple-regression model over mountainous regions *IEEE Trans. Geosci. Remote Sens.* **52** 2436–49

[24] Janeras M, Macau A, Figueras S and Comellas J 2010 Prevention and control of rocky slope instabilities induced by blasting vibration in tunneling – application to Núria rack railway *Rock Mechanics in Civil and Environmental Engineering* (Lausanne: ed J Zhao, V Labiouse, JP Dudt, JF Mathier EUROCK 2010)

Monitoring of unstable slopes by MEMS tilting sensors and its application to early warning

Ikuo Towhata[1], Taro Uchimura[2], Ichiro Seko[3] and Lin Wang[3]

[1] President *of the Japanese Geotechnical Society,* Vice President for Asia *of ISSMGE*
4-38-2, Sengoku, Bunkyo-Ku, Tokyo 112-0011 JAPAN
[2] Associate Professor, *University of Tokyo*
7-3-1, Hongo, Bunkyo-Ku, Tokyo 113-8656 JAPAN
[3] President, *Chuo Kaihatsu Corp.*
3-13-5, Nishi Waseda, Shinjuku-Ku, Tokyo 169-8612 JAPAN
[3] Senior Engineer, *Chuo Kaihatsu Corp.*
3-13-5, Nishi Waseda, Shinjuku-Ku, Tokyo 169-8612 JAPAN

E-mail: towhata.ikuo.ikuo@gmail.com

Abstract. The present paper addresses the newly developed early warning technology that can help mitigate the slope failure disasters during heavy rains. Many studies have been carried out in the recent times on early warning that is based on rainfall records. Although those rainfall criteria of slope failure tells the probability of disaster on a regional scale, it is difficult for them to judge the risk of particular slopes. This is because the rainfall intensity is spatially too variable to forecast and the early warning based on rainfall alone cannot take into account the effects of local geology, hydrology and topography that vary spatially as well. In this regard, the authors developed an alternative technology in which the slope displacement/deformation is monitored and early warning is issued when a new criterion is satisfied. The new MEMS-based sensor monitors the tilting angle of an instrument that is embedded at a very shallow depth and the record of the tilting angle corresponds to the lateral displacement at the slope surface. Thus, the rate of tilting angle that exceeds a new criterion value implies an imminent slope failure. This technology has been validated against several events of slope failures as well as against a field rainfall test. Those validations have made it possible to determine the criterion value of the rate of tilting angle to be 0.1 degree/hour. The advantage of the MEMS tilting sensor lies in its low cost. Hence, it is possible to install many low-cost sensors over a suspected slope in which the precise range of what is going to fall down during the next rainfall is unknown. In addition to the past validations, this paper also introduces a recent application to a failed slope in the Izu Oshima Island where a heavy rainfall-induced slope failure occurred in October, 2013.

1. Introduction

People's concern for safety from natural disasters are increasing significantly in the recent decades and this trend is even more intensified internationally under the initiative of the United Nations conducting many projects. One of the target natural disasters is the slope failure that is induced by heavy rainfall and it is the main theme of the present study.

It is often argued that the probability of extraordinary heavy rain is increasing recently due to the global climate change. Although the authors are not able to judge whether or not this idea is correct, it is certainly true that people are more concerned than before with safety of their lives and properties during heavy rains. Nevertheless, the current situation is very difficult. For example, photo 1 illustrates the post-disaster situation in Hiroshima, Japan, where a residential land at a foot of hill was attacked by slope failure and induced debris flow during a midnight heavy rain on August 20th in 2014. The residents were living on an alluvial fun that was formed by repeated debris flows in the past, but they were not experts of geology and did not appreciate the hazard information issued by public sectors. It is important here that, although people are concerned with safety, they on the other hand decide their places to live on the basis of the convenience in daily life, good landscape, business chance and so on only. Hence, an advice for relocation to safer places is not effective.

Early warning and evacuation to shelters are popular in the recent times. Generally, evacuation order and/or alert is issued by public sectors on the basis of a rainfall criterion which assumes that the possibility of slope disaster is controlled by the intensity of rain. This idea is basically correct and hence rainfall-based early warning is practiced today in many parts of the world.

A possible drawback of this practice is that the rainfall criterion does not take into account the effects of local geology, topography and human actions in a sloping ground. Moreover, the rainfall forecast cannot capture the local variation of rain in a small spatial scale such as hundreds of meters. Therefore, the rainfall-based early warning is not sufficient when a family wants to be prepared for the failure of a particular slope behind its house. To foresee failure of a particular slope that cannot be captured by the regional rainfall-based warning, a supplementary warning system is necessary.

Photo 1 Rainfall-induced disaster in a residential land on an alluvial fun (Hiroshima, August 2014)

2. Principles of slope monitoring by MEMS tiltometer

It has been empirically known that slope failure during heavy rain is preceded by such phenomenon as ground "roaring", muddy water ejection, cracking in the ground surface, smell of methane gas and so on. Although installation of video, sound and smell sensors would capture those precursors in principle, their reliability is too low during heavy rain with possible thunders in dark midnight. The authors, therefore, understand that those precursors are caused by or closely related with minor slope deformation and decided to develop a displacement sensor [1 and 2]. Moreover, an artificial rainfall test to trigger slope failure reported that minor slope deformation started t around 2000 seconds prior to the ultimate failure [3].

Another important point is the cost of a sensor. Although an extensometer has a long and good tradition in landslide monitoring, its price is too high and installation is not very easy for individual citizens who are not engineers but wish to help themselves from natural disaster. The cost is even more important because a rainfall-induced slope failure may occur everywhere and it is difficult to foresee which part of a slope will fall down during the next heavy rain. Thus, many sensors have to be installed in a slope and hence the cost per sensor has to be reduced.

Figure 1 illustrates the MEMS (Micro Electro Mechanical Systems) tiltometer and its operation principle that were developed by the authors. A MEMS tiltometer is stored in a protection box. After punching a rod into a stable base layer that do not fail during heavy rain, this box is attached to the top of the rod. During rain, the surface soft soil may move downwards and push the rod. Because the bottom tip of the rod is penetrated into a stable base layer, the rod starts to rotate or tilt. The tiltometer captures this minor change of the angle prior to the final failure and transmits the signal to the Internet through wireless (figure 2). Because of the use of MEMS sensor, the price per sensor is reduced. Although the precise cost varies with the mass production, it is of the order of hundreds of US Dollars at this moment. Note that this kind of monitoring is mainly interested in shallow failures which are small in scale but significantly more in numbers than more dramatic deep-seated gigantic slope failures.

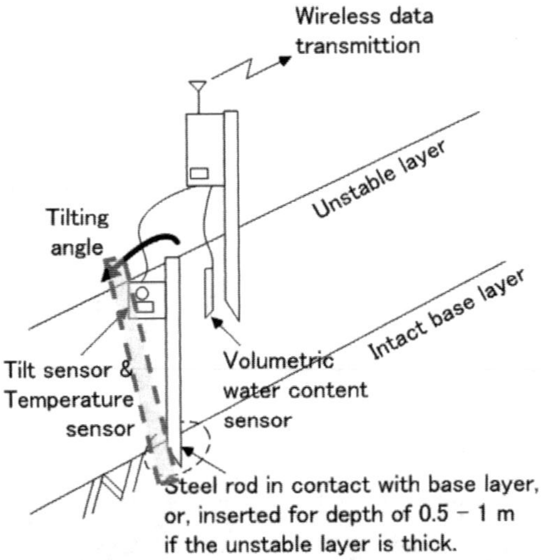

Figure 1 Schematic illustration of MEMS tiltometer sensor for early warning (volumetric water content sensor is optional)

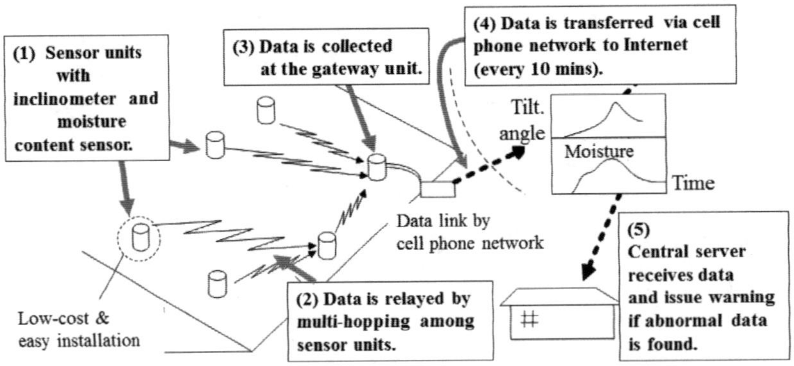

Figure 2 Idea of slope monitoring by multi-displacement sensors connected to internet

To reduce the cost, the following efforts were made so far;

- Water content sensor is optional. Although water content and pore water pressure play major roles in shear failure of soil as soil mechanics states, the use of water information in early warning of slope failure needs to know the exact location of a failure plane. To know this, soil investigation and analysis are required and increase drastically the cost which people cannot afford any more. Consequently, it was decided to issue early warning on the basis of minor displacement precursor.

· The sensor is driven by electricity. This power is provided by dry cells that are substantially cheaper than solar batteries. To reduce the energy consumption, the MEMS sensor works once every 10 minutes, sends the signal through wireless and sleeps again. Accordingly, it has been proven in the field that four batteries can maintain operation for more than one year.

3. Validation of developed technology by artificial rain test

The developed sensor has been installed in many slopes in Japan and China in order to decide the new displacement-based criterion of early warning. To date, the authors propose that an evacuation order should be issued if the rate of the tilting angle exceeds 0.1 deg/hour during rain fall. This deformation criterion was derived by interpreting many field studies during real slope failures and an artificial rainfall test. To begin with, this chapter presents data from an artificial rainfall test that triggered failure in a natural slope.

The artificial rainfall test was conducted in Sichuan Province of China as a joint research project between authors and the Institute of Mountain Hazards and Environment in Chengdu, China. The test slope consists of debris flow deposits that are basically cohesionless, including gravels. Photo 2 illustrates ongoing artificial rain. An unstable cut face was prepared by excavating a part of the slope and water seeped into the upslope side of the face, thus increasing the weight of soil as well as pore water pressure and reducing the factor of safety. The affected cut face, which is of 40 degrees in slope, is shown in photo 3.

Figure 3 shows the locations of installed sensors in the slope. Many tiltometers were embedded both near and at far distance from the cut face and slope failure started from the vicinity of the face and proceeded inwards with time. The intensity of rainfall was 40mm/10 minutes (240 mm/hour) prior to the failure. Figure 4 indicates the time histories of the recorded tilting angles. It may be evident that the rates of tilting angle were 0.15, 0.40 and 0.24 degrees/hour respectively at three sensor locations and then slope failed progressively. Thus, the criterion of 0.1 deg/hour for issuing an evacuation order is reasonable.

. It is of common concern whether or not the new tiltometer is equivalent with the conventional extensometer monitoring. To shed light on this, the surface displacement was assessed at a tiltometer location by (length of vertical rod)*tan(tilting angle) and compared with the displacement obtained by an extensometer. Figure 5 reveals that the two kinds of displacement are similar with minor difference. Thus, it is very possible that more expensive extensometer is replaced by multi-point monitoring by less expensive tiltometers.

(a)

(b)

Photo 2 Ongoing artificial rain to de-stabilize a slope

Photo 3 Cut face to fail during artificial rain

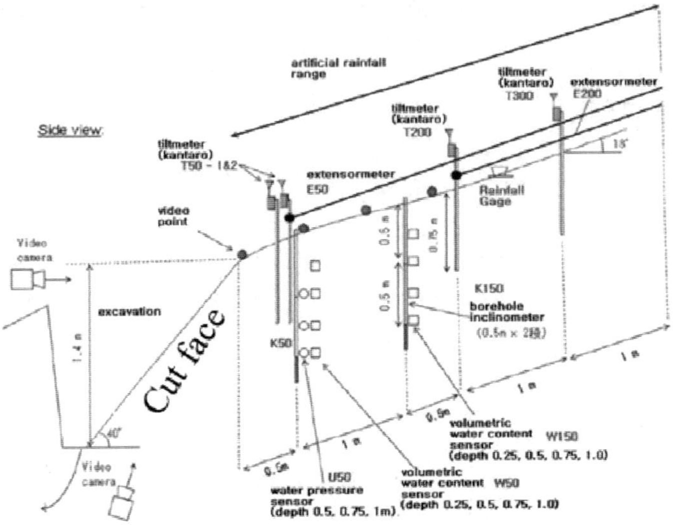

Figure 3 Locations of installed sensors

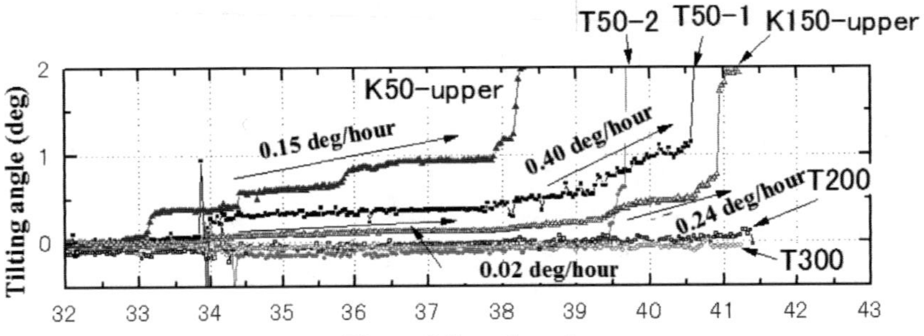

Figure 4 Time history of recorded tilting angles

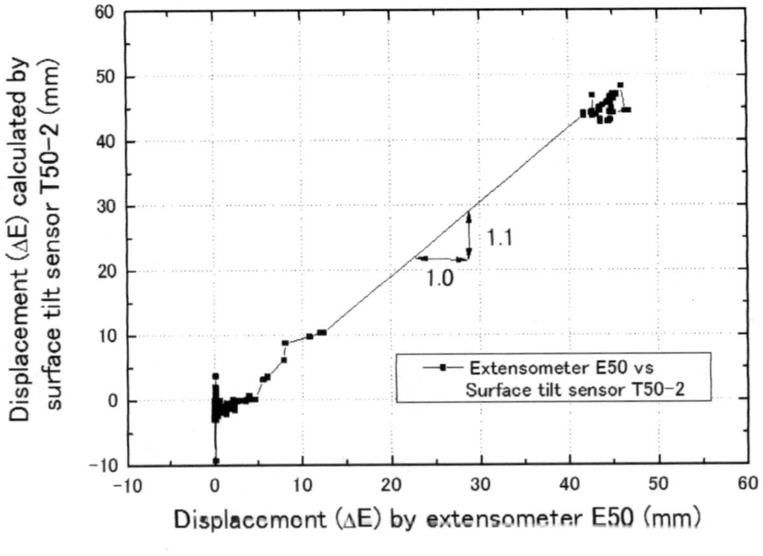

Figure 5 Comparison of extensometer and tiltometer records

4. Validation in failure of natural slope

After the completion of the Three Gorge Dam in China, many slopes along its reservoir, which had been potentially unstable, started to move due to water submergence effects. Photo 4 illustrates one of such sites and, because of this situation, tiltometers were installed for validation of the technology. Photo 5 shows the situation after the slope failure as expectedly happened.

Figure 6 shows the time history of the tilting angle that was recorded by a sensor inside the failed soil mass. No heavy rain happened in this period and the slope failed due to the gravity and water submergence. More than several days before the final failure, the rate of tilting was only 0.008 deg/hour and did not suggest incipient failure. However, precipitations of 15mm and 50mm occurred on June 7th and 8th, respectively, and consequently the rate of tilting was accelerated. As figure 6 shows, the rate became 0.13 deg/hour, exceeding 0.1 deg/hour, and the final failure occurred. Thus, the proposed criterion of 0.1 deg/hour was validated.

It is important that evacuation order is issued well in advance so that people has sufficient time to do so. From this viewpoint, the time histories in figures 4 and 6 are examined to show that the rate of tilting greater than 0.1 deg/hour lasted for hours or even for one day that are sufficiently long for evacuation.

Photo 4 Unstable slope along the Three-Gorge Dam Reservoir

Photo 5 Failure of reservoir slope

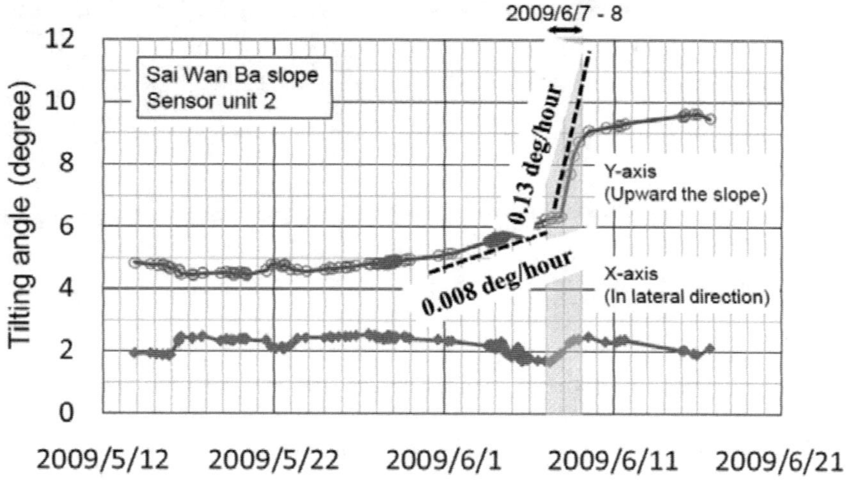

Figure 6 Time history of recorded tilting angles

5. Rainfall-induced slope disaster in Izu-Oshima volcanic island in Tokyo

On October 16th, 2013, a strong typhoon made as much as 800 mm of precipitation in one night in a volcanic island of Izu Oshima, south of Tokyo (figure 7). This rain induced a huge slope failure that claimed 39 lives. The failed slope is shown in photo 6. This slope has been stable for the past 600 years. It was even stable when another typhoon in 1958 produced precipitation of more than 400mm in one day. However, the heavy rain in 2013 exceeded the past record (figure 8) and the disaster was induced. It is noteworthy that the failure in 2013 happened at around 2AM when the accumulated rainfall exceeded the previous record of 400mm. It was very unfortunate that the size of the affected area of was smaller than the target size of the rainfall-criterion strategy of warning. Hence, people were attacked by the disaster without precaution in mid midnight.

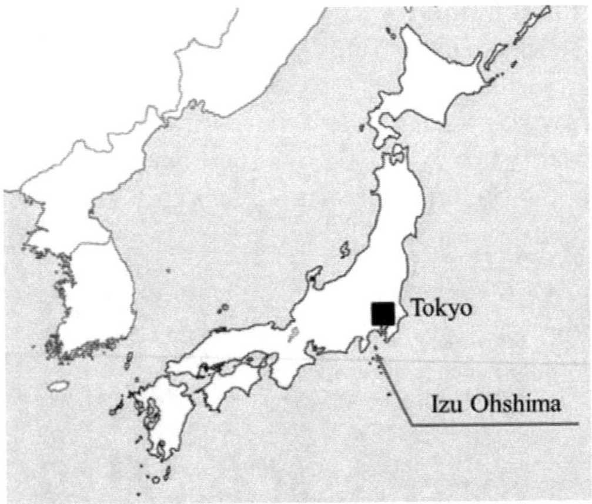

Figure 7 Location of Izu Oshima Island

Photo 6 Failed volcanic slope and soil deposition

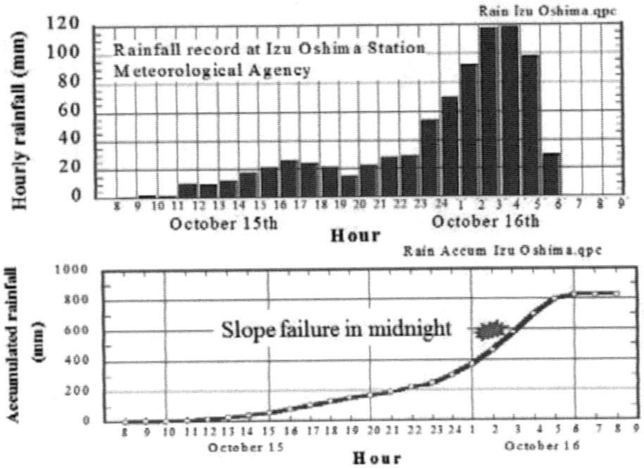

Figure 8 Rainfall record during the disaster in 2013

Photo 7 Piping holes in the failed slope

Failure of a volcanic slope has an interesting characteristic that does not occur in other kind of slopes. It come from the stratified structure which consists of interbedding of more pervious ash deposits and less pervious fine materials as well as lava. Obviously this stratified structure was produced by the history of repeated eruptions in the past. Photo 7 demonstrates one of the most impressive features of the slope failure, which is the occurrence of many piping holes from which ground water was ejected. Certainly, the pervious layers introduced ground water towards the failed slope and the high ground water pressure affected the stability of the surface soil (figure 9). Another important issue is that the failed soil was very thin at the top (photo 7) but the flow of soil thus initiated eroded soils in the middle of the slope to be huge and devastating finally (photo 6).

The fact that the failure occurred in midnight posed several problems in the early-warning and evacuation strategy. Going out of home for evacuation during heavy rain in midnight may be a dangerous action unless good transportation is provided. It is certainly difficult, however, for a driver to visit families and collect people from every house. It may be an alternative idea to designate some streets as evacuation routes and install good illumination. Vertical evacuation, which finds a shelter in the upper floors in the same house or apartment building, can improve the safety. This issue needs further study.

After the disaster, the local administration started to issue evacuation advices well in advance during the daytime, even if heavy rain is expected after midnight. From the safety viewpoint this is a reasonable idea. However, people may not understand the reason why they have to go evacuation when rain has not yet started. Because slopes do not always fail in spite of evacuation advices, people would feel more reluctant to evacuate and spend inconvenient time in a shelter. It is hence being discussed that evacuation should be made more enjoyable by providing light entertainments.

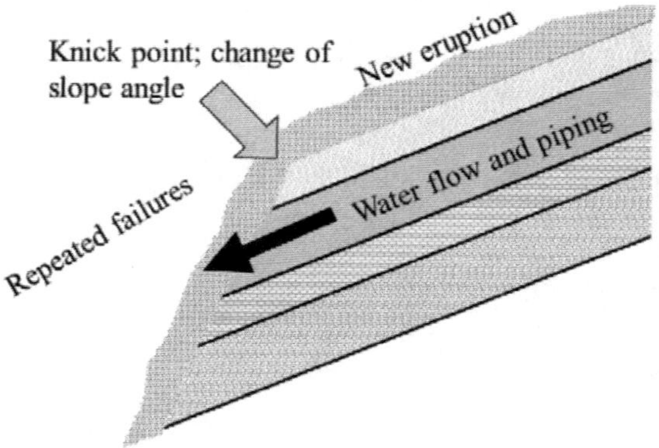

Figure 9 Mechanism of slope failure associated with piping holes

Photo 8 Location of tiltometers in head scarp

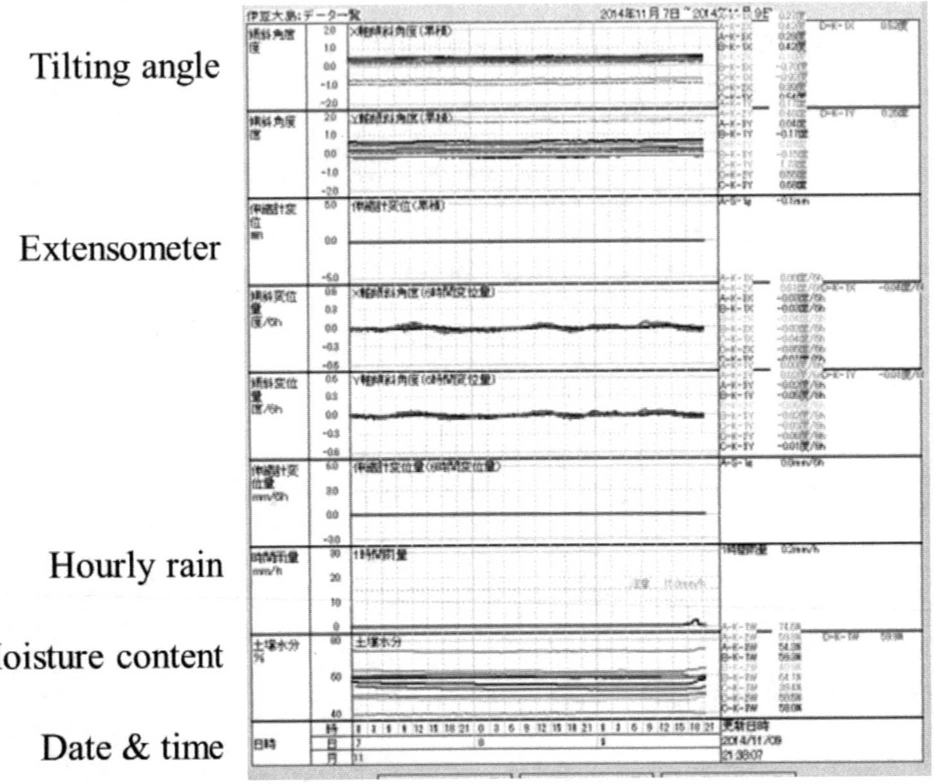

Figure 10 Internet monitoring of slope behaviour in three days in early November, 2014

6. Slope monitoring in Izu Oshima

After the disaster, tiltometers were installed at several head scarps of the failed slope in Izu Oshima where further failure appeared likely (photo 8). It was intended therein to validate the operation of the MEMS tiltometer system through internet. Figure 10 shows a successful example of the record as obtained through internet. Very fortunately no failure occurred during the monitoring period of 2014. Although the issuing of evacuation order can be automatic and computerized, the authors still rely on the judgement of experienced engineers who are watching the record. If the rate of tilting becomes high, a notice is sent to a local government who decides whether or not people should evacuate. Administrators have a dilemma between safety and unnecessary panic of people.

7. Comparison of extensometer and tiltometer

Figure 11 illustrates another monitoring site in Japan where road construction was conducted under the safety control by means of slope monitoring. During rainfalls, the monitored slope developed a certain extent of deformation and they were recorded by tiltometers and extensometers. What is importance is that many sensors were installed here so that different records from different locations and sensors might be compared.

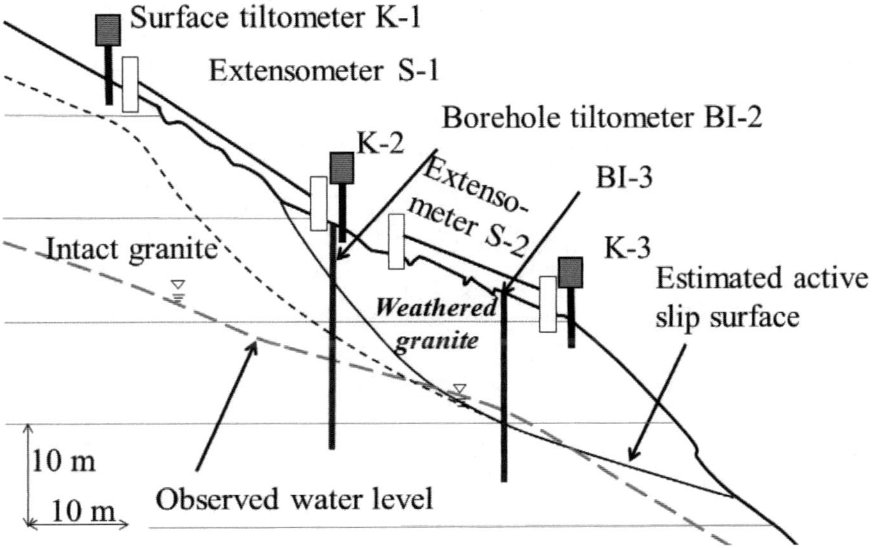

Figure 11 Location of sensors in the studied slope

Figure 12 shows time histories of the records during several rainfall events. It is first noticed that the tilting angle at K-3 near the bottom of the moving soil mass increased faster and more profoundly than those in the upper part of the soil mass. This implies that the soil conditions in the lower part was worse than in the upper part and/or the stress concentration in the lower part played a significant role. In other words, this moving soil mass was pulled down by its bottom. Second, the extensometer record S-1 across the head scarp was greater than the S-2 record within the moving soil mass. This may be consistent with the traditional idea of landslide monitoring by extensometers. However, the S-2 deformation is not negligible and the soil mass was subject to deformation. This finding supports the principle of tiltometer monitoring in which soil deformation induces tilting of the device.

Figure 13 compares the extensometer and tiltometer records in more detail. Herein, records of two extensometers, S-1 across the head of the moving soil mass and S-2 within the middle part of the soil mass, are compared with the tilting angle at K-3 near the bottom. S-1 is always greater than S-2 as stated above. It is noteworthy that each rainfall increases the K-3 record first and then S-1 and S-2 follow. This implies that early warning is more facilitated by a tiltometer than an extensometer because the former responds to the soil mass movement more quickly.

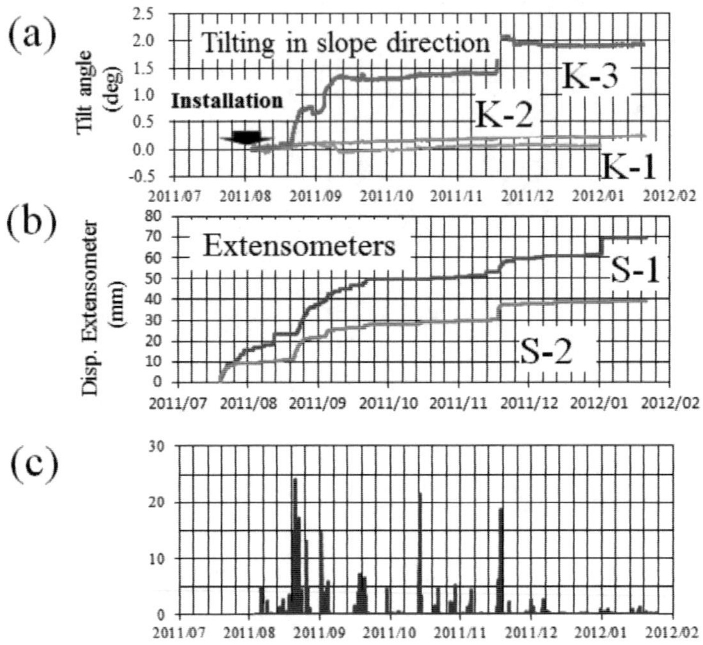

Figure 12 Time histories of recorded slope deformation

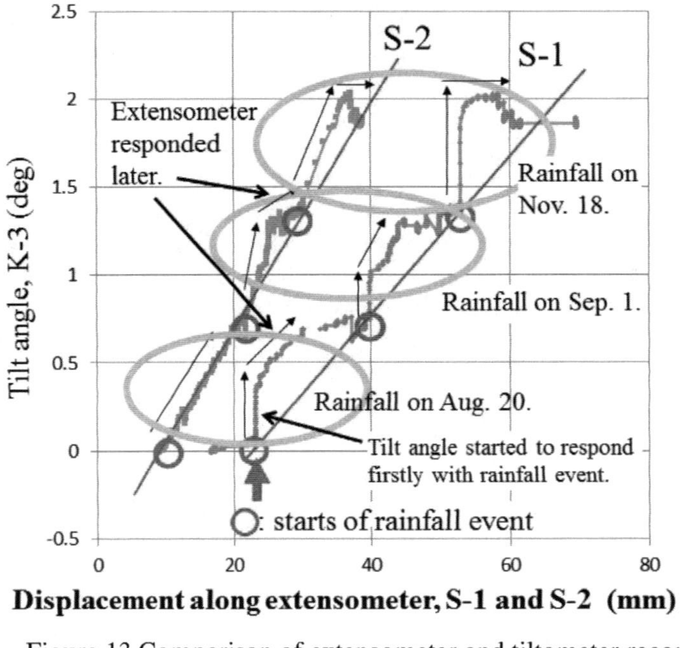

Figure 13 Comparison of extensometer and tiltometer records

8. Conclusion

The authors have been developing an inexpensive early-warning technology that monitors the deformation of unstable slopes undergoing heavy rain. The developed instruments monitor the tilting angle prior to the final failure and, if the rate of tilting exceeds the criterion of 0.1 degree/hour, an evacuation warning is issued. This technology has been validated by several case histories and a field rainfall test, some of which were referred to in this paper. The major conclusions drawn from this study are as what follows.

1. An artificial rainfall test on an unstable slope successfully induced failure.
2. There is a minor slope deformation well in advance of the final failure.
3. A tiltometer sensor was developed and further cost reduction is going on so that early warning may be issued prior to the final failure.
4. The criterion of 0.1 degree/hour is reasonable and there are plenty of time for evacuation after the occurrence of this rate of tilting.
5. Slope monitoring by the newly-developed tiltometer is equivalent with that of the conventional extensometer but much less expensive.
6. Multi-point monitoring, which intends not to miss the slope movement in any part of the moving soil mass, becomes feasible by the use of inexpensive tiltometers.
7. A tiltometer can capture the onset of precursory deformation of a slope earlier than a conventional extensometer.
8. A volcanic slope that failed recently was monitored but no further failure happened in the next year.
9. Rainfall-induced failure of this volcanic slope is characterized by the profound occurrence of water piping that reduced the safety factor of the slope during heavy rain.
10. Evacuation during midnight is not advisable and hence more strategy needs to be studied for successful early warning.

Acknowledgement
The present technology was developed under a financial support from MLIT of the Japanese Government, to which the authors are grateful. The field activities in China were carried out as a collaborative study with the Institute of Mountain Hazards and Environment in Chengdu, China. The authors express their sincere appreciations to Prof. J.P. Qiao and other people of this institute.

References

[1] Uchimura T et al. 2010 *Landslides* **7** 351.
[2] Uchimura T, Towhata I, Wang L, Nishie S, Yamaguchi H, Seko I and Qiao J 2015 accepted by *Soils and Foundations* ***55***.
[3] Sassa K, Fukuoka H, Moriwaki H and Ochiai H 2005 *Annuals of Disaster Prevention Research Institute, Kyoto University* **48C**.

IOP Conf. Series: Earth and Environmental Science **26** (2015) 012059 doi:10.1088/1755-1315/26/1/012059

Modelling of a rapidly evolving rockslide: the Mt. de la Saxe case study

R. Castellanza[1], G.B. Crosta[1], P. Frattini[1], G. Frigerio[2], C. di Prisco[2], G. Dattola[1], D. Bertolo[3]
[1]DISAT – Università di Milano Bicocca, riccardo.castellanza@unimib.it
[2]DICA – Politecnico di Milano, claudio.diprisco@polimi.it
[3]Struttura organizzativa attività geologiche, Regione Valle d'Aosta, Italy

Abstract. To model the temporal evolution of complex landslides, a 1D pseudo-dynamic visco-plastic approach, based on Perzyna's theory, has been conceived. In the original version of the model the viscous nucleus has been assumed to be bi-linear: irreversible deformations develop uniquely for positive yield function values whereas, in a more general case, even for negative values. In this work the model has been enriched by considering: i) an exponential viscous nucleus, ii) a strain-rate softening to reduce friction angle as sliding velocity increases and iii) block interaction forces to cope with complex 3D geometries for the sliding mass. The application of the proposed model to the La Saxe rockslide (Italy) clearly shows how a relatively simple model can be applied to a complex landslide by considering a spatial discretization of the sliding mass.

1. Introduction

In order to model complex sliding masses subject to continuous slow movements related to water table fluctuations it is quite convenient to: i) model the time-dependent mechanical behaviour of the materials by means of a viscous-plastic constitutive law ; ii) assume the water table fluctuation as main input to reproduce time acceleration; iii) consider, the 3D constrains by maintaining a high level of simplicity such to allow implementation into EWS (Early Warning System) for risk management.

In this paper a previously presented 1D pseudo-dynamic visco-plastic model (Secondi et al. 2011), based on Perzyna's theory is applied. The sliding mass is considered as a rigid block subject to its self weight, inertial forces and seepage forces varying with time. All non-linearities are lumped in an interface thin layer positioned between the rigid block and the bedrock. The mechanical response of this interface is assumed to be visco-plastic. The viscous nucleus is assumed to be also exponential, so that irreversible strains develop for both positive and negative values of the yield function; the friction angle is assumed to reduce with strain rates; the sliding mass is discretized in blocks to cope with complex rockslide geometries.

To validate the improvements introduced in this paper one case study is presented. It refers to the recent La Saxe rockslide movements (Aosta valley; Italian Western Alps) widely described in Crosta et al. (2013).

It will be shown that, in its modified version, the mode satisfactorily fits the La Saxe displacements triggered by the fluctuation of the water ground levels induced by snow melting and that the blocks discretization confirms the model suitability in case of complex 3D rockslide.

IOP Conf. Series: Earth and Environmental Science **26** (2015) 012059 doi:10.1088/1755-1315/26/1/012059

2. 1D Viscous plastic model

As is well known, a simple static limit equilibrium analysis is not suitable for modelling and correctly reproducing the time depentent mechanical response of sliding masses triggered by water table fluctuations. A dynamic, or, at least, pseudo-dynamic, analysis should be adopted, and a delayed-plastic constitutive approach employed. The first issue can be addressed by evaluating the inertial effects on the soil mass by using a pseudo-dynamic approach, as that introduced by Newmark (Newmark, 1965) for dams under seismic actions, whereas, viscous components can be described by employing the visco-plasticity theory introduced by Perzyna (1963).

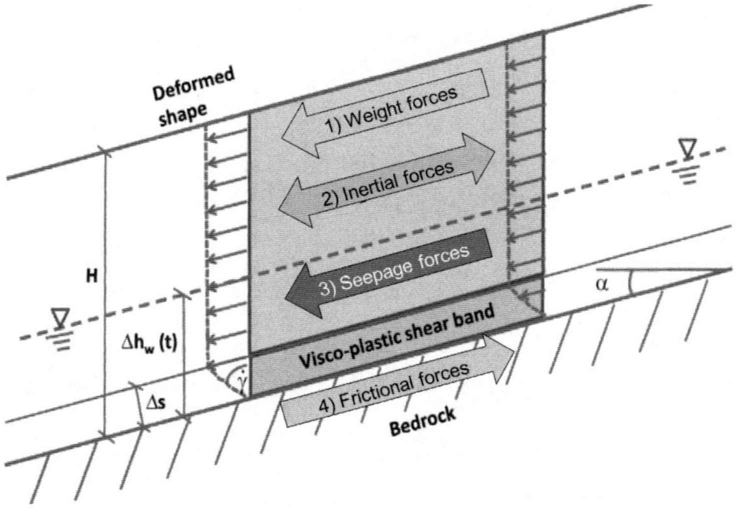

Figure 1. 1D block model with main geometrical elements.

As was previously suggested, according to this approach, the rockslide is interpreted as a rigid block (Figure 1), where the active forces taken into account are (i) the weight, (ii) the inertial forces and (iii) the seepage force deriving from the water table level which is a function of time, $\Delta h_w(t)$. All non-linearities are lumped in a shear band of thickness Δs positioned between the rigid block and the bedrock. All the details and the analytical formulation of this model are reported in Crosta el al. (2013). Thus, the main equations controlling this model are:

Visco-plastic strain rate definition

$$\dot{\gamma}^{vp} = \tilde{\gamma} \cdot \phi(f) \tag{1}$$

Sliding mass displacement rate:

$$\dot{x} = \Delta s \dot{\gamma}^{vp} = \tilde{\gamma}\ \Delta s\ \phi(f) \tag{2}$$

Yield function definition

$$f = \frac{\tau - \tau_{res}}{\sigma'_n} \tag{3}$$

Active shear stress

$$\tau = \tau_{stat} - m^* \ddot{x} \tag{4}$$

$$\tau_{stat} = \left(W\left[\Delta h_w(t) \right] \sin(\alpha) + J \right) / A \tag{5}$$

Failure criterion

$$\tau_{res} = c' + \sigma'_n \tan \phi'_{res} \tag{6}$$

where f is the yield function, $\tilde{\gamma}$ a constitutive viscous parameter, $\phi(f)$ the viscous nucleus, $\dot{x}$ the sliding displacement rate of the mass m, A the total contact area, W' the soil buoyant weight, J the total seepage force, $\Delta h_w(t)$ the groundwater table level varying with time t.

In case f is positive, by substituting equations. (2), (3) and (4) into the expression of the viscous nucleus, and by introducing the geometrical and constitutive parameters, we obtain an equation of the following type:

$$a\ddot{x} + b\dot{x} + c = 0 \tag{7}$$

where coefficients a, b and c depend on time and can be derived from the previous equations. The improvements of the model are presented in the following sections.

3. 1D Viscous nucleous

As is sketched in Figure 2, two different formulation for the viscous nucleus have been considered. The first, proposed in the original version of the model (Secondi et. al. 2011), is a bilinear function (Figure 2a) defined as:

$$\phi(f) = \langle f \rangle \tag{8}$$

where brackets imply the viscous nucleus to coincide with the yield function for $f > 0$ and to be nil for negative values of f (equation 3).

The second expression is an exponential law (Figure 2b) defined: according to the following equation

$$\phi(f) = e^{\alpha \cdot f} \tag{9}$$

The exponential law is also enriched by parameter $\boldsymbol{\alpha}$, governing the shape of the viscous nucleus (Figure 2b).

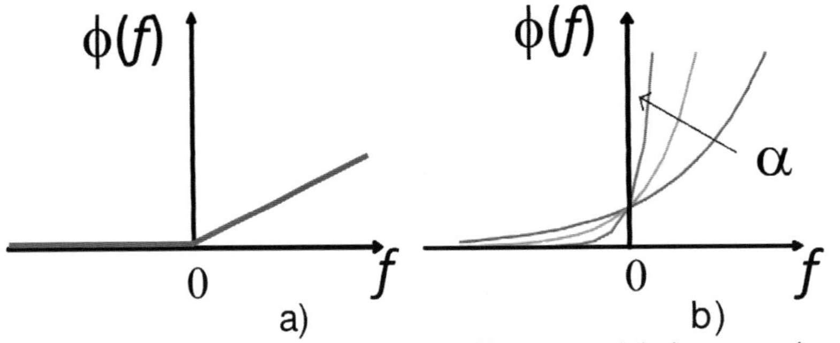

Figure 2. a) bilinear viscous nucleous; b) exponential viscous nucleous

4. Friction weakening

As a preliminary way to simulate the transition from the initiation to the propagation of an unstable rock mass, a friction weakening is here after considered. An increase in the localized shear strain rate (i.e. slip velocity) is thus assumed to produce a slight reduction in the friction angle:

$$\|\dot{x}\| > \dot{x}_w$$

$$\tan\phi'(\dot{x}) = \frac{\tan\phi'_0 - \tan\phi'_w}{\|\dot{x}\| - \dot{x}_w} + \tan\phi'_w \tag{10}$$

otherwise,

$$\tan\phi'(\dot{x}) = \tan\phi'_0$$

where ϕ'_0 and ϕ'_w are the static and the weakened friction angles, respectively, and $\dot{x}_w$ is a threshold characteristic velocity for the weakening to occur.

Many authors (e.g. Rice R. J., 2006, Beeler et al 2008; Kuwano and Hatano, 2011; Vardulakis, 2000; Veveakis et al. 2007) proposed specific studies and theoretical interpretations to justify the frictional weakening as a consequence of multiple mechanisms. Here the authors introduce equation (10) only as one possible ways to simulate, within the adopted 1D model, the acceleration usually recorded before the run-out process. No specific theoretical justification are given; equation (10) should be therefore considered as a macroscopic strength reduction as sliding velocity increases towards catastrophic failure.

5. Discretization

The proposed model can be employed to reproduce the evolution of a non-deformable mass sliding on an inclined rigid plane. A 3D complex landslide a priori does not satisfy all the hypotheses behind the model. To overcome this limitation and to tentatively apply the proposed model to complex 3D landslide settings, a two steps methodology has been developed.

5.1. Splitting the displacement function

In general, the rockslide displacements can be expressed as a vectorial function $\boldsymbol{u}(x, y, z, t)$ defined in space and time. At any instant this function should be properly evaluated from the monitoring. For the sake of simplicity, by assuming:

i) the direction of displacements to be constant within the area considered and to be determined by a unit vector $\boldsymbol{u}'$. This also implies that the dispersion in direction of the velocity within the domain $\hat{\boldsymbol{u}} = \boldsymbol{u}(x, y, z) - \boldsymbol{u}'$ is negligible with respect to $\boldsymbol{u}'$;

ii) the evolution of displacement with time to be independent of space.

The function $\boldsymbol{u}(x, y, z, t) \in C^0$ can be written as the product of three elementary functions:

$$\boldsymbol{u}(x, y, z, t) = \boldsymbol{u}' \bar{f}_1(x, y) \bar{f}_2(z) g(t) \tag{11}$$

where

$$\bar{f}_1(x, y) = f_1(x, y) / u_p(x_p, y_p, z_p, t_0) \tag{12}$$

$$\bar{f}_2(z) = f_2(z) / u_p(x_p, y_p, z_p, t_0) \tag{13}$$

In particular (see Figure. 3):

- $\bar{f}_1(x, y)$ can be considered as a non-dimensional shape function of the spatial superficial distribution of t displacements. This could be evaluated, for example, from GB-InSAR measurements divided by the displacement at a selected point, P, and reference time, t_0;
- $\bar{f}_2(z)$ is the shape function in depth of the displacement at any point on the ground surface (Figure 3) of the rockslide assumed to be time independent. This can be obtained from inclinometer data at a chosen point P by dividing it by the ground displacement measured in P. This function is crucial in order to verify the model assumption of a rigid block sliding on a shear band.

- Function $g(t)$ can be considered as the displacement evolution with time of a representative "pivot point" P.

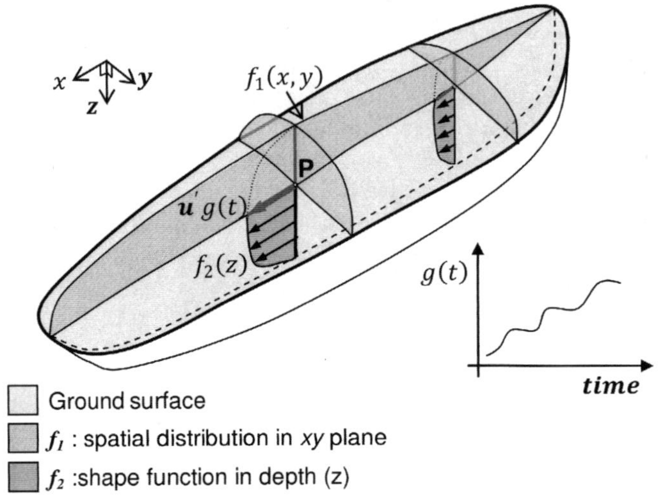

Figure 3. a)Suggested elementary function of $\boldsymbol{u}(x, y, z, t)$ of a rockslide mass.

Once shape functions $\overline{f}_1(x, y)$ and $\overline{f}_2(z)$ are known, it becomes possible to provisionally evaluating the displacement of a complex landslide by simulating or predicting the ground movement at a specific representative point P with respect to time but uniquely defining function $g(t)$.

5.2. Spatial discretization

If the rockslide prevalently slides parallel to the average slope inclination, the outlined model can be a priori employed by assuming for instance both f_1 and f_2 (Heaviside step function) simply describing a rigid block sliding along a plane and variable $\hat{z}$ to coincide with the normal to the average slope inclination (in this case a rotation of the reference frame is suggested, so that $\boldsymbol{u}'^T [1,0,0]$) . For complex 3D rockslides a more refined subdivision of the sliding mass is required. This will be based on a set of criteria employed to identify from monitoring data, geomorphological and geometrical observations, homogeneous displacement zones. Once the rockslide mass is split into subzones, the same subdivision can be applied to the groundwater table, so that for each zone a time dependent oscillation $\Delta h_{w,i}(t)$ can be evaluated. Each zone is then treated has a rigid block resting on a shear band with a viscoplastic behaviour (Figure 1). In order to take into account the interaction between the blocks, equation (5) becomes

$$\tau_{stat,i} = \left(W_i \left[\Delta h_{w,i}(t)\right] \sin(\alpha_i) + J_i + F_i^{rel}\right) / A_i \tag{14}$$

where F_i^{rel} stands for the resultant of both the forces transmitted by the lateral boundaries of each block and the forces transmitted by the uphill blocks to the downhill ones; index i refers to the current block.

6. Case study: the la Saxe Rockslide

During the last decade, a large rockslide along the NW slope of the Mont de la Saxe (Courmayeur, AO - Italy) showed a significant increase in the rate of activity in conjunction with snow melting periods and more recently with important amount of rainfall (see Figure 4).

The 8 Mm^3 rockslide affects a heavily tectonised and intensely fractured mass of black schists, for a surface area of about 150'000 m^2. The rockslide failure surface is typically located at depths of 60 to 80m b.g.l. Because of the highly valuable exposed elements, regional authorities commissioned a comprehensive ground investigation and monitoring campaign, as well as theoretical and numerical analyses, in order to assess slope stability conditions, the possible evolution and the suitable stabilization and mitigation countermeasures (see Crosta et al 2013). Here we briefly present the recent rockslide evolution, recorded in the spring and summer of 2014 and then with the aim to validate the 1D visco-plastic model, the results of a series of simulations considering interacting blocks are presented.

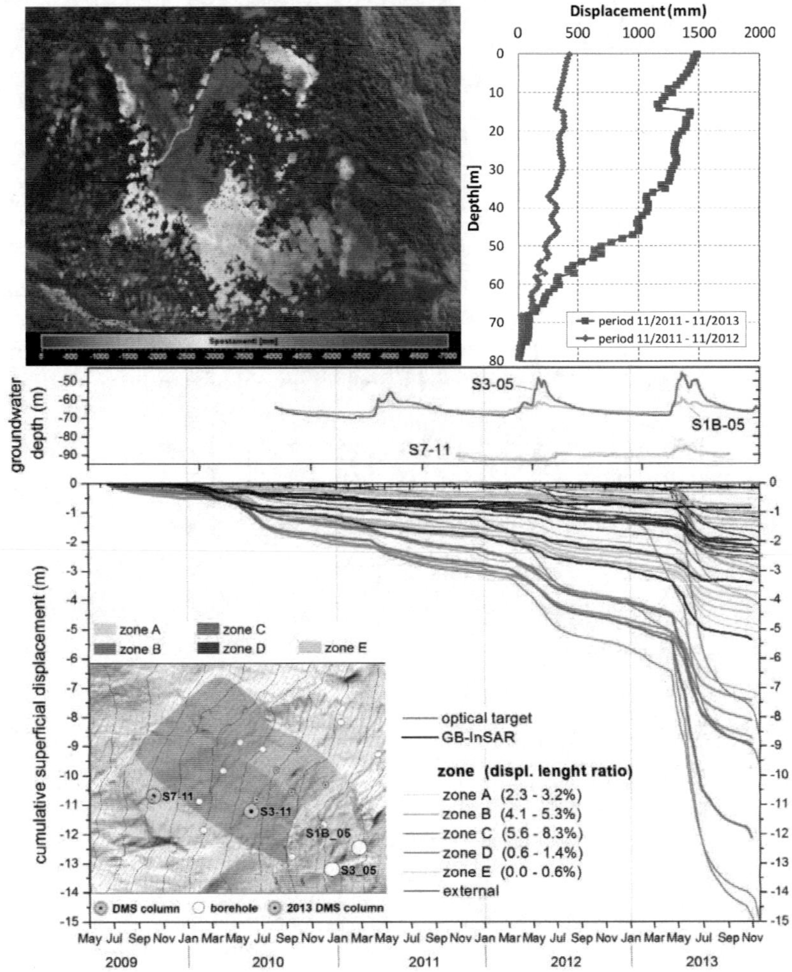

Figure 4. Monitoring data: a) Cumulative GB-InSAR from 2009, b) DMS S3 displacements for two different time intervals; c) GB-streaming points and optical targets data subdivided according to 5 different zones

The rockslide is subject to accelerations every spring due to the combined effect of snow melting and rainfall that implies an increase in the groundwater level (Figure 4). Due to the complexity of the rockslide, this can be subdivided in sub-areas characterized by different geometrical properties (i.e. depth of surface failure, groundwater elevation, Figure 5). This complexity affects the overall rockslide behavior. The 5 blocks discretization (Figure 5b) subdivides the overall rockslide into different sub-areas so that each block has its own basal inclination, height, width length and shear band thickness. The interaction between the blocks has been taken into account by assuming lateral and normal forces (Crosta et al., 2013).

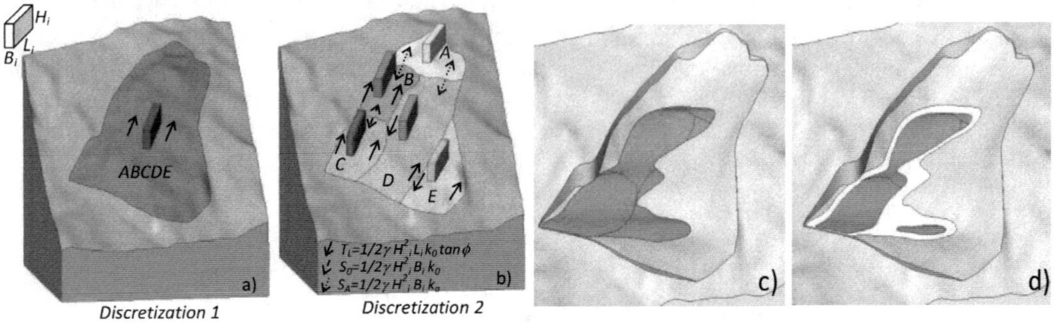

Figure 5. Landslide discretization in 1 (a) and 5 blocks (b). The lower panels show the wet and dry season groundwater table (light blue) above the failure surface.

In Figure 6 the results of the 1D viscoplastic model simulations are reported. The results for the single block discretization (Figure 5a) obtained by assuming averaged geometrical properties for the entire rockslide, are shown in Figure 6b. From this single block discretization the groundwater level results to lay always below the average failure surface with the consequence that the model is unaffected by its oscillation.This clearly suggests that the single block discretization is too poor to reproduce the 3D complex behavior. In contrast, the 5 blocks discretization (Figure 5b) shows the model capability of reproducing the monitored displacements in different subareas of the rockslide (Figure 6c).

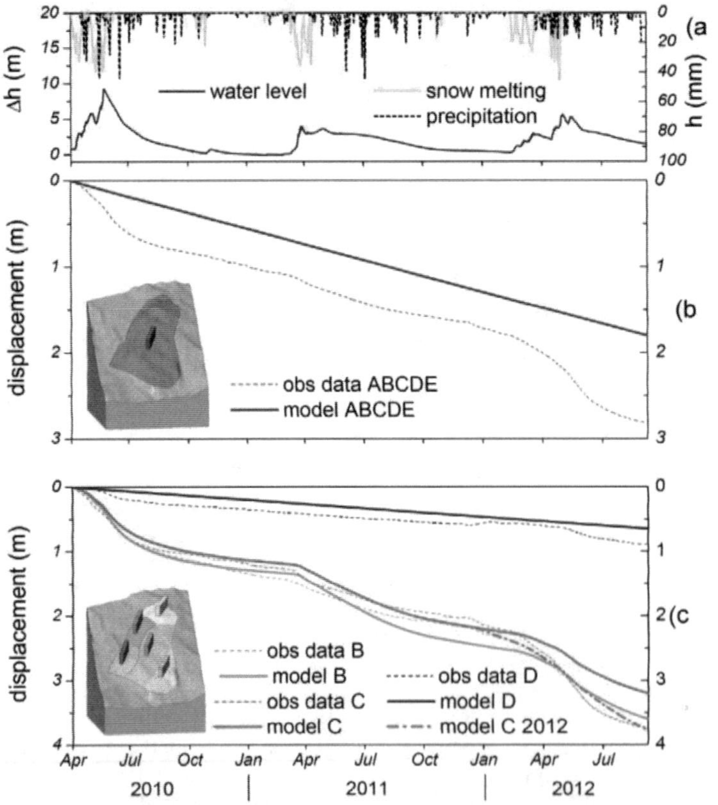

Figure 6. Model simulations in term of *g(t)* functions: a) ground water level fluctuation; b) simulation for Discretization 1; c) simulation for Discretization 2.

Since March 2014 a sector corresponding with Zone C in Figure 5b and involving about 400.000 m3 underwent a strong seasonal reactivation. As a consequence, the landslide mass in this zone reached displacement velocities up to about 0.9m/h and a cumulative displacement of about 30 metres in less than 3 weeks. It is worth noticing that this sector of the landslide, already described as the most sensitive and active one, became every year more and more sensitive to seasonal perturbations.

Among the possible causes and controlling factors for such a behavior and exceptional acceleration, we can list:

a) snow melt contributing to the groundwater recharge favoured by a progressive fracturing of the rock mass nd a consequent increase in permeability;

b) progressive degradation of the rock mass by large displacement, opening of fractures and shear zones till a condition typical of a granular mass;

c) further contribution to groundwater recharge from the neighboring areas.

To understand the landslide behavior and to improve our capabilities to predict landslide displacement we tried to simulate the observed displacements starting from the parameters calibrated on previous time series (2009-2013) and other field observations (Figure 4). In this simulation the depth of the failure surface below the ground surface is reduced to an average value of 30 m.

The triggering causes listed above can be introduced into the 1D model considering the increase in water level above the failure surface associated to the snow melt contribution or lateral inflow, and we also take into account for the seepage force. Rock mass degradation can be simulated by considering an increase in thickness of the shear band and/or a progressive decrease in the average friction angle along the shear zone, due to changes of state characteristics of the material (progressive grain size reduction of black schist and calcschist).

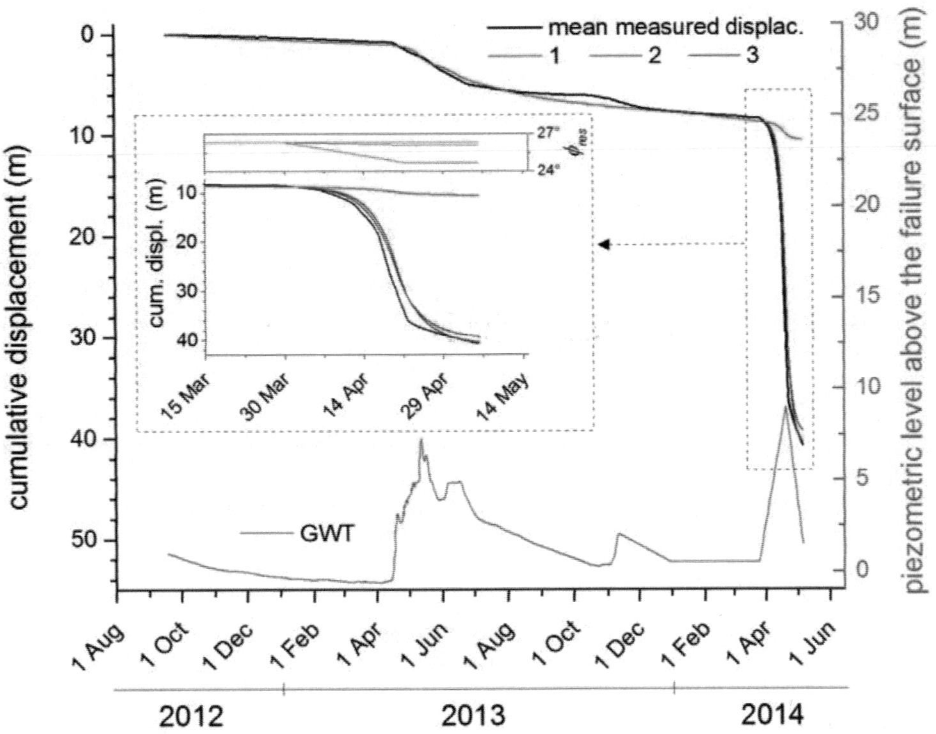

Figure 7. Results of the 1D visco-plastic simulation of the 2013-2014 period. 1) constant friction angle (26.8°, see inset), shear band thickness Δs =1m; 2) friction angle changes linearly (26.8→26°, see inset), shear band thickness Δs =5 m; 3) friction angle changes linearly (26.8→24.7°, see inset), shear band thickness Δs = 1m. Piezometric level (GWT) changes in the same way for all the models.

To accomplish these analyses we carried out a series of simulations by changing different parameters under the above slightly different assumptions:

1. Groundwater recharge with a simplified oscillation comparable to the one observed in the previous year (April-May 2013) and without any change in the material properties (see curve 1 in Figure 7) and shear thickness with respect to previous simulations (□s = 1 m);

2. The change in groundwater level was imposed as in 1) but we progressively decreased the friction angle (26.8→26°, see inset in Figure 7) since the beginning of the acceleration phase and assigned it to a thicker shear band (□s = 5 m);

3. The change in groundwater level was imposed as in 1) but we progressively decreased the friction angle (26.8→24.7°, see inset in Figure 7) since 2013 till the beginning of the 2014 deceleration phase and assigned to a 1 m thick shear band.

Model 1 in figure 7 performs well till the seasonal reactivation occurring on April 2013 but it is unable to replicate the sudden acceleration recorded on April 2014. This last reactivation stage is reproduced by models 2) and 3) thanks to the imposed changes in the mechanical characters (shear zone thickness, friction angle) through simple linear state relationships (see Figure 7) for the friction angle or the increase in shear band thickness. In all the simulations the viscous parameters are maintained constant.

Even if these simulations and their results cannot be used directly for prediction of future evolution, because of the inherent difficulty in collecting the real values of controlling variables, some conclusions can be drawn. In fact it seems possible to state that the groundwater recharge and associated piezometric level oscillation and/or the change in frictional properties associated to material destructuration could be useful predictors to be tested in the laboratory and monitored in situ.

7. Conclusion

A 1D rigid-viscoplastic model has been modified for what concerns the expression of the viscous nucleus, the friction strain-rate weakening and extended to simulate the response of systems of multiple interacting blocks to model 3D complex rockslides . The La Saxe rockslide were chosen in order to show the capability of the proposed model of reproducing the displacement time series triggered by groundwater table fluctuations by adopting an interacting block discretizaton.

References

Beeler N. M., T. E. Tullis, and D. L. Goldsby. (2008) Constitutive relationships and physical basis of fault strength due to flash heating. *J. Geophys. Res.*, **113**, B01401,348, 2008.

Crosta, G. B., di Prisco, C., Frattini, P., Frigerio, G., Castellanza, R., Agliardi, F. (2013) Chasing a complete understanding of the triggering mechanisms of a large rapidly evolving rockslide. *Landslides*, **11**:747–764 DOI 10.1007/s10346-013-0433-1.

Kuwano, O., Hatano, T. (2011) Flash weakening is limited by granular dynamics. *Geophys. Res. Letters*, **38**, L17305, doi: 10.1029/2011GL048530

Newmark, N. M., (1965). Effects of earthquakes on dams and embankments. *Géotechnique*, **15**(2):139-160.

Perzyna P., (1963) The constitutive equations for rate sensitive plastic materials. *Quart.Appl.Math.***20**:321-332.

Rice R. J. Heating and weakening of faults during earthquake slip. *J. Geophys. Res.* **111**,B05311,2006.

Secondi, M., Crosta, G., Di Prisco, C., Frigerio, G., Frattini, P., Agliardi F. (2011) Landslide motion forecasting by a dynamic visco-plastic model. *Proc. WLF 2*, paper WLF2-2011-0571.

Vardoulakis, I. (2000). Catastrophic landslides due to frictional heating of the failure plane. *Int. J. Mechanics of Cohesive and Frictional Materials,* **5**(6), 443-467.

Veveakis E., Vardoulakis I., and Di Toro G., (2007) Thermoporomechanics of creeping landslides: The 1963 Vaiont slide, northern Italy. *Journal of Geophysical Research: Earth Surface* **112**.F3, 2003–2012.

Soil moisture and strength index for earthwork construction quality control

A Sawangsuriya[1], S Wachiraporn[2], and W Sramoon[2]

[1]Department of Highways, Thailand
[2]Mahanakorn University of Technology, Thailand

E-mail: sawangsuriya@gmail.com

Abstract. This paper presents the implementation of soil moisture and strength index measurements for earthwork construction quality control as well as a link between the in situ testing and structural property of earthen materials. Use of the convenient Dynamic Cone Penetrometer (DCP) in conjunction with conventional moisture-density measurements enhances quality control by achieving acceptable level of compaction, more uniform structural properties, and aids developing a controlled design parameter during the earthwork construction. Soil strength in term of DCP index normalized by the deviation of compaction moisture content from the optimum moisture content is proposed as performance criteria for a variety of engineered earth fills and special engineering assessment, prevention, and mitigation of geohazards e.g. earthen flood defense embankments.

1. Introduction

Typical earthwork compaction acceptance criteria are based on specified target dry density of the placed earthen materials achieved through proper moisture content. According to this approach, by achieving a certain dry density using an acceptable level of compaction energy assures attainment of an optimum available level of structural properties and also minimizes the available pore space and thus future moisture changes. Conventional approach is also based on the premise that monitoring dry density as opposed to a structural property is relatively simple and can be applied to generate data for a statistical evaluation of compaction quality.

The difficulty and expense of acquiring quality relevant structural properties have traditionally caused engineers to rely on density tests. The relative compaction alone is not a reliable indicator of the structural properties of compacted earthen materials. In addition, the relative compaction is only a quality index used to judge compaction acceptability and is not the most relevant property for engineering purposes. For engineered earth fills and special engineering assessment, prevention, and mitigation of geohazards e.g. earthen flood defense embankments, the ultimate engineering parameters of interest are often the soil stiffness and strength, which are direct structural properties for determining load support capacity and deformation characteristic in engineering design.

Since the non-uniformity of structural property is directly related to progressive failures and life-cycle cost, a simple, rapid, and direct structural property testing which can be conducted

[1] Corresponding author

independently and in companion with conventional moisture-density testing without interference with the construction process is anticipated to increase test coverage, to improve statistical evaluation, and to reduce variability, thus substantially enhance construction quality control of the entire earth fills. This paper presents the implementation of strength index from the Dynamic Cone Penetrometer (DCP) in conjunction with the conventional compaction control and moisture content measurements for earthwork construction quality control and structural performance of engineered earth fills.

2. Structural property measurement for earthwork quality control

Edil and Sawangsuriya [1] adopted the Soil Stiffness Gauge (SSG) for assessing the soil stiffness of various materials used for the earth fills from different construction sites around the state of Wisconsin, U.S.A. along with the conventional approach of moisture-density control. They investigated the variation of SSG stiffness with the relative compaction (RC) and the deviation of moisture content from the respective optimum moisture content ($w-w_{opt}$) for the natural subgrade soils tested. To account for the effect of moisture content, Edil and Sawangsuriya [1] also proposed a plot of K_{SSG} divided by ($w-w_{opt}$), i.e., normalized stiffness, as a function of RC. They indicated that the normalized stiffness varied very little with RC for compacted soils, while a larger variation was observed for uncompacted soils. The implication of this for compacted soils with the typically rather narrow range of RC was that the effect of compaction levels on stiffness was relatively minor compared to moisture contents [1].

Taesiri *et al.* [2] examined the use of the SSG and the DCP for assessing the in-place stiffness and strength index of pavement layer materials, respectively in the highway construction projects around the city of Bangkok, Thailand. The SSG stiffness and DPI were also correlated with the elastic modulus (E) and California bearing ratio (CBR) of the earthen materials, respectively. A good correlation was obtained between E-value from the SSG and CBR from the DCP [2]. Their studies indicated that both devices exhibited good potential for future implication in the pavement material evaluation. The in-place stiffness and strength properties of various earthen materials can be rapidly and directly monitored in companion with the conventional moisture-density measurements during earthwork construction. Direct monitoring of stiffness and strength of the pavement materials using these two devices appeared to be an effective approach during construction quality control monitoring [2].

According to Thailand Department of Highways (DOH)'s quality control criteria and construction specifications, Wachiraporn *et al.* [3] recommended the DCP for assessing the structural uniformity of layer thickness and routine earthwork quality control evaluation in companion with the conventional moisture-density control test during highway construction in Thailand. Wachiraporn *et al.* [4] suggested an adoption of the DCP as a mechanistic quality control tool during highway construction as well as future development for performance specifications. Based on their studies conducted in the laboratory test box and in the field trial, the DCP provided instantaneous in-place strength index monitoring in term of DCP penetration index (DPI) for earthwork quality control and also exhibited good potential for quality control monitoring of the earthen materials [4]. In addition, Wachiraporn *et al.* [4] studies suggested that the CBRs estimated from the DCP were considerably smaller than the field CBRs (ASTM D4429), while the laboratory unsoaked CBR (ASTM D1883) tended to give the highest value.

3. Test materials and methods

Disturbed earthen materials e.g. natural subgrade, sand embankment, lateritic soil subbase, fine-grained aggregate subbase, and crushed rock base were collected from highway construction sites in Thailand [6]. They included (1) Highway No. 35: Samutsakorn-Amphoe Pakto, (2) Highway No. 351: connection to Sukhapiban 1–eastern outer ring road, and (3) Keharomkroa road development project. A summary of their index properties, soil classification, and compaction characteristics are tabulated in Table 1. The particle size distribution curve is illustrated in Figure 1. Note that according to

Thailand DOH's standard and specifications, the crushed rock base used in the highway construction is generally subdivided into Grade A, Grade B, and Grade C.

Table 1. Material properties and their classification.

Material Classification and Properties	Natural Subgrade	Sand Embankment	Lateritic Soil Subbase	Fine-grained Aggregate Subbase	Crushed Rock Base Grade C	Crushed Rock Base Grade A,B
AASHTO Classification	A-2-7	A-3	A-2-4	A-2-4	A-1-a	A-1-a
50.0 mm (1½")	100	100	100	100	100	100
25.0 mm (1")	100	100	97.7	100	100	98.9
19.0 mm (3/4")	98.7	100	93.0	96.6	-	91.8
9.5 mm (3/8")	88.0	100	68.8	81.3	69.0	63.3
No. 4	72.2	98.5	40.1	63.8	47.0	42.1
No. 10	47.8	97.8	24.8	35.3	30.0	23.6
No. 40	23.2	93.6	16.4	19.6	18.0	10.5
No. 200	6.7	24.1	8.1	12.4	10.0	6.9
D_{10} (mm)	0.12	0.055	0.1	0.051	0.86	0.29
D_{30} (mm)	0.7	0.085	3.0	1.4	2.0	3.0
D_{60} (mm)	3.0	0.18	7.3	4.1	7.5	9.0
LL (%)	44.6	N.P.	23.4	N.P.	N.P.	N.P.
PI (%)	34.8	N.P.	7.6	N.P.	N.P.	N.P.
W_{opt} (%)	14.5	9.5	6.5	5.4	5.7	6.0
$\gamma_{dry,\,max}$ (t/m^3)	1.84	1.94	2.25	2.32	2.31	2.34
CBR (%)	3	26	27	58	83	105
Swell (%)	2.5	-	0.35	-	-	-

N.P. = Non-plastic

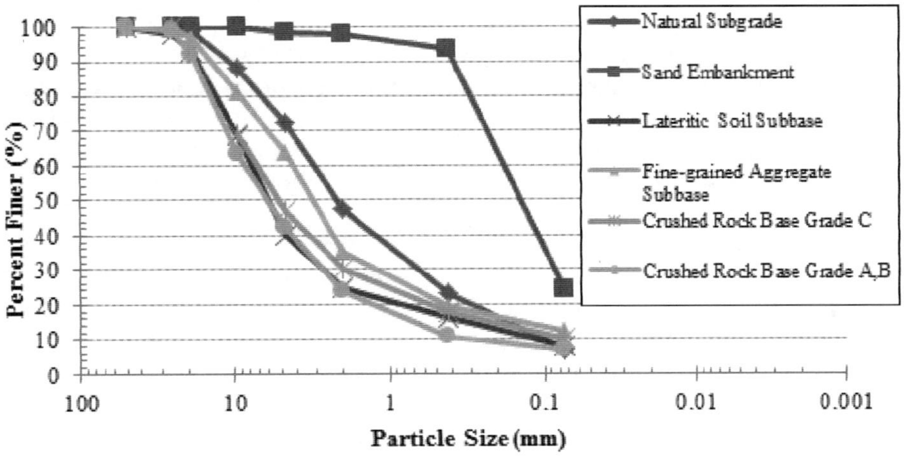

Figure 1. Particle Size Distribution Curve

A DCP was selected to measure the strength index of the earthen materials at these sites. Each field trial had approximately 100-200 m long. The DCP measurements were made at every 10 m depending on the length of field trial. The DCP was simple, rugged, economical, and able to provide a rapid in-place index of strength of earthen materials. It was used for measuring the material resistance to penetration while the cone of the device is being driven into the earth structure. The number of blows during operation was recorded with depth of penetration. The slope of the relationship between number of blows and depth of penetration (in millimeters per blow) at a given linear depth segment was recorded as DCP penetration index (DPI) [5].

Since DCP testing was basically a measure of penetration resistance, expressed as DPI, the analysis of the DCP data must be interpreted to generate a representative value of penetration per blow for the material being tested. In this study, such representative value was obtained by averaging the DPI across the penetration depth of 150 mm. Two methods of calculating the representative DPI value for a penetration depth of 150 mm are: (i) arithmetic average and (ii) weighted average [5]. Since the weighted average method yielded narrower standard deviation for the representative DPI value and provided better correlations to other field tests than the arithmetic average method based on field data available [5]. Thus, the weighted average method was adopted to calculate the representative DPI value in this study. The weighted average method can be obtained as follows:

$$DPI = \frac{1}{H}\sum_{i}^{N}\left[(DPI)_i \cdot (z)_i\right] \tag{1}$$

where z is the penetration distance per blow set and H is the overall penetration depth (e.g. 150 mm). After the compaction procedure, the DCP in companion with conventional moisture-density measurements using the nuclear gauge (ASTM D6938) were conducted instantaneously for each material. Every measurement was made at the adjacent location.

4. Results and Discussion

Earthen materials from three highway construction sites in Thailand were monitored in terms of their DPI_{150mm}, dry unit weight, and moisture content [6]. Figure 2 shows the relationship of the state of density (i.e., relative compaction, RC defined as the ratio of the field dry unit weight divided by the laboratory maximum standard/modified Proctor dry unit weight) to the deviation of moisture content from the respective optimum moisture content ($w\text{-}w_{opt}$) for earthen materials tested. Typical DOH compaction specifications call for RC $\geq$ 95%. Most of the RC of field compacted earthen materials were ranged from 87 to 104% with moisture contents dry of the optimum moisture content, except the lateritic soil subbase having moisture contents wet of optimum moisture content. Furthermore, RC tended to decrease with increasing $w\text{-}w_{opt}$. Figure 3 shows the variation of DPI_{150mm} with $w\text{-}w_{opt}$ for the earthen materials. Excluding the crushed rock base, a strong dependency of DPI_{150mm} on moisture content was evident as DPI_{150mm} varied from 5 to 30 mm/blow for a moisture content deviation of about -6% to +2% of the optimum moisture content. The compacted earthen materials had moisture contents mostly dry of optimum. Of course, there were other factors that affected DPI_{150mm} such as dry unit weight, texture, and soil fabric and they caused the spread in DPI_{150mm} for a given moisture content.

In the case of earthen materials, moisture content and dry unit weight of a test material played significant role on its DPI_{150mm} and their effects are hard to uncouple. To account for the effect of moisture content, DPI_{150mm} was divided by ($w\text{-}w_{opt}$). This normalized DPI_{150mm} was plotted versus RC in Figure 4. The normalized DPI_{150mm} varied very little with RC for compacted earthen materials within a narrow range of $\pm$10. It should be noted that those data set having moisture contents near the optimum moisture content ($w\text{-}w_{opt} \leq 1\%$) were excluded from the plot as the soil was theoretically in the optimum state of density and moisture. Results were consistent with the previous study [1], which indicated that the normalized stiffness (e.g. K_{SSG} divided by ($w\text{-}w_{opt}$)) varied very little with RC for compacted soils. Data from Edil and Sawangsuriya [7] were revisited and re-evaluated to develop a plot of the normalized DPI_{150mm} as a function of RC. The normalized DPI_{150mm} was remarkably constant around -8.4 for compacted soils. Results obtained can be used to establish the performance criteria for a variety of properly compacted engineered earth fills. The implication of this for compacted earth fills with the typically rather narrow range of RC was that the effect of dry unit weight on the strength index was relatively minor compared to moisture content. In addition, those data falling within the proposed envelope indicated proper compaction quality control process and consequently uniform structural properties (e.g. stiffness and strength).

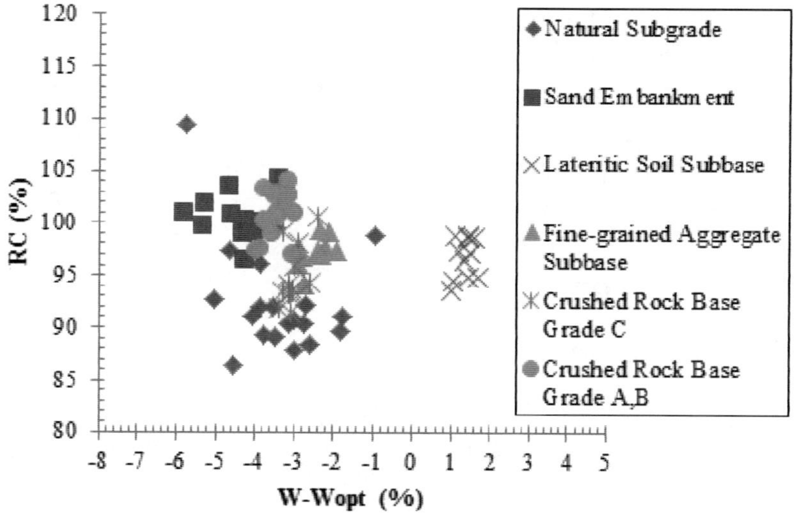

Figure 2. Relative compaction vs. deviation of moisture content from the optimum moisture content for earthen materials tested.

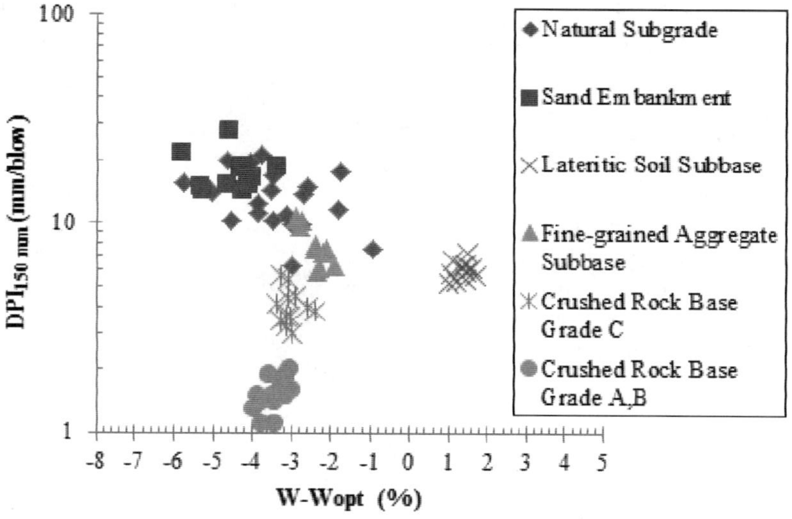

Figure 3. DPI_{150mm} vs. moisture content variance for earthen materials tested.

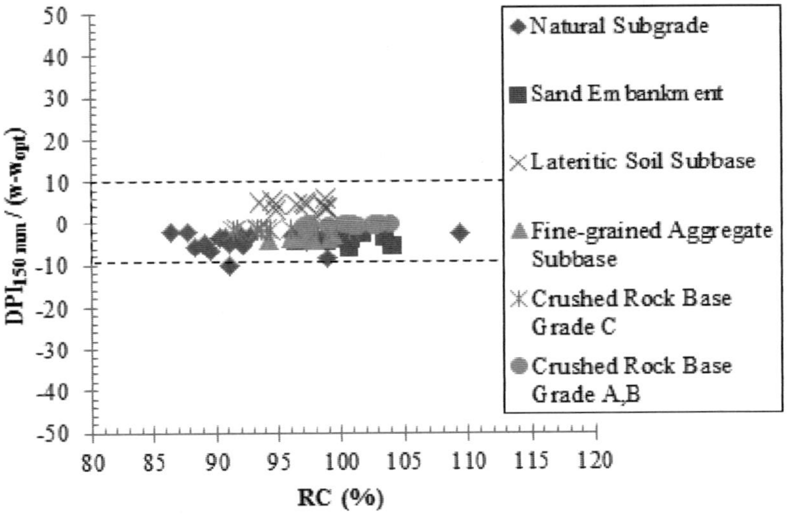

Figure 4. Normalized DPI_{150mm} vs. relative compaction.

5. Conclusions

This paper presents the implementation of soil moisture and strength index from the Dynamic Cone Penetrometer (DCP) measurements for earthwork construction quality control as well as structural performance of engineered earth fills. Use of the convenient DCP in companion with conventional moisture-density measurements enhances quality control by achieving more uniform structural property and aids developing a controlled design parameter during the earthwork construction. DPI_{150mm} normalized by the deviation of compaction moisture content from the optimum moisture content varied within a narrow range of ± 10 for compacted earthen materials and was proposed as performance criteria for a variety of properly compacted engineered earth fills and special engineering assessment, prevention, and mitigation of geohazards e.g. earthen flood defense embankments.

References

[1] Taesiri Y, Sawangsuriya A, Wachiraporn S, and Sramoon W 2009 Assessment of in-situ tests for stiffness and strength characteristic of pavement materials *13th REAAA Conference Korea 2009* Icheon South Korea

[2] Wachiraporn S, Sawangsuriya A, Sunitsakul J, and Sramoon W 2010 Stiffness and strength based in-place evaluation of compacted unbound materials *Proceedings of Sessions of GeoShanghai 2010* Geotechnical Special Publication No. 203 Shanghai China p 347-354

[3] Wachiraporn S, Sawangsuriya A, and Sramoon W 2013 Mechanistic based quality control during highway construction *Proceedings of the 18th Southeast Asian Geotechnical Conference* Singapore p 605–610

[4] Sawangsuriya A and Edil T B 2005 Evaluating stiffness and strength of pavement materials *Proceedings of the Institution of Civil Engineers-Geotechnical Engineering* **158**(4) p 217-230

[5] Edil T B and Sawangsuriya A 2005 Earthwork quality control using soil stiffness *Proceedings of the 16th International Conference on Soil Mechanics and Geotechnical Engineering* Osaka Japan p 1689-1692

[6] Wachiraporn S 2008 *Engineering Property-Based Compaction Control for Highway Construction* M.Eng. Thesis Mahanakorn University of Technology Bangkok Thailand

[7] Edil T B and Sawangsuriya A 2005 *Investigation of the DCP and SSG as alternative methods to determine subgrade stability* Report No. 0092-01-05 Wisconsin Department of Transportation WI

Analytical solutions to assess the stability of rock slopes subject to cracks via limit analysis

W. Wu [1]
School of Engineering, University of Warwick, Coventry, UK

E-mail: weigao.wu@warwick.ac.uk

Abstract. Based on the kinematic approach of limit analysis, a full set of upper bound solutions for the stability of homogeneous rock slopes subjected to cracks are obtained. The generalized Hoek-Brown failure criterion is adopted to describe the non-linear strength envelope of rocks. In this paper, critical failure mechanisms are determined for cracks of known depth but unspecified location, cracks of known location but unknown depth, and cracks of unspecified location and depth. It is shown that there is a nearly up to 50% drop in terms of the stability factors for rock slopes intersected by cracks compared with intact ones. Tables and charts of solutions in dimensionless forms are presented for ease of use by practitioners.

1. Introduction

A substantial range of different numerical tools exist today to investigate slope stability. Traditionally stability analyses of rock slopes have been based on limit equilibrium methods [1, 2]. In the last four decades, numerical methods for continuum mechanics such as the finite element method with strength reduction technique [3, 4] and finite element limit analysis [5] have provided the capability to reliably detect the onset of failure in rock slopes where the presence of joints is accounted for by a so called smear approach. However, in case of moderately fractured rock the smear approach no longer works since the onset of instability is ruled by the behavior of single fractures. In this case, the Discrete Element Method can nowadays be employed for 3D analyses of jointed rock slopes [6]. The development of these analyses within computationally affordable runtimes has been made possible by recent algorithmic advances in terms of contact detection algorithms [7, 8, 9] and rock joint dataset representation [9].

Rock slopes considered in this paper are intact or heavily fractured slopes subject to cracks that can be attributed to tension. These types of cracks tend to be vertical and/or subvertical. A continuum mechanics approach that explicitly considers the presence of cracks is adopted in this paper. Utili [10] and Michalowski [11] investigated the stability of cohesive frictional slopes subject to cracks via the limit analysis upper bound method. The advantage of this method is in providing an analytical solution of general validity that can be used to assess slopes of any geometry and for any value of c, ϕ. But, unlike cohesive frictional materials, rock yield is non-linear so that the Hoek-Brown (H-B) failure criterion [12], is almost universally accepted as the best criterion to express rock yielding. Collins et al. [13] developed a "tangent line" technique to extend the application of the limit analysis upper

[1] W. Wu is with the School of Engineering, University of Warwick, Coventry, UK (Email: weigao.wu@warwick.ac.uk).

IOP Conf. Series: Earth and Environmental Science **26** (2015) 012057 doi:10.1088/1755-1315/26/1/012057

bound method to the stability of rock slopes. Yang et al. [14] improved the analysis of Collins et al. [13] by modifying the generalized Hoek-Brown failure criterion [15]. However, neither Collins et al. [13] nor Yang et al. [14] considered the presence of cracks in rock slopes.

In this paper, an analytical formulation will be provided for the first time to investigate the stability of rock slopes subjected to cracks via the limit analysis upper bound method assuming that the rock slope is uniform and isotropic. Then, preliminary results of the analytical investigation are provided.

2. Hoek-Brown failure criterion

Based on the result of a series of field investigations and triaxial tests on rocks, Hoek & Brown [12] and Hoek et al. [15] introduced the well-known Hoek-Brown failure criterion which has become the most adopted method for the characterization of rock strength by far. The original H-B failure criterion can be expressed as

$$\sigma_1 = \sigma_3 + \sqrt{m\sigma_3\sigma_{ci} + s\sigma_{ci}^2} \tag{1}$$

with σ_1, σ_3 the major and minor principle stress respectively, σ_{ci} uniaxial compression strength for intact rock, m a parameter related to the rock type and s degree of fracturing of the rock mass.

The original Hoek-Brown failure criterion was designed for intact rock with high cohesion. In Hoek et al. [15], a new parameter n is introduced to extend the applicability of the criterion to loose and broken rocks. Thus, the generalized Hoek-Brown failure criterion is written as

$$\sigma_1 = \sigma_3 + \sigma_{ci}\left(m\frac{\sigma_3}{\sigma_{ci}} + s\right)^n \tag{2}$$

with n a parameter accounting for degree of imperfection of the rock mass. The failure envelop for the H-B failure criterion in terms of major and minor principle stresses is shown in Figure 1.

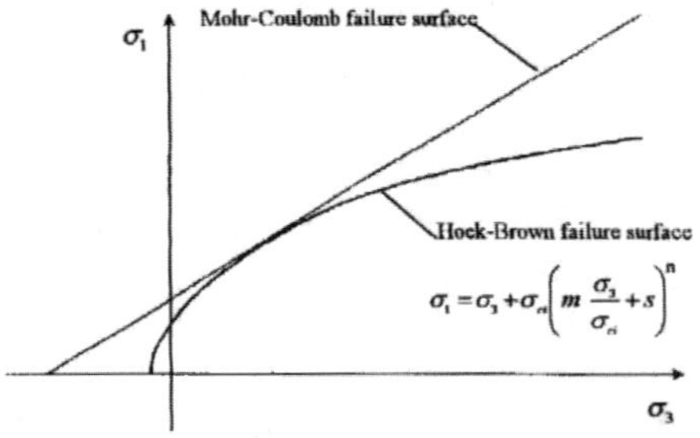

Figure 1. Hoek-Brown failure criterion.

3. Rock slope stability analysis

Although there is no lower/upper bound solution available for the stability of rock slopes with cracks, it was still expected by Li et al. [5] that the rigorous limit analysis results were found to bracket the true slope stability number to within ±9% or better for the intact case. In this paper, the upper bound solutions following the Mohr-Coulomb failure criterion are provided in the first place. The solutions are not directly applicable to materials obeying the H-B criterion. However, in principle, limit analysis can still be adopted as long as the material considered obeys the normality rule. A detailed upper bound solution of a rock slope with a vertical tension crack will be given in this section.

3.1. Limit analysis for slopes following Mohr-Coulomb failure criterion

For illustrative purposes, only a horizontal upper slope with failure line passing through the slope toe is examined (see Figure 2). For more general cases like failure line passing below the slope toe and non-horizontal upper slope are illustrated in Appendix. The slope has a height of H and an inclination of β, with a region of rock $\widehat{EDCB}$ rigidly rotating away about a centre of rotation P. The remaining part is bound by a crack B-C and logarithmic spiral $\widehat{DC}$ with an equation written in polar coordinates with reference to P,

$$r = r_0 \exp\left[\tan\phi(\theta - \theta_0)\right]$$

with r the distance of a generic point of the spiral to its centre P, θ the angle formed by r with a reference horizontal axis (see Figure 2), and θ_0 and γ_0 identifying the angle and distance of a particular point of the spiral F to its centre. According to normality rule [16], the angle between the sliding rate $\dot{u}$ of the rock mass and the failure line $\widehat{DC}$ must always equal to ϕ. However, it is not the case for the angle φ between $\dot{u}$ and the crack B-C, which can be different from ϕ. [10]

In Figure 2, the following geometrical relationships are obtained

$$r_\varsigma = r_\chi \exp\left[\tan\phi(\varsigma - \chi)\right] \tag{3}$$

and

$$r_v = r_\chi \exp\left[\tan\phi(v - \chi)\right] \tag{4}$$

with r_ζ and r_υ the radii of the spiral at the ζ and υ angles respectively, and

$$H = r_\chi \left\{\exp\left[\tan\phi(v - \chi)\right]\sin v - \sin\chi\right\} \tag{5}$$

$$\delta = r_\chi \left\{\exp\left[\tan\phi(\varsigma - \chi)\right]\sin\varsigma - \sin\chi\right\} \tag{6}$$

$$L_1 = r_\chi \left\{\frac{\sin(\chi + \beta)}{\sin\beta} - \exp\left[\tan\phi(v - \chi)\right]\frac{\sin(v + \beta)}{\sin\beta}\right\} \tag{7}$$

$$L_2 = r_\chi \left\{\cos\chi - \exp\left[\tan\phi(\varsigma - \chi)\right]\cos\varsigma\right\} \tag{8}$$

with L_1 and L_2 horizontal lengths as indicated in Figure 2, and δ being the crack depth.

Three different types of mechanisms will be analysed:

(a) slopes with a crack of known depth but unknown location
(b) slopes with a crack of known location but unknown depth
(c) slopes with cracks of unknown location and depth

In conditions (a) and (b), certain geometric constraints are introduced for the selections of χ, υ and ζ. For cracks of known depth, the following constraint is found:

$$\exp(\tan\phi \cdot \varsigma)\sin\varsigma = \exp(\tan\phi \cdot \chi)\sin\chi\left(1 - \frac{\delta}{H}\right) + \frac{\delta}{H}\exp(\tan\phi \cdot v)\sin v \tag{9}$$

Similarly, for cracks of known location, the following constraint is found:

$$\exp(\tan\phi\cdot\chi)\sin\chi=\exp(\tan\phi\cdot v)\sin v+\frac{\exp(\tan\phi\cdot v)\cos v-\exp(\tan\phi\cdot\zeta)\cos\zeta}{x/H}\tag{10}$$

Then, the internally dissipated energy rate $\dot{W}_d$ along the failure line $\widehat{DC}$ and the external work rate $\dot{W}_\gamma$ for the slope are to be detailed respectively.

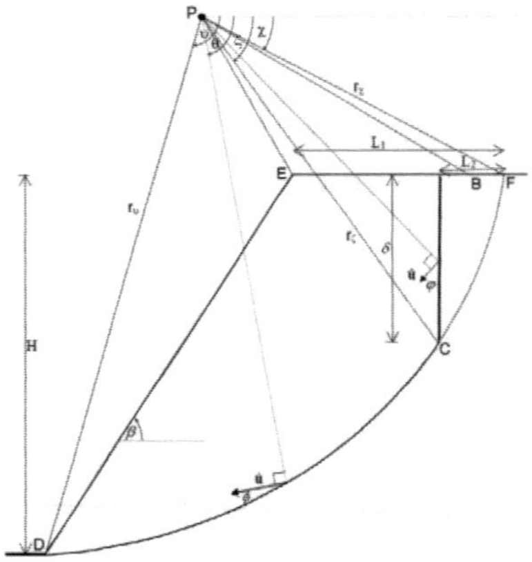

Figure 2. Failure mechanism

According to upper bound theorem and the assumed kinematic mechanism, energy is dissipated only along the logarithmic spiral shaped failure line $\widehat{DC}$, which is indicated as Γ in the integral

$$\dot{W}_d=\int_\Gamma c\dot{u}\cos\phi\frac{rd\theta}{\cos\phi}=c\omega\int_\Gamma r^2d\theta=c\omega r_\zeta^2\int_v^\zeta\exp\left[2\tan\phi(\theta-\zeta)\right]d\theta\tag{11}$$

from which the following expression is obtained

$$\dot{W}_d=c\omega r_\chi^2\exp\left[2\tan\phi(\zeta-\chi)\right]\frac{\exp\left[2\tan\phi(v-\zeta)\right]-1}{2\tan\phi}=c\omega r_\chi^2 f_d(\chi,v,\zeta,\phi)\tag{12}$$

As shown in (Utili and Nova [17]; Utili and Crosta [18,19]), the rate of external work due to the rock weight of region $\widehat{EDCB}$ is computed as the work done by region $\widehat{EFD}$ minus the work of region $\widehat{BFC}$. The rate of external work for region $\widehat{EFD}$ is the result of the work done by region $\widehat{PFD}$ minus P-F-E and P-E-D. Likewise, work done by region $\widehat{BFC}$ can be expressed by the summation of the work done by region $\widehat{PFC}$ minus P-F-B and P-B-C. $\dot{W}_1$, $\dot{W}_2$, $\dot{W}_3$, $\dot{W}_4$, $\dot{W}_5$ and $\dot{W}_6$ indicate the work done by $\widehat{PFD}$, P-F-E, P-E-D, $\widehat{PFC}$, P-F-B and P-B-C respectively. Therefore, the total rate of external work due to the rock weight is given by:

$$\dot{W}_\gamma=\dot{W}_1-\dot{W}_2-\dot{W}_3-\left(\dot{W}_4-\dot{W}_5-\dot{W}_6\right)=\dot{W}_1-\dot{W}_2-\dot{W}_3-\dot{W}_4+\dot{W}_5+\dot{W}_6=\omega\gamma r_\chi^3\left(f_1-f_2-f_3-f_4+f_5+f_6\right)\tag{13}$$

The calculation of the work rates $\dot{f}_1$, $\dot{f}_2$, $\dot{f}_3$, $\dot{f}_4$, $\dot{f}_5$ and $\dot{f}_6$ can be found in Utili [10]. Equating the rate of external work $\dot{W}_\gamma$ to the rate of internal energy dissipation $\dot{W}_d$ gives

$$\dot{W}_\gamma=\dot{W}_d\tag{14}$$

$$\omega\gamma r_{\chi}^{3}\left(f_1 - f_2 - f_3 - f_4 + f_5 + f_6\right) = c\omega r_{\chi}^{2} f_d \tag{15}$$

Dividing by ω and r_χ^2 and rearranging, the stability factor $N_{M-C} = \frac{\gamma H}{c}$ for Mohr-Coulomb failure criterion is obtained as

$$N_{M-C} = \frac{\gamma H}{c} = g\left(\chi,\zeta,\nu\right) = \frac{f_d\left\{\exp\left[\tan\phi\left(\nu-\chi\right)\right]\sin\nu - \sin\chi\right\}}{f_1 - f_2 - f_3 - f_4 + f_5 + f_6} \tag{16}$$

3.2. Limit analysis for slopes following Hoek-Brown failure criterion

Drescher & Christopoulos [20] first proposed a linear failure surface which is a tangent to the actual non-linear failure surface to get an upper bound solution. Meanwhile, the same "tangent line method" was adopted by Collins et al. [13] to linearize the Hoek-Brown failure criterion. Yang et al. [14] introduced a similar method as Collins et al. [13] but more complicated conditions such as sophisticated geometry (e.g., inclined upper slope), different strength parameters (e.g., n for the generalized H-B failure criterion) and pore pressure distribution are considered.

Considering the generalized Hoek-Brown failure criterion, Yang et al. [14] proposed a revised stability factor $N_{H-B} = \frac{\gamma H}{s^n \sigma_c}$, with $s^n\sigma_c$ the uniaxial compressive strength of the rock to be used in the definition of the stability factor N_{H-B} instead of c in equation (16).

The tangential line to the H-B failure criterion is expressed as

$$\tau = \sigma\tan\varphi_t + c_t \tag{17}$$

where τ and σ are the shear and normal stress, c_t is the intercept of the tangential line to τ axes in the (σ, τ) stress space, and φ_t is the angle of the tangential line at the point considered M to the σ axis (see Figure 3.).

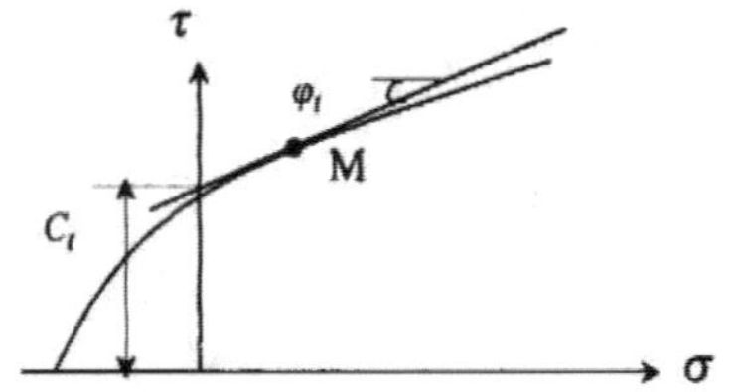

Figure 3. A tangential line for a non-linear failure criterion

According to Yang et al. [14], τ and σ along the failure envelope are determined by the following two equations

$$\frac{\tau}{\sigma_c} = \frac{\cos\varphi_t}{2}\left[\frac{mn\left(1-\sin\varphi_t\right)}{2\sin\varphi_t}\right]^{n/(1-n)} \tag{18}$$

$$\frac{\sigma}{\sigma_c} = \left(\frac{1}{m} + \frac{\sin\varphi_t}{mn}\right)\left[\frac{mn\left(1-\sin\varphi_t\right)}{2\sin\varphi_t}\right]^{1/(1-n)} - \frac{s}{m} \tag{19}$$

From equation (18) and equation (19), c_t is found by

$$\frac{c_t}{\sigma_c}=\frac{\cos\varphi_t}{2}\left[\frac{mn\left(1-\sin\varphi_t\right)}{2\sin\varphi_t}\right]^{n/(1-n)}-\frac{\tan\varphi_t}{m}\left(1+\frac{\sin\varphi_t}{n}\right)\left[\frac{mn\left(1-\sin\varphi_t\right)}{2\sin\varphi_t}\right]^{1/(1-n)}+\frac{s}{m}\tan\varphi_t \tag{20}$$

For the material following the original H-B failure criterion when $n = 0.5$, equation (20) can be simplified in the form

$$\frac{c_t}{\sigma_c}=\frac{m\left(1-\sin\varphi_t\right)^2}{16\sin\varphi_t\cos\varphi_t}+\frac{s}{m}\tan\varphi_t \tag{21}$$

Compared with equation (16), the revised stability factor for H-B failure criterion is defined as

$$N_{H-B}=\frac{\gamma H}{s^n\sigma_c}=\frac{\gamma H}{c_t}\cdot\frac{c_t}{s^n\sigma_c}=N_{M-C}\frac{c_t}{s^n\sigma_c}=\frac{f_d}{f_1-f_2-f_3-f_4+f_5+f_6+p_w}\times\left\{\begin{array}{l}\frac{\cos\varphi_t}{2}\left[\frac{mn\left(1-\sin\varphi_t\right)}{2\sin\varphi_t}\right]^{n/(1-n)}-\frac{\tan\varphi_t}{m}\left(1+\frac{\sin\varphi_t}{n}\right)\left[\frac{mn\left(1-\sin\varphi_t\right)}{2\sin\varphi_t}\right]^{1/(1-n)}\\+\frac{s}{m}\tan\varphi_t\end{array}\right\}\times\frac{\exp\left[\tan\phi_t\left(v-\chi\right)\right]\sin v-\sin\chi}{s^n} \tag{22}$$

The minimum value of the function $N_{H-B}(\chi, v, \zeta, \varphi_t)$ was found by evaluating repeatedly the function over the four variables χ, v, ζ and φ_t in the range of values of engineering interest.

Results

In most cases, no information is available about the depth and the location of cracks, therefore the most critical scenario for the stability of the slope is assumed. This scenario entails the existence of the most critical crack in terms of depth and position for the slope. Obviously this is an unrealistic assumption since in general cracks other than the most critical one will be present. However, this assumption has the advantage of being on the conservative side. In the absence of data a conservative assumption has to be preferred. To search for the upper bound corresponding to the most critical scenario, equation (22) is minimized over all possible χ, v, ζ and φ_t. In Figure 4, it is quite apparent that with the increase of the slope inclination, the values of the stability factors N_{H-B} drop significantly.

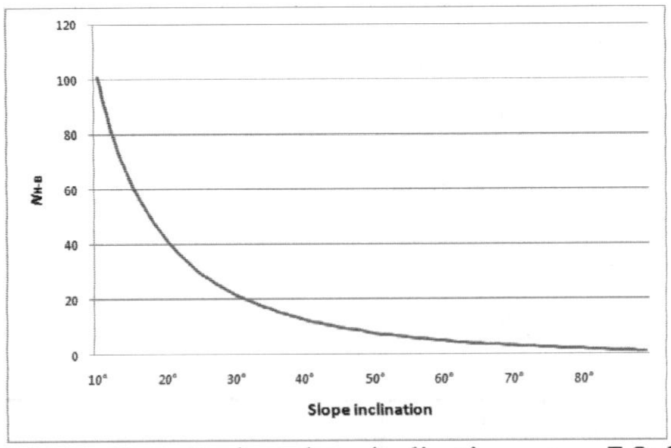

Figure 4. Stability factor N_{H-B} against slope inclination, $m = 7.3$: limestone, $s = 1$.

In practice, the generalized H-B failure criterion accounts for the variability of natural rock mass and the difference of fractured degree with n as a parameter in equation (2). The exponent n varies from 0.5 to 0.7. Sometimes the upper slope has an angle $\alpha \neq 0$ (see Figure 6 in Appendix), which complicates the solutions of the stability factors. Table 1 provides the most critical values of the stability factors $N_{critical}$ for cracked slopes with the slope inclination β varying from 30° to 90°, and α being equal to 5°, 10° and 15°.

From Table I, it is found that the parameter α has little influence on the stability factors. Figure 5 shows the influence of the exponent n (ranging from 0.5 to 0.7) on the stability factors for limestone slopes with $m = 7.3$, $s = 1$ and $\alpha = 5°$. The stability factors $N_{critical}$ increase with n.

Table 1. The stability factors $N_{critical}$ for limestone rock ($m = 7.3$, $s = 1$) with various values of n, α and β.

		Slope inclination β°						
n	α	30°	40°	50°	60°	70°	80°	90°
0.50	5	21.89	13.41	9.65	7.39	5.74	4.33	2.88
	10	21.71	13.35	9.62	7.36	5.70	4.28	2.83
	15	21.02	13.50	9.96	7.71	6.01	4.52	3.00
0.55	5	31.66	17.61	12.15	9.08	6.94	5.17	3.40
	10	31.49	17.58	12.13	9.06	6.90	5.11	3.34
	15	30.75	17.95	12.70	9.60	7.35	5.46	3.57
0.60	5	49.42	24.34	16.00	11.64	8.74	6.43	4.18
	10	49.22	24.33	15.98	11.62	8.69	6.35	4.09
	15	48.35	24.14	15.89	11.54	8.60	6.25	4.01
0.65	5	85.89	38.17	24.42	17.52	13.02	9.51	6.15
	10	85.63	38.17	24.40	17.48	12.96	9.40	6.02
	15	84.61	37.92	24.28	17.37	12.83	9.26	5.90
0.70	5	175.25	64.28	38.66	26.90	19.61	14.12	8.65
	10	175.20	64.34	38.67	26.87	19.52	13.95	8.83
	15	174.29	64.02	38.53	26.73	19.35	13.75	8.65

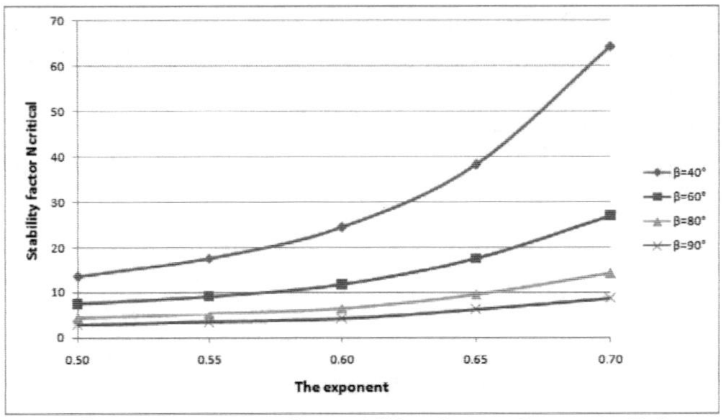

Figure 5. Influence of exponent n on the stability factor for $m = 7.3$, $s = 1$ and $\alpha = 5°$.

4. Conclusions

The kinematic approach of limit analysis and the tangent technique were applied to investigate the stability of rock slopes subjected to tension cracks with rocks obeying the generalized Hoek-Brown failure criterion. From the performed analyses it emerges that rock slopes subject to cracks can suffer a nearly up to 50% drop in stability factor in comparison with the case of intact rock slopes.

Appendix

Case of a non-horizontal upper slope ($\alpha \neq 0$)

The case of $\alpha \neq 0$ is illustrated in Figure 6. In the following, only the equations that assume a different expression from the equations illustrated in the paper for a horizontal upper slope ($\left(1-\frac{N_{k\neq 0}}{N_{k=0}}\right)\times 100\%$) are shown.

Equation (5) becomes

$$H = r_\chi \left\{\exp\left[\tan\phi(\upsilon-\chi)\right]\sin\upsilon - \sin\chi\right\} \tag{14}$$

Equation (6) becomes

$$\delta = r_\chi \left\{\exp\left[\tan\phi(\zeta-\chi)\right]\sin\zeta - \sin\chi\right\} \tag{15}$$

Equation (7) becomes

$$L_1 = r_\chi \left\{\frac{\sin(\upsilon-\chi)}{\sin(\upsilon+\alpha)} - \exp\left[\tan\phi(\upsilon-\chi)\right]\times\sin(\upsilon+\alpha)\frac{\sin(\upsilon+\beta)}{\sin(\upsilon+\alpha)\sin(\beta-\alpha)} - \sin(\chi+\alpha)\right\} \tag{16}$$

Equation (8) becomes

$$L_2 = r_\chi \left\{ \cos\chi - \exp\left[\tan\phi(\zeta - \chi) \right] \cos\zeta \right\} \tag{17}$$

$\dot{W}_2$ becomes

$$\dot{W}_2 = \omega\gamma r_\chi^3 \left[\frac{1}{6}\sin\chi \frac{L_1}{r_\chi}\left(2\cos\chi - \frac{L_1}{r_\chi} \right) \right] = \omega\gamma r_\chi^3 f_2(\chi, \upsilon, \beta, \phi) \tag{18}$$

$\dot{W}_3$ becomes

$$\begin{aligned} \dot{W}_3 &= \omega\gamma r_\chi^3 \left[\begin{array}{l} \frac{1}{6}\exp\left[\tan\phi(\upsilon - \chi) \right]\left[\sin(\upsilon - \chi) - \frac{L_1}{r_\chi}\sin\upsilon \right] \\ \times\left\{ \cos\chi - \frac{L_1}{r_\chi} + \cos\upsilon \exp\left[\tan\phi(\upsilon - \chi) \right] \right\} \end{array} \right] \\ &= \omega\gamma r_\chi^3 f_3(\chi, \upsilon, \beta, \phi) \end{aligned} \tag{19}$$

$\dot{W}_6$ becomes

$$\begin{aligned} \dot{W}_6 &= \omega\gamma r_\chi^3 \left(\frac{1}{3}\exp\left[2\tan\phi(\zeta - \chi) \right](\cos\zeta)^2 \times \left\{ \exp\left[\tan\phi(\zeta - \chi) \right]\sin\zeta - \sin\chi \right\} \right) \\ &= \omega\gamma r_\chi^3 p_3(\chi, \zeta, \phi) \end{aligned} \tag{20}$$

Equation (22) becomes

$$\begin{aligned} N_{H-B} &= \frac{\gamma H}{s^n\sigma_c} = \frac{\gamma H}{c_t}\cdot\frac{c_t}{s^n\sigma_c} = N_{M-C}\frac{c_t}{s^n\sigma_c} = \frac{f_d}{f_1 - f_2 - f_3 - f_4 + f_5 + f_6} \times \frac{\sin\beta}{\sin(\beta - \alpha)} \\ &\left\{ \frac{\cos\varphi_t}{2}\left[\frac{mn(1 - \sin\varphi_t)}{2\sin\varphi_t} \right]^{n/(1-n)} - \frac{\tan\varphi_t}{m}\left(1 + \frac{\sin\varphi_t}{n} \right)\left[\frac{mn(1 - \sin\varphi_t)}{2\sin\varphi_t} \right]^{1/(1-n)} + \frac{s}{m}\tan\varphi_t \right\} \times \\ &\frac{\exp\left[\tan\phi_t(\upsilon - \chi) \right]\sin(\upsilon + \alpha) - \sin(\chi + \alpha)}{s^n} \end{aligned} \tag{21}$$

Equation (9) becomes

$$\exp(\tan\phi\cdot\zeta)\sin(\zeta+\alpha)=$$
$$\exp(\tan\phi\cdot\chi)\sin(\chi+\alpha)\left(1-\frac{\delta}{H}\frac{\sin\beta}{\sin(\beta-\alpha)}\right)$$
$$+\frac{\delta}{H}\frac{\sin\beta}{\sin(\beta-\alpha)}\exp(\tan\phi\cdot\upsilon)\sin(\upsilon+\alpha) \tag{22}$$

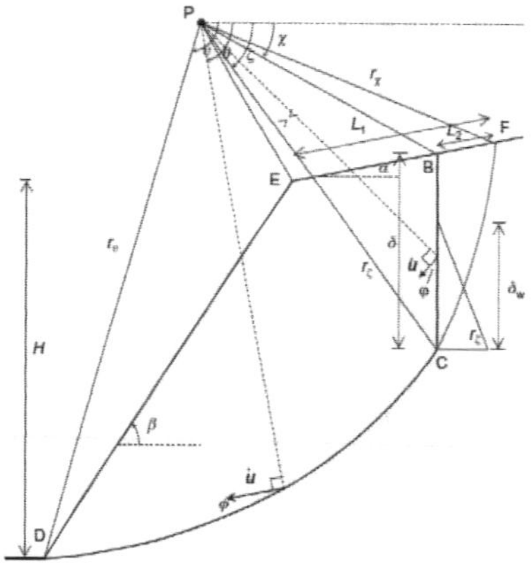

Figure 6. Inclined slope upper surface ($\alpha \neq 0$).

References

[1] Spencer E 1968 Effect of tension on stability of embankments J Soil Mech Foundations **5** 1159-1173

[2] Baker R 1981 Tensile strength, tension cracks and stability of slopes Soils Found **21** 1-17

[3] Dawson E M, Roth W H, Drescher A 1999 Slope stability analysis by strength reduction Géotechnique **49** 835-840

[4] Conte E, Silvestri F, Troncone A 2010 Stability analysis of slopes in soils with strain-softening behaviour Comput Geotech **37** 710-722

[5] Li A J, Merifield R S, Lyamin A V 2008 Stability charts for rock slopes based on the Hoek-Brown failure criterion Int J Rock Mech Min Sci **45** 689-700

[6] Boon C W, Houlsby G T, Utili S 2014 New insights in the 1963 Vajont slide using 2D and 3D Distinct Element Method analyses Géotechnique **64** 800-816

[7] Boon C W, Houlsby G T, Utili S 2012 A new algorithm for contact detection between convex polygonal and polyhedral particles in the discrete element method Comput Geotech **44** 73-82

[8] Boon C W, Houlsby G T, Utili S 2013 A new contact detection algorithm for three dimensional non-spherical particles Powder Technology **248** 94-102

[9] Boon C W, Houlsby G T, Utili S 2015 A new rock slicing algorithm based on linear programming Comput Geotech **65** 12–29

[10] Utili S 2013 Investigation by limit analysis on the stability of slopes with cracks Géotechnique **63** 140-54

[11] Michalowski R L 2013 Stability assessment of slopes with cracks using limit analysis Can Geotech J **50** 1011-1021

[12] Hoek E, Brown E T 1980 Empirical strength criterion for rock masses J Geotech Eng ASCE **106** 1013-35

[13] Collins I F, Gunn CIM, Pender M J, Yan W 1998 Slope stability analyses for materials with a non-linear failure envelope In J Numer and Anal Meth Geomech **12** 533-50

[14] Yang X L, Li L, Yin J H 2004 Stability analysis of rock slopes with a modified Hoek-Brown failure criterion In J Numer and Anal Meth Geomech **28** 181-90

[15] Hoek E, Carranza-Torres C, Corkum B 2002 Hoek–Brown failure criterion-2002 edition In: Proceedings of the North American rock mechanics society meeting, Toronto

[16] Chen W F 1975 Limit analysis and soil plasticity (New York: Elsevier)

[17] Utili S, Nova R 2007 On the optimal profile of a slope Soils Found **47** 717-729

[18] Utili S, Crosta G B 2011 Modelling the evolution of natural slopes subject to weathering: Part I. Limit analysis approach J Geophys Res **116** F01016

[19] Utili S, Crosta G B 2011 Modelling the evolution of natural slopes subject to weathering: Part II. Discrete element approach J Geophys Res **116** F01017

[20] Drescher A, Christopoulos C 1998 Limit analysis slope stability with nonlinear yield condition J Numer and Anal Meth Geomech **12** 341-345

Investigation of the slope stability problem using the Material Point Method

F Fatemizadeh[1] and C Moormann[2]

[1]PhD candidate, Institute of Geotechnical Engineering (IGS), University of Stuttgart, Germany
[2]University professor, Institute of Geotechnical Engineering (IGS), University of Stuttgart, Germany

E-mail: farzad.fatemizadeh@igs.uni-stuttgart.de

Abstract. The Finite Element Method (FEM) has become a standard tool in engineering, although its shortcoming in large deformation analysis is apparent. Mesh distortions are common in this area. In order to be able to investigate applications including large deformations or displacements of the material, advanced numerical methods are needed. In this study we introduce the Material Point Method (MPM) as a powerful tool to simulate applications including large deformations. This capability of MPM plays an important role in the study of land and rockslides, avalanches, mudflow, etc. Here we will present a brief introduction on the governing equations and the solution procedure of MPM. Based on the presented formulation and using a two dimensional program developed, a slope stability problem is solved to show the capability of this method to be used in studying geohazard related phenomena. At the end some concluding remarks and recommendations are presented.

1. Introduction

The Finite Element Method (FEM) is a standard numerical tool in different fields of engineering especially in the field of geotechnical engineering. Some applications in this field include large deformations or displacements of the soil. FEM is not able to simulate these large deformations or displacements and suffers from what is called the “mesh distortion”. Examples of such applications are land and rockslides, avalanches, mudflow, etc. During the simulation of such applications the finite element mesh becomes so distorted that the determinant of the Jacobian matrix becomes negative. Remedies like using updated Lagrangian formulation introduce errors during the mapping process [1] and remeshing is computationally expensive.

In order to study these applications other suitable numerical methods should be used. Available numerical methods for the simulation of large deformations can be categorized into three major groups. In the first group, the methods use the advantages of the both Lagrangian and Eulerian discriptions of motion while avoiding their drawbacks. The Arbitrary Lagrangian Eulerian (ALE) method [2] and the Coupled Eulerian Lagrangian (CEL) method [3] are categorized in this group. Second group includes the mesh free methods like the Element Free Galerkin (EFG) method [4] and the Smooth Particle Hydrodynamic (SPH) method [5]. The last group contains the mesh based particle methods. The Particle in Cell (PIC) method [6], Fluid Implicit Particle (FLIP) method [7] and the Material Point Method (MPM) [8,9] are amongst others in this group.

MPM discretizes the continuum using the material points or particles and discretizes the space using an Eulerian background fixed mesh on which the equations of motion are solved (figure 1). This mesh should cover the whole space where the material may go during the simulation process. Material points are the integration points which can move during the simulation and using this property MPM is able to analyze large deformations or displacements of the material. Particles carry all the physical information (e.g. stresses, strain, etc.) during the simulation process and no permanent data are stored on the mesh.

Working procedure of MPM for one time step consists of three phases. First is the initialization phase where all the data are mapped from particles to the nodes of the background mesh using the nodal shape functions. Then is the Lagrangian phase in which the equations of motion are solved on the mesh. At the end is the convective phase. In this phase information are mapped back from the mesh to the material points and updates the data of the particles. Background mesh goes back to its original position (figure 2).

For the first time Sulsky et al. [8] applied the PIC method from fluid to solid mechanics and called it the material point method [9]. Bardenhagen et al. [10] introduced a frictional contact algorithm to the method based on the Coulomb friction law. Bardenhagen and Kober [11] presented the Generalized Interpolation Material Point Method (GIMP) to avoid the grid crossing error presented in the framework of original MPM. Quasi static formulation of MPM was first developed by Beuth et al. [12]. The abovementioned formulations use the explicit time integration scheme whereas Cummins and Brackbill [13], Guilkey and Weiss [14], Sulsky and Kaul [15] and Stolle et al. [16] presented unconditionally stable implicit time integration schemes for MPM. As MPM showed its capability to simulate large deformations especially in the field of geotechnical engineering, this method is used to study several applications in this area [17,18].

After the short introduction in chapter 1, an example of a one dimensional spring and mass system is solved in chapter 2 as a validation for the MPM formulation which is thoroughly discussed in appendix A. Next an example of a box which slides on a steep surface is simulated in chapter 2 as a validation for the contact algorithm which is discussed in appendix B. The problem of slope stability is analyzed using the developed code by the author based on the explained algorithms in chapter 3. At the end some concluding remarks are presented in chapter 4.

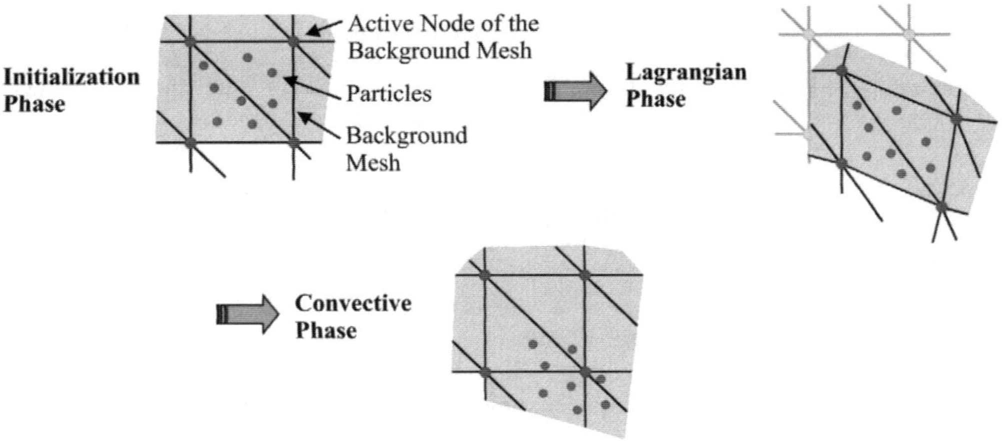

Figure 2. MPM working procedure for one time step.

2. One dimensional spring and mass system

To validate the complete formulation of MPM presented in appendix A, a one dimensional spring and mass example is simulated in this chapter using the code developed by the author. The spring is $1\ m$ long and has a cross section of $0.1\ m^2$ and elastic modulus of $5 \times 10^3\ kN\ /\ m^2$. The mass has a

IOP Conf. Series: Earth and Environmental Science **26** (2015) 012019 doi:10.1088/1755-1315/26/1/012019

length of 5 cm and the same cross section as the spring with a density of 1000 kg / m^3. Linear elastic material behavior is assumed for the spring and mass. Ten particles per element were used and the gravitational acceleration is assumed to be $g = 10\ m / s^2$. Density of the spring is small in comparison with the mass to represent the massless spring.

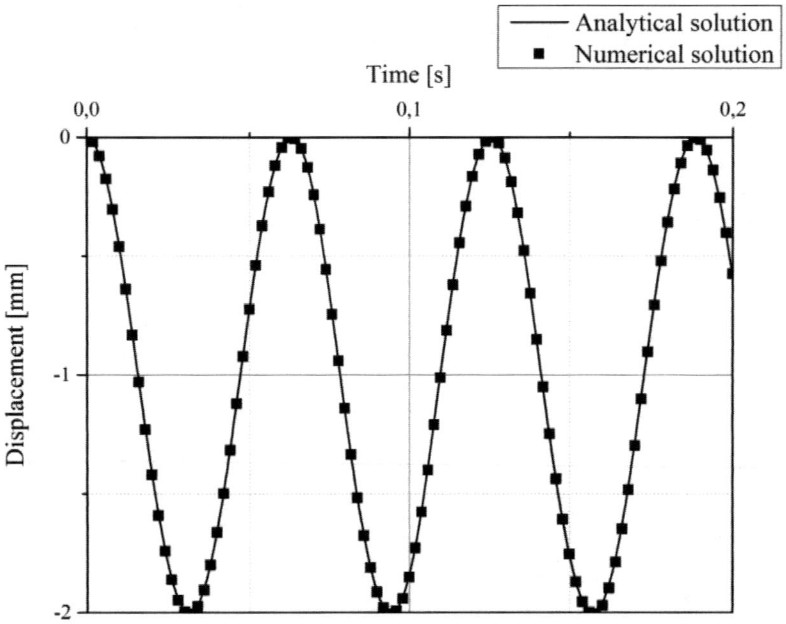

Figure 3. Analytical and numerical solution of the one dimensional spring and mass system.

For this problem the analytical solution exists which can be written as

$$y_{(t)} = \frac{F_0}{k}[1 - cos(\omega t)]. \quad (1)$$

By introducing the corresponding values it can be written that

$$y_{(t)} = 10^{-3}[1 - cos(100t)]. \quad (2)$$

The simulation results are shown in figure 3 together with the analytical results. A good agreement can be seen between the curves and MPM can catch the analytical results very well.

3. Sliding of a box on the steep surface

In order to show the effectivity of the described contact algorithm from appendix B, a simple example of a box sliding on a steep surface on which the coefficient of friction is defined, is solved using the code developed by the author and the results are compared with the analytical ones.

Linear elastic material behavior is assumed for both the box and the surface. Ten particles per element were used for the MPM simulation. The surface is assumed to be 10 cm thick and 1 m long with an inclination angle of 30°. The box has a length of 15 cm and a height of 10 cm with a density of 200 kg/m^3. For this problem the analytical solution exists by assuming the coefficient of friction between the box and the surface to be 0.4 and the initial velocity and displacement to be zero which states

$$x_{(t)} = 0.768\ t^2 \quad (3)$$

and for the coefficient of friction of 0.3 and the same initial conditions

$$x_{(t)} = 1.2\ t^2. \quad (4)$$

The results from the MPM simulation together with the analytical results are shown in figure 4, which shows a good agreement between them. With the coefficient of friction of 0.3 larger displacements are obtained than the case with the coefficient of friction of 0.4 which was expected from the analytical solution as well. Figure 5 shows the position of the box after one second of displacement before the box hits the wall for the case with the coefficient of friction equal to 0.4.

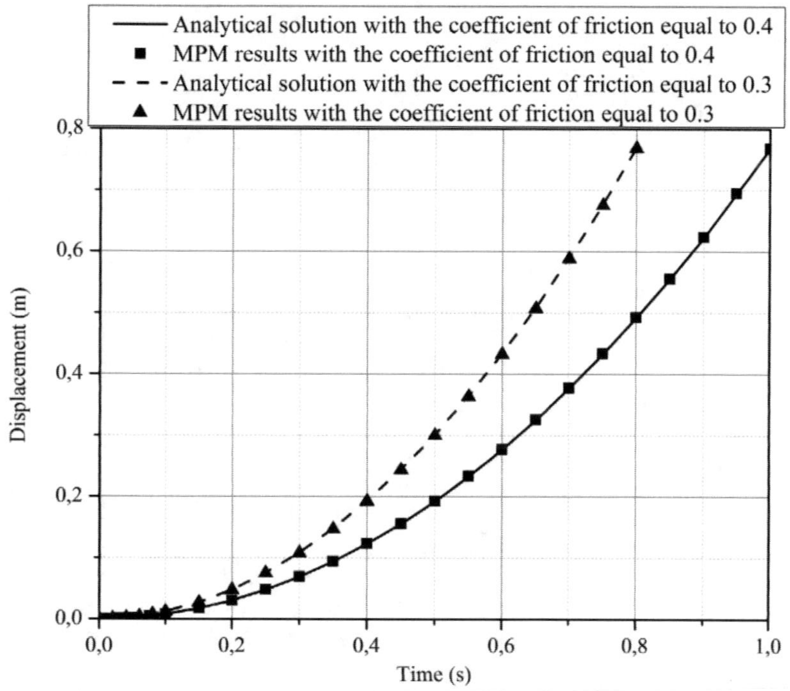

Figure 4. Analytical and numerical solution of the box sliding on the steep surface example.

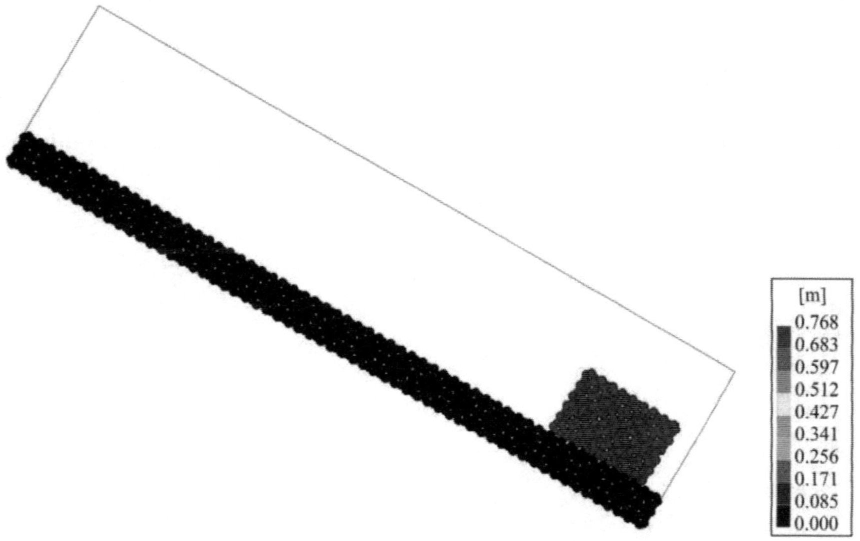

Figure 5. Position of the box after one second of movement.

4. Stability of the slope

To show the applicability of the MPM to be used in the field of geotechnical engineering for the simulation of large deformations or displacements of the soil, the problem of slope stability is analyzed using the code developed by the author based on the described methods in the previous chapters. In order to overcome the volumetric locking that happens in low order elements (here linear triangular elements) the enhanced volumetric strain method suggested by Detournay & Dzik [19] is used. This method is based on averaging the volumetric part of the strain on the nodes then averaging these values to be used for the particles inside the elements. During the averaging process volumetric part of the strain is smeared using the values from neighboring elements and this method helps to relax the elements from volumetric locking.

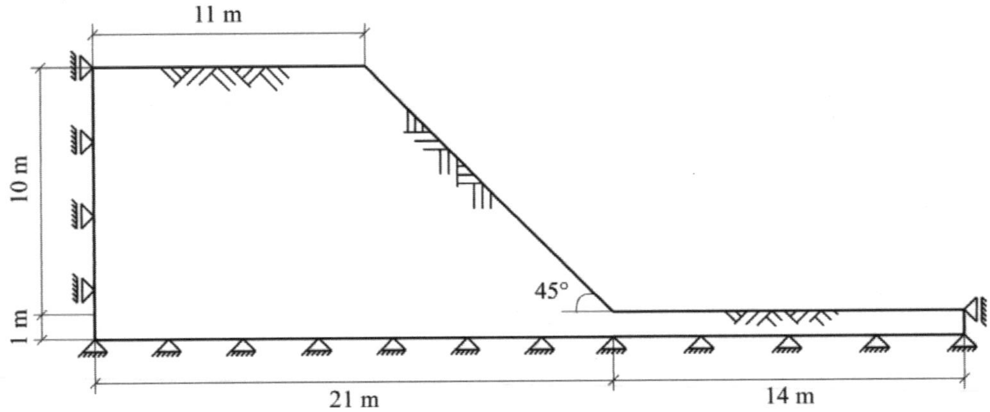

Figure 6. Geometry and boundary conditions of the soil.

Geometry and boundary conditions of the slope are shown in figure 6. The elastic-perfectly plastic Mohr-Coulomb material model is used to describe the behavior of the soil. In order to focus on the capabilities of MPM and exclude the effects of different material models on the final results, the simple failure criteria of Mohr-Coulomb is adopted. The soil of the slope is assumed to be sand with the friction angle of 30°, cohesion of 1 kpa, module of elasticity of $4 \times 10^4\ kN/m^2$, Poisson ratio of 0.3 and density of 1800 kg/m^3. Four particles per element were used for the MPM simulation. Total displacement of the particles is shown in figure 7. As can be seen in figure 7 the slope failure happens due to the self-weight of the slope under gravitational acceleration. The failure zone starts from the toe of the slope and goes up to the surface in a circular shape, an attitude which can also be seen in simulations and lab tests from other publications and case studies [20]. The failed soil slides down the slope and stops in the downhill.

In the nature during this sliding, the failed soil mass causes damage and destroys the buildings and structures in its way. How far the sliding soil will travel, can be predicted to prevent the damages caused during this phenomenon.

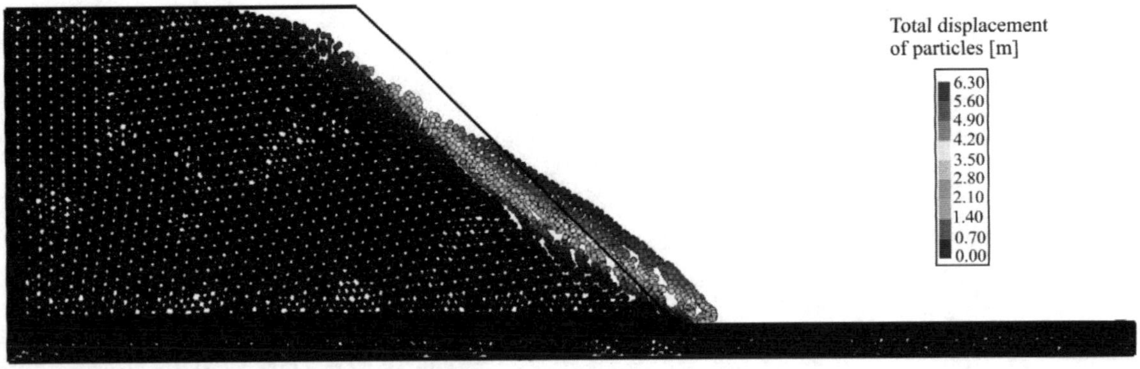

Figure 7. Total displacement of the particles.

5. Summery and conclusion

In the present work MPM is introduced as a powerful numerical tool which is able to simulate large deformations or displacements of the material without the mesh tangling problem which is present in Lagrangian FEM formulations. The governing equations of the method are presented briefly in appendix A and validated using a simple one dimensional spring and mass example. The contact algorithm based on the Coulomb friction law is described in appendix B and validated using a simple example of a box which slides on a steep surface for which a coefficient of friction is defined. Good agreement between the simulation results and the analytical solutions can be seen in the examples. To show the applicability of the method in the field of geotechnical engineering the problem of slope stability is analyzed. The simulation results show the potential of this numerical method to be used in order to study the large deformations and displacements of continuums in the field of geotechnical engineering and scrutiny the geohazards related phenomena like land and rockslides, avalanches, mudflow, etc.

Further research on the development of the MPM is currently done at the Institute of Geotechnical Engineering (IGS) of the University of Stuttgart.

Appendix A: Governing equations of MPM

Starting point is the momentum balance equation

$$\rho\ddot{\boldsymbol{u}} = \nabla \cdot \boldsymbol{\sigma} + \rho\boldsymbol{g} \quad and \quad \boldsymbol{t} = \boldsymbol{\sigma} \cdot \boldsymbol{n} \tag{A.1}$$

where ρ is the material density, $\boldsymbol{u}$ is the displacement vector, a superposed dot declares differentiation with respect to time, $\boldsymbol{\sigma}$ is the Cauchy stress tensor and $\boldsymbol{g}$ is the gravitational acceleration vector. The surface traction acting on the external boundary is denoted by $\boldsymbol{t}$ and $\boldsymbol{n}$ is the outward unit normal of the boundary.

Applying the virtual work principle to a domain of volume V which is surrounded by boundary S and using the integration by parts and the divergence theorem, the weak form of the momentum equation can be written as

$$\int_V \delta\boldsymbol{u}^T\rho\ddot{\boldsymbol{u}}dV = -\int_V \delta\boldsymbol{\varepsilon}^T\boldsymbol{\sigma}dV + \int_V \delta\boldsymbol{u}^T\rho\boldsymbol{g}dV + \int_S \delta\boldsymbol{u}^T\boldsymbol{t}dS \tag{A.2}$$

where δ denotes a virtual quantity, $\boldsymbol{\varepsilon}$ is the strain tensor and the superscript T specifies the transpose.

The displacement field is discretized using the nodal shape functions $\boldsymbol{N}$ and the nodal displacements $\boldsymbol{a}$

$$\boldsymbol{u} = \boldsymbol{N}\boldsymbol{a} \tag{A.3}$$

which leads to the definition of the strain tensor

$$\boldsymbol{\varepsilon} = \boldsymbol{B}\boldsymbol{a} \qquad \boldsymbol{B} = \boldsymbol{L}\boldsymbol{N} \tag{A.4}$$

where B is the strain displacement matrix and L is the linear differential operator. Substituting equation (3) and equation (4) into equation (2) and doing some mathematical operations, the well-known equilibrium equation will be derived

$$\boldsymbol{M}\ddot{a} = \left(\boldsymbol{F}^{ext} - \boldsymbol{F}^{int}\right) \tag{A.5}$$

where $\boldsymbol{M}$ is the mass matrix

$$\boldsymbol{M} = \int_V \rho \boldsymbol{N}^T \boldsymbol{N} dV \tag{A.6}$$

$\boldsymbol{F}^{ext}$ is the external force vector

$$\boldsymbol{F}^{ext} = \int_V \rho \boldsymbol{N}^T \boldsymbol{g} dV + \int_S \boldsymbol{N}^T \boldsymbol{t} dS \tag{A.7}$$

and $\boldsymbol{F}^{int}$ is the internal force vector

$$\boldsymbol{F}^{int} = \int_V \boldsymbol{B}^T \boldsymbol{\sigma} dV. \tag{A.8}$$

To increase the computational efficiency lumped mass matrix is used which is the row some of the entries of the consistent mass matrix. In the original formulation of MPM the density field is discretized using the Dirac delta function i.e.

$$\rho_{(x_i)} = \sum_{P=1}^{n_p} m_p \delta\left(x_i - x_i^p\right) \tag{A.9}$$

where x_i is an arbitrary position vector and x_i^p is the position vector at particle p. Using Dirac delta function to discretize the density field indicates that the mass of the continuum is approximated to be in the particles and as long as the mass of the particles do not change during the simulation, conservation of mass is considered automatically.

Here the explicit time integration scheme of the discretized momentum equation is adopted via applying an Euler forward integration scheme

$$\ddot{\boldsymbol{a}}^t = [\boldsymbol{M}_L^t]^{-1} \boldsymbol{F}^t \quad , \quad \dot{\boldsymbol{a}}_p{}^{t+\Delta t} = \dot{\boldsymbol{a}}_p{}^t + \Delta t \boldsymbol{N}_p \ddot{\boldsymbol{a}}^t \tag{A.10}$$

where Δt is the time increment, $\dot{\boldsymbol{a}}_p{}^{t+\Delta t}$ and $\dot{\boldsymbol{a}}_p{}^t$ are the particle velocities at time t and $t + \Delta t$ respectively and $\ddot{\boldsymbol{a}}^t$ is the nodal accelerations at time t.

The nodal velocities $\dot{\boldsymbol{a}}^{t+\Delta t}$ at time $t + \Delta t$ are then calculated from the updated particles velocities, solving the following equation

$$\boldsymbol{M}_L^t \dot{\boldsymbol{a}}^{t+\Delta t} \approx \sum_{p=1}^{n_p} m_p \boldsymbol{N}_p^T \dot{\boldsymbol{a}}_p{}^{t+\Delta t}. \tag{A.11}$$

Vector of nodal displacements is calculated by using an Euler backward integration method. Then the position of particles are updated

$$\Delta \boldsymbol{a}^{t+\Delta t} = \Delta t \dot{\boldsymbol{a}}^{t+\Delta t} \quad , \quad \boldsymbol{X}_P^{t+\Delta t} = \boldsymbol{X}_P^t + \boldsymbol{N}_P \Delta \boldsymbol{a}^{t+\Delta t}. \tag{A.12}$$

The time step size of the explicit time integration scheme is restricted by the CFL condition in which

$$\Delta t = \alpha_c \Delta t_{cr} \quad , \quad \Delta t_{cr} = {}^{h_{min}}/{}_{c_d} \tag{A.13}$$

where h_{min} is the minimum representative distance in an element, c_d is the compression wave speed and α_c is called the Courant's number and is a value between one and zero.

Appendix B: Contact algorithm for the material point method

Due to the single valued velocity field in the framework of the original MPM formulation, interpenetration of the particles are excluded automatically. York et al. [21] proposed that when two bodies are colliding each other conventional formulation of MPM is used to avoid interpenetration of the particles and when the two bodies are separating from each other no restrictions should be applied in order to allow the free separation. Bardenhagen et al. [10] proposed a frictional contact algorithm based on the Coulomb friction law for MPM. His formulation is widely used and will be employed in this work as well. Since the formulation of frictional contact algorithm is proposed, many modifications to this algorithm are introduced by other researchers [20,22].

Contact algorithm of Bardenhagen et al. [10] for MPM is formulated on the nodes of the background mesh. First equations of motion for each body and for the whole system should be solved separately. Contact node is detected by comparing the nodal velocities of each body and the nodal velocities of the whole system. In figure 1 node p is a contact node while for this node states

$$\dot{\boldsymbol{a}}_{A,p}^{t+\Delta t} \neq \dot{\boldsymbol{a}}_{p}^{t+\Delta t} \tag{B.1}$$

where $\dot{\boldsymbol{a}}_{A,p}^{t+\Delta t}$ is the velocity of node p from body A at time $t + \Delta t$ and $\dot{\boldsymbol{a}}_{p}^{t+\Delta t}$ is the velocity of node p from the whole system (bodies A and B in figure 1) at time $t + \Delta t$ (whereas point q is not a contact node). Contacting nodes on which the bodies are separating from each other do not need any corrections and the separation without any constraint should be allowed. For the contacting nodes on which the bodies are approaching each other it should be checked if the bodies are sliding on each other or not. This is done by comparing the tangential force in the contact node between the bodies and the maximum allowed tangential force by the Coulomb friction law. If the tangential force between bodies is less than the maximum force, the bodies are sticking to each other and no restriction is needed to be applied. But when the tangential force is more than the maximum value, the bodies slide on top of each other and two restrictions are needed to be applied. First restriction is applied on the normal velocities of each body to prevent the interpenetrations of the particles and the second one is applied on the tangential velocities of each body to satisfy the Coulomb friction law.

The same restrictions can also be applied by introducing forces into the equation of motion instead of correcting the velocities. In this work the correction of predicted velocities is adopted.

After correcting the velocities on the nodes of the background mesh, acceleration vector for each body is calculated and the conventional solution process of MPM is followed.

References

[1] Więckowski Z 2001 Analysis of granular flow by the material point method *European Conf. on Computational Mechanics*, Cracow, Poland

[2] Hirt C W, Amsden A A and Cook J L 1974 An arbitrary Lagrangian-Eulerian computing method for all flow speeds *J. Comput. Phys.* **14(3)** 227-53

[3] Noh W F 1964 CEL: A time-dependent, two-space-dimensional, coupled Eulerian-Lagrangian code *Methods in Computational Physics* **3** 117-79

[4] Belytschko T, Lu Y Y and Gu L 1994 Element-free Galerkin methods *Int. J. Numer. Meth. Eng.*

37(2) 229–56

[5] Monaghan J J 1988 An introduction to SPH *Comput. Phys. Commun.* **48(1)** 89–96

[6] Harlow F H 1964 The particle-in-cell computing method for fluid dynamics *Methods in Computational Physics* **3** 319–43

[7] Brackbill J U and Ruppel H M 1986 FLIP: A low dissipation particle-in-cell calculations of fluid flows in two dimensions *J. Comput. Phys.* **65(2)** 314–43

[8] Sulsky D, Zhou S J and Schreyer H L 1995 Application of a particle-in-cell method to solid mechanics *Comput. Phys. Commun.* **87** 236–52

[9] Sulsky D and Schreyer H L 1996 Axisymmetric form of the material point method with applications to upsetting and Taylor impact problems *J. Comput. Method. Appl. M.* **139** 409–29

[10] Bardenhagen S G, Brackbill J U and Sulsky D 2000 The material-point method for granular materials *Comput. Method. Appl. M.* **187** 529-41

[11] Bardenhagen S G and Kober E M 2004 The generalized interpolation material point method *Computer modelling in engineering and sciences* **5(6)** 477-96

[12] Beuth L, Benz T, Vermeer P A, Coetzee C J, Bonnier P and Van Den Berg P 2007 Formulation and validation of a quasi-static material point method *Proc. of the 10th International Symp. on Numerical Methods in Geomechanics* **10** 189-95

[13] Cummins S J and Brackbill J U 2002 An implicit particle-in-cell method for granular materials *J. Comput. Phys.* **180** 506-48

[14] Guilkey J E and Weiss J A 2003 Implicit time integration for the material point method: Quantitative and algorithmic comparisons with the finite element method *Int. J. Numer. Meth. Eng.* **57** 1323-38

[15] Sulsky D and Kaul A 2004 Implicit dynamics in the material-point method *Comput. Method. Appl. M.* **193** 1137-70

[16] Stolle D, Jassim I and Vermeer P 2009 Accurate simulation of incompressible problems in geomechanics *Computer Methods in Mechanics* 347-61

[17] Coetzee C J, Vermeer P A and Basson A H 2005 The modeling of anchors using the material point method *Int. J. Numer. Anal. Met.* **29** 879-95

[18] Jassim I K, Hamad F M and Vermeer P A 2011 Dynamic material point method with applications in geomechanics *Proc. 2nd Int. Symp. on Computational Geomechanics* Croatia

[19] Detournay C and Dzik E 2006 Nodal mixed discretization for tetrahedral elements *4th Int. FLAC Symp. on Numerical Modeling in Geomechanics* Madrid Spain

[20] Jassim I K 2013 Formulation of a Dynamic Material Point Method (MPM) for Geomechanical Problems *PhD thesis* Institute of Geotechnical Engineering, University of Stuttgart

[21] York A R, Sulsky D and Schreyer H L 1999 The material point method for simulation of thin membranes, *Int. J. Numer. Meth Eng.* **44** 1429-56

[22] Bardenhagen S G, Guilkey J E, Roessig K M, Brackbill J U, Witzel W M and Foster J C 2011 An improved contact algorithm for the material point method and application to stress propagation in granular material, *Computer modeling in engineering and sciences* **2** 509-22

Earthquake-induced displacement of cohesive-frictional slopes subject to cracks

Akram H Abd [1,2]
[1] School of Engineering, University of Warwick, CV4 7AL, Coventry, UK.
[2] Civil Engineering Dept., College of Engineering, Tikrit University, Tikrit, Iraq.

E-mail: akramciv22@gmail.com

Abstract. The upper bound theorem of limit analysis together with Newmark's method are employed to evaluate the displacement of soil slopes subject to cracks. The pseudo static approach has been routinely used in the literature to estimate the seismic displacement of soil slopes. However, the effect of cracks on the slope displacement has yet to be tackled. In this paper, a new technique is proposed to estimate the horizontal displacement at the slope toe due to a given earthquake postulating rough estimation of real time crack formation. Rotational failure mechanisms for intact slopes exhibiting the formation of cracks as part of the failure process and the case of cracks which are pre-existing in the slope were considered. On the basis of Newmark's method, the seismic-induced displacement is calculated by incorporating a stepwise yield acceleration corresponding to the cracks occurring in the slope. Results of the proposed technique can reasonably bridge the gap between the conservatism of assuming the slopes subject to the most detrimental cracks, and the overestimation of slope stability resulted from the neglect of crack formation. An example illustrating the procedure for a given earthquake is presented. Also, charts providing the values needed to calculate the stepwise yield accelerations are proposed.

1. Introduction

Cracks can be found in soil slopes and embankments due to tensile stresses such as seismic action or external static loading, and/or due to desiccation and cycles of wetting and drying. Pre-existing cracks can cause significant reduction in the stability of soil slopes [1-3], especially if these slopes are subjected to seismic action [4]. The presence of a vertical crack can reduce the safety factor of the slope depending mainly on its location and depth, and it does not only form a potential part of the slip surface, but also an easy flow channel for rainfall water which reduces the soil strength and exerts a lateral stress, inducing the failure when the crack is filled with water.

Methods for assessing the seismic stability of slopes have been developed during the last century. The Mononobe-Okabe method is one of the first published works that address the stability of retaining walls and dams during earthquake incorporating dynamic earth pressure [5, 6]. Thereafter, several limit equilibrium methods were developed for this purpose (e.g. [7-9]) which remain the most commonly used by practitioners. More recently, numerical methods for continuum mechanics, such as the finite element method with strength reduction technique (e.g. [10, 11]) and finite element limit analysis [12] have provided the capability to reliably detect the onset of failure in slopes according to the approach of continuum mechanics. However, if cracks are present, a continuum approach no longer works since the onset of instability is ruled by the behaviour of single fractures. In this case, the Discrete Element Method can nowadays be employed for 3D analyses of slopes with cracks [13].

Recent algorithmic advances in terms of contact detection algorithms [14, 15] have substantially reduced the runtime of these analyses. However, when little information on the presence of cracks is available, extensive parametric analyses requiring large computational times are necessary. In this case, an analytical approach is very desirable so that numerical analyses would be run only for the case(s) identified by the analytical approach as the most critical.

Newmark's analytical method [16] is popular among practitioners where a pseudo-static force is used instead of the dynamic excitation to calculate earthquake-induced displacements. The analytical solution for earthquake-induced displacements undergone by intact slopes subject to a rotational failure mechanism is presented by [17]. One of the main limitations of Newmark's method is the neglect of the earthquake induced strength degradation of the soil, i.e. it assumes a constant yield acceleration throughout the analysis [18]. In this paper, however, earthquake induced crack occurrence and the consequent reduction of yield acceleration are accounted for. Seismic induced displacements are calculated based on a stepwise time varying yield acceleration.

2. Formulation of the Problem

The kinematic approach of limit analysis will be used to calculate the least upper bound on the yield (critical) coefficient of acceleration K_y for a given uniform c, ϕ slope. The yield acceleration can be defined as the minimum level of horizontal acceleration (vertical acceleration being proportional to the horizontal acceleration) that brings the slope to failure (i.e. safety factor =1). According to Newmark's method [16], slope displacements start to accumulate whenever the seismic induced acceleration exceeds the yield acceleration. Then, displacements occurring during the earthquake can be obtained by double integrating the differences between the applied accelerations and the yield one during the time intervals when the ground velocity is larger than zero.

Pre-existing cracks, i.e. cracks exiting in the slope before any seismic excitation occurs, can significantly reduce the yield acceleration for a given slope, depending on their locations and depths [4]. Here, we shall consider an initially intact slope subject to the formation of tension cracks as a result of the earthquake. In this case, the cracks are formed as part of the failure process at the first time the slope yield acceleration is exceeded. Then, in order to calculate the slope displacements generated by the earthquake, a new yield acceleration, accounting for the presence of the cracks formed the first time the yield acceleration of the intact slope was exceeded, needs to be calculated for all the subsequent steps. Four cases are considered in this paper:

I. Slopes made of rocks / cohesive soils of unlimited tensile strength, hence not subject to tension cracks.
II. Slopes made of rocks / cohesive soils of limited tensile strength.
III. Slopes made of rocks / cohesive soils of zero tensile strength.
IV. Slopes subject to the most unfavourable crack from a stability point of view pre-existing the onset of the earthquake.

The procedure for calculating the stepwise time varying yield acceleration is outlined as follows:

1- Determine the yield acceleration for an initially intact slope subject to the formation of tension cracks $K^{1}_{y(c)}$. Vertical tension cracks are formed as part of the occurring failure mechanism since energy is needed to form any crack [3]. Therefore the yield acceleration of a slope subject to the formation of tension cracks is lower than the yield acceleration of a slope of unlimited tensile strength $K^{\text{int}}_{y(c)}$, i.e. $K^{1}_{y(c)} \leq K^{\text{int}}_{y(c)}$. This acceleration is used to calculate the displacements at the first time that the seismic acceleration exceeds the yield acceleration $K_{y(in)}$.

2- Determine the yield acceleration for the same slope but accounting for the presence of the crack generated in step 1, treated now as a pre-existing crack (i.e. the crack is already present so that no energy is dissipated for crack formation). This new value of yield acceleration, $K^2_{y(c)}$ is used to calculate the displacements in all the subsequent steps.

3- Determine the accumulated wedge displacement D_i with respect to the ground surface, at each time step (i) when the seismic acceleration exceeds $K^2_{y(c)}$.
4- Calculate the dimensionless coefficient C that relates the displacement of the slope toe to the integral of the earthquake acceleration record above the level of yield acceleration.
5- Determine the accumulated horizontal displacement at the slope toe D_{xi}, where $D_{xi} = C \times D_i$, and then the total horizontal displacement D_t is to be found.

It should be noted that, although several tension cracks at different locations in the slope may form during an earthquake, only the crack which has the worst detrimental effect on slope stability needs to be considered in the calculation, since according to the kinematic approach of limit analysis, the failure mechanism taking place is the most critical mechanism for the stability of the slope among all the kinematically feasible mechanisms.

3. Calculation of Yield Acceleration

The upper bound theorem of limit analysis is employed here to calculate the yield acceleration for both intact and cracked slopes. The analytical expressions for the calculation of the external work done by soil masses sliding along composite log-spiral failure surfaces, which requires the use of fictitious wedges bordered by a log-spiral, were first presented in [19, 20] for the case of slopes with horizontal upper part subject to a sequence of landslides, and for more general case of slopes with an inclined upper part, see [21]. Note that these calculations apply to slopes made of bonded granulates [22, 23] as well. In [24], the calculation of the work done by a wedge enclosed by two log-spirals was first presented. The analytical solution is derived here for the case of a horizontal upper slope surface and vertical pre-existing cracks from the upper slope (see figure 1). However, the solution can be straightforwardly extended to the case of a non-horizontal upper slope and that of cracks departing from the slope face, such an extension is reported in [2].

According to the upper bound theorem of limit analysis, the failing wedge E-D-C-B rotates rigidly and log-spirally around the centre of rotation P, as yet undefined, with the ground lying on the right of the log-spiral piece D-C and of the vertical crack C-B remaining at rest. The equation of log-spiral D-C is:

$$r = r_\chi \exp\left[\tan\phi(\upsilon - \chi)\right] \tag{1}$$

with r being the distance of a generic point of the spiral to its centre, ϕ being the angle of internal friction, r_χ identifying the distance of point F of the spiral to its centre, and χ, υ being the angles made by segment P-F and segment P-D with the horizontal, respectively (see figure 1).

The upper bound on the yield acceleration K_y will be derived imposing energy balance for the failing wedge E-D-C-B:

$$\dot{D} = \dot{W} \tag{2}$$

where $\dot{D}$ and $\dot{W}$ are the rate of dissipated energy and of external work, respectively. The dissipated energy $\dot{D}$ has two terms as follow:

$$\dot{D} = \dot{D}_{B-C} + \dot{D}_{C-D} \tag{3}$$

where $\dot{D}_{B-C}$ and $\dot{D}_{C-D}$ are the rates of dissipated energy along the crack (B-C) and the log-spiral segment (D-C), respectively. The analytical expression for the energy dissipated to form a new crack is given in [3]:

$$\dot{D}_{B-C} = \dot{\theta} r_\chi^2 \left(\frac{\sin\chi}{\tan\theta_c} \right)^2 \left[\frac{f_c}{2} \int_{\theta_c}^{\zeta} \frac{1-\sin\theta}{\cos^3\theta} d\theta + \frac{f_t}{1-\sin\phi} \int_{\theta_c}^{\zeta} \frac{\sin\theta - \sin\phi}{\cos^3\theta} d\theta \right] \tag{4}$$

with θ_c: is the angle made by the segment P-B with the horizontal (see Fig. 1), f_c and f_t are the unconfined compressive and tensile strength of the geo-material, respectively. According to the Mohr-Coulomb failure criteria (see figure 2), they can be expressed as:

$$f_c = 2c \frac{\cos\phi}{1-\sin\phi} \tag{5}$$

$$f_t = 2\alpha c \frac{\cos\phi}{1+\sin\phi} \tag{6}$$

where α is a dimensionless coefficient introduced here to express the tensile strength of the soil(i.e. $0 \le \alpha \le 1$). In this Paper, values of α of 0, 0.5 and 1 will be considered. Substituting equation (5) and equation (6) into equation (4), the following expression is obtained:

$$\dot{D}_{B-C} = c\dot{\theta} r_\chi^2 f_{B-C}(\chi,\upsilon,\zeta,\phi,\alpha) = c\dot{\theta} r_\chi^2 \left(\frac{\sin\chi}{\tan\theta_c} \right)^2 \left[\frac{\cos\phi}{1-\sin\phi} \int_{\theta_c}^{\zeta} \frac{1-\sin\theta}{\cos^3\theta} d\theta + \frac{2\alpha\cos\phi}{1-\sin^2\phi} \int_{\theta_c}^{\zeta} \frac{\sin\theta-\sin\phi}{\cos^3\theta} d\theta \right]$$

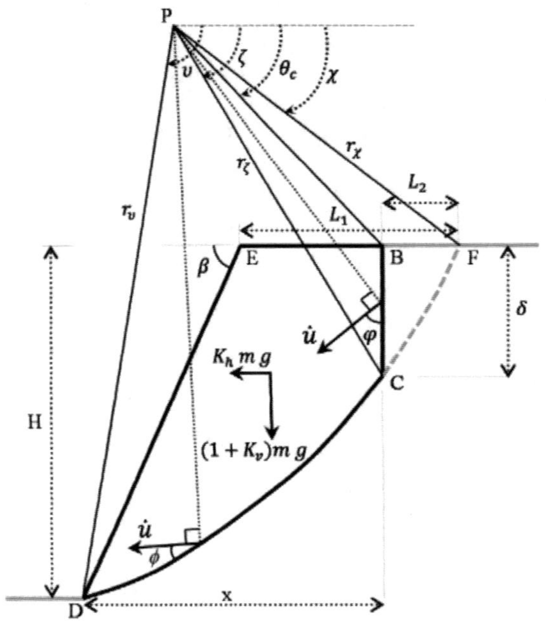

Figure 1. Failure mechanism. Note that $\varphi \neq \phi$. The wedge of soil enclosed by black lines D-C (logarithmic spiral failure line), B-C (pre-existing crack), B-E (upper surface of the slope) and E-D (slope face) rotates around point P.

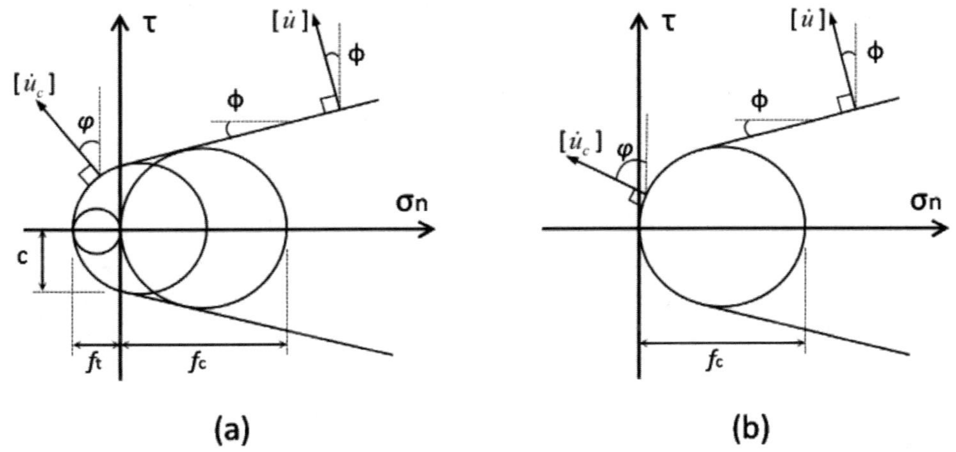

Figure 2. Yield stress criteria for (a) soil of limited tensile strength. (b) soil of zero tensile strength (tension cut-off), after [3].

Also energy is dissipated along the log-spiral segment (D-C). Its analytical expression, reported in, is:

$$\dot{D}_{C-D} = c\dot{\theta}r_\chi^2 f_{C-D}(\chi,\upsilon,\zeta,\phi) = c\dot{\theta}r_\chi^2 \exp\left[2\tan\phi(\zeta-\chi)\right]\frac{\exp\left[2\tan\phi(\upsilon-\zeta)\right]-1}{2\tan\phi} \quad (6).$$

The rate of external work $\dot{W}$ for the sliding wedge E-B-C-D can be calculated as the work of block E-D-F minus the work of block B-C-F. The work of block E-D-F is calculated by algebraic summation of the work of blocks P-D-F, P-E-F and P-D-E that are here called $\dot{W}_1$, $\dot{W}_2$ and $\dot{W}_3$, respectively. The work of block B-C-F is calculated by algebraic summation of the work of blocks P-C-F, P-B-F and P-C-B that are called $\dot{W}_4$, $\dot{W}_5$ and $\dot{W}_6$, respectively. So $\dot{W}$ can be calculated from the following summation:

$$\dot{W} = \dot{W}_1 - \dot{W}_2 - \dot{W}_3 - \left(\dot{W}_4 - \dot{W}_5 - \dot{W}_6\right) = \dot{W}_1 - \dot{W}_2 - \dot{W}_3 - \dot{W}_4 + \dot{W}_5 + \dot{W}_6 \quad (7).$$

The expressions for $\dot{W}_i$ are derived for each block by calculation of the vectorial product of the displacement rate u of the block (see figure 1) times their weight force [2]. Here instead, in addition to the weight force, a horizontal pseudo-static force, $F_h = mK_h g = \gamma K_h A$ with g being the gravitational acceleration and m the mass of the wedge, and a vertical one, $F_v = mK_v g = \gamma K_v A$, are added to account for seismic action. For the sake of space, only the final expressions of the yield acceleration are reported here:

$$\dot{W}_1 = \dot{\theta}\gamma r_\chi^{\ 3}\left[(1+k_v)f_{1v}(\chi,\upsilon,\phi) + k_h f_{1h}(\chi,\upsilon,\phi)\right] \quad (8)$$

$$= \dot{\theta}\gamma r_\chi^{\ 3}\left[\begin{array}{l}(1+K_v)\dfrac{\exp\left[3\tan\phi(\upsilon-\chi)\right](3\tan\phi\cos\upsilon+\sin\upsilon)-3\tan\phi\cos\chi-\sin\chi}{3\left(1+9\tan\phi^2\right)}+\\ K_h\dfrac{\exp\left[3\tan\phi(\upsilon-\chi)\right](3\tan\phi\sin\upsilon-\cos\upsilon)-3\tan\phi\sin\chi+\cos\chi}{3\left(1+9\tan\phi^2\right)}\end{array}\right]$$

with $\dot{\theta}$ being the rate of angular displacement of the failing wedge E-B-C-D. For block P-E-F instead:

$$\dot{W}_2 = \dot{\theta}\gamma r_\chi^{\ 3}\left[(1+K_v)f_{2v}(\chi,\upsilon,\phi) + K_h f_{2h}(\chi,\upsilon,\phi)\right] \quad (9)$$

$$= \dot{\theta}\gamma r_\chi^{\ 3}\left[(1+K_{\rm v})\frac{L_1}{6r_\chi}sin\chi\left(2\cos\chi-\frac{L_1}{r_\chi}\right)+K_{\rm h}\frac{L_1}{3r_\chi}\sin^2\chi\right]$$

for block P-D-E:

$$\dot{W}_3 = \dot{\theta}\gamma r_\chi^{\ 3}\left[(1+K_{\rm v})f_{3{\rm v}}(\chi,\upsilon,\phi)+K_{\rm h}f_{3{\rm h}}(\chi,\upsilon,\phi)\right] \tag{10}$$

$$= \dot{\theta}\gamma r_\chi^{\ 3}\left[\begin{array}{l}\frac{(1+K_{\rm v})}{6}\exp[\tan\phi(\upsilon-\chi)]\left(\sin(\upsilon-\chi)-\frac{L_1}{r_\chi}sin\upsilon\right)\left(cos\chi-\frac{L_1}{r_\chi}+\exp[\tan\phi(\upsilon-\chi)]cos\upsilon\right)+\\ \frac{K_{\rm h}}{6}\exp[\tan\phi(\upsilon-\chi)]\left(\sin(\upsilon-\chi)-\frac{L_1}{r_\chi}sin\upsilon\right)\left(\sin\chi+\exp[\tan\phi(\upsilon-\chi)]\sin\upsilon\right)\end{array}\right]$$

for block P-C-F:

$$\dot{W}_4 = \dot{\theta}\gamma r_\chi^{\ 3}\left[(1+K_{\rm v})f_{4{\rm v}}(\chi,\upsilon,\phi)+K_{\rm h}f_{4{\rm h}}(\chi,\upsilon,\phi)\right] \tag{11}$$

$$= \dot{\theta}\gamma r_\chi^{\ 3}\left[\begin{array}{l}(1+K_{\rm v})\frac{\exp[3\tan\phi(\zeta-\chi)](3tan\phi cos\zeta+sin\zeta)-3tan\phi cos\chi-sin\chi}{3(1+9tan^2\phi)}+\\ K_{\rm h}\frac{\exp[3\tan\phi(\zeta-\chi)](3tan\phi\sin\zeta-\cos\zeta)-3tan\phi\sin\chi+\cos\chi}{3(1+9tan^2\phi)}\end{array}\right]$$

for block P-B-F:

$$\dot{W}_5 = \dot{\theta}\gamma r_\chi^{\ 3}\left[(1+K_{\rm v})f_{5{\rm v}}(\chi,\upsilon,\phi)+K_{\rm h}f_{5{\rm h}}(\chi,\upsilon,\phi)\right] \tag{12}$$

$$= \dot{\theta}\gamma r_\chi^{\ 3}\left[(1+K_{\rm v})\frac{L_2}{6r_\chi}sin\chi\left(2cos\chi-\frac{L_2}{r_\chi}\right)+K_{\rm h}\frac{L_2}{3r_\chi}\sin^2\chi\right]$$

for block P-C-B:

$$\dot{W}_6 = \dot{\theta}\gamma r_\chi^{\ 3}\left[(1+K_{\rm v})f_{6{\rm v}}(\chi,\upsilon,\phi)+K_{\rm h}f_{6{\rm h}}(\chi,\upsilon,\phi)\right] \tag{13}$$

$$= \dot{\theta}\gamma r_\chi^{\ 3}\left[\begin{array}{l}\frac{(1+K_{\rm v})}{3}\exp[2\tan\phi(\zeta-\chi)]\cos^2\zeta\left(\exp[\tan\phi(\zeta-\chi)]sin\zeta-\sin\chi\right)+\\ \frac{K_{\rm h}}{6}\exp[\tan\phi(\zeta-\chi)]\cos\zeta\left(\exp[2\tan\phi(\zeta-\chi)]\sin^2\zeta-\sin^2\chi\right)\end{array}\right]$$

Substituting equation (3) and equation (7) into equation (2), the following expression is obtained:

$$c\dot{\theta}r_\chi^2\left(f_{B-C}+f_{C-D}\right)=\dot{\theta}\gamma r_\chi^3\left[\begin{array}{l}(1+K_{\rm v})(f_{1{\rm v}}-f_{2{\rm v}}-f_{3{\rm v}}-f_{4{\rm v}}+f_{5{\rm v}}+f_{6{\rm v}})\\ +K_{\rm h}(f_{1{\rm h}}-f_{2{\rm h}}-f_{3{\rm h}}-f_{4{\rm h}}+f_{5{\rm h}}+f_{6{\rm h}})\end{array}\right] \tag{14}$$

The vertical coefficient of acceleration is introduced by λ as a ratio of the horizontal acceleration, where $\lambda = K_v/K_h$. Consistently with figure 1, the + sign indicates vertical downward acceleration, whereas the – sign indicates vertical upward acceleration. An upper bound on the coefficient of yield acceleration K_y is obtained by solving equation (14) with respect to K_h as follow:

$$K_y = \frac{\left[\begin{array}{l}\frac{c}{\gamma H}\left(f_{B-C}+f_{C-D}\right)\left(\exp\left[\tan\phi(\upsilon-\chi)\right]\sin\upsilon-\sin\chi\right)- \\ \left(f_{1v}-f_{2v}-f_{3v}-f_{4v}+f_{5v}+f_{6v}\right)\end{array}\right]}{\lambda\left(f_{1v}-f_{2v}-f_{3v}-f_{4v}+f_{5v}+f_{6v}\right)+\left(f_{1h}-f_{2h}-f_{3h}-f_{4h}+f_{5h}+f_{6h}\right)} \quad (15)$$

$$K_y = f_y\left(\chi,\upsilon,\zeta,\phi,c/\gamma H,\beta,\lambda,\alpha\right)$$

The global minimum of $f_y\left(\chi,\upsilon,\zeta,\phi,c/\gamma H,\beta,\lambda,\alpha\right)$ over the three geometrical variables χ,υ,ζ provides the least upper bound on the coefficient of yield acceleration, assuming that the most unfavourable crack for the slope is present. However, mathematical constraints could be applied to equation (15) in order model a slope with a prescribed crack depth or crack location. Using equation (15), charts providing the yield acceleration for soil slopes with either zero tensile strength (i.e. $\alpha = 0$) or half the Mohr-Coulomb tensile strength (i.e. $\alpha = 0.5$) are presented in figure 3. It should be noted that the solid lines in figure 3 refer to soil slopes that are initially intact but they undergo crack formation as part of the developing failure mechanism. The dashed lines, however, are for soil slopes with an earthquake induced crack, since in this case the crack is already formed and treated here as an open crack. The stepwise yield acceleration, proposed in this paper, can be found using these two lines (i.e. solid and dashed), where for a given soil slope properties, two values of yield acceleration are obtained. The one obtained from the solid line represents the starting value of the yield acceleration which steps down to the value obtained from the dashed line as soon as exceeded for the first time by the applied acceleration, given by the earthquake record. The definition of the stepwise yield acceleration is detailed by an illustrative example later in this paper. It should also be noted that failure passes below the slope toe were not permitted during the calculations for these charts. This type of failure might occur for gentle slope with low angle of internal friction [25].

4. Calculation of Seismic Displacements

The maximum horizontal displacement of the slope face occurs at the slope toe [17]. This displacement is denoted here as δu_x (see figure 4). Based on Newmark's method, the rate of δu_x can be calculated as follow [25]:

$$\delta u_x = r_\upsilon \sin\upsilon\delta\theta = r_\upsilon \sin\upsilon \int_{\delta t}\int_{\delta t} \ddot{\theta}\, dt\, dt = C\int_{\delta t}\int_{\delta t}\left(K_i - K_y\right)g\, dt\, dt \quad (16)$$

where $\delta\theta$ is the angular displacement $\ddot{\theta}$ is the angular acceleration K_i is the applied horizontal coefficient of acceleration at step *i*, and *C* is a dimensionless coefficient that relates the displacement of the slope toe to the integral of the earthquake acceleration record above the level of yield acceleration. Performing the calculations, this coefficient can be expressed as:

$$C = \frac{\gamma r_\chi^4 \exp\left[\tan\phi(\upsilon-\chi)\right]\sin\upsilon\left[\begin{array}{l}\lambda\left(f_{1v}-f_{2v}-f_{3v}-f_{4v}+f_{5v}+f_{6v}\right)+ \\ \left(f_{1h}-f_{2h}-f_{3h}-f_{4h}+f_{5h}+f_{6h}\right)\end{array}\right]}{Gl^2} \quad (17)$$

with *G* being the weight of the potential sliding mass and *l* is the distance from point P to the centre of gravity of that mass. The calculations for *G* and *l* are listed in Appendix A. To this end, the seismic induced displacements can be calculated using equation (16) by assigning an earthquake record and calculating the yield acceleration for the slope of interest.

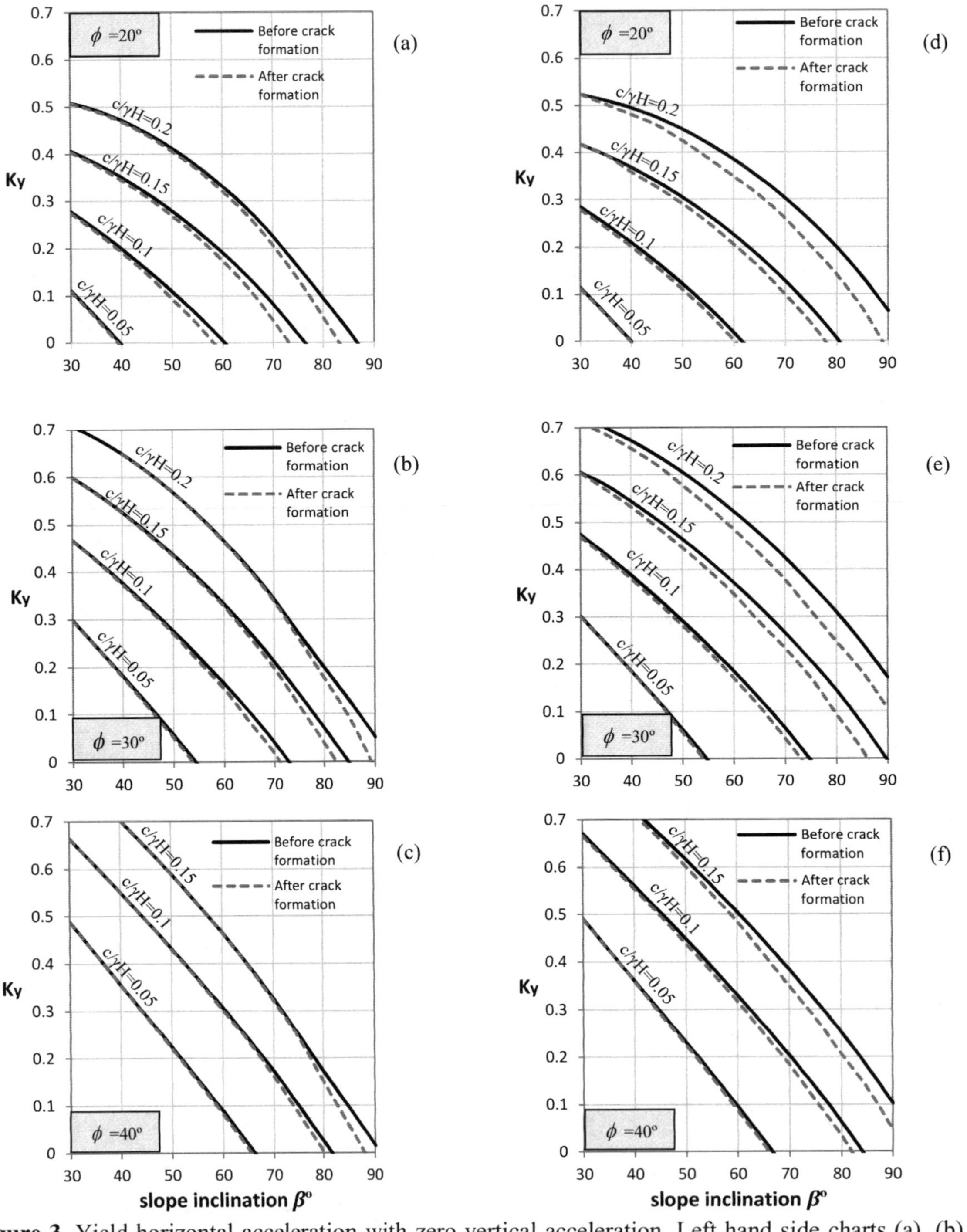

Figure 3. Yield horizontal acceleration with zero vertical acceleration. Left hand side charts (a), (b) and (c) are for soil slopes with zero tensile strength while (d), (e) and (f) are for soil slopes with limited tensile strength (i.e. half of Mohr-Coulomb's tensile strength).

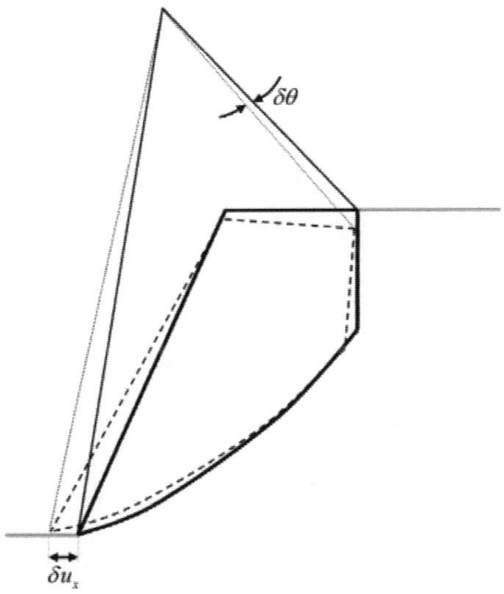

Figure 4. Illustration of the horizontal displacement at the slope toe δu_x and the angular displacement $\delta\theta$.

5. Illustrative Example

Here we shall consider a slope with $\beta = 70°$, $\phi = 20°$ $\lambda = 0$ and $c/\gamma H = 0.15$ subjected to the Northridge earthquake (1994), whose main characteristics are listed in table 1. Four cases are considered: case (I) soil slope with full tensile strength, (i.e. $\alpha = 1$), therefore not subject to tension cracks, case (II) soil slope of limited tensile strength, in this case $\alpha = 0.5$, case (III) soil slope of zero tensile strength, (i.e. $\alpha = 0$), and case (IV) soil slope subjected to the most adverse pre-existing crack.

Now, according to the procedure mentioned earlier in this paper, the stepwise yield acceleration for soil with limited or zero tensile strength is illustrated in figure 5(a). It can be noticed that the yield acceleration for a soil slope with limited tensile strength is reduced significantly when it is exceeded for the first time by the applied acceleration. This is because the crack formed as part of the failure at that instance, is then treated as a pre-existing one. Consequently, this could increase the estimated displacement as shown in figure 5(b).

As the displacement corresponding to a slope with the most detrimental pre-existing crack seems over conservative, at the same time, assuming an intact slope that remains intact during the earthquake may underestimate the displacement. However, assuming a limited tensile strength for the soil slope seems to reasonably bridge the gap between the conservatism, corresponding to a slope with the most detrimental pre-existing crack, and the underestimation of the displacement when ignoring the crack formation (i.e. intact slope). Figure 5(c) provides an insight as to the way the limited tensile strength could change the crack properties and the orientation of the failure mechanism.

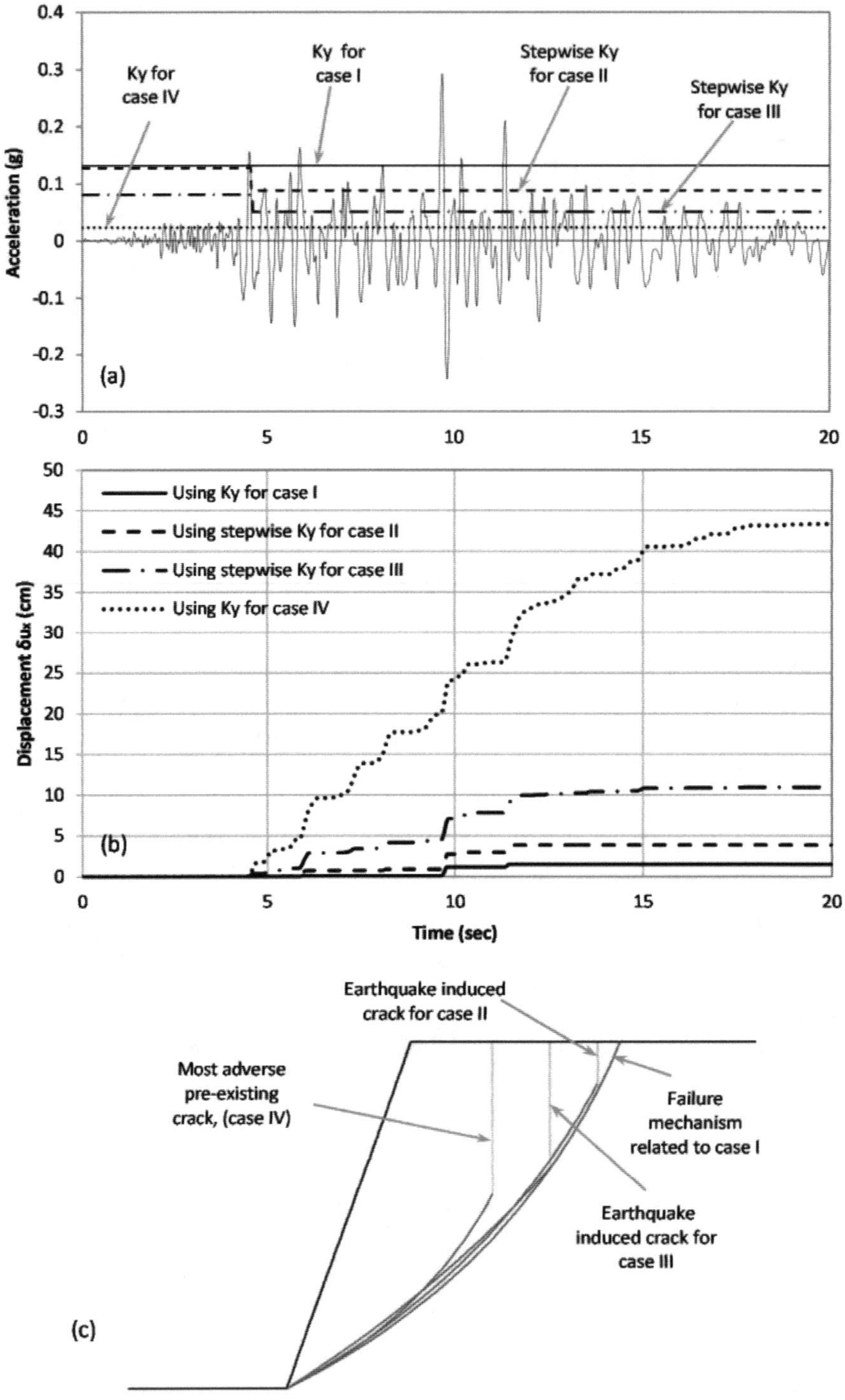

Figure 5. Slope with β=70, ϕ=20, c/γH=0.15. (a) Illustration of the calculated yield for the four cases considered employing the Northridge earthquake (1994) as seismic input. (b) Accumulated horizontal displacement at the slope toe, for the four cases considered. (c) Failure mechanisms associated with the calculated yield accelerations.

Table 1. Main characteristics of the earthquakes considered in the example case.

Date	17/1/1994
Station	24283 Moorpark - Fire Sta.
Magnitude	6.7
Direction	180°
Peak acceleration (g)	0.292
Epicentre distance (km)	23

6. Conclusions

The upper bound theorem of limit analysis together with the pseudo static approach were employed to evaluate the displacements of cohesive frictional slopes subject to the formation of tension cracks. The formation of earthquake-induced tension cracks and their effect on the displacements were considered. The assumption of the stepwise yield acceleration can be used to reasonably bridge the gap between the conservatism corresponding to a slope with the most detrimental pre-existing crack, and the underestimation of the displacement when ignoring the crack formation throughout the analysis. Four cases were considered here: intact slopes of unlimited tensile strength, intact slopes of limited tensile strength, intact slopes with no tensile strength, and slopes subject to cracks pre-existing the seismic event. Charts providing the values needed to calculate the stepwise yield acceleration are presented.

Acknowledgements

The results presented in this Paper are based on research supported mainly by HCED (Higher Committee of Education Development in Iraq) and partially by Tikrit University. This support is greatly appreciated.

Appendix A

The calculations of the weight of the sliding mass, called G in the manuscript, are detailed as follow:

$$G = \gamma A \tag{A.1}$$

with $A = A_1 - A_2 - A_3 - A_4 + A_5 + A_6$

$$A_1 = \frac{1}{2} r_\chi^2 \left[\frac{\exp\left[2\tan\phi(\upsilon - \chi)\right] - 1}{2\tan\phi} \right] \tag{A.2}$$

$$A_2 = \frac{1}{2} r_\chi L_1 \sin\chi \tag{A.3}$$

$$A_3 = \frac{1}{2} r_\chi H \frac{\exp[\tan\phi(\upsilon - \chi)]\sin(\beta + \upsilon)}{\sin\beta} \tag{A.4}$$

$$A_4 = \frac{1}{2} r_\chi^2 \left[\frac{\exp[2\tan\phi(\zeta - \chi)] - 1}{2\tan\phi} \right] \tag{A.5}$$

$$A_5 = \frac{1}{2} r_\chi L_2 \sin\chi \tag{A.6}$$

$$A_6 = \frac{1}{2}\delta r_\zeta \cos\zeta = \frac{1}{2} r_\chi \delta\left[\exp[\tan\phi(\zeta - \chi)]\cos\zeta\right] \tag{A.7}$$

The arm length of the weight, l, is given by:

$$l = \frac{\sqrt{\left(\gamma r_\chi^3 (f_1 - f_2 - f_3 - f_4 + f_5 + f_6)\right)^2 + \left(\gamma r_\chi^3 (f_{h1} - f_{h2} - f_{h3} - f_{h4} + f_{h5} + f_{h6})\right)^2}}{G} \tag{A.8}$$

References

[1] Baker R 1981 Tensile strength, tension cracks, and stability of slopes *Soils and Foundation* **21** 1-17

[2] Utili S 2013 Investigation by limit analysis on the stability of slopes with cracks *Geotechnique* **63** 140-154

[3] Michalowski R L 2013 Stability assessment of slopes with cracks using limit analysis *Can. Geotech. J.* **50** 1011-1021

[4] Utili S and Abd A H 2015 On the stability of fissured slopes subject to seismic action *Int. J. Num. Analy. Meth. Geo.* under review

[5] Okabe S 1924 General theory of earth pressure and seismic stability of retaining wall and dam *J. Japan Soc. Civ Engrs, Tokyo, Japan* **12**

[6] Mononobe N a M, H. 1929 On the determination of earth pressures during earthquakes *Proc. world Engrg, Congr. Tokyo, Japan* **9**

[7] Janbu N 1954 Application of composite slip surfaces for stability analysis *Proc. European Conf. on Stability of Earth Slopes, Stockholm, 1954* 3 43-49

[8] Little A and Price V 1958 The use of an electronic computer for slope stability analysis *Geotechnique* **8** 113-116

[9] Morgenstern N and Price V E 1965 The analysis of the stability of general slip surfaces *Geotechnique* **15** 79-93

[10] Dawson E, Roth W, and Drescher A 1999 Slope stability analysis by strength reduction *Geotechnique* **49** 835-840

[11] Conte E, Silvestri F, and Troncone A 2010 Stability analysis of slopes in soils with strain-softening behaviour *Computers and Geotechnics* **37** 710-722

[12] Li A-J, Merifield R S, and Lyamin A V 2010 Three-dimensional stability charts for slopes based on limit analysis methods *Can. Geotech. J.* **47** 1316-1334

[13] Boon C, Houlsby G, and Utili S 2014 New insights into the 1963 Vajont slide using 2D and 3D distinct-element method analyses *Geotechnique* **64** 800-816

[14] Boon C, Houlsby G, and Utili S 2012 A new algorithm for contact detection between convex polygonal and polyhedral particles in the discrete element method *Computers and Geotechnics* **44** 73-82

[15] Boon C, Houlsby G, and Utili S 2013 A new contact detection algorithm for three-dimensional non-spherical particles *Powder Technology* **248** 94-102

[16] Newmark N M 1965 Effects of earthquakes on dams and embankments *Geotechnique* **15** 139-159

[17] Chang C-J, Chen W F, and Yao J T 1984 Seismic displacements in slopes by limit analysis *J. Geotech. Engrng.* **110** 860-874

[18] Jibson R W 2011 Methods for assessing the stability of slopes during earthquakes—A retrospective *Engineering Geology* **122** 43-50

[19] Utili S 2005 An analytical relationship for weathering induced slope retrogression: a benchmark *RIG (Italian Geotechnical Journal)* **39** 9-30

[20] Utili S and Nova R 2008 DEM analysis of bonded granular geomaterials *Int. J. Num. Analy. Meth. Geo.* **32** 1997-2031

[21] Utili S and Crosta G 2011 Modelling the evolution of natural slopes subject to weathering: Part I. Limit analysis approach *Journal of Geophysical Research - Earth Surface* **116**

[22] Utili S and Crosta G 2011 Modelling the evolution of natural slopes subject to weathering: Part II. Discrete element approach *Journal of Geophysical Research - Earth Surface* **116**

[23] Jiang M, Zhu F, Liu F, and Utili S 2014 A bond contact model for methane hydrate- bearing sediments with interparticle cementation *Int. J. Num. Analy. Meth. Geo.* **38** 1823-1854

[24] Utili S and Nova R 2007 On the optimal profile of a slope *Soils and Foundations* **47** 717-729

[25] You L and Michalowski R L 1999 Displacement charts for slopes subjected to seismic loads *Computers and Geotechnics* **25** 45-55

IOP Conf. Series: Earth and Environmental Science **26** (2015) 012040 doi:10.1088/1755-1315/26/1/012040

Forensic analysis of Malin landslide in India

Pinom Ering[1], **Ramesh Kulkarni**[2], **Yashwant Kolekar**[3], **Satyanarayana Murty Dasaka**[4] and **G. L. Sivakumar Babu**[5]

[1]Research scholar, *Department of Civil Engineering*, IISc Bangalore, Karnataka-India. E-mail: pinom432@gmail.com

[2]Engineer, Soil Tech Private Limited, Ground Engineers and Consultants, Pune-India. E-mail: rkulkarni@soiltech.in

[3]Associate professor, Civil Engineering Department, College of Engineering, Pune-India. E-mail: yashwantlolekar@gmail.com

[4]Associate professor, Civil Engineering Department, Indian Institute of Technology, Bombay,Maharashtra-India. E-mail: dasaka@civil.iitb.ac.in

[5]Professor, *Department of Civil Engineering*, IISc Bangalore, Karnataka – India E-mail: gls@civil.iisc.ernet.in

Abstract: A devastating landslide occurred on 30th July 2014, resulting in the burial of a village of about 40 houses called Malin, in western India and also led to about 160 deaths. The landslide was triggered by heavy rainfall in the area and mass movement of debris. The paper investigates slope failure in the Malin area using back analysis and numerical methods. Site investigation was conducted to obtain representative information of the area. Finite difference analyses using FLAC 2D is performed for the failed slope to determine the possible cause of failure. Analysis results show that slope failure occurred due to the loss of suction strength at the interface between rock and local soil.

1. Introduction

Landslides constitute a major threat to both lives and property worldwide. One of the most common triggering mechanisms for landslides is rainfall and the consequent water infiltration. Rainfall induced landslides are the results of a change in the groundwater conditions, especially in the unsaturated zone. The location of groundwater table in most of the high slopes is deep below the ground surface and the pore-water pressures in the soil above ground water table are negative. This negative pore-water pressure is called matric suction when referenced to pore-air pressure which contributes to the stability of unsaturated soil slopes (Fredlund and Rahardjo, 1993; Rahardjo et al. 1995; Griffiths and Lu, 2005). Hence the effect of rainfall infiltration on slope could result in changing soil suction and positive pore pressure, or main water table, as well as raising soil unit weight, reducing shear strength of rock and soil.

Deep-seated rotational and shallow translational slides can often be observed in slopes which used to be stable under normal rainfall conditions due to extreme rainfall infiltration. In these cases deep-seated slides are generated by a rise in the groundwater level and consequently an increase in pore-water pressures and a lowering of the effective stresses in the soil. The failure surface often is located below the phreatic line. This type of sliding occurs in fine-grained clayey soils. On the other hand, shallow translational slides are mainly triggered in the zone above the groundwater level in silty/sandy soils. The thickness of this zone depends much on the climatic conditions of the area. It can be of the order of a meter in temperate regions and up to tens of meters in tropical and subtropical regions. Once the rain-water starts infiltrating this vadose zone of the soil, the negative pore-water pressures will tend to dissipate due to an increase in the soil water content. This process contributes to lowering the shear strength of the soil layers close to the surface. Under critical conditions, the shear strength of the soil can be reduced below the mobilized shear stress on a potential slip surface and failure of the slope may occur. The majority of hill slopes can be considered unsaturated during normal conditions. Due to this unsaturated state, the slopes can often be found at angles steeper than what would be "theoretically possible" by using the common saturated soil mechanics. In practice, most of the slope stability calculations neglect any suction present above the groundwater table. The difficulties associated with the measurement of negative pore-water pressures and their incorporation into the slope stability analysis is the primary reason for this practice. It is difficult to predict the stability of a natural slope subjected to environmental changes due to the many factors involved in the process such as the soil hydraulic properties, climatic data and the initial groundwater conditions within the slope. Neglecting the negative pore-water pressures can be reasonable in many situations where the slip surface is mainly lying below the phreatic line. However, for situations where the groundwater level is deep and/or shallow sliding is of concern, negative pore-water pressures should not be ignored.

In this study, a rainfall induced landslide which occurred along the upslope of Malin village, Ambegaon Tehsil in Pune district of Maharashtra (India) is examined. The landslide occurred in the early hours of July 30th 2014 claiming 151 human lives and damaging 45 houses. Heavy rainfall was reported in the area on that day and over the previous one week. Hence rainfall infiltration is assumed to be the direct cause of landslide in our analyses. Back analysis of the failed slope is performed to determine the shear strength parameters and other conditions at the time of failure. Generally, back analysis is an effective approach to provide an insight into the underlying failure mechanism and improve the understanding regarding the factors controlling the stability of slopes. One of the advantages of back analysis is that it can account for important factors that may not be well represented in laboratory and in-situ tests such as the presence of cracks and pre-existing shear planes within the soil mass. Finally, the cause of slope failure/landslide is explained based on the basic concepts of unsaturated soil mechanics.

2. Site description

Location: The Malin village is located in AmbegaonTaluka of Pune District of western Maharashtra, India. It is located at an elevation of 760 m and about 95 kms away from Pune city. Geographical coordinates of the village are: Latitude N19°09'34.40", Longitude E73°41'19". Around 70 dwelling units existed before landslide. The hill slopes at levels above and below the village were partially converted to terraces for paddy cultivation.

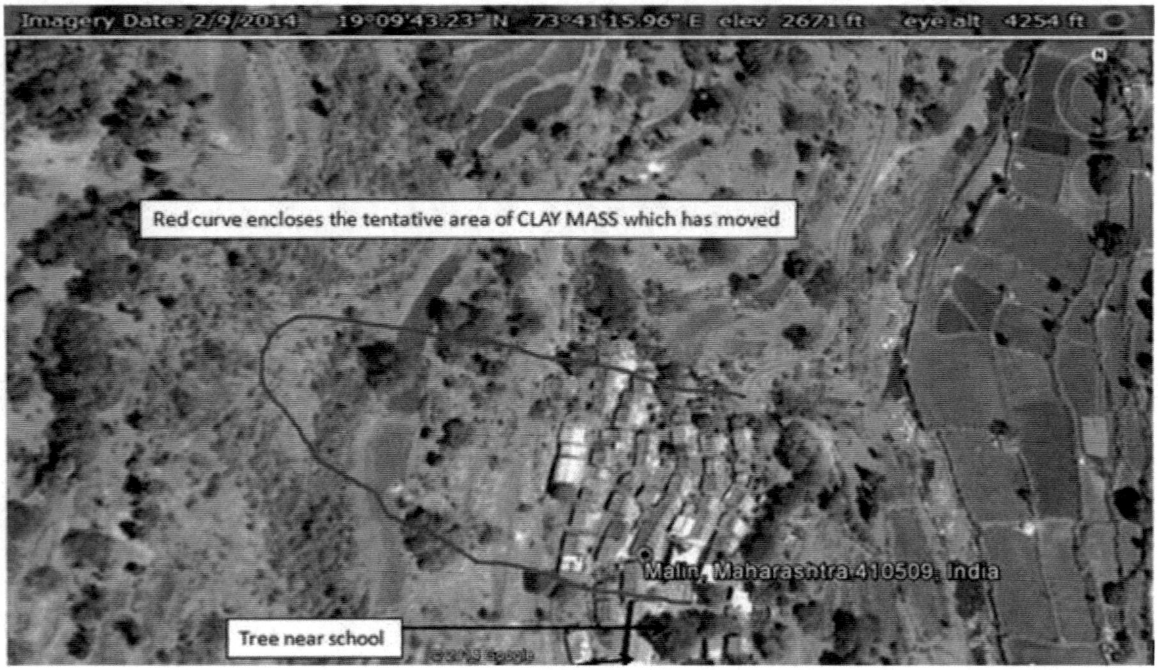

Figure 1: Malin landslide area

Physiography and geology of the area around Malin village

The Malin landslide area is located on the eastern slope of a roughly N-S trending hill. The area is drained by the Ghod River and its tributaries. The area around Malin village is represented by the basaltic lava flows. The seismo-tectonic map of the area revealed the occurrence of two low magnitudes seismic activities (4.0 to 4.9) within 100 km radius of the landslide location. However, these seismic events do not appear to have played any significant role in inducing slope instability in Malin.

Topography

The topography of Malin hill slope is broadly uniform. The entire hill slope where the landslide occurred can be divided into four zones:
Zone 1: from Nala bed level to the road level, which is relatively flatter in its lower part followed by gentler slope in upper part.
Zone 2: from road level to the 2nd slope break/level, which is relatively gentler
Zone 3: from 2nd major slope break/level to the 3rd major slope break, this is moderately inclined.
Zone 4: from 3rd level till the plateau edge/apex of hill slope, this is moderately inclined.

The overall slope inclination as measured from the base topographic map is 25°. Considering the similarity of slope configuration for the entire Malin hill slope, a longitudinal section A-A' is drawn in the southern side of the slide from across the river up to just above the crown level of Malin landslide. This section roughly reveals the pre-slide slope configuration of slide portion. Another section B-B' is drawn almost through the middle of the landslide. This reveals the slope modified due

IOP Conf. Series: Earth and Environmental Science **26** (2015) 012040 doi:10.1088/1755-1315/26/1/012040

to landslide. Slope angles at various zones along the two sections are given in Table 1. Figure 2 shows the longitudinal topographical section along A-A' line of the slope.

Table 1: Slope angles at different slope zones along sections AA' and BB'

Section	1st Zone		2nd Zone		3rd Zone		4th Zone		
A-A'	20°		10°	25°	30°		18°	25°	50°
B-B'	10°	28°	10°	25°	30°	20°	25°	30°	40°

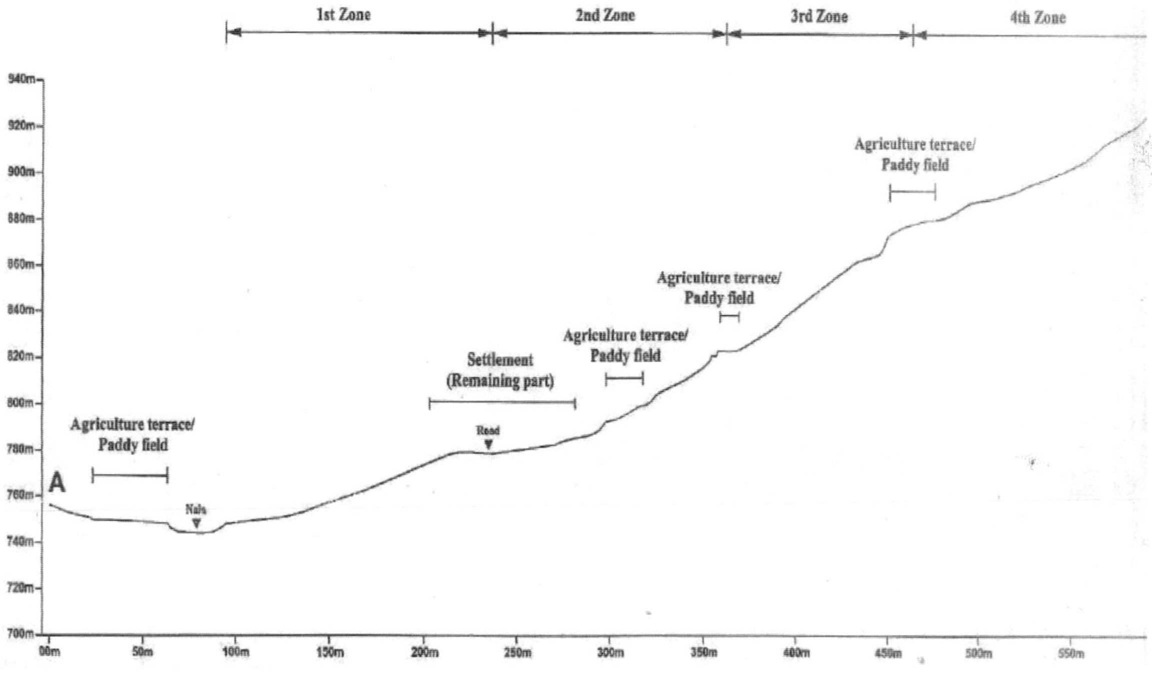

Figure 2: Longitudinal topographical section along A-A'

Features of the Malin Landslide

The height of the landslide is roughly estimated as 190 m while the width of the slide varies from 45 m to 134 m. The entire length of the slide from the crown to toe is 514 m and the area affected is 44245 m^2. The crown of the landslide is at a height of 936 m and marks the top of 4th zone. The width of the landslide in this zone varies from 45 m to 134 m. The Landslide depleted part of 4th zone, entire 3rd zone and maximum part of 2nd zone. Lowest part of 2nd zone and 1st zone form the zone of accumulation. Unfortunately, this zone of accumulation is the settlement area of the Malin village. The maximum thickness of the material slided during landslide could not be established accurately and approximated as 7 m because when the detailed mapping work was carried out, most of the failed slope material has been removed from the place as part of rescue effort.

Representative soil samples were collected from Malin landslide area. These samples were collected from three different levels in main landslide. Table 2 shows the results of geotechnical tests performed on soil samples from failed slope.

Table 2: Results of geotechnical tests on soil samples

Parameter	From slided area		
Zone	2	3	4
Moisture content (%)	36.41	27.92	38.70
Dry density (g/cc)	1.32	1.36	1.24
Liquid limit (%)	53	51	46
Plastic limit (%)	31	29	28
Plasticity Index (PI)	23	23	18
Specific Gravity	2.57	2.51	2.58
Grain Size analysis (Silty clay, %)	57	78	31
Cohesion (Kg/cm^2)	0.125	0.391	0.339
Friction angle (°)	4.06	10.31	41.41

The grain size analysis indicated that the samples collected from slide area belong to silty-clay group. The plasticity index values vary from 18 to 23 %. The average bulk density of the slope material is taken as 1560 kg/m^3.

Rainfall process

In the Malin village there are no rain gauge stations; therefore rainfall data were recorded from different rain gauge stations located around Malin. Figure 3 shows the rainfall data of Malin village from 22nd July to 30th July. In the figure it can be noted that during previous few days i.e from July 22 to 28, the antecedent rainfall was nothing extraordinary. However the rainfall record for 29th July i.e after 168 hours shows high amount of rainfall (108 mm). This may have played a significant role in slope instability.

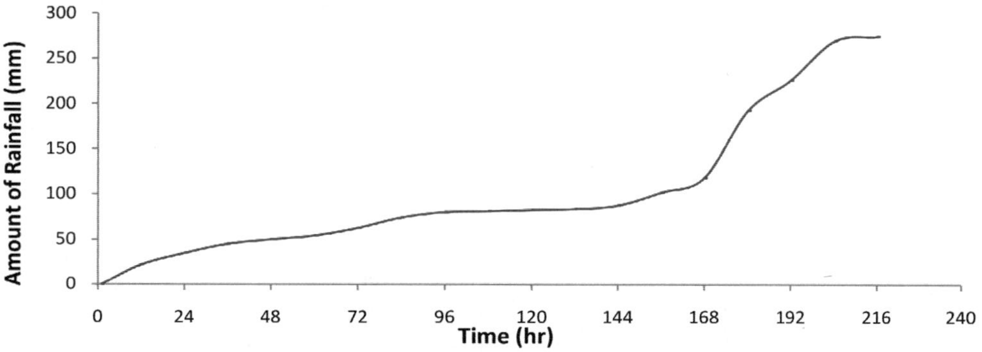

Figure 3: Rainfall data of Malin village

3. Method of Analysis
Back Analysis of the failed slope

A slope failure implies that the factor of safety of the slope at the moment of failure is unity. Based on this information, back analysis is often carried out to improve knowledge on slope stability parameters. Stability parameters may include both soil strength parameters and pore water pressure parameters at the moment of slope failure. The landslide in Malin village provides us with information that factor of safety of the slope at the time of failure is unity and the length of failure surface is 514 m. To achieve a safety value of unity, the soil resistance can be varied which is usually represented by a combination of c' and φ'. Back analysis requires determining appropriate values of cohesion and friction angle which controls failure in the original slope. To determine the shear strength parameters of the slope at failure, the method proposed by Duncan and Stark (1992) is used. This method enables the estimation of friction angle from the known values of plasticity index and hence one can back-calculate the cohesion parameter at the moment of failure. Known parameters: Factor of safety =1, Length of failure surface = 514 m, Plasticity Index = 23. Unknown parameters: Friction angle, cohesion.

Table 3: Typical values of φ' at different values of plasticity index

Plasticity Index	Values of φ' (in degrees)	
	Fully softened	Residual
0-10	30-40	18-30
10-20	25-35	12-25
20-40	20-30	10-20
40-80	15-25	7-15

For the given value of plasticity index (PI = 23), friction angle is expected to in the range of 20-30 degree. Hence a value of friction angle value of 22° for the slope considered in the study. The rotational model is adopted in the analysis. The circular failure surface extends down to the weathered/unweathered interface and has a long radius to represent long shallow failure surface as observed in the failed slope. Bishop's method of slices is applied to evaluate the shear strength parameters of the slope at failure. Knowing the length of failure surface, a slip circle is drawn along the slided slope from the crown in the 4th zone till the 2nd zone. The soil mass above the assumed slip circle is divided into a number of vertical slices of equal width. The number of slices taken is 14 in this case. Failure length = 514 m, Radius of arc taken = 295 m, Number of slices = 14, width of each slice = 25 m.

Considering the whole length of slip surface L as 514 m, the total driving and resisting forces are:

Driving forces = $\sum T$; Resisting force = $\sum c'L + \tan\varphi' \sum N$

Hence, the factor of safety against sliding is

$$F = (c'L + \tan\varphi' \sum N) / \sum T \quad (1)$$

Where, N is the normal component of weight and T is the tangential component of weight. Normal force and Tangential component are calculated as$N = W*\cos\alpha$ and $T = W*\sin\alpha$.

From the analysis: $\varphi' = 22°$, $\sum N = 86452.17$kN/m, $\sum T = 201075.1$kN/m, L = 514 m, F = 1

Since factor of safety is one, the left hand side of equation (1) equals the right hand side. Assuming the value of φ' as 22° and solving equation (1), we obtained the value of cohesion mobilized along the slip/failure surface as 10.14 kPa.

To validate the results obtained using this limit equilibrium method of slope stability, the stability of same slope is examined using finite difference program FLAC 2D (Fast Langrangian analysis of continua). The geometry of the slope with finite difference grid is shown in figure 4.Mohr coulomb model is used as material model in the analysis. The cohesion and friction angle values considered are 10.14 kPa and 22°. Other properties of soil are taken from table 2. Boundary conditions are such that both the vertical boundaries are fixed in x-direction and the bottom boundary is fixed in both x and y directions. FLAC organizes its zones in a row and column fashion. However, the grid can be distorted to fit complicated shapes. As a general rule, the aspect ratio (i.e height to width ratio) of the zone should be close to unity and the ratio of areas between adjacent zones should be less than 4:1. Considering the above rules, different mesh sizes from coarser to finer grids were used in the problem. It was noted that due to the complexity of slope geometry in our problem, finer grids got distorted under the influence of gravity conditions and could not simulate the physical system. Hence coarser grids with gradual variation in gird size from zone 2 to zone 4 were used to solve the problem. Finer grids were used in zone 4 to ensure maximum accuracy in the displacement and stress values. The number of zones used is 830. The analysis is performed in gravity conditions only. The entire slided slope is divided into 2nd, 3rd and 4th zones, each having different slope inclinations as given in Table 1. Different zones are depicted in different colour shades. FLAC evaluates factor of safety using strength reduction method. It is found that for the combination of shear strength parameters considered, factor of safety evaluated is one. Also the modelled slip surface agrees well with the observed one as the failure surface covers part of 4th zone, whole of 3rd zone and part of 2nd zone as observed in the site. Figure 5 shows the Malin slope at the time of failure.

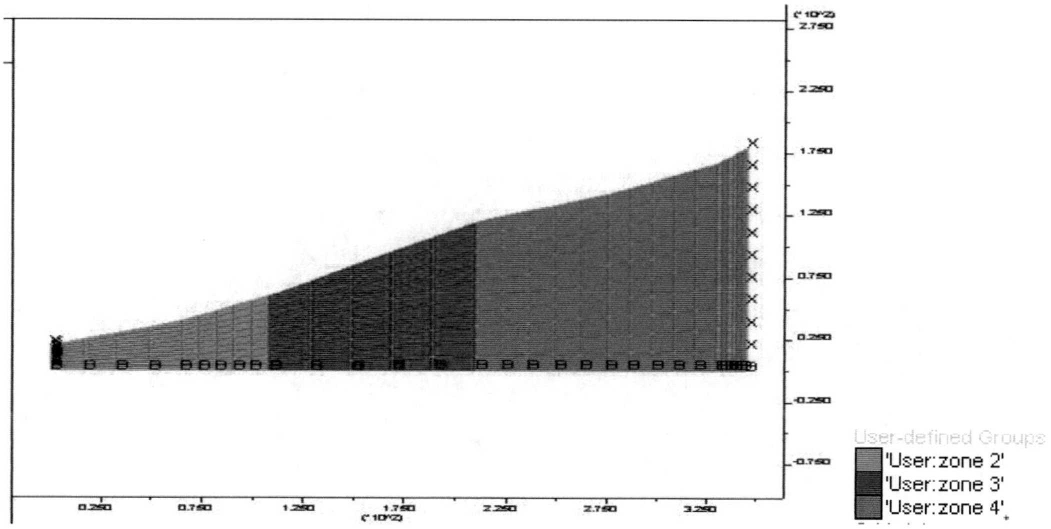

Figure 4: Finite difference grid

Figure 5: Malin slope at the time of failure (Factor of safety = 1.00)

Analysis performed using FLAC confirms the accuracy of back analysis performed on the failed slope. Figure 6 shows the displacement contours of the slope and it is found that the maximum thickness of slip surface is 10 m. During the site investigation, the thickness of material depleted from various zones during landslide was approximated as 7 m although the value was not measured accurately because when the detailed mapping work was carried out, most of the failed slope material was removed from the place as part of rescue efforts.

Figure 6: Displacements contours at the time of failure

Using FLAC, one can determine the pre-slide shear strength parameters of the slope. It is understood that stable slopes have factor of safety greater than 1 and in most natural soil slopes the conventional factor of safety assumed for stable slope is 1.5 or greater than 1.5. In FLAC one can perform an analysis to determine the combination of cohesion and friction angle which gives a factor of safety of 1.50 or above. For the same slided slope we have carried out the stability analysis. Mohr coulomb model is used as material model in the analysis. Boundary conditions are same as the previous problem and the analysis is performed in gravity conditions only. The entire slided slope is divided into three zones, each having different slope inclinations and is depicted in different colour shades. From the analysis it is found that for values of cohesion = 36 kPa and friction angle = 22°, the factor of safety of slope is 1.50. Hence the pre-slide strength parameters of the slope are considered as 36 kPa and 22°.

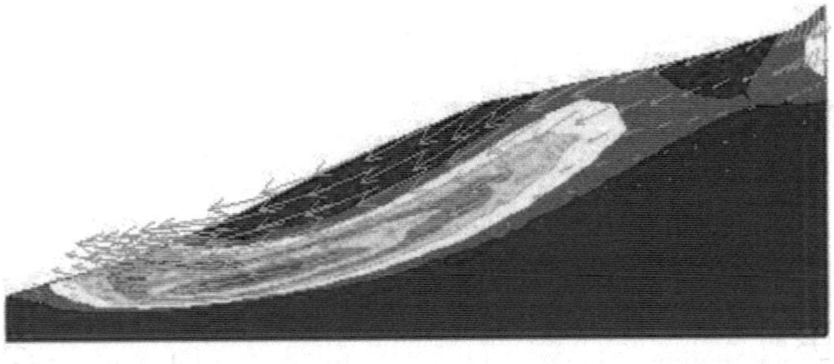

Figure 7: Pre-slide Malin slope (Factor of safety = 1.52)

Figure 7 shows that the pre-slide slope suggested deep rotational slip circle, however the rainfall infiltration into the soil profile decreased the strength parameters and caused shallow slide. Table 4 compares the strength parameters of pre-slide, failure state and post slide state of the same slope where landslide occurred. It is observed that for the same friction angle value of 22° there is a considerable decrease in the value of cohesion.

Table 4: Pre-slide and failure state strength parameters

Parameter	Pre-slide	Failure state	Post-slide
Cohesion (kPa)	36	10.14	38.3
Friction angle (°)	22	22	10.31

It is evident that there has been a decrease in strength parameters from pre-slide state due to failure of slope (parameters of post-slide are taken from Table 2). Excessive rainfall in the area for more than a week has been identified as the triggering mechanism for the failure. Chowdhury and Flentje (2002) observed that high pore water pressures that are generated after prolonged and intense rainfall trigger most cases of significant landsliding. The mobilized shear strength as low as c' = 0 kPa and φ' = 15° have been determined from back analyses and for deep-seated landslides, the shear strength parameters determined are c' = 0 and *φ'* = 9°.SivakumarBabu and Dasaka (2005) showed that decrease of matric suction with time is a time dependent process and evaluated the reliability of a typical landslide in Himalayan terrain. Zhang et al. (2009) performed back analysis of slope failure using probabilistic methods. A cut slope failed due to heavy rainstorm was back-analyzed in a probabilistic way. The mean value of strength parameters determined from the analysis were c' = 8kPa and φ' = 38°. Soon Min Ng et al. (2014) investigated the slope failure triggered by rainfall using numerical back analysis method. Back analyses were performed via finite element shear strength reduction method. The results showed that the shear strength parameters at failure were c' = 11 kPa and φ' = 20°. The slope consisted of silty soil in large amounts. Hence the results obtained by back analysis based on limit equilibrium and finite difference methods are comparable with the results from Soon Min Ng et al. (2014).

This decrease in strength parameters which resulted in failure of slope can be attributed to the loss of matric suction in the unsaturated part of the slope due to excessive rainfall infiltration. Once the rain-water starts infiltrating the vadose zone, the negative pore-water pressures will tend to dissipate due to an increase in the soil water content. This process contributes to lowering the shear strength of the soil layers close to the surface. Under normal conditions the negative pore water pressure creates suction in the vadose zone and increases the strength of the soil.

4. Role of matric suction in the stability analysis of unsaturated soil slopes

To study the stability of natural slopes subjected to extreme rainfall, one needs knowledge of the pore pressures in normal conditions in addition to those during and after the event. The pore-water pressures, together with the shear strength parameters, govern the stability of a slope. In unsaturated soils, pore-water pressures are dependent upon the flux of water infiltration and upon the hydraulic properties of the soil. The most important hydraulic parameters for unsaturated soils are the storage capacity characterized by the soil-water characteristic curve and the water coefficient of permeability of the soil.

In a saturated soil, the pore space is entirely filled with water i.e the degree of saturation is 100%. But in an unsaturated soil, the volume of water stored within the soil depends upon the negative pressure or suction within the pores. This negative pressure can be varying with time and space in the soil. A function is therefore required to describe the changes in water content related to different suction pressures in the soil. The soil-water characteristic curve (SWCC) can be viewed as the continuous sigmoidal function describing the amount of water in the soil as it is subjected to changes in soil matric suction. The shear strength of a soil is an essential parameter for numerous types of stability analyses. In a saturated soil, the stress state variable is called the effective stress and is often represented by the following:

$$\sigma' = \sigma - \mu_w \tag{2}$$

Where σ′ = effective normal stress, σ = total normal stress, μ_w = pore water pressure. The shear strength of a saturated soil is most often expressed using the Mohr-Coulomb failure criteria and the effective stress concept as:

$$\tau_{ff} = c' + (\sigma_n - \mu_w)_f \tan \varphi' \tag{3}$$

Where τ = shear stress, c' = cohesion, $(\sigma_n - u_w)$ = effective normal stress, φ' = friction angle. Unlike saturated soils, the shear strength of unsaturated soils cannot be described by a singlestress-state variable. Fredlund et al. (1978) justified the need for two independent stress state variables known as the net normal stress (σ-μ_a) and the matric suction (μ_a-μ_w). They proposed that the shear strength of an unsaturated soil could be expressed as:

$$\tau_{ff} = c' + (\sigma_n - \mu_a) \tan \varphi' + (\mu_a - \mu_w) \tan \varphi^b \qquad (4)$$

Where φ^b = angle indicating the increase in shear strength relative to the matric suction and all other parameters have the same meanings. The parameter φ^b is determined experimentally, and is normally found to be somewhere between φ' and $\varphi'/2$ (Fredlund and Rahardjo 1993). This angle describes the increase of shear strength due to an increase in matric suction, while the friction angle (φ') describes an increase in shear strength caused by an increase in net normal stress. Hence from equation (4), an unsaturated soil tends to give larger shear strength than that of the same saturated soil. Because it gets strength from three components: cohesion, frictional strength and suction strength.

In this study, the cause of Malin landslide was the rainfall over the previous one week. This antecedent rainfall may have played a role in saturating the slope. Due to increase in saturation, the matric suction in the soil decreased which is clearly evident in the analysis as for the same friction angle value, the value of cohesion decreased from 36 kPa in pre-slide state to 10.14 kPa at the moment of failure. The most convenient form for expressing the decrease in shear strength due to suction is a decrease in the cohesion.

$$c = c' + (\mu_a - \mu_w) \tan \varphi^b \qquad (5)$$

Where c = total cohesion of the soil and all other parameters are the same. Considering that the matric suction has the same effect as that of confining pressure and hence the friction angle with respect to matric suction is the same friction angle. Solving the above equation for c = 36 kPa and 10.14 kPa, the matric suction in the soil is calculated and found that it decreased from 90kPa to 25 kPa. This decrease in matric suction is attributed to the rainfall infiltration. Hence it is important to study the stability of normal slopes affected by rainfall using unsaturated soil mechanics concept which may help in understanding many problems of rainfall induced landslides. Further analysis of the problem stated in this paper is carried out using the concepts of unsaturated soil mechanics and intensity-duration-frequency analysis of rainfall.

5. Conclusions

The catastrophic landslide which occurred along the upslope of Malin village in Pune district of Maharashtra, India in the early hours of July 30th 2014 and claimed the lives of 151 humans and damaged 45 houses was studied. Heavy rainfall over the previous one week was identified as the triggering mechanism of slope failure. Back analysis is performed on the failed slope to identify the cause of slope failure/landslide. Using Bishop's method of slices, the shear strength parameters at the time of failure is determined by assuming factor of safety of unity and known length of failure surface. A combination of cohesion and friction angle values of 10.14 kPa and 22° is reported as the shear strength parameters at the time of failure. To validate these values a numerical analysis was performed using FLAC2D. FLAC2D program is used to analyze the stability of the pre-slide Malin slope. For pre-slide state, cohesion and friction angle values is evaluated as 36 kPa and 22°. Hence there has been considerable decrease in soil strength due to rainfall infiltration. This is due to the loss of matric suction in the unsaturated zone of soil slope which causes lowering of strength of soil. Hence stability of soil slopes subjected to rainfall must incorporate the concepts of unsaturated soil mechanics and rainfall intensity-frequency analysis.

Acknowledgements

The reconnaissance efforts and detailed study of Malin landslide were completed by Geological Survey of India (GSI). The detailed mapping of the slide area was carried out. The soil samples collected from the Malin landslide area were tested for geotechnical parameters by M/s Soiltech, Pune.

References

1) Chowdhury, R.N. and Flentje, P. (2002). "Uncertainties in rainfall-induced landslide hazard", *Quarterly Journal of Engineering Geology and Hydrogeology,**35**, 61–70*
2) Duncan, J.M and Stark, T.D (1992). "Soil strengths from back analysis of slope failures", *Proc. of speciality conf. Stability and performance of slopes and embankments, II, ASCE, Berkeley, CA, June, **Vol. 1**, pp 890-904*
3) Fredlund, D.G. and Rahardjo, H. (1993). "Soil Mechanics for Unsaturated Soils". *John Wiley and Sons, Inc., pp. 517.*
4) FLAC 2D Version 7.0. *ITASCA Consulting group,Inc.*
5) Griffths, D.V and Lu, N. (2005). "Unsaturated slope stability analysis with steady infiltration or evaporation using elastic-plastic finite element", *Int.J. Numer. Anal. Math.Geomech., **Vol.29**, pp: 249-267.*
6) Morgenstern, N.R. and Widger, R.A. (1978). "The shear strength of unsaturated soils". *Canadian Geotechnical J., **15(3)**: 313-321.*
7) Rahardjo, H., Lim T.T., Chang, M.F. and Fredlund, D.G. (1995). "Shear strength characteristics of a residual soil", *Canadian Geotechnical Journal, **Vol.32**, pp: 60-77.*
8) Ng, Soon Min; Mohamad Ismail, Mohd Ashraf and Abustan, Ismail (2014). "Back analysis of slope failure using finite element with point estimate method", *Journal of Civil Engineering Research, **4(3A)**:31-35.*
9) Sivakumar Babu, G.L.,and Dasaka,S.M., (2005). "Reliability analysis of unsaturated soil slopes". *Journal of Geotechnical and Geoenvironmental Engineering, ASCE, **Vol. 131**, No.11, pp. 1423-1428.*
10) Zhang, J., Wilson H. Tang, and L. M. Zhang (2009). "Efficient probabilistic back-analysis of slope stability model parameters." *Journal of Geotechnical and Geoenvironmental Engineering **136.1**: 99-109.*

Evolution of rock falls in the Northern part of the Peloponnese, Greece

V Zygouri, I K Koukouvelas

Department of Geology, University of Patras, 26500 Patras, Greece

E-mail address: zygouri@upatras.gr

Abstract Rock falls are a common fast – moving type of slope failures. Earthquake triggered rock falls attracted widespread attention since they represent serious hazard during strong earthquakes, causing severe damages and even fatalities. Strong earthquakes and their associated rock falls give rise to a sudden change in landscape evolution in tectonically active areas. The associated risk can be high both to communities and to critical infrastructures even far away from the active source slopes. Distinguishing between climatic induced and tectonically induced rock falls triggered by past earthquakes is a challenging task based on the development and the fault related discontinuities of a rock slope. We chose two case studies located in the Northern part of the Peloponnese (in Ilia and Corinthia prefecture), the Skolis Mountain and the Acrocorinthos area, in order to establish the rock fall susceptibility for each case study through the implementation of shadow angle β. The proposed methodology is based on the integrated analysis of the recurrence of rock falls, their spatial distribution and their mapping through field survey and aerial photography. Our mapping is integrated through Geographic Information System taking into account also the catalogue of historical and recent recorded seismicity in an attempt to examine triggering mechanisms and causes including the effects of climatic conditions for each case study. After the analysis of the spatial relationships between rock falls and the distribution of seismic epicentres and active faults as seismogenic sources, we conclude that both studied areas have suffered extensive rock fall phenomena induced by shallow seismicity and that the relationship between geomorphologic parameters and rock fall occurrence is strong. The research steps are described, namely, the recognition, identification, mapping and evolution of rock fall phenomena through time. Our results propose a critical threshold value of 24° for shadow angle β as the worst case scenario, suggesting that isolated boulders pose the greater risk on the associated communities.

1. Introduction

Rock fall is defined as individuals rocks are removed from a slope by sliding, toppling or falling and proceeding downslope [1]. The study of earthquake – induced rock fall distribution has a major importance for a better understanding of the relationship between rock fall density, type and size and the causal mechanisms. The increasing availability and precision of Digital Elevation Models improve the assessment of rock fall prone areas where only few data are available [2, 3, 4, 5, 6]. Two simple ways are usually applied to construct rockfall maps: the first is using the fahrböschung or reach angle [7, 8, 9] and the second one is using the shadow angle [8, 10, 11, 12]. The difference between them is the starting point of the travel distance estimation of a fallen block. The reach angle introduces the definition of the release point of a rock fall, whereas the shadow angle defines the talus apex as the starting point of the talus slope travel path (figure 1). Thus, determining the shadow angle β is a common geometrical problem, calculating the arctangent β as the result of the relationship Ht/Lt where the Ht and Lt are defined by the intersection of the talus cone starting beneath a steep cliff with the maximum horizontal distance that a rock fall can travel to (see figure 1). In this way, the maximum travel distance of a block can be described by the shàdow angle β [12 and references therein]. According to Evans and Hungr [8] the shadow angle method is better suited for rock falls smaller than 10^5 m^3 which is the scenario for both case studies here considered.

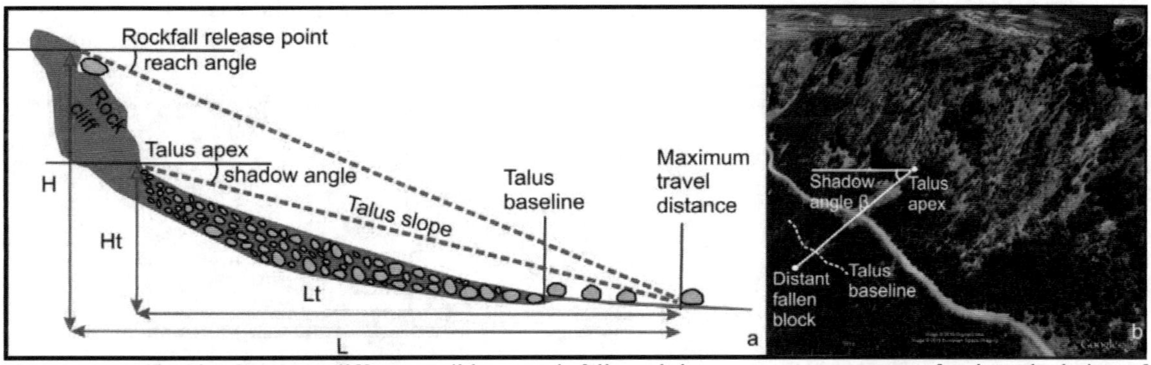

Figure 1 (a) Sketch of a steep cliff susceptible to rock falls and the necessary parameters for the calculation of reach and shadow angle (modified by Copons et al. [12]). **(b)** Example of key elements needed for the shadow angle calculation in the Skolis area near Santameri.

This paper presents the preliminary results of a rock fall mapping survey carried out through a large set of optical high resolution photo archives and air photo archives for pre and post earthquake times. This approach is performed in 4 main steps which include: DEM creation, identification of geomorphologic features, mapping of lithology and preliminary rock fall assessment estimation of large unstable volumes. The current study aims to contribute to the understanding of the causal mechanisms of rock falls triggered by earthquakes by preparing detailed rock fall inventory maps and correlating them with seismic and morphological parameters. For this reason we concentrated and tried to analyze simultaneously, both geomorphological, lithological and climatic data. The employed rainfall data were acquired from meteorological stations at Portai and Velos villages in Ilia and Corinthia prefectures respectively (figure 2). These rainfall data show that the period from April to September is almost dry in both areas. The annual precipitation ranges between 800 and 1000 mm for the Ilia area and between 580 and 650 mm for the Corinthos area (figure 2). The study areas have little vegetation on the steep slopes, mainly bushes, but Skolis area displays few small tree clusters and Acrocorithos presents cultivated land. From the geomorphological point of view there are few surface streams draining both areas.

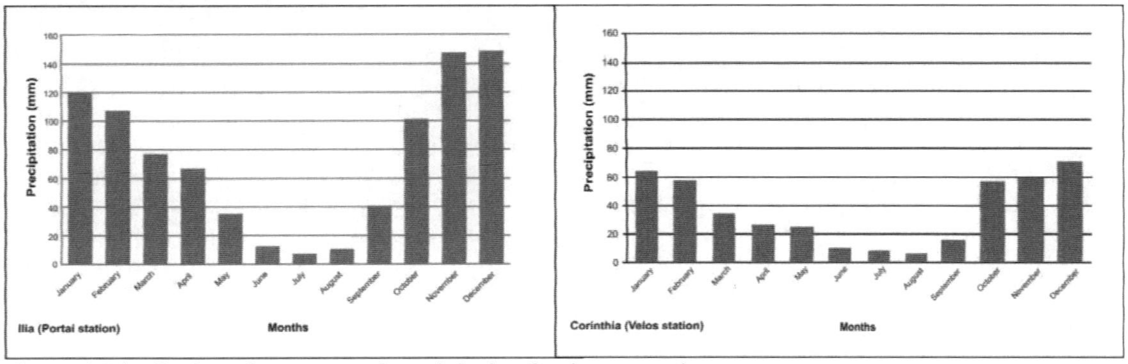

Figure 2 Distribution of mean annual precipitation for the two study areas.

2. Geologic setting

According to Heim [7] one of the main factors necessary for rock fall initiation is the steepness of the slope. Gravity driven surface processes in mountainous regions are principally correlated to the steepness of the topography and the relief morphology which therefore reflect these instabilities [13, 14]. Apart from topography, Keefer [15, 16] suggested that moderate to strong earthquakes can also trigger landslides. Earthquake triggered landslides can be either located on active faults or in the affected epicentral area [15, 17].The studied areas are extended in the Northern part of the Peloponnese. Both studied areas present rock fall distribution related to the main faults that are responsible for the mountainous relief and separate limestone bedrock from the Neogene deposits (figure 3).

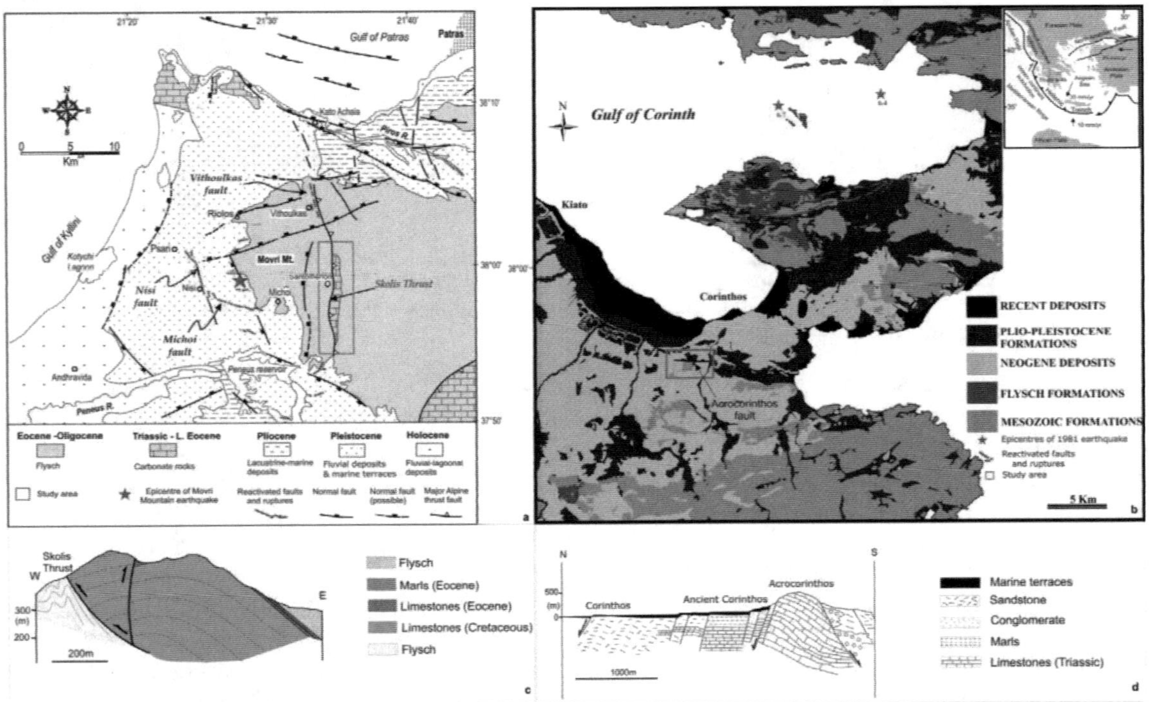

Figure 3 (a, b) Geological maps of the studied areas. The red stars show the epicentres of the most recent strong earthquakes and the red lines the associated coseismic ruptures. **(c, d)** Cross sections of the studied areas.

2.1. The case of Skolis Mountain

The Skolis Mountain is a fault related open asymmetric anticline 9 km long and 1.5 km wide (figure 3a, c). The mountain consists of late Cretaceous – Eocene carbonates and late Eocene – Oligocene flysch. The concentration of rock falls, in Skolis Mountain, is related to the structurally damaged rockwalls [18]. On the 8th of June 2008, a strong earthquake of Mw=6.4 occurred in northwestern Peloponnese. North Peloponnese is located in the most tectonically and seismically active region of Greece [19, 20, 21, 22]. The 2008 event was the largest earthquake occurred in north western Peloponnese during the past 30 years [23]. The secondary effects following the Mw=6.4 earthquake, were landslides and liquefaction phenomena near the epicentral area [23, 24, 25, 26, 27, 28]. Santomerion village was greatly affected due to its proximity to the steep slopes of Skolis Mountain and its high elevation. As a result a large volume of rock falls and boulders, hit the village (figure 4a).

2.2. The case of the Acrocorinthos in the gulf of Corinth

The gulf of Corinth represents a sedimentary basin trending WNW-ESE across the Hellenic mountain range. Sediments of the basin are exposed at an elevation of almost 2 km approximately suggesting high post – orogenetic uplift rate. In combination with the fast uplift rate, current rates of extension in

N-S direction reach up to a rate of 15 mm/yr [30], while at the eastern end of the gulf this is reduced to 6 mm/yr [31]. The sedimentary processes are controlled by the tectonic activity creating high relief through seismic ruptures and suitable conditions for high sediment supply [22]. Historical and instrumental seismic catalogues in the eastern part of the gulf of Corinth include records of at least twelve rupture events from 426 BC [32, 33] up to the last destructive Alkyonides event in 1981 AD [34]. In many cases these events triggered significant rock falls and/or liquefactions in coastal areas as recorded in 1858 [35], in 1928 [19] and in 1981 event [11, 36] (figure 4b).

Figure 4 Rock falls after a): the 2008 Movri Mountain earthquake and b): the 1981 Alkyonides earthquake (a. [23], b. [29]).

The Acrocorinthos represents a tectonic horst related to the activity of the north facing active normal fault called hereinafter Acrocorinthos fault. The Acrocorinthos fault juxtaposes fractured Mesozoic limestone from Neogene marls and conglomerate deposits (figures 3b, d). It has an almost E-W direction and its length is almost 5 km. The fault related damage zone has caused dense fracturing in the carbonate rocks which led to slope instability phenomena.

3. Methodology

Analog and digital air photos were used to analyze landslides evolution of the Skolis Mountain over the period from 1945 until 2008. Classical analog airphotos were available for the years of 1945 and 1996, digital air photos from the Greek Cadastral of 2008 and Quickbird satellite images (GoogleEarth) of 2007 and 2008. From the 1945 and 1996 orthorectified airphotos and from the 2007 GoogleEarth and 2008 cadastral digital mosaics the rock fall points and polygons were digitized. The onscreen digitations were performed at 1/1000 scale (figure 5). For the detailed description of the methodology see Koukouvelas et al. [18]. The same procedure was followed for the area of Acrocorinthos. Classical analog air photos were available for the years 1965 and 1987, while Quickbird satellite images (GoogleEarth) of 2006 to 2013 were also used. The interpretation of the rock falls was carried out using the air photo archives and the DEM derived for each study area. The DEM of the Skolis area and the Acrocorinthos area are generated using digitized contour lines from 1:5000 scale topographic maps with contour intervals 4m. An example of orthorectified air photos for both areas, is illustrated in figure 5 along with the rockfall dispersion in relation to different slopes.

Both study areas comprise a widespread rock fall terrain (figures 5, 6). The Skolis Mountain area shows 42 rock fall sites derived by the orthorectified air photos of 1945, 72 rock fall sites are noticed in 1996 orthorectified air photos, 75 are from the 2013 GoogleEarth images and 89 are recorded immediately after the earthquake. In correspondance 26 rock fall sites are present in the 1965 orthorectified air photos of the Acrocorinthos area, 38 are derived from the 1987 orthorectified air photos and 35 from the 2013 GoogleEarth Images.

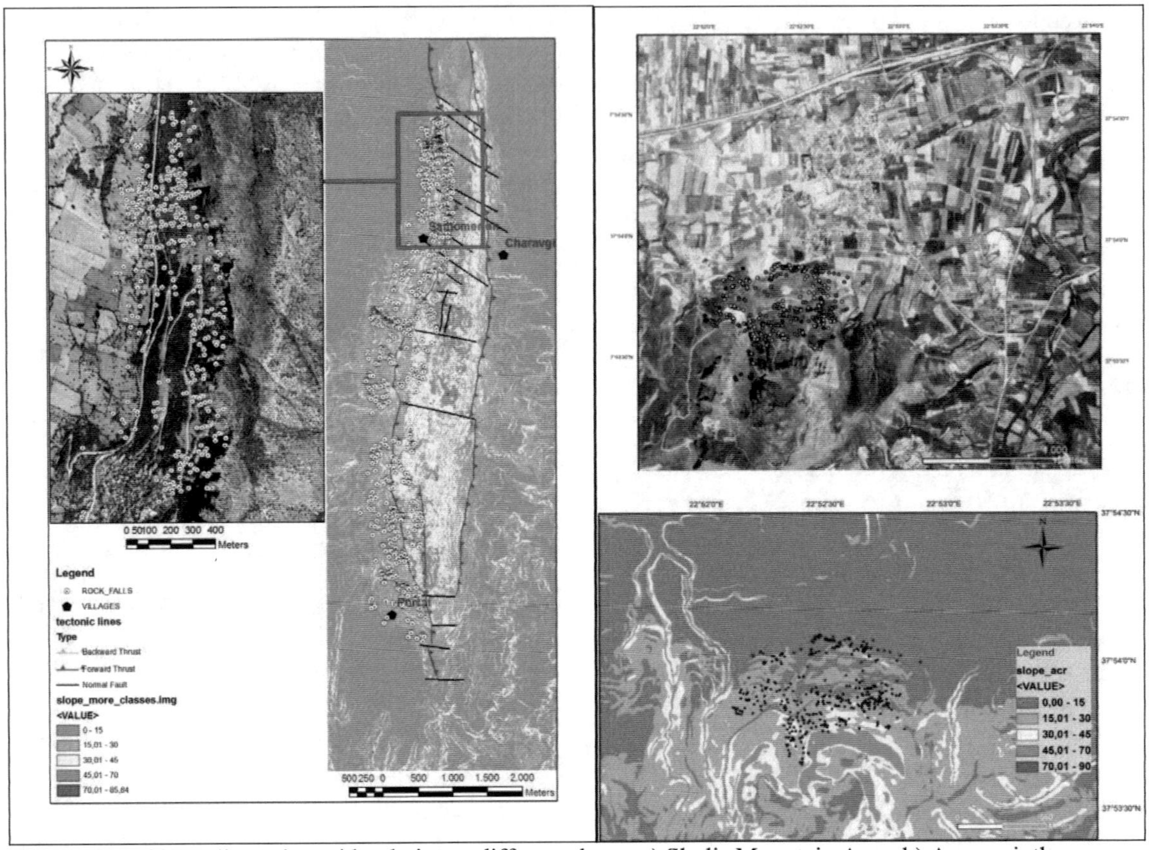

Figure 5 Boulders' dispersion with relation to different slopes a) Skolis Mountain Area, b) Acrocorinthos area.

Figure 6 Examples of the widespread rock fall terrains in the study areas. The photo on the left is from Skolis Mountain and on the right from Acrocorinthos.

4. Characteristics and distribution of rock fall sites

Having identified and mapped the hazardous areas we can produce a rockfall inventory map for each case study based on field observations, structural mapping and orthorectified air photo analysis. The Skolis Mountain was analyzed for sixty three years before the 2008 earthquake, while the north slope of Acrocorinthos was analyzed for half a century. Both areas have been affected by at least one earthquake during this time span. Thus, we can try to distinguish between climatologically and seismically induced rock falls by comparison.

4.1. The Skolis Mountain inventory map

The mapping of rock falls along the western slope of Skolis Mountain indicated that pre-existing instabilities have been increased in number immediately after the earthquake activity. Air photo mapping results suggest that the 2008 earthquake reactivated 72 taluses and triggered 17 new rock fall sites (figure 7). In addition a reactivation took place in almost all rock fall paths. To estimate the rock fall susceptibility, we identified the location of rock fall release points. These release points are recognized in a zone with elevation ranging from 450 m to 970 m. From these release points large boulders bounced or rolled on a slope surface ranging between 70-30° (figure 5), concentrated in talus cones with talus apex ranging between 650 m to 400 m elevation height. Typically the maximum spread of a block is described by the shadow angle β [8]. Pre earthquake mapping of rock falls show that boulders are dispersed at shadow angles between 30°-33°. On the contrary, post earthquake mapping of boulders reveals a change in this angle, calculated in 24° (figure 7).

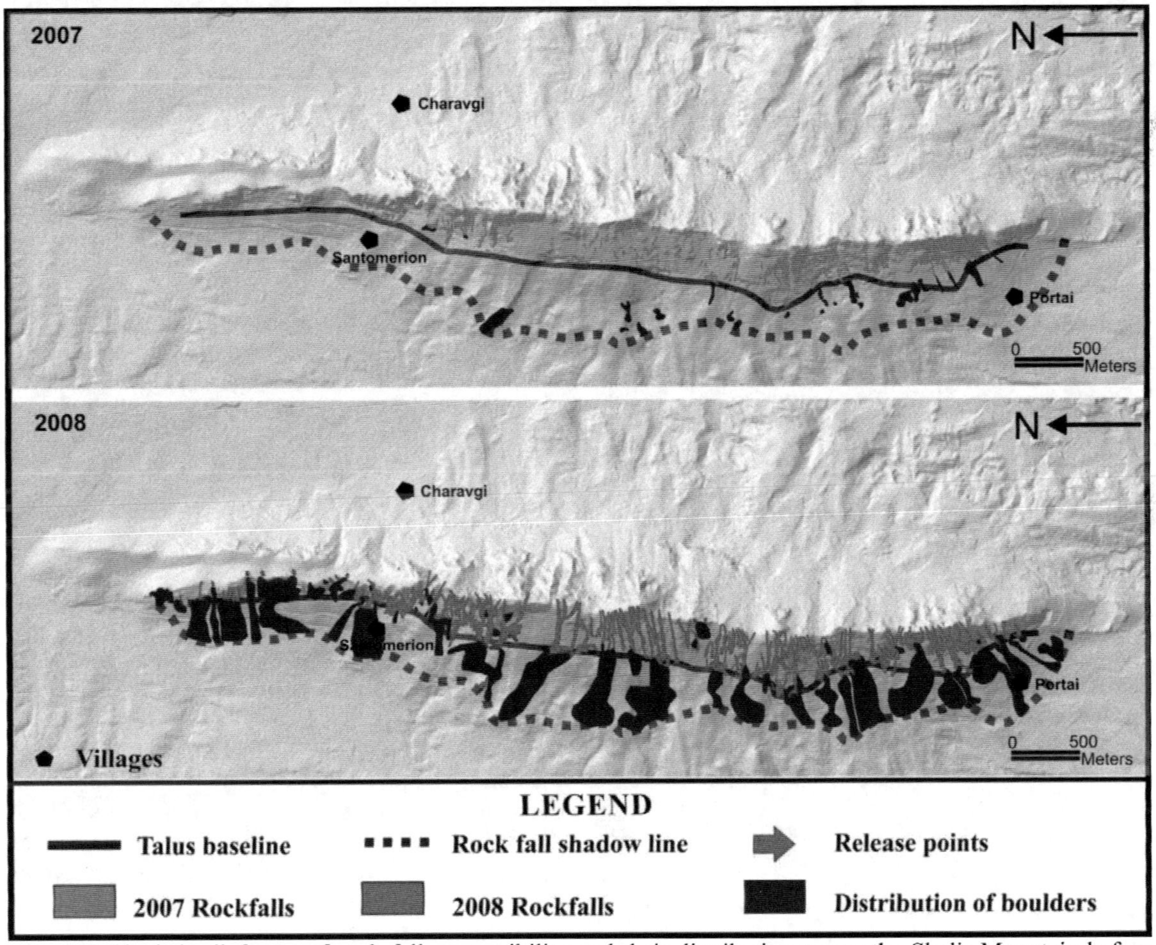

Figure 7 Shaded relief map of rock fall susceptibility and their distribution across the Skolis Mountain before and after the earthquake of 8th June 2008.

4.2. The Acrocorinthos inventory map

In the Acrocorinthos horst the interpretation and evaluation of the air photo archive of the last 50 years show the susceptibility of the area to extensive rock falls (figure 5b). During field mapping and air photo mapping, two separate populations of boulders were recognized based on their lithology. The first one is composed of Mesozoic limestone and originates from the steep slope of the Acrocorinthos cliff, whereas the second one is composed of cemented conglomerate (figure 6). The second cluster corresponds to the dismembered Neogene marine terraces (see figure 3d) and it is a common feature through the southern part of the gulf of Corinth. It originates from the toppling of marine caprock lying over marls. We do not take into consideration these boulders for the calculation of the shadow angle β since our field observations show that these are not subject to large movements over time. Thus, we do not consider them as possible hazard even though they are in close proximity to the village of ancient Corinthos (figure 8 right photo).

Figure 8 Typical boulders in Acrocorinthos area. Photo on the left shows the Mesozoic limestone boulders, while photo on the right, the Neogene conglomerate boulders.

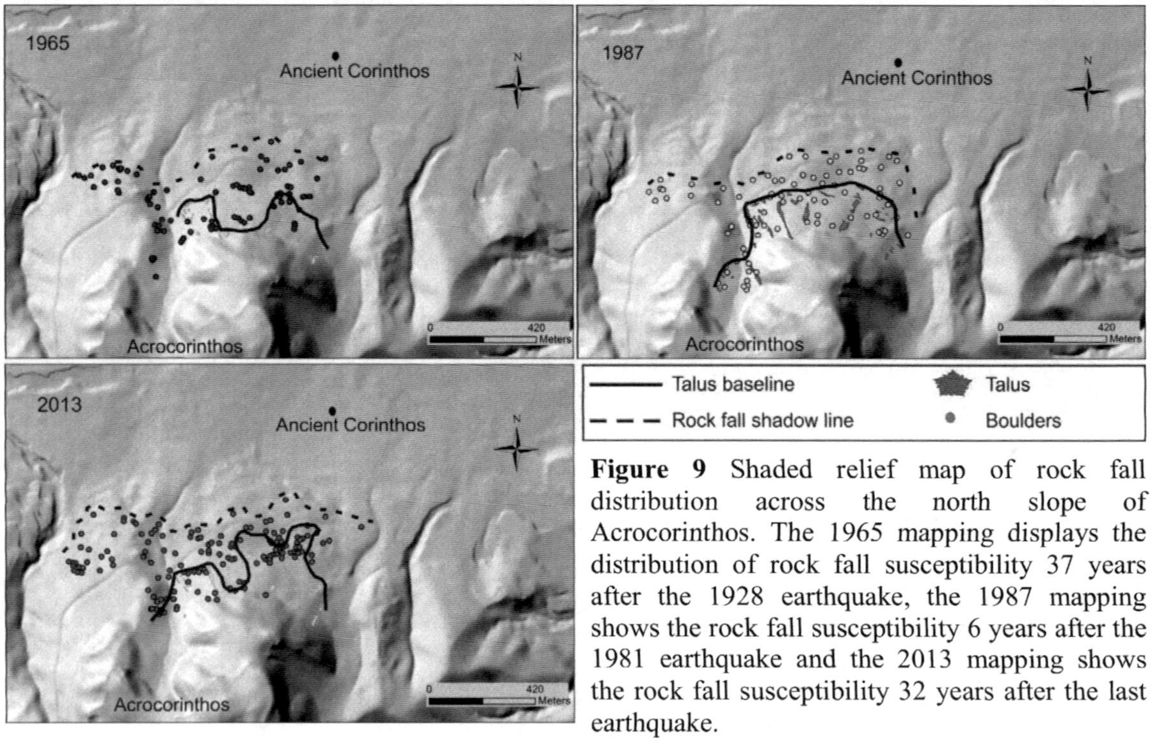

Figure 9 Shaded relief map of rock fall distribution across the north slope of Acrocorinthos. The 1965 mapping displays the distribution of rock fall susceptibility 37 years after the 1928 earthquake, the 1987 mapping shows the rock fall susceptibility 6 years after the 1981 earthquake and the 2013 mapping shows the rock fall susceptibility 32 years after the last earthquake.

The 1965 air photo mapping suggests a number of 26 rock fall sites. After the 1981 earthquake, the 1987 air photo mapping suggests that the identified rock falls are 38. Thus, they did not present a spectacular increase as in the case of Skolis Mountain. This increase is attributed to the 1981 earthquake swarm. However, the 26 km distance from the 1981 earthquake epicenter does not seem to have such a great impact on the Acrocorinthos area. Closest to the 1981 earthquake epicentres extensive rock falls have been mapped at the Skinos – Alepochori area by Koukis and Rozos [36] and Marinos et al. [29]. In 2013 GoogleEarth images 35 rock fall sites can be identified. It seems that 30 years after the 1981 earthquake a talus cone can redistribute its fragments on much gentler slopes. If this is the case then the 1965 air photo archive mapping suggests a similar distribution after the 1928 earthquake that according to historical and recorded data had a great effect in the area. Field survey showed that boulders that bounced on the steep slopes and rolled over the Acrocorinthos northern slope rest at a distance of 300 m from the talus apex. The zone in which the apex of the talus can be identified is located at an elevation ranging from 250 to 350 m. Mapping of boulders shows that boulders are dispersed at shadow angle with threshold value of 24° after the 1981 earthquake, while the same angle is estimated at 27° from the 1965 and 2013 photo archives (figure 9).

5. Discussion

This study aimed to generate a comprehensive data base of rock falls triggered by moderate to strong earthquakes in two areas that are prone to slope instabilities. For this reason we used air photo archives and tried to correlate them with climatic and lithologic factors. In Skolis Mountain, the west dipping stratification in limestone and the joints affecting it seem to generate favourable conditions for producing rock fall hazard. The rainfall pattern seems to promote these instabilities in a rather slow way over long periods of time. On the contrary, the 2008 Movri Mountain earthquake accelerates and enhances this process, according to our observations derived from field survey and photo archive. Thus, seismicity events turn out to be more important than rainfall in the initiation of slope failures. Therefore, we consider seismicity as a substantial factor that influences the rock mass strength. Keeping in mind that the newly formed or reactivated rock falls of 2008 are the consequence of recent seismicity, distant boulders recognised in photos taken prior to 2007 can be attributed to past earthquakes.

Similarly, the Acrocorinthos area appears as triggering rock falls throughout time. The tectonic pattern and the fractured carbonate rocks create predisposing conditions in favor of slope instability, not only associated with strong earthquakes but also with heavy rainfalls that sometimes occur as flash floods (flash flood event of 1997, [37]). The maximum distance travelled by rock fall blocks from the talus apex reaches 300 m, while in Skolis Mountain reached 200 m beyond the active taluses (talus baseline). However, some scarce boulders rest at a distance of 600 m from the base of the steep slope. We suggest that this difference is associated to the slope gradient which is smoother and lower respectively in Acrocorinthos than in Skolis. The talus cones in the Acrocorinthos area accumulate on a slope ranging from 40° to 30°, whereas the slope angle in Skolis Mountain is between 45° to 30° (the Ht factor is higher in Skolis case). This variation reflects the difference in pre-earthquake shadow angle. Even though the kinetic energy of the fallen block is largely decreased after the first impacts on the talus surface, suggesting an unaffected shadow angle by the rock fall height (elevation height of the release zone which is substantial in the calculation of reach angle), a block is able to travel further away over a long inclined talus slope due to its friction coefficient. Thus, smaller shadow angles sometimes can be estimated in places where the talus and/or ground surface is relatively smooth and flatter, or large boulders roll over finer taluses [8 and references therein]. Fine grained taluses seem to accumulate in the Acrocorinthos area affecting the shadow angle envelope (figures 4, 10 left photo). Nevertheless, the post earthquake – induced rock fall distribution appears similar with the one described in Skolis Mountain. The worst case scenario of 24° minimum shadow angle is valid for both the Skolis Mountain and the Acrocorinthos area. Moreover, the difference on pre earthquake estimations of angle β between the two areas could also be influenced by the plant cover [38, 39, 40] of each area. In Acrocorinthos area the plant cover is mostly rural with olive trees and vineyards just below the talus baseline (as defined in figure 1), while the Skolis Mountain slopes display tree clusters. These tree clusters offer some protection during heavy rainfalls that limit the travel distance of a fallen block (figure 10, right photo).

Figure 10 View of characteristic talus accumulation in the Acrocorinthos area on left. In the right photo, view of a boulder constrained by trees in the Skolis Mountain active slope.

The 1987 mapping of Acrocorinthos shows that the taluses have grown bigger in size and number since 1965 mapping. Thus, after the earthquake pre existing rock fall sites reactivated along with the development of new rock fall sites and caused reduced shadow angle. However, the 2013 data analysis seems alike the 1965 one. Given that both maps describe the rock fall distribution 30 years after the occurrence of a strong earthquake it seems that boulders can be dispersed even further in a long period after the earthquake. As the boulders are released from the carbonate rockwalls they usually follow trajectory paths. These trajectories may be or not obvious but it seems that boulders are concentrated immediately after the earthquake. After a heavy rainfall or a flash flood some of these boulders especially the ones at the base of the talus cones of the Acrocorinthos area that lack protective forest plant cover may have been set in motion again to their final resting place [41, 42] For this reason a protective metal - wired fence has been installed on the northern slope of the Acrocorinthos cliff.

6. Conclusions

In the current study we mapped and compared two areas hosting moderate earthquakes and prone to rock falls. Based on our results we summarize the following preliminary conclusions of the rock fall hazard in Skolis Mountain and Acrocorinthos area.

1) Both areas shows that the slope failures are related to fractured carbonate rocks and fault related structures.
2) Earthquake induced rock falls provoke the growth of pre existing rock fall sites or create new ones.
3) Isolated rock falls climatic or earthquake triggered, pose the greatest hazard for the residents of the area. However in both cases the road that crosses the rock fall terrain can act as a barrier for boulders travelling down slope.
4) The use of shadow angle β shows that a 24° envelope captures the farthest travelled boulder and highlights the zone of potential rock fall hazard for both the Skolis Mountain and the Acrocorinthos area.

Acknowledgements

This work was supported by Grant E078 to V Zygouri and I Koukouvelas from the Research Committee of the University of Patras (Programme K. Karatheodori).

References

[1] Varnes D J 1978 *Landslides, Analysis and Control* Special Report **176** (Transportation Research Board) (Washington) 11-33
[2] Guzzetti F, Reichenbach P and Wieczorek G F 2003 *Nat. Hazards Earth Sys. Sci.* **3** 491-03
[3] Derron M H, Jaboyedoff M and Blikra L H 2005 *Nat. Hazards Earth Sys. Sci.* **5** 285-92

[4] Frattini P, Crosta G, Carrara A and Agliardi F 2008 *Geomorhology* **94** 419-37
[5] Mihmoud C, Derron M H, Horton P, Jaboyedoff M, Baillifard F J, Loye A, Nicolet P, Pedrazzini A and Queyrel A 2012 *Nat. Hazards Earth Sys. Sci.* **12** 615-29
[6] Frattini P and Crosta G 2013 *Earth Planet. Sci. Lett.* **361** 310-19
[7] Heim A 1932 *Bergsturz und Menschenleben* Fretz and Wasmuth Verlag (Zurich) 218
[8] Evans S G and Hungr O 1993 *Can. Geotech. J.* **30** 620-36
[9] Corominas J 1996 *Can Geotech J* **33**, 260-71
[10] Jaboyedoff M, Dudt J P and Labiouse V 2005 *Nat. Hazards Earth Sys. Sci.* **5** 621-32
[11] Wieczoreck G F, Stock G M, Reichenbach P, Snyder J B, Borchers J W and Godt J W 2008 *Nat. Hazards Earth Sys. Sci.* **8** 421-32
[12] Copons R, Vilaplana J M and Linares R 2009 *Nat. Hazards Earth Sys. Sci.* **9** 2107-18
[13] Montgomery D R and Brandon M T 2002 *Earth Planet. Sci. Lett.* **201** 481-89
[14] Agliardi F, Crosta G B, Meloni F, Valle C and Rivolta C 2013 *Tectonophysics* **605** 34-47
[15] Keefer D K 1984 *Geol. Soc. Am. Bull.* **95** 406-21
[16] Keefer D K 2002 *Surveys in Geophysics* **23** 473-510
[17] Keefer D K 1984 *Science* **223** 1288-90
[18] Koukouvelas I, Litoseliti A, Nikolakopoulos K and Zygouri V 2015 *Eng. Geol.* **191** 71-85
[19] Papazachos B and Papazachou C 1997 *The Earthquakes of Greece* (Ziti and Co, Thessaloniki) 304
[20] Ambraseys N 2009 *Earthquakes in the Mediterranean and Middle East: a multidisciplinary study of seismicity up to 1900* (Cambridge University Press) 947
[21] Koukouvelas I K, Katsonopoulou D, Soter S and Xypolias P 2005 *Terra Nova* **17** 158-64
[22] Zygouri V, Verroios S, Kokkalas S, Xypolias P and Koukouvelas I K 2008 *Tectonophysics* **453** 193-210
[23] Koukouvelas I, Kokkalas S and Xypolias P 2010 *Int. Geol. Rev.* **52** 249-68
[24] Kokkalas S, Xypolias P and Koukouvelas I 2008 *Proc 31st General Assembly of the European Seismological Commission* (Crete)
[25] Koukouvelas I 2008 *Proc 31st General Assembly of the European Seismological Commission* (Crete)
[26] Pavlides S, Valkaniotis S, Papathanassiou G, Sboras S, Chatzipetros A 2008 *Proc 3rd Panhellenic Conference of Earthquake Engineering and Engineering Seismology* (Athens)
[27] Papadopoulos G A, Karastathis V, Kontoes C, Papoutsis I, Daskalaki E, Diakogianni G and Fokaefs A 2008 *Proc 31st General Assembly of the European Seismological Commission* (Crete)
[28] Papathanassiou G 2012 *Bull. Eng. Geol. Env.* **71** 187-94
[29] Marinos P, Koukis G, Stournaras and Skia S 1986 *Bulletin of KEDE* **3-4** 93-105
[30] Briole P, Rigo A, Lyon-Caen H, Ruegg J, Papazissi K, Mistakaki C, Balodimou A, Veis G, Hatzfeld D and Deschamps A 2000 *J. Geophys. Res.* **B105**, 25605-25
[31] Avallone A, Briole P, Agatza-Balodimou A M, Billiris H, Charade O, Mitsakaki C, Nercessian A, Papazissi K, Paradissis P and Veis G 2004 *Comptes Rendus Geoscience* **336** 301-11
[32] Ambraseys N and Jackson J 1990 *Geophys. J. Int.* **101** 663-708
[33] Tsapanos T, Koravos G, Zygouri V, Tsapanos M, Kortsari N, Kijko A and Kalogirou E 2011 *J. Balkan Geophys. Soc.* **14** 1-14
[34] King G C P, Ouyang Z X, Papadimitriou P, Deschamps A, Gagnepain L, Houseman G, Jackson J A Soufleris C and Virieux J 1985 Geophys. *J. R. Astron, Soc* **80** 677-93
[35] Koustas G 1858 Corinthos Earthquake (Pandora) 225-29
[36] Koukis G and Rozos D 1985 *Bull. Int. Assoc. Eng. Geol.* **32** 91-95
[37] Karymbalis E, Katsafados P, Chalkias C and Gaki-Papanastassiou K 2012 *Z. Geomorphol.* **56** 45-67
[38] Berger F and Dorren L 2007 *Schweiz. Zeitsch. für Forstwesen* **158** 157-65.
[39] Clouet N and Berger F 2010 *International Symposium Interpraevent 382-90.*

[40] Rammer W, Brauner M, Dorren L, Berger F and Lexer M J 2010 *Nat. Hazards Earth Sys. Sci.* **10** 699-11
[41] Wieczorek G F, Morrissey M M, Iovine G and Godt J 1999 *U.S. Geological Open File Report* 99-578
[42] Brideau M A, Yan M and Stead D 2009 *Geomorphology* **103** 30-49

IOP Conf. Series: Earth and Environmental Science **26** (2015) 012045 doi:10.1088/1755-1315/26/1/012045

Seismic displacement of geosynthetic-reinforced slopes subject to cracks

Akram H Abd [1,2]
[1] School of Engineering, University of Warwick, CV4 7AL, Coventry, UK.
[2] Civil Engineering Dept., College of Engineering, Tikrit University, Tikrit, Iraq.

E-mail: akramciv22@gmail.com

Abstract. The kinematical approach of limit analysis associated with pseudo static assumption is employed to evaluate the displacement of geosynthetically reinforced soil slopes subject to cracks. According to existing literature, the seismic displacements for soil slopes have been calculated with the effect of possible cracking being neglected, such cracking is likely to emerge due to an earthquake with even moderately large motion. In this paper, a new technique is proposed to estimate the horizontal displacement of the slope toe for geosynthetically reinforced slopes resulting from a given earthquake postulating a rough estimation of real time crack propagation. The effect of crack formation as part of the failure process during the earthquake on the horizontal displacement of the slope toe is specifically tackled. The seismic displacement is estimated by incorporating a stepwise yield acceleration corresponding to postulated crack propagation. Rotational failure mechanisms accounting for either intact reinforced slopes that can show cracks or reinforced slopes with pre-existing cracks are considered. Two types of reinforcement layouts are employed here; uniformly distributed reinforcement along the slope height and linearly increasing distribution (i.e. the spacing between layers decreases linearly with depth). An example illustrating the procedure for a given earthquake is presented. Results show that the horizontal displacement of the slope toe calculated using the stepwise yield acceleration for both uniform distribution of reinforcement and for linearly increasing distribution can provide a reasonable estimation of the slope displacement. Furthermore, in terms of the slope displacement, linearly increasing distribution yields better results than the uniform layout.

1. Introduction

Cracks can cause significant reduction in the stability of unreinforced soil slopes [1, 2], especially if these slopes are subjected to seismic action [3]. Although, geosynthetics have been used successfully and effectively as soil reinforcement material for the last thirty years [4, 5], pre-existing cracks can be detrimental for geosynthetically reinforced slopes [6]. The presence of a vertical crack can reduce the safety factor of the slope depending mainly on its location and depth. Cracks form not only potential parts of the slip surface, but also they form easy flow channels for rainfall water which reduces the soil strength and exerts a lateral stress inducing the failure when these cracks are filled with water. Cracks can be found in soil slopes and embankments due to tensile stresses such as seismic action or external static loading, and/or due to desiccation and cycles of wetting and drying.

Methods for assessing the seismic stability of slopes have been developed during the last century. The Mononobe-Okabe method might be the first published work that addresses the stability of retaining walls and dams during earthquake incorporating dynamic earth pressure [7, 8]. Then, several limit equilibrium methods were developed for this purpose [9-11] which are still the most commonly used by practitioners. More recently, numerical methods for continuum mechanics such as the finite element method with strength reduction technique [12, 13] and finite element limit analysis [14] have provided the capability to reliably detect the onset of failure in slopes according to the approach of continuum mechanics. However, if cracks are present, a continuum approach no longer works since the onset of instability is ruled by the behaviour of single fractures. In this case, the Discrete Element Method can nowadays be employed for 3D analyses of slopes with cracks [15]. Recent algorithmic advances in terms of contact detection algorithms [16, 17] have substantially reduced the runtime of these analyses. However, when little information on the presence of cracks is known, extensive parametric analyses requiring large computational times are necessary. In this case, an analytical approach is very desirable so that numerical analyses would be run only for the case(s) identified by the analytical approach as the most critical.

Newmark's method [18] is an analytical method which is very popular among practitioners where a pseudo-static force is used instead of the dynamic excitation to calculate earthquake induced displacements. Analytical solution for earthquake induced displacements undergone by slopes subject to a rotational failure mechanism is presented by [19]. One of the main limitations of Newmark's method is the neglect of the earthquake induced strength degradation of the soil, i.e. it assumes a constant yield acceleration throughout the analysis [20]. In this paper, earthquake induced crack occurrence and the consequent reduction of yield acceleration will be accounted for geo-reinforced slopes. Seismic induced displacements will be calculated based on a stepwise time varying yield acceleration.

2. Formulation of the Problem

The kinematical approach of limit analysis will be used to calculate the least upper bound on the yield (critical) acceleration K_y for uniform c-ϕ geo-reinforced slopes. Two different distributions of reinforcement are considered, namely uniform distribution (UD) and linearly increasing distribution with depth (LID), both illustrated in figure 1. Let us introduce the dimensionless variable K_t as the average tensile strength of reinforcement per unit height of the slope. Where K_t is a function of the tensile strength of the reinforcement layer per unit width, T, and of the spacing between reinforcement layers, S, and can be written as:

$$K_t = \frac{T}{S} \tag{1}$$

the two cases of reinforcement distributions have been analysed by assigning the corresponding value of K_t for uniform distribution (UD):

$$K = K_t \tag{2}$$

and for linearly increasing distribution (LID):

$$K = 2K_t \frac{\exp[\tan\phi(\theta - \chi)]\sin\theta - \sin\chi}{\exp[\tan\phi(\upsilon - \chi)]\sin\upsilon - \sin\chi} \tag{3}$$

where θ, χ and υ are the angles made by r (or r_c), r_χ and r_υ respectively with a reference axis, (see figure 2), r is the distance between point P and any point on the log-spiral slip surface, r_c is the distance between point P and any point on the crack, r_χ and r_υ are the lengths of the chords P-F and P-D respectively, and ϕ is the angle of internal friction of the soil.

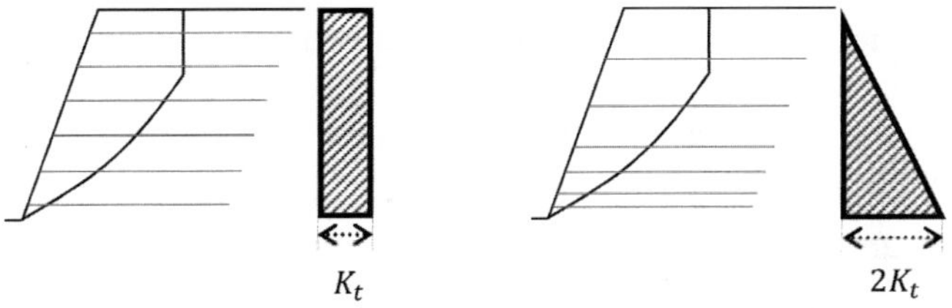

Figure 1. Geosynthetic-reinforcement layouts. (a) Uniformly distribution (UD), and (b) Linearly increasing distribution (LID).

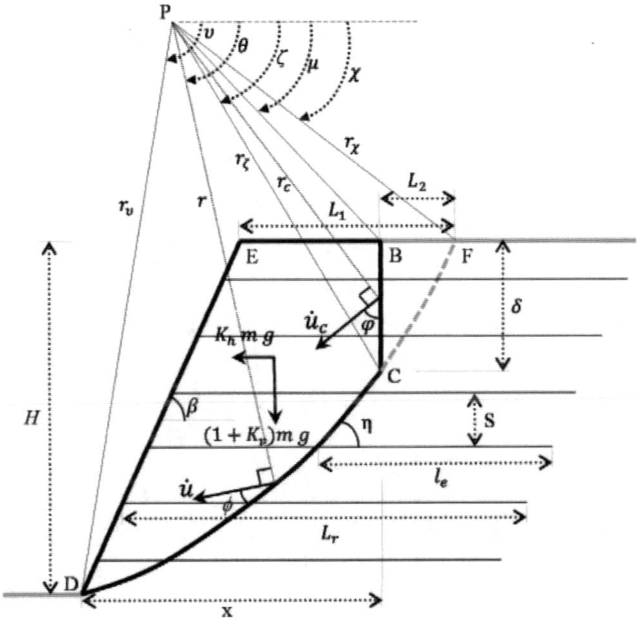

Figure 2. Failure mechanism. Note that $\varphi \neq \phi$. The wedge of soil enclosed by black lines D-C (logarithmic spiral failure line), B-C (pre-existing crack), B-E (upper surface of the slope) and E-D (slope face) rotates around point P.

The yield acceleration can be defined as the minimum level of horizontal acceleration (vertical acceleration being proportional to the horizontal acceleration) that brings the slope to failure (i.e. safety factor =1). According to Newmark's method [18], slope displacements start to occur whenever the seismic induced acceleration exceeds the yield acceleration. Then, the displacements occurring during the earthquake can be obtained by double integrating the differences between the applied accelerations and the yield one during the time intervals when the ground velocity is larger than zero.

In the following, an initially intact slope subject to the formation of tension cracks as result of the earthquake is considered. In this case, the cracks are formed as part of the failure process the first time the slope yield acceleration is exceeded. Then, in order to calculate the slope displacements generated by the earthquake, a new yield acceleration, accounting for the presence of the cracks formed the first time the yield acceleration of the intact slope was exceeded, needs to be calculated for all the subsequent steps. Four cases are considered in this paper:

a) Slopes made of cohesive soils of unlimited tensile strength, hence not subject to tension cracks.
b) Slopes made of cohesive soils of limited tensile strength.
c) Slopes made of cohesive soils of zero tensile strength.
d) Slopes subject to the most unfavourable crack from a stability point of view pre-existing the onset of the earthquake.

The procedure for calculating the stepwise time varying yield acceleration is outlined as follows:

1- Determine the yield acceleration for an initially intact slope subject to the formation of tension cracks $K^1_{y(c)}$. Vertical tension cracks are formed as part of the occurring failure mechanism [21] since energy is needed to form any crack. Therefore, the yield acceleration of a slope subject to the formation of tension cracks is lower than the yield acceleration of a slope of unlimited tensile strength $K^{\text{int}}_{y(c)}$, i.e. not subject to tension crack formation: $K^1_{y(c)} \le K^{\text{int}}_{y(c)}$. This acceleration is used to calculate the displacements, the first time that the seismic acceleration exceeds the yield acceleration $K_{y(in)}$
2- Determine the yield acceleration for the same slope, but accounting for the presence of the crack generated in step 1, treated now as a pre-existing crack, i.e. the crack is already present so that no energy is dissipated for crack formation). This new value of yield acceleration, $K^2_{y(c)}$, is used to calculate the displacements in all subsequent steps.
3- Determine the accumulated wedge displacement D_i with respect to the ground surface, at each time step (i) when the seismic acceleration exceeds $K^2_{y(c)}$.
4- Calculate the dimensionless coefficient C that relates the displacement of the slope toe to the integral of the earthquake acceleration record above the level of yield acceleration.
5- Determine the accumulated horizontal displacement at the slope toe D_{xi}, where $D_{xi} = C \times D_i$, and then the total horizontal displacement D_t is to be found.

It should be noted that, although several tension cracks at different locations in the slope may form during the earthquake, only the crack which has the worst detrimental effect on slope stability needs to be considered in the calculation. This is because, according to the kinematic approach of limit analysis, the failure mechanism taking place is the most critical mechanism for the stability of the slope among all the kinematically feasible mechanisms.

3. Calculation of Yield Acceleration

The upper bound theorem of limit analysis will be here employed to calculate the yield acceleration for both intact and cracked slopes. The analytical expressions for the calculation of the external work done by soil masses sliding along composite log-spiral failure surfaces, which requires the use of fictitious wedges bordered by a log-spiral, was first presented in [22, 23] for the case of slopes with horizontal upper part subject to a sequence of landslides, and in [24] for the more general case of slopes with an inclined upper part. Note that these calculations apply to slopes made of bonded granulates [25, 26] as well. In [27], the calculation of the work done by a wedge enclosed by two log-spirals was first presented. The analytical solution is derived here for the case of a horizontal upper slope surface and vertical pre-existing cracks from the upper slope (see figure 1). However, the solution can be straightforwardly extended to the case of a non-horizontal upper slope and the case of cracks departing from the slope face. Such an extension is reported in [2] for the static case.

Let us consider the failing wedge E-D-C-B which is about to rotate rigidly around the centre of rotation P, as yet undefined, with the ground lying on the right of the log-spiral piece D-C and of the vertical crack C-B remaining at rest. The equation of log-spiral D-C is:

$$r = r_\chi \exp\left[\tan\phi(\upsilon - \chi)\right] \quad (4).$$

The upper bound on the yield acceleration K_y will be derived imposing energy balance for the failing wedge E-D-C-B:

$$\dot{D} = \dot{W} \tag{5}$$

where $\dot{D}$ and $\dot{W}$ are the rate of dissipated energy and of external work respectively. In this paper $\dot{D}$ has four terms as follow:

$$\dot{D} = \dot{D}s_{C-D} + \dot{D}s_{B-C} + \dot{D}r_{C-D} + \dot{D}r_{B-C} \tag{6}$$

where $\dot{D}s_{C-D}$ and $\dot{D}s_{B-C}$ are the rates of dissipated energy within the soil along the log-spiral segment (D-C) and along the crack (B-C) respectively. While $\dot{D}r_{C-D}$ and $\dot{D}r_{B-C}$ are the rates of dissipated energy within the geosynthetic reinforcement along the log-spiral segment (D-C) and the crack (B-C), respectively. The energy dissipated within the soil, $\dot{D}s_{C-D}$, along the log-spiral segment (D-C) can be written as:

$$\dot{D}s_{C-D} = c\dot{\theta}r_\chi^2 f_{C-D}\left(\chi,\upsilon,\zeta,\phi\right) \tag{7}$$

$$\dot{D}s_{C-D} = c\dot{\theta}r_\chi^2 \exp\left[2\tan\phi\left(\zeta-\chi\right)\right]\frac{\exp\left[2\tan\phi\left(\upsilon-\zeta\right)\right]-1}{2\tan\phi} \tag{8}$$

the dissipated energy within the soil $\dot{D}s_{B-C}$ along the crack, as it forms as part of the failure [21], can be written as:

$$\dot{D}s_{B-C} = \dot{\theta}r_\chi^2\left(\frac{\sin\chi}{\tan\theta_c}\right)^2\left[\frac{f_c}{2}\int_{\theta_c}^{\zeta}\frac{1-\sin\theta}{\cos^3\theta}\mathrm{d}\theta + \frac{f_t}{1-\sin\phi}\int_{\theta_c}^{\zeta}\frac{\sin\theta-\sin\phi}{\cos^3\theta}\mathrm{d}\theta\right] \tag{9}$$

with θ_c :is the angle made by the segment P-B with the horizontal (see figure 2), f_c and f_t are the unconfined compressive and tensile strength of the geo-material, respectively. According to the Mohr-Coulomb failure criteria, they can be expressed as:

$$f_c = 2c\frac{\cos\phi}{1-\sin\phi} \tag{10}$$

$$f_t = 2\alpha c\frac{\cos\phi}{1+\sin\phi} \tag{11}$$

where α is a dimensionless coefficient introduced here to express the amount of tensile strength, with $0 \le \alpha \le 1$. Now substituting equations (10 and 11) into equation (9), the following expression is obtained:

$$\dot{D}s_{B-C} = c\dot{\theta}r_\chi^2 f_{B-C}\left(\chi,\upsilon,\zeta,\phi,\alpha\right) \tag{12}$$

$$\dot{D}s_{B-C} = c\dot{\theta}r_\chi^2\left(\frac{\sin\chi}{\tan\theta_c}\right)^2\left[\frac{\cos\phi}{1-\sin\phi}\int_{\theta_c}^{\zeta}\frac{1-\sin\theta}{\cos^3\theta}\mathrm{d}\theta + \frac{2\alpha\cos\phi}{1-\sin^2\phi}\int_{\theta_c}^{\zeta}\frac{\sin\theta-\sin\phi}{\cos^3\theta}\mathrm{d}\theta\right] \tag{13}.$$

The third term of the dissipated energy is the one that occurs within the geosynthetic reinforcement along the log-spiral part, $\dot{D}r_{(\text{C-D})}$. This can be calculated by integrating the product of the infinitesimal increment of strain rate undergone by the reinforcement and the tensile strength of the reinforcement T averaged over the spacing S between consecutive layers of reinforcement [28]. For the sake of space, the calculations here are only for uniform distribution (UD) of reinforcement (i.e. $K = K_t$):

$$dr = \int_0^{t/\sin\eta} K \sin\eta\, \dot{\varepsilon}_x \,\mathrm{d}x = K \sin\eta\, \dot{u} \cos(\eta - \phi) \tag{14}$$

with $\dot{\varepsilon}_x$: strain rate in the direction of reinforcement, t : thickness of the discontinuity layer, and η : angle made by the reinforcement layer with discontinuity surface, which can be written as:

$$\eta = \frac{\pi}{2} - \theta + \phi \tag{15}$$

now, by integrating equation (14) over the log-spiral part (C-D),

$$\dot{D}r_{(\text{C-D})} = \int_{\text{C-D}} K \sin\eta\, \dot{u} \cos(\eta - \phi) \frac{r\,\mathrm{d}\theta}{\cos\phi} \tag{16}$$

after substituting and simplifying, the following expression is obtained:

$$\dot{D}r_{(\text{C-D})} = r_\chi^2 \dot{\theta} \int_\zeta^\upsilon K\left(\exp[2\tan\phi(\theta - \chi)]\sin\theta\cos\theta + \sin^2\theta\tan\phi\exp[2\tan\phi(\theta - \chi)]\right)\mathrm{d}\theta \tag{17}$$

for uniform distribution of reinforcement (UD), $K = K_t$, then:

$$\dot{D}r_{(\text{C-D})} = K_t r_\chi^2 \dot{\theta} g_2(\chi, \upsilon, \zeta, \phi) \tag{18}$$

$$\text{with } g_2 = \frac{1}{2}\left[\exp[2\tan\phi(\upsilon - \chi)]\sin^2\upsilon - \exp[2\tan\phi(\zeta - \chi)]\sin^2\zeta\right] \tag{19}$$

the dissipated energy by the reinforcement along the crack B-C has been reported by [6]. An analytical formula similar to that one presented by [28] to calculate the energy dissipated along the log-spiral part C-D is employed. Note that here the angle made by the velocity vector of the ground mass slipping away and the crack, φ, is different from the soil friction angle, ϕ, (see figure 2). Now using equation (14), but with vertical crack (i.e. $\eta = \frac{\pi}{2}$) the following expression can be obtained:

$$\dot{D}r_{(\text{B-C})} = \int_{\text{B-C}} K \dot{u}_c \sin\phi\, \mathrm{d}\delta \tag{20}$$

$$\text{with } \dot{u}_c = r_c \dot{\theta} = \left(\frac{r_\zeta \cos\zeta}{\cos\theta}\right)\dot{\theta} \tag{21}$$

$$\mathrm{d}\delta = \frac{r_c\,\mathrm{d}\theta}{\cos\theta} \tag{22}$$

with $\dot{u}_c$ the velocity vector along the crack B-C, r_c the distance between point P and any point along the crack B-C, then,

$$\dot{D}r_{(\text{B-C})} = \dot{\theta}\int_{\mu}^{\zeta} K r_c \sin\theta \frac{r_c\, \mathrm{d}\theta}{\cos\theta} \tag{23}$$

with μ is the angle made by the line P-B and a horizontal reference.

$$\dot{D}r_{(\text{B-C})} = r_\chi^2 \dot{\theta} \exp[2\tan\phi(\zeta-\chi)]\cos^2\zeta \int_{\mu}^{\zeta} K \frac{\sin\theta}{\cos^3\theta}\mathrm{d}\theta \tag{24}$$

for uniform distribution of reinforcement (UD), $K = K_t$, then, integration leads to

$$\dot{D}r_{(\text{B-C})} = K_t r_\chi^2 \dot{\theta}\, g_3(\chi,\upsilon,\zeta,\phi)$$

$$g_3(\chi,\upsilon,\zeta,\phi) = \frac{1}{2}\left(\exp[2\tan\phi(\zeta-\chi)]\sin^2\zeta - \sin^2\chi\right) \tag{25}$$

The rate of external work for the sliding wedge E-B-C-D, (i.e $\dot{W}$), is calculated as the work of block E-D-F minus the work of block B-C-F [29]. The work of block E-D-F will be calculated by algebraic summation of the work of blocks P-D-F, P-E-F and P-D-E called here $\dot{W}_1$, $\dot{W}_2$ and $\dot{W}_3$ respectively. The work of block B-C-F will be calculated by algebraic summation of the work of blocks P-C-F, P-B-F and P-C-B called here $\dot{W}_4$, $\dot{W}_5$ and $\dot{W}_6$ respectively. So, $\dot{W}$ can be calculated from the following summation:

$$\dot{W} = \dot{W}_1 - \dot{W}_2 - \dot{W}_3 - \left(\dot{W}_4 - \dot{W}_5 - \dot{W}_6\right) = \dot{W}_1 - \dot{W}_2 - \dot{W}_3 - \dot{W}_4 + \dot{W}_5 + \dot{W}_6 \tag{26}$$

The expressions for $\dot{W}_i$ are derived for each block by calculation of the vectorial product of the displacement rate, u, of the block (see figure 2) times its weight force. Here instead, in addition to the weight force, a horizontal pseudo-static force, $F_h = mK_h g = \gamma K_h A$, with g being the gravitational acceleration and m the mass of the wedge, and a vertical one, $F_v = mK_v g = \gamma K_v A$, are added to account for seismic action. For the sake of space, only the final expressions are reported here:

$$\dot{W}_1 = \dot{\theta}\gamma r_\chi^{\,3}\left[(1+k_\text{v})f_{1\text{v}}(\chi,\upsilon,\phi) + k_\text{h} f_{1\text{h}}(\chi,\upsilon,\phi)\right] \tag{27}$$

$$= \dot{\theta}\gamma r_\chi^{\,3}\left[\begin{array}{l}(1+K_\text{v})\dfrac{\exp\left[3\tan\phi(\upsilon-\chi)\right](3\tan\phi\cos\upsilon+\sin\upsilon) - 3\tan\phi\cos\chi - \sin\chi}{3(1+9\tan\phi^2)} + \\ K_\text{h}\dfrac{\exp\left[3\tan\phi(\upsilon-\chi)\right](3\tan\phi\sin\upsilon-\cos\upsilon) - 3\tan\phi\sin\chi + \cos\chi}{3(1+9\tan\phi^2)}\end{array}\right]$$

with $\dot{\theta}$ being the rate of angular displacement of the failing wedge E-B-C-D. For block P-E-F instead:

$$\dot{W}_2 = \dot{\theta}\gamma r_\chi^{\,3}\left[(1+K_\text{v})f_{2\text{v}}(\chi,\upsilon,\phi) + K_\text{h} f_{2\text{h}}(\chi,\upsilon,\phi)\right] \tag{28}$$

$$=\dot{\theta}\gamma r_\chi^{\ 3}\left[(1+K_\mathrm{v})\frac{L_1}{6r_\chi}sin\chi\left(2\cos\chi-\frac{L_1}{r_\chi}\right)+K_\mathrm{h}\frac{L_1}{3r_\chi}\sin^2\chi\right]$$

for block P-D-E:

$$\dot{W}_3=\dot{\theta}\gamma r_\chi^{\ 3}\left[(1+K_\mathrm{v})f_{3\mathrm{v}}(\chi,\upsilon,\phi)+K_\mathrm{h}f_{3\mathrm{h}}(\chi,\upsilon,\phi)\right] \tag{29}$$

$$=\dot{\theta}\gamma r_\chi^{\ 3}\begin{bmatrix}\frac{(1+K_\mathrm{v})}{6}\exp[\tan\phi(\upsilon-\chi)]\left(\sin(\upsilon-\chi)-\frac{L_1}{r_\chi}sin\upsilon\right)\left(cos\chi-\frac{L_1}{r_\chi}+\exp[\tan\phi(\upsilon-\chi)]cos\upsilon\right)+\\ \frac{K_\mathrm{h}}{6}\exp[\tan\phi(\upsilon-\chi)]\left(\sin(\upsilon-\chi)-\frac{L_1}{r_\chi}sin\upsilon\right)(\sin\chi+\exp[\tan\phi(\upsilon-\chi)]\sin\upsilon)\end{bmatrix}$$

for block P-C-F:

$$\dot{W}_4=\dot{\theta}\gamma r_\chi^{\ 3}\left[(1+K_\mathrm{v})f_{4\mathrm{v}}(\chi,\upsilon,\phi)+K_\mathrm{h}f_{4\mathrm{h}}(\chi,\upsilon,\phi)\right] \tag{30}$$

$$=\dot{\theta}\gamma r_\chi^{\ 3}\begin{bmatrix}(1+K_\mathrm{v})\frac{\exp[3\tan\phi(\zeta-\chi)](3tan\phi cos\zeta+sin\zeta)-3tan\phi cos\chi-sin\chi}{3(1+9tan^2\phi)}+\\ K_\mathrm{h}\frac{\exp[3\tan\phi(\zeta-\chi)](3tan\phi\sin\zeta-\cos\zeta)-3tan\phi\sin\chi+\cos\chi}{3(1+9tan^2\phi)}\end{bmatrix}$$

for block P-B-F:

$$\dot{W}_5=\dot{\theta}\gamma r_\chi^{\ 3}\left[(1+K_\mathrm{v})f_{5\mathrm{v}}(\chi,\upsilon,\phi)+K_\mathrm{h}f_{5\mathrm{h}}(\chi,\upsilon,\phi)\right] \tag{31}$$

$$=\dot{\theta}\gamma r_\chi^{\ 3}\left[(1+K_\mathrm{v})\frac{L_2}{6r_\chi}sin\chi\left(2cos\chi-\frac{L_2}{r_\chi}\right)+K_\mathrm{h}\frac{L_2}{3r_\chi}\sin^2\chi\right]$$

for block P-C-B:

$$\dot{W}_6=\dot{\theta}\gamma r_\chi^{\ 3}\left[(1+K_\mathrm{v})f_{6\mathrm{v}}(\chi,\upsilon,\phi)+K_\mathrm{h}f_{6\mathrm{h}}(\chi,\upsilon,\phi)\right] \tag{32}$$

$$=\dot{\theta}\gamma r_\chi^{\ 3}\begin{bmatrix}\frac{(1+K_\mathrm{v})}{3}\exp[2\tan\phi(\zeta-\chi)]\cos^2\zeta\left(\exp[\tan\phi(\zeta-\chi)]sin\zeta-\sin\chi\right)+\\ \frac{K_\mathrm{h}}{6}\exp[\tan\phi(\zeta-\chi)]\cos\zeta\left(\exp[2\tan\phi(\zeta-\chi)]\sin^2\zeta-\sin^2\chi\right)\end{bmatrix}$$

substituting equation (6) and equation (26) with their components into equation (5), the following expression is obtained:

$$c\dot{\theta}r_\chi^2(f_{B-C}+f_{C-D})+K_t r_\chi^2\dot{\theta}(g_2+g_3)=\dot{\theta}\gamma r_\chi^3\begin{bmatrix}(1+K_\mathrm{v})(f_{1\mathrm{v}}-f_{2\mathrm{v}}-f_{3\mathrm{v}}-f_{4\mathrm{v}}+f_{5\mathrm{v}}+f_{6\mathrm{v}})\\ +K_\mathrm{h}(f_{1\mathrm{h}}-f_{2\mathrm{h}}-f_{3\mathrm{h}}-f_{4\mathrm{h}}+f_{5\mathrm{h}}+f_{6\mathrm{h}})\end{bmatrix} \tag{33}$$

Now, let us introduce the ratio of vertical to horizontal acceleration, λ as $\lambda=K_v/K_h$. Consistently with figure 2, the + sign indicates vertical downward acceleration, while the – sign indicates vertical upward acceleration. An upper bound on the coefficient of yield acceleration K_y is obtained by solving equation (33) with respect to K_h :

$$K_y = \frac{\left[\frac{c}{\gamma H}\left(f_{B-C} + f_{C-D}\right) + \frac{K_t}{\gamma H}\left(g_2 + g_3\right)\right] - \frac{r_\chi}{H}\left(f_{1v} - f_{2v} - f_{3v} - f_{4v} + f_{5v} + f_{6v}\right)}{\frac{r_\chi}{H}\left[\begin{array}{l}\lambda\left(f_{1v} - f_{2v} - f_{3v} - f_{4v} + f_{5v} + f_{6v}\right)\\ +\left(f_{1h} - f_{2h} - f_{3h} - f_{4h} + f_{5h} + f_{6h}\right)\end{array}\right]} \tag{34}$$

$$K_y = f_y\left(\chi, \upsilon, \zeta, \phi, \frac{c}{\gamma H}, \frac{K_t}{\gamma H}, \beta, \lambda, \alpha\right) \tag{35}$$

The global minimum of $f_y\left(\chi, \upsilon, \zeta, \phi, c/\gamma H, \beta, \lambda, \alpha\right)$ over the three geometrical variables χ, υ, ζ provides the least upper bound on the coefficient of yield acceleration, assuming the most unfavourable crack to be present. Results obtained using equation (34) are presented in figure 3, providing the two terms needed to find the proposed stepwise yield acceleration for soil slopes with either zero tensile strength (i.e. $\alpha = 0$) or half the full tensile strength of the Mohr-Coulomb criterion (i.e. $\alpha = 0.5$). The solid lines in figure 3 refer to slopes that are initially intact but they can exhibit cracks forming as part of the incipient failure mechanism, while the dashed lines are for slopes with earthquake induced cracks (i.e. the formed crack is treated here as an open crack). The stepwise yield acceleration, proposed in this paper, can be found using these two lines (i.e. solid and dashed), where for a given soil slope properties, two values of yield acceleration are obtained. The one obtained from the solid line represents the starting value of the yield acceleration which steps down to the value obtained from the dashed line as soon as it is exceeded for the first time by the applied acceleration, given by the earthquake record. It can be observed that the two lines (solid and dashed) in figure 3 are close to each other. This is because the depth of the formed crack within reinforced soil is relatively shallow, especially for gentle slopes. The definition of the stepwise yield acceleration is detailed by an illustrative example later in this paper. It also should be noted that failure passes below the slope toe were not permitted during the calculations for these charts. This type of failure might occur for gentle slope with a low angle of internal friction [30].

4. Seismic Displacement

Based on Newmark's method, and following [19, 30], the maximum horizontal displacement of the slope face that occurs at the slope toe, denoted here as δu_x (see figure 4), is calculated as follow:

$$\delta u_x = r_\upsilon \sin\upsilon\,\delta\theta = r_\upsilon \sin\upsilon \int_{\delta t}\int_{\delta t} \ddot{\theta}\, \mathrm{d}t\, \mathrm{d}t = C \int_{\delta t}\int_{\delta t} \left(K_i - K_y\right) g\, \mathrm{d}t\, \mathrm{d}t \tag{36}$$

where $\delta\theta$ is the angular displacement, $\ddot{\theta}$ is the angular acceleration, K_i is the applied horizontal acceleration at step *i*, and *C* is a dimensionless coefficient relates the displacement of the slope toe to the integral of the earthquake acceleration record above the level of yield acceleration. This coefficient depends on the slope geometrical features and the ground strength parameters and can be expressed as:

$$C = \frac{\gamma r_\chi^4 \exp\left[\tan\phi\left(\upsilon - \chi\right)\right] \sin\upsilon \left[\begin{array}{l}\lambda\left(f_{1v} - f_{2v} - f_{3v} - f_{4v} + f_{5v} + f_{6v}\right) + \\ \left(f_{1h} - f_{2h} - f_{3h} - f_{4h} + f_{5h} + f_{6h}\right)\end{array}\right]}{G l^2} \tag{37}$$

with *G* being the weight of the potential sliding mass and *l* is the distance from point P to the centre of gravity of that mass. The calculations for *G* and *l* are listed in Appendix A. To this end, the seismic induced displacements can be calculated using equation (36) by assigning an earthquake record and calculating the yield acceleration for the slope of interest.

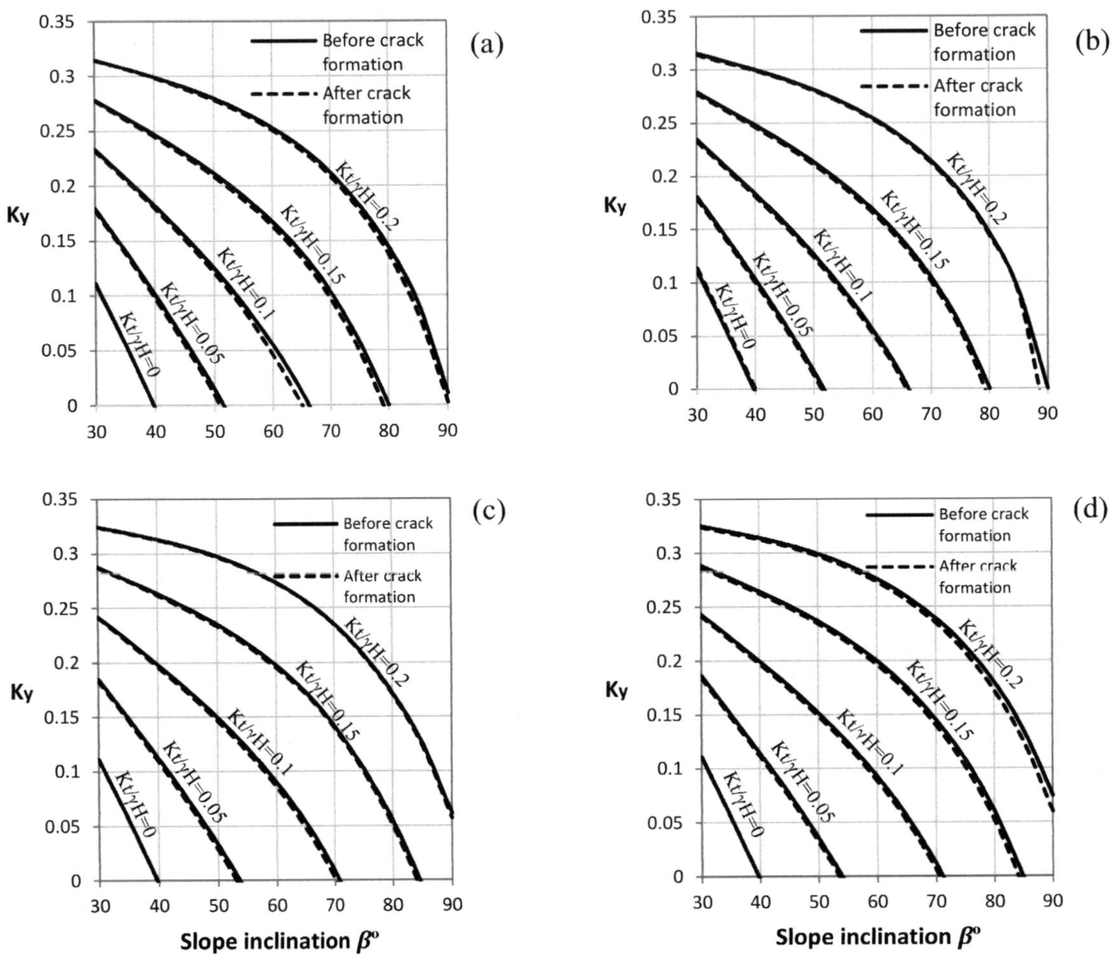

Figure 3. Yield horizontal acceleration for uniform distribution of reinforcement with $\phi = 20$, $c/\gamma H = 0.05$ and $\lambda = 0$. (a and b) for uniform distribution of reinforcement and (c and d) for linearly increasing distribution. Left hand side charts are for soil slopes with zero tensile strength while the right hand side are for soil slopes with limited tensile strength (i.e. half of Mohr-Coulomb's tensile strength).

5. Illustrative Example

The two types of reinforcement distributions, used in this paper, are compared here along with this illustrative example. A soil slope with $\beta = 75°$, $\phi = 20°$, $c/\gamma H = 0.1$, $\lambda = 0$, and $K_t/\gamma H = 0.1$,is considered and assumed to be subjected to the Northridge earthquake (1994), whose main characterises are listed in table 1. Four cases are analysed: case (a) soil slope with full tensile strength, (i.e. $\alpha = 0$), therefore not subject to tension cracks, case (b) soil slope of limited tensile strength, in this case $\alpha = 0.5$, case (c) soil slope of zero tensile strength, (i.e. $\alpha = 0$), and case (d) soil slope subjected to the most adverse pre-existing crack.

Now, according to the procedure mentioned earlier in this paper, the stepwise yield acceleration for soil with limited or zero tensile strength is illustrated in figure 5(a). It can be noticed that the yield acceleration for a soil slope with limited or zero tensile strength is reduced significantly when it is exceeded for the first time by the applied acceleration, because once the crack is formed as part of the failure at that instance, it is then treated as an earthquake-induced (pre-existing) crack. Consequently, this increases the estimated displacement as shown in figure 5(b). It can be seen that the displacement corresponding to a slope with the most detrimental pre-existing crack seems over conservative, at the same time, assuming an intact slope that remains intact during the earthquake may underestimate the slope displacement. Hence, assuming a limited tensile strength for the soil slope seems to reasonably bridge the gap between the conservatism, corresponding to a slope with the most detrimental pre-existing crack, and the underestimation of the displacement when ignoring the crack formation (i.e. intact slope). Figure 5(c) provides an insight as to the way the limited tensile strength can change the crack properties and the orientation of the failure mechanism.

Comparing the left hand side of figure 4 which is related to a slope with UD of reinforcement, with the right hand side of the same figure related to the same slope and with the same amount and strength of reinforcement but with the LID of reinforcement. It is noticed that slope with LID has better performance than the same slope with UD. The yield acceleration seems less affected by the presence of crack, whether earthquake-induced or pre-existing crack. Consequently, the total horizontal displacement of the slope reinforced with LID is reduced.

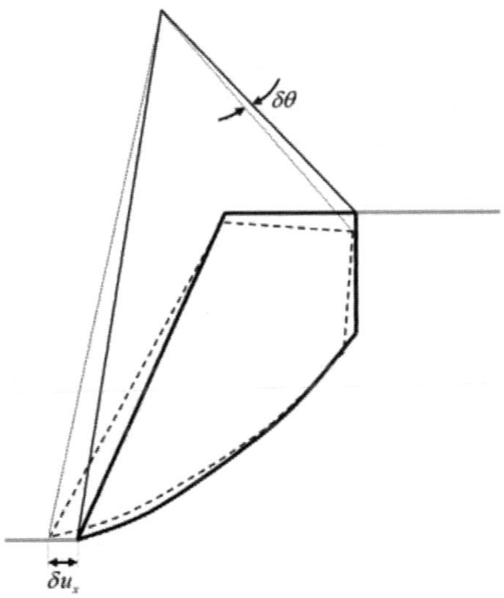

Figure 4. Illustration of the horizontal displacement at the slope toe δu_x and the angular displacement $\delta\theta$.

Table 1. Main characteristics of the earthquakes considered in the example case.

Date	17/1/1994
Station	24283 Moorpark - Fire Sta.
Magnitude	6.7
Direction	180°
Peak acceleration (g)	0.292
Epicentre distance (km)	23

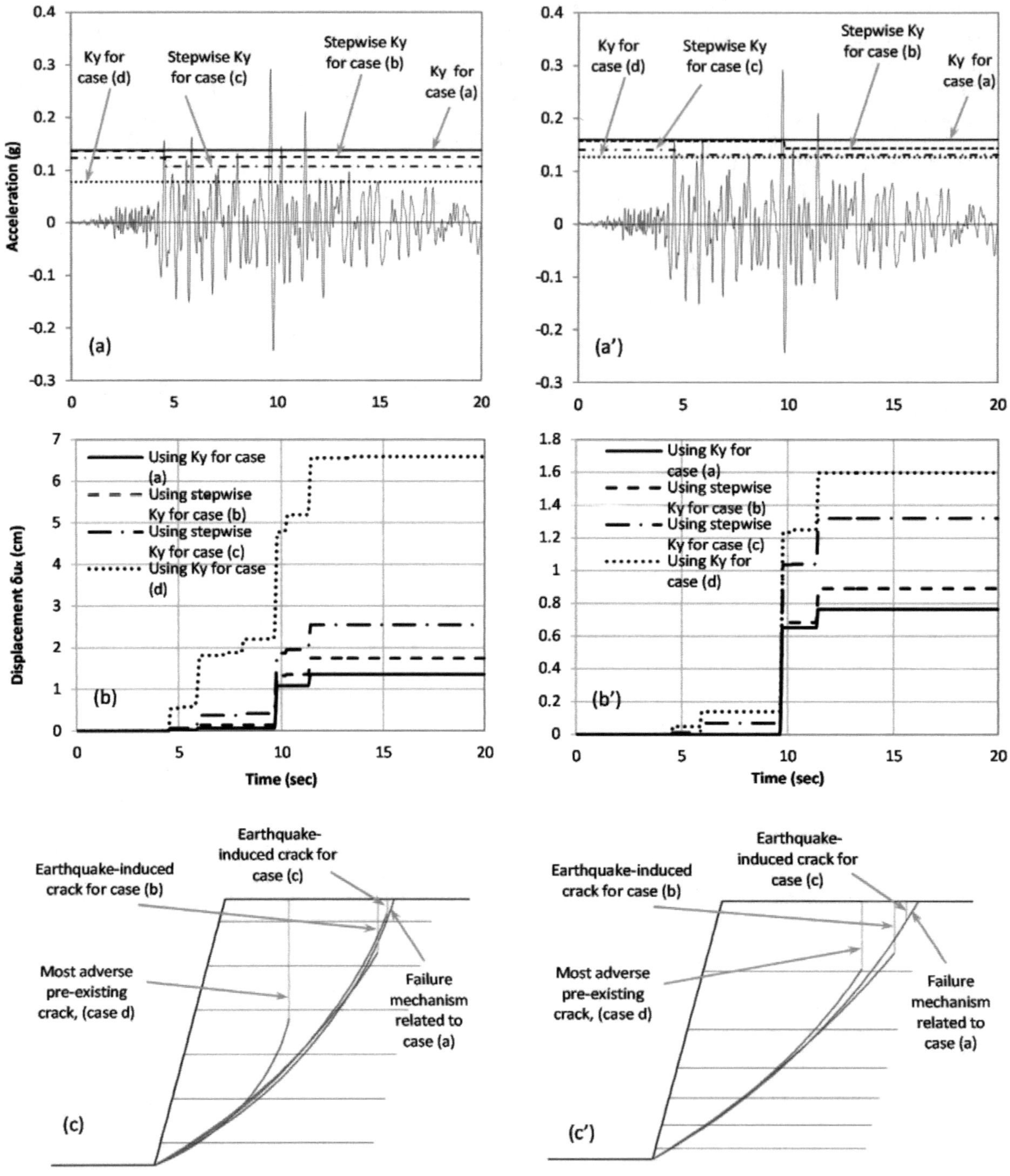

Figure 5. Yield accelerations, horizontal displacement (δux) and failure mechanisms for a slope with β=75, φ=20, c/γH=0.1, Kt/γH=0.1. Left hand side is for uniform distribution of reinforcement while the right hand side is for linearly increasing distribution of reinforcement. (a and a') Yield accelerations corresponding to the four cases explained earlier employing Northridge earthquake (1994). (b and b') Comparison of the accumulated horizontal displacement of the slope toe. (c and c') Failure mechanisms related to the yield accelerations illustrated in (a and a').

6. Conclusions

The upper bound theorem of limit analysis associated with pseudo static approach is employed to evaluate the displacement of soil slopes subject to cracks. A new technique is proposed to estimate the horizontal displacement of geosynthetic reinforced soil slopes at the slope toe, due to a given earthquake postulating a rough estimation of real time crack propagation. It should be pointed out that, in reality, during a moderate or strong earthquake, the soil slopes whether reinforced or not, are likely to show cracks even when assuming that the soil has full tensile strength. So, it is reasonable to assume that the soil has limited tensile strength when designing soil slopes subject to seismic actions.

The procedure for the new technique is described herein with an illustrative example, according to which, soil slopes with LID have better performance than the same slope with UD. Where using LID, the yield acceleration seems less affected by the presence of crack, whether earthquake-induced or pre-existing crack. Consequently, the total horizontal displacement of the slope reinforced with LID is less than the UD's one.

Acknowledgement

The results presented in this paper are based on research supported mainly by HCED (Higher Committee of Education Development in Iraq) and partially by Tikrit University. This support is greatly appreciated.

Appendix A

The calculations of the weight of the sliding mass B-C-E-D, called *G* in the manuscript is detailed below.

$$G=\gamma A \tag{A.1}$$

$$\text{with } A = A_1 - A_2 - A_3 - A_4 + A_5 + A_6 \tag{A.2}$$

$$A_1 = \frac{1}{2} r_\chi^2 \left[\frac{\exp\left[2\tan\phi(\upsilon-\chi)\right]-1}{2\tan\phi} \right] \tag{A.3}$$

$$A_2 = \frac{1}{2} r_\chi L_1 \sin\chi \tag{A.4}$$

$$A_3 = \frac{1}{2} r_\chi H \frac{\exp\left[\tan\phi(\upsilon-\chi)\right]\sin(\beta+\upsilon)}{\sin\beta} \tag{A.5}$$

$$A_4 = \frac{1}{2} r_\chi^2 \left[\frac{\exp\left[2\tan\phi(\zeta-\chi)\right]-1}{2\tan\phi} \right] \tag{A.6}$$

$$A_5 = \frac{1}{2} r_\chi L_2 \sin\chi \tag{A.7}$$

$$A_6 = \frac{1}{2}\delta r_\zeta \cos\zeta = \frac{1}{2} r_\chi \delta\left[\exp\left[\tan\phi(\zeta-\chi)\right]\cos\zeta\right] \tag{A.8}$$

The arm length of the weight, *l*, is given by:

$$l = \frac{\sqrt{\left(\gamma r_\chi^3 (f_1 - f_2 - f_3 - f_4 + f_5 + f_6)\right)^2 + \left(\gamma r_\chi^3 (f_{h1} - f_{h2} - f_{h3} - f_{h4} + f_{h5} + f_{h6})\right)^2}}{G} \tag{A.9}$$

References

[1] Baker R 1981 Tensile strength, tension cracks, and stability of slopes *Soils and Foundation* **21** 1-17

[2] Utili S 2013 Investigation by limit analysis on the stability of slopes with cracks *Geotechnique* **63** 140-154

[3] Utili S and Abd A H 2015 On the stability of fissured slopes subject to seismic action *Int. J. Num. Analy. Meth. Geo. -under review-*

[4] Michalowski R L 1998 Limit analysis in stability calculations of reinforced soil structures *Geotextiles and Geomembranes* **16** 311-331

[5] Ausilio E, Conte E, and Dente G 2000 Seismic stability analysis of reinforced slopes *Soil Dynamics and Earthquake Engineering* **19** 159-172

[6] Abd A H and Utili S 2015 Accounting for the presence of cracks in the design of geosynthetic-reinforced slopes *Geotext. Geomemb. -under reveiw-*

[7] Okabe S 1924 General theory of earth pressure and seismic stability of retaining wall and dam *J. Japan Soc. Civ Engrs, Tokyo, Japan* **12**

[8] Mononobe N a M, H. 1929 On the determination of earth pressures during earthquakes *Proc. world Engrg, Congr. Tokyo, Japan* **9**

[9] Janbu N 1954 Application of composite slip surfaces for stability analysis *Proc. European Conf. on Stability of Earth Slopes, Stockholm, 1954* 3 43-49

[10] Little A and Price V 1958 The use of an electronic computer for slope stability analysis *Geotechnique* **8** 113-116

[11] Morgenstern N and Price V E 1965 The analysis of the stability of general slip surfaces *Geotechnique* **15** 79-93

[12] Dawson E, Roth W, and Drescher A 1999 Slope stability analysis by strength reduction *Geotechnique* **49** 835-840

[13] Conte E, Silvestri F, and Troncone A 2010 Stability analysis of slopes in soils with strain-softening behaviour *Computers and Geotechnics* **37** 710-722

[14] Li A-J, Merifield R S, and Lyamin A V 2010 Three-dimensional stability charts for slopes based on limit analysis methods *Can. Geotech. J.* **47** 1316-1334

[15] Boon C, Houlsby G, and Utili S 2014 New insights into the 1963 Vajont slide using 2D and 3D distinct-element method analyses *Geotechnique* **64** 800-816

[16] Boon C, Houlsby G, and Utili S 2012 A new algorithm for contact detection between convex polygonal and polyhedral particles in the discrete element method *Computers and Geotechnics* **44** 73-82

[17] Boon C, Houlsby G, and Utili S 2013 A new contact detection algorithm for three-dimensional non-spherical particles *Powder Technology* **248** 94-102

[18] Newmark N M 1965 Effects of earthquakes on dams and embankments *Geotechnique* **15** 139-159

[19] Chang C-J, Chen W F, and Yao J T 1984 Seismic displacements in slopes by limit analysis *J. Geotech. Engrng.* **110** 860-874

[20] Jibson R W 2011 Methods for assessing the stability of slopes during earthquakes—A retrospective *Engineering Geology* **122** 43-50

[21] Michalowski R L 2013 Stability assessment of slopes with cracks using limit analysis *Can. Geotech. J.* **50** 1011-1021

[22] Utili S 2005 An analytical relationship for weathering induced slope retrogression: a benchmark *RIG (Italian Geotechnical Journal)* **39** 9-30

[23] Utili S and Nova R 2008 DEM analysis of bonded granular geomaterials *Int. J. Num. Analy. Meth. Geo.* **32** 1997-2031

[24] Utili S and Crosta G 2011 Modelling the evolution of natural slopes subject to weathering: Part I. Limit analysis approach *Journal of Geophysical Research - Earth Surface* **116**

[25] Utili S and Crosta G 2011 Modelling the evolution of natural slopes subject to weathering: Part II. Discrete element approach *Journal of Geophysical Research - Earth Surface* **116**

[26] Jiang M, Zhu F, Liu F, and Utili S 2014 A bond contact model for methane hydrate- bearing sediments with interparticle cementation *Int. J. Num. Analy. Meth. Geo.* **38** 1823-1854

[27] Utili S and Nova R 2007 On the optimal profile of a slope *Soils and Foundations* **47** 717-729

[28] Michalowski R L 1997 Stability of uniformly reinforced slopes *Journal of Geotechnical and geoenvironmental Engineering* **123** 546-556

[29] Chen W F 1975 Limit analysis and soil plasticity *Elsevier, Amsterdam, Netherlands*

[30] You L and Michalowski R L 1999 Displacement charts for slopes subjected to seismic loads *Computers and Geotechnics* **25** 45-55

Effect of hysteresis on the stability of an embankment under transient seepage

Kang Liu, Philip J. Vardon, Patrick Arnold and Michael A. Hicks

Section of Geo-Engineering, Department of Geoscience & Engineering, Delft University of Technology, Delft, The Netherlands

E-mail: k.liu@tudelft.nl

Abstract. Hysteresis is a well-known phenomenon that exists in the soil water retention behaviour of unsaturated soils. However, there is little research on the effects of hysteresis on slope stability. If included in slope stability analyses, commonly the suction in the unsaturated zone is taken as non-hysteretic. In this paper, the authors investigate the effect of hysteresis on the stability of an embankment under transient seepage. A scenario of water level fluctuation has been assessed, in which a cyclic external water level fluctuates between a low and high level. It was found that the factor of safety (FOS), the volumetric water content and the suction in the unsaturated zone are significantly affected by hysteresis. It was also found that, when the period of water level fluctuation in one cycle is relatively small, there is little difference in the FOS between the hysteretic case and non-hysteretic case. However, when the period exceeds a certain threshold value, significant differences between these two cases can be observed. Compared to the case in which hysteresis is considered, the FOS is higher in the case which does not consider hysteresis. This suggests that the non-hysteretic case may overestimate slope stability, leading to a potentially dangerous situation. Moreover, the period under which there emerge large differences between the hysteretic and non-hysteretic case is strongly related to the magnitude of hydraulic conductivity and the period of the cyclic water level fluctuation.

1. Introduction

Wetting-drying hysteresis in the soil water retention curve (SWRC) of unsaturated soils has been recognised for a long time. The relationships between water content and suction for the wetting path and the drying path are not identical. The reason for hysteresis is, in the main, attributed to the "ink-bottle" effect, i.e. entrapped air in the voids and the contact angle of the meniscus [1, 2]. Normally, the hysteresis in water retention behaviour is ignored in seepage analysis [3, 4]. Yang et al. [5, 6] investigated the influence of hysteresis on the seepage in a soil column by comparing the numerical results to three infiltration tests. The authors showed that the difference in the variation of suction profiles using hysteretic and non-hysteretic formulations was significant. They found that, when hysteresis was considered, the computed suction and volumetric water content in the soil column were closer to the experimental results. Tsai and Chen [7] and Ma et al. [8] studied the influence of hysteresis on the stability of infinite slopes subjected to rainfall and pointed out that hysteresis behaviour affects the distribution of water content and suction, and thus the slope stability.

This paper investigates the effect of wetting-drying hysteresis on the stability of an embankment slope under transient seepage conditions. The authors first introduce the formulations and numerical implementation for analysing transient seepage in an embankment considering the hysteresis in water retention behaviour, and the computation of factor of safety (FOS). Then, the authors presents a comparison between hysteretic and non-hysteretic cases in terms of matric suction profiles and the variation of FOS. They also present a brief analysis of the parameters which can affect the emergence of the differences between non-hysteretic and hysteretic cases.

This paper follows on from recent work undertaken in this research group. Arnold and Hicks [9, 10] investigated the effect of rainfall on the stability of heterogeneous soil slopes under steady state and transient conditions, respectively. Li et al. [11], Li and Hicks [12] and Hicks et al. [13] focused on slope stability computations in 3D, investigating the impact of material property variability, and trying to improve the realism of slope stability computation in comparison to 2D. Wang et al. [14] extended the research to the propagation of slope failure using the material point method and de Gast et al. [15] has investigated how analyses are used to determine the FOS in the Netherlands.

2. Formulation of transient seepage considering hysteresis and slope stability

2.1. Governing flow equation

The governing equation of 2D transient flow is based on mass conservation. In this paper, the soil skeleton is considered rigid for the flow analaysis, which implies that any volume change potentially affecting the flow is not accounted for. Therefore, the hydraulic governing equation is given as follows,

$$\frac{\partial}{\partial x}\left(k_x\frac{\partial h}{\partial x}\right)+\frac{\partial}{\partial y}\left(k_y\frac{\partial h}{\partial y}+k_y\right)=C(h)\frac{\partial h}{\partial t} \tag{1}$$

where h is the pore water pressure head, and k_x and k_y are the hydraulic conductivities in the x and y directions, respectively, $C(h)=\frac{\partial\theta}{\partial h}$ is the specific moisture capacity function, θ is the volumetric water content and t is time.

2.2. Water retention model

The soil water retention curve is a function relating the suction head, h, with the volumetric water content, θ. In this paper, the suction head is,

$$h=-s/\gamma_w=-(u_a-u_w)/\gamma_w \tag{2}$$

where s is the matric suction, u_w is the pore water pressure, u_a is the pore air pressure, which in this paper is assumed to be atmospheric, and γ_w is the unit weight of water.

The van Genuchten model [16], in combination with the pore size distribution relationship proposed by Mualem [17], is used:

$$\begin{aligned} S &= \tfrac{\theta-\theta_r}{\theta_s-\theta_r} = \tfrac{1}{[1+(\alpha|h|)^n]^{1-\frac{1}{n}}} && h<0 \\ S &= 1 && h\geq 0 \end{aligned} \tag{3}$$

where S is effective degree of saturation, α is the inverse of the air-entry suction head and θ_s and θ_r are the saturated and residual volumetric water contents, respectively (see Figure 1). Due to the hysteretic behaviour of the water retention curve, the main drying and wetting curves have different values of α, i.e. $\alpha_w \neq \alpha_d$. The model parameter n defines the slope of the water retention curve and is assumed to be identical for the main wetting and drying curves.

The function of hydraulic conductivity is derived from the SWRC, which is proposed by van Genuchten [16]:

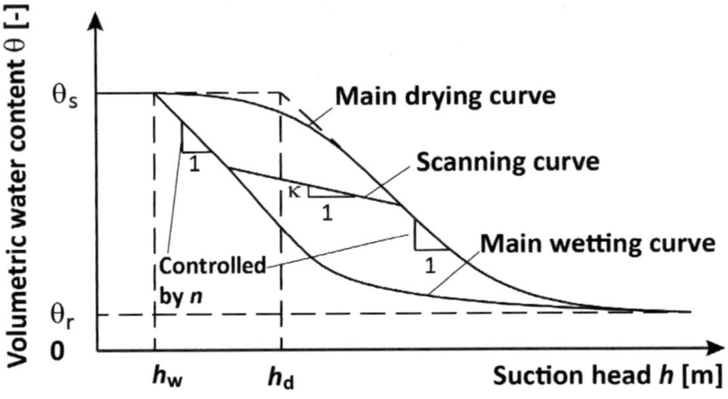

Figure 1. Soil water retention curve.

$$k = k_{sat}\sqrt{S}\left[1 - \left(1 - S^{\left(\frac{n}{n-1}\right)}\right)\right]^2 \tag{4}$$

in which k_{sat} is the saturated hydraulic conductivity of the soil.

In this paper, the authors use linear scanning curves to model the transition from drying to wetting and vice versa (Figure 1) [5, 6]. The slope of the scanning curve, κ, is constant and defined as:

$$-\kappa = \frac{d\theta}{dh} \tag{5}$$

2.3. Slope stability assessment

In this work, Bishop's effective stress combined with the extended Mohr-Coulomb failure criterion has been used to calculate the shear strength [9, 10]:

$$\tau = c' + \sigma \tan\phi' + \chi s \tan\phi' \tag{6}$$

where τ is the shear strength, c' and ϕ' are effective cohesion and friction angle, χ is a scalar parameter defining the suction, s, induced effective stress and σ is the normal stress to the shear strength plane due to gravitational loading. In this paper, the suction stress, χs, is defined as,

$$\begin{array}{ll} \chi s = \frac{\theta-\theta_r}{\theta_s-\theta_r} s & s > 0 \\ \chi s = 0 & s \leq 0 \end{array} \tag{7}$$

The unit weight, γ, of the unsaturated soil is influenced by the volumetric water content, θ, [7] and can be expressed as,

$$\gamma = [(1 - \theta_s)\rho_w G_s + \rho_w \theta]g \tag{8}$$

where ρ_w is the density of water, G_s is the specific gravity of the soil solid and g is the gravitational acceleration.

3. Numerical implementation

In the flow analysis, 4-node quadrilateral elements are used in the finite element method (FEM) program based on Smith and Griffiths [19]. The governing flow equation has been solved by use of the modified Picard iteration method [20, 21]. In the flow analysis code, the volumetric water content for the main drying and wetting curves is computed via Equation 3. The water retention behaviour along the linear scanning curve is computed via an algorithm which determines the current volumetric water content as a function of the variation in the suction head, Δh, and the previous volumetric water content $\theta^{t+1} = f(\Delta h, \theta^t)$ [5].

In the FEM slope stability code, 8-node quadrilateral elements are used. After the flow analysis, the suction stress and volumetric water content are imported into the stability analysis and mapped to the Gaussian points of the 8-node elements. If the pore water pressure at the Gaussian point is negative, the unit weight is calculated by Equation 8; otherwise, the unit weight is the saturated unit weight. The initial stresses are based on the suction stress and gravitational loading. The strength reduction method [22] is applied here to compute the FOS of the slope. The FOS is defined as the factor by which the original shear strength is reduced in order to cause the slope to fail:

$$\begin{aligned} c'_f &= c'/FOS \\ \phi'_f &= \arctan\left(\frac{tan\phi'}{FOS}\right) \end{aligned} \tag{9}$$

where c'_f and ϕ'_f are the factored shear strength parameters.

4. Example set-up

In this paper, an embankment under water level fluctuation is analysed to show the influence of hysteresis on slope stability. The geometry of the embankment is shown in Figure 2. The height of the earth embankment is 12m. The width of the embankment crest and foundation are 4m and 52m, respectively. The upstream and downstream side slopes are both 1:2. The upstream water level is changing with time and the downstream water level remains at foundation level (z=0m). The upstream water level fluctuation is simulated by the superposition of two sinusoidal curves (Figure 3).

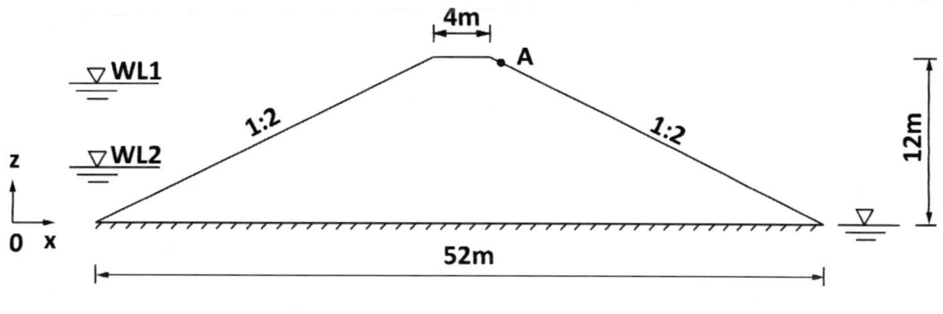

Figure 2. Geometry of the embankment under water level fluctuation.

In Figures 2 and 3, $WL1$ and $WL2$ are the highest and lowest water levels in the process of water level fluctuation. The two sinusoidal curves have different periods, i.e. T = 10d and $T_1 = 0.2T$ = 2d, but the same amplitude. The saturated hydraulic conductivity, k_{sat}, of the soil has been set to be 0.0864m/d. The saturated and residual volumetric water contents, θ_s and θ_r, are 0.38 and 0.0038, and the retention curve parameter, n, is 1.226. The inverse of the air-entry suction head of the main drying and wetting curves, α_d and α_w, are 0.1m^{-1} and 0.2m^{-1}, respectively, with $\alpha = \alpha_d$ being used for the non-hysteretic case. The slope of the scanning curves is κ = 0.00006. The high and low water levels, $WL1$ and $WL2$, are 10m and 4m, respectively.

The transient seepage starts from the result of a steady state analysis using the drying curve when the water level is 10m. The bottom boundary of the embankment is impermeable and fixed.

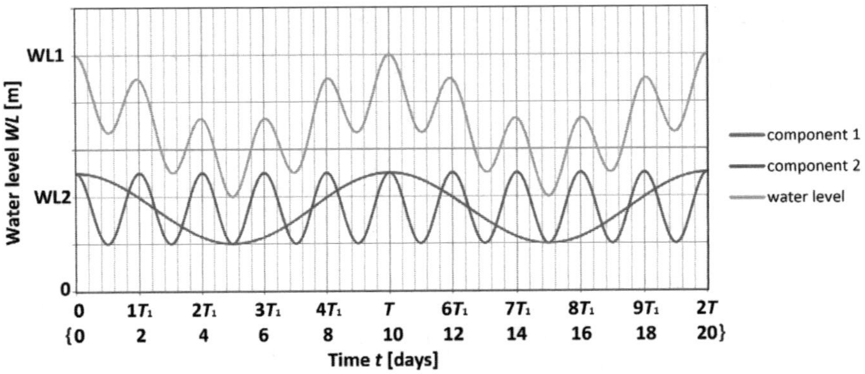

Figure 3. Two cyclic fluctuations of the simulated water level.

5. Results

5.1. Influence of hysteresis on seepage behaviour

The hysteresis in water retention behaviour implies that the variation of volumetric water content in soil becomes more complex. In Figure 4, the suction and water content have been tracked at position A shown in Figure 2. It can be seen from Figure 4 (a) that, when the soil is considered as non-hysteretic, the variation of volumetric water content are always on the main drying curve. However, if the hysteretic behaviour is taken into account (Figure 4 (b)), the variation of volumetric water content with suction can follow the scanning curve and can even reach the main wetting curve (although not in this case at this point).

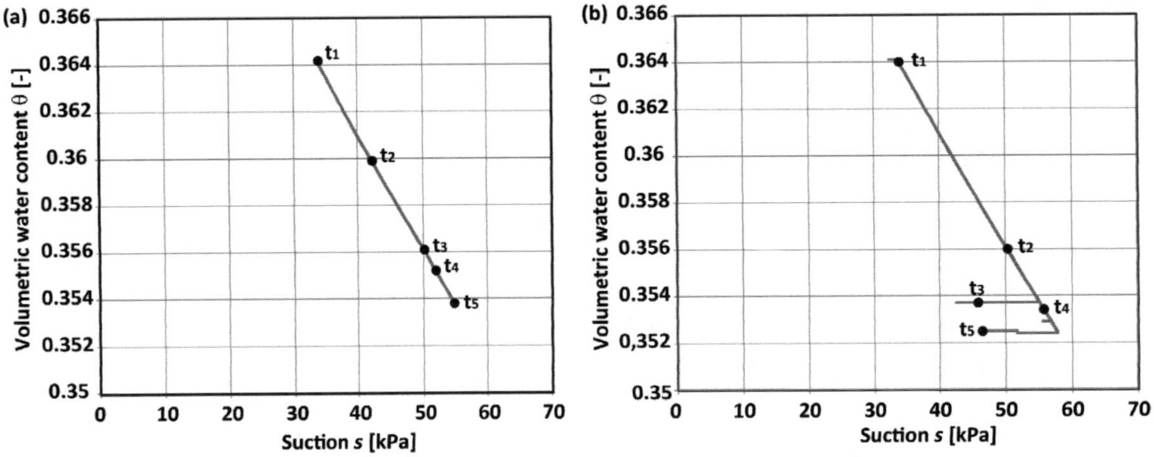

Figure 4. Volumetric water content vs. suction at point A for: (a) non-hysteretic case, and (b) hysteretic case, with t_1=0d, t_2=5d, t_3=10d, t_4=15d and t_5=20d.

Figure 5 shows the comparison of pore water pressure and volumetric water content, between non-hysteretic and hysteretic cases, at the end of the first cycle, i.e. $t = T = 10$d. The internal gradients of pore pressure and volumetric water content for the hysteretic case less steep than those for the non-hysteretic case.

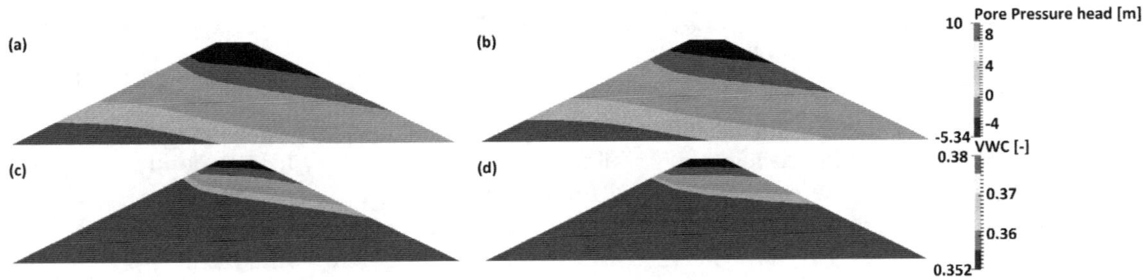

Figure 5. Pore water pressure and volumetric water content at t=10d, for non-hysteretic case ((a) and (c)) and hysteretic case ((b) and (d)).

5.2. Influence of hysteresis on slope stability

The influence of hysteresis on the pore water pressure and the volumetric water content results in an impact on the shear strength (Equation 6), and thus on the slope stability. The variation of the FOS during the transient process of water level fluctuation is shown in Figure 6.

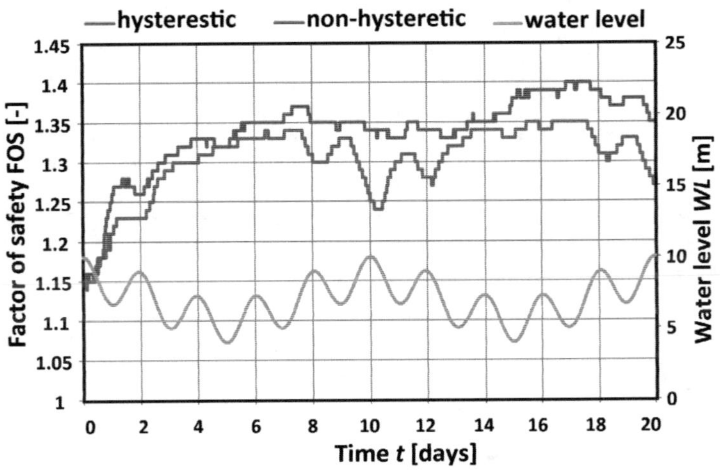

Figure 6. FOS variation with time.

In Figure 6, there is a significant difference in FOS between the non-hysteretic and hysteretic cases. The FOS results of the non-hysteretic case overestimate the slope stability compared to the hysteretic case. This means that the FOS computed without considering the hysteretic water retention behaviour is non-conservative and could lead to a higher risk in design. The higher FOS for the non-hysteretic case and the lower FOS for the hysteretic case are due to the combined effects of gravitational loading, overturning force and suction. For example, in the drying process of the first cycle, the response of the pore water pressure and volumetric water content in the non-hysteretic case is slower than that of the hysteretic case. The combined effect of those three factors results in a slightly higher FOS. In addition, during the wetting process in the hysteretic case, the variation of volumetric water content with suction follows the scanning curve; there is a rapid decrease in suction, while the increase of volumetric water content is insignificant, which means that the gravitational loading and overturning force do not change much. The rapidly reduced suction results in a significant decrease in the effective stress, which is the reason for the decrease in FOS for the hysteretic case at the end of the first cycle. Moreover, the highest FOS does not occur at the same time as the water level reaches the lowest point; it occurs at a slightly later time.

5.3. Parameters which affect the difference between non-hysteretic and hysteretic cases

In this section, the authors discuss the influence of the saturated hydraulic conductivity and the period of water level fluctuation on the final FOS. Figure 7 shows the evolution of FOS corresponding to four different periods using the same soil property values. To simplify the variation of water level, only a single sinusoidal curve is used here, which is component 1 in Figure 3. $WL1$ and $WL2$ are now 12m and 6m, respectively.

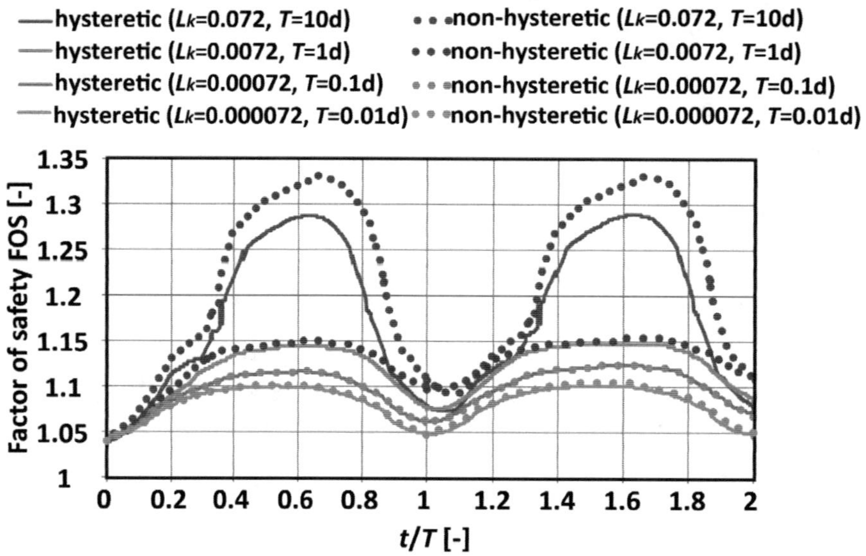

Figure 7. Differences between non-hysteretic and hysteretic cases corresponding to four different values of L_k.

The difference in the results is due to the interplay between the period of the water level fluctuation and the saturated hydraulic conductivity. The authors proposed a non-dimensional variable to explore the result of this interplay, $L_k = TK_{sat}/H$ (where H is the height of the embankment). Although the amplitude of the water level change and the width of the embankment are also important, for this specific embankment these two features are ignored in the expression of L_k. It can be seen from Figure 7 that there are only negligible differences between the non-hysteretic and hysteretic cases when L_k is relatively small. This is because, when the period is small compared to the saturated hydraulic conductivity, there is little time for the pore water pressure and volumetric water content changes to propagate and their variations are limited into a small range for both cases. Therefore, the variation of FOS for each case during the transient process is small, meanwhile the difference between these two cases is not apparent. The authors found that when the variable L_k increases, there is a greater difference between the non-hysteretic and hysteretic cases. In Figure 7, the approximate minimum period for which there is a noticeable difference is 1d (L_k =0.0072), although it is still small, but it increases further with a 10d period (L_k =0.072).

In Figure 8, the variation of the maximum difference between the non-hysteretic and hysteretic cases with the non-dimensional variable L_k is presented. It can be seen from Figure 8 that the maximum difference in FOS between the two cases in the transient seepage process increases with increasing L_k.

This non-dimensional variable could be a good indication for design purposes. As to a specific type of embankment or slope, the engineer could find a critical value of L_k and quickly decide whether it is necessary to compute the slope stability by considering hysteresis or not.

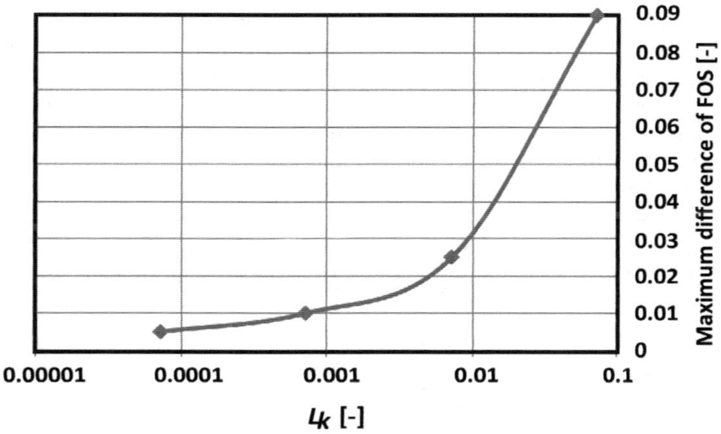

Figure 8. Maximum difference of FOS vs. L_k.

6. Conclusions

This paper investigates the effects of hysteresis in the water retention behaviour of unsaturated soil on the slope stability of an embankment under transient seepage. Due to the existence of hysteresis, the distribution of pore water pressure and volumetric water content are different from the non-hysteretic case. The differences in pore water pressure and volumetric water content result in differences in the slope stability between the non-hysteretic and hysteretic cases. In all cases, the hysteretic soil has a lower FOS.

The authors found that the differences can be related to a non-dimensional variable which shows the interplay between period, saturated hydraulic conductivity and slope dimensions. When the saturated hydraulic conductivity is large, the period for which there is a difference between non-hysteretic and hysteretic cases is small, and vice versa.

References

[1] Pham H Q, Fredlund D G, and Barbour S L 2005 A study of hysteresis models for soil-water characteristic curves *Canadian Geotechnical Journal* **42(6)** 1548-1568
[2] Laloui L 2013 *Mechanics of Unsaturated Geomaterials* (London: John Wiley & Sons)
[3] Tsaparas I, Rahardjo H, Toll D G and Leong E C 2002 Controlling parameters for rainfall-induced landslides *Computers and Geotechnics* **29(1)** 1-27
[4] Le T M H, Gallipoli D, Sanchez M, Wheeler, S J 2012 Stochastic analysis of unsaturated seepage through randomly heterogeneous earth embankments *International Journal for Numerical and Analytical Methods in Geomechanics* **36(8)** 1056-1076
[5] Yang C, Sheng D and Carter J P 2011 Hysteretic seepage analysis in unsaturated soil covers *Proceedings of the 5th Asia-Pacific Conference on Unsaturated Soils* ed Jotisankasa A (Bangkok: Kasetsart University) pp 541-546
[6] Yang C, Sheng D and Carter J P 2012 Effect of hydraulic hysteresis on seepage analysis for unsaturated soils *Computers and Geotechnics* **41** 36-56
[7] Tsai T L and Chen H F 2010 Effects of degree of saturation on shallow landslides triggered by rainfall *Environmental Earth Sciences* **59(6)** 1285-1295
[8] Ma K C, Tan Y C and Chen C H 2005 The influence of water retention curve hysteresis on the stability of unsaturated soil slopes *Hydrological Processes* **25(23)** 3563-3574

[9] Arnold P and Hicks M A 2010 Stochastic modelling of unsaturated slope stability *Proc.: Fifth International Conference on Unsaturated Soils (UNSAT)* ed Alonso E and Gens A. (Barcelona: Taylor & Francis) pp 1237-1242
[10] Arnold P and Hicks M A 2011 A stochastic approach to rainfall-induced slope failure *Proceedings of 3rd Int Symp Safety and Risk* ed Vogt N (Munich: BAW) pp 107-115
[11] Li Y, Hicks M A and Nuttall J D 2013 Probabilistic analysis of a benchmark problem for slope stability in 3D *Proceedings of the 3rd International Symposium on Computational Geomechanics (COMGEO III)* ed Pietruszczak S and Pande G N (Krakow: International Centre for Computational Engineering) pp 641-648
[12] Li Y and Hicks M A 2014 Comparative study of embankment reliability in three dimensions *The 8th European Conference on Numerical Methods in Geotechnical Engineering (NUMGE)* ed Hicks M A, Brinkgreve R and Rohe A (Delft: CRC Press) pp 467-472
[13] Hicks M A, Nuttall J D and Chen J 2014 Influence of heterogeneity on 3D slope reliability and failure consequence *Computers and Geotechnics* **61** 198-208
[14] Wang B, Vardon P J and Hicks M A 2013 Implementation of a quasistatic material point method for geotechnical applications *Proceedings of the 3rd International Symposium on Computational Geomechanics (COMGEO III)* ed Pietruszczak S and Pande G N (Krakow: International Centre for Computational Engineering) pp 305-313
[15] de Gast T, Vardon P J, Jommi C and Hicks M A 2015 The history of safety factors for Dutch regional dykes *Proceedings of the 5th International Symposium on Geotechnical Safety and Risk (ISGSR)* Submitted
[16] van Genuchten M T 1980 A closed-form equation for predicting the hydraulic conductivity of unsaturated soils *Soil Science Society of America Journal* **44(5)** 892-898
[17] Mualem Y 1976 A new model for predicting the hydraulic conductivity of unsaturated porous media *Water Resources Research* **12(3)** 513-522
[18] Cai F and Ugai K 2004 Numerical analysis of rainfall effects on slope stability *International Journal of Geomechanics* **4(2)** 69-78
[19] Smith I M, Griffiths D V and Margetts L 2013 *Programming the Finite Element Method* (London: John Wiley & Sons)
[20] Celia M A, Bouloutas E T and Zarba R L 1990 A general massconservative numerical solution for the unsaturated flow equation *Water Resources Research* **26(7)** 1483-1496
[21] Lehmann F and Ackerer P H 1998 Comparison of iterative methods for improved solutions of the fluid flow equation in partially saturated porous media *Transport in Porous Media* **31(3)** 275-292
[22] Griffiths D V and Lane P A 1999 Slope stability analysis by finite elements *Géotechnique* **49(3)** 387-403

IOP Conf. Series: Earth and Environmental Science **26** (2015) 012048 doi:10.1088/1755-1315/26/1/012048

Modelling weathering induced retreat of c-φ cliffs with limited tensile strength

C Voulgari
PhD Student, *University of Warwick*,
School of Engineering, Library Rd, CV4 7AL, Coventry

E-mail: C.Voulgari@warwick.ac.uk

Abstract. Natural cliffs subject to weathering induced retreat are typically made of hard soils and / or weak rocks exhibiting limited tensile strength. In this paper, the morphologic evolution of uniform c, φ slopes subject to weathering is investigated for a range of values of tensile strengths employing the limit analysis upper bound method. This paper extends the analytical framework set up in [1, 2] by accounting for the limited tensile strength of the ground which was previously disregarded. The solutions were obtained by employing the kinematic method of limit analysis providing rigorous upper bounds to the true collapse values. The inclusion of tension cracks leads to modified analytical expressions of the energy balance equation (the balance between external work and dissipated energy) and as a consequence, of the function whose minimum provides the solution in terms of failure mechanisms and associated values of soil strength. Pre-existing cracks are considered, as well as cracks that form as part of the failure mechanism. It turns out that the presence of tension cracks may significantly alter the size of each landslide contributing to the retrogression of the slope. Results in the form of dimensionless ready-to-use charts are produced for any value of engineering interest of friction angle and slope inclination for the case of dry cracks. Moreover, upper bounds for values not included in the charts can be achieved either by interpolation from the charts or by running the minimisation of the analytical functions provided in the paper.

1. Introduction

Morphological evolution of cliffs (natural and excavated slopes) is a traditional subject in engineering geology and geomorphology. Modelling of the progressive retreat of cliffs has recently received considerable attention by the engineering community due to increasing coastal erosive processes caused by climate change and amplified environmental awareness at national and European level [1, 3]. The insurance industry needs reliable models for the prediction of the amount of cliff retreat over time for residential buildings located in exposed areas, whereas local authorities and decision makers need to know the level of risk faced by the public infrastructure (e.g. coastal roads, pedestrian footpaths, car parks, etc.).

Cracks or fissures are widely present in soil slopes and can cause a significant decrease in their stability [2, 4, 5], as they provide preferential flow channels which increase the soil permeability and decrease the soil strength, they can form a part of the critical slip surface with no shear strength and when it is water filled an additional driving force is applied on the slope. They can be the result of a variety of phenomena, for instance low tensile resistance, cycles of wetting-drying [6, 7], desiccation [8-10] and weathering [11].

Weathering tend to turn hard rocks into soft rocks characterised by higher void ratios and reduced bond strengths; then soft rocks are transformed into granular residual soils typically by destroying the bonds keeping the rock grains together [12]. Note that also other natural processes, e.g. methane hydrate dissociation in methane hydrate bearing sands [13], can be conceptualized as weathering. Therefore, a common feature of weathering processes is the degradation of the mechanical properties of the geomaterials (rocks and soils) of which the slope face is made; which in turn leads to the occurrence of successive landslides. This paper aims at investigating and developing an engineering model of the morphological evolution of natural cliffs with limited tensile strength subject to weathering induced landslides.

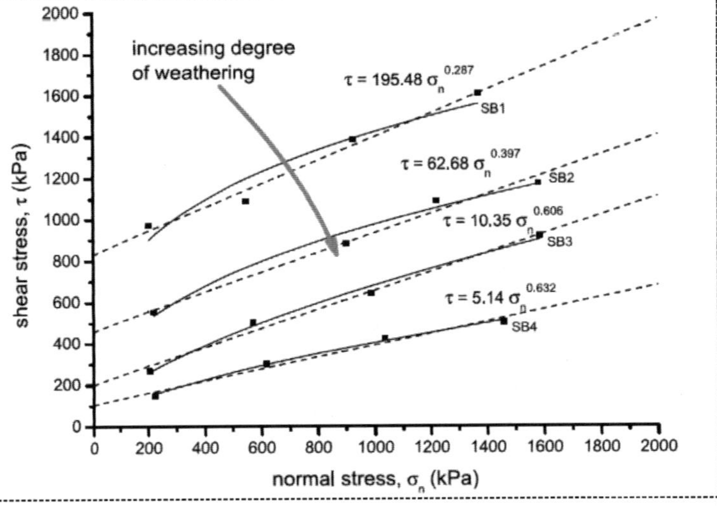

Figure 1. Weathering of granite after (Kimmance 1988)

Most stability analyses of slopes are mainly based on the conventional limit equilibrium method. The weakness with limit equilibrium methods, when it comes to studying slopes with cracks, is that they are not rigorous methods and are limited in their capacity of analysis, since they usually require the user to assume a crack depth and location in the slope. Physical modelling has also been used through the years to simulate the behaviour of slopes subject to weathering, the results of which can be used for the validation of numerical models [14, 15]. The use of numerical methods such as Finite Element Method [16, 17] and Material Point Method [18] to provide approximate solutions to the slope stability problems is also increased in the latest decades. The Discrete Element Method has also been employed [19], more recently for 3D analysis of the stability of rock slopes [20]. This has been made possible by computational advances in the DEM contact detection algorithms to deal with polyhedral blocks [21, 22]. However, in cases of rather uniform slopes subject to a few cracks, such a numerical approach is not justified. In this paper, the limit analysis upper bound method is used to derive the analytical law describing the evolution of slopes subject to weathering and to provide rigorous solutions. Soil strength is characterised by the Mohr-Coulomb failure criterion therefore only three parameters are needed to describe the soil properties (unit weight, friction angle and cohesion).

Soil strength can be expressed by two parameters, cohesion c and internal friction angle φ, according to the Mohr-Coulomb yield criterion (Figure 1). Most of the models developed in the studies mentioned above, assume that the normal and shear stresses along the slip surface complies this criterion, which is the most used criterion for slope stability problems in cohesive soils. Thus, weathering can be taken into consideration by assuming that cohesion (and friction angle) reduces by time. In Figure 1, failure envelopes obtained from tests on a granitic rock subject to various degrees of weathering are illustrated [23]. As shown in Figure 1, weathering causes mainly a decrease in the cohesion and a lower decrease in the friction angle. In this work the case of cohesion only decrease is tackled.

2. Limit analysis model

In order to study the influence due to the presence of cracks to the stability of a slope with uniform cohesion c and internal friction angle φ subject to weathering, the kinematic approach of limit analysis upper bound method has been adopted. As described by [24], an upper-bound solution can be obtained by considering a rotational discontinuity (Figure 2). A homogeneous slope, with zero pore pressure and constant unit weight is presented, with H and β being the height and the inclination of the slope respectively. Basic elements of the slope under study are illustrated in Figure 2. In this part, detailed calculations for the upper-bound limit analysis, considering dry cracks of any possible depth and location, departing from the upper surface of the slope are illustrated.

The weathering of the slope can cause decrease in the cohesion and the friction angle of the slope's material. In the following, only the case of decreasing cohesion is illustrated. The calculations presented below refer to the case of a horizontal slope crest, α=0. Since the limit analysis method is not able to give any information about the final geometry of the debris propagation after each landslide, the material accumulated at the slope toe cannot be taken into account in the model. It is therefore, assumed that the accumulated debris is taken away by atmospheric agents or erosion.

2.1. First failure mechanism

Concerning the occurrence of the first landslide and according to the failure mechanism assumed above, the region of soil B-C-D-E (Figure 2) rotates rigidly around a centre of rotation P_1, as yet undefined, with the material below the logarithmic spiral C-D and right of the vertical crack B-C remaining at rest. The calculations for the first failure mechanisms are obtained from [2], the mechanism is defined by three variables x_1, z_1, y_1 (Figure 2). For the needs of this paper, only the analytical expressions for the occurrence of the second failure mechanism will be illustrated, as the analytical expressions for the first failure can be found in [2].

2.2. Second (successive) failure mechanism

The analytical expressions for the second failure also appeal to every successive failure that will take place. As stated by [24], the stability number Ns=γH/c can be derived from the work rate balance equation:

$$\dot{W}_{ext} = \dot{W}_d \qquad [1]$$

where $\dot{W}_{ext}$ and $\dot{W}_d$ are the external work rate and the internally dissipated energy along the failure line I-D respectively. Note that this equation only appeals for dry cracks, where the rate of external work is due to the soil weight only, in the case of water-filled cracks there is also the contribution of the hydrostatic water pressure filling the crack. According to the limit equilibrium, the angle between the displacement rate (velocity) vector u of the soil mass sliding away and the failure line I-D must be always equal to φ [24]. This condition is met by the adoption of a logarithmic spiral line, where:

$$r = r_0 \exp[\tan\varphi(\theta - \theta_0)] \qquad [2]$$

Some of the geometrical relationships, as arising from Figure 2, will be used for the calculations of the various work rates are given below. Where r_x, r_y and r_z are the radii of the spiral at the angles x, y and z respectively, L_1, l_1 L_2, l_2 are the horizontal lengths as indicated in Figure 2 and H is the height of the slope.

$$r_{y1} = r_{x1}\exp[\tan\varphi(y_1 - x_1)] \quad [3]$$

$$r_{z1} = r_{x1}\exp[\tan\varphi(z_1 - x_1)] \quad [4]$$

$$r_{y2} = r_{x2}\exp[\tan\varphi(y_2 - x_2)] \quad [5]$$

$$r_{z2} = r_{x2}\exp[\tan\varphi(z_2 - x_2)] \quad [6]$$

$$H = r_x \exp[\tan\varphi(y - x)]\sin y_1 - \sin x_1 \quad [7]$$

$$L_1 = r_{x1}\cos x_1 - \cos y_1 \exp[\tan\varphi(y_1 - x_1)] \quad [8]$$

$$l_1 = r_{x1}\cos x_1 - \cos z_1 \exp[\tan\varphi(z_1 - x_1)] \quad [9]$$

$$L_2 = r_{x2}\cos x_2 - \cos y_2 \exp[\tan\varphi(y_2 - x_2)] \quad [10]$$

$$l_2 = r_{x2}\cos x_2 - \cos z_2 \exp[\tan\varphi(z_2 - x_2)] \quad [11]$$

The analytical expression of the stability factor Ns=γH/c, where γ is the unit weight of soil, H is the slope height and c the cohesion of the soil, can be obtained by calculating the rate of the dissipated energy along the logarithmic spiral shaped failure line I-D. Energy is dissipated only along the failure line I-D, according to the assumed rigid rotational mechanism:

$$\dot{W}_d = c\omega\int_{y_2}^{z_2} r^2 d\theta$$

$$= \exp[2\tan\varphi(z_2 - x_2)]\frac{\exp[2\tan\varphi(y_2 - z_2)] - 1}{2\tan\varphi} \quad [12]$$

$$= \omega\gamma r_{x2}^{\ 2} f_d(x_2, y_2, z_2, \varphi)$$

where ω is the angular velocity.

After the region B-C-D-E has slipped away, further weathering will at some point induce a second landslide. The double logarithm spiral shaped area G-I-D-C-B will rigidly rotate around the center of rotation P_2, as yet undefined, with the material below the logarithmic spiral I-D and right of the vertical crack G–I remaining at rest. The equation for the energy balance for a double logarithmic region rigidly rotating away was first presented in [25]. Here, the mechanism is defined by six variables x_1, z_1, y_1, x_2, z_2, y_2. The rate of external work due to the soil weight of region G-I-D-C-B is calculated as the work done by the region J-D-E subtracting the work done by the region B-C-D-E and the work done by the region J-I-J. A direct integration of the rate of the external work due to the soil of the regions mentioned above is very complicated. Instead the method of superposition by first finding the rates of works $\dot{W}_1^o$, $\dot{W}_2^o$, $\dot{W}_3^o$, $\dot{W}_4^o$, $\dot{W}_5^o$, $\dot{W}_6^o$, $\dot{W}_1^n$, $\dot{W}_2^n$, $\dot{W}_3^n$, $\dot{W}_4^n$, $\dot{W}_5^n$ and $\dot{W}_6^n$ will be used. The rate of the work of G-I-E-J is given by six contributors $\dot{W}_1^o$-$\dot{W}_2^o$-$\dot{W}_3^o$-($\dot{W}_4^o$- $\dot{W}_5^o$- $\dot{W}_6^o$), where $\dot{W}_1^o$, $\dot{W}_2^o$, $\dot{W}_3^o$, $\dot{W}_4^o$, $\dot{W}_5^o$, $\dot{W}_6^o$ are the rates of work done by regions P_1FC, P_1FE, P_1ED, P_1FC, P_1FB and P_1BC respectively. The rate of the work of J-D-E is given by three contributors $\dot{W}_1^n$- $\dot{W}_2^n$- $\dot{W}_3^n$, where $\dot{W}_1^n$, $\dot{W}_2^n$ and $\dot{W}_3^n$ are the rates of work done by regions P_2JC, P_2JE and P_2ED respectively. And, finally the rate of the region J-I-G is also given by three contributors $\dot{W}_4^n$- $\dot{W}_5^n$- $\dot{W}_6^n$, where $\dot{W}_4^n$, $\dot{W}_5^n$ and $\dot{W}_6^n$ are the works done by regions P_2JI, P_2JG and P_2GI respectively.

Therefore, it can be summed up that the rate of external work due to the weight of the soil is a total of twelve different contributors, each one referring to the work done by the soil weight of the regions that are shown in Figure 2, given by the following equation:

$$\dot{W}_{ext} = \dot{W}_1^{\,n} - \dot{W}_2^{\,n} - \dot{W}_3^{\,n} - \dot{W}_4^{\,n} + \dot{W}_5^{\,n} + \dot{W}_6^{\,n} - \dot{W}_1^{\,o} + \dot{W}_2^{\,o} + \dot{W}_3^{\,o} + \dot{W}_4^{\,o} - \dot{W}_5^{\,o} - \dot{W}_6^{\,o} \quad [13]$$

with n and o referring to the new and the old landslide respectively.

To find a solution to the problem, it is necessary to express all the various work contributions in terms of a common variable, here chosen as r_θ, so that the equation obtained will be linear in H. The calculation of the work rates $\dot{W}_1^n$, $\dot{W}_2^n$ and $\dot{W}_3^n$ can be found in Chen (1975). For the demands of this paper only the first and the final expressions of all calculations are given; detailed calculations for each of the following expressions are available.

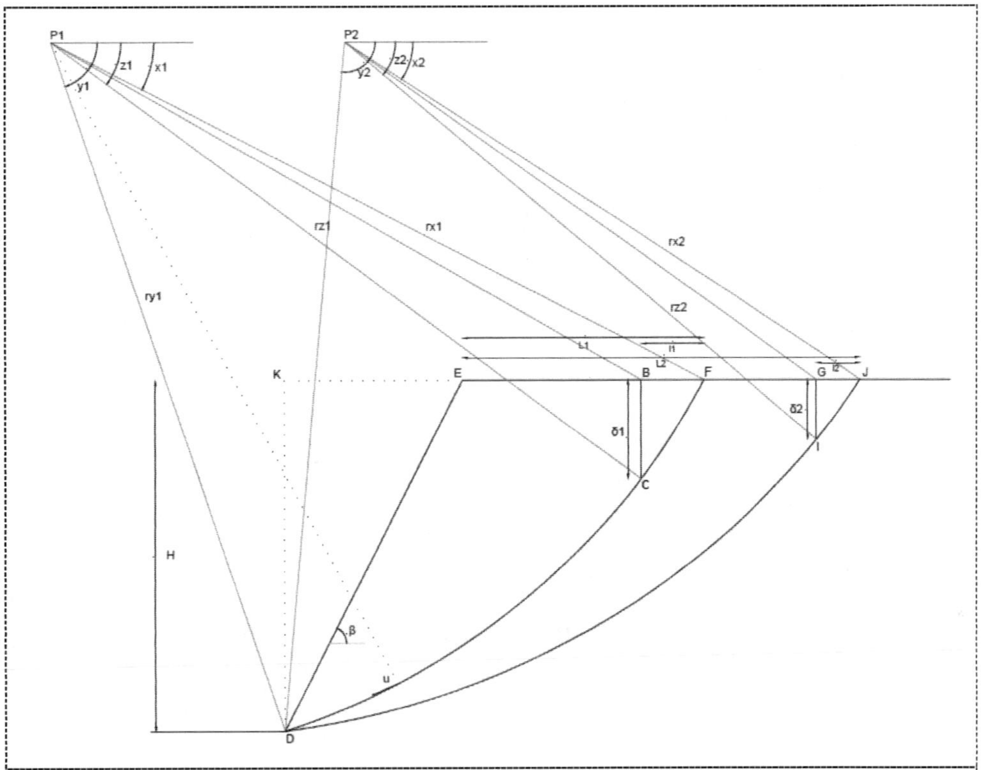

Figure 2. Second Failure mechanism: Region of soil G-I-D-C-B by rotates around point P2. Region of soil B-C-D-E has already slipped away when the first failure occurred.

Considering the region P_2JD:

$$\dot{W}_1^n = \dot{W}_{P_2JD} = (\omega \frac{2}{3} r_\theta \cos\theta)(\frac{1}{2}\gamma r_\theta^2 d\theta)$$

$$= \omega\gamma r_{x2}^{\ 3} \frac{\exp[3\tan\varphi(y_2 - x_2)](3\tan\varphi\cos y_2 + \sin y_2) - 3\tan\varphi\cos x_2 - \sin x_2}{3(1+9\tan^2\varphi)} \quad [14]$$

$$= \omega\gamma r_{x2}^{\ 3} f_1^n(x_2, y_2, \varphi)$$

Considering the region P_2JE:

$$
\begin{aligned}
\dot{W}_2{}^n &= \dot{W}_{P_2JE} = [\omega\frac{1}{3}(2r_{x_2}\cos x_2 - L_2)(\frac{1}{2}\gamma L_2 r_{x_2}\sin x_2) \\
&= \omega\gamma r_{x2}{}^3\frac{1}{6}\frac{L_2}{r_{x_2}}(2\cos x_2 - \frac{L_2}{r_{x_2}})\sin x_2 \\
&= \omega\gamma r_{x2}{}^3 f_2{}^n(x_2, y_2, \varphi)
\end{aligned}
\qquad [15]
$$

Considering the region P_2ED:

$$
\begin{aligned}
\dot{W}_3{}^n &= \dot{W}_{P_2ED} = (\omega\frac{2}{3}r_{y_2}\cos y_2)(\frac{1}{2}\gamma H r_{y_2}\cos y_2) \\
&= \omega\gamma r_{x2}{}^3\frac{1}{3}[\exp(\tan\varphi(y_2 - x_2)\sin y_2 - \sin x_2]\cos^2 y_2 \exp(2\tan\varphi(y_2 - x_2) \\
&= \omega\gamma r_{x2}{}^3 f_3{}^n(x_2, y_2, \varphi)
\end{aligned}
\qquad [16]
$$

Considering the region P_2JI, the rate of the external work is calculated for an infinitesimal slice, as illustrated in Figure 3:

$$
d\dot{W}_4{}^n = u_4 dF_4 = \omega\,|\,x_{G4} - x_P\,|\,\gamma dA = \omega\frac{1}{3}r^3\gamma\cos\theta d\theta
$$

where u_4 is the displacement rate of the infinitesimal region, dF_4 is the weight force, and x_{G4} and x_P are the x coordinates of the gravity center of the soil region and of the center of rotation P_2 respectively. Integrating over the whole region, the rate of the work of the region P_2JI is calculated:

$$
\begin{aligned}
\dot{W}_4{}^n &= \dot{W}_{P_2JI} = (\omega\frac{2}{3}r_\theta\cos\theta)(\frac{1}{2}\gamma r_\theta{}^2 d\theta) \\
&= \omega\gamma r_{x2}{}^3\frac{\exp[3\tan\varphi(z_2 - x_2)](3\tan\varphi\cos z_2 + \sin x_2) - 3\tan\varphi\cos x_2 - \sin x_2}{3(1 + 9\tan^2\varphi)} \\
&= \omega\gamma r_{x2}{}^3 f_4{}^n(x_2, z_2, \varphi)
\end{aligned}
\qquad [17]
$$

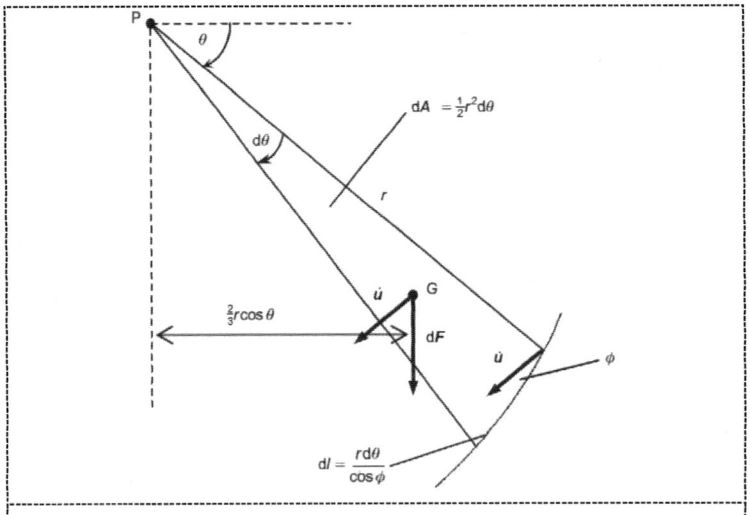

Figure 3. Infinitesimal slice of logarithmic spiral regions.

Considering the region P_2JG:

$$\dot{W}_5^n = \dot{W}_{P_2AB} = [\omega \frac{1}{3}(2r_{x_2}\cos x_2 - l_2)(\frac{1}{2}\gamma l_2 r_{x_2}\sin x_2)$$

$$= \omega\gamma r_{x2}^{\ 3}\frac{1}{6}\frac{l_2}{r_{x_2}}(2\cos x_2 - \frac{l_2}{r_{x_2}})\sin x_2 \qquad [18]$$

$$= \omega\gamma r_{x2}^{\ 3} f_5^n(x_2, z_2, \varphi)$$

Finally, considering the region P_2GI:

$$\dot{W}_6^n = \dot{W}_{P_2GI} = (\omega\frac{2}{3}r_{y_2}\cos z_2)(\frac{1}{2}\gamma H r_{z_2}\cos z_2)$$

$$= \omega\gamma r_{x2}^{\ 3}\frac{1}{3}[\exp(\tan\varphi(z_2 - x_2)\sin z_2 - \sin x_2]\cos^2 z_2 \exp(2\tan\varphi(z_2 - x_2) \qquad [19]$$

$$= \omega\gamma r_{x2}^{\ 3} f_6^n(x_2, z_2, \varphi)$$

All calculations of the external works mentioned above, refer to the occurrence at the second landslide and have been achieved by calculating the moment of each soil region around the point P_2. The calculations for the remaining six regions that refer to the occurrence of the first landslide should also be achieved by calculating the moment of each soil region around the point P_2. Note that for the calculation of the rates of the external work $\dot{W}_1^o$, $\dot{W}_2^o$ and $\dot{W}_3^o$, the front of the slope is considered vertical for the simplicity of the calculations and that is the reason why the inclination of the slope β does not appear in the calculations. After subtracting $\dot{W}_2^o$ and $\dot{W}_3^o$ from $\dot{W}_1^o$ the result is the same.

For the region P_1FD, the rate of the external work is calculated for an infinitesimal slice, similar to $\dot{W}_4^n$, by considering this time the point P_2 as the center of rotation:

$$d\dot{W}_1^o = (\omega\frac{2}{3}r_\theta\cos\theta - r_{y_1}\cos y_1 + r_{y_2}\cos y_2)(\frac{1}{2}\gamma r_\theta^2\, d\theta)$$

And after integration by parts, manipulations and substitutions ([3][5]) of the obtained expression:

$$\dot{W}_1^o = \dot{W}_{P_1FD}$$

$$= \omega\gamma r_{x2}^{\ 3}\left[\begin{array}{l} \left(\dfrac{r_{x1}}{r_{x2}}\right)^2 \dfrac{[\exp(\tan\varphi(y_2 - x_2)\cos y_2(\exp(2\tan\varphi(y_1 - x_1) - 1)}{4\tan\varphi} \\ +\left(\dfrac{r_{x1}}{r_{x2}}\right)^3 \dfrac{\exp(3\tan\varphi(y_1 - x_1))(\sin y_1 + 3\tan\varphi\cos y_1) - \sin x_1 - 3\tan\varphi\cos x_1}{3(1+9\tan^2\varphi)} \\ -\left(\dfrac{r_{x1}}{r_{x2}}\right)^3 \dfrac{[\exp(\tan\varphi(y_1 - x_1)\cos y_1[\exp(2\tan\varphi(y_1 - x_1) - 1]}{4\tan\varphi)} \end{array}\right] \qquad [20]$$

$$= \omega\gamma r_{x2}^{\ 3} f_1^o(x_1, y_1, x_2, y_2, \varphi)$$

Considering the region P_1FE:

$$\dot{W}_2^o = \dot{W}_{P_1FE} = \left[\frac{1}{3}(2r_{x_1}\cos x_1 - L_1 + r_{y_2}\cos y_2 - r_{y_1}\cos y_1\right]\left(\frac{1}{2}\gamma L_1 r_{x_1}\sin x_1\right)$$

And after manipulations and substitutions ([3][5][8]) of the obtained expression, the following expression is obtained:

$$\dot{W}_2^o = \omega\gamma r_{x2}^{\ 3}\frac{1}{2}\frac{L_1}{r_{x1}}\left(\frac{r_{x1}}{r_{x2}}\right)^2 \sin x_1\left[\frac{1}{3}\frac{r_{x1}}{r_{x2}}(\cos x_1 - 2\exp(\tan\varphi(y_1 - x_1)\cos y_1 + \exp(\tan\varphi(y_2 - x_2)\cos y_2\right] \qquad [21]$$
$$= \omega\gamma r_{x2}^{\ 3} f_2^{\ o}(x_1, y_1, x_2, y_2, \varphi)$$

Considering the region P_1ED:

$$\dot{W}_3^o = \dot{W}_{P1ED} = (\omega\frac{2}{3}r_{y_1}\cos y_1 - r_{y_1}\cos y_1 + r_{y_2}\cos y_2)(\frac{1}{2}\gamma H r_{y_1}\cos y_1)$$

And after manipulations and substitutions ([3][5][7]) of the obtained expression:

$$\dot{W}_3^{\ o} = \omega\gamma r_{x2}^{\ 3}\frac{1}{2}\left(\frac{r_{x1}}{r_{x2}}\right)^2\left[\exp(\tan\varphi(y_1 - x_1)\sin y_1 - \sin x_1\right]\exp(\tan\varphi(y_1 - x_1)\cos y_1$$
$$\left[[\exp(\tan\varphi(y_2 - x_2)\sin y_2 - \frac{1}{3}\frac{r_{x1}}{r_{x2}}[\exp(\tan\varphi(y_1 - x_1)\cos y_1]\right] \qquad [22]$$
$$= \omega\gamma r_{x2}^{\ 3} f_3^{\ o}(x_1, y_1, x_2, y_2, \varphi)$$

The external work for the region P_1FC is calculated similar to that of $\dot{W}_1^{\ o}$ and:

$$d\dot{W}_4^{\ o} = (\omega\frac{2}{3}r_\theta\cos\theta - r_{y_1}\cos y_1 + r_{y_2}\cos y_2)(\frac{1}{2}\gamma r_\theta^{\ 2}\, d\theta)$$

And after integration by parts, manipulations and substitutions ([4][6]) of the obtained expression:

$$\dot{W}_4^{\ o} = \dot{W}_{P1FC}$$
$$= \omega\gamma r_{x2}^{\ 3}\left[\begin{array}{l}\left(\dfrac{r_{x1}}{r_{x2}}\right)^2\dfrac{[\exp(\tan\varphi(y_2 - x_2)\cos y_2(\exp(2\tan\varphi(z_1 - x_1) - 1)}{4\tan\varphi} \\ +\left(\dfrac{r_{x1}}{r_{x2}}\right)^3\dfrac{\exp(3\tan\varphi(z_1 - x_1))(\sin z_1 + 3\tan\varphi\cos z_1) - \sin x_1 - 3\tan\varphi\cos x_1}{3(1 + 9\tan^2\varphi)} \\ -\left(\dfrac{r_{x1}}{r_{x2}}\right)^3\dfrac{[\exp(\tan\varphi(y_1 - x_1)\cos y_1[\exp(2\tan\varphi(z_1 - x_1) - 1]}{4\tan\varphi}\end{array}\right] \qquad [23]$$
$$= \omega\gamma r_{x2}^{\ 3} f_4^{\ o}(x_1, z_1, x_2, z_2, \varphi)$$

Considering the region P_1FB:

$$\dot{W}_5^o = \dot{W}_{P1FB} = \left[\frac{1}{3}(2r_{x_1}\cos x_1 - l_1 - r_{y_1}\cos y_1 + r_{y_2}\cos y_2\right]\left(\frac{1}{2}\gamma l_1 r_{x_1}\sin x_1\right)$$

And after manipulations and substitutions ([4][6][9]) of the obtained expression and dividing by $r_{x2}^{\ 3}$, the following expression is obtained:

$$\dot{W}_5^o = \omega\gamma r_{x2}^{\ 3}\frac{1}{2}\frac{l_1}{r_{x1}}\left(\frac{r_{x1}}{r_{x2}}\right)^2\sin x_1\left[\frac{r_{x1}}{r_{x2}}(\frac{1}{3}(2\cos x_1 - \frac{l_1}{r_{x1}}) - \exp(\tan\varphi(y_1 - x_1)\cos y_1) + \exp(\tan\varphi(y_2 - x_2)\cos y_2\right] \qquad [24]$$
$$= \omega\gamma r_{x2}^{\ 3} f_2^{\ o}(x_1, y_1, z_1, x_2, y_2, z_2, \varphi)$$

Finally, considering the region P_1BC:

$$\dot{W}_6^o = \dot{W}_{P_1BC} = (\omega \frac{2}{3} r_{z_1} \cos z_1 - r_{y_1} \cos y_1 + r_{y_2} \cos y_2)(\frac{1}{2} \gamma H r_{z_1} \cos z_1)$$

And after manipulations and substitutions [4][6] of the obtained expression and dividing by r_{x2}^3, the following expression is obtained:

$$\dot{W}_6^o = \omega\gamma r_{x2}^3 \frac{1}{2}\left(\frac{r_{x1}}{r_{x2}}\right)^2 \left[\exp(\tan\varphi(z_1 - x_1)\sin z_1 - \sin x_1\right]\exp(\tan\varphi(z_1 - x_1)\cos z_1$$

$$\left[\exp(\tan\varphi(y_2 - x_2))\cos y_2 + \frac{r_{x1}}{r_{x2}}(\frac{2}{3}\exp(\tan\varphi(z_1 - x_1)\cos z_1) - \exp(\tan\varphi(y_1 - x_1)\cos y_1))\right] \quad [25]$$

$$= \omega\gamma r_{x2}^3 f_3^o(x_1, y_1, z_1, x_2, y_2, z_2, \varphi)$$

Substituting equations [7][12][13][14][15][16][17][18][19][20][21][22][23][24][25] into equation [1] and dividing by ω and r_{x2} and rearranging, the stability factor, Ns=γH/c, is obtained as

$$\frac{\gamma H}{c} = g(x_1, y_1, z_1, x_2, y_2, z_2, \varphi)$$

$$= \frac{fd}{(f_1^o - f_2^n - f_3^n - f_4^n + f_5^n + f_6^n - f_1^o + f_2^o + f_3^o + f_4^o - f_5^o - f_6^o)}[\exp(\tan\varphi(y_2 - x_2))\sin y_2 - \sin x_2] \quad [26]$$

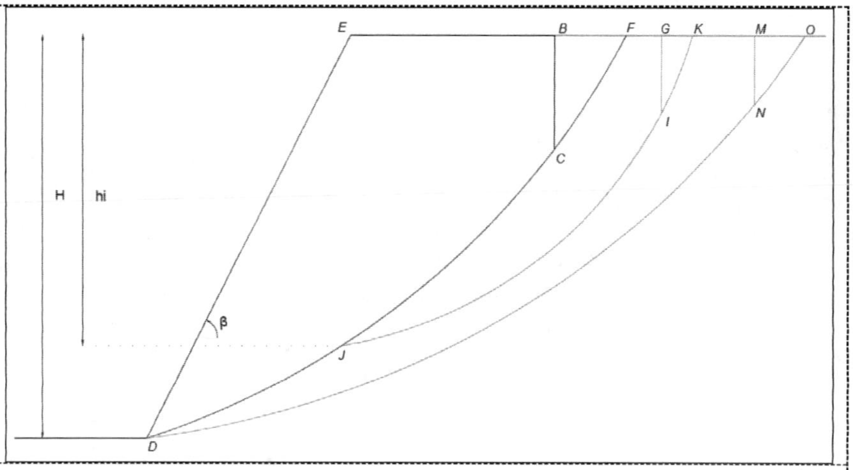

Figure 4. Failure lines relative to the different mechanisms considered.

This second (and every latter) mechanism could pass through any point since the current slope profile is no longer straight, as presented in Figure 4. To take this possibility into consideration, the slope that was created after the occurrence of the first landslide, has been divided into a discrete number of points (n) and each point has been assumed as the toe of a slope with height, h_i. In order to find the most critical mechanism among all the possible failure mechanisms the minimum stability number must be obtained. The minimum of $g(x_1, y_1, z_1, x_2, y_2, z_2, \varphi)$ [26], provides the upper bound for the stability factor, which corresponds to the maximum value of the dimensionless cohesion c/(γH). The critical cohesion values, c_i, and angles, x_i, y_i and z_i, associated with the critical log spiral, should be determined for all n slopes of different height, hi, where the parameter y_i assumes a different value associated with each of the n slopes assumed. The most critical failure mechanism among the n potential mechanisms is the one with the highest cohesion value.

2.3. Failure mechanism including crack formation

Apart from the case of pre-existing cracks, the case of slopes with a crack forming in the soil during failure, for a tension cut-off (where f_t=0) and slopes with soil tensile strength limited to half of that described by the classical Mohr–Coulomb yield condition are tackled.

For any failure mechanism including the crack-opening the expenditure of energy needed to open the crack should be added to the energy dissipation along the logarithmic spiral. The final expression of this energy is given in [5]:

$$\dot{W}_{cf} = r_{x2}^{\ 2}\omega\int_{\theta_2}^{z_2} \frac{\sin(x_2)}{\tan(\theta_2)\cos\theta}\left(f_c\frac{1-\sin\theta}{2} + f_t\frac{\sin\theta-\sin\varphi}{1-\sin\varphi}\right)\frac{d\theta}{\cos\theta} \quad [27]$$

$$=\omega\gamma r_{x2}^{\ 2} f_d(x_2, z_2, \theta_2, \varphi)$$

where: $f_c = \dfrac{2c\cos\varphi}{1-\sin\varphi}$ and $f_t = \dfrac{2c\cos\varphi}{1+\sin\varphi}$, the unconfined compressive strength and the tensile strength of the soil respectively, as described by the classical Mohr-Coulomb yield condition.

The final expression to calculate the stability number Ns=γH/c, including the energy required for the formation of the crack is:

$$\frac{\gamma H}{c} = g(x_1, y_1, z_1, x_2, y_2, z_2, \theta_2, \varphi)$$

$$= \frac{f_d + f_{cf}}{(f_1^o - f_2^n - f_3^n - f_4^n + f_5^n + f_6^n - f_1^o + f_2^o + f_3^o + f_4^o - f_5^o - f_6^o)}[\exp(\tan\varphi(y_2 - x_2))\sin y_2 - \sin x_2] \quad [28]$$

3. Cohesion decrease

As stated by [23] it is reasonable, as a first approximation, to assume that weathering causes mainly a decrease of cohesion, and to a much lesser extent of the friction angle. For the needs of this paper only the cohesion decrease is tackled and the slopes are assumed to be homogeneous.

Computations were carried out for a wide range of parameters (friction angle φ and initial slope inclination β) for slopes with pre-existing cracks, slopes with no tensile strength (tension cut-off) where crack forming requires work to open and slopes with soil tensile strength limited to half of that described by the classical Mohr–Coulomb yield condition. Results for the evolution of the stability number from the analysis for ten successful failures for initial slope inclination β=60° and friction angles φ=20° and 40° are illustrated in Figure 5.

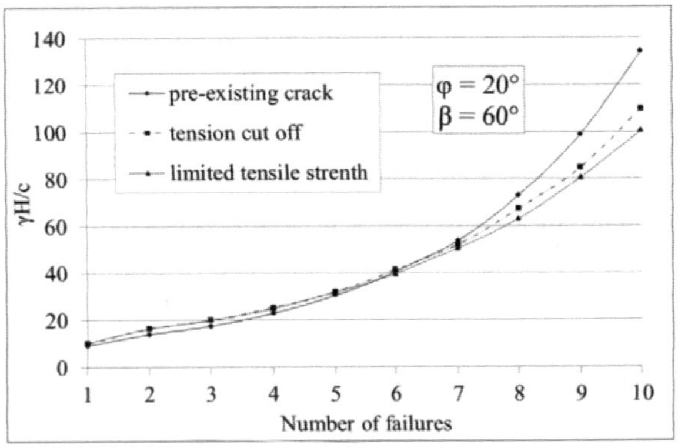

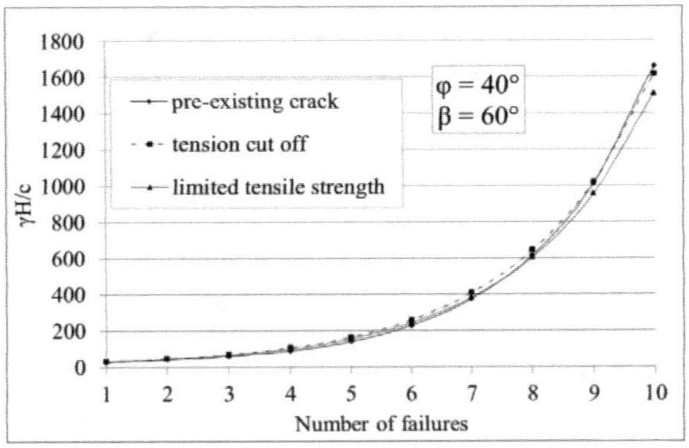

Figure 5. Preliminary results for the evolution on the stability number for 10 successive failures for initial slope angles β=60° and friction angle φ=20° and 40°.

To make meaningful comparisons, the outcome of calculations in terms of the critical height for selected geometrical parameters is also presented in Table 1. The values of the crest retreat normalised by the initial slope height, the sliding area normalised by the square of the initial height and the associated crack depth normalised by the height are listed for initial slope inclination β=60° and friction angles φ=20° and 40° for the three different cases of the appearance of the crack.

Table 1. Preliminary Results for Associated Crest Retreat, Sliding Area and Crack Depth for Slopes With Initial Inclination β and Friction Angle φ, for slopes with pre-existing cracks, slopes with no tensile strength where crack forming requires work to open and slopes with soil tensile strength limited to half of that described by the classical Mohr–Coulomb yield condition

	β = 60°					
	φ = 20°			φ = 40°		
	CR/H	A/H^2	d/H	CR/H	A/H^2	d/H
Failure	**Pre-existing crack**					
1	0.316	0.335	0.339	0.133	0.163	0.167
2	0.630	0.218	0.238	0.254	0.058	0.119
3	0.864	0.115	0.189	0.330	0.021	0.084
4	1.038	0.061	0.145	0.377	0.007	0.055
5	1.167	0.034	0.110	0.406	0.003	0.036
6	1.262	0.018	0.083	0.424	0.001	0.022
7	1.333	0.010	0.060	0.435	0.000	0.013
8	1.385	0.005	0.045	0.442	0.000	0.008
9	1.423	0.003	0.033	0.446	0.000	0.005
10	1.451	0.002	0.025	0.448	0.000	0.003
Failure	**Tension cut-off**					
1	0.431	0.359	0.219	0.148	0.163	0.115
2	0.770	0.280	0.113	0.276	0.062	0.064
3	1.072	0.171	0.102	0.359	0.023	0.046
4	1.301	0.103	0.078	0.412	0.009	0.028
5	1.485	0.062	0.062	0.446	0.004	0.018
6	1.625	0.039	0.048	0.468	0.001	0.011
7	1.738	0.023	0.038	0.482	0.001	0.007
8	1.824	0.015	0.029	0.490	0.000	0.004
9	1.893	0.009	0.023	0.496	0.000	0.003
10	1.946	0.005	0.018	0.499	0.000	0.002
Failure	**Limited tensile strength**					
1	0.378	0.349	0.062	0.137	0.159	0.044
2	0.753	0.287	0.031	0.269	0.068	0.024
3	1.054	0.175	0.024	0.360	0.027	0.015
4	1.297	0.108	0.019	0.418	0.011	0.009
5	1.486	0.069	0.015	0.456	0.004	0.006
6	1.639	0.043	0.012	0.479	0.002	0.004
7	1.757	0.027	0.009	0.494	0.001	0.002
8	1.853	0.017	0.008	0.503	0.000	0.002
9	1.927	0.011	0.006	0.509	0.000	0.001
10	1.987	0.007	0.005	0.513	0.513	0.000

Note that for the case of slopes with pre-existing crack the most critical mechanism for the second failure is the one that passes through point C (Figure 6, mechanism G-C) with a crack almost as deep as the one from the previous mechanism. The evolution of the slope then will follow the line C-I (Figure 6 - dashed gray lines illustrate the successive log-spirals), until a point where the mechanism J-K will become more critical. Since the analytical formulation for a slope profile made up of more than one log-spiral lines becomes prohibitive, two profiles for the initiation of the second failure were compared. The profiles were F-C-D and I-C-D and the failure mechanisms starting from these two profiles were calculated, giving very close results. Slightly more critical solutions in terms of stability numbers are obtained by assuming that the initial profile for the second failure is F-C-D and was adopted for all the results presented in this paper.

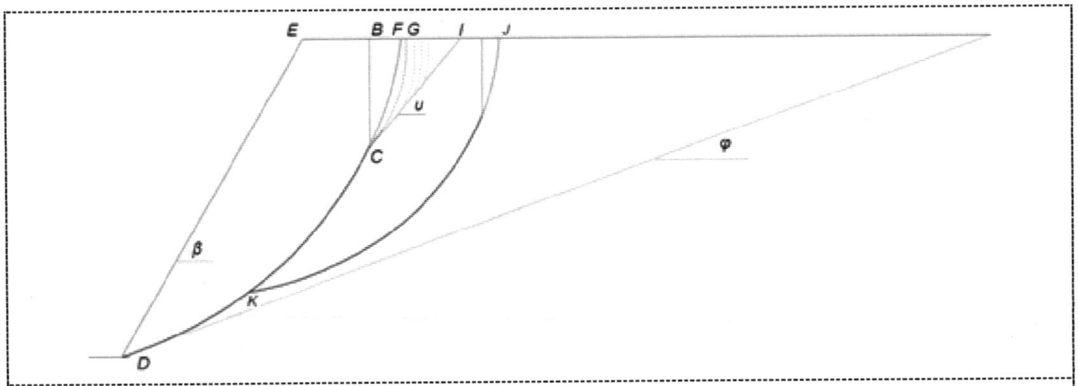

Figure 6. Failure lines relative to the different mechanisms considered for β=60°.

4. Conclusions

The presence of cracks may heavily affect the evolution of slope subject to weathering. Consideration of crack formation as part of the failure mechanism in the slope is also important. In this paper, a model based on the kinematic approach of limit analysis to predict evolution of slopes with limited tensile strength has been proposed. The location and the depth of the cracks as well as the most critical failure mechanism for every failure were calculated via an optimisation procedure. Solutions were provided for three different cases: determination of the evolution of slopes exhibiting pre-existing cracks, slopes with no tensile strength (tension cut-off), and slopes of limited soil tensile strength. With the presented model it is possible to account for the presence of cracks into the prediction of the evolution of natural slopes subject to a series of successive failures caused by the progressive degradation of soil strength over time.

5. References

[1] Utili S. and Crosta G., "Modeling the evolution of natural cliffs subject to weathering: 1. Limit analysis approach," *Journal of Geophysical Research: Earth Surface (2003–2012),* vol. **116**, p. F01016, 2011.

[2] Utili S., "Investigation by limit analysis on the stability of slopes with cracks," *Geotechnique,* vol. **63**, pp. 140-154, 2013.

[3] Bray M. J. and Hooke J. M., "Prediction of soft-cliff retreat with accelerating sea-level rise," *Journal of Coastal Research,* vol. **13**, pp. 453-467, 1997.

[4] Baker R., "Tensile strength, tension cracks, and stability of slopes," *Soils and foundations,* vol. **21**, pp. 1-17, 1981.

[5] Michalowski R. L., "Stability assessment of slopes with cracks using limit analysis," *Canadian Geotechnical Journal,* vol. **50**, pp. 1011-1021, 2013.

[6] Yesiller N., Miller C., Inci G., and Yaldo K., "Desiccation and cracking behavior of three compacted landfill liner soils," *Engineering Geology,* vol. **57**, pp. 105-121, 2000.

[7] Morris P. H., Graham J., and Williams D. J. , "Cracking in drying soils," *Canadian Geotechnical Journal,* vol. **29**, pp. 263-277, 1992.

[8] Péron H., Hueckel T., Laloui L., and Hu L., "Fundamentals of desiccation cracking of fine-grained soils: experimental characterisation and mechanisms identification," *Canadian Geotechnical Journal,* vol. **46**, pp. 1177-1201, 2009.

[9] Dyer M., Utili S., and Zielinski M., "Field survey of desiccation fissuring of flood embankments," *Proceedings of the ICE-Water Management,* vol. **162**, pp. 221-232, 2009.

[10] Konrad J. M.and Ayad R., "A idealized framework for the analysis of cohesive soils undergoing desiccation," *Canadian Geotechnical Journal,* vol. **34**, pp. 477-488, 1997.

[11] Hales T. and Roering J. J., "Climatic controls on frost cracking and implications for the evolution of bedrock landscapes," *Journal of Geophysical Research: Earth Surface (2003–2012),* vol. **112**, 2007.

[12] Utili S., "Evolution of natural slopes subject to weathering: An analytical and numerical study," Ph. D thesis, Politecnico di Milano, Milan, Italy, 2004.

[13] Jiang M., Zhu F., Liu F., and Utili S., "A bond contact model for methane hydrate-bearing sediments with interparticle cementation," *International Journal for Numerical and Analytical Methods in Geomechanics,* vol. **38**, pp. 1823-1854, 2014.

[14] Le Cossec J., Duperret A., Vendeville B. C., and Taibi S., "Numerical and physical modelling of coastal cliff retreat processes between La Hève and Antifer capes, Normandy (NW France)," *Tectonophysics,* vol. **510**, pp. 104-123, 2011.

[15] Gabrieli F., Lambert P., Cola S., and Calvetti F., "Micromechanical modelling of erosion due to evaporation in a partially wet granular slope," *International journal for numerical and analytical methods in geomechanics,* vol. **36**, pp. 918-943, 2012.

[16] Huang M. and Jia C. Q., "Strength reduction FEM in stability analysis of soil slopes subjected to transient unsaturated seepage," *Computers and Geotechnics,* vol. **36**, pp. 93-101, 2009.

[17] Zheng H., Liu D., and Li C., "Slope stability analysis based on elasto-plastic finite element method," *International Journal for Numerical Methods in Engineering,* vol. **64**, pp. 1871-1888, 2005.

[18] Yerro A., Alonso E., and Pinyol N., "The material point method for unsaturated soils," *Geotechnique,* vol. **65**, pp. 201-217, 2015.

[19] Utili S. and Crosta G., "Modeling the evolution of natural cliffs subject to weathering: 2. Discrete element approach," *Journal of Geophysical Research-Part F-Earth Surface,* vol. **116**, p. F01017, 2011.

[20] Boon C., Houlsby G., and Utili S., "New insights into the 1963 Vajont slide using 2D and 3D distinct-element method analyses," *Geotechnique,* vol. **64**, pp. 800-816, 2014.

[21] Boon C., Houlsby G., and Utili S., "A new algorithm for contact detection between convex polygonal and polyhedral particles in the discrete element method," *Computers and Geotechnics,* vol. **44**, pp. 73-82, 2012.

[22] Boon C., Houlsby G., and Utili S., "A new contact detection algorithm for three-dimensional non-spherical particles," *Powder Technology,* vol. **248**, pp. 94-102, 2013.

[23] Kimmance G., "Computer aided risk analysis of open pit mine slopes in kaolin mined deposits," Ph. D thesis, Imperial College London (University of London), 1988.

[24] Chen W., "Limit analysis and soil plasticity, 1975," ed: Elsevier, 1975.

[25] Utili S., Nova R., "On the optimal profile of a slope," *Soils and foundations*, vol. **47**, pp. 717-729, 2007.

A surface and subsurface model for the simulation of rainfall infiltration in slopes

H Zhang, F Zhang, K Shen and M Yuan
China Three Gorges University, Collaborative Innovation Center for Geo-hazards and Eco-environment in Three Gorges Area, Yichang, 443002, China

E-mail: cezhua@qq.com

Abstract. Rainfall infiltration is one of the major triggering factors leading to slope failures in geotechnical engineering. Numerical investigation on rainfall infiltration is often based on Richards' equation, which ignores the surface water effects and simplifies the boundary conditions. In reality, rainfall, infiltration, and surface runoff are interacted simultaneously. In this paper a new conjunctive one-dimensional surface flow and two-dimensional subsurface flow model for geotechnical slope is developed. The interaction between surface and subsurface flow is the interface infiltration rate, which is obtained by iterations. The results of comparisons between coupled and uncoupled models show that the surface water depth rises up as runoff increases and it tends to a dynamic balance state with a steady surface water depth. Interaction between surface and subsurface flow has remarkable effects on infiltration process. According to the results of coupled model, more rainwaters infiltrate into the slope. Therefore, pore water pressure changes faster and the wetting front moves deeper into the soil. Under initial drier condition, the capacity of infiltration is higher and more rainfall can be absorbed into slope, thus the differences of infiltration rate and pore water pressure between the coupled and uncoupled model are more significant.

1. Introduction

Rainfall is within a part of the hydrological cycle. Generally, when rainwater falls on a slope, part of the amount of rainfall will infiltrate into soil mass, while the others will run off along the slope surface if the rainfall intensity is larger than the infiltration capacity of soil. American Petroleum Institute [1] produces a thorough review on numerous estimation methods of infiltration rate through unsaturated zone. Based on the above research, Ravi and Williams [2] classify those estimation methods into three categories, i.e. (a) empirical models; (b) Green-Ampt models; and (c) Richards equation models. The empirical models only depict cumulative infiltration and infiltration rate changes with time, but can not provide pore water pressure and water content information. Green-Ampt models describe one-dimensional infiltration process by assuming water advancing downward like a piston, but those still can not provide pore water pressure distribution information. Richards equation is the mathematical governing equation of physical flow in saturated-unsaturated soils. After applying proper initial and boundary conditions and solving the partial differential equation, the pore water pressure distribution is obtained.

Since pore water pressure directly contributes to soil shear strength and volumetric change, it is the key issue considered by geotechnical engineers in slope engineering. The mechanism of rain-induced slope failure has been studied by early researchers [3-6]. Neuman [7] successfully introduces finite element method to solve Richards equation. Lam [8] successfully solves some saturated-unsaturated

IOP Conf. Series: Earth and Environmental Science **26** (2015) 012025 doi:10.1088/1755-1315/26/1/012025

seepage problems in geotechnical engineering by finite element method. Fredlund [9-10] presents extended shear strength and volumetric change theories for unsaturated soils, which takes the pore water pressure into consideration. Many numerical studies on infiltration in slopes has presented in references [11-17].

The above numerical rainfall infiltration simulations assume the surface water depth is zero when runoff occurs. But, in fact, the physical processes of rainfall infiltration and surface runoff interact simultaneously. To reflect this interaction, conjunctive surface and subsurface flow model should be adopted. Akan and Yen [18-19] develop a conjunctive one-dimensional surface flow and two-dimensional subsurface flow model. Morita and Yen [20-21] extend it to a conjunctive two-dimensional surface and three-dimensional subsurface flow model. Recently, other researchers [22-26] present new models for solving conjunctive surface and subsurface flow.

In this paper, a new conjunctive one-dimensional surface flow and two-dimensional subsurface flow model is developed. The governing equations of one-dimensional surface water flow on steep slope are modified Saint Venant equations [27] where the effect factor of slope angle is introduced into momentum equation. The governing equation of two-dimensional subsurface flow equation is Richards equation. It is solved by finite element method. The interaction between surface and subsurface flow is the interface infiltration, which is determined by iterations between surface and subsurface flow models. To verify the validity and illustrate the characteristics of conjunctive surface and subsurface flow model, four numerical study cases are performed. A summary of those numerical study cases is presented in Table 1.

Table 1. Summary of numerical simulation cases.

Case No	1	2	3	4
SWCC	See Figure 5	See Figure 5	See Figure 5	See Figure 5
Ks (m/s)	2.0×10^{-5}	1.23×10^{-5}	1.23×10^{-5}	1.23×10^{-5}
Dimension	1D	1D	2D	2D
Boundary type	Head	Flux	Head/Flux automatically specified	Head/Flux automatically specified
Rainfall intensity (m/s)		2.5×10^{-6}	2.74×10^{-5}	2.74×10^{-5}
Initial condition	Constant saturation S=0.1	Negative hydrostatic distribution	Steady state solution with low groundwater table	Steady state solution with high groundwater table
Remark	For model verification	For model verification	For model comparison	For model comparison

2. Conjunctive surface-subsurface flow model

2.1. Theory of surface flow model

The governing equations for the gradually varied unsteady shallow water flow are commonly known as the Saint Venant equations [28]. In one-dimensional condition, Yen [27] extended the Saint Venant equations to accommodate the rapid varied unsteady flow in steep slope condition as shown in Equations (1) and (2).

$$\frac{\partial h}{\partial t}+\frac{\partial(uh)}{\partial x}=r-f \tag{1}$$

$$\frac{\partial(uh)}{\partial t}+\frac{\partial(u^2h)}{\partial x}+gh\cos\theta\frac{\partial h}{\partial x}=gh(S_o-S_f) \quad (2)$$

A sketch of variables definition of the modified Saint Venant equations is presented in Figure 1. $S_0=\sin\theta$ is the bottom slope; and S_f is the friction slope. The main difference between modified and original Saint Venant equations is the pressure force term in the momentum equation. When the slope angle is close to zero, factor $\cos\theta$ tends to 1.0 and modified Saint Venant equations return to its original form.

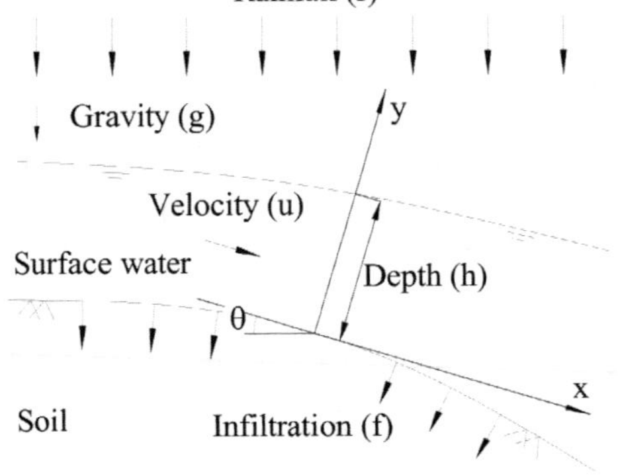

Figure 1. Description of variables in Saint Venant equations

The friction slope S_f is derived according to the Darcy-Weisbach [28] formula

$$S_f=f_d\frac{u^2}{8gh} \quad (3)$$

where f_d is the frictional resistance coefficient, which is calculated from an approximated form of the Moody diagram [29]. Under laminar, transitional, and turbulent flows conditions, f_d is calculated as follows respectively:

$$f_d=\frac{C_L}{R_e} \qquad \text{For laminar flow} \quad (4)$$

$$f_d=\frac{0.223}{R_e^{0.25}} \qquad \text{For transition flow} \quad (5)$$

$$f_d=\left[2\log\left(\frac{2h}{k}\right)+1.74\right]^{-2} \qquad \text{For turbulent flow} \quad (6)$$

where R_e is the Reynolds number, which is the ratio of inertia force to viscous force ($R_e=uh/\nu$) with ν being the kinematic viscosity; k is a length measure of surface roughness; C_L is a coefficient which is a function of rainfall intensity. If the effect of infiltration on the resistance is assumed negligible, C_L is equal to 24 [27-28].

2.2. Numerical scheme for surface flow model

The traditional Preissmann four point implicit finite difference scheme is used to solve Equations (1) and (2), which has the advantage of producing unconditionally stable solutions. Newton iteration technique is used to deal with the non-linear problem. Since the flow regime may become supercritical instead of subcritical [28], the boundary conditions for the Saint Venant equations will be changed from one upstream and one downstream to two upstream boundary conditions automatically. A schematic flow chart of surface flow is illustrated in Figure 2.

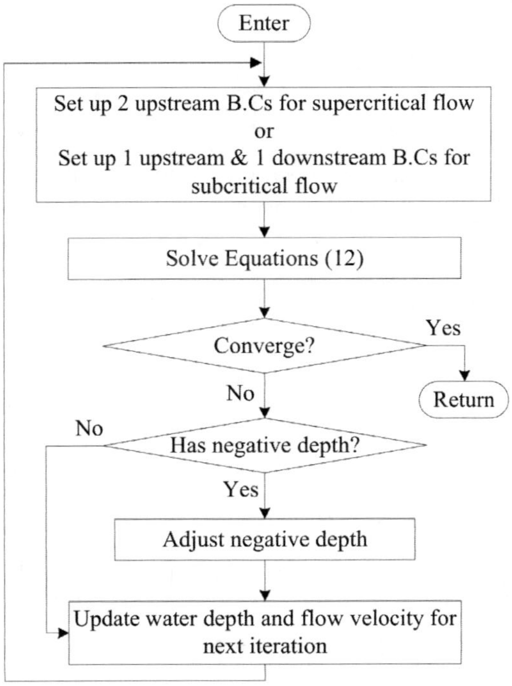

Figure 2. Flow chart of surface flow model

2.3. *Theory of subsurface flow model*

Two-dimensional Richards equation and the flux boundary condition without consideration of source term are presented as follows

$$m_w\gamma_w\frac{\partial H_w}{\partial t}-\frac{\partial}{\partial x}\left(k_x\frac{\partial H_w}{\partial x}\right)-\frac{\partial}{\partial y}\left(k_y\frac{\partial H_w}{\partial y}\right)=0 \qquad \text{on } \Omega \tag{7}$$

$$(\mathbf{k}\cdot\nabla H_w)\cdot\mathbf{n}-\bar{q}=0 \qquad \text{on } \Gamma \tag{8}$$

where H_w is total head; k_x and k_y are hydraulic conductivity in the x-direction and y-direction respectively; $mw=\partial\theta_w/\partial u_w$, which is the slope of soil-water characteristic curve, where θ_w is volumetric water content and u_w is pore water pressure; γ_w is the unit weight of water; t is the time; **k** is hydraulic conductivity tensor; ∇Hw is the gradient of total head at boundary surface; **n** is the unit normal direction vector; and $\bar{q}$ is the boundary flux.

2.4. *Numerical scheme for subsurface flow model*

The Galerkin method is used to solve the finite element equation (7) and (8). Set **N** as the interpolating function vector and **H** as the total head vector of the element. Let

$$\mathbf{C}=\int_\Omega \mathbf{N}^T\mathbf{N}(m_w\gamma_w)d\Omega \tag{9}$$

$$\mathbf{K}=\int_\Omega\left[\left(\frac{\partial\mathbf{N}}{\partial x}\right)^T k_x\left(\frac{\partial\mathbf{N}}{\partial x}\right)+\left(\frac{\partial\mathbf{N}}{\partial y}\right)^T k_y\left(\frac{\partial\mathbf{N}}{\partial y}\right)\right]d\Omega \tag{10}$$

$$\mathbf{Q} = \int_{\Gamma} \mathbf{N}\bar{q}\, d\Gamma \tag{11}$$

Then, the finite element equation can be expressed symbolically as follows:

$$\mathbf{C} \cdot \dot{\mathbf{H}} + \mathbf{K} \cdot \mathbf{H} = \mathbf{Q} \tag{12}$$

Using the backward difference method to discretize the time term leads to the following equation:

$$\left[\frac{\mathbf{C}}{\Delta t} + \mathbf{K}\right]\mathbf{H}_{n+1} = \left[\frac{\mathbf{C}}{\Delta t}\right]\mathbf{H}_{n} + \mathbf{Q}_{n+1} \tag{13}$$

Equation (13) is the matrix form of finite element equation usually used to solve the saturated-unsaturated seepage problem. Based on the subsurface model presented above, a FORTRAN module is written to solve two-dimensional saturated-unsaturated subsurface flow. Lumped mass matrix method is adopted for mass matrix **C** to enhance the stability of solution. A schematic flow chart of subsurface flow is illustrated in Figure 3.

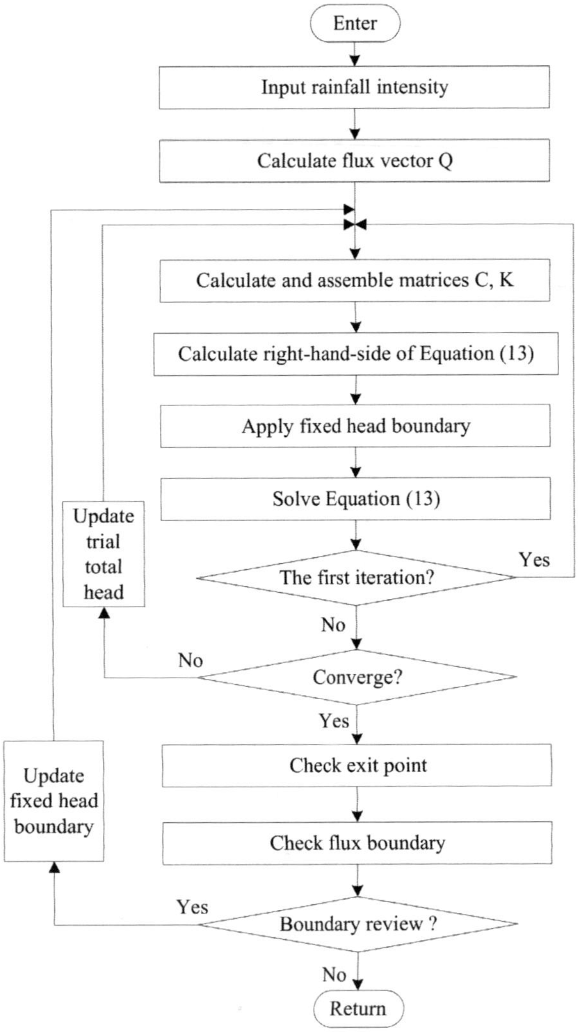

Figure 3. Flow chart of subsurface flow model

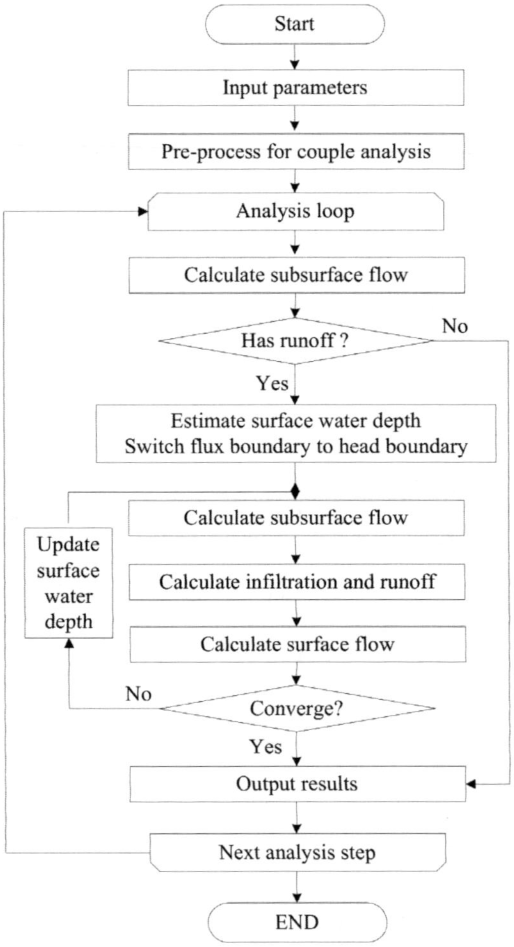

Figure 4. Flow chart of conjunctive flow model

2.5. Judgment of runoff generation

A trial and error procedure is used to judge runoff generation on slope surface. First assume all the rainfall supplies at that time interval could infiltrate into soil under its own gravity and matric suction gradients. Then calculate the subsurface flow. If positive water pressure occurs somewhere, it means that extra pressure is needed to force the given rainfall infiltrate into slope, which is against the previous assumption of no extra forces. Thus, runoff will generate at that position. In contrast, Morita and Yen [20-21] use the infiltrability of soil to judge runoff generation, but it is hard to be estimated.

2.6. Calculate infiltration and runoff

Infiltration is calculated after solving subsurface flow. It is a backward procedure of solving total head or pore water pressure distribution under a given boundary condition. The right hand side term $\mathbf{Q}_{n+1}$ in Equation (13) denotes the equivalent nodal flux. When the total head vector at n+1 time is obtained, it can be calculated as follows:

$$\mathbf{Q}_{n+1} = \mathbf{K}\mathbf{H}_{n+1} + \left[\frac{\mathbf{C}}{\Delta t}\right](\mathbf{H}_{n+1} - \mathbf{H}_{n}) \tag{14}$$

Since the nodal flux is obtained, the unit flux can be easily estimated by dividing the length of its edge. Thus, infiltration and runoff can be obtained.

2.7. Conjunctive surface and subsurface flow model flow chart

The main flow chart of conjunctive surface and subsurface flow model is shown in Figure 4. First, prepared data and parameters are read and a series of pre-processes for conjunctive model analysis are done. Then, it enters into the analysis loop and conducts simulation one time step by one time step. In each time step, firstly run subsurface flow model to check if runoff occurs on slope. Then, estimate surface water depth as trial value and switch flux boundary to head boundary to calculate subsurface flow again. When the convergence is obtained, calculate the infiltration and runoff respectively. Then use the runoff as input to conduct the surface flow modeling and calculate surface water depth along the slope. Finally, compare the trial and calculated surface water depths. If the differences of those depths tend to an acceptable convergence, it will end the current time step, output corresponding results, and return to next analysis step, or else, it will update the trial surface water depth and modify the corresponding head boundary to iterate again.

3. Model verification

3.1. One-dimensional infiltration under head boundary

In case 1, the height of soil column is 4 m. At the top of soil column, the total head boundary is applied with the value of 4 m, which means the pressure head is zero. The saturated and residual volumetric water contents of soil are 0.2 and 0.02 respectively. The porosity is 0.2; saturated conductivity is 2.0×10^{-5} m/s. Initial saturation of whole soil column is set to be 0.1. The soil-water characteristic curve and hydraulic conductivity function are illustrated in Figure 5. In numerical simulation, finite element mesh space in vertical direction dz is equal to 0.1 m and the time step dt is equal to 6 seconds. Infiltration simulation in 3 hours duration is performed.

The comparison of saturation profiles among conjunctive model, analytical solution [30], and SEEP/W at 1, 2, and 3 hours respectively are presented in Figure 6. Figure 6 shows that the degree of saturation profiles of conjunctive model are overlapped with that of analytical solution and SEEP/W. The relative errors among those three models are within 0.2%. In this case, the soil column is 4 times deeper and saturated conductivity is about 3 times larger than that of Morita and Yen [20]. But both the differences between numerical and analytical solution are rather small.

3.2. One-dimensional infiltration under head boundary

In case 2, the height of soil column is 4m. A kind of homogeneous fine sand [17] is used with the Ks is 1.23×10^{-5} m/s (44.3mm/hr) and the porosity is 0.347. Its soil-water characteristic curve and hydraulic conductivity function are shown in Figure 7. The residual volumetric water content is about 0.184. The rainfall duration is 12 hours. The initial pore water pressure distribution in soil column is in negative hydrostatic distribution. The rainfall is applied at the top surface and the bottom of soil column is closed and impermeable.

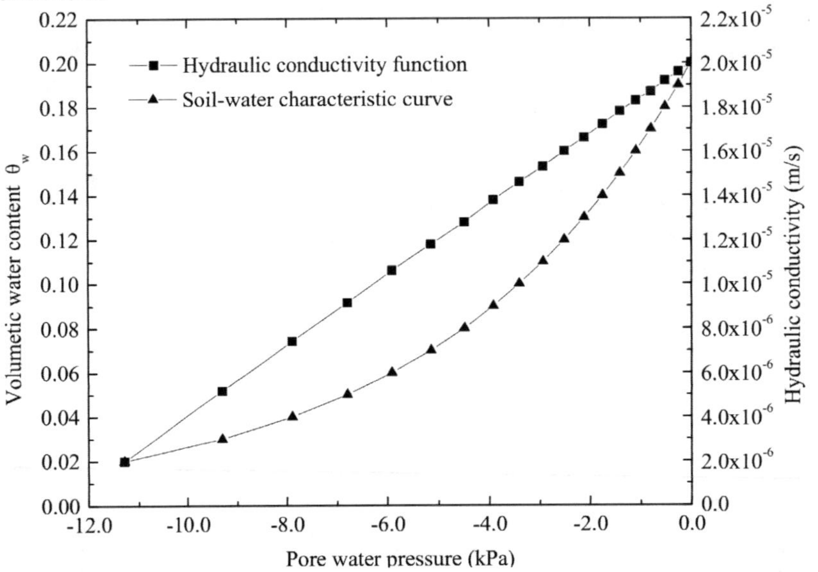

Figure 5. Soil-water characteristic curve and hydraulic conductivity function used in analytical model

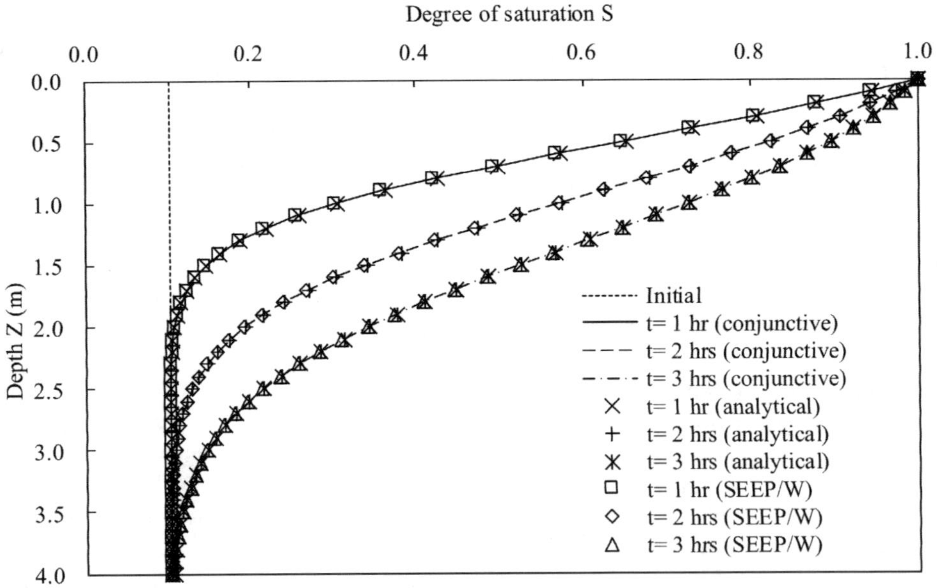

Figure 6. Comparisons of saturation profile among conjunctive model, analytical solution, and SEEP/W under head boundary

The calculated results of conjunctive model and SEEP/W are illustrated in Figure 8. At 4 hours rainfall, the wetting front affects to about 0.7 m in depth. The maximum difference of pore water pressure at wetting front between conjunctive model and SEEP/W is about 3 kPa. At the end of 8 and 12 hours rainfall, the wetting front advances to about 1.1 m and 1.6 m in depth respectively and the maximum difference of pore water pressure at wetting front between two models is still about 3 kPa. While, in the nearly saturated zone, the pore water pressure changes mildly against that on wetting front and the pore water pressure result of conjunctive model is highly consistent with that of SEEP/W. The pore water pressure of nearly saturated zone is converged to a value of about -7 kPa, which is consistent with the pore water pressure value in hydraulic conductivity function with respect to conductivity of 2.5×10^{-6} m/s. The difference of pore water pressure distribution between conjunctive model and SEEP/W is small and stable within 12 hours rainfall duration. It shows that the result of conjunctive model is valid and reliable in solving flux boundary problem.

As a result, the subsurface model in conjunctive surface and subsurface flow model is verified under both head boundary and flux boundary. The results show that it is robust in solving highly non-linear saturated-unsaturated flow.

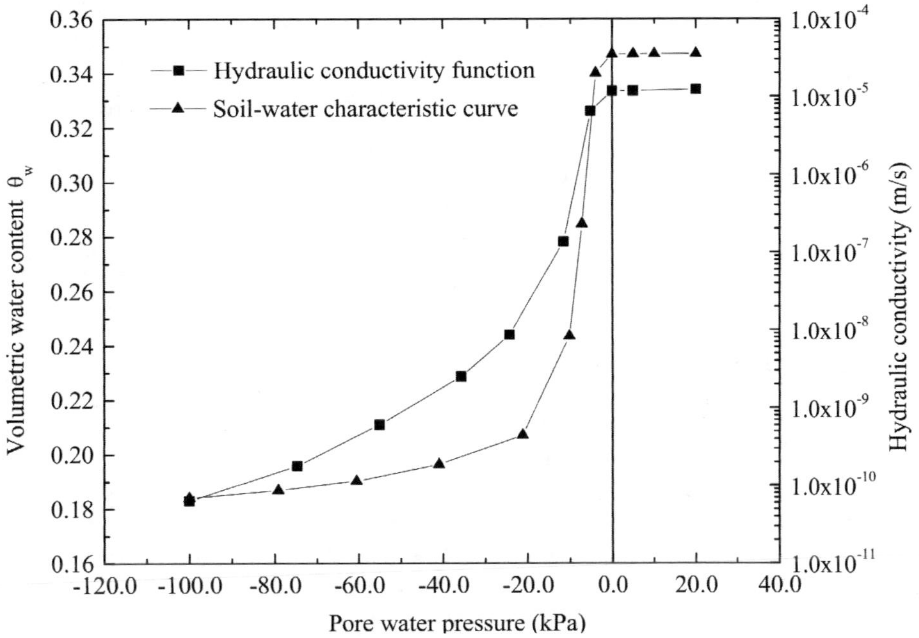

Figure 7. Soil-water characteristic curve and hydraulic conductivity function of fine sand

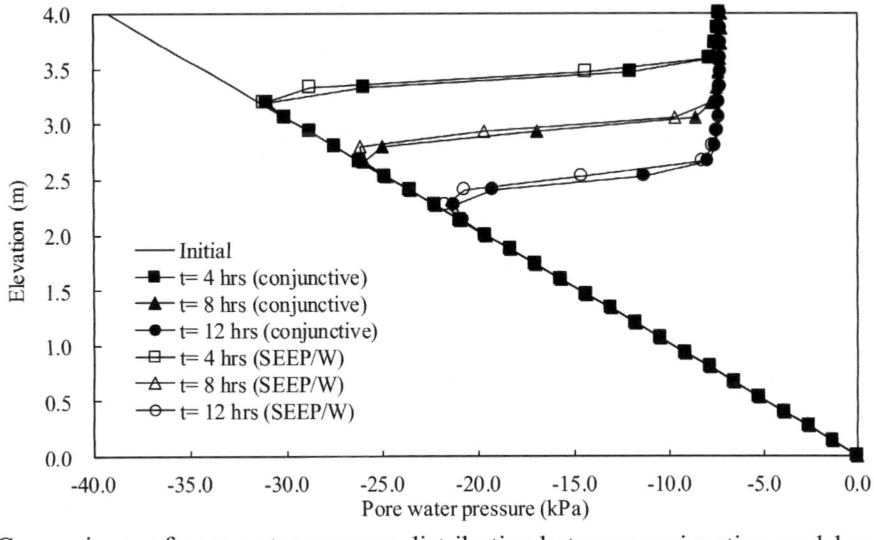

Figure 8. Comparisons of pore water pressure distribution between conjunctive model and SEEP/W under flux boundary

4. NUMERICAL SIMULATION

4.1. Study cases

In case 3, the initial groundwater table is low and the slope soil is dry, while in case 4 the initial groundwater table is high and the slope soil is relative wet. The soil properties are the same as that of case 2. 4 hours uniform rainfall with intensity of 2.74×10^{-5} m/s (98.6 mm/hour) is adopted. The numerical time step is about 6 seconds. Total simulation time is 48 hours. The total rainfall amount is close to a high daily rainfall 394 mm that is presented in Ng and Shi's study [11], but the hourly rainfall intensity is several times larger.

4.2. Geometry and meshes

The geometry and finite element mesh for subsurface flow analysis are presented in Figure 9. The length of the slope is 50 m; the slope angle is 30°; and the depth of calculation zone is 15 m. The whole subsurface flow domain is divided into 2500 quadrilateral elements with 2626 nodes.

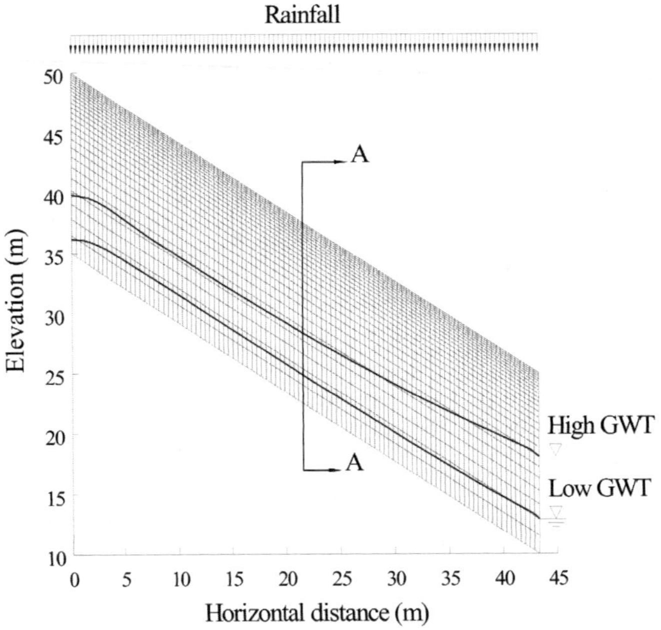

Figure 9. Slope geometry, mesh, and initial groundwater tables

4.3. Initial and Boundary conditions

Initial groundwater table and pore water pressure are estimated by running steady state flow under proper boundary conditions. A small rainfall intensity with the value of 1.0×10^{-9} m/s (0.0864 mm/day) is applied on the slope surface. A river is assumed at the toe of slope with the elevation of 12.9 m in case 3 and 18.0 m in case 4. The positions of two initial groundwater tables are illustrated in Figure 9. As for surface flow, the initial condition is assumed that the slope surface is dry.

For subsurface flow, the slope surface is a flux boundary subjected to rainfall. When the surface is saturated, the flux boundary will switch to head boundary. The proportion of right side boundary above the river level is specified as potential seepage surface and the exit point will be checked during iterations. Besides, the other boundaries are impermeable. For surface flow, the surface water depth at upstream is equal to the base flow water depth, i.e. 0.15 mm. If surface flow is subcritical, the downstream water depth is specified to be equal to normal depth. If surface flow is supercritical, the upstream inflow velocity is specified according to base flow.

4.4. Results and discussions

In conjunctive model, the upstream surface water depth is fixed as base flow water depth, when runoff discharges from upstream to downstream, the surface water depth increases gradually and the downstream water depth is the highest which is sensitive with time. The comparisons of downstream surface water depth hydrograph within the first 5 hours between two investigated cases are shown in Figure 10. It shows that at about 0.35 hour the downstream point begins to have remarkable surface water under low groundwater table condition. While under high groundwater table condition, the remarkable surface water generates earlier at about 0.22 hour. This is because the slope soil with high groundwater table is wetter than that with low groundwater. And the initially drier soil slope can absorb more water than that of wetter soil slope. So under the same rainfall pattern, the occurrence of surface runoff in initially drier slope will delay a short time. When larger amount of runoff occurs and discharges to downstream, the downstream water depth rises up prominently. From 0.22 to about 1.0 hour, the downstream water depth increases rapidly. During this time, the downstream water depth under high groundwater table condition is larger than that under low groundwater table condition. At about 1.0 hour, the difference of surface water depth tends to quiet small, which shows that the influence of initial condition begins to disappear. At about 1.5 hours, the surface water depth reaches to a limit value of about 5.2 mm, which means that at the coupled processes among rainfall, infiltration, and surface runoff have reached to a relative steady state. At the end of 4 hours, the downstream water depth reaches to about 5.3 mm. After rainfall ceases at 4 hours, the surface water on slope discharges rapidly. The water depth at downstream drops from about 5.3 mm to zero in 7 minutes. In contrast, a similar hydrograph on mild open channel slope is presented by Morita and Yen [21]. The slope length is about 1/5 of our study; uniform rainfall intensity is about 1/2 of our study; and duration is 1/12 of our study. But the downstream surface water depth is about 10 mm which is about twice of those cases.

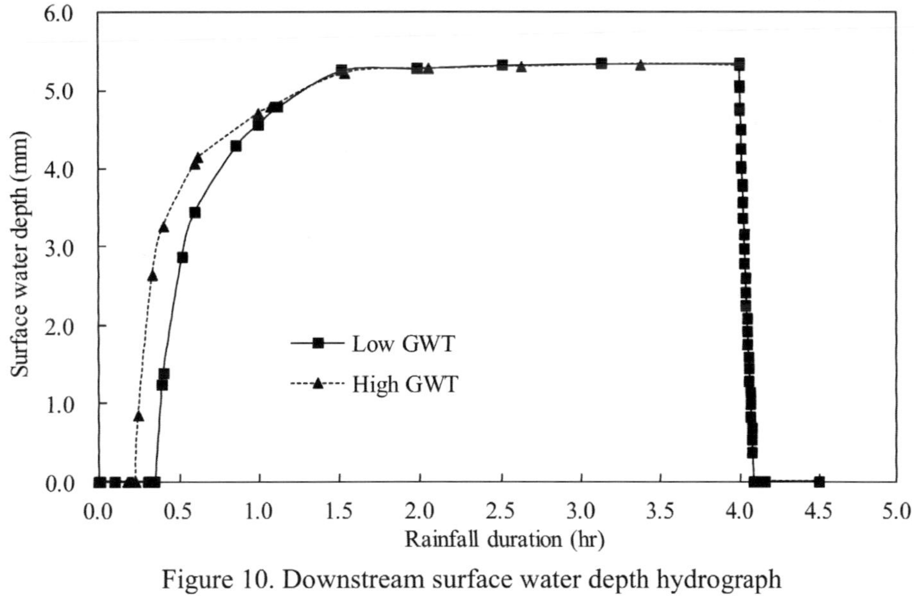

Figure 10. Downstream surface water depth hydrograph

The results of surface water depth profile under different initial conditions are compared at selected times, which are illustrated in Figure 11. Because the upstream boundary of surface flow is fixed, the surface water depth at distance of zero along slope surface, i.e. upstream point, is equal to 0.15 mm. 4 observing times are selected to illustrate the surface water depth profile, i.e. 0.4, 0.6, 1.0, and 4 hours. Figure 11 shows that at 0.4 hour the surface water is shallow under low groundwater table condition with the maximum water depth of about 1.4 mm. The surface water depth increases gradually along the slope from upstream to downstream. While, under high groundwater table condition the surface water depth profile rises up to higher position with the maximum water depth of about 3.2 mm. At 0.6 hour, the surface water depth increases quickly under low groundwater table. The maximum water depth reaches to about 3.4 mm, which is 2 mm higher than that at 0.4 hour. While, under high groundwater table the surface water profile increases about 0.8 mm. At this time, the difference of maximum water depth between different initial conditions is about 0.6 mm, which is much smaller than that at 0.4 hour. Then, at 1 hour, the difference of the maximum surface water depth reduces to about 0.15 mm and the difference of surface water profile between different initially conditions becomes to disappear. Finally, when rainfall ceases at 4 hours, the difference of maximum water depth is about 0.01 mm which is too small to distinguish and two profiles are nearly superposed.

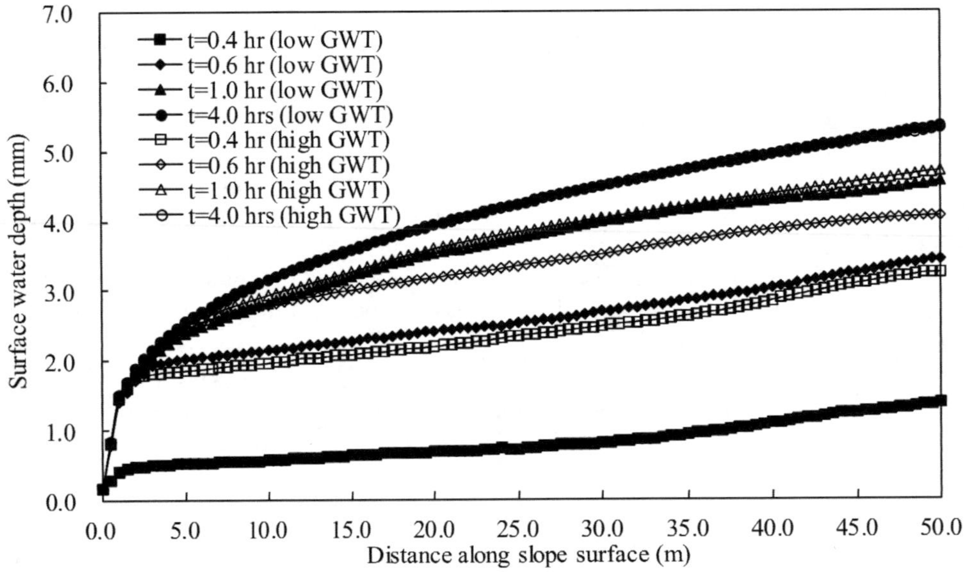

Figure 11. Surface water depth profiles at different times

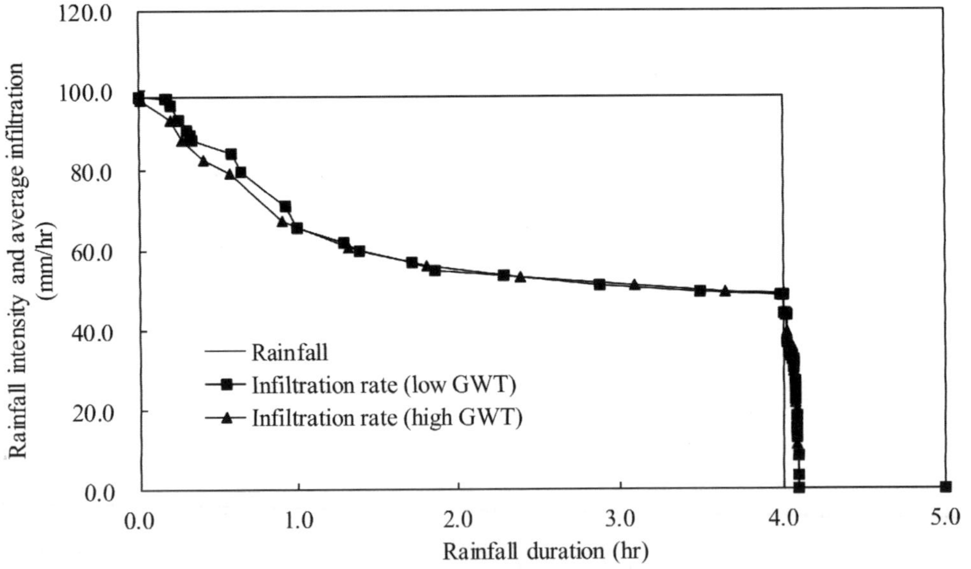

Figure 12. Comparisons of rainfall intensity and infiltration rate in different initial conditions

The comparisons of infiltration rate between different initial conditions are presented in Figure 12. Hillel [31] shows that the infiltration capacity of soil increases as soil becomes dry, thus under initial lower groundwater table condition, the slope soil can fully absorb higher rainfall intensity at beginning. As shown in Figure 12, in the first 0.15 hour, almost all the rainfall water is absorbed into slope soil under initial lower groundwater table condition, while under high groundwater table condition the surface becomes saturated sooner after rainfall applies on slope. Morita and Yen [21] show a similar process. That is, when the infiltration capacity of soil is greater than the rainfall intensity, all rainwater enters into soil. Else, parts of the slope surface become saturated and runoff occurs on it. Then, infiltration rate decreases gradually and tends to the constant value of saturated hydraulic conductivity [31]. At about 1 hour, the infiltration rates under different initial conditions tend to be equal, which is closely consistent with the changes of surface water depth as shown in Figure 11. After about 1.5 hours, the infiltration rate changes slightly and the coupled processes of infiltration and runoff reach to a relative steady state. When rainfall stops, the infiltration continues for about 7 minutes as surface water exists on slope.

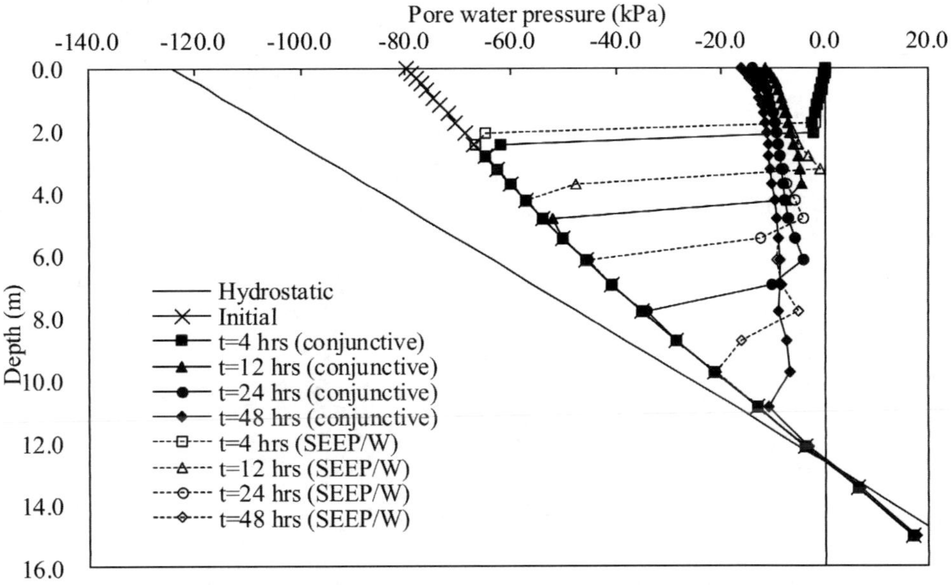

Figure 13. Comparisons of pore water pressure distribution between conjunctive model and SEEP/W at section A-A

The comparisons of pore water pressure distributions in different initial conditions at A-A section is illustrated in Figure 13. Figure 13 shows that under high groundwater table condition, the pore water pressure at slope surface is about -60 kPa, which is 20 kPa greater than that under low groundwater table condition. At 4 hours, the difference of wetting front between conjunctive model and SEEP/W is about 0.4 m under low groundwater table condition, while under high groundwater table condition, that difference is nearly zero. Although the wetting front of coupled model reaches to nearly the same depth under different initial conditions, the infiltration amount is different. Then, at the end of 24 hours, the pore water pressure has redistributed for about 20 hours. The difference of wetting front between conjunctive model and SEEP/W under low groundwater table condition is about 1.6 m, which is about two times of that under high groundwater table condition. Finally, at the end of 48 hours, under low groundwater table condition, the difference of wetting front between conjunctive model and SEEP/W is still large, which is about 2.0 m. While under high groundwater table condition, the wetting front of both conjunctive model and SEEP/W reaches to groundwater table, which causes groundwater table raising 1.2 m and 0.4 m for coupled and uncoupled results respectively. According to the comparisons, in a word, the difference of pore water pressure distribution between conjunctive model and SEEP/W under initial dry condition is more significant than that under initial wet condition. The comparison of pore water pressure profiles between conjunctive model and SEEP/W at A-A section under low groundwater table condition is shown in Figure 14. It shows that the wetting front of conjunctive model reaches to about 2.4 m in depth from the ground surface after 4 hours rainfall, which is about 0.4 m deeper than that of SEEP/W. The difference reflects that under the positive pressure head of surface water and interaction between infiltration and runoff, the infiltration in each time step is greater than that of uncoupled model SEEP/W. So, during rainfall time, more water infiltrates into slope and the wetting front moves deeper as shown in conjunctive model result. When rainfall stops at 4 hours, in conjunctive model the surface water keeps on affect the infiltration for about 7 minutes. While in SEEP/W, infiltration stops with the cease of rainfall simultaneously. When runoff disappears from the slope surface, pore water pressure begins to redistribute within the slope. The pore water pressure at surface decreases with time. The wetting front continues to move downward. At the end of 12 hours, the wetting front of conjunctive model reaches to about 4.8 m in depth, which is about 1.2 m deeper than that of SEEP/W. In addition, at the end of 24 hours, the wetting front of conjunctive model moves to about 7.6 m in depth, which is about 1.6 m deeper than that of SEEP/W. Finally, at 48 hours, the wetting front is close to groundwater table. As water discharges into groundwater, the difference of pore water pressure distribution between conjunctive model and SEEP/W tends to lessen.

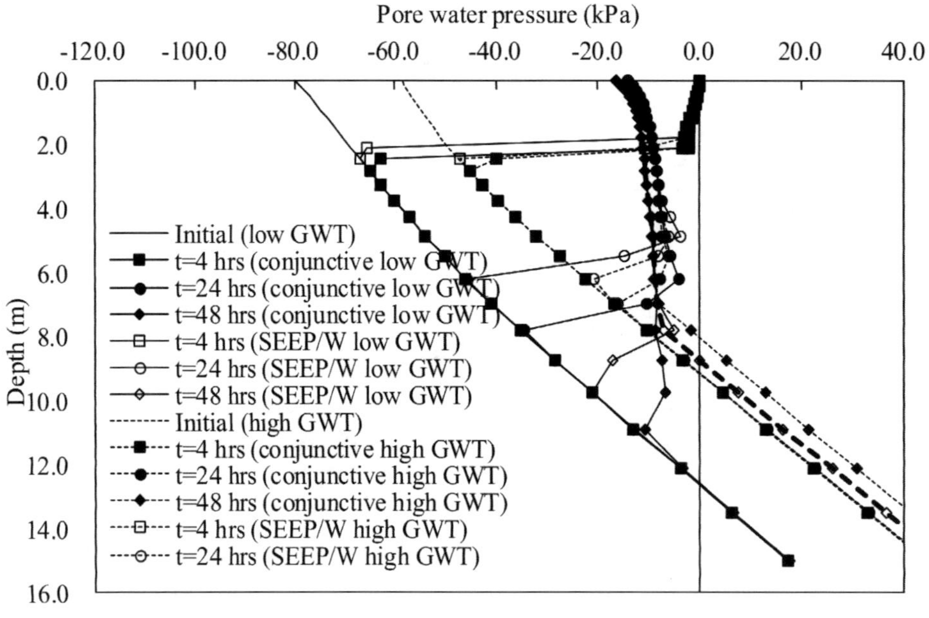

Figure 14. Comparisons of pore water pressure distribution in different initial conditions at section A-A

5. Conclusions

A new conjunctive one-dimensional surface flow and two-dimensional subsurface flow model is developed to analyze infiltration a nd runoff simultaneously in steep slope. The governing equations of one-dimensional surface flow are modified Saint Venant equations. The governing equation of two-dimensional subsurface flow is Richards equation. The subsurface model in conjunctive model is verified by head boundary problem and flux boundary problem respectively. Then, the characteristics of conjunctive model are illustrated. Typical surface water depth and infiltration rate changes with rainfall duration are presented. Furthermore, the pore water pressure distributions are compared and discussed between conjunctive model and SEEP/W, and the effect of initial condition is studied. Based on the above studies, the following conclusions can be drawn:

(1) Using new developed conjunctive surface and subsurface flow model, surface flow on steep slope can be solved; the infiltration rate versus time can be determined; and the actual infiltration and runoff processes can be simulated.

(2) When runoff occurs on slope, the surface water depth can be calculated. As runoff increases, the surface water depth rises up. When infiltration and runoff tend to a relative steady state, the surface water depth also rises to a relative steady level. Under 98.6 mm/hour rainfall intensity, the maximum downstream water depth over a 30° slope after 4 hours duration is about 5.3 mm. When rainfall stops, the surface water discharges rapidly within about 7 minutes.

(3)Surface water depth and the interaction between infiltration and runoff have remarkable effects on rainfall infiltration process. According to the results of conjunctive surface and subsurface flow model, a greater amount of rainwater infiltrates into the slope. Therefore, the pore water pressure changes faster and the wetting front moves deeper into the soil.

(4) Under initial drier condition, the capacity of infiltration is higher and more rainfall can be absorbed into slope. The difference of pore water pressure distribution between conjunctive model and SEEP/W under initial dry condition is more significant than that under initial wet condition.

Acknowledgments
This research is supported by the project 51178251 provided by National Natural Science Foundation of China and research grants RGC-HKUST6108/99E provided by the Research Grants Council of the Hong Kong Government of the Special Administrate Region. The authors are grateful to Prof. Ng, CWW and Prof. Tung, YK from HKUST.

References

[1] American Petroleum Institute. Review of methods to estimate moisture infiltration, recharge, and contaminant migration rates in the vadose zone for site risk assessment. API Publication, No. 4643, Washington, D.C., 1996.

[2] Ravi V, Williams JR. Estimation of infiltration rate in the vadose zone: compilation of simple mathematical models. Environmental Protection Technology Series, EPA/600/R-97/128a, vol. I, 1998.

[3] Lumb, PB. Effects of rain storms on slope stability. Sym. on Hong Kong Soils., Hong Kong, 1962; 73-87.

[4] Fukuoka, M. Landslides associated with rainfall. Geotechnical Engineering, Journal of the Southeast Asia Society of Soil Engineering 1980; 11:1-29.

[5] Anderson, MG, Pope, RG. The incorporation of soil water physics models into geotechnical studies of landslide behaviour. Proc. 4th. Int. Sym. on Landslides, vol. 4, 1984; 349-353.

[6] Wolle, CM, Hachich, W. Rain-induced landslides in south-eastern Brazil. Proc. 12th Int. Conf. Soil Mech. & Fnt Engg, Rio de Janeiro, vol. 3, 1989; 1639-1644.

[7] Neuman, SP. Galerkin approach to saturated-unsaturated flow in porous media. Finite Elements in Fluids. vol. 1, viscous flow and hydrodynamical, Wiley: London, 1974; 201-217.

[8] Lam, L, Fredlund, DG, Barbour, SL. Transient seepage model for saturated-unsaturated soil systems: a geotechnical engineering approach. Canadian Geotechnical Journal 1987; 24:565-580.

[9] Fredlund DG, Morgenstern, NR, Widger, RA. The shear strength of unsaturated soils. Canadian Geotechnical Journal 1978; 15(3):313-321.

[10] Fredlund, DG, Rahardjo, H. Soil Mechanics for Unsaturated Soils. Wiley: Chichester, NY, 1993.

[11] Ng, CWW, Shi, Q. A numerical investigation of the stability of unsaturated soil slopes subjected to transient seepage. Computers and Geotechnics 1998; 22(1):1-28.

[12] Rahardjo, H, Leong, EC, Gasmo, JM, Deutscher, MS. Rainfall-induced slope failures in Singapore: investigation and repairs. Proceedings of the 13th Southeast Asian Conference, vol. 1, 1998; 147-152.

[13] Ng, CWW, Pang, YW. Influence of stress state on soil-water characteristics and slope stability. Journal of Geotechnical and Geo-environmental Engineering, ASCE 2000; 126(2):157-166.

[14] Ng, CWW, Wang, B, Tung, YK. Three-dimensional numerical investigations of groundwater responses in an unsaturated slope subjected to various rainfall patterns. Canadian Geotechnical Journal 2001; 38(5):1049-1062.

[15] Tung, YK, Ng, CWW, Liu, JK. Lai Ping road landslide investigation- Three dimensional groundwater flow computation. Technical Report, part I, Geotechnical Engineering Office of the Hong Kong Special Administrative Region, Hong Kong, 1999.

[16] Tsaparas, I, Rahardjo, H, Toll, DG, Leong, EC. Controlling parameters for rainfall-induced landslides. Computers & Geotechnics 2002; 29(1):1-27.

[17] Geo-Slope international Ltd. SEEP/W for Finite Element Seepage Analysis (version 5) User's Guide, Geo-Slope International, Canada, 2002.

[18] Akan AO. Unsteady shallow water flow on porous media. Thesis of Doctor of Philosophy, University of Illinois at Urbana-Champaign, U.S., 1976.

[19] Akan, AO, Yen, BC. Mathematical model of shallow water flow over porous media. Journal of Hydraulic Engineering, ASCE 1981; 107(4):479-494.

[20] Morita, M, Yen, BC. Numerical methods for conjunctive two-dimensional surface and three-dimensional sub-surface flows, International Journal for Numerical. Methods in Fluids 2000; 32:921-957.

[21] Morita, M, Yen, BC. Modeling of conjunctive two-dimensional surface-three-dimensional subsurface flows. Journal of Hydraulic Engineering, ASCE 2002; 128(2):184-201.

[22] Kwok, YF. Stochastic analysis of coupled surface and subsurface flow model in steep slopes for slope stability analysis, Thesis of Master of Philosophy, Civil Engineering, Hong Kong University of Science and Technology, Hong Kong, 2003.

[23] Tung, Y.K., Zhang, H. Ng, C.W.W. & Kwok, Y.F. Transient seepage analysis of rainfall infiltration using a new conjunctive surface-subsurface flow model. Proc. 57th Can. Geot. Conf. and 5th joint CGS-IAH Conf. 24-27 Oct. 2004, Quebec City, Canada. Session 7C, 17-22.

[24] He, Z.G., W.M. Wu, S.S.Y. Wang. Integrated Two-Dimensional Surface and Three-Dimensional Subsurface Contaminant Transport Model Considering Soil Erosion and Sorption. Journal of Hydraulic Engineering-Asce， 2009， 135(12): 1028-1040.

[25] Rihani, J.F., R.M. Maxwell, F.K. Chow. Coupling groundwater and land surface processes: Idealized simulations to identify effects of terrain and subsurface heterogeneity on land surface energy fluxes. Water Resources Research， 2010， 46.

[26] Tian, D.F., D.F. Liu. A new integrated surface and subsurface flows model and its verification. Applied Mathematical Modelling， 2011， 35(7): 3574-3586.

[27] Yen, BC. Stormwater runoff on urban areas of steep slope. Environmental Protection Technology Series. EPA/600/2-77-168, 1977.

[28] Kareliotis, SJ, Chow, VT. Computer solution of a hydrodynamic watershed model (IHW Model II), Civil Engineering Studies, Hydraulic Engineering Series No. 25, University of Illinois at Urbana-Champaign, U.S., 1971.

[29] Chaudhry, MH. Open-Channel Flow. Prentice-Hall: Englewood Cliffs, NJ, 1993.

[30] Philip, JR. Theory of infiltration, Advances in Hydroscience, Chow, VT (ed.), vol. 5, Academic: New York, 1969; 215-296.

[31] Hillel, D. Introduction to Soil Physics. Academic: New York, 1982.

IOP Conf. Series: Earth and Environmental Science **26** (2015) 012027 doi:10.1088/1755-1315/26/1/012027

Macro-Level Assessment of Seismically Induced Landslide Hazard for the State of Sikkim, India Based On GIS Technique

Naveen James[1] and T G Sitharam[2]

[1]Assistant Professor, Department of Civil Engineering, Birla Institute of Technology and Science Pilani, Hyderabad Campus, Hyderabad, 500078, India
[2] Professor, Department of Civil Engineering, Indian Institute of Science, Bangalore, 560012, India
E-mail: naveenjames7@gmail.com

Abstract. This paper presents a macro-level seismic landslide hazard assessment for the entire state of Sikkim, India, based on the Newmark's methodology. The slope map of Sikkim was derived from ASTER Global Digital Elevation Model (GDEM). Seismic shaking in terms of peak horizontal acceleration (PHA) at bedrock level was estimated from deterministic seismic hazard analysis (DSHA), considering point source model. Peak horizontal acceleration at the surface level for the study area was estimated based on nonlinear site amplification technique, considering B-type NEHRP site class. The PHA at surface was considered to induce driving forces on slopes, thus causing landslides. Knowing the surface level PHA and slope angle, the seismic landslide hazard assessment for each grid point was carried out using Newmark's analysis. The critical static factor of safety required to resist landslide for the PHA (obtained from deterministic analysis) was evaluated and its spatial variation throughout the study area is presented. For any slope in the study area, if the in-situ (available) static factor of safety is greater than the static factor of safety required to resist landslide as predicted in the present study, that slope is considered to be safe.

1. Introduction

Landslide hazard poses major threat to most of the hilly regions in the world. In India, about 15% of the total land area (approximately 0.49 million km^2) is susceptible to landslides. Regions which are highly susceptible to landslide are the Himalayas, North-east India and the Western Ghats NDMA, 2009). Mountainous topography, very high regional seismicity and high rainfall, make the Himalayan region and Northeast India highly susceptible to landslides. The present study makes an attempt to assess seismically induced landslide hazard at a macro-level for Sikkim, a state in India, situated in the Himalayas (fig.1). Several studies have showed that the Himalayan belt, which is an active plate boundary due to the convergent movement between Indian plate and Eurasian plate, is one the most seismically active regions in the world. The 2011 earthquake of magnitude (M_w) 6.9 along India-Nepal border, has triggered more than 300 landslides in Sikkim, affecting transportation and power infrastructures in the state (Chakraborty et al., 2011). The recent devastating earthquake of Nepal with a magnitude of 7.8, has also triggered landslides in the state of Sikkim. All this suggests a pressing need for a quantitative, macro-level seismic landslide hazard assessment for the state of Sikkim, in order to identify the regions with high hazard. The landslide hazard mapping provides decisive information to identify the landslide vulnerable regions, thereby effectively implementing the appropriate mitigation works.

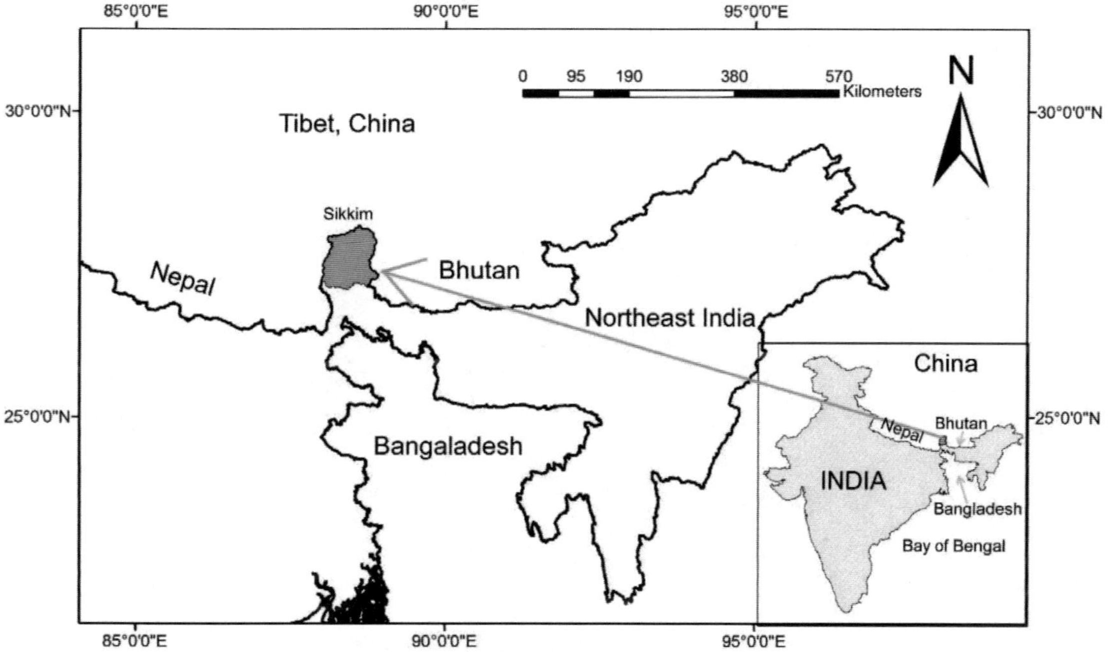

Figure 1. Location of the Study Area

2. Study Area and Seismicity

The state of Sikkim lies in a high seismic zone along the Himalayan belt. It is situated very close to the rupture zone of the 1934 M=8.4 Bihar-Nepal earthquake which is one of the most devastating earthquakes in the region. All the earthquake events within 500 km radius from the boundary of the study area were collected along with epicentral coordinates, focal depth, magnitude, date, time, and year of occurrence. Figure 2 presents the declustred earthquake event map within 500 km from the boundary of the state. Declustering was carried out in order to remove any repetitive events, foreshocks, aftershocks. After declustring, there are about 2830 seismic events occurred in the study area from 1901 to 2014, out of which 943 events are of magnitude 4 and above. There are about 11 events in the study area, having magnitude above 7. And from these, it is very clear that the Sikkim state is located in very high seismically active region Himalayas, where large magnitude earthquakes have occurred in the past.

3. Deterministic Seismic Hazard Analysis

As the literature suggests, there are two methodologies for estimating the seismic hazards at a site they are, Probabilistic Seismic Hazard Analysis (PSHA) and Deterministic Seismic Hazard Analysis (DSHA). The DSHA considers maximum magnitude-distance scenario at a shortest possible distance to the site for quantifying seismic hazard. Since it only considers the critical scenario, the DSHA gives an upper bound value for the ground motion when compared with that of predicted by PSHA at lower return periods. Hence DSHA is more suitable for assessing seismic ground shaking for the critical structures like nuclear power plants, big dams, bridges, hazardous waste contaminant facilities etc. In the present study, the DSHA has been adopted for the evaluation of seismic hazard.

Seismic hazard assessment using a deterministic methodology typically involves the following steps.

i. Identification and characterization of all earthquake sources in the study area.
ii. Selection of source to site distance. Generally the shortest possible distance between the source and the site (hypocentral distance) is considered for hazard estimation.

iii. Estimation of seismic hazard at a site from all known seismic sources using appropriate attenuation models or predictive relations.
iv. Selecting the critical value of ground shaking produced from all the sources as the seismic hazard of that site

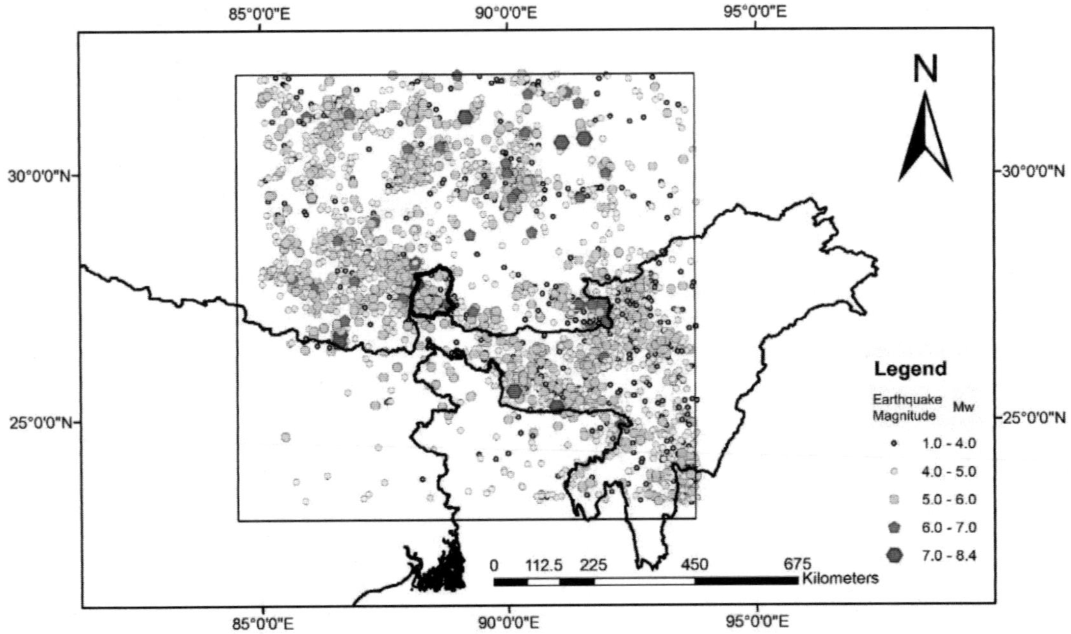

Figure 2. Earthquake events within 500 Km from the Sikkim's boundary

The seismic hazard at location is also dependent on the seismic source model used for the analysis. A seismic source model primarily defines the spatial characteristics of a source and the earthquake distribution within it. In the current study, the hazard analysis was carried out using the point source model. In a point source model, the location of an earthquake event itself is considered as a seismic source, thus, characterizing its geometry as a point. The point source model is simple and can very well characterize the earthquakes associated with a small fault or volcanoes. In spite of its simplicity, the point has been observed that the conventional point source model is always associated with some location error. Hence a smoothing procedure is applied to account for the source dimension of the earthquake events and for location errors (Costa et al., 1993; Panza et al., 1999).

The attenuation relation proposed by Sharma et al. (2009) for the Himalayan region (as in equation 1) has been used in the present study for evaluating PHA. Attenuation models or predictive relationships usually express the ground motion parameter as functions of magnitude, distance etc., which are generally developed from a strong motion data set based on regression analysis.

$$\log(y) = b_1 + b_2 M + b_3 \log\left[\sqrt{R_{JB}^2 - b_4^2}\right] - b_5 S + b_6 H \quad (1)$$

Where b_1, b_2, b_3, b_4, b_5 and b_6 are the regression coefficients, y is spectral acceleration in terms of m/s^2 and S is 1 for rock site, H is 1 for strike-slip and 0 for reverse mechanism M is the earthquake magnitude and R_{JB} is Joyner-Boore distance.

3.1. Hazard Assessment
A MATLAB program was developed to compute hazard parameter mainly PHA for the study areas. The major steps involved in the deterministic hazard assessment are presented below.

IOP Conf. Series: Earth and Environmental Science **26** (2015) 012027 doi:10.1088/1755-1315/26/1/012027

i. All the earthquake sources within 500 km radius from a particular grid point were considered in the program for evaluating seismic hazard.
ii. The minimum distance from the selected grid point to each of the sources was calculated. The focal depth was taken as 15 km.
iii. Considering the maximum magnitude and the minimum possible hypocentral distance, the PHA values for the given grid point, were calculated using attenuation relation for each of the sources.
iv. The maximum of these PHA values was taken as the PHA for that particular grid point.
v. Similar analysis was done for all the grid points and the PHA values were obtained.

Figure 3 presents the spatial variation of PHA values at the bedrock level, throughout the state of Sikkim. The PHA value at ground surface for each grid point was estimated using a non-linear site amplification technique (Stewart et al., 2003; Raghu Kanth and Iyengar, 2007). The surface level PHA was estimated for grid points which are having slope value greater than or equal to 10 degrees. The grid points having slope angle less than 10 degree were considered to be flat and hence neglected. The site class for the grid points having slope angle 10 degree and above (gradient $\geq$ 0.176 m/m) was found to be B-type (Wald and Allen, 2007) for which the average shear wave velocity for top 30 m overburden (V_{s30}) is more than 760 m/s (BSSC, 2003) The PHA at the ground surface Y_s for a particular grid was evaluated by multiplying suitable amplification factor F_s by the bedrock level PHA (Y_{br}) as shown in equation 2.

$$Y_s = Y_{br} \times F_s \tag{2}$$

Raghu Kanth and Iyengar (2007) have developed a regression relation (eqn.3) for determining amplification factor F_s.

$$\ln(F_s) = a_1 Y_{br} + a_2 + \ln(\delta_s) \tag{3}$$

For site class B, $a_1 = 0$ and $a_2 = 0.49$ are site class dependent regression coefficients, Y_{br} is the spectral acceleration at bedrock level and $\delta_s = 0.08$ is the error term as per Raghu Kanth and Iyengar (2007).

4. Development of Slope Map

Digital elevation models (DEMs) are raster data files containing terrain information for a specified region in form of pixels (fig. 4(a)). The resolution of the DEM refers to the distance between two adjacent pixels. The ASTER Global Digital Elevation Model (ASTER GDEM) was developed and released by the Ministry of Economy, Trade, and Industry (METI) of Japan, with the collaboration of National Aeronautics and Space Administration (NASA) United States, covering about 99% of the total land area with a resolution of 1 arc-second (30 m). The slope map of the study area was developed from ASTER DEM using ArcGIS 10 (fig. 4(b)).

5. Landslide Hazard Assessment

Several studies have showed that the Newmark's sliding-block model is the most viable method to assess seismic landslides on natural slopes (Wilson, 1983; Wilson and Keefer, 1985; Wieczorek et al., 1985; Jibson, 1993). Wilson and Keefer (1985) showed that the Newmark's method results in a less conservative and more cost effective design than the pseudostatic methods and thus most appropriate for predicting landslide in a natural slope. In the Newmark's method, the soil mass at failure is modelled as a rigid block sliding along an inclined plan, where the resisting forces is just been exceeded by driving forces. Hence the factor of safety drops below 1.0 (Kramer, 1996) and the block requires a critical acceleration (a_c) as per equation 4 which is the threshold acceleration to initiate sliding (Jibson et al., 2000).

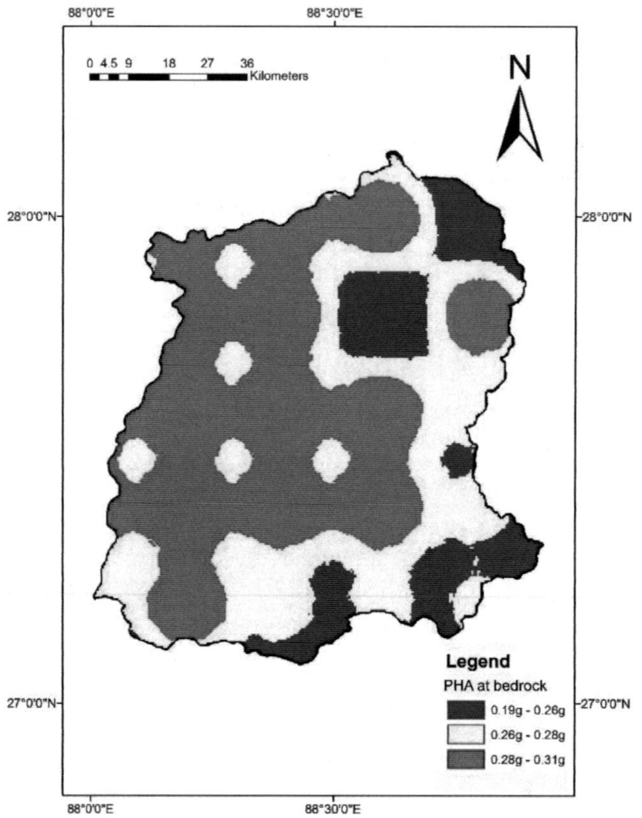

Figure 3. Spatial variation of PHA values throughout Sikkim using DSHA

$$a_c = (FS - 1)g \sin \alpha \quad (4)$$

Here FS is the static factor of safety, which depends on the in-situ soil parameters and slope angle (α). In the present study, landslide hazard is predicted in terms critical value of factor of safety (FS_c) required to resist landslide was evaluated as per equation 5 against a destabilizing force generated by surface level shaking obtained from the DSHA. The slope is considered to be safe, if the in-situ static factor of safety is greater than the computed critical static factor of safety (FS_c)

$$FS_c = (PHA/ \sin \alpha) + 1 \quad (5)$$

The study area is first divided into 4 districts and the seismic landslide hazard map of each district was developed considering a grid size of 30 m × 30 m size. The landslide hazard maps of all 4 district are then merged, to obtain the landslide hazard map of Sikkim. The slope angle value for each grid point was derived from ASTER DEM, using GIS. Knowing the seismic acceleration and the slope angle, the static factor of safety (FS_c) required to resist landslide is evaluated for each grid point. The static factor of safety for any slope depends upon the in-situ soil parameters such as shear strength, density and slope angle etc., which can be evaluated using conventional slope stability analysis. A high value of static factor of safety for a slope implies a higher stability of the slope against earthquake shaking. Thus if the in-situ static factor of safety is greater than the computed static factor of safety (FS_c) required to resist landslide, the slope is considered to be safe. Figure 5 presents the landslide hazard map of Sikkim, showing the spatial variations of the critical static factor of safety required to resist landslide throughout the state.

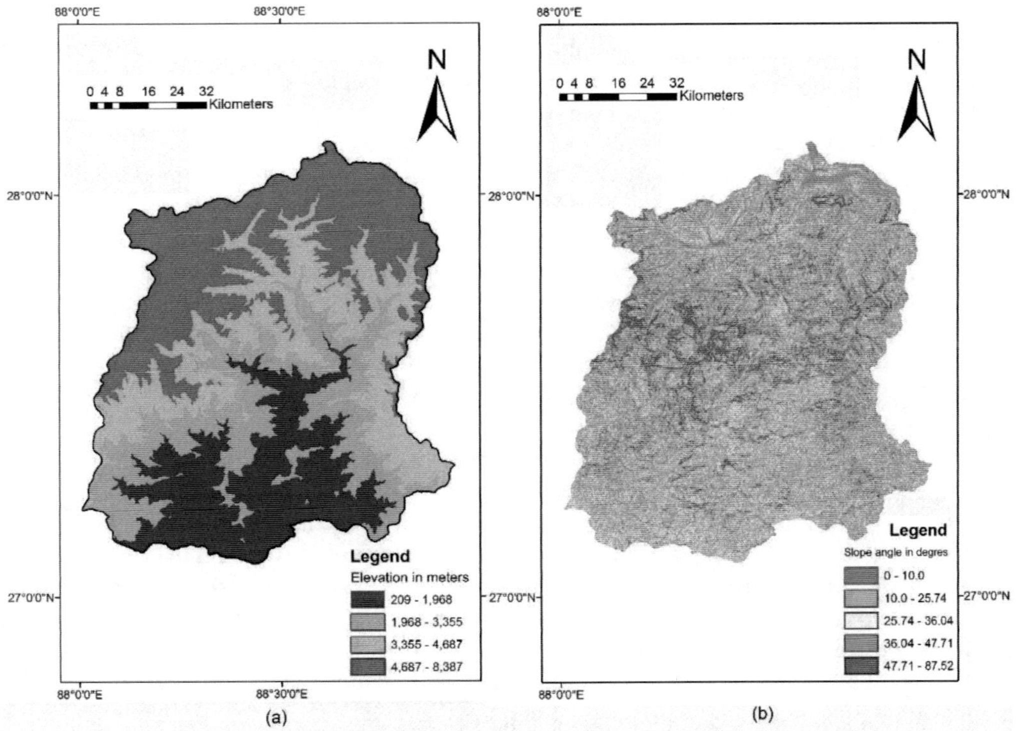

Figure 4. (a) Digital Elevation Model and (b) Slope map of the study area

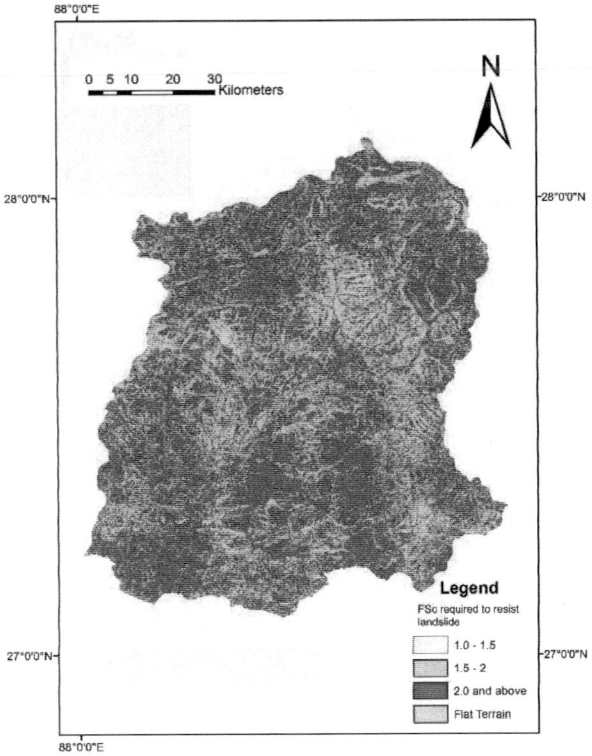

Figure 5. Seismic landslide hazard map of Sikkim

6. Conclusions

The present study describes the quantitative evaluation of seismic landslide hazard for the entire state of Sikkim in terms of static factor of safety required to resist landslide, based on the Newmark's method. Topographic slope map for the entire state of Sikkim was derived from the ASTER GDEM. Peak horizontal acceleration (PHA) at ground surface, obtained from DSHA, is used as critical acceleration for evaluating static factor of safety to resist landslide. This aper presents a landslide hazard map of Sikkim, showing the spatial distribution static factor of safety required to resist landslide. For a given natural slope, located within the state of Sikkim, the insitu static factor of safety for that slope can be computed using conventional slope stability analysis. The static factor of safety based on dynamic finite element method is recommended for evaluating dynamic slope stability. Thus the map developed in the present study is an excellent first level landslide inventory database, which is helpful for a quick assessment of seismically induced landslides while planning big projects like new road/rail routes, setting up power generation/transmission structures, tunnels etc.

The present study classifies Sikkim into 3 hazard zones based on the range of static factor of safety required to resist landslide. They are, low hazard region ($1.0< FS_c <1.5$), moderate hazard ($1.5< FS_c <2$) and high hazard (FS_c 2.0 and above). From fig. 5, it can be observed that majority of Sikkim falls is high hazard zone, where a minimum static factor of safety of 2.0 and above against an expected earthquake ground shaking, to resist landslide. Such a high value of seismic landslide hazard is due to high level of expected ground shaking ranging from 0.2g to 0.3g at bedrock level and steep terrain. Hence it is highly recommended to carry out a rigorous micro-level landslide hazard assessment for hilly region in the state were human settlement is high. The current landslide hazard map for the state of Sikkim has been developed based on 110 years of earthquake data, hence it requires revision after any major earthquake, occurred within the radius 500km from the boundary of Sikkim.

References

[1] BSSC 2003 NEHRP recommended provisions for seismic regulations for new buildings and other structures (FEMA 450), Part 1: Provisions. Building Seismic Safety Council for the Federal Emergency Management Agency, Washington, D.C., USA.

[2] Chakraborty I, Ghosh S, Bhattacharya D, and Bora A 2011 Earthquake induced landslides in the Sikkim-Darjeeling Himalayas-An aftermath of the 18th September 2011 Sikkim earthquake. *Geological Survey of India*, Kolkata.

[3] Costa G, Panza G F, Suhadolc P, and Vaccari F 1993 Zoning of the Italian territory in terms of expected peak ground acceleration derived from complete synthetic seismograms Journal of *Applied Geophysics* 30(1):149–160.

[4] Jibson R W 1993 Predicting earthquake-induced landslide displacements using Newmark's sliding block analysis *Transportation Research Record* pp. 9–9.

[5] Jibson R W, Harp E L, and Michael J A 2000 A method for producing digital probabilistic seismic landslide hazard maps *Engineering Geology* 58(3):271–289.

[6] Kramer S L 1996 *Geotechnical earthquake engineering*. Pearson Education Pvt. Ltd, Reprinted 2003, Delhi, India.

[7] NDMA 2009 *National Disaster Management Guidelines - Management of Landslides and Snow Avalanches*. National Disaster Management Authority, Government of India.

[8] Panza G F, Vaccari F, and Cazzaro R 1999 Deterministic seismic hazard assessment Vrancea earthquakes: *Tectonics, Hazard and Risk Mitigation* pp. 269–286.

[9] Raghu Kanth S T G and Iyengar R N 2007 Estimation of seismic spectral acceleration in peninsular India *Journal of Earth System Science* 116(3):199–214.

[10] Sharma M L, Douglas J, Bungum H, and Kotadia J, 2009 Ground-motion prediction equations based on data from the Himalayan and Zagros regions *Journal of Earthquake Engineering* 13(8):1191–1210.

[11] Stewart J P, Liu A H, and Choi Y 2003 Amplification factors for spectral acceleration in tectonically active regions *Bulletin of the Seismological Society of America* 93(1):332–352.

[12] Wald D J, and Allen T I, 2007 Topographic slope as a proxy for seismic site conditions and amplification *Bulletin of the Seismological Society of America* 97(5):1379–1395.

[13] Wieczorek G F, Wilson R C, and Harp E L, 1985 Map showing slope stability during earthquakes in San Mateo County, California. *US Geological Survey*.

[14] Wilson R C, and Keefer D K 1983 Dynamic analysis of a slope failure from the 6 august 1979 Coyote lake, California, earthquake *Bulletin of the Seismological Society of America* 73(3):863–877.

[15] Wilson R C, and Keefer D K, 1985 Predicting areal limits of earthquake-induced landsliding Evaluating Earthquake Hazards in the Los Angeles Regine, *US Geol. Surv. Profess. Pap 1360:317–345.Strite S and Morkoc H 1992 J. Vac. Sci. Technol.* B 10 1237

Susceptibility analysis of rapid flowslides in southern Italy

M C Mandaglio[1], N Moraci[1], D Gioffrè[1] and A Pitasi[1]

[1]Department of Civil, Energy, Environmental and Materials Engineering, "Mediterranea" University of Reggio Calabria, Via Graziella - Loc. Feo di Vito - 89122 Reggio Calabria (RC)

E-mail: nicola.moraci@unirc.it

Abstract. Qualitative and quantitative estimation of distribution of the existing landslides or that could occur inside a determined area is named landslide susceptibility. The zoning of landslide susceptibility consists of inventory of landslides occurred in the past and identification of the areas where the landsliding could occur in the future. The paper shows the results of a study of fast flowslides susceptibility performed on an area placed between Scilla and Favazzina (RC) regularly and historically interested by rainfall and in some cases by earthquake induced landslides. The trend of inclination and lithological features of the potential detachment zones has been analyzed at 1:5.000 scale. Classes of inclinations, where the most part of detachment zones are clustered, have been identified and the distribution of these classes for each lithological type has been graphed. Among all surveyed events, the inclinations and lithological features of the detachment zones similar to those of the events occurred in the 2001 and 2005 have been found. Finally, in order to evaluate the susceptibility of the area, the paths of flowslides have been evaluated by numerical simulations using the SPH model.

1. Introduction

Flowslides are landslides consisting of events of mass transports constituted of water and solid material triggered by natural factors, mainly heavy rains and earthquakes, but also by human activities. They occur periodically on established paths, usually gullies and first or second order drainage channels. Thus, flowslide vulnerability is specific to a given path and a deposition area ("debris fan"). This, and the periodicity of occurrence at the same location, influences the methodology of vulnerability studies and contrasts with related phenomena, such as debris avalanches, whose occurrence is not bound to an established path [1]. The interaction between fluid and solid component plays a key role in the physics of the phenomena and makes them particularly destructive. In these phenomena the two main components of the mixture, water and solid particles, can be present in different proportions and the behaviour can belong to the movements of the slope or the flooding depending on the concentration of the mixture. The separation between the two types of phenomena is very difficult. The most widely used classification is the system proposed by Varnes [2], [3] and modified by Cruden and Varnes [4]. Hungr et al. [1] proposed an update of the Varnes classification to modify the definition of landslide-forming materials and to correlate the terminology of landslides with the mechanical behaviour.

Nowadays flowslides are responsible for some of the largest landslide disasters [5]. Furthermore, population growth has driven development further into debris-flow-prone areas, increasing risk from this hazard [5]. For these reasons, a large part of research on risk of flowslides has been performed to in - depth understand of the susceptibility analysis. Many of these studies have focused on the physical characteristics and processes of flowslides. These included geologic settings, triggering mechanisms, transport processes, deposition characteristics, and mitigation options in order to identify the most appropriate strategies for the mitigation of landslide risk to which people and properties (structural and infrastructural facilities) are exposed.

The paper shows the results of a study of rapid flowslides susceptibility performed at large scale on a wide area placed between Scilla and Favazzina (RC), regularly and historically interested by rainfall-induced landslides and sometimes also by earthquake landslides. The paper is developed on the base of geomorphologic data, slope gradients, lithological maps, and numerical approach. In particular, the aim of the research is to provide susceptibility maps drawn for a specific rheological behavior of the material involved in the landslides and a specific erosion coefficient on the bottom of the channel.

The study is a further development of a research on analysis of areal scale of landslides induced by rainfall that the Department DICEAM has been carried out for a long period [6], [7], [8], [9].

2. Analysis of the events classified as fast flowslides in the south of Italy

The analysis of the flow-like landslides has been performed using the data provided by the Basin Authority of the Calabria Region, updated to 2006 and data available on the National Geoportal. These data have been catalogued according to IFFI Project (Inventory of landslides in Italy) whose target is to create a single information system for the landslides in Italy and the homogenization of the national data.

The area chosen as study case is the zone between Scilla and Favazzina, affected by several events of flow-like landslides. In particular, two events, occurred in May 2001 and in March 2005 (figure 1), produced extensive damage involving various lifelines. The flowslides in the area are generally translational landslides involving the superficial and weathered part of the metamorphic bedrock; they occur upslope to the channels, concomitant with very intense rainfalls. The landslides mass is channelled and fluidized by stream waters. Previous reconstructions have shown that the volumes at the end of the propagation are tripled respect to detachment volumes [10].

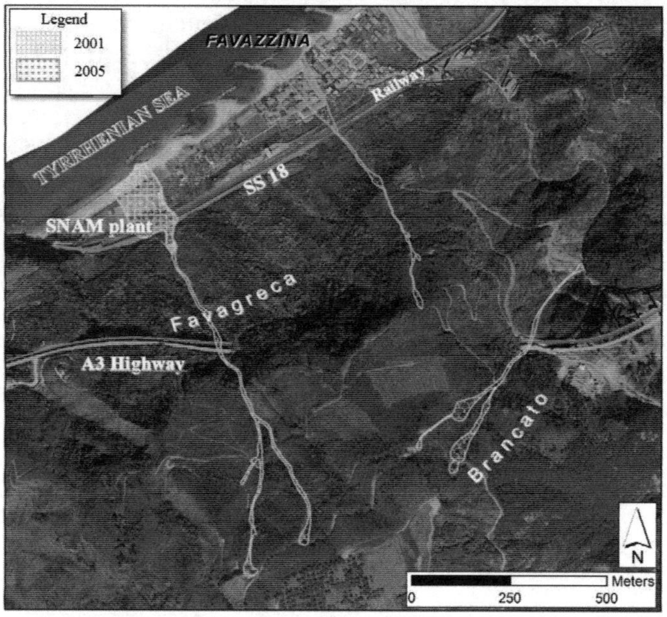

Figure 1 – 2001 and 2005 flow-like landslides events in Favazzina.

IOP Conf. Series: Earth and Environmental Science **26** (2015) 012029 doi:10.1088/1755-1315/26/1/012029

Figure 1 shows the detachment zones of the flowslides, propagation and accumulation areas and the infrastructures involved during the events. On 12 May 2001, two translational landslides were triggered by heavy rainfalls at the head of the Favagreca River, respectively at 570 and 550 m above sea level in correspondence of two incisions that are joined at about 300 m a. s. l. The movement of the translational landslides turned into a rapid flow-like landslide that hit the SNAM station of the methane pipeline, the main road SS 18 and the railway causing the derailment of the intercity train Turin - Reggio Calabria. A second smaller flow slide triggered in an adjacent valley above the A3 highway. On 31 March 2005, a similar phenomenon has been activated in the valley near to Favazzina village, causing several damages to the highway and the railway, causing the derailment of the intercity train ICN Reggio Calabria – Milan.

The soil involved are weathered covers of metamorphic bedrock that have thickness ranging from centimetres to one meter (Figure 2).

Figure 2 – Typical weathered covers of metamorphic bedrock.

Because of the vegetation cover and the inaccessibility of the place, the weathered covers are not easy to observe. The geomorphic characteristics of the area have been determined by in situ surveys and by photo-interpretation of flight Calabria 2001 stereoscopic frames (scale 1: 15.000 and 1: 18.000).

The high relief energy make possible the gravitational sliding that mark the slope with frequent episodes, mainly present at the head of streams. The landslides affect the weathered covers of metamorphic bedrock, and often involve in their movement also the man-made terraces for agricultural purposes. The instabilities turn into debris flows that, channelling into valley, reach the coast at the base of the slope, involving traffic road and railway. In the area, the most part of sediments that reach the coast and those that forms the area behind the beach were originated by these processes. From hydraulic point of view, it must be pointed out the strong erosive and transport capacity of the drained water of the streams that are able to transport disruptive amount of water and debris.

The total number of real events occurred in the area, obtained using the data provided by the Basin Authority of the Calabria Region and data available on the National Geoportal, is equal to 20. In order to characterize the collected events, inclinations and lithological features of the detachment zones have been used as susceptibility descriptors. The inclinations has been analysed at 1:5.000 scale and their values, and relative lithological features are reported in table 1.

Table 1 – Inclinations and lithological features of the real flow-like landslides.

Landslide N°	i [°]	Lithological features
1	42	Paragneiss, Gneiss, Biotitic schists
2	41	Paragneiss, Gneiss, Biotitic schists

3	42	Paragneiss, Gneiss, Biotitic schists
4	41	Debris
5	40	Debris
6	40	Paragneiss, Gneiss, Biotitic schists
7	38	Gneiss occhiadini
8	42	Gneiss occhiadini
9	38	Gneiss occhiadini
10	42	Gneiss occhiadini
11	44	Gneiss occhiadini
12	39	Paragneiss, Gneiss, Biotitic schists
13	37	Paragneiss, Gneiss, Biotitic schists
14	42	Paragneiss, Gneiss, Biotitic schists
15	41	Paragneiss, Gneiss, Biotitic schists
16	40	Paragneiss, Gneiss, Biotitic schists
17	41	Paragneiss, Gneiss, Biotitic schists
18	40	Paragneiss, Gneiss, Biotitic schists
19	41	Paragneiss, Gneiss, Biotitic schists
20	38	Paragneiss, Gneiss, Biotitic schists

The variation of the frequency of the events versus the slopes and the classes of slopes are shown in figure 3a and 3b. In particular, the figure 3b shows the slopes clustered in four classes: ≤ 37°, 38°-40°, 41°-43° and >43°.

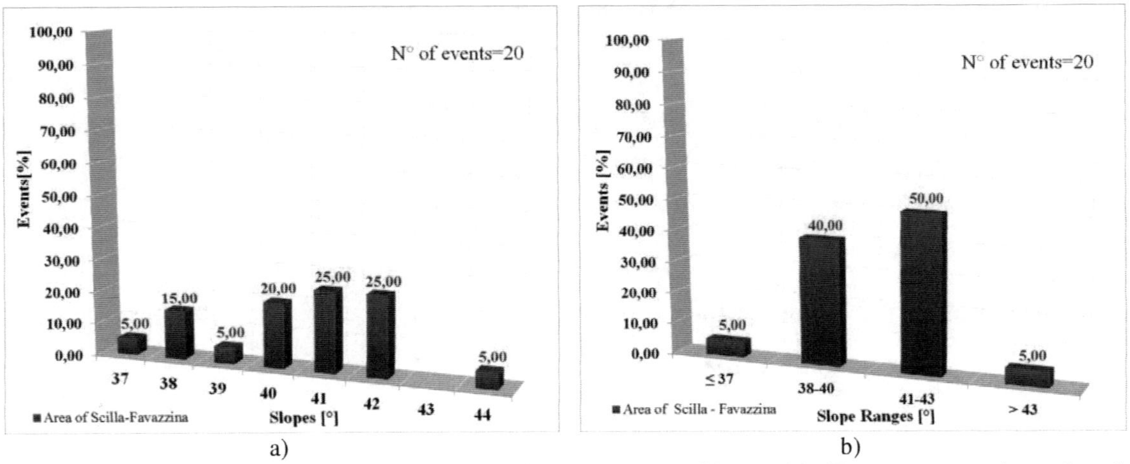

Figure 3 – Histograms Frequency number of real events – Slopes (a); Frequency number of real events – Slope Ranges.

It can be noticed that the most part of the events occurs on slopes of 41° and 42° (Figure 3a) and in the classes 38° - 40° and 41° - 43° (Figure 3b). In areas with slopes less than 37° and greater than 43° the number of events is small and equal to 5%.

The second parameter used as susceptibility descriptor of the flowslides has been the lithology of the areas affected by landslides. The statistical distribution of lithologies affected by the events of flowslides is shown in figure 4.

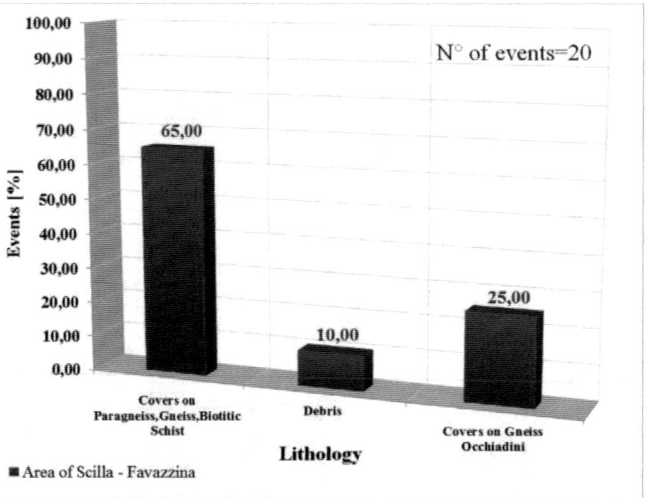

Figure 4 – Histogram of frequency of the real events per lithology.

It can be noticed that paragneiss, gneiss and biotitic schists are involved for the 65% of the events; the 25% of the events occurred on gneiss occhiadini and only the 10% of the events occurred on debris.

It has also been determined the areal incidence of each type of lithology compared to the total area, as a function of the frequency of the number of events. This analysis was conducted by means of a lithological chart at scale 1:5000.

The area is approximately equal to 4,80 km^2. The areal distribution of each lithology affected by events of flowslides is shown in the table 2.

Table 2 – Areal distribution of lithologies.

Total area [km^2]	Lithology	N° of events	Frequency of the events [%]	Lithology area [km^2]	Areal distribution [%]
4,76	Debris	2	14	0,48	10,03
	Covers on Paragneiss–Gneiss–Biotitic schists	12	86	3,68	77,31
	Covers on Gneiss Occhiadini	0	0	0,08	1,78

It can be noticed that paragneiss, gneiss and biotitic schists cover the 3.68% of the examined area.

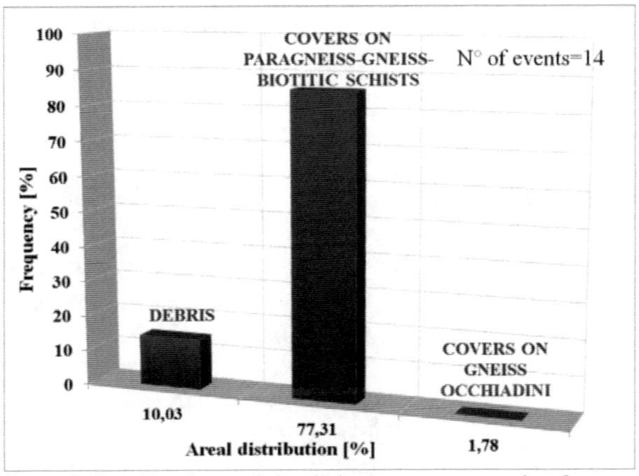

Figure 5 – Histogram of areal distribution of the lithology versus the frequency of the events.

Figure 5 shows the areal distribution versus the frequency of events. As expected, the areal distribution is equal to 77,31% on the paragneiss, gneiss and biotitic schists with 86% of frequency of the events, while it is equal to 10,03% on debris with events frequency equal to 14% and the areal distribution is equal to 1,78% on gneiss occhiadini but no events occurred here.

3. Susceptibility analysis of the representative area of Favazzina

Analysis of the real events in the study area allowed to obtain the inclination values and lithological features that may be affected by flow-like landslides. Then, hypothetical detachment zones have been identified (Figure 6) taking into account points with inclinations and lithological features similar to those of the real events, and geomorphological evidences such as crowns, deep eroded gullies, and steep slope. The total number of so obtained points was 65 but only the points with inclinations between 37° and 44° degrees, like those of the real events, were considered.

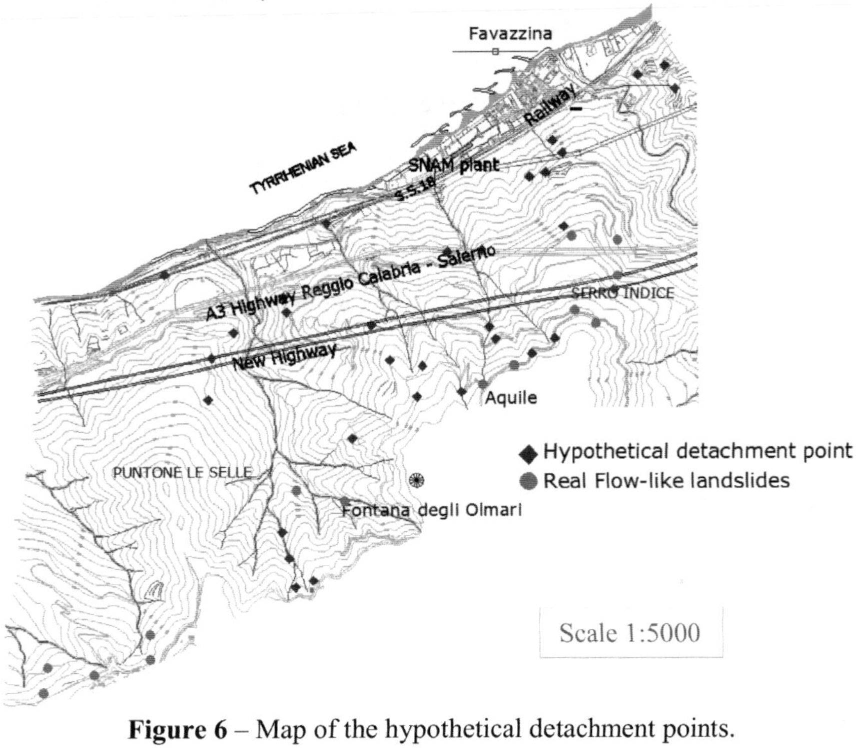

Figure 6 – Map of the hypothetical detachment points.

In figure 6, the blue squares represent the hypothetical detachment points with the same inclinations and lithological features of the real events, while the red circles represent the real flow-like landslides. 32 hypothetical detachment points have been individuated.

3.1. Statistical analysis

The slope values, the lithological features and the geomorphological evidences of each detachment point is listed in table 3, where the paragneiss, gneiss and biotitic schists have been clustered and called paragneiss because the covers on these rocks are the same and they are weakly clay sandy silts.

Table 3 – Slopes, lithological features and geomorphological evidences of the hypothetical detachment points.

Detach. point	i [°]	Lithological features	Geomorphological evidences
1	37	Covers on Paragneiss	Crown
2	43	Covers on Paragneiss	Crown
3	37	Covers on Paragneiss	Crown
4	42	Covers on Paragneiss	Deep eroded gully
5	40	Covers on Paragneiss	Deep eroded gully
6	44	Covers on Paragneiss	Deep eroded gully
7	37	Debris	Crown
8	39	Debris	Crown
9	39	Covers on Paragneiss	Collapse slope
10	41	Debris	Deep eroded gully
11	41	Covers on Paragneiss	Collapse slope
12	37	Covers on Paragneiss	Deep eroded gully
13	43	Covers on Paragneiss	Steep slope
14	44	Covers on Paragneiss	Crown
15	38	Covers on Paragneiss	Collapse slope
16	42	Covers on Paragneiss	Deep eroded gully
17	39	Covers on Paragneiss	Crown
18	40	Covers on Paragneiss	Crown
19	38	Continental deposits	Crown
20	37	Continental deposits	Crown
21	44	Covers on Paragneiss	Crown
22	37	Covers on Paragneiss	Crown
23	37	Covers on Gneiss occhiadini	Deep eroded gully
24	41	Covers on Gneiss occhiadini	Deep eroded gully
25	39	Covers on Paragneiss	Crown
26	43	Covers on Paragneiss	Crown
27	41	Covers on Paragneiss	Steep slope
28	42	Covers on Paragneiss	Crown
29	39	Covers on Paragneiss	Crown
30	42	Covers on Paragneiss	Crown
31	44	Covers on Paragneiss	Crown
32	40	Covers on Paragneiss	Crown

As for the real events, the statistical distributions of the frequency of the hypothetical events versus the slopes (a) and classes of slopes (b) have been graphed.

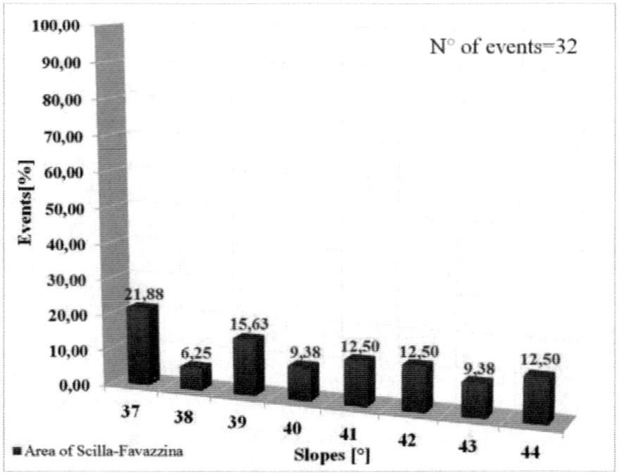

a)

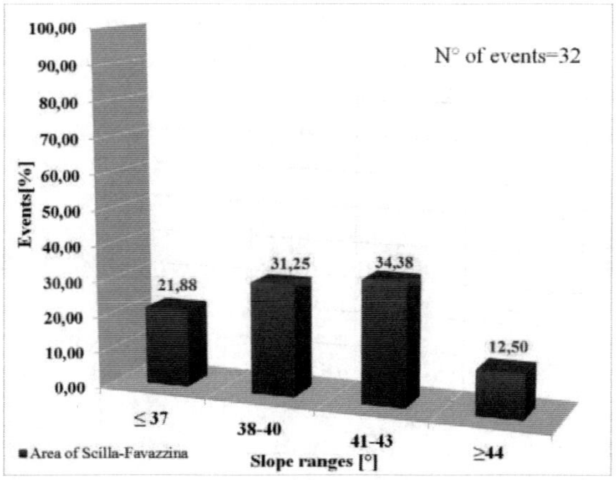

b)

Figure 7 – Histograms Frequency number of hypothetical events – Slopes (a); Frequency number of hypothetical events – Slope Ranges.

Figure 7 (a) shows that greatest number of events (21,88% and 15,63%) occurred on slopes equal to 37° and 39°. The histogram in figure 7 (b) indicates, clearly, the most part of the possible events occurred on slopes in the ranges 38°-40° (31,25%) and 41°-43° (34,38%).

Also in this case, the statistical distribution of lithologies has been graphed in function of the frequency of the number of events (Figure 8).

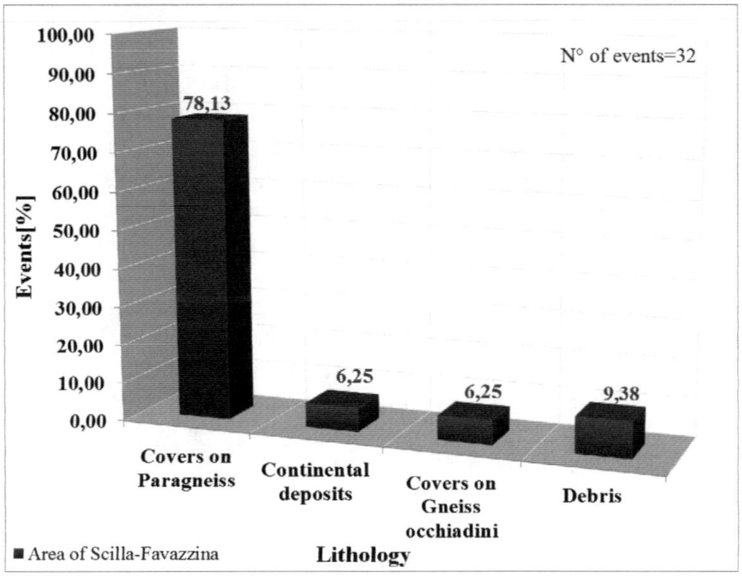

Figure 8 – Histogram of the frequency of the hypothetical events per lithology.

The most involved lithology has been covers on paragneiss with frequency of the events equal to 78,13%; the debris have been affected only by 9,38% of events. A few number of events occurred in other lithologies. This happened because in the area especially paragneiss outcrop (Table 2).

3.2. Numerical simulation using SPH model

As mentioned previously, the analysis of the representative area of Favazzina aims to define the parameters to use in order to define the susceptibility of the area. In addition to the inclination and lithology parameters representing the common indices for the trigger phase of the flow-like landslides, it is important to focus on the propagation phase which identifies the runout area and the flow velocity. The prediction of both runout distances and velocity of fast flowslides by means of mathematical modelling allows to identify the elements exposed at risk. It provides a mean to estimate the vulnerability and it provides the suggestions for the choice of appropriate mitigation measures.

Most of the current numerical models, available in literature, are based on grids, like finite differences or finite elements and volumes. An interesting and powerful alternative to these methods is provided by 'meshless' numerical methods that have been developed in the past decades. The key idea of the meshfree methods is to provide accurate and stable numerical solutions for integral equations or partial differential equations with all kinds of possible boundary conditions with a set of arbitrarily distributed nodes (or particles) without using any mesh that provides the connectivity of these nodes or particles [11]. Smoothed particle hydrodynamics (SPH) [12] [13], as a meshfree and particle method, was originally invented for modelling astrophysical phenomena, and later widely extended for applications to problems of continuum solid and fluid mechanics because of relatively strong ability to incorporate complicated physical effects into the SPH formulations. The earliest applications of SPH were mainly focused on fluid dynamics related areas and has also been applied to model the propagation of catastrophic landslides. In this paper a model, based on the SPH method and proposed by Pastor [14], able to simulate the propagation of fast debris flows is presented.

The mathematical model is based on the mixtures theory and the numerical model is defined depth integrated model because its equations are integrated along the vertical axis. Finally, a rheological model has added in order to describe the fluidized geo material [14].

Like previously mentioned in the study area two particularly damaging landslides occurred: 2001 and 2005 events. In particular the numerical analysis was performed by the use of the results of PhD thesis on "Numerical modelling of the phenomena of debris flows" [15] that analyses the real flowslide occurred on May 2001. One of the main difficulties encountered when modelling run out of landslides is the selection of the rheological model and its material parameters. In some cases, as in the present study, it will be possible to obtain the soil properties with rheometer or viscometer tests, like in the case of muddy materials without large granular particles. Effectively, the analysis carried out on the soil where hypothetical points are located showed that it is the same of the involved soil by 2001 flowslide, a sandy silt with a high content of fine-grained material (60% of silt and clay).

In the PhD thesis was carried out a calibration of rheological parameter and the viscous Bingham law was the rheological model that better approximated the real event. The rheological parameters of water-sandy silt mixtures used in the PhD was carried out by laboratory test and the simulation results was compared with real evidences. The numerical analysis carried out in the present study take into account some difficulties such problems related to initial conditions of the mobilized geomaterials, problems related to the behavior of the fluidized material, problems related to interaction between propagating material and the basal surface (erosion). After defining the rheological model, considering the results of PhD thesis, in the present study was carried out different calibration analyzes regarding: sensitivity to initial mass, sensitivity to the number of points describing landsliding mass, sensitivity to DTM grid spacing and calibration of erosion factor.

In particular, the analysis carried out on the 2001 landslide indicated that the solid concentration by volume was equal to 60% and the total volume of the mass was equal to 1130 m^3 [10]; a calibration analysis was carried out to determine the best rheological parameters values, varying the critical shear stress τ_c between 19 and 190 [Pa] and the viscosity coefficient μ between 4 and 40 [Pa·s]. In the numerical model, the Hungr equation [16] is implemented depending on E_s, defined as the "growth rate". Also this erosion coefficient was calibrated varying his values between 0.0001 and 0.002 m^{-1}.

In the present paper these calibration analysis weren't reported because the principal aim is to describe and draft the susceptibility map.

IOP Conf. Series: Earth and Environmental Science **26** (2015) 012029 doi:10.1088/1755-1315/26/1/012029

The mathematical relations which represent Bingham model have been the following:

$$\tau_c = 0{,}2959\exp(0{,}1293 \cdot C_v) \quad (1)$$
$$\mu = 0{,}0090\exp(0{,}1666 \cdot C_v) \quad (2)$$

τ_c is the critical shear stress, C_v is the solid concentration by volume and μ is the viscosity coefficient. The values used in the following numerical simulation of these parameters are listed in table 4.

Table 4 – Bingham parameters.

C_v [%]	τ_c [Pa]	μ [Pa·s]
60	692,5	197,4

While the erosion rate E_s is equal to 0,002 m^{-1}, according to the best result of the calibration analysis.

Figure 9 shows the digital terrain model of the whole studied area of Favazzina and a hypothetical detachment point is pointed out.

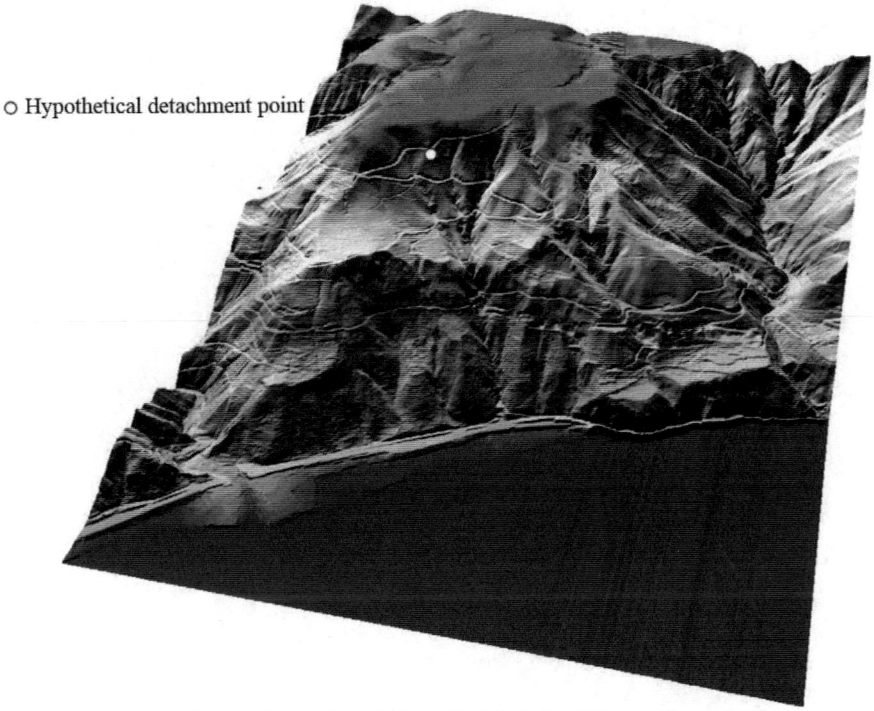

Figure 9 – Digital Terrain Model of Favazzina.

The modelling results of the hypothetical detachment point is shown in figure 10. The simulation time is equal to 180 seconds, assuming a propagation velocity, approximately, equal to 6 m/s; the behaviour is undrained and the bed is erodible. The figures 10 a, b, c and d show the height soil in 4 different times: 0, 30, 60 and 180 seconds.

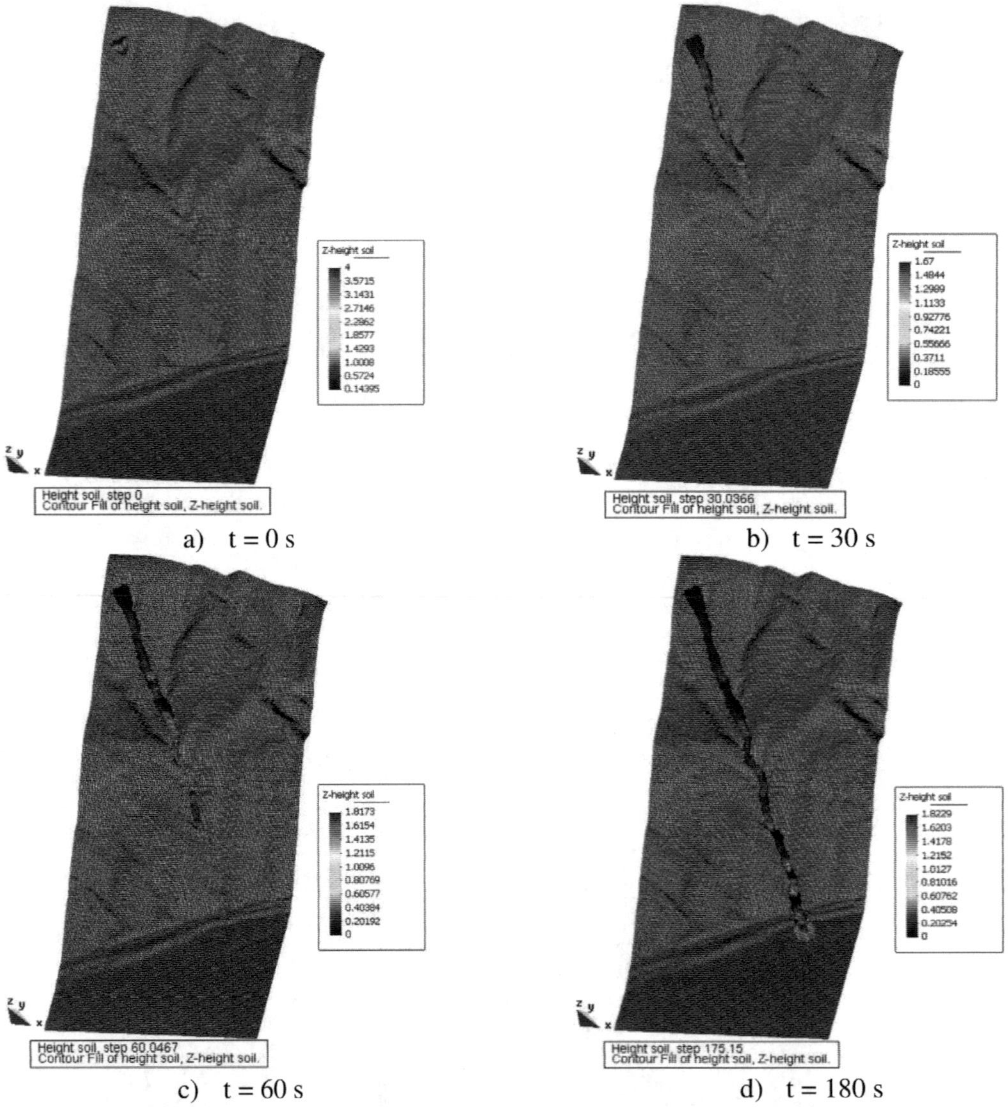

Figure 10 – Simulation results of hypothetical detachment point. a) Landslide position at 0 s; b) landslide position at 30 s; c) landslide position at 60 s; d) landslide position at 180 s.

It can be noticed that the topography has been reproduced in detail; the landslide is well channelled along the path and the accumulation area is well defined.

The maximum height and velocity of the flow at the end of the path and the final volume are listed in table 5.

Table 5 – Simulation results.

$Height\_soil_{max}$ [m]	v_{max} [m/s]	V_{final} [m^3]
1,83	10,05	5422,226

The final volume is four times the initial one according to the values of the real events of the area. In figure 11 is represented the landslide contour along AA section in the accumulation zone.

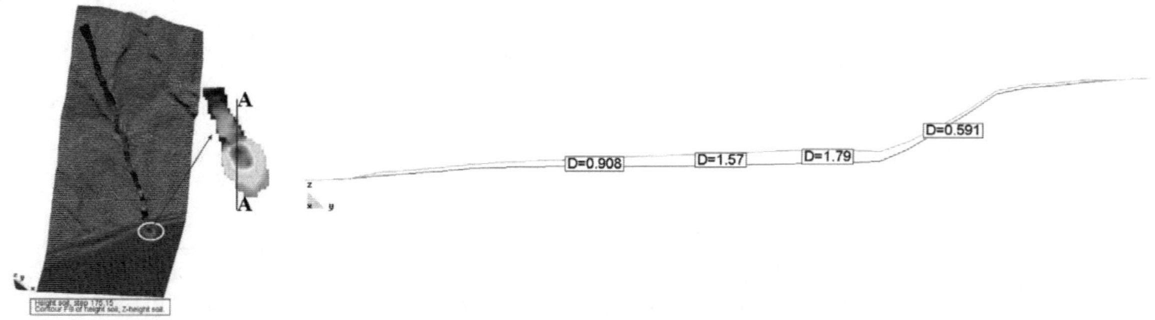

Figure 11 – Height soil values along AA section.

Figure 11 shows the thicknesses (D) of the flow-like landslide along the section AA in the accumulation area. These values are important to identify appropriate mitigation measures.

The previous numerical analysis was performed on more hypothetical detachment points allowing so the drafting of a susceptibility map of the whole studied area, represented in Figure 12.

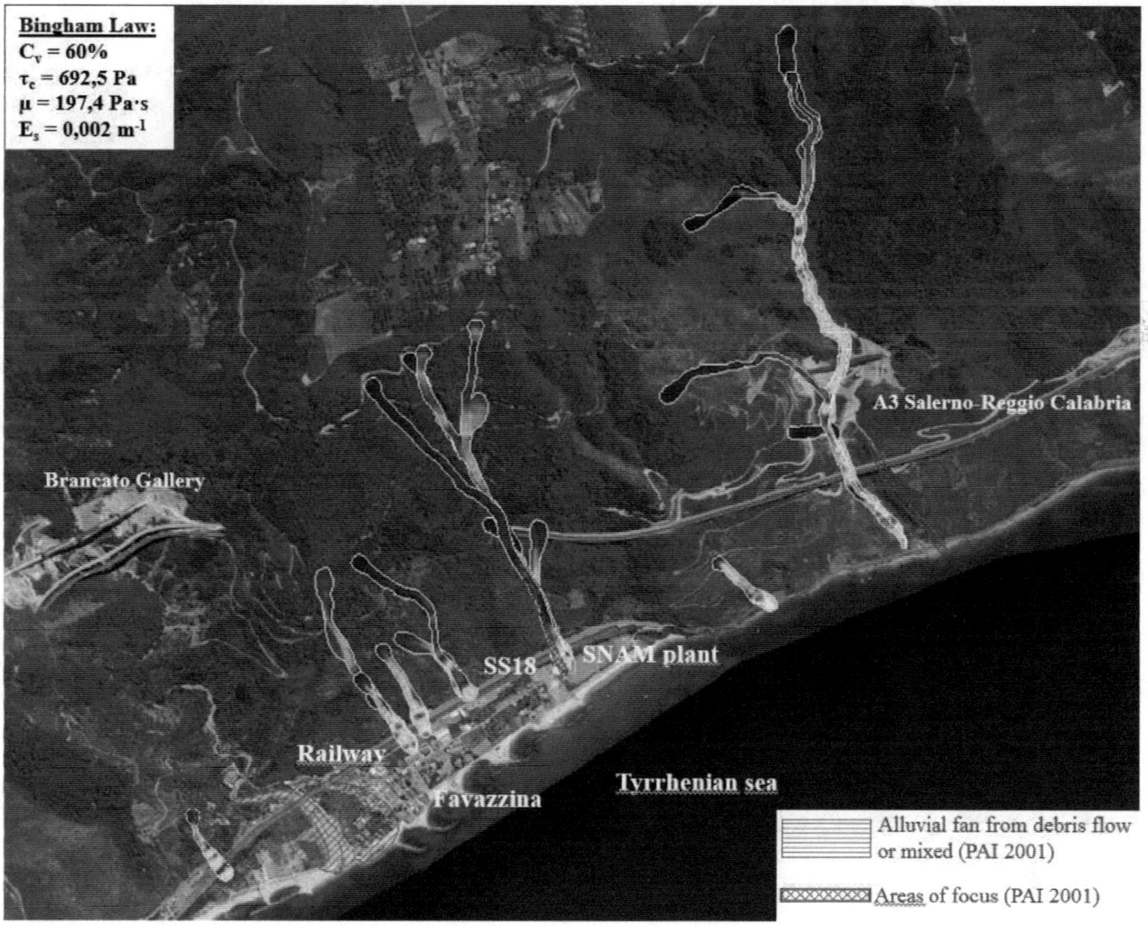

Figure 12 – Susceptibility map of the Favazzina area.

The achieved results demonstrate that, for the study area, a viscous rheology with the parameters shown at the top right of the figure and in table 4 is adequate to simulate this type of landslide.

IOP Conf. Series: Earth and Environmental Science **26** (2015) 012029 doi:10.1088/1755-1315/26/1/012029

4. Conclusions

The analysis of the events classified as flow-like landslides occurred in the southern Italy has been carried out; in particular the area between Scilla and Favazzina has been chosen. Two susceptibility descriptors have been used in order to characterize these events, natural slopes and lithological features of the detachment zones of the landslide mass. The investigation has shown the largest number of the events has been occurred in the slope ranges of 38°-40° and 41°-43°. Lithology mainly affected by the flowlike landslides has been paragneiss, gneiss and biotitic schists, which covered almost all the examined area.

The results achieved have allowed to individuate hypothetical detachment zones, in the study area, with slopes and lithological features similar to those of real events. In this case, the most involved slopes have been 41°-43°, whereas lithology more affected has been paragneiss as previously checked.
In order to evaluate the runout distances and the velocity of the landslide, numerical simulations have been carried out to show the susceptibility of the elements at risk. A depth-integrated, coupled SPH method has been used, considering a Bingham viscous law as rheological model, taking into account the erosion phenomenon. The achieved results have demonstrated the capability of the model to simulate the propagation phase and have allowed to obtain essential data to design appropriate prevention and mitigation measures for the elements at risk.

Acknowledgment
All authors have contributed in equal manner to the development of the paper.

5. References

[1] Hungr O, Leroueil S and Picarelli L 2014 The Varnes classification of landslide types, an update *Landslides* **11** 167-194
[2] Varnes DJ 1954 Landslide types and processes *Landslides and engineering practice, special report 28*, ed E B Eckel (Washington, DC: Highway research board. National Academ of Sciences) pp. 20–47
[3] Varnes DJ 1978 Slope movement types and processes. In: Schuster RL, Krizek RJ (eds) *Landslides, analysis and control, special report 176*: Transportation research board, National Academy of Sciences, Washington, DC., pp. 11–33
[4] Cruden DM, Varnes DJ 1996 Landslide types and processes. In: Turner AK, Schuster RL (eds) *Landslides investigation and mitigation*. Transportation research board, US National Research Council. Special Report 247, Washington, DC, Chapter 3, pp. 36–75
[5] Jakob M and Hungr O 2005 Introduction. In: Hungr O, Jakob M (eds) *Debris-flow hazards and related phenomena*. Springer-Praxis, Berlin, pp 1–7
[6] Gullà G, Mandaglio M C, Moraci N and Sorriso-Valvo M 2004 Definizione degli elementi generali di dei modelli geotecnici per l'analisi delle instabilità superficiali per scorrimento-colata in Calabria jonica. *Atti del XXII Convegno Nazionale di Geotecnica, Palermo 22-24 settembre 2004*, 127-134.
[7] Gullà G, Mandaglio M C, Moraci N 2006 - Effect of weathering on the compressibility and shear strength of a natural clay.*Canadian Geotechnical Journal* **43**: 618-625
[8] Gullà G., Mandaglio M.C., Moraci N. 2005 - Influence of degradation cycles on the mechanical characteristics of natural clays. *Proceedings of 16th International Conference on Soil Mechanics and Geotechnical Engineering, Osaka (Japan), 12-16 September 2005*: 2521-2524.
[9] Borrelli L., Gioffrè D., Gullà G., Moraci N. 2012 - Susceptibility to shallow and rapid landslides in ground alteration: A possible contribution of propagation modeling - Suscettibilità alle frane superficiali e veloci in terreni di alterazione: un possibile contributo della modellazione della propagazione. *Rend. Online Soc. Geol. It., Vol. 21 (2012), pp. 534-536, Società Geologica Italiana, Roma 2012.*

[10] Bonavina M, Bozzano F, Martino S, Pellegrino A, Prestininzi A and Scandurra R 2005 Le colate di fango e detrito lungo il versante costiero tra Bagnara Calabra e Scilla (Reggio Calabria): valutazioni di suscettibilità. *Giornale di Geologia Applicata 2* 65-74

[11] Liu G R and Liu M B 2003 Smoothed Particle Hydrodynamics: A Meshfree Particle Method World Scientific (ed)

[12] Lucy L B 1977 Numerical approach to testing the fission hypothesis, *Astronomical Journal*, **82**:1013-1024

[13] Gingold R and Monaghan J J 1977 Smoothed Particle Hydrodynamics: Theory and Application to Non-spherical stars, *Monthly Notices of the Royal Astronomical Society*, **181**:375-389

[14] Pastor M, Haddad B, Sorbino G, Cuomo S and Drempetic V 2009 A depth-integrated, coupled SPH model for flow-like landslides and related phenomena, *Int. J. Numer. Anal. Meth. Geomech.,* **33**:143–172

[15] Gioffre' D 2005 Modellazione numerica dei fenomeni di frane di colata. PhD Thesis in Geothecnical Engineering (XVII cycle) "Mediterranea" University of Reggio Calabria

[16] Hungr O 1995 A model for the runout analysis of rapid flow slides, debris flows and avalanches. *Canadian Geotechnical Journal*, **32**, 610-623

IOP Conf. Series: Earth and Environmental Science **26** (2015) 012031 doi:10.1088/1755-1315/26/1/012031

GIS-aided Statistical Landslide Susceptibility Modeling And Mapping Of Antipolo Rizal (Philippines)

A J Dumlao, J A Victor
Institute of Civil Engineering
University of the Philippines Diliman, Quezon City, Philippines

E-mail: jaimeangelovictor@yahoo.com

Abstract. Slope instability associated with heavy rainfall or earthquake is a familiar geotechnical problem in the Philippines. The main objective of this study is to perform a detailed landslide susceptibility assessment of Antipolo City. The statistical method of assessment used was logistic regression. Landslide inventory was done through interpretation of aerial photographs and satellite images with corresponding field verification. In this study, morphologic and non-morphologic factors contributing to landslide occurrence and their corresponding spatial relationships were considered. The analysis of landslide susceptibility was implemented in a Geographic Information System (GIS). The 17320 randomly selected datasets were divided into training and test data sets. K- cross fold validation is done with k= 5. The subsamples are then fitted five times with k-1 training data set and the remaining fold as the validation data set. The AUROC of each model is validated using each corresponding data set. The AUROC of the five models are; 0.978, 0.977, 0.977, 0.974, and 0.979 respectively, implying that the models are effective in correctly predicting the occurrence and non-occurrence of landslide activity. Field verification was also done. The landslide susceptibility map was then generated from the model. It is classified into four categories; low, moderate, high and very high susceptibility. The study also shows that almost 40% of Antipolo City has been assessed to be potentially dangerous areas in terms of landslide occurrence.

1. Background

Landslide is one of the natural hazards that is unexpected and with high magnitude that threatens humans and properties. It is a downward movement of rocks and soil, which includes rock, falls, and deep failure of slopes, shallow debris flows and avalanches. The movement is triggered by gravity due to other hazards like earthquake and high amount of rainfall. Other factors such as geology, morphology, elevation and human activities affect the slope stability of an area.

The stu dy will be conducted in Antipolo City due to its increase in population and establishments on its steep slope covering its 306 km^2 area. As of 2009, government records show that 545,157 are residing in the city but an unofficial count estimates the population to reach 800,000 due to informal settlers. The number of business establishments also continues to rise with 6,903. The subdivisions in the city are also increasing with a total of 405 and 10 are still under development.

Although Antipolo City has experienced several landslide events in the past years, preparedness and mitigation in the area are lacking. It is because there are not enough data, and hazard experts available to guide the concerned authorities.

2. Research Objectives

The object of this study is to investigate and apply multivariate statistical analysis (specifically logistic regression) within a GIS environment to determine which independent intrinsic variables significantly influence the occurrence of contemporary landslide activity within the study area. Results will be utilized to create a susceptibility map and analyzed to assess the viability of utilizing GIS modeling as part of a broader decision support system pertaining to slope instability monitoring, mitigation and indeed planning for natural hazards.

3. Logistic Regression

Landslide susceptibility analysis may be approached with the use logistic regression as method of mathematical modeling. This method relies on the estimation of future landslides based on the current presence and absence of landslides associated with the values of a set of predictor variables. (Lee, 2005).

In logistic regression, the normal distribution of the dependent variable is not required. When compare with multiple linear regression, one of the main advantages of logistic regression is that it does not assume that there is a linear relationship between the dependent and independent variables. It assumes a linear relationship between the outcome and the logit of the independent variables. Also, there are no assumptions required with regards to the homogeneity of the variance and normally distributed error terms. (Nandi et. al, 2008). The dependent variable can also have 2 values, the event occurring and the event not occurring. (Dai and Lee, 2002)

In landslide susceptibility assessment, the outcome demands to be predicted is the occurrence/ non- occurrence of landslide in a given area. The independent variables are the physical controlling factors that affect the instability of an area and the dependent variable is the presence or absence of landslides. Because of the binary response, two alternative groups are established; mapping units free of landslides, and mapping units having landslides.

The relationship of the occurrence of landslide and the preparatory factors affecting it is expressed as (Hosmer and Lemeshow, 2000):

$$P_r = \frac{1}{1 + e^{-z}} \tag{1}$$

$$Z_n = b_o + b_1x_1 + b_2x_2 + \cdots b_nx_n \tag{2}$$

Where P_r is the estimated probability of landslide event with values varying from 0 to 1, b_0 is the intercept of the model, n is the number of independent variables or predictors, $b_{(i-n)}$ is the ith coefficient of the model, $x_{(i-n)}$ is the xth independent variable or predictor

The data analysis was utilized using Statistical Package for Social Sciences (SPSS). The raster map from the ArcGIS was used to obtain the values that will be needed for the data analysis. Each raster map was divided into cells, which comprised the whole study area. The DEM was used to define the location of the cells using the sample tool in ArcGIS 10.0. A total of 3527765 pixels were extracted to each raster map. The values obtained in ArcGIS 10.0 were saved as a dbf file for compatibility in the SPSS. A total of 41 independent variables were used including six continuous and 35 categorical variables. The independent set consists of the Aspect [*Flat, North(0-22.5), Northeast, East, Southeast, South, Southwest, West, Northwest, North(337.5-360)*], Slope angle, Elevation, Distance to River, Distance to Road, TWI, SPI, Planform Curvature [*Convex, Flat, Concave*], Profile Curvature [Convex,

Flat, Concave], Soil Type [*Mountain soil, Antipolo soil, Antipolo clay, Binangonan clay, Marikina silt loam, Antipolo clay loam, Binangonan lowland clay, Marikina clay loam*], and Geology [*Cretaceous, Oligocene-miocene, Paleocene-eocene, Oligocene-miocene (sedimentary and metamorphic), Recent, Oligocene, Neogene, Upper miocene-pliocene, Pliocene-quaternary, Pliocene-pleistocene, Cretaceous-paleocene*]

The dependent variable, which is the observed landslide occurrence (1580 landslide events), is coded as 1 and 0 for presence or absence of landslide respectively. The statistical analysis was performed using randomly selected training data sets with equal number of presence and absence of landslide value. A total of 17320 cases were chosen as subsamples for determining the relationship of the dependent to the independent variables.

Before performing the regression analysis of the datasets, dummy variables for the categorical variables were created. Since some of the variables will be insignificant in the model it is much easier to remove it by means of the generation of dummy variables. Dummy variables were also used as binary representation of categorical data (e.g. soil type, lithology) based on the presence (1) and absence (0).

The logistic regression analysis was performed using the binary logistic tool under the analyze tab in SPSS. The output in the SPSS produced different summary of the model. It includes Homer and Lemeshow Test, -2 Log likelihood, Cox and Snell R Square, and Nagelkerke R Square. The analysis was then again performed as the insignificant variables were removed. Insignificant variables were determined as variables exceeding the threshold value of 0.05 (level of significance). Variables exceeding the tolerance value accept the null hypothesis thus affecting the model insignificantly. The coefficients of the variables in the model were identified after all the variables portray significant influence in the occurrence of the event.

4. Validation

To have scientific significance, LSI prediction models should have validation measures (Chung and Fabbri, 2003). The models generated were validated using k- fold cross-validation. The randomly selected dataset were randomly divided into five folds in the SPSS. The performance of each model was measured using the corresponding test data set by creating a confusion matrix. The coefficients of the model were used to manually compute the predicted probability of each case of the test data set. A cut- off value of 0.5 was used for the predicted probability. The predicted probability of each case is then compared to the observed outcome of the landslide occurrence. The correctly predicted and wrongly predicted cases of occurrence and non- occurrence of landslide were tabulated in a table to compute for the overall correct percentage of each model.

The AUROC are also performed using the a separate independent validation data set of each model (approximately 20% of the total input data).

5. Thematic Data Layers

There are several factors that can cause a landslide event. The selection and preparation of these factors are one of the crucial components for modeling a landslide susceptibility map. The dataset that will be selected must be available all over the investigated area and categorical data must be defined at a comparable spatial accuracy. The factors selected were namely: elevation, slope, curvature, aspect, topographic wetness index (TWI), stream power index (SPI), lithology, distance from roads and rivers. Most of the data were obtained from Department of Science and Technology (DOST) specifically from Project Nationwide

Operational Assessment of Hazards (NOAH) and Disaster Risk and Exposure Assessment for Mitigation (DREAM) Program while others were available from online resource databases. The DEM used was obtained from Synthetic Aperture Radar (SAR), which were post- processed by the UP Training Center for Applied Geodesy and Photogrammetry (UP-TCAGP). It has a resolution of 10 m, tile size of 10 x 10 km, and an absolute vertical map accuracy of LE90

5.1 Morphometric Preparatory Factors

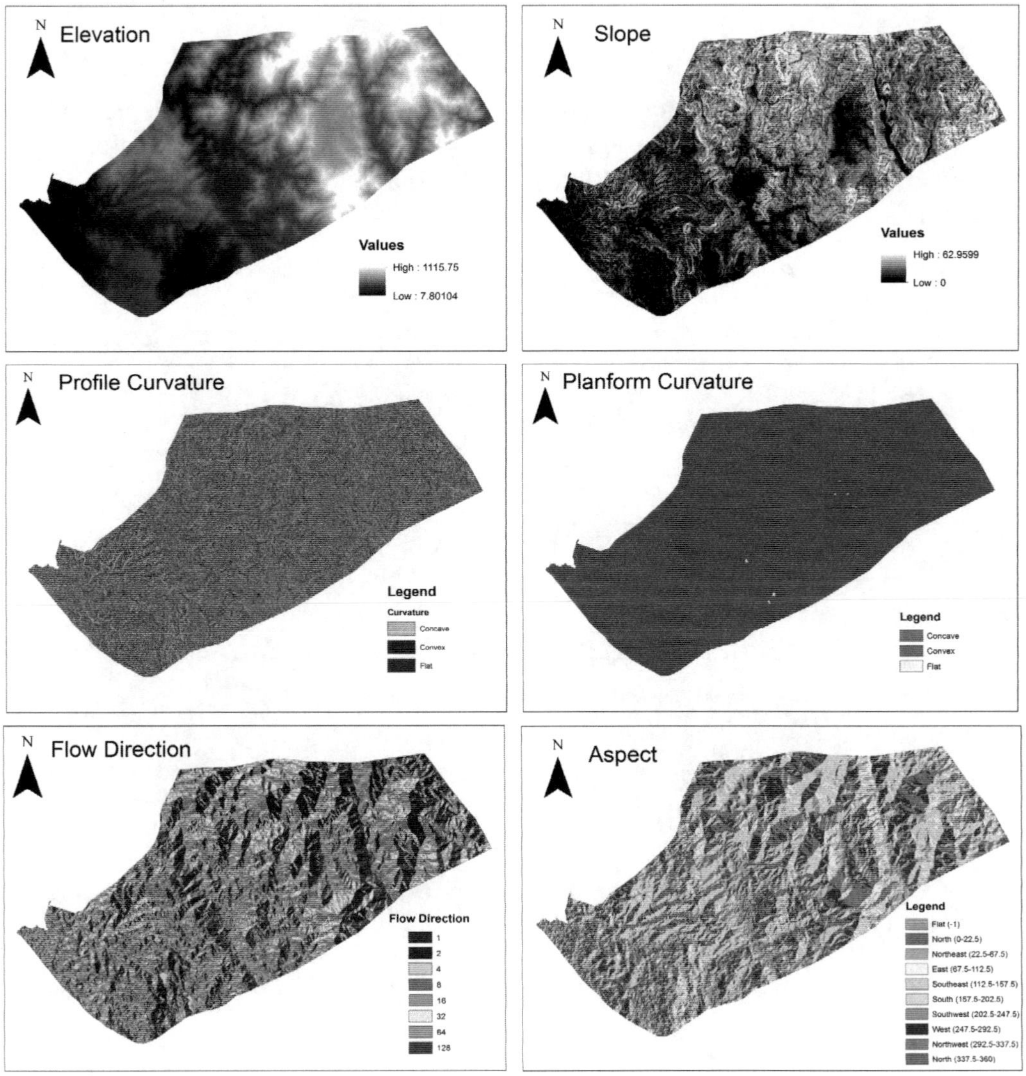

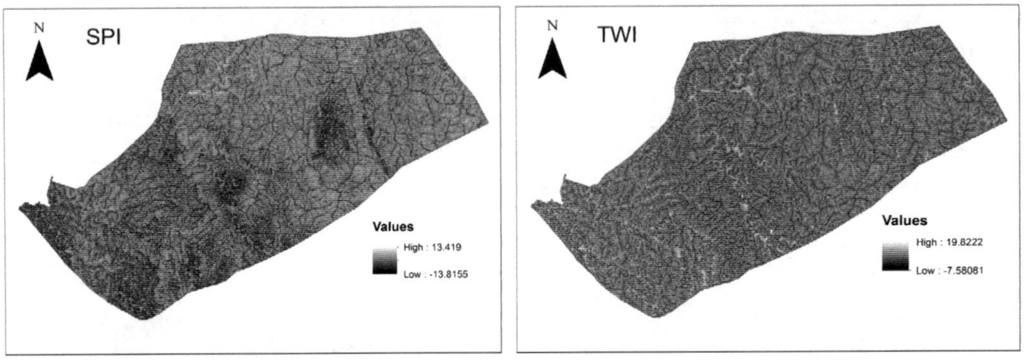

Figure 1. Thematic factor maps of morphometric variables related to landslide occurrence: Slope, Aspect, Elevation, Flow direction, Planform curvature, Profile Curvature, SPI, and TWI raster maps of Antipolo City.

5.2 Non- morphometric Preparatory Factors

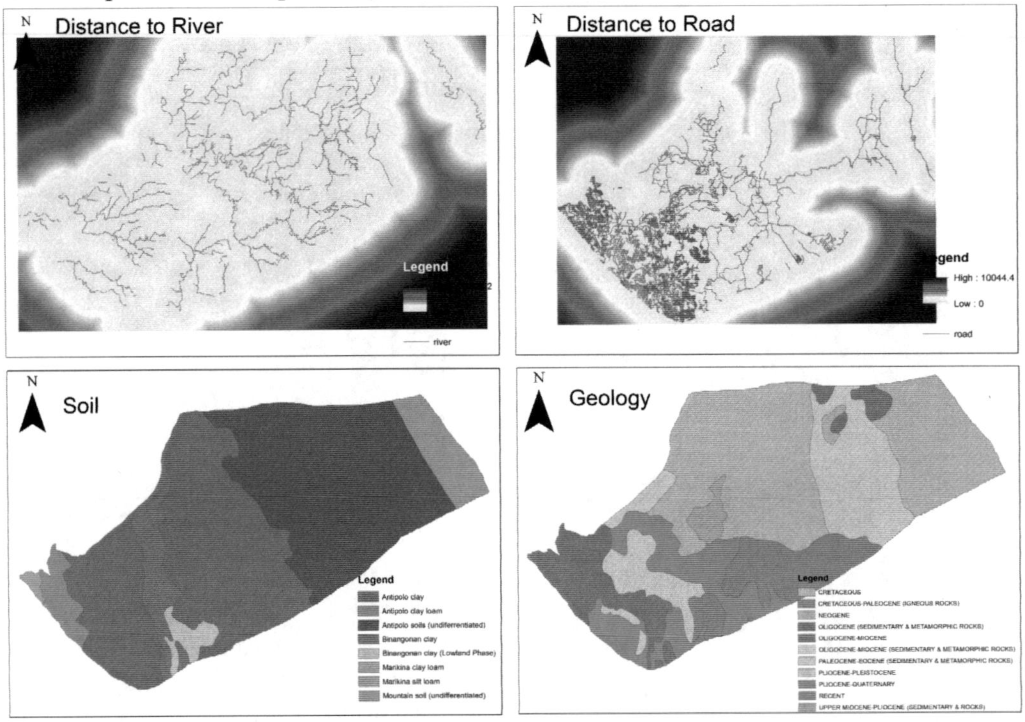

Figure 2. Non-morphometric factors considered in this study: Distance to river, Distance to road, Soil and Geology raster map.

6. Tests for Multicollinearity

The results of the multicollinearity tests done on the set of independent variables are summarized in table 1. An evidence of presence of multicollinearity is tolerance value of less than 0.2, and consequently, a variance inflation factor of greater than 5 (Menard, 1995). Logistic regression can proceed with the chosen set because multicollinearity among the variable is not evident in the set presented by tolerance values greater than 0.2 and consequently, variance inflation factor values less than 5.

Table 1: Results of Multicollinearity Tests.

Model	Coefficient	Collinearity Statistics	
		Tolerance	VIF
1	(Constant)		
	Distance to river	.807	1.239
	Aspect	.983	1.018
	Distance to road	.555	1.801
	Elevation	.461	2.169
	Geology	.795	1.258
	Plan curvature	.800	1.250
	Profile curvature	.939	1.065
	Slope	.873	1.145
	SPI	.752	1.329
	TWI	.552	1.811
	Soil	.483	2.071

Several runs of logistic regression were performed until all significant values of the variables were under the threshold value. Despite the good fit statistics show that each model accurately predict the outcome of the dependent variable, it is still not clear which variables have an active influence to the landslide occurrence. So the usual threshold of significance value of 0.05 is set to identify those variables that will be effective in susceptibility assessment. It shows in the table that removal of those variables that deemed to be insignificant mostly gives positive effect on the significance value of the remaining variables.

The coefficients derived from the logistic regression indicate the relationship of each independent variable to the probability of occurrence of the event. The exponential value of the coefficient represent the log- odds. Thus positive coefficients tend to increase the probability of a landslide activity and decrease it otherwise. The constant or the intercept of the models represent the odds of an outcome happening with all the variables removed.

From all the models, the geology 3 yields the highest value followed by aspect 9. The aspect 9, are slopes facing northwest, and geol 3, Paleocene-eocene rock, implies that these variables have the most significant value in explaining the occurrence of landslide. Other factors such as the slope, TWI, SPI, and curvature always yield lower values than the other aspect and geology factors hence does not explained well the occurrence of landslides.

Table 2. Summary of the coefficients of each training model.

Paremeters/ Coefficients	Model 1	Model 2	Model 3	Model 4	Model 5
Slope	0.37	0.40	0.32	0.33	0.34
SPI	2.33	2.10	2.72	2.67	2.54
TWI	-2.31	-2.06	-2.69	-2.63	-2.52
Flat	-2.91	-2.61	-4.64	-2.86	-2.69

North (0-22.5)	-0.73	-0.73	-0.90	-0.65	-0.72
Northeast	-0.72	-0.90	-1.03	-0.77	-0.83
East	-0.49	-0.54	-0.66	-0.55	-0.44
South	0.00	-0.31	-0.38	-0.32	0.00
West	-0.66	-0.69	-0.87	-0.77	-0.63
Northwest	-1.54	-1.89	-1.88	-1.83	-1.66
North (337.5-360)	-1.25	-1.27	-1.32	-1.37	-1.30
Oligocene-miocene	-2.28	-1.95	-2.07	-2.30	-2.00
Paleocene-eocene	-1.91	-1.77	-1.69	-1.78	-1.85
Oligocene	-2.55	-2.64	-2.40	-2.38	-2.12
Neogene	-2.77	-2.47	2.72	-2.16	-2.56
Upper miocene-pliocene	-0.92	0.00	-0.60	-0.99	-0.57
Pliocene-quaternary	0.39	0.79	0.80	0.72	0.79
Pliocene-pleistocene	-2.10	-1.87	-1.76	-1.71	-1.70
Cretaceous-paleocene	0.00	0.85	0.73	0.79	0.83
Plan convex	0.19	0.00	0.00	0.00	0.20
Antipolo clay	0.00	-0.24	-0.20	0.00	-0.24
Binangonan clay	-0.43	-0.61	-0.67	-0.48	-0.60
Binangonan clay (lowland phase)	0.00	-2.24	-1.30	-1.73	-1.16
Constant	-0.40	-1.42	1.47	1.01	0.52

The summary shown on table 2 presents that not all variables were effective in determining the outcome of the landslide occurrence. It can be observed that for all the models; the slope, SPI, Pliocene-quaternary, and Cretaceous-paleocene showed positive values of coefficients. Thus, regions with high slope and SPI are more probable to landslide as well as areas lying in bedrocks composed of Pliocene-quaternary, and Cretaceous-paleocene rocks. For models 1 and 5 convex planform curvature also have positive coefficient hence also increase the instability of the area.

Table 3. Statistics and accuracy of the each model.

		Model 1	Model 2	Model 3	Model 4	Model 5
Hosmer and Lemeshow test	Chi - square	130.926	105.452	91.033	84.556	100.437
	df	8	8	8	8	8
	Sig	0	0	0	0	0
-2 Log likelihood		5224.280	5143.240	5134.659	5029.286	5151.012
Cox & Snell R Square		0.635	0.638	0.638	0.640	0.638
Nagelkerke R Square		0.847	0.851	0.850	0.853	0.850
Accuracy		90.2 %	90.7 %	90.8 %	91.0 %	90.7 %

IOP Conf. Series: Earth and Environmental Science **26** (2015) 012031 doi:10.1088/1755-1315/26/1/012031

The summary of the statistics in the table 3 shows goodness of fit of the model to the data and its accuracy. The Hosmer and Lemeshow test yields very small significance values implying that the models were statistically significant. The Cox & Snell and Nagelkerke R squares show a reliable value which indicates the effectiveness of the model. The accuracy confirms the value of the R squares yielding more than 90% of the training data sets were correctly predicted.

7. Results Validation

The AUROC of each model is validated using each corresponding data set. The AUROC of the five models are; 0.978, 0.977, 0.977, 0.974, and 0.979 respectively. The value of the AUROC implies that the models are effective in correctly predicting the occurrence and non- occurrence of landslide activity. The data used in evaluating the AUROC is the independent validation set that is not used in the susceptibility model calibration. Confusion matrices were also done to determine the predictive capability of the models, as show in tables 4 to 8.

Table 4. Confusion matrix of validation data set 1 evaluated using model 1.

observed		predicted		
		landslide		percentage correct
		absence	presence	
landslide	absence	1568	165	90.48
	presence	153	1581	91.18
Overall Percentage				**90.83**

Table 5. Confusion matrix of validation data set 2 evaluated using model 2.

observed		predicted		
		landslide		percentage correct
		absence	presence	
landslide	absence	1635	102	94.13
	presence	228	1455	86.45
Overall Percentage				**90.29**

Table 6. Confusion matrix of validation data set 3 evaluated using model 3.

observed		predicted		
		landslide		percentage correct
		absence	presence	
landslide	absence	1547	185	89.32
	presence	168	1570	90.33
Overall Percentage				**89.83**

IOP Conf. Series: Earth and Environmental Science **26** (2015) 012031 doi:10.1088/1755-1315/26/1/012031

Table 7. Confusion matrix of validation data set 4 evaluated using model 4.

observed		predicted		
		landslide		percentage correct
		absence	presence	
landslide	absence	1533	210	87.95
	presence	148	1629	91.67
Overall Percentage				**89.81**

Table 8. Confusion matrix of validation data set 5 evaluated using model 5.

observed		predicted		
		landslide		percentage correct
		absence	presence	
landslide	absence	1537	178	89.62
	presence	147	1581	91.49
Overall Percentage				**90.56**

8. Final Landslide Susceptibility Map

The landslide susceptibility map generated from model 1 is shown in figure 3. It is classified into four categories; low, moderate, high and very high susceptibility. Most of the areas with low susceptibility to landslide are lying in the low altitude areas with gentle slopes. The susceptibility increases as it traverses eastward because of the mountain ranges that covers the terrain. It can also be observed that small portions with low susceptibility are river networks.

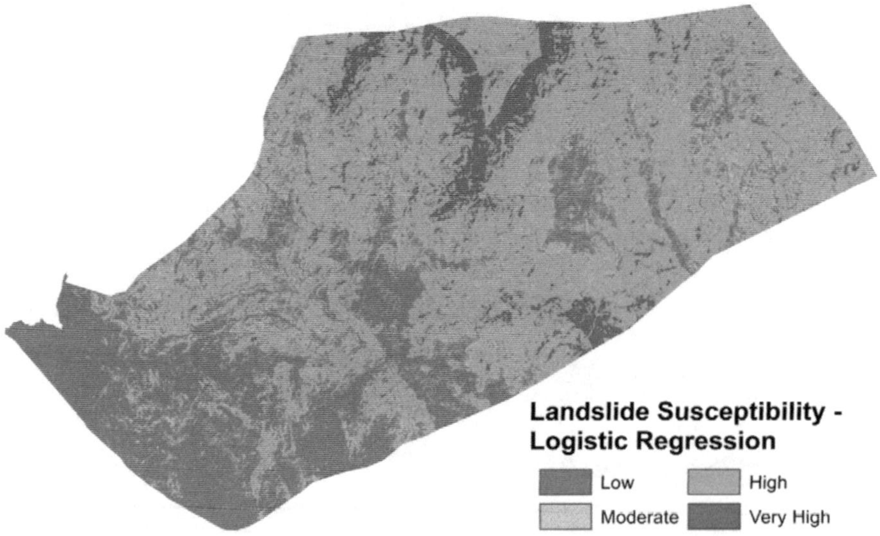

Figure 3: Landslide susceptibility map of Antipolo City.

9. Conclusions

Based on the results of the logistic regression analysis of the landslide susceptibility of Antipolo City, the objectives of the study had been attained. The study shows the ability of the model to delineate areas of instability.

From the study, it shows that area with more than 15° of steepness and facing the southwest direction have high probability of failure. The lithology of the area also contributes to the occurrence of landslide. It was found out that area lying in bedrocks composed of Cretaceous rocks and topsoil cover of Antipolo- origin soil was more prone to landslide compared to other bedrocks and top soil covers.

The study also shows that almost 40% of Antipolo City has been assessed to be potentially dangerous areas in terms of landslide occurrence. The landslide susceptibility of each barangays was also examined. Brgy. San Juan yields the highest percentage of highly susceptible land area with more than 65%. Other barangays such as Bagong Nayon, Calawis, and Inarawan should also be aware of instability with more than 40% of their areas were situated in high landslide susceptibility.

The model generated also shows high reliability based on the several statistical tests applied using test data set. It can be concluded that the model can accurately predict landslide occurrence given the physical factors that are present in the area. Thus it can be used for further studies such as landslide hazard and risk assessments. The generated map can also be used by concerned authorities to properly mitigate areas that were found to be unstable. Residences that were lying in highly susceptible areas should also be force evacuated to avoid further property damages and loss. As well as the establishments developed should also be removed or converted to other kind of recreational area. Based on the map, an extensive zoning should also be conducted since the population, business establishments, and subdivisions are increasing rapidly in the city.

10. References

1. Lee S (2005) Application of logistic regression model and its validation for landslide susceptibility mapping using GIS and remote sensing data journals. Int J Remote Sens 26(7):1477–1491.

2. Nandi A, Shakoor A (2008) Application of logistic regression model for slope instability prediction in Cuyahoga River Watershed, Ohio, USA. Georisk 1:12

3. F.C. Dai, and S. Lee, "Landslide characteristics and slope instability modeling using GIS, Lantau Island, Hong Kong", Geomorphology, Vol. 42, 2002, PP. 213-227.

4. Hosmer D, Lemeshow S (2000). Applied Logistic Regression, Second Edition, John Wiley & Sons, Inc., Print ISBN: 9780471356325

5. C. J. F. Chung and A. G. Fabbri, Validation of spatial prediction models for landslide hazard mapping, Natural Hazards, vol. 30, pp. 451-472, 2003.

6. Menard SW (1995) Applied logistic regression analysis. SAGE, Thousand Oaks

Risk assessment and management of unstable slopes on the national forest estate in Scotland

M Humphreys[1], I Nettelton[1] and K Leech[2]

[1] Coffey Geotechnics UK Ltd, Atlantic House, Atlas Park, Simonsway, Manchester M22 5PR, UK
[2] Forestry Commission, Forest Enterprise Scotland Head Office, 1 Highlander Way, Inverness, IV2 7GB; UK

Abstract. The National Forest Estate in Scotland has a wide range of geotechnical hazards present, primarily landslides, which may cause a significant risk to people and key infrastructure. UK land owners are increasingly required to understand the risks associated with their land and how their activities may affect landsliding and, in particular, where landslides originating from their land may impact third party assets. A Geographic Information System (GIS) based landslide susceptibility assessment by the British Geological Survey (BGS) identified a number of sites in the National Forest Estate as being susceptible to landslide hazards. Coffey Geotechnics Ltd and the BGS are currently undertaking "ground-truthing" of selected sites to identify and characterise the hazards, pathways and elements at risk. A "Slope Stability Appraisal of Risk" system was used to assign a risk category to areas identified during the "ground-truthing" phase which need to be managed. This system is based on the combination of hazards, receptor type, vulnerability and pathway in a similar manner to that of the Australian Geomechanics Society. A long term strategy for risk management of unstable slopes is under development by Forestry Commission Scotland to provide strategic guidance on future land management and guidance for existing felled sites. Options for risk management include: maintenance of existing systems i.e. drainage; silviculture where establishment of woodland can assist in slope stabilisation; and engineering works such as barrier systems, retaining structures and rock remedial works.

1. Introduction

The National Forest Estate (NFE) in Scotland currently extends to 650,000 hectares of land. The terrain is variable across the estate, but includes a high proportion of mountainous landform, steep slopes and geological diversity. [1].

Compared with many countries, the United Kingdom (UK) has a low level of landslide hazard; however, due to the high population in the UK and the density of critical infrastructure, the tolerance for landslides is considered to be low. In 2010 Forestry Commission Scotland (FCS) commissioned an ambitious programme of work – 'The Slope Stability Project'. The project was targeted at identification and quantification of the landslide hazard and development of risk mitigation measures for the NFE and any adjacent third party assets. As part of this the British Geological Survey (BGS) was commissioned by FCS to undertake a desk based study using Geographical Information Systems (GIS) to target sites for further assessment [2].This work is detailed in 'Landslide hazard assessment and characterisation for forestry asset management in Great Britain' [3]. The total area of forest on steep slopes amounts to some 40,000 hectares, and of these approximately 17,500 hectares has been identified as being potentially unstable [1].

2. Hazard Assessment

The 17,500 hectares of potentially unstable land are currently being subject to an on-site 'ground truthing' geotechnical assessment, by Coffey Geotechnics Ltd and the BGS, to identify and characterise the types of hazards, triggers and propagating factors. There are two main categories of hazards identified during the assessments; natural hazards and man-made hazards.

Natural hazards observed in the NFE include: formation of debris dams in stream beds and at culvert inlets due to gully erosion sediment, landslides and trees debris (figure 1); landside failures in the forest and above the tree line; debris flows both channelised and open slope (primarily channelised, figure 2); rock falls due to weathering, discontinuity controlled failures, root jacking (figure 3) and windthrow, and boulders and rock material resting on slopes and against trees (figure 4). Each of these hazards has the potential to adversely affect the surrounding land whether that is on or outside the NFE.

Figure 1. Erosion of channel in forest.

Figure 2. Debris flow onto Trunk Road.

Figure 3. Root jacking of a rock crag.

Figure 4. Boulders caught up against trees.

Man-made hazards observed in the NFE are commonly associated with construction of forest roads. Forest roads are typically built for timber extraction and as such, must be relatively low cost to construct [4]. Historically, constructing cost effective roads across steep and challenging terrain led to practices such as building roads out onto trees dropped against tree stumps ('high stumps'). Whilst this method is acceptable in the short term, over time the trees and stumps rot away leading to stability issues for the outer edge of road embankments and the presence of loose and poorly compacted fill.

Good drainage is fundamental to any forest road especially in areas of high relief where landsliding is a common problem. In such areas the drainage can be affected by minor instability, which may lead to drainage issues and subsequent larger scale instability. A forest road should not inhibit the flow of existing watercourses and it should be designed to cope with foreseen issues such as transportation of sediment, debris (e.g. brash) and storm events.

3. Risk Assessment

As an aid to risk assessment and risk management each site, once assessed, was split into discrete Areas based on the hazards and exposures identified. The Slope Stability Appraisal of Risk (STAR) system developed in 2008 [5] was used to assign a risk value, risk category and action to each Area. The STAR system is based on the combination of hazards, receptor type, receptor vulnerability and pathway in a similar manner to that of the Australian Geomechanics Society [6]. Risk (STAR Value) = Hazard x Receptor Vulnerability x Receptor Type x Pathway

The definitions and values for the Hazard, Receptor Vulnerability and Receptor Type are presented in Table 1. The Hazard is an estimation of the significance of the slope feature with the potential to do harm with a value of 5 having the maximum potential for harm. The Receptor Vulnerability is an estimation of the damage that could be caused by the hazard with a value of 5 being the maximum damage possible. The Receptor Type is the kind of receptor which the hazard could potentially affect. An undeveloped area where people are not present has been assigned a value of 1 (minimum) and an area in which numerous people could be present has been assigned a value of 5 (maximum). The Pathway is an estimation of the likelihood of the Hazard interacting with the Receptor Type with a value of 1 being almost certain and a value of 0 being barely credible.

Table 1. Definitions and Values of Hazards and Receptors.

Values	Hazard Description	Receptor Vulnerability	Receptor Type
1	Minor failure / erosion / weathering - typically slopes < 25°.	Little or no effect	Undeveloped land
2	Moderate slope failure - typically slopes < 25°. Small rock fall individual blocks <$0.1m^3$.	Nuisance or minor damage	Unoccupied / Infrequently visited building / Public rights of way
3	Substantial slope failure – typically slopes 25° to 40°. Moderate rock fall <$10m^3$.	Significant damage	Roads / Footpaths
4	Large slope failure – typically slopes at 25° to 40°. Large rock fall 10 to $50m^3$.	Major damage / Major Injury	Residential property / Commercial buildings (single to few lives)
5	Major slope failure – typically slopes at > 40°. Major rock fall >$50m^3$.	Total loss / Loss of life	Major public buildings (many lives)

The STAR values therefore range between a minimum of 0 and a maximum of 125. The STAR values, the corresponding risk category and actions are given in Table 2.

Table 2. STAR Values.

Risk Value	Risk Category	Action
<10	Low	Normally accepted
10 to <25	Low to Moderate	Significant Risk - need to be made aware of hazards and monitor
25 to <75	Moderate to High	Significant Risk requiring remedial measures / risk management actions
75 to <100	High	Significant Risk requiring major remedial measures
100 or >	Very High	Significant Risk requiring urgent action

4. Risk Management

4.1. General

There are different levels of risk management and remedial works which can be applied to the sites depending on the outcome of the risk assessment. Low risk sites can typically be managed with general inspections and maintenance of infrastructure. Low to Moderate risk sites can typically be managed with maintenance and refurbishment of existing infrastructure and drainage along with implementing control measures, such as preventing public access to forest roads during foresting operations. Moderate to High, High and Very High risk sites are likely to require further inspection to fully characterise the hazards. It is likely that these will require remedial works to reduce the risk to a level that can be managed effectively.

Studies in steep forested areas throughout the world have noted an increase in landsliding after vegetation is removed from the slopes; Endo and Tsuruta [7]; Megahan et al. [8]; Wu and Swanston [9]; Bathurst et al. [10]; Preston and Crozier [11]; Jakob [12]; Dhakal and Sidle [13]. This is important to consider when planning foresting operations and subsequent risk management strategies. FCS are currently developing a Guidance Report for Sustainable Management on Steep Slopes. The guidance will cover two main areas of concern; steep ground working and slope stability. These are not mutually dependant; however, they are linked. The guidance will outline underlying principles of slope stability; provide strategic and tactical guidance on future land-use management in respect to steep slopes; provide guidance for existing felled sites on steep slopes; identify good practice in planning and management on steep slopes and identify appropriate silvicultural and geotechnical management systems to meet site specific challenges [1].

4.2. Silvicultural

4.2.1. Land Management. Most forests on the NFE are already well-established. Many of these are in the process of being re-structured and as this takes place it will be important to establish what the objectives of management should be given the nature of the site. On unstable sites where there is a clear threat to third party assets, public health and safety, or environmentally sensitive areas, then slope stability will be the principal objective. In these cases foresting will be driven by risk management rather than commercial pressures.

4.2.2. Protection Forests. The role of forests as protection systems has seen increased research over the last decade; Motta & Haudemand [14]; Brang [15]; Hurand & Berger [16]; Dorren et al. [17]; Stokes [18]. In the European Alps, and in parts of the Scottish Highlands, rockfalls tend to be low magnitude / high frequency events with single rocks (<5m3) being displaced [19 & 20]. In such cases trees are effective as protection systems and can assist in reducing the velocity and the rebound heights of falling rock [21]. This has been seen throughout the site assessments, even where the trees offer only a relatively thin protection band. The NFE acts, to some extent, as a natural protection system to the elements at risk by creating a barrier against falling rock debris, and also acts to stabilise superficial deposits and reduce surface water run-off intensity. On such critical sites, as the trees are removed this barrier is lost, which coupled with the potential for foresting operations to disturb boulders and rock debris on slopes and rock material from crags, may lead to an increased potential for hazard and receptor interaction. Hence, in critical locations specific management actions and even artificial protection systems may be required to manage the risk.

4.2.3. Tree Removal. The removal of trees from a site can be the trigger for the onset of instability, but it is also the opportunity to influence the stability of a site for the future. Once trees have been removed from a site there is a period of time during which the roots of the pre-existing trees will exert a beneficial influence on the soil. The deterioration of the old root system occurs over a number of years but the flexing action of the stems in the wind will be removed. This offers a window of opportunity to establish a crop of successor trees either through natural regeneration or by planting. Newly established trees contribute little in the way of rooting strength to the soil until around eight or nine years of age. It is important that on potentially unstable sites a successor crop is established before the integrity of the old root system becomes severe [1].

4.2.4. Mixed Root Development. The establishment of woodland has been identified as having a positive effect on slope stability with a growing body of scientific literature supporting it [22]. This has long been recognised in European countries where the protective capacity of forest cover has underpinned silvicultural practice. Soil protection is provided by grasses and herbaceous species, but this is greatly enhanced by the addition of shrubs and trees with woody root systems [23]. Tree roots can help to bind loose friable material and loose rocks on slopes, but can lead to 'root jacking' and rockfall hazards on rock crags or cuttings. They can also assist in intercepting water and controlling its movement through the soil.

The development of a rooting system on unstable sites can be beneficial in reducing the opportunity of soil movement; however, if carried out in the wrong way this can also exacerbate the problem. Root systems that are shallow and form an extensive plate can create a shearing plane. Add to this the tall thin stems that are prone to excessive movement in wind, can create the opportunity for windthrow and the subsequent exposure of unstable slopes. A variety of different tree species with various rooting structures binds the soil and rock horizons both horizontally and vertically (figure 5) and a varied canopy protects the forest from catastrophic windthrow [1].

Figure 5. Mixed root development to assist with slope stabilisation.

4.3. Geotechnical

For risk management and remedial works there are four principal methods available: avoidance; removal; control and containment; and strengthening. Avoidance, i.e. moving the infrastructure away from the unstable slope is typically not a practical option in the NFE, especially where third party land is involved.

4.3.1. Removal. Where there are a limited number of well-defined hazardous features then removal is considered appropriate. Based on past experience in the NFE the biggest cause of instability is due to water on the slopes. In the majority of cases the water could have been controlled by implementing an adequate drainage system or in the cases where drainage is already present, in having sufficient inspection and maintenance regimes in place. This will form part of the Guidance Report for Sustainable Management on Steep Slope [1]. Removal of material build-up in the culverts and ditches can ensure that water is allowed to flow through the sites freely, preventing water backing up in the slopes which could lead to a build-up of pore pressures and possibly, landslides. Removal of tree and rock debris from the gully beds can prevent the creation of debris dams. Material can back up behind such dams until eventually the dam fails suddenly releasing significant quantities of debris and water downslope. Such debris dam failure can lead to significant downstream erosion, obstruction of culverts and potentially debris flows.

With regards to rock crags and boulders, where failures do not have the potential to significantly undermine material or exacerbate other potential failures then removal of key hazard areas is recommended, otherwise control and containment should be considered. Removal of trees from the crest of rock crags within the NFE can prevent 'root jacking'.

4.3.2. Control and Containment. Forest Research, the research agency of the Forestry Commission, has been investigating protective barrier systems to assist with natural hazards and foresting on steep slopes [24]. Due to the proximity of major roads to the NFE, installation of protective barrier systems will need to be in conjunction with traffic management. Forestry Commission have been working closely with Transport Scotland (governing body in charge of Scottish road network) in some areas to enable effective traffic management with as little disruption as possible to drivers. Protective barrier systems considered by Forest Research include rolling blockades of Heavy Goods Vehicles (HGV), Roman Army style 'Armadillo' rolling barrier system and catch fence systems (figures 6 & 7). The rolling blockade using HGVs would consist of a row of HGV trailers modified and strengthened to provide barriers or catch fences. The trailers would also be 'weighted' to reduce movement or overturning during any debris strike. Such systems would be positioned in the lane of the road nearest to the forest area, allowing the far lane to remain open to traffic. Any expected rock or landslide within design parameters would be contained by the barrier [24].

Due to the hazards and risks associated with access and construction activities on steep slopes, it has been identified that reducing the construction activity on the slope would be desirable and reduce costs. Forest Research has been undertaking research on the stability of trees on steep slopes [25] and the anchorage capacity of coniferous trees [26], to develop a temporary catch fence design utilising trees as substitute 'fence posts' and upslope fence post support anchorages. This would reduce the need for foundations, posts, erection of posts and the installation of anchorages.

4.3.3. Strengthening. Where localised failures have the potential to undermine the slope or infrastructure in the NFE then strengthening is considered to be appropriate. In past experience on the NFE, strengthening is typically required for embankment failures along forest roads where slopes below are too steep to build out onto. In these cases strengthening options such as gabion walls and soil nailing have been utilised (figures 8 & 9).

Figure 6. Barrier allowing road to remain open.

Figure 7. Catch fences using trees as posts.

Figure 8. Soil nailing below forest road.

Figure 9. Gabion wall landslide reinstatement.

5. Conclusions

Following an initial GIS based screening of the landslide susceptibility across the NFE Scotland a 'ground truthing' exercise is currently under way to validate and further characterise the landslide hazard and risk associated with the higher susceptibility sites identified, thereby enabling prioritisation and budgeting to be undertaken.

Based on the terrain, the hazards and the elements at risk encountered during the 'ground truthing' it is clear that management of the NFE need to consider long term management not just the completion of the next phase in the foresting cycle. Guidance is currently being developed and updated by FCS to inform planning and foresting operations for steep and potentially landslide susceptible terrain.

In many of the sites most susceptible to landsliding manipulation of natural processes followed by land use management is likely to be more successful and cost effective than intrusive engineering solutions. However, there are a number of critical sites which require engineering works to mitigate or manage the risks.

Acknowledgements

This paper is published with the permission of the Forestry Commission and special thanks go to Kim Leech (Project Manager) and Ben Lennon for their work and support.

References

[1] Forestry Commission Scotland. 2014. *Sustainable Management on Steep Slopes Guidance Report*. Draft.

[2] Hobb, P R N, Foster C, Smit, J, Pennington C, Plim J, Winter M G, Reeves H J and Culshaw M G 2011 *Landslide Potential & Characterisation of Landslides within Forestry Commission Scotland Managed Land*. British Geological Survey COMMERCIAL Report, CR/11/006. 62pp.

[3] Foster C, Reeves H J, Pennington C, Hobbs P R N, Culshaw M G, Plim J, Smith J T, MacCulloch F, Queen G and Winter M G 2012 *Landslide Hazard Assessment and Characterisation for Forestry Asset Management in Great Britain*. In: Eberhart, E., (ed.) Landslides and Engineered Slopes: Protecting Society through Improved Understanding. CRC Press, p6.

[4] Hay R 1996 *Forest Road Design*. Proc. Seminar on Environmentally Sound Forest Roads and Wood Transport. Romania.

[5] Tonks D M, Nettleton I M and Denney R M 2008 *Some Developments in Rock Slope Hazard Assessments for Road and Rail*. In Advances in Transportation Geotechnics. Ellis, Yu, McDowell & Thorn (eds). (Taylor & Francis Group: London).

[6] Australian Geomechanics Society *2007 Guideline for Landslide Susceptibility, Hazard and Risk Zoning for Land Use Management*. AGS Landslide Taskforce. Landslide Zoning Working Group. Australian Geomechanics. **Vol 42(1)**.

[7] Endo T and Tsuruta T 1969 *The Effect of the Tree's Roots on the Shear Strength of Soil*. In Annual Report, 1968. (Hokkaido Branch Forest Experimental Station: Sapporo) pp 167–182.

[8] Megahan W F, Day N F and Bliss T M 1978 *Landslide Occurrence in the Western and Central Northern Rocky Mountain Physiographic Province in Idaho*. In Proc. 5th North American Forest Soils Conference. (Colorado State University: Fort Collins) pp 116–139.

[9] Wu T H, and Swanston D N 1980 Risk of landslides in shallow soils and its relation to clearcutting in southeastern Alaska. *Forest Science* **26**: pp 495–510.

[10] Bathurst J C et al 1996 *Modelling the Impacts of Climate and Land Use Change on Basin Hydrology and Soil Erosion in Mediterranean Europe*. In Mediterranean Desertification and Land Use, Brandt CJ, Thornes JB (eds). (John Wiley: Chichester) pp 355–387.

[11] Preston N J and Crozier M J 1999 Resistance to shallow landslide failure through root-derived cohesion in East Coast hill country soils, North Island, New Zealand. *Earth Surface Processes and Landforms* **24**: pp 665–675.

[12] Jakob M 2000 The impacts of logging on landslide activity at Clayoquot Sound, British Columbia. *Catena* **38(4)**: pp 279–300.

[13] Dhakal A S and Sidle R C 2003 Long-term modelling of landslides for different forest management practices. *Earth Surf Process Landforms*, **28**: pp 853–868. doi: 10.1002/esp.499

[14] Motta R and Haudemand J C 2000 *Protective Forests and Silvicultural Stability. An Example of Planning in the Aosta Valley*. Mt Res. Dev. p 20

[15] Brang P 2001 Resistance and elasticity: promising concepts for the management of protection forests in the European Alps. *For Ecol Manage*. p 145.

[16] Hurand A and Berger F 2002. Forets et risqué naturels. Protection contre l'erosion. les mouvements de terrain et les avalanches. *Houille Blanche* p 3.

[17] Dorren L, Maier B, Seijmonsbergen H and Berger F 2004 *Rockfall, Forest and Human Interactions in the European Alps*. Int. Symp. Interpraevent 2004. Riva/Trient.

[18] Stokes A 2006. Selecting tree species for use in rockfall-protection forests. For. *Snow Landsc Res*. p 80.

[19] Berger F, Quetel C and Dorren L K A 2002 *Forest: a Natural Protection Mean Against Rockfalls, but with which Efficiency?* Int. Cong. Interpraevent 2002 in the Pacific Rim, Matsumoto, Japan. Congress publication, volume 2. pp. 815-826.

[20] Stoffel M, Lievre I, Monbaron M and Perret S 2005 Seasonal timing of rockfall activity on a forested slope at Taschgufer (Swiss Alps) – A dendrochronological approach. Z. *Geomorphol*. p 59.

[21] Dorren L K A, Berger F, Jonnson M, Krautblatter M, Moelk M, Stoffel M and Wehrli A 2007

State of the art in rockfall - forest interactions. *Schweizerische Zeitschrift für Forstwesen.* **158(6)**: pp 128-141.

[22] Stokes A, Norris J E, Van Beek L P H, Bogaard T, Cammaerat E, Mickovski S B, Jenner A, Di Idorio A and Fourcaud T 2008 *How Vegetation Reinforces Soil on Slopes.* In Slope Stability and Erosion Control: Ecotechnological Solutions. (Springer: Dordrecht, NL) pp 65-118 .

[23] Raynor B and Nicoll B 2012 *Potential for Woodland Restoration above the A83 in Glen Croe to Reduce the Incidence of Water Erosion and Debris Flows.* Forest Research Report.

[24] Forest Research 2008 *Felling and Extracting Large Trees on Steep Ground Adjacent to Major Roads – a Case Study on the A82.* Internal Project Information Note 13/07.

[25] Nicoll B C, Achim A, Mochan S and Gardiner B A 2005 Does steep terrain influence tree stability? A Field Investigation. *Canadian Journal for Forestry Research*, **35**, p 2360 to 67.

[26] Nicoll B C, Gardiner B A, Rayner B and Peace AJ 2006 Anchorage of coniferous trees in relation to species, soil type, and rooting depth. *Canadian Journal for Forestry Research*, **36**, pp 1871 to 83.

IOP Conf. Series: Earth and Environmental Science **26** (2015) 012055 doi:10.1088/1755-1315/26/1/012055

System reliability analysis of granular filter for protection against piping in dams

Amit Srivastava[1]
Associate Professor, *Department of Civil & Environmental Engineering*, ITM University, Sector – 23A, HUDA, Gurgaon, Haryana – India

E-mail: amitsrivastava@itmindia.edu

G. L. Sivakumar Babu[2]
Professor, *Department of Civil Engineering*, Indian Institute of Science, Bangalore, Karnataka – India 560012

E-mail: gls@civil.iisc.ernet.in

Abstract. Granular filters are provided for the safety of water retaining structure for protection against piping failure. The phenomenon of piping triggers when the base soil to be protected starts migrating in the direction of seepage flow under the influence of seepage force. To protect base soil from migration, the voids in the filter media should be small enough but it should not also be too small to block smooth passage of seeping water. Fulfilling these two contradictory design requirements at the same time is a major concern for the successful performance of granular filter media. Since Terzaghi era, conventionally, particle size distribution (PSD) of granular filters is designed based on particle size distribution characteristics of the base soil to be protected. The design approach provides a range of D_{15f} value in which the PSD of granular filter media should fall and there exist infinite possibilities. Further, safety against the two critical design requirements cannot be ensured. Although used successfully for many decades, the existing filter design guidelines are purely empirical in nature accompanied with experience and good engineering judgment. In the present study, analytical solutions for obtaining the factor of safety with respect to base soil particle migration and soil permeability consideration as proposed by the authors are first discussed. The solution takes into consideration the basic geotechnical properties of base soil and filter media as well as existing hydraulic conditions and provides a comprehensive solution to the granular filter design with ability to assess the stability in terms of factor of safety. Considering the fact that geotechnical properties are variable in nature, probabilistic analysis is further suggested to evaluate the system reliability of the filter media that may help in risk assessment and risk management for decision making.

1. Introduction

Seepage and piping is one of the major sources of concern for the safety of the dams [1, 2, and 3]. Piping is a phenomenon in which the hydraulic gradient at the exit point exceeds the critical hydraulic gradient value. The critical exit point is normally at the toe of the structure where seeping water flowing through foundation exits. Under the influence of seepage flow, soil at the exit point has the tendency to get eroded. Once eroded, the erosion continues and moves in the backward direction, ultimately, leading to the failure of the whole structure [4]. Conventionally, granular filters are provided to protect the dam from failure against piping. Figure 1(a) depicts typical piping failure in a concrete dam and Figure 1(b) shows provision of granular filter provided at the downstream end for protection against piping. Similarly, in earth dams, granular filters are provided as toe filter, blanket drains and (or) chimney drains as per design requirements as shown in Figures 1(c) & 1(d), respectively. Detailed discussions on the topic "granular filters for dams" are available in FEMA [5] document. It is well understood that the critical location for the occurrence of piping in any water retaining structure is at the toe of the structure where the seeping water exits and the flow is in the vertically upward direction resulting in the reduction of vertical effective stress due to upward seepage and a condition very much favourable for development of piping.

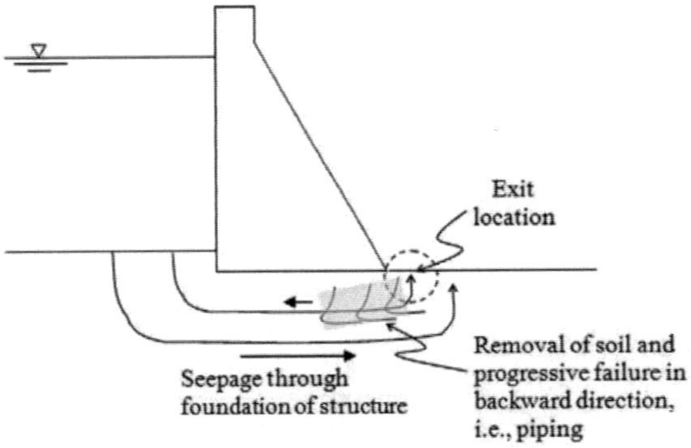

Figure 1(a) Seepage through the foundation of a water retaining structure and phenomenon of piping in dams

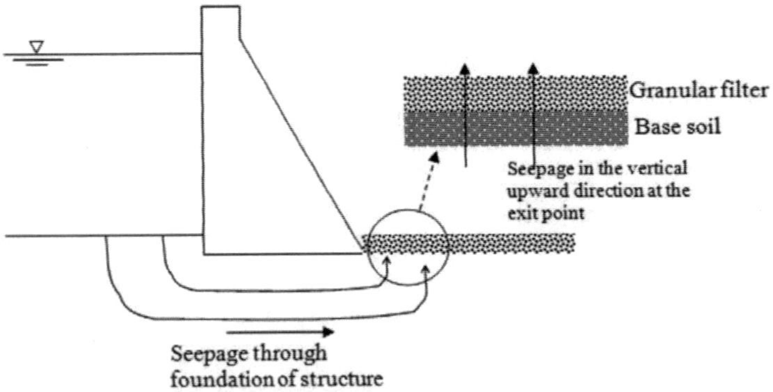

Figure 1(b) Provision of granular filters for protection of dam against piping

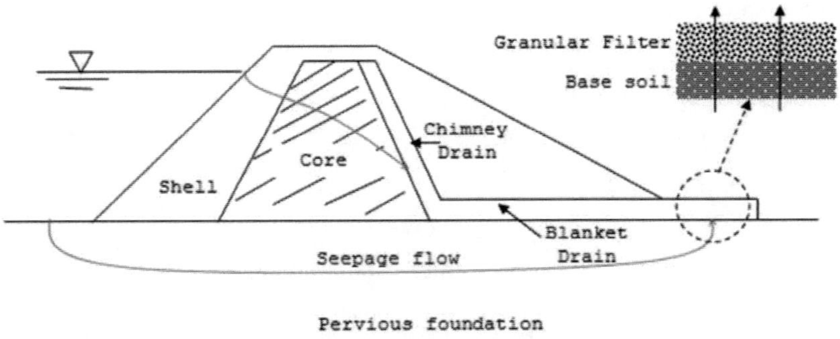

Figure 1(c) Provision of granular filters (chimney drain and blanket drain) for protection of earth dam

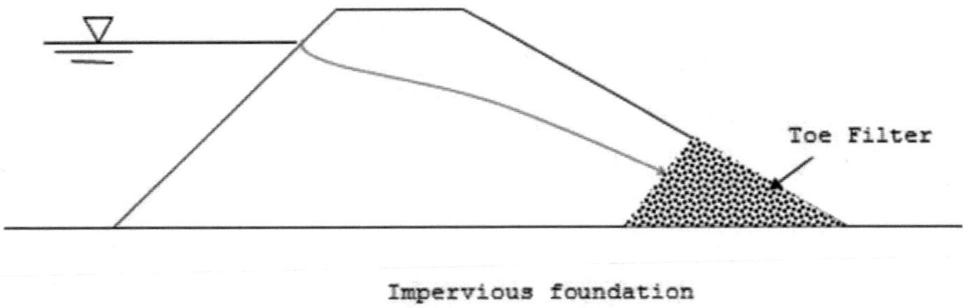

Figure 1(d) Provision of toe (granular) filters for protection of earth dam

Classical examples of seepage and piping failures in water retaining structures are reported in literature [2, 6]. Few examples are failure of right abutment of Fontenelle Dam (1965), Teton dam failure (1976) and Cowlitz County dam failure (2002).

2. Granular filter Design criteria

Granular filters protect the base soil from migration or being washed away under the influence of seepage forces and at the same time it allows the smooth passage of seeping water to avoid development of excess pore water pressure and loss of shear strength in the base soil., The size of voids in the filter media should be small enough to fulfil the first requirement, and, at the same time it should be large enough to fulfil the second requirement. Satisfying these two conflicting requirements at the same time is a challenging task for the design of granular filter. Terzaghi [7] suggested design criteria for obtaining the particle size distribution (PSD) characteristics of the granular filter media based on the PSD characteristics of the base soil. Later, several researchers [8, 9, 10, 11, 12, 13, 14, 15, and 16] established filter design guidelines on the same line of thoughts depending on soil types, experimental procedure, experience and specific design requirements, such as, self healing and internal stability. In the existing filter design guidelines, consideration of geotechnical properties of the base soil as well as filter media and existing hydraulic conditions is lacking. A few theoretical studies on the soil transport phenomenon [10, 17] help in understanding the physical mechanisms of the soil particle transport and retention. Recent works into piping phenomenon focused on the development of predictive mathematical models for particle transport and filter clogging [18, 19, 20, 21, 22 and 23] conducted study on the estimation of the time of development of piping for prediction purposes. Researchers like [24, 25] provided excellent literature review on critical appraisal of granular filter design. Terzaghi [7] in his original work provided the following criteria for the PSD design of granular filters:

Soil retention criterion

$$\frac{D_{15f}}{D_{85b}} < 4.0 \tag{1}$$

Permeability criterion

$$\frac{D_{15f}}{D_{15b}} > 4.0 \tag{2}$$

where, D_{15f} and D_{15b} are the diameters of particles at 15% passing for filter material and base soil, respectively; D_{85b} is the diameter of particles at 85% passing for base soil. The design criteria provide a range of D_{15f} values and there exists an infinite number of options for the choice of appropriate particle size distribution characteristics of the filter media. Although, these guidelines are well established and have been successfully used through decades of experiences, the safety of the filter media with respect to the two design criteria cannot be quantified. Authors [26, 27, and 28] proposed the following analytical solutions for quantifying the factor of safety with respect to soil particle migration and permeability criteria in granular filter design:

Soil retention criterion

$$FS_m = \frac{\left[\frac{d_s^3}{6} + \frac{1}{2\sqrt{2}} d_s^3 \tan\phi'\right](G_s - 1)}{\left[\frac{3}{32} d_s n_f D_{of}^2 + \frac{1}{4} d_s^3\right] i} \tag{3}$$

Permeability criterion

$$FS_p = \frac{(G_s - 1)}{mpH_t i}\left[\left(\frac{D_{15f}}{D_{15b}}\right)^2 \frac{(mp-1)^3}{m(p-1)^3} H_b + H_f\right] \tag{4}$$

Where,

$$m = \frac{(\gamma_d)_b}{(\gamma_d)_f} \tag{5}$$

$$p = \frac{G_s \gamma_w}{(\gamma_d)_b} \tag{6}$$

In the above equations, ϕ' is the effective angle of internal friction of the base soil, n_f is the porosity of filter media, D_{of} is the size of pore channels in the filter media with the assumption that the filter media can be represented as a bundle of tubes, d_s is the weighted average particles size of the base soil, H_t is the total thickness of base soil (H_b) and filter media (H_f), G_s is the specific gravity of the soil particles, $(\gamma_d)_b$ and $(\gamma_d)_f$ are the dry unit weights of the base soil and filter material, respectively, and i is the existing hydraulic gradient. It can be observed that the proposed analytical equations takes into consideration some of the basic geotechnical properties like effective angle of internal friction, unit weight, representative particle sizes, pore size, and existing hydraulic conditions and enable carrying out a stability assessment as well as for the design of granular filter media in terms of factor of safety.

3. Uncertainties Issues: Factor of safety vs. probabilistic approach

The input parameters required in equations (3) & (4) are obtained through various methods, such as, field or laboratory testing, correlation, or mathematical modelling. Soil being a natural material, it has inherent variation in its properties. These factors contribute to the uncertainty in the estimation of the soil input parameters. Conventionally, uncertainties are taken care by use of *factors of safety* in the design philosophy. In geotechnical design and practices, a factor of safety in the range of 1.5 to 3.0 is *generally* accepted [28]. Selection of appropriate values of *factor of safety* comes from past experience, good engineering judgment and the confidence level of the designer. It lacks rationality and there is no mathematical basis. Hence, a question that often arises in practice is to know "how safe is safe?" or to what extent the question of safety can be addressed adequately.

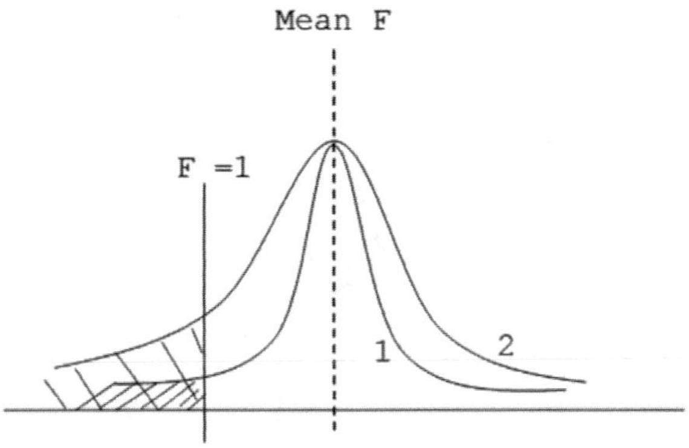

Figure 2 Comparison of two cases with same system response (mean factor of safety) but different failure probabilities or associated risks

Considering the fact that variation in the input parameters introduces variation in the output response, in the present study, the output response, i.e., stability of the system, is measured in terms *of factor of safety (F) and its variations*. The system fails when $F < 1.0$. Figure 2 compares the two cases in which the mean system response (i.e., mean F) is same but the extent of variation in F is different due to the extent of variation in the input parameters being different. The area below the *frequency curve* for F < 1.0 defines the probability of failure or the risk associated with the system. It can be observed that although, the mean system response is although same in the two cases, a higher risk (or probability of failure) is associated with case 2. Such rationality is never achieved in conventional factor of safety approaches and there is no means of identifying the factors for risk reduction and making the system more reliable. The extent of variation and its influence on the system response can be captured in a probabilistic framework. In the probabilistic approach, input parameters are, generally, assumed to be normally or log-normally distributed continuous random variables and parameters of distribution are related to unbiased estimate of mean and variance of the measured data set [29, 30].

Sample Mean (μ)

$$\mu_{FS} = \frac{1}{N}\sum_{i=1}^{N} FS_i \tag{7}$$

Variance (σ^2)

It is a measure of dispersion of data about the mean value. The square root of variance is defined as standard deviation (σ).

$$\sigma_{FS}^2 = \frac{1}{(N-1)}\sum_{i=1}^{N}\left(FS_i - \overline{FS}\right)^2 \tag{8}$$

The coefficient of variation (δ), which is defined as the ratio of sample standard deviation (σ) and the sample mean (μ), is commonly used to quantify geotechnical uncertainties. The advantage of using the coefficient of variation is that it is dimensionless and provides a better estimate of relative dispersion of data around mean. Where site-specific data are not available, typical values of coefficients of variation measured in soil properties are taken from published literature [31, 32, and 33]. In the probabilistic analysis, performance of the system is assessed in terms of an index called *reliability index* (β).USACE [34] made specific recommendations on target reliability indices (β) in geo-technical and infrastructure projects.

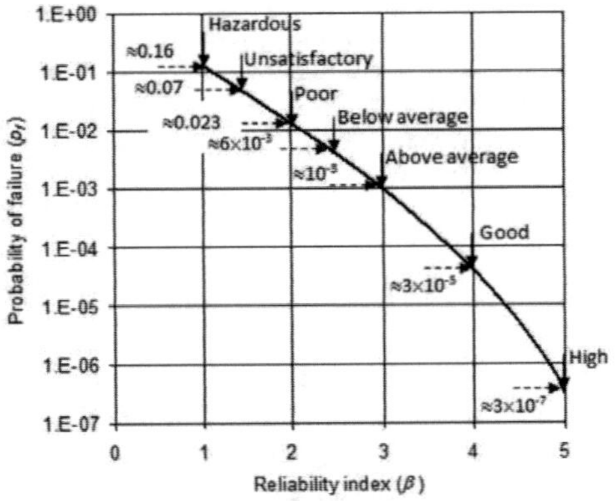

Figure 3. USACE (1997) guidelines for reliability index (β) and corresponding probability of failure (p_f) (adapted from Table B-1) (Phoon, 2004)

The suggested guidelines state that a reliability index (β) value of at least 3.0 is expected for an average performance of the system and 5.0 for excellent performance. There are many methods of estimation of reliability index (β) available in literature [30, 35, and 36].

3.1 System Reliability

The two conflicting design requirements for granular filters introduce multiple failure modes in the base soil- filter media system. Hence, occurrence of anyone of the failure mode will involve failure or non-performance of the granular filter media system. In the estimation of system reliability, evaluation of reliability of system (in terms of reliability indices, β) with respect to individual failure modes, i.e., component reliability is first obtained. A detailed discussion on system reliability approach is provided in [36, 37]. It can be noted that the two failure modes in granular filter are not statically independent as the analytical equations (3) and (4) share several input parameters. If the correlation coefficient between two failure modes is taken as (ρ), the system reliability can be evaluated. For correlated normal variates, Ditlevsen [38] suggested following bounds (equation 9) for joint probability of failure Pr[MP], which is an approximation as the exact solution is cumbersome to determine. Here, *M* (m) is used for particle migration and *P* (p) is used for permeability consideration.

$$max\,(p_m, p_p) \le pr[MP] \le p_m + p_p \quad (9)$$

Where

$$p_m = \Phi\,(-\beta_m)\Phi\left(-\frac{\beta_m-\rho\beta_p}{\sqrt{1-\rho^2}}\right) \quad (10)$$

$$p_p = \Phi\,(-\beta_p)\Phi\left(-\frac{\beta_p-\rho\beta_m}{\sqrt{1-\rho^2}}\right) \quad (11)$$

Where p_m and p_p are the probabilities of failure with respect to particle migration and permeability consideration, respectively. β_m and β_p are the reliability index values from particle migration and permeability consideration and these individual components of reliability indices are first evaluated.

3.2 Reliability analysis of individual component (β_i)

If the performance function g(x) is defined as g(x) = FS-1, where FS is factor of safety with respect to particle migration and permeability consideration (as the case may be), the reliability index (β) through First Order Reliability Method (FORM) for log-normally distributed FS is obtained from the following expression [30]

$$\beta = \frac{\ln\left[\left(\overline{FS}\right)\sqrt{\left(1+\delta_{FS}^2\right)}\right]}{\sqrt{\ln\left[\left(1+\delta_{FS}^2\right)\right]}} \quad (12)$$

where, $\overline{FS}$ is the mean factor of safety and δ_{FS} is the coefficient of variation in *FS* whose values are obtained through Monte Carlo Simulation (*MCS*). For MCS, N numbers of random data (sample) are generated for each set of input parameters that follows the characteristics of probability density function (*pdf*) defined for that input parameters. Factor of safety with respect to soil particle migration $(FS_m)_i$ and permeability $(FS_p)_i$ are evaluated with the help of equations (3) and (4), respectively for each set of the input parameters. It provides *N* number of data set for FS_m and FS_p and the mean and variance of factor of safety are then evaluated using equations (7) and (8). Reliability index values with respect to particle migration (β_m) and permeability criteria (β_p) are then estimated using equation (10). For system reliability evaluation, the sample correlation coefficient (r) between FS_m and FS_p is to be obtained from the following expression:

$$r = \frac{\sum_{i=1}^{N}(FS_i-\overline{FS}_m)(FS_i-\overline{FS}_p)}{\sqrt{\sum_{i=1}^{N}(FS_i-\overline{FS}_m)^2(FS_i-\overline{FS}_p)^2}} \quad (13)$$

4. Illustrative Example

In the present section, evaluation of system reliability of a base soil-filter media system is presented by means of an example application. From the particle size distribution (*PSD*) of the base soil, the weighted average particle size (d_s) is assumed to be obtained as 0.716 mm. To protect this base soil from particle migration and the onset of piping, granular filter layer is provided. The *PSD* of granular filter layer is such that the D_{of} (pore size of filter media) value is obtained as 0.519mm. Other geotechnical properties of base soil and filter media are provided in Table 1.

Table 1 (a). Geotechnical properties of base soil

Properties of base soil	Numerical values
Dry unit weight $[(\gamma_d)_b]$	13.0kN/m^3
Angle of internal friction (ϕ')	32°
D_{15b}	0.750 mm
Weighted average particle size (d_s)	0.716 mm
Parameter "p"	2.00
Parameter "m"	0.93

Table 1 (b). Geotechnical properties of filter media

Properties of filter media	Numerical values
Dry unit weight $[(\gamma_d)_f]$	14.0 kN/m^3
D_{15f}	2.8 mm
Void ratio (e_f)	0.857
Porosity (n_f)	0.461

For the reliability analysis, coefficients of variation in the input soil parameters are taken from the range of values suggested in the literature (Table 2). It is worth mentioning here that the applied hydraulic gradient (i) is assumed as a deterministic parameter.

Table 2. Coefficient of variation (δ%) considered in the input soil parameters and representative particle sizes

Input soil parameters	Range CoV%	Selected CoV%
d_s, D_{of}, D_{15b}, D_{15f}	10-15% [39]	15%
Dry unit weight (γ_d)	3-7% [33, 34]	7%
Effective friction angle (ϕ')	2-13% [33, 34]	13%

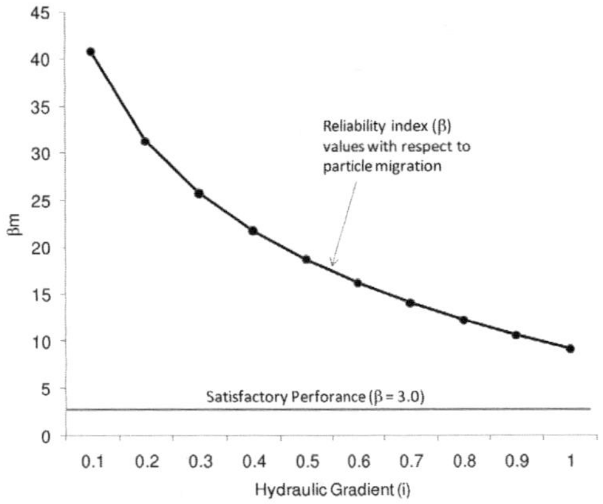

Figure 4(a) Reliability index values with respect to particle migration

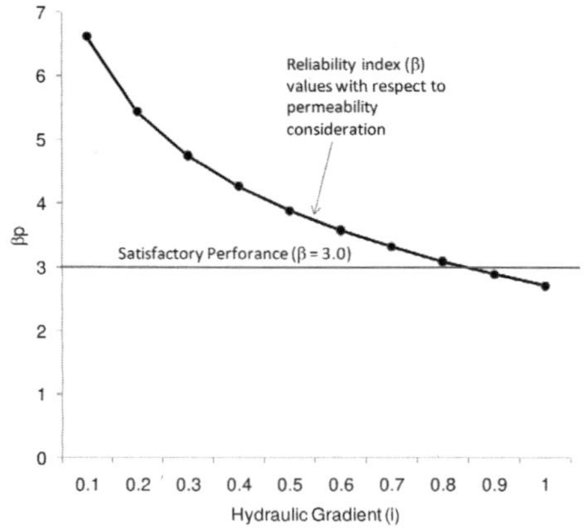

Figure 4(b) reliability index values with respect to permeability

Evaluation of reliability index with respect to two failure modes, i.e., particle migration (β_m) and permeability criteria (β_p) require information on mean and variance of the factor of safety with respect to particle migration (FS_m) and permeability (FS_p), respectively. The mean ($\overline{FS}_m$ *or* $\overline{FS}_p$) and coefficient of variation of the factor of safety $[(\delta_{FS})_m$ or $(\delta_{FS})_p]$ with respect to these two criteria can be evaluated through Monte Carlo Simulations (sample size N) with the help of the proposed analytical solutions. The reliability index values from the two failure modes (β_m & β_p) were estimated by utilizing commercially available software tool STRUREL [COMREL 8 -TI (Time invariant analysis)] [40] and results are compared in the following Figures 4a & 4b.

It can be noted that the granular base soil filter media system indicates acceptable values of reliability indices with respect to particle migration for all the ranges of hydraulic gradient values considered. On the other hand, with respect to permeability, filter performance is satisfactory up to a hydraulic gradient value of 0.8.

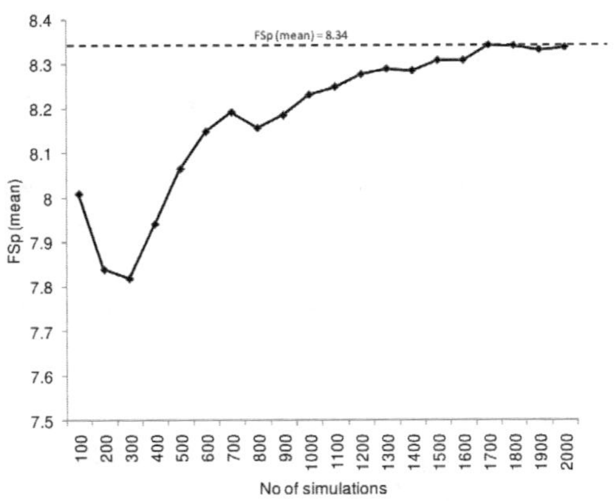

Figure 5 (a) Estimate mean FS_m with sample size

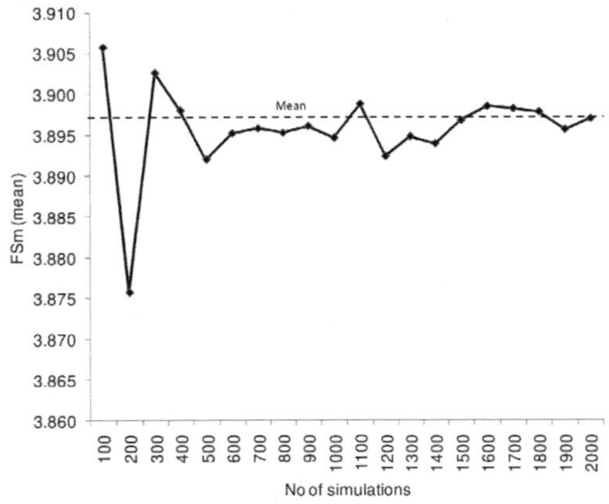

Figure 5 (b) Estimate mean FS_p with sample size

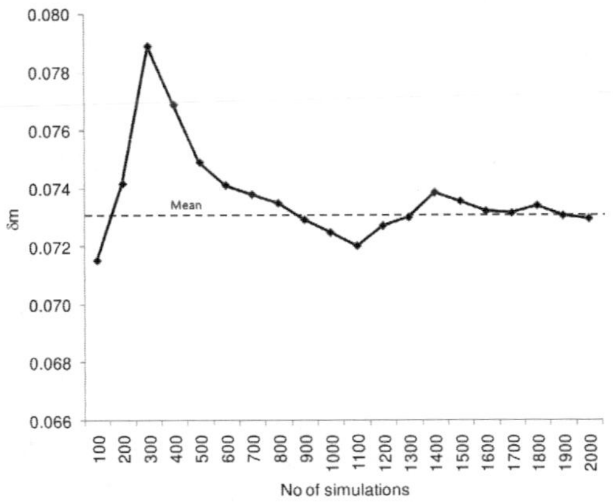

Figure 5 (c) Estimate δ_m with sample size

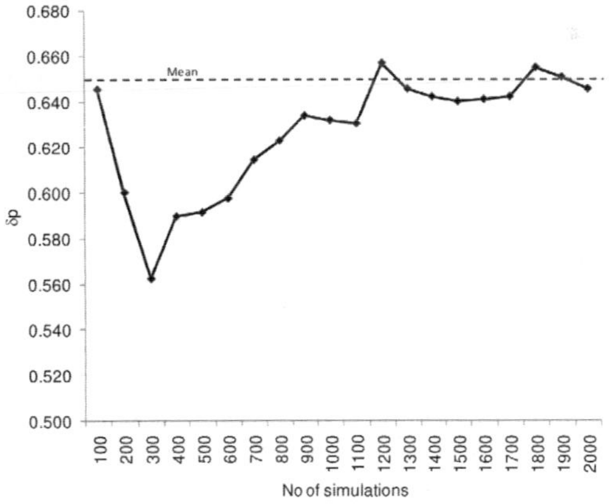

Figure 5 (d) Estimate δ_p with sample size

5.2 Sample size (N) for Monte Carlo simulations

Sample size (N) for Monte Carlo simulation determines both accuracy of the results and time for the analysis. A higher sample size would provide more accurate results but it is time consuming. On the other hand, a smaller sample size although would provide faster results but would compromise the accuracy of the results. To obtain the sample size (N) for Monte Carlo simulations, mean and coefficient of variation of output response, i.e., FS_m and FS_p were obtained for increasing numbers of simulations as shown in the Figure 5. It can be noted that after 1600-1800 simulations, the estimated mean and coefficient of variation stabilize. Hence, in the present study, a sample size of 2000 is considered to be sufficient for the Monte Carlo Simulation.

5.3 System reliability evaluation

For system reliability evaluation of base soil – filter media system, the (sample) correlation coefficient (r) between FS_m and FS_p is obtained using equation (13). FS_i is the factor of safety value obtained for each set of input parameters through Monte Carlo simulations and the value of r is obtained as +0.83. It can be noted that both the factor of safety with respect to particle migration (FS_m) and the one with respect to permeability (FS_p) are positively correlated. It is noted that reliability index values at lower hydraulic gradients are quite high and it can be safely assumed that the system reliability is high for low values of hydraulic gradient. In Table 3, the component reliability indices (β_m , β_p) at high hydraulic gradients are reported.

Table 3. Component reliability indices at higher hydraulic gradients

Hydraulic gradient (i)	Reliability index (β_m)	Reliability index (β_p)
1.1	7.563	2.513
1.2	5.726	2.286
1.3	3.604	2.024
1.4	1.092	1.714

Figure 6 shows the system reliability indices (lower bound) values for the base soil – filter media system at higher hydraulic gradients. It is noted that the base soil- filter media system is likely to have better system performance level up to a hydraulic gradient of 1.3, beyond which, it decreases.

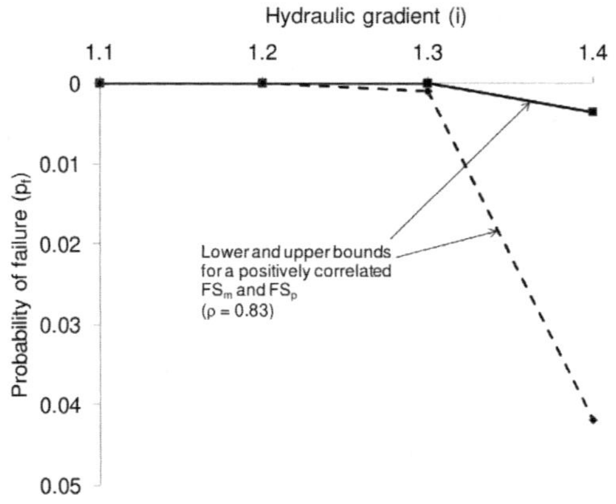

Figure 6. Lower and upper bound for positively correlated FS_m and FS_p

5. Conclusion

- The proposed analytical solutions with respect to soil particles migration and permeability criteria takes into consideration some of the basic geotechnical properties of the base soil and filter media as well as the existing hydraulic conditions.
- The variability in the geotechnical parameters and its influence on the system response can be captured through reliability analysis using presented analytical solutions.
- Reliability analysis when used in conjunction with the conventional factor of safety approach provides a sound mathematical basis for the decision making process.
- Mutually dependent failure modes in a base soil-filter media system, FS_m and FS_p are positively correlated with correlation coefficient +0.83.
- System reliability analysis provides an overall assessment of the system response where there exist multiple failure modes and the failure of one component affects the performance of the other ones.

References

[1] FEMA 1987 *Dam safety: an owner's guidance manual* Federal Emergency Management Agency manual (last access dated 9/04/2015) (http://www.damsafety.org/media/Documents/FEMA/145_GuidanceManual_87.pdf)

[2] Foster M, Fell R, and Spannagle M 2000 A method for assessing the relative likelihood of failure of embankment dams by piping *Canadian Geotechnical Journal*, **37(5)**, 1025-61.

[3] USBR 2001 *Safety Evaluation of Existing Dams (SEED) Manual*, United States Bureau of Reclamation, Denver, Colorado (last accessed 9/04/2015) (http://www.usbr.gov/pmts/instrumentation/policies/SEED.pdf)

[4] Robin Fell, Patrick MacGregor, David Stapledon, Graeme Bell, Mark Foster. Sellmeijer, H., López de la Cruz, J., Beek, V. and van Knoeff, H. 2011 *Geotechnical Engineering of Dams* CRC press, Taylors & Francis Groups.

[5] FEMA 2011 Filters for Embankment Dams – Best Practices for Design and Construction, Federal Emergency Management Agency. http://www.fema.gov/media-library-data/20130726-1813-25045-0568/fema_filter_manual_links.pdf (access date 20/6/2015)

[6] ICOLD 1995 *Dam failures statistical analysis*. International Commission on Large Dams (ICOLD), Bulletin 99.

[7] Terzaghi K 1943 *Theoretical Soil Mechanics*, John Wiley and Sons, New York.

[8] Sherard J L, Dunnigan, L P and Talbot J R 1984a Basic properties of sand and gravel filters. *Journal of Geotechnical Engineering*, ASCE, **110 (6)**, 684-700.

[9] Sherard J L, Dunnigan L P and Talbot J R 1984b Filters for silts and clays, *Journal of Geotechnical Engineering*, ASCE, **110 (6)**, 701-718.

[10] Kenny T C and Lau D 1985 Internal stability of granular filters *Canadian geotechnical journal*, **22**, 215-225.

[11] Honjo Y and Veneziano D 1989 Improved filter criterion for cohesionless soil *Journal of Geotechnical Engineering*, ASCE, **115(1)**, 75-95.

[12] Sherard J L and Dunnigan L P 1989 Critical filters for impervious soils *Journal of Geotechnical Engineering*, ASCE, **115(7)**, 927–947.

[13] Khor C H and Woo H K 1989 Investigation of crushed rock filters for dam embankment *Journal of Geotechnical Engineering*, ASCE, **115(3)**, 399-412.

[14] NRCS1994 *National Engineering Handbook* National Resources Conservation Services (NRCS) U.S. Department of Agriculture, Washington, D.C., Chapter 26, Part 633.

[15] Indraratna B 1997 Analytical model for particle migration with in base soil-filter system *Journal of Geotechnical and Geoenvironmental Engineering*, ASCE, **123(2)**, 100-109.

[16] Locke M and Indraratna B 2002 Filtration of broadly graded soils: the reduced PSD method *Geotechnique*, **52(4)**, 285-287.

[17] Vanmarcke E H and Honjo Y 1985 Probabilistic description of the void phase of soil with reference to assessing risk of internal erosion *Proc. 4th International conference on structure safety and Reliability*, Kobe, Japan, **III**, 321-330.

[18] Reddi L N, Ming X, Hajra M G and Lee I M 2000 Permeability reduction of soil filters due to physical clogging *Journal of Geotechnical and Geoenvironmental Engineering*, ASCE, **126(3)**, 236–246.

[19] Locke M, Indraratna B and Adikari G 2001 Time-dependent particle transport through granular filters Journal of Geotechnical and Geoenvironmental Engineering, ASCE, **127(6)**, 521–529.

[20] Reddi L N and Kakuturu S P 2006a Evaluation of the parameters influencing self-healing in earth dams *Journal of Geotechnical and Geoenvironmental Engg*, ASCE, **132(7)**, 879-889.

[21] Reddi L N and Kakuturu S P 2006b Mechanistic model for self-healing of core cracks in earth dams *Journal of Geotechnical and Geoenvironmental Engineering*, ASCE, **132(7)**, 890-901.

[22] Fell R, Wan C F, Cyganiewics J and Foster M 2003 Time for development of internal erosion and piping in embankment dams *Journal of Geotechnical and Geoenvironmental Engineering*, ASCE, **129(4)**, 307–314.

[23] Annandale G W, Murthy R and Beckwith G 2004 Simulation of dam breach initiated by cracks and earth fissures in semi-arid and arid environments *In: Proceedings ASDSO Dam safety* Phoenix, Arizona.

[24] Richards K and Reddy K R 2007 Critical Appraisal of Piping Phenomena in Earth Dams *Bulletin of Engineering Geology and the Environment*, **66 (4)**, 381-402.

[25] Indraratna B, Laricar Dominic O. Trani and Hadi K 2008 A critical review on granular dam filter behaviour – from particle sizes to constriction-based design criteria *Geomechanics and Geoengineering Journal*, **3(4)**, 279 – 290.

[26] Sivakumar Babu G L and Srivastava A 2007 A procedure for the design of protective filters *Canadian Geotechnical Journal*, 44, 490-495.

[27] Srivastava A. and Sivakumar Babu G L 2013 Reliability analysis-based selection procedure for granular filters with respect to soil-boiling condition, *International Journal of Geotechnical Engineering*, **7(2)**, 193-198.

[28] Srivastava A and Sivakumar Babu G L 2011 Analytical solutions for the protective filters based on soil retention and permeability criteria with respect to phenomenon of soil boiling *Canadian Geotechnical Journal*, **48**, 956–969.

[29] Harr M E 1987 *Reliability-based design in civil engineering* McGraw-Hill, NY.

[30] Baecher G B and Christian J T 2003 *Reliability and statistics in geotechnical engineering* John Wiley Publications, NY.

[31] Lacasse S and Nadim F 1996 Uncertainties in characterizing soil properties *In Shackleford, C. D., Nelson, P. P., and Roth, M. J. S. (Eds.), Uncertainty in the Geologic Environment* (GSP 58), ASCE, New York, 49-75.

[32] Duncan J M 2000 Factors of safety and reliability in geotechnical engineering *Journal of Geotechnical and Geoenvironmental Engineering*, ASCE, **126**, 307–316.

[33] Uzielli M, Lacasse S, Nadim F and Phoon K K 2007 *Soil variability analysis for geotechnical practice*. Characterization and engineering properties of natural soils – Tan, Phoon, Hight& Leroueil (eds), Taylor & Francis group, London, ISBN 978-0-415-42691-6.

[34] USACE 1997 *Risk-based analysis in Geotechnical Engineering for Support of Planning Studies, Engineering and Design* US Army Corps of Engineers, Department of Army, Washington, DC, 1997. pp. 20314-100.

[35] Hasofer A M and Lind N C 1974 Exact and invariant second moment code format *Journal of Engineering Mechanics*, ASCE, **100(EMl)**, 111-121.

[36] Ang AH.-S and Tang W H 1984 *Probability concepts in engineering planning and design: Decision, Risk, and Reliability*, **Vol. 2**, John Wiley & Sons, New York.

[37] Minguez R, Delgado F, Escuder I. and De Membrillera G. 2006 Reliability assessment of granular filters in embankment dams *International Journal of Numerical and Analytical Methods in Geomechanics*, **30**, 1019-37.

[38] Ditlevesen O D 1981 *Uncertainty modelling with applications multidimensional civil engineering systems*. McGraw-Hill, New York.

[39] Haldar A and Mahadevan S 2000 *Probability, reliability and statistical methods in engineering design*. NY: John Wiley & Sons.

[40] STRUREL 2012 *Structural Reliability Software* Users Manual, RCP, Munich, Germany.

IOP Conf. Series: Earth and Environmental Science **26** (2015) 012050 doi:10.1088/1755-1315/26/1/012050

Geohazard assessment lifecycle for a natural gas pipeline project

D Lekkakis[1], M D Boone[2], E Strassburger[3], Z Li[1] and W P Duffy[1]

[1] Bechtel Ltd, 2 Lakeside Drive, Park Royal, London NW10 7FQ, UK
[2] Bechtel Corporation, 50 Beale St, San Francisco, CA 94105, USA
[3] Bechtel Corporation, 3000 Post Oak Boulevard, Houston, TX 77056, USA

Email: Lekkakis@Bechtel.com

Abstract. This paper is a walkthrough of the geohazard risk assessment performed for the Front End Engineering Design (FEED) of a planned large-diameter natural gas pipeline, extending from Eastern Europe to Western Asia for a total length of approximately 1,850 km. The geohazards discussed herein include liquefaction-induced pipe buoyancy, cyclic softening, lateral spreading, slope instability, groundwater rise-induced pipe buoyancy, and karst. The geohazard risk assessment lifecycle was comprised of 4 stages: initially a desktop study was carried out to describe the geologic setting along the alignment and to conduct a preliminary assessment of the geohazards. The development of a comprehensive Digital Terrain Model topography and aerial photography data were fundamental in this process. Subsequently, field geohazard mapping was conducted with the deployment of 8 teams of geoprofessionals, to investigate the proposed major reroutes and delve into areas of poor or questionable data. During the third stage, a geotechnical subsurface site investigation was then executed based on the results of the above study and mapping efforts in order to obtain sufficient data tailored for risk quantification. Lastly, all gathered and processed information was overlain into a Geographical Information database towards a final determination of the critical reaches of the pipeline alignment. Input from Subject Matter Experts (SME) in the fields of landslides, karst and fluvial geomorphology was incorporated during the second and fourth stages of the assessment. Their experience in that particular geographical region was key to making appropriate decisions based on engineering judgment. As the design evolved through the above stages, the pipeline corridor was narrowed from a 2-km wide corridor, to a 500-m corridor and finally to a fixed alignment. Where the geohazard risk was high, rerouting of the pipeline was generally selected as a mitigation measure. In some cases of high uncertainty in the assessment, further exploration was proposed. In cases where rerouting was constrained, mitigation via structural measures was proposed. This paper further discusses the cost, schedule and resource challenges of planning and executing such a large-scale geotechnical investigation, the interfaces between the various disciplines involved during the assessment, the innovative tools employed for the field mapping, the classifications developed for mapping landslides, karst geology, and trench excavatability, determining liquefaction stretches and the process for the site localization of the Above Ground Installations (AGI). It finally discusses the objectives of the FEED study in terms of providing a route, a ± 20% project cost estimate and a schedule, and the additional engineering work foreseen to take place in the detailed engineering phase of the project.

IOP Conf. Series: Earth and Environmental Science **26** (2015) 012050 doi:10.1088/1755-1315/26/1/012050

1. Introduction

The envisioned pipeline project has an approximate length of 1,800 km and a diameter of 48-to-56 inches and comprises of a buried natural gas pipeline system including all above ground facilities. The above ground facilities include temporary facilities such as laydown areas and construction camps as well as permanent facilities including compressor stations, metering stations, off-take stations, block valve stations, telecom and control systems, and other related equipment and installations. The pipeline also includes an offshore portion, the details of which will not be discussed in the present paper. The onshore pipeline is predominately buried within a 3-m deep trench, with the exception of the various crossings (river, road, utility) where environmental, soil erosion or construction constraints may potentially dictate a greater burial depth.

2. Pipeline alignment evolution

The initial route identified was a 2-km wide corridor, routed to satisfy basic geographical constraints such as the gas metering stations at the bordering countries, in-country fixities (off-take stations), known archaeological sites and populated areas, and to accommodate the offshore crossing fixities (landfalls) and to provide the shortest alignment as possible. This first revision was largely a product of experience, running parallel to existing pipelines. As more data became available, the route was updated and the corridor reduced to 500m, within which an alignment was fixed. Once the alignment was fixed, all subsequent changes were recorded in "route change files", recording the reason for the change, the economic benefits (shortening of alignment, reducing inflection points and thus bends, etc.), whether the change was within the identified 500m or 2km corridors and whether it potentially clashed with constraints such as constructability, geohazards, socioeconomic (settlements and public safety), archaeology and cultural heritage, environmental and authority/stakeholders. The project underwent extensive route development since inception. The finalized "FEED" route was the 8th revision.

3. Geohazard risk assessment lifecycle

3.1. Desktop study

A desk study was performed to describe the geologic setting along the alignment and to conduct a preliminary assessment of the geohazards. The main inputs to the desktop study included aerial photography and Digital Terrain Modelling topography data over a 2-km wide pipeline corridor, geologic maps from government agencies and local authorities, as well as data from open sources (i.e. province boundaries, district boundaries, roads, rivers, etc.). The study output was a preliminary route selection report and a surficial geology report, on which the initial route revisions were based.

3.2. Field geohazard mapping

The field geohazard mapping was conducted with the deployment of 8 teams of geoprofessionals, who spatially covered the 500-m pipeline corridor. State-of-the-art tools such as rugged Tablet PC's were employed for data collection. These added both to the productivity of the teams and to the accuracy of the data collection, as its embedded GPS functionality allowed swift positioning and orientation and its preloaded GIS layers such as topo and geology maps provided a holistic view of the area to the field personnel. The standardization provided by use of these tools further minimized the post-processing effort and allowed for seamless exchange of information between the field teams and the design office (basemap and route revision updates, embedding field photos). At critical areas, the field teams were supported by Subject Matter Experts (SME) primarily in the fields of Karst and Landslides, with further information provided in Sections 4.2 and 4.3. The field mapping efforts were completed within 2.5 months, including reporting.

IOP Conf. Series: Earth and Environmental Science **26** (2015) 012050 doi:10.1088/1755-1315/26/1/012050

3.3. Planning and execution of the geotechnical exploration program

Once the pipeline had reached a fixed alignment in the 5th revision of the route (500-m corridor), the geotechnical investigation program could be developed. Due to the size and linear nature of the project, consideration had to be given to the cost and schedule of the investigation, the purpose of the investigations and with a view to what data will be used and in which aspects of the pipeline design. The data would be used to:

- Primarily, identify the trench excavatability. To this end, 3-m deep trial pits were assigned at intervals of approximately 1.5 km via use of conventional rubber tired backhoe. For a proper excavatability assessment, the recommendation is that the trial pit excavator be of similar characteristics (tonnage, bucket size, horsepower, output) as the equipment foreseen to be mobilized during construction. If not specified, subcontractors will tend to mobilize lightweight equipment to achieve faster completion of the investigation works thus skewing the excavatability estimates.
- Determine the subsurface characteristics of potential landslides identified during field reconnaissance (geohazard mapping) that could not be rerouted, their impact on the pipeline in case of failure, and the extent of the civil measures that would have to be employed towards mitigation and/or monitoring (depth of failure surface, lateral extent). Boreholes for landslide investigation were typically deeper than the explorations for trench excavatability (up to 50m depth) and the soil column would be continuously cored to enable identification of thin low strength bands or development of shear surfaces.
- Assess the subsoil conditions at crossing locations; primarily at major river crossings, where the depth of the pipeline invert would increase due to soil erosion/scour considerations. This was done via boreholes and trial pits with the aim of identifying conditions that would impact horizontal directional drilling (HDD) design.
- Investigate alignment sections in which potentially high groundwater table in combination with loose sand deposits could lead to liquefaction. This was performed with a combination of boreholes and piezocone penetration tests (CPTu). Due to the CPTu being limited to soft and loose soils, CPTu's were only advanced when boreholes and Trial Pits in close proximity revealed presence of loose sandy materials and high groundwater table. A first CPTu was advanced adjacent to an existing borehole or trial pit and the results were immediately evaluated using pre-developed liquefaction evaluation charts. If the resulting analysis indicated significant liquefaction, additional CPTu's were advanced on each side on the alignment, until a negative result (no significant liquefaction) was obtained or until it was evident in the field or from aerial photography that conditions had changed to preclude liquefaction.
- Investigate the potential Above Ground Installation (AGI) sites, primarily on a high level to assist with site selection, and determine groundwater level and approximate foundation parameters. This was evaluated via conventional boreholes (typically 2 boreholes per candidate site). Refer to Section 4.5 below for further information on AGI site selection process.
- Map the soil electrical resistivity along the alignment for purposes of pipeline corrosion protection design.

As expected for such a large-scale investigation spanning over a wide geographical area, there were severe access and weather constraints to overcome. As the investigation commenced in the summer, the most remote parts of the pipeline, i.e. in mountainous terrain which would become inaccessible during winter, were first investigated. Successful completion of the entire alignment, not just accessible areas, was integral to the overall success of the investigation program. The geotechnical spread mobilized as well as the duration and productivities of the investigations are presented in table 1 below for reference:

Table 1. Geotechnical investigation metrics.

Investigation type	Avg. Nos rigs mobilized	Nos. of investigations	Duration (months)	Productivity[a]	Cost (percent of overall programme cost)
Trial Pits	2	1400	8	4 Nos/rig/day	23%
Boreholes	10	430 (7,000m drilled)	8	6 m/rig/day	67%
Cone Penetration Tests	1	230	4	2.5 Nos/rig/day	10%

[a] Includes downtime due to equipment breakdown, transportation, contractual stoppages, access constraints, inclement weather, time-off. Based on 6-day workweek, 1 shift.

3.4. Data compilation into GIS database

The project utilized GIS to manage the project spatial data, and to provide spatial analysis and mapping support. The GIS consisted of a folder based file geodatabase (.gdb) system to store spatial data in a logical collection of folders. Data are held in various formats to enable the data to be used by various software packages, including CAD, Google Earth, Global Mapper and ArcGIS. All data is projected and stored in file geodatabases using the grid-based WGS84 coordinate system. Following is an indicative list of the layers available to the route engineers:

- Aerial photography, digital terrain model (DTM), 1:2000 scale photogrammetric vector topographic features which could produce 2-m elevation contours,
- Parcel boundary and owner information, authority data (third party pipelines, local government boundaries etc.)
- Physical (topographic), biological, socio-economic, and archaeological field survey and ministry data.
- Data from open sources (i.e. all province boundaries, district boundaries, roads, rivers, geologic maps, etc.)
- Geologic mapping (from public sources as well as from independent consultants)
- Desktop geohazard mapping data of the 500m wide route corridor including field photo records
- Geotechnical exploration data
- Seismicity contours (regulator's, project preliminary Probabilistic Seismic Hazard Assessment, independent studies)
- Pipeline corridor and alignment revisions

All data were subject to rigorous quality checks to ensure consistency and completeness, documented in terms of their source and receipt date, and released for use on a weekly basis.

4. Study deliverables

4.1. Trench excavatability and dewatering

Two major cost and constructability items affect the pipeline, namely trench excavatability and trench dewatering. The following ground conditions data along the pipeline route were considered:

- aerial photography including vegetation and local surface characteristics

- topography
- surficial geology
- geotechnical exploration data
- groundwater measurements

Using the above datasets, a set of criteria was developed, with input from construction personnel which allowed subdivision of the pipeline alignment into a finite number of classes, in terms of the expected excavatability and trench dewatering needs. For each available and relevant investigation along the pipeline route, the soil profile down to 3m depth was subdivided into two predominant material types. Excavatability and dewatering classes were then assigned to each material type. Guidelines and criteria used for classifying each material into dewatering and excavatability classes are presented in the following Sections 4.1.1 and 4.1.2. The exploration as a whole was then assigned a class chosen as the most unfavorable class of the predominant materials within the reach. Finally, the pipeline reach corresponding to each investigation point analyzed was derived based on review of the area topography, orthophotos and route geology report, rather than a simple midpoint to midpoint length selection between investigation locations.

4.1.1. Criteria for dewatering classifications. The developed dewatering classification scheme is shown in table 2 below.

Table 2. Trench dewatering classification.

Class	Representative equipment	Representative groundwater and soil conditions
1	High Dewatering needs (combination perimeter dewatering system and/or cutoff wall).	If the groundwater table is between 0 and 2m below ground surface, combined with presence of coarse grained, permeable soils (GW, GP, GW-GM, GW-GC, GP-GM, GP-GC, SW, SP, SW-SM, SW-SC, SP-SM, SP-SC) or highly weathered rock in the upper 3m of soil stratification. Exceptions: Bedrock that exhibit low permeability in its weathered state is not considered herein and may be considered in subsequent project phases.
2	Medium Dewatering needs (bottom dewatering e.g. centrifugal trash pump)	If the groundwater table is between 2 and 4m below ground surface, as well as presence of coarse grained, permeable soils (sands, gravel) or highly weathered rock in the top 3m of soil stratification. Otherwise, if the groundwater table is between 0 and 3m with the presence of medium permeability soils (GM, GC, SM, SC) in the same depth range.
3	No Dewatering required	If the groundwater table is below 4m from ground surface, or if the top 3m of soil stratification is comprised solely of impermeable soils (CL, ML, CH, MH, OL/OH) or intact or fresh rocks.

4.1.2. Criteria for excavatability classifications. The developed excavatability classification scheme is shown in table 3 below.

Table 3. Trench excavatability classification.

Class	Representative equipment	Representative soil conditions
1	Trench excavation is carried out by means of conventional excavator only (15-20 ton).	All soils, including boulders/blocks of up to 1000mm maximum dimension as well as soft rocks.
2	Trench excavation is carried out by combination of bulldozer ripper and excavator. Hydraulic jackhammer may also be used for trenching beyond ripping depth.	Heavily fractured strong rocks, or intact weak rocks (i.e. claystone, sandstone, tuffites, conglomerates, anhydrite) of up to 20MPa Uniaxial Compressive Strength, or soils including boulders/blocks of greater than 1000mm maximum dimension.
3	Trench excavation is carried out by rock trencher. Hydraulic jackhammer may also be used as an aid.	Intact or fractured rocks of up to 130 MPa Uniaxial Compressive Strength (i.e. obsidian, limestone, marl, agglomerate, marble, shale, gypsum).
4	Trench excavation is carried out by means of drilling, installation of charges, blasting and finally use of excavator for trench clean-up.	Intact rocks of Uniaxial Compressive Strength greater than 130 MPa (i.e. most igneous rocks such as basalt, andesite, granite, gabbro)

4.1.3. Recommendations and design. The evaluation concluded that approximately 2/3 of the pipeline had to be excavated by conventional means, whereas an approximate 20% required the aid of a ripper (e.g. such as the one attached to the Caterpillar D9 bulldozer). Heavier means of excavation such as rock trencher and/or blasting will be required for an estimated 15% of the alignment. Regarding dewatering, it is expected that approximately 90% of the alignment will be situated in dry trench or trench with low dewatering requirements. The remaining length will require some form of dewatering, depending on the proximity to the water bodies, groundwater level, construction methods and type of soils in which the trench is excavated. These recommendations were formalized in a report which specified, by kilometer post (KP), what type of excavation and dewatering, if any, was likely required.

4.2. Karst characterization

4.2.1. Karst identification and assessment. The characterization of karst features utilized the classification scheme of Waltham [1], in which surface observations are used to assign karst landscapes into five morphological assemblages ranging from kI – Undeveloped Karst to kV – Extreme Karst. Much of the karst along the Project alignment is a gypsum karst and with the aid of the project SME, a site-specific classification scheme was developed for the gypsum karst encountered. This scheme is summarized in table 4.

4.2.2. Recommendations and design. The key recommendation in Karst areas is to control drainage so as not to change existing drainage patterns. Such changes will lead to new dissolution and impacts to the existing conditions. As with most geohazards, if the hazard could be reasonably avoided through rerouting, the pipeline alignment was rerouted around the hazard. However, of all geohazards encountered along the pipeline alignment, the karstic zones were the most widespread over a large geographic area which made rerouting impractical. Where crossing of karstic areas was required, a detailed analysis was performed and the pipeline was assumed to be unsupported with the length of the unsupported span given by the length of the intersected karstic area. For example, a karst crossing with low hazard of ground movement, designing for an unsupported section of 25 meters may be appropriate while for a karst crossing with high hazard of ground movement, designing for an unsupported section of 50 meters may be inappropriate.

The result is a generic classification scheme for much of the alignment, in terms of kilometers, but a site-specific classification scheme with detailed descriptions for morphology, the hazard to the pipeline, and mitigation measures, for the region where the karst risk is highest. The general classification scheme from literature and the site-specific classification scheme allowed having higher confidence in the determination of risk to the pipeline and provided a transparent process between observation of morphology to hazard and mitigation.

Table 4. Classification scheme developed for gypsum karst.

Class	Morphology	Karst Geohazard	Mitigation Measures
kg1	**Karst margin**, narrow zone along cliff line boundary	Caves and notches undercut steep slopes, increasing slope failure hazard	Cross at right angles to minimize length within hazard zone
kg2	**Polygonal karst**, upland with network of low bedrock ridges between closely-spaced, soil-floored solution dolines, and may contain large, old, collapse dolines	Settlement and suffosional loss of soil on doline floors, possible new subsidence sinkholes a few metres across, very small hazard of collapses 20m across	Avoid crossing doline floors where possible; control the drainage during construction, and do not let the trench fill become a new conduit
kg3	**Plateau karst**, upland and high ground with no surface drainage, with scattered, large, old, collapse dolines	Small but measurable hazard of collapses up to 20 metres across	Control the drainage; possibly thicken the pipe wall to allow spanning a new collapse
kg4	**Immature karst**, generally on mixed rock sequences	Minor, rare small subsidence sinkholes	Not on route

kg5	**Mantled karst**, upland areas with thick soils over the gypsum, also some basins on kg3 plateaus	Small hazard of suffosion sinkholes that start small but may grow to 10m across	Minimal geohazard
kg6	**Polje karst**, alluviated poljes and lowland valley floors, many with strips of kg1 along cliff margins	Minor in alluviated centers, enhanced solution features along margins	Minimize the lengths across steep marginal zones with undercuts and collapse; Control the drainage

4.3. Landslide identification and risk assessment

The landslide identification methodology generally followed the classification scheme of Cruden and Varnes [2], slightly modified for the use of the project. The following were recorded during the desktop study and field mapping:

- material type (rock, debris, or soil)
- movement mechanism (fall/topple, spread/flow, slide/complex)
- state (active, and inactive (latent or dormant))
- depth (shallow, <1m; intermediate; 1-6m; and deep-seated; >6m)
- size (very small, <0.1ha; small, 0.1-1ha; medium, 1-10ha; or large, >10ha)

As a measure of reference, the FEED effort increased the previously mapped landslide database by 30%, totalling approximately 300 sites. The majority of the features comprised slides (85%) whereas falls and flows only occupied 15% of the sites. The material type was primarily debris (50%) followed by sites in soil (35%) and lastly sites within rock (15%).

Subsequently, the hazard level is defined as the potential for structural failure of the pipeline. The developed criteria leading to the subject scenario include:

- Lateral displacement – horizontal and/or vertical movement of the pipeline may occur. Generally, welded steel pipelines can accommodate some movement. If the soil moves more than 5m, failure was assumed.
- Spanning –pipeline failure was assumed if landsliding resulted in an unsupported pipe section greater than 30m.
- Loading – pipe rupture as a result of landslide loading by burial or impact is evaluated.
- Exposure and impact by landslide debris – the sequence of flooding, scour, and debris flow during or after a storm can cause pipeline failure. The risk due to this sequence of events was evaluated for landslide crossings.

Finally, the risk of landslide reactivation causing structural failure to the pipeline within the typical 50-year project design life is assessed. Reactivation during construction may occur by removal of material at the toe, equipment-induced vibrations, by increasing the driving load at the slide crown, and/or by poor construction surface water retention. Other reactivation scenarios include increasing the pore-pressures by rainfall, and seismic ground shaking. The risk was assessed via a risk matrix scheme (based on the proposals of Lee, Jones and Charman [3], [4], suitably modified for the project), depending on proximity to the pipeline, position of the pipeline relative to the landslide, activity state of the landslide, and the thickness of the landslide deposits crossed by the pipeline (refer to tables 5 and 6). The qualitative classification scheme then determines the landslide as one of the following risk categories: none, negligible, low, medium, and high.

Table 5. Landslide hazard assessment matrix for active slides. Risk level corresponds to 50-yr design life.

Pipeline position relative to slide	0 to 5m	5 to 10m	10 to 25m	25 to 50m	Greater than 50m
Upslope of backscarp	High risk	Low risk	Low risk	Negligible	No risk
Main body or beyond its lateral flanks	High risk	Low risk	Negligible	No risk	No risk
Downslope of toe	High risk	Low risk	Negligible	No risk	No risk
Run-out apron	Low risk	Negligible	No risk	No risk	No risk

Table 6. Landslide hazard assessment matrix for non-active slides. Risk level corresponds to 50-yr design life.

Pipeline position relative to slide	0 to 5m	5 to 10m	10 to 25m	Greater than 25m
Upslope of backscarp	Medium risk	Low risk	Negligible	No risk
Main body or beyond its lateral flanks	Medium risk	Low risk	Negligible	No risk
Downslope of toe	Medium risk	Negligible	Negligible	No risk
Run-out apron	Negligible	Negligible	No risk	No risk

The risk identification exercise revealed that 2% of the sites were in the high risk category, 4% in medium risk, 14% in low risk, 26% in negligible risk and 54% in the no-risk category.

4.3.1. Recommendations and design. Nyman et al. [5] recommend that, as a basic mitigation strategy, the pipelines be rerouted to avoid mapped landslides. State of practice, as delineated by Sweeney [6] further enforces routing along ridges as opposed to sideslopes to minimize exposure to landslides. This recommendation was adopted for the project via rerouting around features of high and medium risk levels. Where rerouting is impractical and therefore crossing a landslide was required, other mitigation options including civil measures (retaining structures or ground stabilization) or locally increasing the pipeline burial depth are to be considered during subsequent project design phases. Individual suspect features may also require additional geotechnical exploration or instrumentation monitoring but no such cases were identified during the particular project design phase. Heavy rainfall events and seismic shaking may also generate new, unmapped features which should be identified and risk-characterized during inspections by geoprofessionals throughout the project operational life.

4.4. Liquefaction

4.4.1. Liquefaction identification and assessment. An initial liquefaction screening was made by evaluating all subsurface explorations (borings, piezocone penetration tests [CPTu], and machine excavated trial pits) to determine if potentially liquefiable soil was present at or below the alignment and if groundwater was within 1.5 meters of the ground surface. In floodplains, the groundwater table was conservatively assumed at the ground surface. The trial pits were helpful to determine locations where rock was encountered at shallow depths (>2m) or where clearly non-liquefiable soils were present, thus used in a qualitative manner. Where the initial screening of liquefaction indicated a potential for liquefaction (soil and relatively shallow groundwater) liquefaction triggering calculations using borehole and CPTu data were performed. One problem encountered in the liquefaction analyses was the very high design peak horizontal ground acceleration (PGA) used in the simplified method estimation of seismic cyclic shear stress (CSR) through the soil column. See Seed and Idriss [7] for a detailed explanation of the simplified method for estimating CSR. The Client provided a preliminary probabilistic seismic hazard analysis (PSHA) indicated regions of PGA equal to 1.6g. At the ground surface this results in a CSR using the simplified method of 1.04 assuming a magnitude 7.5, CSRM=7.5. A recent database of field observations of liquefaction by Boulanger et al. [9] and Boulanger and Idriss [8] report a maximum CSRM=7.5 of 0.686 making the design CSRM=7.5 approximately 50% higher than the largest field observation of liquefaction. Therefore current liquefaction methods require significant extrapolation from field observations to access liquefaction potential for the design seismic hazard. Cyclic laboratory and centrifuge tests summarized by Giannakou et al. [10] indicate that the extrapolation of current liquefaction triggering analyses appears to properly estimate the onset of liquefaction, defined as an excess pore pressure ratio, r_u, defined as the ratio of excess pore water pressure to total vertical stress, equal to 1. Figure 1 presents the cyclic resistance ratio (CRR), or the soil's resistance to liquefaction, versus clean sand normalized blow count for two liquefaction triggering methodologies along with the case history database from Boulanger et al. [9] and Boulanger and Idriss [8]. The methodology of NCEER [16], Youd et al. [17], and Youd et al. [18] is referred to herein as NCEER. Note that the NCEER method considers any soils with a normalized clean sand blowcount of equal to or greater than 30 too dense to liquefy.

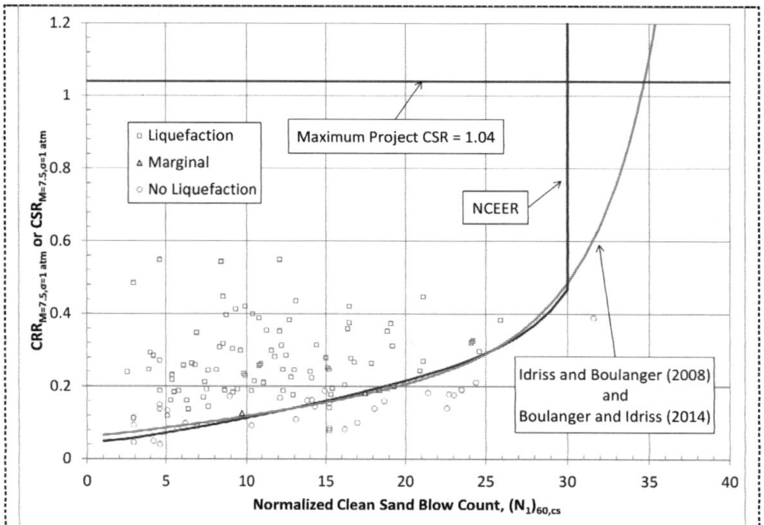

Figure 1. Comparison of liquefaction triggering methodologies and reported case histories.

4.4.2. Recommendations and design. The hazards from liquefaction are generally pipe buoyancy and lateral spread, leading to lateral deformation of the pipeline. In areas of high groundwater, the pipe is generally designed to be neutrally buoyant meaning that the unit weight of the pipeline in cross-section is near, or slightly above, the hydrostatic pressure. As an added safeguard, tie-downs were recommended in liquefaction areas. Tie-downs mechanically connect the pipeline to more competent material below the pipeline to resist uplift. In areas of lateral spread, the pipeline should generally be buried beneath the zone of lateral spreading.

For dense sands where liquefaction analyses indicate liquefaction, due to high seismic design inputs, the consequences of liquefaction must be considered. While shaking may cause liquefaction in terms of excess pore pressure ratio, Robertson [11] and Cox [12] show that, in terms of strength, soil state dictates soil strength with loose sands (looser than the Critical State Line,CSL) exhibiting strain softening, moderate density sands (slightly loose of the CSL) tending to exhibit limited strain softening, and dense sands (dense of the CSL) tending to exhibit strain hardening. This process is shown graphically on figure 2. Soils loose of the CSL have a liquefied strength, or shear strength of the liquefied soil, less than that of the static drained strength but as density increases and the soil state crosses the CSL the static drained strength governs. This occurs at approximately a normalized blowcount, $(N_1)_{60}$, for clean sands of 19 to 35 depending on drainage characteristics. Therefore, while liquefaction calculations indicate liquefaction may occur, where the soil is dense enough, or dense of CSL, no mitigation measures are necessary.

Waterman et al. [15] present preferred crossing orientations where lateral spreading is expected for the design seismic input. Generally, the pipeline alignment is adjusted to cross the area of lateral spread perpendicularly to reduce the crossing length and make the lateral movement along the axis of the pipe, or to increase the depth of the pipeline such that the alignment is below the expected depth of the zone subject to spreading.

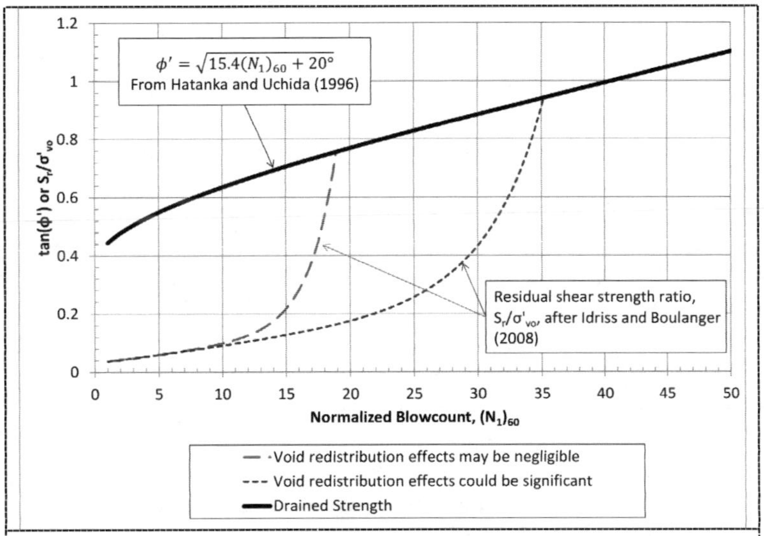

Figure 2. Static drained strength and residual strengths assuming clean sands.

4.5. Above ground installations site localization

The term AGI refers broadly to all above ground installations and facilities that support the pipeline. The present section refers solely to the Compressor Stations and Metering Stations. The site localization study is essentially the process for determining the exact location of each AGI which is subject to engineering, authority and socio-environmental constraints. Initially, the pipeline hydraulic flow study is used to determine the required general distances along the pipeline at which compressor stations are needed. This study utilizes the pipeline elevation and length data, pipeline diameter(s) and the required flow and outputs the number of compressor stations required to assure the flow, as well as their approximate Kilometric Stations (typically in 50-km ranges). The engineering team would then superpose aerial photography and topography data to identify 5 to 7 candidate sites within the above

IOP Conf. Series: Earth and Environmental Science **26** (2015) 012050 doi:10.1088/1755-1315/26/1/012050

range for hosting the AGI in question. The land plots required by the various AGIs were 700m square for the compressor stations and 300m square for the metering stations. These sites would then be ranked according to a series of criteria such as: topography, water and electricity, access, environmental/land use, proximity to settlements, hydrology, and mapped geohazards. The top 3 ranked sites are selected and assigned geotechnical investigations (typically 2-3 boreholes each). A site visit of a multi-disciplinary team follows, comprising of construction, civil, and geo-professionals to confirm the findings of the desktop study and identify any additional hazards/concerns. The sites are then re-ranked according to the findings of the site visit and the borehole findings. Finally, permitting and land acquisition processed take place on the selected site.

5. Scope foreseen to be performed in subsequent engineering phases

While the FEED study is fundamental in providing a project cost estimate (thus aiding the Final Investment Decision) and allowing for an early placement of orders for the long-lead items, the subsequent phase of Detailed Engineering will ultimately finalize the route, provide closure to any outstanding regulatory requirements and be used for the remaining material requisitions. On the geotechnical front, the following studies and exploration programs are foreseen to take place in the Detailed Engineering design:

- Finalizing identification of and performing additional exploration of areas deemed to require further detailed exploration. These included in particular:
    - Reaches of significant realignment of the pipeline, new pipeline segments, and reaches in which exploration could not be performed due to external constraints (ground conditions, military zones, etc.);
    - River crossings where horizontal directional drilling is proposed and the existing data is inadequate for such constructions;
    - Landslide areas which could not be readily routed around, and in which boring under the slide was a consideration;
    - Areas in which the extent of liquefiable material along the pipeline alignment could not be determined from existing explorations and surface features;
    - Fault crossings, which require very specific fault geometry and other design information, and;
    - Above Ground Installations, particularly the compressor stations, which required design level geotechnical investigations.
- Further collection of groundwater levels to provide data over at least a one year duration.
- An update to the initial probabilistic seismic hazard analysis (PSHA) will be required for detailed engineering. The update to the PSHA will use the latest seismic-hazard source model and ground motion prediction equations (GMPE) to calculate the seismic requirements for design along the pipeline alignment for both the pipeline itself and AGIs.
- Fault trenching of 13 active faults along the alignment will be required to identify the precise location, activity, orientation of movement, and distribution of shear for each fault. These data will be used in designing the pipeline fault crossings.

6. Lessons learnt, summary and conclusions

- In order to timely deliver a successful FEED, the engineering party must be very clear regarding the FEED goals, as those are delineated in the contract documents. Usually these goals are defined in terms of a design effort target (approximately 30% design), as well as a cost estimate target (+/-20% cost estimate). Therefore, on the geohazard-driven routing front, attention should be drawn on those elements that may yield major reroutes between FEED and Detailed Engineering such as large landslides, karst landscapes, and less on elements of minor impact on the overall cost, i.e. crossing locations and design depths or adjusting the geotechnical investigation program whenever the route gets revised.

- Execution of the geotechnical investigation as early as possible will prove beneficial to the project as it provides valuable excavatability and geohazard information, addressing most of the uncertainty regarding route selection. Route changes that may occur after the investigation has been planned can be easily addressed while the crews are still on site.
- Mobilization of expatriate teams and experts was not a straightforward task and required planning ahead. Also, availability and tight schedules resulted in many of the subcontracts to be awarded sole-sourced.
- Geotechnical subcontractors, based on their domestic competition and frequency of large-scale projects materializing in their areas of operation, do not follow the same standards in all parts of the world. This became particularly evident in the initial stages of the project, where efforts were concentrated in technical supervision. It is important that unresolved and continuing quality concerns are documented and addressed in the design stage, by increasing factors of safety and looking at the data from a critical viewpoint.
- In this project, the Owner hired a separate geotechnical firm to provide field investigation and field quality control, but the contracts were set up to be mutually exclusive, each with the Owner independently. Thus the engineering contractor had no contractual control over the investigation contractor, even though his job was to control the quality of the work produced. Needless to say, this produced significant conflicts between the parties. Additionally, both companies hired independent consultants as their field staff for much of the field works, leading to erratic consistency.
- Towards the aim of economizing, the Owner's contract also stipulated that the investigation contractor would have one field engineer on each of 2 drill rigs, as long as the rigs were within an hour's travel of each other. Continuous soil core was also stipulated from each boring, which reduced average production to about 6m per day. Blow counts were recorded by the driller, but due to absence of the field engineer much of the time, the quality of the logs did not include field observations normally recorded during drilling. In some cases, work was found to have progressed for 3 days without an engineer's log having been prepared. Thus, the quality of the final log was suspect and inconsistencies were found in many logs. Sample preservation was also questionable, particularly in adverse weather.

References

[1] Waltham T 2002 The engineering classification of karst with respect to the role and influence of caves *International Journal of Speleology* **31** 19-35
[2] Cruden D M and Varnes D J 1996 Landslides: Investigation and Mitigation *Transportation Research Board Special Report (Washington, USA)* vol 247 ed Turner A K and Schuster R L (National Academy Press) pp 36-75
[3] Lee E M and Jones D K C 2013 *Landslide Risk Assessment* (London, UK: Thomas Telford)
[4] Lee E M and Charman J 2005 Geohazards and risk assessment for pipeline route selection *Terrain and Geohazard Challenges Facing Oinshore Oil and Gas pipelines* ed M. Sweeney, (Thomas Telford) pp 95-116.
[5] Nyman D J, Hall W J and Cluff L S. 2006 Observations and Lessons Learned from Three Decades of Seismic Hazard Mitigation for Major Oil and Gas Pipeline Projects *Proc. of the 8th U.S. National Conference on Earthquake Enginering (San Francisco, USA,18-22 April 2006)* Paper No. 1016.
[6] Sweeney M 2005 *Terrain and geohazard challenges facing onshore oil and gas pipelines* (London, UK: Thomas Telford)
[7] Seed H B and Idriss I M 1971 Simplified procedure for evaluating soil liquefaction potential *J. Geotech. Engrg. Div. ASCE* **97** 1249–73
[8] Boulanger R W and Idriss I M 2014 CPT and SPT based liquefaction triggering procedures *Report No. UCD/CGM-14/01* Center for Geotechnical Modeling, Department of Civil and Environmental Engineering Report, University of California, Davis (Davis, USA) p 134

[9] Boulanger R W, Wilson D W and Idriss I M 2012 Examination and re-evaluation of SPT-based liquefaction triggering case histories *Journal of Geotechnical and Geoenvironmental Engineering, ASCE* **138** 898-909

[10] Giannakou A, Travasarou T, Ugalde J, Chacko J and Byrne P 2011 Calibration methodology for liquefaction problems considering level and sloping groundconditions *5th International Conferenceon Earthquake Geotechnical Engineering (Santiago, Chile, 10-13 January 2011)* Paper No. CMFGI

[11] Robertson P K 1994 Suggested termninology for liquefaction *Proc., 47th Canadian Geotechnical Conference (Halifax, Nova Scotia)* pp. 277-286.

[12] Cox B R 2006 Development of a direct test method for dynamically assessing the liquefaction resistance of soils in situ *PhD. Thesis (University of Texas, Austin).*

[13] Hatanaka M and Uchida A 1996 Empirical Correlation Between Penetration Resistance and Internal Friction Angle of Sandy Soils *Soils and Foundations* **36-4** 1-9

[14] Idriss I M and Boulanger R W 2008 Soil liquefaction during earthquakes *Monograph MNO-12,* (Oakland, USA: Earthquake Engineering Research Institute) p 261

[15] Waterman M, Hundl J D, Wiggs J E, Boone M D, Kottke A K and Sholley, M G 2014 Geohazard mapping and evaluation for a natural gas pipeline project *Poster presentation, GSA 2014.*

[16] Youd T L and Idriss I M, National Center for Earthquake Engineering Research (NCEER) 1997 *Proc. of the NCEER Workshop on Evaluation of Liquefaction Resistance of Soils* ed Technical Report NCEER-97-022 p 302

[17] Youd T L, Idriss I M, Andrus R D, Arango I, Castro G, Christian J T, Dobry R, Finn W D L, Harder Jr L F, Hynes M E, Ishihara K, Koestor J P, Liao S S C, Marcuson III W F, Martin G R, Mitchell J K, Moriwaki Y, Power M S, Robertson P K, Seed R B and Stokoe II K H 2001 Liquefaction Resistance of Soils: Summary Report from the 1996 NCEER and 1998 NCEER/NSF Workshops on Evaluation of Liquefaction Resistance of Soils, *Journal of Geotechnical and Geoenvironmental Engineering, ASCE* **127(10)** 817-33

[18] Youd T L, Idriss I M, Andrus R D, Arango I, Castro G, Christian J T, Dobry R, Finn W D L, Harder Jr L F, Hynes M E, Ishihara K, Koestor J P, Liao S S C, Marcuson III W F, Martin G R, Mitchell J K, Moriwaki Y, Power M S, Robertson P K, Seed R B and Stokoe II K H 2003 Closure to "Liquefaction Resistance of Soils: Summary Report from the 1996 NCEER and 1998 NCEER/NSF Workshops on Evaluation of Liquefaction Resistance of Soils," *Journal of Geotechnical and Geoenvironmental Engineering, ASCE* **129(3)** 284-286

Advancement of liquefaction assessment in Chinese building codes

Haoyu Sun[1,2,3], **Fang Liu**[1,2,3] **and Minjing Jiang**[1,2,3]

[1] State Key Laboratory of Disaster Reduction in Civil Engineering, Tongji University, 1239 Siping Road, Shanghai 200092, China
[2] Key Laboratory of Geotechnical and Underground Engineering (Tongji University), Ministry of Education, Shanghai 200092, China
[3] Department of Geotechnical Engineering, College of Civil Engineering, Tongji University, 1239 Siping Road, Shanghai 200092, China

Corresponding author: Fang Liu (liufang@tongji.edu.cn)

Abstract. China has suffered extensive liquefaction hazards in destructive earthquakes. The post-earthquake reconnaissance effort in the country largely advances the methodology of liquefaction assessment distinct from other countries. This paper reviews the evolution of the specifications regarding liquefaction assessment in the seismic design building code of mainland China, which first appeared in 1974, came into shape in 1989, and received major amendments in 2001 and 2010 as a result of accumulated knowledge on liquefaction phenomenon. The current version of the code requires a detailed assessment of liquefaction based on in situ test results if liquefaction concern cannot be eliminated by a preliminary assessment based on descriptive information with respect to site characterization. In addition, a liquefaction index is evaluated to recognize liquefaction severity, and to choose the most appropriate engineering measures for liquefaction mitigation at a site being considered.

Key words: liquefaction assessment; seismic design code; anti-seismic design

1. Introduction

Liquefaction often occurs in saturated granular soils due to earthquake shaking, and causes significant damages on buildings and other urban infrastructure systems as a result of dramatic reduction in strength of liquefied soils. As one of the most important subjects in the field of geotechnical earthquake engineering, the phenomenon of liquefaction has attracted intensive studies in the world after two major earthquakes in 1964 [1]. Liquefaction assessment has become part of seismic design of any important project located in seismically affected areas prone to liquefaction. Like other seismically active regions, China has also suffered from several devastating earthquakes with over one hundred thousand casualties in the past century [2] including 1976 Tangshan earthquake, in which extensive examples of liquefaction have been observed. With accumulated knowledge on liquefaction, the specifications related to liquefaction assessment appeared in the national seismic building code in mainland China in 1974 and received three major amendments according to lessons learnt from historical strong earthquakes. Although the blow count of a standard penetration test (SPT) is also employed in Chinese building codes to feature liquefaction resistance of a site, the detailed procedure is less acquainted elsewhere in the world and distinct from the simplified procedure originally proposed by Seed and his colleagues [1, 3, 4]. Chen [5] provided a good review in the local language on the revolution of the Chinese seismic design code before 2001. Yuan and Sun [6] proposed further improvement based on an overview of the state of practice of liquefaction assessment specified in the Chinese building code. However, the readership of their works was confined locally due to language, and a comprehensive review on the upgrade in the past 15 years remains insufficient. This constitutes the motivation of this paper.

2. Timeline of major earthquakes and amendments in Chinese building codes

Earthquake-induced liquefaction has occurred in almost every strong earthquake in China. A comprehensive overview of typical liquefaction phenomena and their characteristics during recent major earthquakes in China can be found in [2]. Some new phenomena, such as liquefaction of gravelly soil and deep-level sandy soils, and re-liquefaction, have been also observed in these recent earthquakes, drawing increasing attention in further study [2]. Among other strong earthquakes, 1970 Tonghai earthquake, 1975 Haicheng earthquake, and 1976 Tangshan earthquake provided particularly important data resulting in improvements in the liquefaction assessment procedure adapted in Chinese building codes [6].

Figure 1 illustrates the timeline of a number of major earthquakes in mainland China and the milestones leading to evolution of Chinese seismic building codes particularly regarding liquefaction assessment. The efforts for launching a national seismic code began in early 1920s by conducting intensive site investigations after 1918 Nanao earthquake in Guangdong province with the magnitude of 7.3 and 1920 Haiyuan earthquake in Ningxia Province with the magnitude of 8.6. These efforts resulted in the first seismic intensity map in China, but further studies were unfortunately terminated due in large part to the wars. Because of an increasing demand of seismic disaster mitigation after the civil war in China, a revisit of previous work gave birth to the first draft of Chinese seismic design code *Building Regulations in Seismic Area* in 1959, in which the seismic design is based on the response spectrum theory in compliance with the *Building Regulations in Seismic Area CH-8-57* published by the former Soviet Union. The draft version of *Seismic Code of Building* was accomplished and available in 1964. This draft introduced the concept of site classification, which paved the way for the first specification regarding liquefaction concern in *Seismic Code for Industrial and Civil Buildings (TJ11-74)* available in 1974 [7]. This code implements liquefaction assessment by comparing the field measurement of SPT blow count and the critical blow count, below which liquefaction likely occurs. As a milestone, the code *TJ11-74* introduced a quantitative approach to evaluate liquefaction potential of a site, and accordingly provided detailed guidelines on selecting appropriate engineering measures for liquefaction mitigation. Since then, liquefaction assessment becomes a routine practice of seismic design of buildings in China. The code *TJ11-74* was revised four years later in order to implement lessons learnt from the disastrous earthquake at Tangshan, and an updated code *Seismic Code for Industrial and Civil Buildings (TJ11-78)* was available to public in 1978 [8]. However, the terms regarding liquefaction assessment remain largely unchanged in the updated version. By looking back at the lessons learnt from earthquakes taking place during 1970 to 1989, a further update was made and *Code for Seismic Design of Buildings (GBJ11-89)* entered into force in 1989 [9]. The major revisions regarding liquefaction assessment in *GBJ11-89* is the first usage of the liquefaction index, which indicates liquefaction severity and guides the anti-liquefaction design. Another amendment was made in 2001 [10] to extend the depth limit of liquefaction assessment from 15 m to 20 m, considering liquefaction observed in deep-level sandy soils in the historical earthquakes. This revision was implemented into *Code of Seismic Design of Buildings (GB50011-2001)*, and was continuously used in the latest version of national code in mainland China, *Code of Seismic Design of Buildings (GB50011-2010)* [11], which incorporated a number of new findings from strong motions taking place in 2000s including the catastrophic shake in 2008 at Wenchuan. In summary, the country was frequently attacked by strong earthquakes, which prompted advancement in the seismic building codes particularly regarding liquefaction assessment.

3. Methodology of liquefaction assessment adapted in Chinese seismic codes and its revolution

3.1 The Basic Principle

Figure 2 schematically illustrates the flow chart of liquefaction assessment adapted in the current version of Chinese seismic design code of buildings [11], which remains largely unchanged from the earlier version *TJ11-89* [8]. As shown in the figure, the general procedure includes site characterization, preliminary assessment (qualitative), detailed assessment (quantitative) and liquefaction severity recognition.

In the phase of site characterization, a comprehensive set of information about the site is collected, including the building classification, hydraulic conditions of the site, seismic characteristics of the site, and the site classification. A preliminary assessment is then conducted based on the information in a qualitative manner by simply looking up the matrix of site characterization. A further quantitative assessment in details is necessitated if the site is recognized as liquefiable in the step of preliminary assessment. If the detailed assessment indicates the site is prone to liquefaction, liquefaction severity is identified using the liquefaction index and guides the anti-liquefaction design. The general framework of liquefaction assessment was preserved in the past nearly 30 years, while the detailed procedure in each step has been evolved with growing knowledge on liquefaction obtained from post-earthquake reconnaissance. The remaining of this section is to review the major revisions of each step.

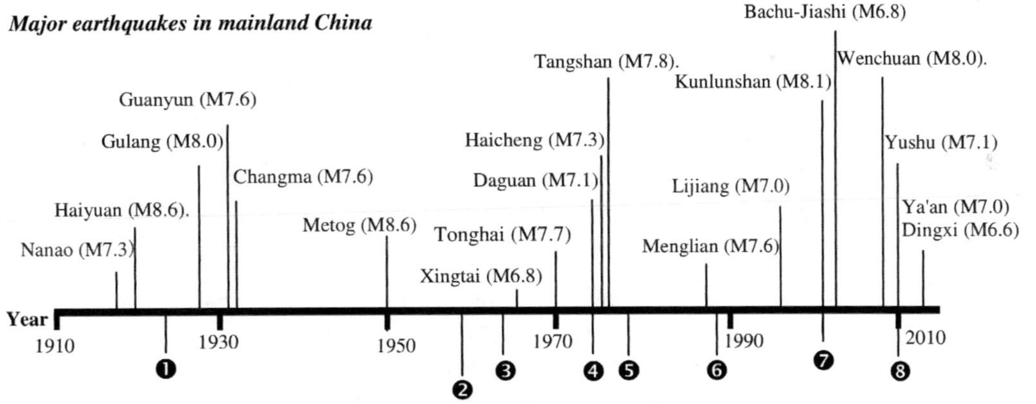

No.	Year	Events
❶	1920s	The first seismic intensity map
❷	1959	*Building Regulations in Seismic Area (draft)*: the first version of building regulation at seismic areas in mainland China
❸	1964	*Seismic Code of Building (draft)*: implementing the concept of site ranking, which paved the way for liquefaction assessment
❹	1974	*Seismic Code for Industrial and Civil Buildings (TJ11-74)*: the first national seismic code with consideration of liquefaction
❺	1978	*Seismic Code for Industrial and Civil Buildings (TJ11-78)*: revisions after 1976 Tangshan earthquake.
❻	1989	*Code for Seismic Design of Buildings (GBJ11-89)*: usage of liquefaction index
❼	2001	*Code of Seismic Design of Buildings (GB50011-2001)*: extending the depth limit from 15 m to 20 m in liquefaction assessment
❽	2010	*Code of Seismic Design of Buildings (GB50011-2010)*

Figure 1. Major amendments of Chinese seismic building codes regarding liquefaction assessment inspired by destructive earthquakes in mainland China

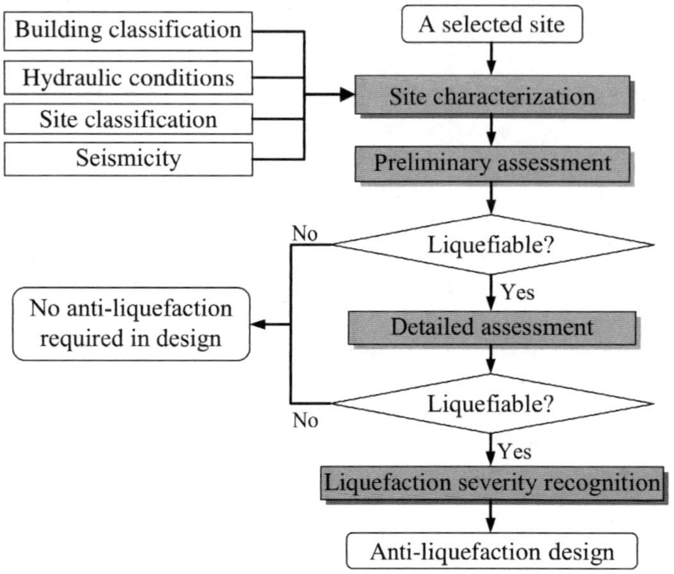

Figure 2. The procedure of liquefaction assessment adapted in the Chinese seismic building code

3.2 Amendments

3.2.1 Site characterization

The information collected in the phase of site characterization provides necessary data or parameters for liquefaction assessment. Building classification indicates different levels of importance of buildings according to their types and functions, and hydraulic conditions are mainly about the ground water table and its seasonal variation. The methods used to feature these two groups of information have not been significantly altered, while several major amendments have been made for seismicity characterization and site characterization attributed to a rapid advancement of geotechnical field testing techniques and seismic theory.

The national seismicity zonation is the prior information for site characterization. Table 1 summarizes the evolved approaches for seismicity zonation in China. The earliest seismic map of China was published in 1920s, and however it only covered eastern China. A complete seismic map of the entire country was not available until 1956 after the civil war. This map use Chinese seismic intensity scale to classify seismic impacts into 12 degrees of intensity in Roman numerals from I for insensible to XII for landscape reshaping, in a similar manner of the 1992 version of the European macroseismic scale (EMS-92). This approach had been continuously used in the subsequent two upgraded versions until the 2001 version, in which a subjective evaluation was developed based on the peak ground acceleration and the characteristic period of the response spectra. Additionally, a probabilistic approach was implemented. This framework continues in the latest 2010 version, which is still under development. Note that the upgrade of the seismic evaluation has affected the procedure of liquefaction assessment adapted in the Chinese building codes.

Table 1. Evolution of seismic zonation assessment maps.

Version	Seismic parameter(s)
1956 Version	Seismic intensity
1977 Version	Seismic intensity
1990 Version	Current standard issue by using seismic risk analysis of comprehensive probability method
2001 Version	The peak ground acceleration and the period of response spectra
2011 Version	The peak ground acceleration and the period of response spectra

IOP Conf. Series: Earth and Environmental Science **26** (2015) 012056 doi:10.1088/1755-1315/26/1/012056

Table 2 lists the evolution of site characterization in different versions of the Chinese seismic building code. Initially, a very rough categorization was used in the 1964 draft of *Seismic Code of Building* by considering types of soils underlying the site. The descriptive approach remained in *TJ11-74* (see Table 3) and the site categories were reduced from four to three: rock, soil and soft soil. With the development of in situ testing tools, a more precise approach based on the shear wave velocity was proposed in *GB50011-2001* for characterizing a site in the quantitative manner (see Table 4). This approach remains in the latest version *GB50011-2010*, while the class I is divided into two sub-classes (i.e., I_0 and I_1) to further distinguish a site with equivalent shear wave velocity over 800 m/s (see Table 5).

Table 2. Evolved methods for site classification

Code version	Method
1964 Draft	Descriptive terms (qualitative)
TJ11-74	Descriptive terms (qualitative with more accurate statement)
GBJ11-89	Descriptive terms (qualitative) and consideration of rigidity of soils and overburden thickness (quantitative)
GB50011-2001	Consideration of equivalent shear wave velocities (quantitative)
GB50011-2010	Consideration of equivalent shear wave velocities in a wide range (quantitative)

Table 3. Qualitative site classification in *TJ11-74*

Site category	Description
I	*Rocks* consisting of strongly cemented rocks.
II	*General soils* consisting of gravels, sands, clays soil, collapsible loess and other geomaterials excluded in I and III.
III	*Soft soils* consisting of saturated loose sands, high plastic light loam, muck and mucky soil, soft soils with high compressibility, and fills.

Table 4. Quantitative site classification in *GB50011-2001*

Equivalent shear wave velocity V_{se} (m/s)	Averaged thickness of overlaying soils (m)			
	I	II	IV	IV
$V_{se} > 500$	0			
$500 \geq V_{se} > 250$	<5	≥5		
$250 \geq V_{se} > 140$	<3	3~50	>50	
$V_{se} \leq 140$	<3	3~15	15~80	>80

Table 5. Quantitative site classification in *GB50011-2010*

Equivalent shear wave velocity V_{se} (m/s)	Averaged thickness of overlaying soils (m)				
	I_0	I_1	II	III	IV
$V_{se} > 800$	0				
$800 \geqslant V_{se} > 250$		0			
$500 \geqslant V_{se} > 250$		<5	≥5		
$250 \geqslant V_{se} > 150$		<3	3~50	>50	
$V_{se} \leqslant 150$		<3	3~15	15~80	>80

3.2.2 The preliminary assessment

In general, liquefaction assessment is not required at a site with the seismic intensity of 6. To be effective, *TJ11-89* started to recommend a preliminary assessment of liquefaction potential of a site before a detailed evaluation. The preliminary assessment is qualitatively conducted in a rapid manner simply based on the descriptive information collected in the phase of site characterization. According to *TJ11-89*, liquefaction concern can be eliminated in any of the following circumstances:

1) Soil deposits at the site are the Late Pleistocene of Quaternary (Q_3) deposits or older;

2) The fine content of the soils (with the particle size less than 0.005 mm) at the site is more than 10%, 13% or 16% if a site is designed at the seismic intensity of 7, 8, or 9 at the China scale, respectively;

3) The unliquefiable regime is inferred based on Figure 3 for a shallow foundation embedded at a depth less than 2 m.

These terms remains largely unchanged in *GB50011-2010* except two minor amendments: (1) The first term above was amended as "Soil deposits at the site are the Late Pleistocene of Quaternary (Q_3) deposits or older if the seismic intensity is 7 or 8"; (2) It explicitly states that the preliminary assessment was not suitable for loess.

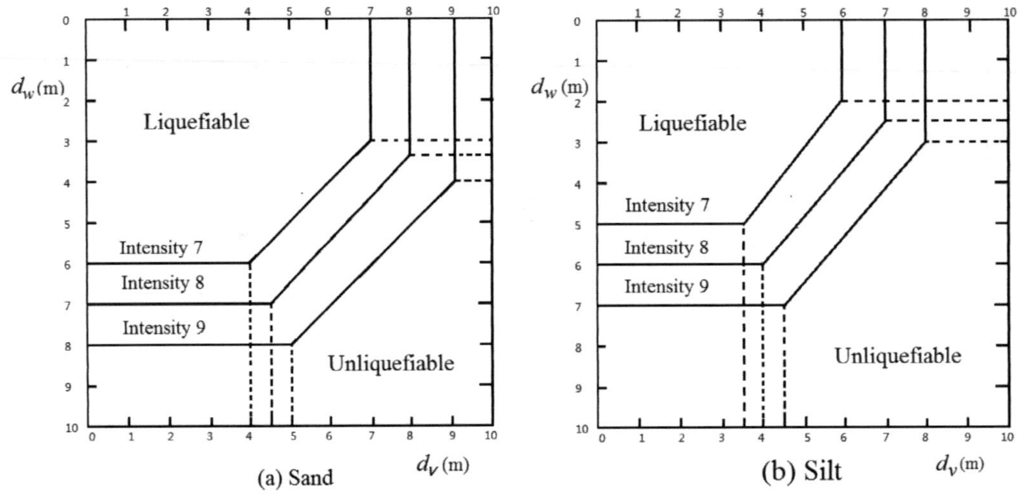

Figure 3. Liquefaction preliminary assessment chart (d_w and d_v are the depth of the underground water table and thickness of the overburden unliquefiable layer, respectively. Note that this chart is also applicable to foundations with embedded depth of 2~5 m by replacing d_w and d_v with corrected values)

3.2.3 The detailed assessment

A detailed quantitative assessment of liquefaction potential is required if liquefaction concern cannot be eliminated in the preliminary assessment. In general, liquefaction is inferred by the following inequality:

$$N_{63.5} \leq N_{cr}, \text{ liquefiable} \quad (1)$$

in which the basic principle is to compare the SPT blow count measured at the site without length correction ($N_{63.5}$) and the critical SPT blow count (N_{cr}), below which the site being considered likely liquefies. This differs from the simplified procedure proposed by Seed and Idriss [1], which compares the cyclic stress ratio representing the earthquake demand and the cyclic resistance ratio inferred from in situ tests such as SPT, cone penetration tests and shear wave velocity test [4].The SPT method is the only quantitative liquefaction assessment method mentioned in *GB50011-2010*. Note that SPT is not applicable to some particular soils such as gravelly soils [12,13,14,15], and thus *GB50011-2010* cannot deal with liquefaction assessment for such soils. Nevertheless, some other methods are included in some local building codes and speciality codes. For example, the shear wave velocity method was used in local building code of Tianjing City (*TBJI-88*) [16], and the method based on cone penetration test (CPT) was also suggested in *Code for Investigation of Geotechnical Engineering* (*GB50021-2001*)[17].

Although Eq. (1) has not been changed in the subsequent upgrades since it first appeared in *TJ11-74*, the procedure to determine N_{cr} has been updated several times as summarized in Table 6. Considering the formula for computing N_{cr} in *TJ11-74* being less reliable for clayey soils, *TJ11-89* took the effect of the fine content into account and provided a unified formula for sandy and clayey soils. Due to increasing demand of deep foundations, the impact depth of liquefaction was extended from 15 m to 20 m. The formula of N_{cr} was accordingly modified in *GB50011-2001* to manipulate the liquefaction assessment in deep soils. In the current version *GB50011-2010*, a unified formula was developed for an entire depth profile and a corrected factor β was introduced to account for the effect of seismic levels. In general, β is equal to 0.80, 0.95 and 1.05 at the first, second and third types of earthquake, respectively. The type of design earthquake is defined according to Table 7.

Regardless of changes in the formula for computing N_{cr}, the reference blow count (N_0) is always a necessary parameter related to seismic input, and its suggested values have been changed in accordance with the evolution of seismic zonation in China. In the earliest version *TJ11-74*, N_0 is assigned based on seismic intensity at the China scale. The updated version *GBJ11-89* suggests distinguishing different earthquake types (i.e., near-field earthquake and far-field earthquake). Since 2001, the peak ground acceleration has been used as one of the parameters in seismic zonation in China, and accordingly N_0 is assigned based on the peak ground acceleration instead of seismic intensity. Table 8 provides suggested values of N_0 in the latest code *GB50011-2010* for the peak ground acceleration no more than 0.4g, above which *GB50011-2010* provides no detailed guidance unfortunately.

Table 6. Formula for calculating N_{cr} in different versions of national seismic code in China

Code version	Formula for the critical SPT blow count *	Improvement
TJ11-74	$N_{cr} = N_0[1 + 0.125(d_s - 3) - 0.05(d_w - 2)]$	Corrected SPT blow count according to depth and ground water table
GBJ11-89	$N_{cr} = N_0[0.9 + 0.1(d_s - d_w)]\sqrt{3/\rho_c}$	Effect of fine content
GB50011-2001	$N_{cr} = \begin{cases} N_0[0.9 + 0.1(d_s - d_w)]\sqrt{3/\rho_c}, & d_s \le 15 \\ N_0(2.4 - 0.1d_s)\sqrt{3/\rho_c}, & 15 < d_s \le 20 \end{cases}$	Impact depth extended from 15 m to 20 m
GB50011-2010	$N_{cr} = N_0\beta[\ln(0.6d_s+1.5)\text{-}0.1d_w)]\sqrt{3/\rho_c}$	A correction factor β

* Note: d_s – SPT depth (m); d_w – underground water level (m); N_0 – reference SPT blow count; ρ_c – fine content in percentage (ρ_c=3 if fine content is less than 3%); β – correction factor depending on the seismic level.

Table 7. Characteristic periods of earthquake in *GB50011-2010*

Type of design earthquake	Site type				
	I_0	I_1	II	III	IV
First type	0.20	0.25	0.35	0.45	0.65
Second type	0.25	0.30	0.40	0.55	0.70
Third type	0.30	0.35	0.45	0.65	0.90

Table 8. The reference blow count of SPT in *GB50011-2010*

The basic design earthquake acceleration (g)	0.10	0.15	0.20	0.30	0.40
The reference value of SPT, N_0	7	10	12	16	19

3.2.4 Liquefaction severity

Assessment at an individual depth is inadequate to define the severity of liquefaction of an entire profile of a site. Accordingly, the liquefaction index (I_{lE}) appeared in *GBJ11-89* to indicate the severity of liquefaction at a site. This index can be computed as follows:

$$I_{lE} = \sum(1 - N_i/N_{cri})d_i\omega_i \quad (2)$$

where the subscript i denotes the number of the layer, d is the thickness in meter, and ω is the weight factor decreasing with the increase of the depth of the layer as illustrated in Figure 4. This definition is similar to the liquefaction potential index proposed by Iwasaki et al. [18,19] by considering the blow count ratio as a safety factor against liquefaction.

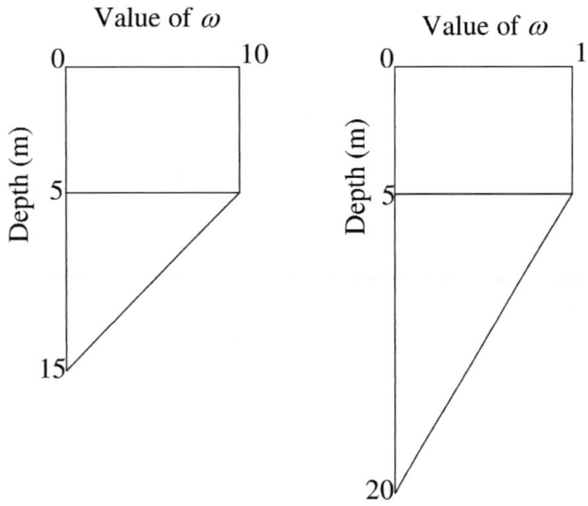

a) *GBJ11-89* b) *GB50011-2010*

Figure 4. The definition of weight factor ω in different versions of Chinese codes (not for scale)

The usage of I_{lE} provides a way to distinguish different levels of liquefaction severity. Table 9 provides the categories of liquefaction severity based on the liquefaction index in different versions of the code. Note that the border values for each category have been slightly justified to account for the change of impact depth of liquefaction. Another benefit of I_{lE} is to group routine design options into different levels of liquefaction severity, providing guidance on appropriate engineering measure to mitigate liquefaction hazards for different building types defined in Table 10. For the significant buildings falling into Category I, particular in-depth anti-liquefaction design should be conducted. Table 11 provides general guidelines on engineering measures of liquefaction mitigation for other buildings.

Table 9. Liquefaction severity based on the liquefaction index

Code	Light	Moderate	Serious
GBJ11-89	$0<I_{lE}\leq 5$	$5< I_{lE}\leq 15$	$I_{lE} >15$
GB50011-2010	$0<I_{lE}\leq 6$	$6< I_{lE}\leq 18$	$I_{lE} >18$

Table 10. Classification of building types in *GB50011-2010*

Building type	Description
Category I	Special buildings, buildings with high seismic vulnerability and etc.
Category II	Lifeline system in the national key cities and buildings with relatively high level seismic vulnerability.
Category III	Others except category I, II and IV.
Category IV	Buildings with low population density and low seismic vulnerability.

Table 11. General guidelines of anti-liquefaction design in *GB50011-2010*

Building type	Liquefaction severity		
	Light	Moderate	Serious
Category II	EPLS or FSM	EALS or EPLS + FSM	EALS
Category III	FSM or no measures	at least FSM	EALS or EPLS + FSM
Category IV	no measures	no measures	FSM or SM

* Note: EALS—eliminate all liquefaction-induced settlement; EPLS—eliminate partial liquefaction-induced settlement; FSM — foundation and structural measures; SM — supplemental measures.

4. Summary

This paper reviews major advancement of the specifications on liquefaction assessment in the seismic design building code in mainland China. The country suffered extensive liquefaction hazards during a large number of destructive earthquakes in the past century. The post-earthquake reconnaissance efforts gave birth to a methodology of liquefaction assessment in 1974, which later became mature in 1989 and continuously received amendments after every strong earthquake as a result of increasing knowledge on liquefaction phenomenon. In the current version of the code, a detailed assessment of liquefaction potential is required based on SPT blow count without depth correction, if liquefaction concern cannot be eliminated by a preliminary assessment based on descriptive information collected in the phase of site characterization. A liquefaction index is evaluated to recognize liquefaction severity considering an entire profile of the site being considered, and also to filter the most appropriate engineering measures for liquefaction mitigation.

Acknowledgement

The authors acknowledge the support from National Nature Science Foundation of China (with grant No. 41102173), the Ministry of Science and Technology of China (with grant No. SLDRCE 14-B-11), the Fundamental Research Funds for the Central Universities, and the Scientific Research Foundation for the Returned Overseas Chinese Scholars, State Education Ministry.

References

[1] Seed H B and Idriss M I 1971 *J. Soil Mech. Found. Div.* **97** 1249-73.
[2] Huang Y, Yu M and Bhattacharya S 2014 *Indian Geotech J* **44**(2):218–24.
[3] Seed H B, Tokimatsu K, Harder L F and Chung R M 1985 *Journal of Geotechnical Engineering, ASCE,* **111**(12) 1425-45.
[4] Youd T, Idriss I, Andrus R, Arango I, Castro G, Christian J, Dobry R, Finn W, Harder L, Jr. Hynes M, Ishihara K, Koester J, Liao S, Marcuson W III, Martin G, Mitchell J, Moriwaki Y, Power M, Robertson P, Seed R and Stokoe K II 2001 *Journal of Geotechnical and Geoenvironmental Engineering, ASCE,* **127**(10) 817-33.
[5] Chen G X 2003 *J. Disaster Prevention and Mitigation Engineering* **23**(1) 102-13 (in Chinese)
[6] Yuan X M, Sun R 2011 *Rock and Soil Mechanics* **32**(S2) 351-8 (In Chinese).
[7] The State Commission of Seismicity of the People's Republic of China 1974 *Code for Industrial and Civil Buildings (trial version), TJ11-74* (Beijing: China Building Industry Press) (in Chinese).
[8] The State Commission of Basic Building of the People's Republic of China 1978 *Code for Industrial and Civil Buildings, TJ11-78* (Beijing: China Building Industry Press) (in Chinese).
[9] The Ministry of Construction of the People's Republic of China 1989 C*ode for Seismic Design of Buildings, GBJ11-89* (Beijing: China Architecture and Building Press) (in Chinese).
[10] The Ministry of Construction of the People's Republic of China 2001 *Code for Seismic Design of Buildings, GB50011-2001* (Beijing: China Architecture and Building Press) (in Chinese).

[11] Ministry of Housing and Urban-Rural Construction of the People's Republic of China 2010 *Code for Seismic Design of Buildings, GB50011-2010* (Beijing: China Architecture and Building Press) (in Chinese).
[12] Cao Z Z, Youd T and Yuan X M 2008 *Soil Dynamics and Earthquake Engineering* **31**(8) 1132-43 (in Chinese).
[13] Yuan X M and Cao Z Z 2011 *Earthquake Engineering and Engineering Vibration* **10**(3) 339-47 (in Chinese).
[14] Lin P S, Chang C W and Chang W J 2004 *Soil Dynamics and Earthquake Engineering* **24** 675–87 (in Chinese).
[15] Zhao X L and Cai G J 2014 *Marine Georesources & Geotechnology* **33**(3) 272-81 (in Chinese).
[16] Tianjin Urban & Rural Construction Commission 1988 *Code for design of building foundation, TBJI-88* (Tianjin: Tianjin People Press) (in Chinese).
[17] The Ministry of Construction of the People's Republic of China 2001 *Code for investigation of geotechnical engineering, GB50021-2001* (Beijing: China Architecture and Building Press) (in Chinese).
[18] Iwasaki T, Tatsuoka F, Tokida K, and Yasuda S 1978 *Proc. 2nd Inter. Conf. on Microzonation (San Francisco)* p 885.
[19] Sonmez H 2003 *Environ. Geol.* **44** 862-71.

IOP Conf. Series: Earth and Environmental Science **26** (2015) 012022 doi:10.1088/1755-1315/26/1/012022

Flood control and loss estimation for paddy field at midstream of Chao Phraya River Basin, Thailand.

T C Cham[1] and Y Mitani[2]

[1 2]Graduate School of Civil and Structural Engineering, Kyushu University, Motooka 744, Fukuoka, Japan

Email: coreyhut@gmail.com/ cham@doc.kyushu-u.ac.jp

Abstract. 2011 Thailand flood has brought serious impact to downstream of Chao Phraya River Basin. The flood peak period started from August, 2011 to the end of October, 2011. This research focuses on midstream of Chao Phraya River Basin, which is Nakhon Sawan area includes confluence of Nan River and Yom River, also confluence of Ping River and Nan River. The main purpose of this research is to understand the flood generation, estimate the flood volume and loss of paddy field, also recommends applicable flood counter measurement to ease the flood condition at downstream of Chao Phraya River Basin. In order to understand the flood condition, post-analysis is conducted at Nakhon Sawan. The post-analysis consists of field survey to measure the flood marks remained and interview with residents to understand living condition during flood. The 2011 Thailand flood generation at midstream is simulated using coupling of 1D and 2D hydrodynamic model to understand the flood generation during flood peak period. It is calibrated and validated using flood marks measured and streamflow data received from Royal Irrigation Department (RID). Validation of results shows good agreement between simulated result and actual condition. Subsequently, 3 scenarios of flood control are simulated and Geographic Information System (GIS) is used to assess the spatial distribution of flood extent and reduction of loss estimation at paddy field. In addition, loss estimation for paddy field at midstream is evaluated using GIS with the calculated inundation depth. Results show the proposed flood control at midstream able to minimize 5% of the loss of paddy field in 26 provinces.

1. Introduction

The Chao Phraya River Basin is normally divided into upstream (Northern Thailand), midstream (Nakhon Sawan) and downstream (Chao Phraya Delta) according to hydrological features (fig. 1). In the midstream of the Chao Phraya River Basin, Ping River, Yom River and Nan River flow down from the mountainous areas of Northern Thailand and converge at Nakhon Sawan. Bhumibol Dam (reservoir capacity of 13.5 billion m^3) was constructed on Ping River and Sirikit Dam (reservoir capacity of 9.5 billion m^3) was constructed on Nan River for irrigation, power generation and flood control purposes. A total of five dams were constructed for Ping River, Wang River and Nan River watersheds and contributes to 25 billion m^3 reservoirs capacity [13]. Komori et al. (2012) concluded the 2011 flood was caused by increased rainfall by 143% over doubled runoff during rainy season rainfall during 1982 – 2001. Moreover, 5 typhoons made landfall in Thailand, which strongly influenced the rainfall. The flood destroyed the water gates and broke levees, especially at the left bank of the upper Chao Phraya Dam, which led to uncontrollable flooding. Thai government intends to build Kaeng Suea Ten Dam (reservoir capacity of 1.15 billion m^3) on Yom River to increase the total capacity in the country. However, the plan was against by the residents and to be constructed. Limited studies were conducted on the midstream of Chao Phraya River Basin. Oki (2012) clarified and analyzed on the flood damage in the downstream of Chao Phraya River Basin. However, flood condition of midstream was not clarified in details. Kwak et al. (2012) estimated total flood volume in Chao Phraya River Basin by using modified remote sensing indices to near-real-time Moderate Resolution Imaging Spectroradiometer (MODIS) images and flood inundation level (FIL) model with the Digital Elevation Model (DEM). It studied flood condition using static method; however, this research will use hydrodynamic simulation method to understand the flood condition.

The Ministry of Finance (2012) reported the total damage and losses from the 2011 floods in Thailand amounted to USD 46.5 billion. Table 1 shows the damage and losses for agriculture sector. The Ministry of Finance (2012) defines crop damages as damage to farm machinery such as tractors, plows, threshers and farm tools. The damage to the farm machinery and losses in 26 provinces for crops were USD 0.2 billion and USD 1 billion, respectively. From table 2, it is able to understand the total losses to rice of 26 assessed provinces are estimated to be roughly USD 0.8 billion. The total loss includes 3 components, which are total production losses, higher production costs and losses due to forced early sale. Flood damage is divided into 2 categories, tangible damage and intangible damage. Intangible damage includes negative psychological impact such as depression and health. Tangible damage is subdivided in to direct damage and indirect damage. Direct damage is measureable and often referred to as the damage. Indirect damage is not physical but subsequent negative effect on economic activities (JICA). This study will use hydrodynamic simulation method to examine the flood condition at the midstream area of Chao Phraya River Basin, particularly paddy field by GIS. The aim is to understand the economic impact of the flood on the paddy field in midstream region. The direct damage for paddy field is calculated by formula (1). Different scenario of flood control at 3 rivers – Nan River, Yom River and Ping River were simulated and the loss estimation of paddy field was analysed. Ultimately, some effective mitigation measures are recommended.

$$one\ type\ of\ damage = area * yield * farm\ gate\ price * damage\ rate \quad (1)$$

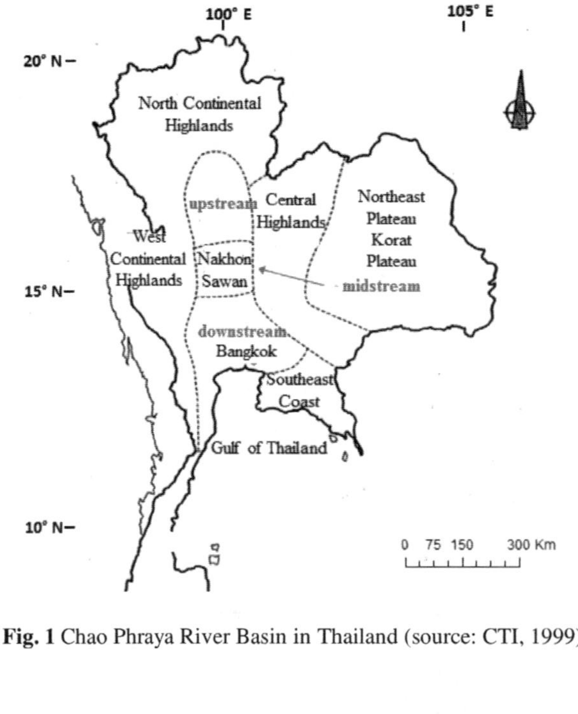

Fig. 1 Chao Phraya River Basin in Thailand (source: CTI, 1999)

Table 1 Agriculture sector - Damage, losses and needs in USD, millions (source from Ministry of Agriculture and Cooperatives, Thailand)

Sub-sector	Disaster Effects			Ownership		Needs		
	Damage	Losses	Total	Public	Private	~6 mths	6-24 mths	>24 mths
Agriculture	154	927	1,081	0	1,081	230	30	0
- Food crops	150	808	957	0	957	-	-	-
- Perennial crops	5	119	124	0	124	-	-	-
Livestock	10	83	93	0	93	5	2	0
Fisheries	4	25	28	0	28	1.6	0.8	0
Cross cutting sector						16	15	0.6
Total	323	1,962	2,285	0	2,283	253	48	0.6

Table 2 Agricultural Sector - Estimated losses in 2011-2014 by commodity in USD, millions (source from Ministry of Agriculture and Cooperatives, Thailand)

	2011	2012	2013	2014
Crop	820	57	36	14
Rice	794	0	0	0
Sugar cane	9	0	0	0
Fruit trees	7	57	36	14
Flowers	4	0	0	0
Maize	0	0	0	0
other	6	0	0	0
Livestock	60	23	0	0
Poultry	35	0	0	0
Swine	1	0	0	0
Cattle	22	23	0	0
Other	2	0	0	0
Fisheries	25	0	0	0
Tilapia	10	0	0	0
Catfish	8	0	0	0
Shrimp	7	0	0	0
Striped snakehead fish	0	0	0	0
Total	905	80	36	14

2. Simulation of Flood Scenario at Study Area

Fig 2 shows the location of midstream of Chao Phraya River Basin, Nakhon Sawan. Those labeled in red are gauging stations. Nan River and Yom River converge at gauging station N. 67. Ping River and Nan River converge before reaching gauging station C. 2. Streamflow data of these gauge stations are received from Royal Irrigation Department (RID). Data of gauging stations at upper reach are used as input parameters for model, such as P. 7A, Y. 17 and N. 7. Data of gauging stations at lower reach are used for validation, such as N. 67 and C. 2. Simulation of 2011 Thailand flood event is done with MIKE Flood, a Hydraulic Model from Denmark Hydraulic Institute (DHI). The hydrodynamic and rainfall parameters data are obtained from RID. Cross section data are provided by The International

Centre for Water Hazard and Risk Management (ICHARM) and source of bathymetry data is Shuttle Radar Topography Mission (SRTM) with the resolution of 90m.

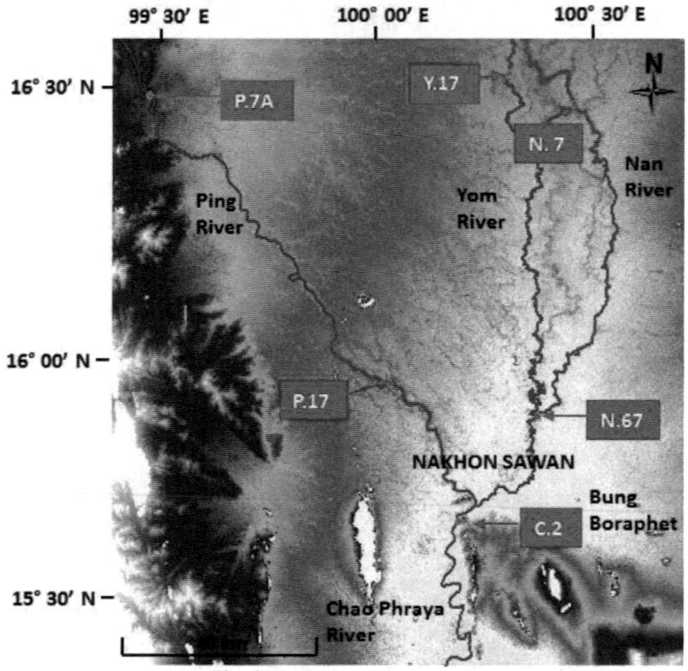

Fig. 2 Study area of research

2.1. *Field Survey at Nakhon Sawan*

Flood marks investigation was conducted in the area of Nakhon Sawan from September 27 to 29, 2012, by measuring rod. The targeted areas were the riverbanks, natural banks, back marshes, narrow segments and others. Some measured flood marks are out of study area, as a result, a total of 8 flood marks are used for validation (fig. 4).Information such as starting date of floods, the flow direction of floods, living condition during floods and awareness of floods were collected through the interview with the residents.

2.2. *Calibration of Modelling*

The flowchart of modelling is shown on fig. 5. Nedbør-Afrstrømnings-Model (NAM) Rainfall-Runoff model in MIKE 11 (DHI.co) is used for simulating 1D hydraulic modelling of rivers condition in study area. Rainfall data in year 2011 for upper reach of Nan River, Ping River and Yom River are used. Streamflow data for gauging stations Y. 17, N. 7 and P. 7A are used as inflow boundary type for upper reach in model. Projection of the model has defined as Universal Transverse Mercator (UTM) North 47 and cross sections are defined in each river with at least 7 cross sections on it. The simulation period for MIKE 11 is from April 1 to December 31, 2011 as limited streamflow data obtained from RID. SRTM is used as bathymetry for MIKE 21 in 2D modelling. It was interpolated into resolution of 100m from 90m to have a smooth running of simulation. Finally, both 1D and 2D are coupled in MIKE FLOOD to simulate the flood generation. In the consideration of running time and quality of result, simulation period is focused from August 1 to December 30, 2011 with time step interval of 15 seconds. In this model, Manning M is used as bed resistance and only parameters from main rivers (Nan River, Yom River, Ping River and Chao Phraya River) are included as limited data obtained. Simulation model did not include the consideration of small streams around those main rivers.

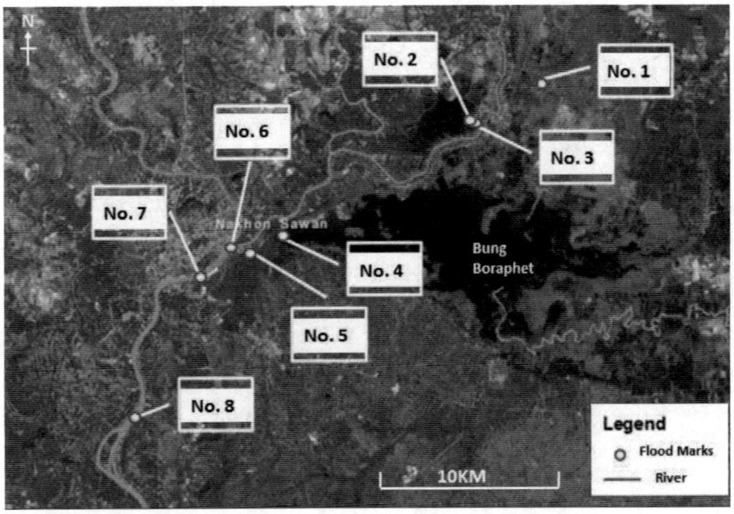

Fig. 4 Location of flood marks at study area, which are used for data validation

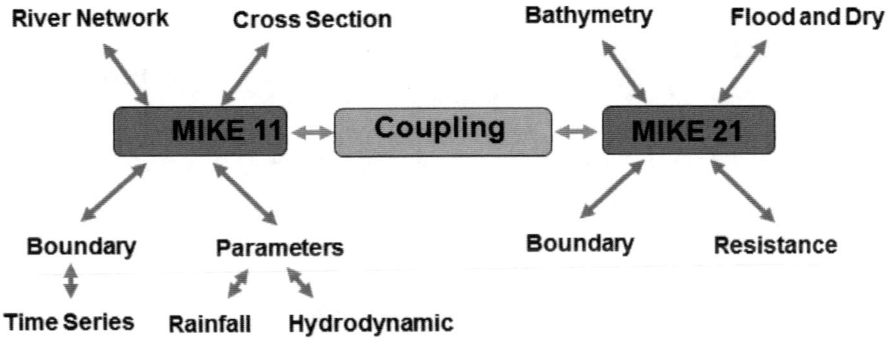

Fig. 5 Flow chart of Simulation

2.3. Validation and Discussion of Simulation Result

The result of simulated streamflow is validated with data received from RID such as data at gauging stations N. 67 and C. 2. Fig 6 shows the differences of streamflow before and after confluence of Nan River and Yom River by the comparing the combination data at gauging stations Y. 17 and N. 7, and N. 67. The simulated result at gauging station N. 67 was validated with RID data. From RID data, the flood condition before gauging station N. 67 is more serious then after as it showed a difference of nearly 63 hundred million m^3 occurred. These flood amount flows into lower area, small streams and flood plain before reaching gauging station N. 67 and caused the flood condition eased after gauging station N. 67 at a streamflow less than 2000 m^3/s. Dotted line in fig 6 represents the result of simulation at gauging station N. 67 and it showed about 10 hundred million m^3 differences compared with RID data from August 1 to October 31, 2011. This study focused on the main river and the flood volume could flow into the small streams of the Nan River which may contribute to the observed differences.

Harmless streamflow limit for gauging station C.2 is $4000m^3/s$ [5]. In fig.7, RID data at gauging station C. 2 shows streamflow increases gradually from August to October 2011 and reaches flood peak which over the harmless streamflow limit after September 21.The total flood amount at gauging station C. 2 shows a big difference compared with gauging station N. 67. This is caused by flood amount before gauging station N. 67 returned and flowed back into Chao Phraya River (gauging station C. 2). It is the main reason of the flood peak occurred from September 21 to the end of

October, 2011. When compared the simulated result from MIKE Flood with the RID data, it shows the same pattern of flow although it does not rise smoothly as RID data. Simulated result presented pulsed undulations wave due to limitation of data and variable parameters, which is discussed in the end of this section. The different flood volume between simulated result and RID data is about 3 hundred million m^3. In other words, flood volume at gauging station C. 2 has only 3 hundred million m^3 which were affected by small streams surrounding it. The flood volume at gauging station C. 2 could be mitigated under the harmless limit if the flood returned volume is control at the upper reaches of Nan River and Yom River.

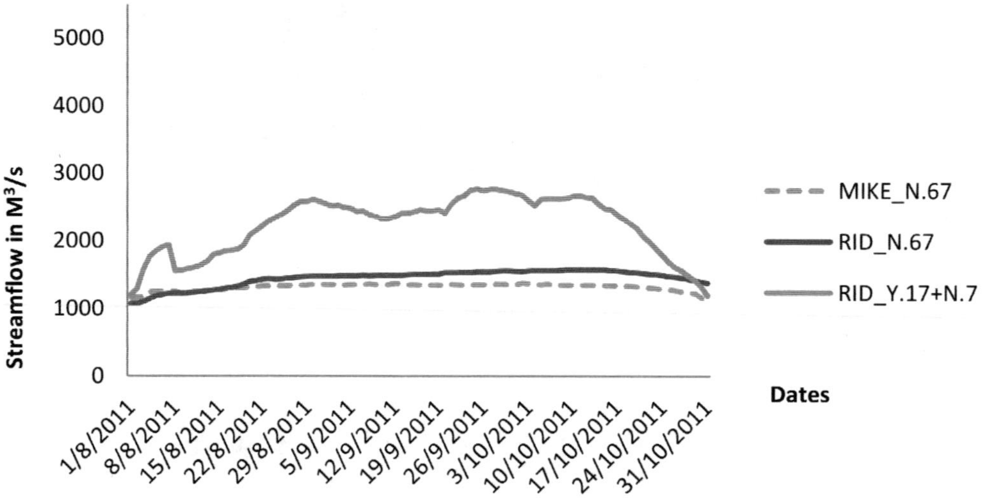

Fig. 6 Streamflow result before and gauging station N. 67

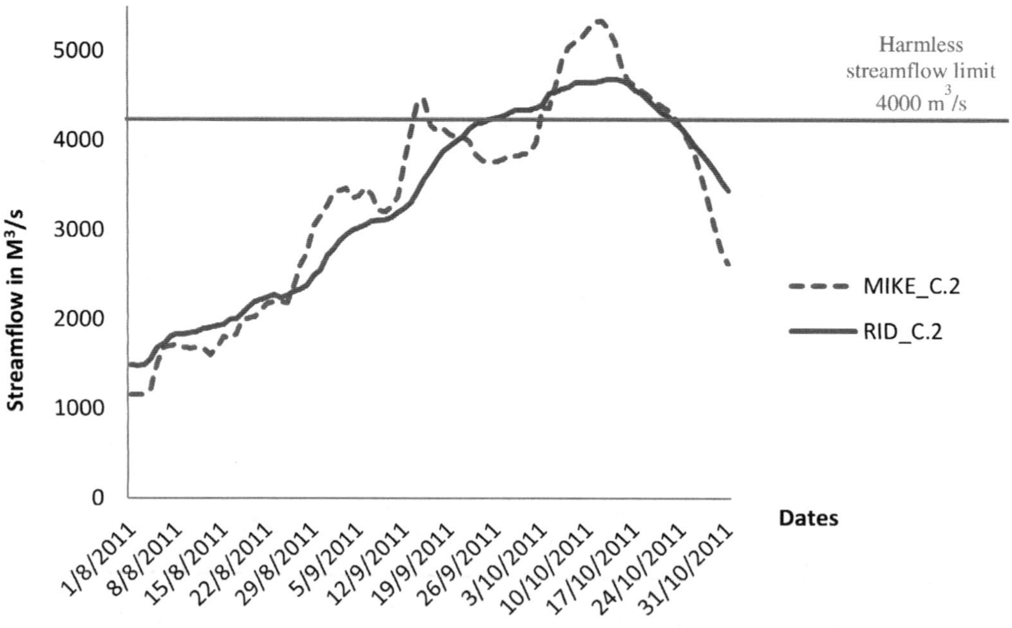

Fig. 7 Streamflow after confluence of Ping River and Nan River, gauging station C. 2

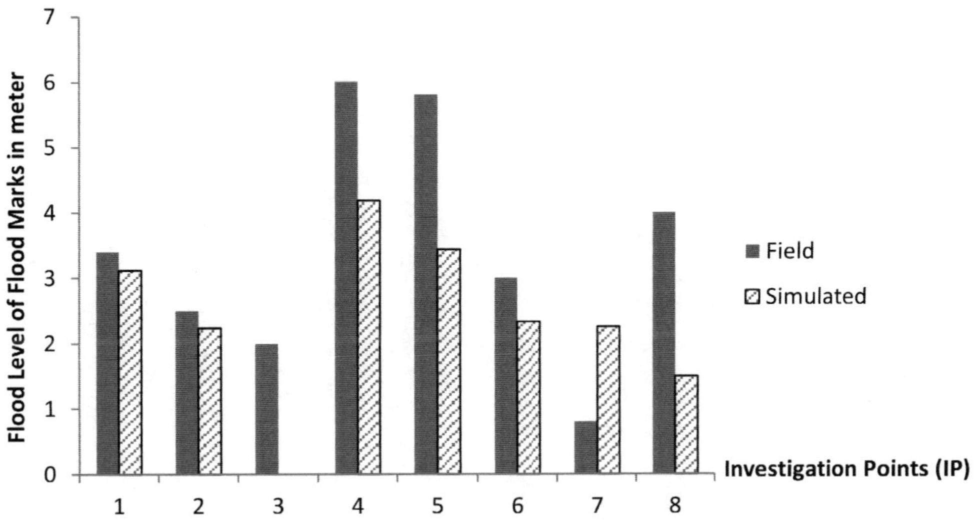

Fig. 8 Comparison of flood level in between field investigations and simulated results.

The flood inundation from the simulated result has validated with measurement of flood marks at Nakhon Sawan (fig 8). Among 8 investigation points (IP) (fig 4), the IP 1, IP 2 and IP 6 are matched with difference of + 0.5m. However, the resolution of SRTM has limited accuracy and could affect the simulated result. IP 3 falls into dry area in this model because the location is near to dike as in SRTM. IP 7 showed a difference of + 1m near to the Chao Phraya River. Therefore, the location of IP 7 in this model might be the other side of the warehouse. Furthermore, the IP 4 and IP 5 at Bung Boraphet area have a difference of + 2m. It might be affected by the water level of Bung Boraphet in SRTM. As this simulation model did not include the consideration of small streams around those main rivers. Therefore, the result of flood extent is smaller than RID in this study. Total flood volume calculated in simulated result is about 3600 million m^3. The estimation of flood volume of RID by spatial analyst in ArcGIS is about 4300 million m^3. Thus, the small streams of the main river may contribute to the differences of 700 million m^3.

2.4 Error Propagation

RRMSE for streamflow data at Obs. C2 was calculated using formula (1) with the period of every ten days from August 1 to October 20 using the streamflow generated from simulation. The result is shown as fig. 9. Although chart was not smooth, however all of value was less than 1.4 m. The most accurate period of data happened in the end of September and this might be caused by similar flood distribution generated from simulation with the RID data.

There are some error propagation effects in this study. Differences could be seen in flood volume, streamflow, pulsed undulation flood wave, and flood inundation depth between observed and simulated flood marks. These differences between observed and simulated results could be induced by various artificial structures such as man-made structures, micro topography, variables of rainfall volume, lacking of cross sections along rivers, which have possibilities to affect the dynamics of flood wave, particularly when the flood depth became low. Due to data constraints, small streams, which played crucial role to affect flood condition, are excluded in this model. Therefore, the simulated result tends to be underestimated, compare to actual extent. The simulation could be enhanced if more details of data obtained such as streamflow of small streams, details field survey of cross sections, better resolution imagery for bathymetry and including details artificial structures along rivers such as weirs and culverts.

$$RRMSE = \sqrt{E\left[\left(\frac{(Q_s(t) - Q_o(t))}{\overline{Q_o(t)}}\right)^2\right]} \quad (1)$$

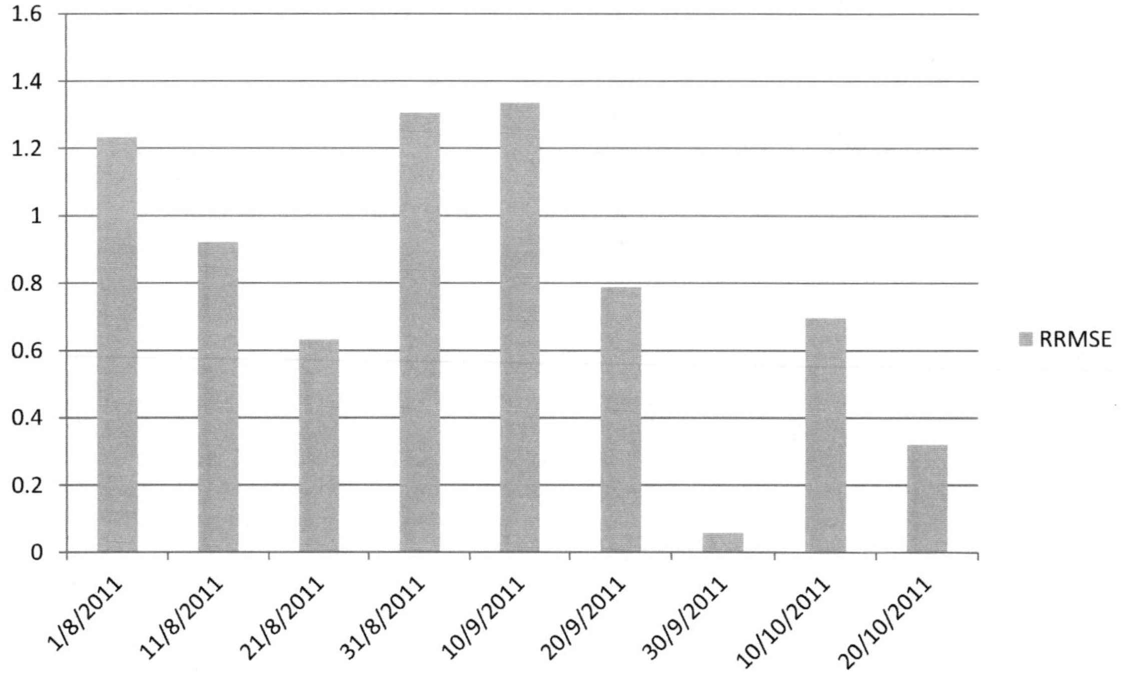

Fig. 9 RRMSE for streamflow data at Obs. C2 in the period of every 10 days.

3. Economic Impact Analysis for Rice field

The loss estimation on paddy field is evaluated using ArcGIS after validation of the simulation result of 2011 flood is conducted. Spatial distribution of flood extent area in simulated result is smaller than the real flood condition because of model's variable uncertainty, simulated result is compared with RID flood coverage and the different of loss estimation will be used in section 5 to include flood coverage affected by small streams and flood returned.

CTI. (1999) stated upstream of Chao Phraya is dominant by traditional rice field with some High Yielding Variety (HYV). Therefore, the calculation of the loss analysis will be based on 70% of traditional variety and 30% of HYV. The farm gate price is calculated referring to farm price index stated in bank of Thailand and land use data of Year 2010 is used for loss analysis. This study used Japan flood mitigation manual (table 3), which is used by CTI. (1999) to estimate loss rate in Japan International Cooperation Agency (JICA report).

Table 3 Flood Damage Rate of Rice (Values are for inundation of 7 days or more) Unit (%), (source from flood mitigation manual, Ministry of construction, Japan)

	Inundation Depth (m)						
Variety	x<0.3	0.3<x<0.5	0.5<x<1.0	<1.0<x<1.5	1.5<x		
High Yielding Variety(HYV)	0	50	71	100	100		
Traditonal Variety	0	0	51	71	100		
Deep Water Rice	0	0	51	71	100		
	Inundation Depth (m)						
	x<1.0	<1.0<x<1.5	1.5<x<2.0	2.0<x<2.5	2.5<x<3.0	3.0<x<3.5	3.5<x
Floating Rice	0	6	18	30	42	54	60

4.0 Result and Discussion of Loss Estimation

4.1 Comparison with the Loss Estimation in rice field with World Bank Report

The original result of simulation is used to calculate the losses in rice field. Fig.10 shows the flood area with inundation depth value in August 6, September 21 and October 15 for the original result from simulation. The starting of flood event, which is August 6, has flood extent area about 574 km^2, and 40% is covered on rice field. When the flood expands till September 21, the flood extent area is about 1844 km^2 and 35% of total spatial extent is covered on rice field. The widest flood extent happened on October 15, which is about 36% covered on rice field while the total flood area is about 2217 km^2. Within these 3 months, more than 60% of inundated area is covered on rice field. The losses in rice field for October is near to USD 0.1 billion. World Bank report shows total losses is about USD 794 million for 26 provinces. Thus, the percentage of losses in study area is nearly 25% of total losses in 26 provinces.

When the flood spatial distribution of simulated result and RID are compared as shown in fig. 11, rice field area covered by RID is about 25% more than simulated result. When calculation of loss estimation is done, it shows RID has 7.4% more than simulated result on October 15. Thus, 7.4% is added for loss estimation in different scenarios of simulated result, which is considered loss in flood area caused by small streams and flood returned. Result of loss estimation will be discussed in section 4.2.

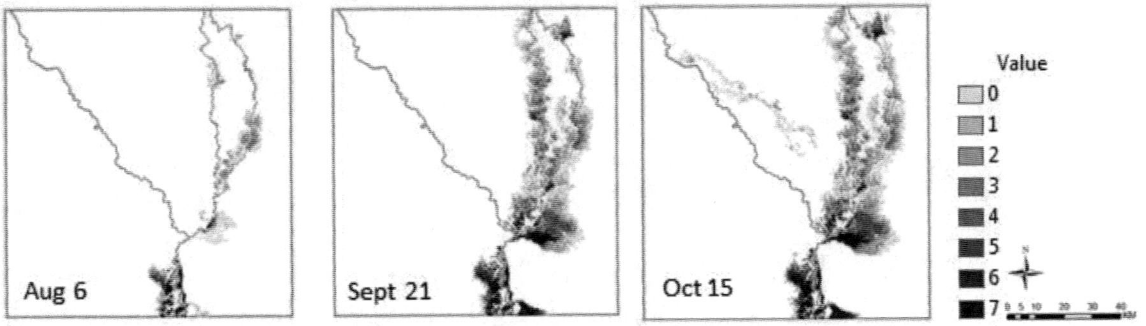

Fig. 10 Spatial distribution of flood extent area in 3 months from original result of simulation

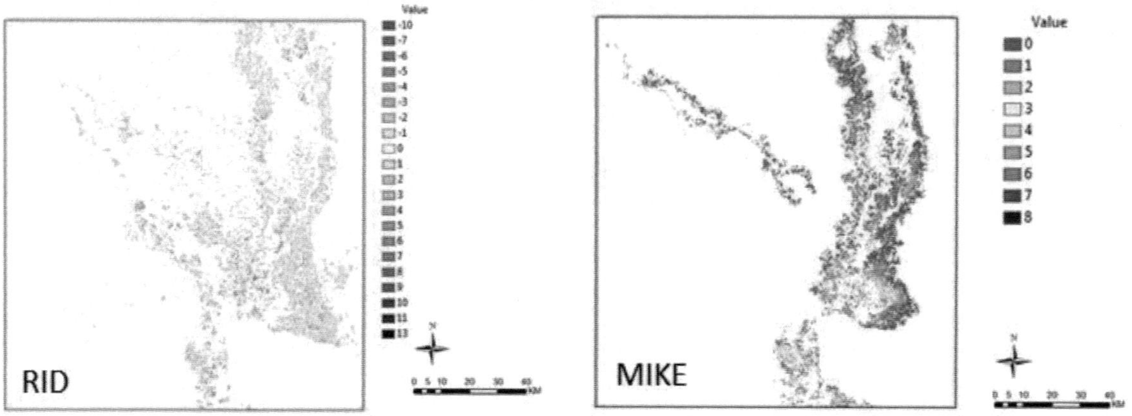

Fig. 11 Difference of area covered by rice field for RID and simulated result on October 15

4.2 Comparison with different scenario of flood condition

In this research, 3 scenarios quantified the benefits of flood control operation on existing reservoir and its potential to minimize flood risk. Parameters of streamflow data are modified in each case of scenario to predict conditions and effects when flood is under controlled at midstream. The hydraulic model is run repeatedly for each of the scenario cases with variable parameters. Condition of each scenario describes in table 4, each scenario has 3 cases. The baseline of model (original) for comparison is validated model as discussed in section 2.3. In order to choose the most applicable measures to control the flood condition, changes of streamflow along rivers and flood distribution are compared.

4.2.1. Comparison on 3 Scenarios of flood condition for losses in paddy field

Since October 15 has widest spatial distribution of flood extent, it is used to compare difference of loss estimation in rice field for scenario 1, scenario 2 and scenario 3. Fig. 12 shows the percentage of loss estimated in study area. The red line is the percentage of baseline. Based on the graph, loss is estimated in scenario 1(Case 1a – 1c) has decreased more than 2%, which is highest compare to other scenarios. Loss estimated in scenario 2(case 2a – 2c) has decreased more than 1% while loss estimated in scenarios 3(case 3a – 3c) only decreased less than 1%. Therefore, controlling flood at upper reach of Nan River (scenario 1) able to reduce losses more effectively than other rivers. In contrast, controlling flood at upstream of Yom River (scenario 3) only contribute minor effect.

Table 4 List of cases in 3 scenarios for flood controlling

Scenario	**Case**	**River Controlled**	**Condition**
Scenario 1	1a	Nan	Reduced 10% streamflow
	1b	Nan	Reduced 20% streamflow
	1c	Nan	Reduced 30% streamflow
Scenario 2	2a	Ping	Reduced 10% streamflow
	2b	Ping	Reduced 20% streamflow
	2c	Ping	Reduced 30% streamflow
Scenario 3	3a	Yom	Reduced 10% streamflow
	3b	Yom	Reduced 20% streamflow
	3c	Yom	Reduced 30% streamflow

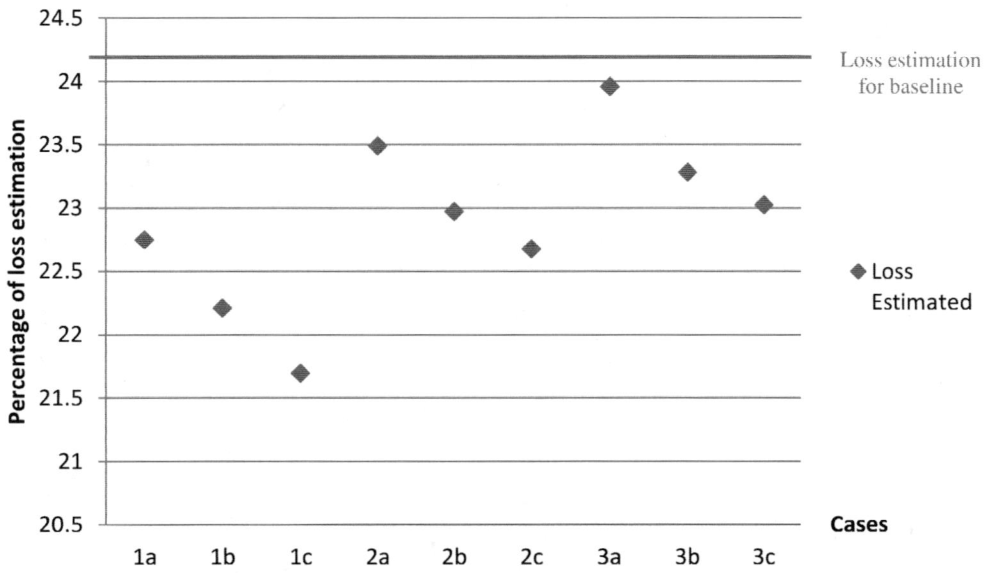

Fig. 12 Comparison of loss estimation for cases in 3 scenarios

4.2.2. Comparison on 3 Scenarios of flood distribution in Paddy Field

Spatial distribution of flood extent for these 3 scenarios are used to compare before proposed a suitable flood control. Fig. 13 shows the difference of 3 scenarios in each case with different result of flood distribution. In scenario 1, the changes of flood distribution can be seen reduced 16% of coverage in case 1a .In addition, the coverage is reduced to 18% and 21% respectively in case 1b and case 1c. Thus, the more percentage of flood volume controlled at upper reach of Nan River could help to reduce 2-3% of coverage on rice field. In scenario 2, the changes of flood distribution are much wider than scenario 1. The flood coverage has 12% reduced in case 2a, 15% in case 2b and 17% in case 2c. Difference of 3 cases in scenario 2 is about 2-3%, which is similar to scenario 1. In scenario 3, the flood distribution changes is smallest compared to scenario 1 and 2. In case 3a, there is reduced 9% of coverage, which is lesser than both case 1a and 2a. Case 3b and 3c only reduced 12% and 13% of coverage respectively. Difference of each case in scenario 3 is about 1-2% only. Based on the results, Nan River is able to mitigate flood more than Ping River and Yom River. In contrast, Yom River is less effective as the river width is narrower than the other 2 rivers.

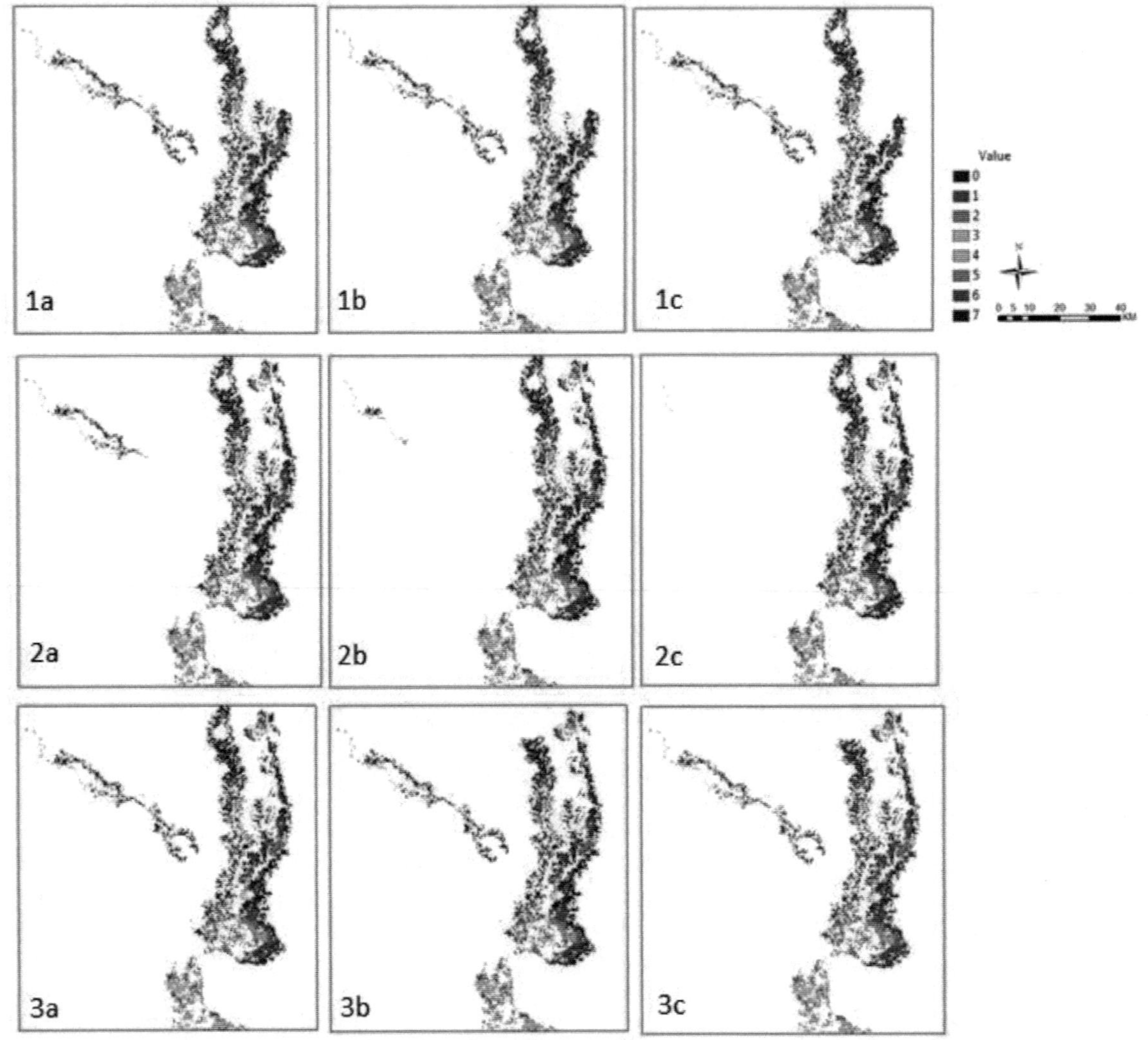

Fig. 13 Changes of Flood extent coverage for 3 scenarios

5. Proposed Flood Controlling

In the consideration of estimated loss and flood distribution on rice field, a combination of flood volume control at upstream of Nan River (20%), Yom River (20%) and Ping River (10%) is recommended. During 2011 Thailand flood, streamflow at Nan River and Yom River exceeded its capacity and flow into small streams around them, result in gauging station N.67 has lower streamflow. When flood volume is controlled at upper reaches of Nan River and Yom River, the streamflow would not exceed its capacity during the period from August to October and it flows down to downstream smoothly. As a result, higher streamflow occurred at gauging station N.67 from September and October.

Although there is about 200 million m^3 more than original in the period of September 21 to October 24 happened at gauging station N.67, which is location of Bung Boraphet, however the daily streamflow did not over harmless streamflow limit for N.67, which is $1500m^3/s$. The most important is the flood returned volume from small streams to C2 reduced tremendously. When flow out from upstream to small streams decreased, the flood volume that over harmless stream flow limit also reduced indirectly.

6. Conclusions

By controlling flood volume at upper reach of Nan River helps to minimize the loss of rice field and flood extent more effectively rather than controlling flood volume at upstream of Yom River or Ping River. If flood volume is controlled at upper reaches of Nan River, Ping River and Yom River, it minimizes about 5% of total loss and reduces nearly 30% of flood extent on rice field. Result of this research shows work plan for management of major water reservoirs and formulation of water management in Thailand master plan workable, although there are some weakness such as lack of concrete proposal to compensate farmers in flood retention area and inadequate attention to the complex long-term issues of fragmented water management.

Error propagation effects between observed and simulated results are caused by limitation of data and variable parameters such as various artificial structures, micro topography, variables of rainfall volume, lacking of cross sections along rivers, which have possibilities to affect the dynamics of flood wave, particularly when the flood depth became low. The impervious surface is closely related with surface runoff. It is suggested to conduct detail field survey on cross sections along rivers and artificial structures along rivers is needed in order to improve the result of simulation. Furthermore, streamflow data for small streams in study area is needed for enhancing the result in future.

References

[1] Bates P D and DeRoo A P J 2000 A simple raster-based model for flood inundation simulation *J. Hydrology* **236** pp 54–77

[2] Champathong A, Komori D, Kiguchi M, Sukhapunnaphan T, Oki T, Nakaegawa T 2013 Future Projection of Mean River Discharge Climatology for The Chao Phraya River Basin *Hydrological Research Lett* **7** 2 pp 36-41

[3] Cham T C, Mitani Y, Ikemi H, Fujii K 2013 Flood Hazard Assessment in Upstream Region of Chao Phraya River with GIS *Proc. of Asia Geospatial Conf* (Kuala Lumpur, Malaysia)

[4] Chang T J, Hsu M H, Teng W H, Huang C J 2000 A GIS assisted distributed watershed model for simulating flooding and inundation *J. the American Water Resources Association* **36** 5 pp 975–988

[5] CTI Engineering International CO LTD, INA Corporation 1999 *The Study on Integrated Plan for Flood Mitigation in Chao Phraya River Basin* pp 2.1-2.11

[6] Nkeki F N, Henah P J, Ojeh V N 2013 Geospatial Techniques for the Assessment and Analysis of Flood Risk along the Niger-Benue Basin in Nigeria *J. Geographic Information System* **5** pp123-135

[7] Fujii K, Tsukahara K, Hayashi H, Mitani Y, Ikemi H, Cham T C, Shimatani Y 2013 Investigation Report on the Flooding Condition in the Midstream Area of Chao Phraya River During the Thai Flooding in 2011 *J. Disaster Research* **8** pp 424-431

[8] Gül G, Harmancıoğlu N, Gül A 2010 A combined hydrologic and hydraulic modeling approach for testing efficiency of structural flood control measures. *Natural Hazards* **54** 2 pp 245-260

[9] Higuchi S 2012 *Integrated Flood Mitigation Plan Chao Praya Basin* (Japan: Agricutural Development consultants Association)

[10] International Centre for Water Hazard and Risk Management (ICHARM), http://www.icharm.pwri.go.jp/

[11] Mateo C M, Hanasaki N, Komori D, Tanaka K., Kiguchi M, Champathong A, Sukhapunnaphan T, Yamazaki D, Oki T 2014 Assessing the Impacts of Reservoir Operation to Floodplain Inundation by Combining Hydrological, Reservoir Management, and Hydrodynamic Models *Water Resources Research* **50** 9 pp 7246 – 7266

[12] Ministry of Agriculture and Cooperatives, Thailand http://eng.moac.go.th/main.php?filename=index

[13] Komori D, Nakamura S, Kigushi M, Nishijima A, Yamazaki D, Suzuki S, Kawasaki A, Oki K

and Oki T 2012 Characteristics of the 2011 Chao Phraya River flood in Central Thailand *Hydrological Research Lett* **6** pp 41-46

[14] Kourgialas N, & Karatzas G 2011 Flood management and a GIS modelling method to assess flood-hazard areas—a case study *J. Hydrological Sciences* **56** 2 pp 212-225

[15] Kwak Y J, Park J G, Yorozuya A., Fukami K. 2012 Estimation of Flood Volume in Chao Phraya River Basin, Thailand, from MODIS Images Couppled with Flood Inundation Level. *Proc. the 32nd annual IGARSS Symp* (Germany: IEEE Geoscience and Remote Sensing Society)

[16] Oki T, Komori D 2012 *Facing the floods in Thailand* (Japan:Japan Science and Technology Agency) (In Japanese)

[17] Oki T 2012 *The Chaophraya flood 2011 and flood disaster of Thailand* (Japan: The University of Tokyo) (In Japanese)

[18] Patro S, Chatterjee C, Mohanty S, Singh R, Raghuwanshi N S 2009 Flood Inundation Modeling using MIKE FLOOD and Remote Sensing Data *J. Indian Society Remote Sensing* **37** pp107-118

[19] Poaponsakorn N and Meethom P 2013 Impact of the 2011 Floods, and Flood Management in Thailand ERIA Discussion Paper Series (Indonesia: Economic Research Institute for ASEAN and East Asia)

[20] Prajamwong S, Suppataratarn P 2009 Integrated: Flood Mitigation Management in the Lower Chao Phraya River Basin. *Expert Group Meeting on Innovative Strategies Towards Flood Resilient Cities in Asia-Pacific 2009 Meeting Documents*

[21] Remo J, Carlson M, Pinter N 2012 Hydraulic and flood-loss modeling of levee, floodplain, and river management strategies, Middle Mississippi River, USA *Natural Hazards* **61** 2 pp 551-575

[22] Sayama T, Tatebe Y, Tanaka S 2015 An Emergency Response-Type Rainfall-Runoff-Inundation Simulation for 2011 Thailand Floods *J. Flood Risk Management* pp 1-14

[23] Shimakage J, Yamazaki F 2012 Detection of Flooded Areas Following the 2011 Thailand Floods using ASTER Images *The Proc.33rd Asian Conf. on Remote Sensing 2012* (Thailand: Asian Association on Remote Sensing)

[24] Strategic Formulation Committee for Water Resources Management (SCWRM) 2012 *Long Term Strategy for Flood Prevention and Mitigation in the Chao Phraya River Basin* (Thailand: Office of the National Economic and Social Development Board)

[25] Takeya K 2012 *JICA's Support to Toward Safe & Resilient Thailand" through Revising the Comprehensive Flood Management Plan for the Chao Phraya River Basin* (Japan: JICA)

[26] The Ministry of Finance 2012 *Thai Flood 2011: Rapid Assessment for Resilient Recovery and Reconstruction Planning* (Thailand: Royal Thai Governmental and The World Bank)

[27] Wongsa S 2013 2011 Thailand Flood *J. Disaster Research* **8** 3 pp 380-385

[28] Kwak Y J, Park J, Yorozuya A, Fukami K 2012 Estimation of Flood Volume in Chao Praya River Basin, Thailand, from MODIS Images Coupled with Flood Inundation Level *Proc. Int. Geoscience and Remote Sensing Symp* pp 887-890

The costs of housing developments on sites with elevated landslide risk in the UK

K Barclay[1] and A Heath[2]
[1] Graduate Engineer, Interserve Construction Limited, UK
[2] Senior Lecturer (Associate Professor), BRE Centre for Innovative Construction Materials, Department of Architecture and Civil Engineering, University of Bath, UK

E-mail: A.Heath@bath.ac.uk

Abstract. New housing targets are being set for local planning authorities resulting in more areas being zoned for development. There is currently no requirement for a landslide assessment prior to this zoning, and sites at elevated risk of landslides are being put forward for development without consideration of the additional costs and other impacts of building on these higher risk sites. This study aimed to reveal the increased financial, economic, social and environmental costs associated with these decisions. Case studies were focused on the city of Bath, an area of increasing population and "*one of the most intensely landslipped areas in Britain*". The case studies found the financial costs associated with building in a landslide risk area to be significantly higher than the equivalent construction in areas of greater geological stability. Furthermore, it was found that uncertainty in cost when developing in unstable areas exacerbates this problem as the final cost cannot be accurately predicted before construction.

1. Introduction

With a growing population, strict new housing targets are being set for UK local planning authorities resulting in more areas being zoned for development [1]. There is currently no requirement for a landslide assessment prior to this zoning, and sites at elevated risk of landslides are being put forward for development without consideration of the additional costs and other impacts of building on these higher risk sites.

There have been a numerous studies into landslides, particularly into the damage that they do to buildings and infrastructure and the casualties that result. There have been some studies into the costs of specific landslide events [2,3,4]. Rather than investigating the costs of landslide events as in these previous studies, this paper will explore the costs associated with building in a landslide area. These costs may be financial, economic, social or environmental affecting just individuals, a community or the whole economy. As well as being very damaging to the population, landslides can have a very detrimental impact upon buildings and infrastructure. Unfortunately, the occurrence of landslides in the UK is rising, with a total of 177 events in the UK in 2012 and thousands more worldwide [5].

The British Geological Survey (BGS) have analysed the data from the Met Office and concluded that the increased average rainfall and extreme rainfall events have led to a rise in the amount of landslides occurring in the UK [5]. The occurrence of landslides in the UK during 2012 is compared to rainfall during the same period in Figure 1, showing there is a clear increase in landslides during or in the month following increased rainfall.

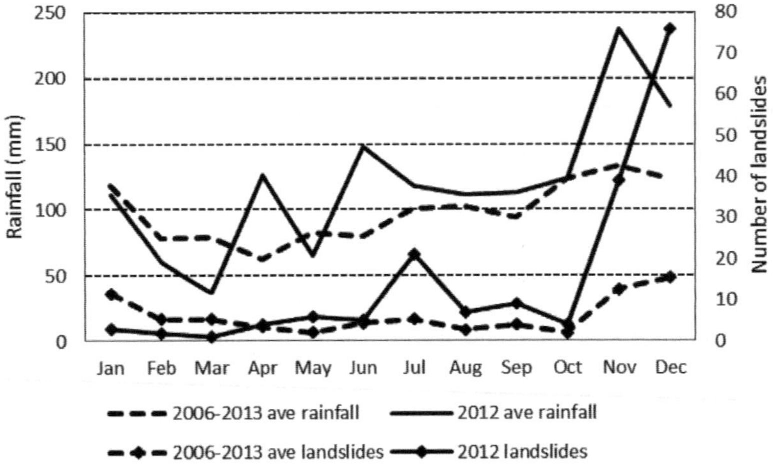

Figure 1. Rainfall and landslides in the UK (data from [5])

While the links between rainfall and landslides, and between landslides and damage to buildings and infrastructure are reasonably well established, data surrounding the increased costs of building in a landslide area is very limited. For this study, the increased costs were investigated through a series of case studies focused on the city of Bath, an area of increasing population and "*one of the most intensely landslipped areas in Britain*" [6]. Numerous options, including surveys, insurance quotes and case studies, were considered as potential methods to determine the costs that building housing in landslide prone areas had on costs. With surveys relying on responses from the general public who may not be aware of the issues, and insurance quotes requiring a high number of input variables to be assumed, it was decided that using case studies would be the most suitable approach. These case studies were based on the results from a series of interviews with industry experts and from data in geotechnical reports.

2. Development pressures and planning guidance

With rising UK Government pressure to build more housing there is also an increase in unsuitable sites being put forward for development and local councils may not have the knowledge and expertise to assess these sites. The main guidance paper surrounding the topic of developing in an area of mass movement was Planning Policy Guidance note 14 (PPG14) [7] which was removed from the public domain in 2012. Annex 1 of PPG14 further explored the guidance given and specialised it to the problems associated with building in a landslide area. It was therefore used by landowners, developers and Local Planning Authorities requiring assistance in identifying landslide areas. By restricting development in areas highlighted as a potential landslide risk, the impact of any mass movement in these regions would have been mitigated.

As already mentioned, PPG14 Annex 1 is no longer in the public domain yet it has also not been replaced hence leaving a large gap. Previous research warned of this in a discussion of the potential outcome of the planning system reformation and predicted that Planning Policy Statements (PPS) and PPGs with particular reference to PPG14, may be lost [8]. As a consequence it was suggested that this loss of guidance would *"further diminish the degree to which planning controls can be used as a mechanism to control landslide hazards"* [2]. This has resulted in local authorities, landowners and developers having less certainty than initially anticipated. While it is possible for landowners, developers and local councils to obtain a landslide hazard report from the British Geological Survey (BGS) before any planning decision is taken, this is not required or even commonly undertaken even in some landslide prone areas. This leads to inappropriate planning applications, detailed studies and hearings which cost far in excess of the minimal cost of a landslide hazard report.

The Building Regulations for England and Wales specify that *"The building shall be constructed so that the ground movement caused by: (a) swelling, shrinkage or freezing of the subsoil; or (b) land-slip or subsidence (other than subsidence arising from shrinkage), in so far as the risk can be reasonably foreseen, will not impair the stability of any part of the building."* [9] This clearly relates to construction in areas of slope instability however, it is not preceded by an avoidance strategy. It is obvious that if construction in a landslide risk area is avoided, the associated problems of building in this area are mitigated. However, neither this, nor how to identify potentially problematic areas prior to a development proposal are discussed. Therefore, by the time construction is being considered or initiated it is normally too late for avoidance of unstable areas.

3. Geological context

The geology of Bath lends itself to be very susceptible to mass movement and in particular, the mixed nature of the limestone/mudrock sequence is of high engineering interest. There are three core groups of strata: Great Oolite, Inferior Oolite and Lias. These consist of strata types with varying soil properties and formation patterns, but the general section through the valley slopes in Bath are as in Table 1. The formations are gently dipping at 1-2 degrees for areas away from the valley slopes, although there is a major fault running through the Avon valley in Bath.

Table 1. Major geological units from top of hills to valley floor in the City of Bath [10,11]

Group	Formation	Approximate thickness (m)	Name
Great Oolite	Chalfield Oolite	10	Bath Oolite Twinhoe Beds Coombe Down Oolite
	Fuller's Earth	60	Upper Fuller's Earth Fuller's Earth Rock Lower Fuller's Earth
Inferior Oolite	Salperton Limestone	10	Inferior Oolite
Lias	Bridport Sand	60	Midford Sands
	Charmouth Mudstone	100	Lower Lias Clay

While there is some cambering within the Oolitic Limestones in Bath, the majority of slope stability problems are due to two of the weaker materials. The Fuller's Earth formation clays are of high plasticity and low shear strength [12] making this strata likely to move under local changes such as excavation or applied loads from construction. The Lias Group has also been widely studied due to its relation to landslides [13]. Predominately, this group comprises of sandstones and mudstones with varying intermittence of limestone layers. This makes the group as a whole of much lower permeability than the Inferior and Great Oolite Groups sitting above it. This causes weathering of its upper layers, creating a region that is highly susceptible to slope stability problems.

4. Construction case studies

Bath is considered a desirable place to live which naturally increases the demand for housing and has resulted in an average house price well above the national or regional average. As a result, there is additional financial incentive for developers to construct in areas at risk of landslides. The following case studies demonstrate the unpredictability of building in a landslide risk area and the accumulation of costs that slope instability problems entail. These case studies are considered on an individual basis prior to comparison and discussion. Due to a lack of ground investigations on the historic landslides and because the areas are now built up, it was not possible to obtain accurate cross sections through any of the landslides showing the location of the slip surface.

In all cases the principle was to mitigate the effect of the construction on local slope stability, rather than attempting to stabilise the slopes against future failures. In other words, the geotechnical design was to prevent the construction from inducing a landslide, rather than to stabilise an existing slope with marginal stability which is impossible in most cases because of the large marginally stable areas. The locations on the case study buildings are illustrated in Figure 2.

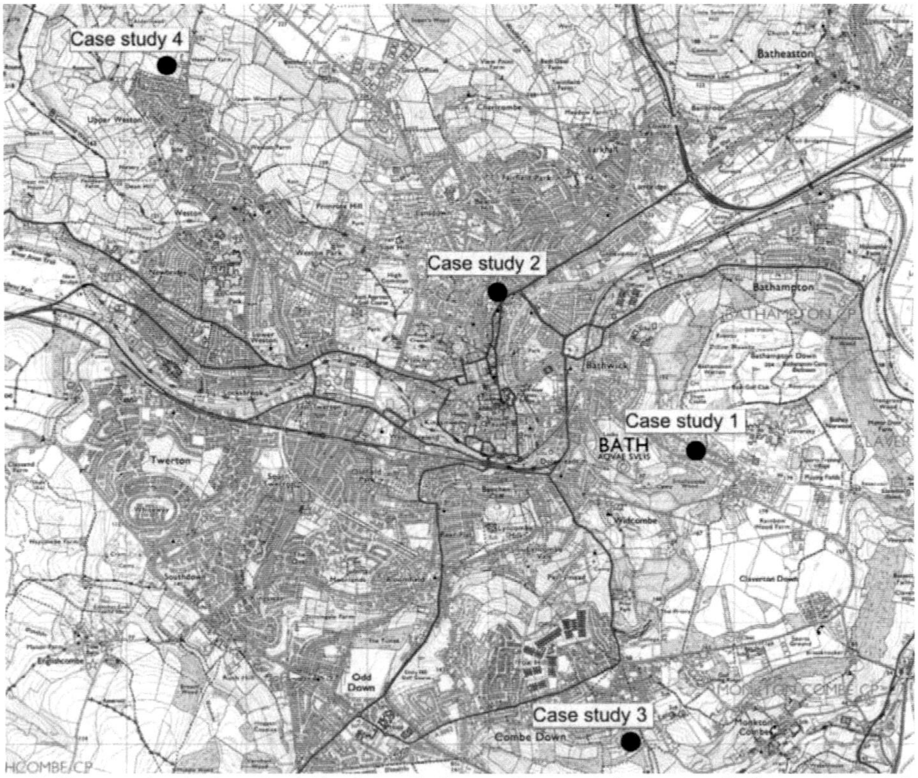

Figure 2. Case study site locations (contains Ordnance Survey information © Crown Copyright and Database Right 2015. Ordnance Survey (Digimap Licence))

4.1. Case study 1

Constructed in 2008, and featured in the television series "Grand Designs", this is a 5-bedroom home with internal floor area of 427m^2. Numerous problems were encountered due to slope instability, typical of a building in a landslide risk area. Due to the potential slope stability problems, a raft slab was used as the foundation to form a platform for the house to sit on. Behind this a retaining wall structure was built using 60, 12m deep, bored piles in order to hold back the unstable slope above the site. The water presence between the permeable limestone and impermeable mudrock layers in the ground caused the drilled holes to collapse before the piles could even be installed. In order to overcome this problem, a casing had to be added to divert the underground water which added a significant cost that had not been anticipated. In addition the boundary wall with a neighbouring property had to be reconstructed as a reinforced concrete retaining wall, further reinforced with pipe nails. The final design is illustrated in Figure 4.

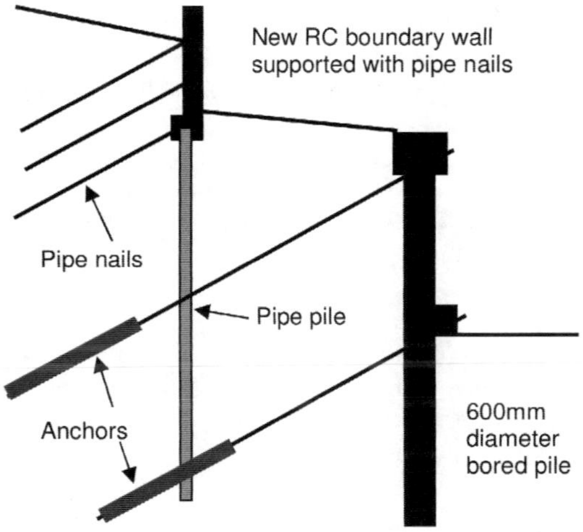

Figure 4. As built retaining wall for case study 1 (not to scale).

The initial cost estimations for the groundworks were approximately £300,000 but the actual cost was £385,000. The retaining wall required the installation of anchors as well as a deeper, more heavily reinforced structure than would be required in an area not suffering from ground instability. On top of this, a large capping beam was required which would not be required in a non-landslide area. Based on information from the groundworks contractor, the groundworks cost £265,000 more than in an equivalent construction in a non-landslide area [14]. The final groundworks have been stable to date.

4.2. Case study 2

Case study 2 was for a separate extension to a furniture store showroom rather than a dwelling, but was comparable in size to a large dwelling (389m^2 internal floor area) and was across the road from the main furniture store location. The new extension was built in 2012 at the bottom of Hedgemead Park, an area well known for its landslide risk. There were a number of previous landslides in the location as shown in Figure 5, with the most recent landslides being the Camden Crescent slip occurring during the construction of Camden Crescent in the 18th century, and the Hedgemead slip occurring during the 19th century. The extent of the historic Beacon Hill is difficult to determine because of the later landslides and because of the extensive construction over it, but this has been inferred from the mass movement deposits in the most recent BGS geological map [11].

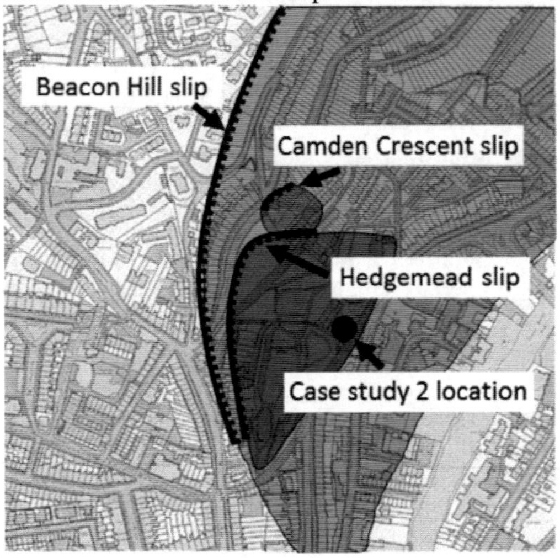

Figure 5. Location of case study 2 with approximate location of nearby historic landslips [11,15] (contains Ordnance Survey information © Crown Copyright and Database Right 2015. Ordnance Survey (Digimap Licence))

As with Case Study 1, a reinforced concrete retaining wall and piled foundation structure with anchors was installed. The total costs for the groundworks was approximately £126,000 which the groundworks contractor estimated to be £81,000 more than would be the case if the construction in an area with low landslide hazard [14].

As with Case Study 1, the majority of the additional costs were from the construction of a retaining wall to provide a level base for the building and the installation of piles and ground beams instead of a shallow foundation system.

4.3. Case study 3

The house for Case Study 3 was also featured in the television series "Grand Designs" and was constructed in 2009 as a redevelopment of an existing building with an internal floor area of approximately £400m^2. It is set in the steep valley area of Combe Down and is an old industrial building on a plot of two acres. It took three years to achieve planning permission due to the area being of outstanding natural beauty, part of the green belt zone and a grade II listed property within the Bath World Heritage boundary. The original building consisted of two barns located on an ancient landslip which required a network of deep piles and steel reinforced concrete ring beams in order to stabilise both the slope and the structure. In order to alleviate the problem of land movement underpinning and consolidation of the footings of the building was also required.

The total cost for these groundworks was approximately £70,000 which was an increase of approximately £30,000 compared to an equivalent new construction in a more geologically stable area, As this particular project was a retrofit rather than new construction, there is a possibility that no remedial foundation work would have been required had it been in a geologically stable area (as commonly the case for barn to housing conversions).

4.4. Case study 4

Case study 4 is the only case study which has not been constructed, but relates to a proposed large area for development. The case studies previously discussed focus on the development of one building therefore it needs to be investigated as to whether, if conducted on a larger scale, it is feasible to develop landslide areas and include the required amount of affordable housing (typically 30% or more depending on the development).

Development of the site was originally to include 300 homes however, this was subsequently reduced to 150 [16]. Reasons for this may have included a change in site location and the slope instability problems highlighted by a geotechnical desk study of the area [17] which was only commissioned after objections related to slope instability were raised during the consultation phase for the Bath Core Strategy. It can be assumed that, although property size and style will vary, these homes will average three bedrooms in size. For a two storey, three bedroom home the recommended minimum floor area is 96m^2 [18].

The site is in an area of slope instability and has nearby areas where mass movement has occurred, including some landslips in the past 50 years, as shown in Figure 6.

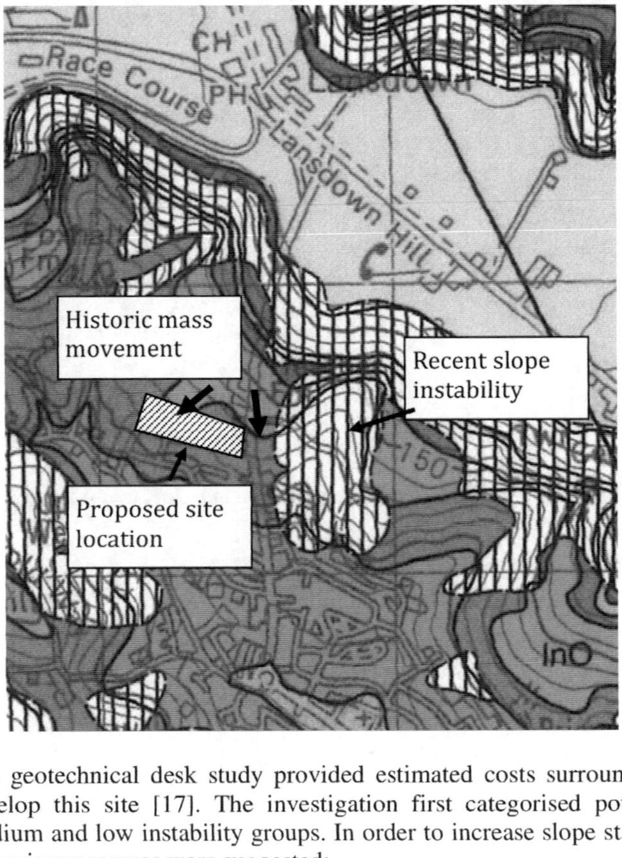

Figure 6. Location of case study 4 with nearby mass movement (contains British Geological Survey materials © NERC 2011 and Ordnance Survey information © Crown Copyright 2011.)

The geotechnical desk study provided estimated costs surrounding the remedial works required to develop this site [17]. The investigation first categorised potential development areas into high, medium and low instability groups. In order to increase slope stability for development of the site the following measures were suggested:

- The installation of gravel filled shear trenches;
- terracing of the steeper sloped areas;
- retaining walls, most likely contiguous piled walls with a 5m retention height;
- piles in areas of medium and high risk.

Based on costs provided for these measures and taking the number of dwellings and area to be developed into account, the development of each dwelling in the low and high risk areas of the site will require between £27,000 and £37,000 extra groundwork costs in order to provide sufficient slope stabilisation. Prior to construction, a detailed site investigation would need to be undertaken to confirm the results of this desk study. The costs associated with this detailed site investigation would have to be added to the cost per dwelling.

This particular site was eventually rejected as suitable for development by the Planning Inspector, and it is therefore unlikely that this site will be developed in the near future. The reasoning by the Planning Inspector was related to visual impact and did not mention issues of slope instability.

5. Discussion

The case studies reveal a consistent increase of the construction cost in areas prone to slope stability failure, however the extent of this increase varies mainly due to site specific factors such as slope angle and immediate geology. Table 2 highlights this increase in cost through a comparison of the groundwork costs for the construction case studies detailed above.

Table 2. Additional groundworks costs for building in area at risk of landslides

Case study	Additional cost (£)	Internal floor area (m^2)	Additional costs (£/m^2)	Cost increase from construction in stable area (%)
Case study 1	265,000	427	620	220
Case study 2	81,000	389	208	180
Case study 3	30,000-70,000	400	75-175	75+
Case study 4*	26,000-37,000	96	271-385	271-385**

* Each dwelling

**Based on estimate £100/m^2 for foundations in non-landslide areas

This clearly indicates that construction in a landslide area is far more expensive than in a non-landslide area, but is still considerably less than the minimum of £600,000 for repair and stabilisation costs for the smallest landslide reported in one of the few previous studies on costs of landslides [2]. These results are particularly interesting, as they clearly demonstrate the consistent extra finances required to build on top of this natural hazard, whilst also highlighting the variation of this increase between the case studies. The individual case studies shows that there is a high level of variability and uncertainty in the final groundworks costs which could deter some developers. There was insufficient data to conclude whether the variability in cost was mainly due to different geotechnical design approaches or different landslide risk (ground conditions, slope angles and other contributing factors).

With the exception of case study 4 which involves a number of dwellings, the others were what could be considered high value applications (large houses / commercial premises with value around or over £1 million). A key reason for this is due to the most substantial cost being getting the required expertise and construction methods, such as a piling rig to site. Once the equipment is on-site replicating the process for more than one build is likely to be lower than for a single building.

It costs between £867.60/m^2 and £1323.60/m^2 to build a home, depending on the desired quality [19] and it can therefore be seen that the increase in groundwork costs experienced by the case studies is a large proportion of these overall construction costs. Specifically, the increase of up to £385 per m^2 for the large development (case study 4) is significant for an area where affordable housing was proposed.

This value is more than 75% above the approximately £220/m^2 extra over cost to take a 96m^2 3 bedroom semi-detached home from the building regulations minimum to the Code for Sustainable Homes Level 5 requirements (the second highest level available) [20]. It was initially proposed that the development in the area for case study 4 and other areas proposed for redevelopment as part of the Bath Core Strategy was at the Code for Sustainable Homes Level 5, but the Planning Inspector removed this requirement as too onerous on developers.

6. Conclusions

Pressure from the Government to increase the housing stock in the UK is rising and this is shown through the extensive list of initiatives which continues to grow. These initiatives strongly encourage local authorities to increase housing in their respective areas. However, this is leading to unsuitable sites being zoned for future development as the previous guidance on developing in areas of slope instability in the UK was removed in 2012.

Although the final property value often outweighs the increased construction costs from building in a landslide prone area, the results from this study clearly show that, even if it is technically feasible, it is much more expensive to develop in a landslide prone area. This increase is due to a combination of extra design costs, installation of slope stability measures, the impact of mass movement during construction and reduced development area within a site as a result of the slope stabilisation structures and required spacing between homes.

There is a large variation between individual sites and the reason for this variation is primarily based on site specific factors which amongst others include the angle of the slope, surrounding properties, immediate geology (and geological features) as well as the weather during the construction process.

The construction case studies highlighted that the development of individual buildings in areas of mass movement may not be feasible for affordable housing however, for larger projects it may be possible to incorporate some affordable housing.

The costs of the construction in landslide prone areas can far exceed the extra over costs for building houses to the Code for Sustainable Homes Level 5 standard, and a decision should be taken on whether it is more in the national interest to construct in areas with increased risk of natural hazard, or build more sustainable buildings, particularly when the remedial measures for slope stability to not guarantee landslides will not occur in the future.

7. Acknowledgements
The information and assistance provided by John Peterson, Technical Director of Foundation Piling Ltd. Is gratefully acknowledged.

8. References

[1]HM Government, 2011, *Laying the foundations: a housing strategy for England.* ISBN 9781409832348, London: HMSO.

[2] Gibson, A.D., Culshaw, M.G., Dashwood, C., Pennington, C.V.L., 2012. Landslide management in the UK – the problem of managing hazards in a 'low-risk' environment. *Landslides*, **10**, 599-610

[3] Isle of Wight Centre for the Coastal Environment, 2006. *The economic impacts of natural hazards in coastal zones, taking account of the consequences of climate change*: LIFE Environment Project 2003–2007 'RESPONSE': Life 03 ENV/UK/000611. Isle of Wight Council, UK

[4] Clements, M., 1994. The Scarborough experience—Holbeck landslide, 3/4 June 1993, *Municipal Engineer*, **103**, June, 63–70.

[5] British Geological Survey, 2013. *Increased incidence of landslides in 2012* [Online]. Nottingham: British Geological Survey. Available from: http://www.bgs.ac.uk/landslides/November2012.html [Accessed 23/4/15]

[6]Kellaway, G.A. & Taylor, J.H. 1968. The influence of landslipping on the development of the City of Bath, England. *Proc. 23rd Int. Geol. Conf.* **12**, 65–76.

[7] Department for Communities and Local Government, 1996. *Planning Policy Guidance 14:* Development on Unstable Land – Annex 1: Landslides and Planning. London: HMSO.

[8] Brook, D. and Marker, B., 2008. Planning for development on land that is potentially prone to subsidence in England. *Quart Jnl Eng Geo Hydro*, **41**, 403–408

[9] HM Government, 2013. *Approved Document A - Structure (2004 Edition incorporating 2010 and 2013 amendments)* ISBN: 978 1 85946 508 0 London: HMSO.

[10] British Geological Survey, 2008. *Bath's 'foundered strata'- a reinterpretation.* UK: British Geological Survey Research Report.

[11] British Geological Survey, 2011. *Bath: Bedrock and Superficial Deposits, England and Wales* Sheet 265, 1:50 000 series, BGS, Nottingham.

[12] Hawkins, A.B. Lawrence, M.S. Privett, K.D., 1988. Implications of weathering on the engineering properties of the Fuller's Earth formation. Géotechnique, **38** (4), 517-532.

[13] Jones, D.C.K. and Lee, E.M., 1994. *Landsliding in Great Britain.* London: HMSO

[14] Peterson, J, 2013. *Pers.comm*, Foundation Piling Ltd.

[15] Kellaway, G.A. & Taylor, J.H. 1968. The influence of landslipping on the development of the City of Bath, England. *Proc. 23rd Int. Geol. Conf.* **12**, 65–76.

[16] Bath and North East Somerset Council, 2013. *Schedule of Core Strategy Amendments.* Bath: B&NES

[17] ARUP, 2013. *Strategic Greenfield Allocation at Weston, Bath: Water Infrastructure and Geotechnical Prioritisation Report.* UK: ARUP

[18] Hughes, 2011. *The Case for Space: the size of England's new homes.* UK: RIBA

[19] EC Harris, 2006. *Indicative Building Costs: Costs updated to 3rd Quarter 2006.* UK: EC Harris

[20] Element Energy and Davis Langdon, 2011. *Cost of building to the Code for Sustainable Homes: Updated cost review,* UK: DCLG

Can we reduce debris flow to an equivalent one-phase flow?

Bruno Chareyre*, Donia Marzougui*, Julien Chauchat†

*Univ. Grenoble Alpes, 3SR, F-38000 Grenoble, France and
CNRS, 3SR, F-38000 Grenoble, France
†Univ. Grenoble Alpes, LEGI, F-38000 Grenoble, France and
CNRS, LEGI, F-38000 Grenoble, France

E-mail: bruno.chareyre@3sr-grenoble.fr

Abstract. A recent extension of the discrete element method is reported for the simulation of dense mixtures of non-colloidal particles and viscous fluids in the non-inertial regime. As an application, we examine the interplay between rate dependent dilatancy and hydro-mechanical coupling which can be expected in debris flow. The numerical model includes sphere-sphere contacts using a soft contact approach [2], short range hydrodynamic interactions defined by frame-invariant expressions of forces and torques in the lubrication approximation, and drag forces resulting from the poromechanical coupling computed with the DEM-PFV technique [3]. The bulk shear stress is decomposed into contact stress and hydrodynamic stress. Both contributions are shown to be increasing functions of a dimensionless shear rate Iv, in agreement with experimental results [4]. Statistics of microstructural variables highlight a complex interplay between solid contacts and hydrodynamic interactions. In contrast with a popular idea, the results suggest that lubrication may not necessarily reduce the contribution of contact forces to the bulk shear stress. The proposed model is general and applies directly to sheared satured granular media in which pore pressure feedback plays a key role. We argue that it can be the case for debris fow, especially during the triggering phase, when run-out include transitional phases, and when the flow is stopped. It is then concluded that debris cannot be computed by assuming solely the rheological properties of an equivalent mixture.

1. Introduction

Simulating gravity driven flows of saturated geomaterials such as debris flow and rapid landslides raises two challenging questions: 1) how to reflect the rheological properties of the moving materials in relevant constitutive laws and, 2) how to solve a boundary value problem when this constitutive law is used combined with very large deformations and a complex morphology of the run-out slope. Advances are needed on both points in order to provide the engineers with better tools, enabling numerical studies of gravitational flows from the triggering to the possible impact on structures and infrastructures.

Regarding the second question, the conventional FEM methods are helpless due to (namely) the moving free surface of the flow. The methods usually used for fluid dynamics (such as direct Navier-Stokes solvers or the Lattice-Boltzman method) may have less problems with the free surface but they make the implementation of complex visco-elasto-plastic models extremely difficult. To overcome these difficulties dedicated techniques have been developed in the recent years, using e.g. the SPH[1] (Smoothed Particle Hydrodynamics) or the FEMLIP[2] (FEM with lagrangian integration points). In this paper we focus on the first question: the rheology of flowing geomaterials, and we propose an approach of this question through micromechanical modeling as in [3]. Recent developments of the discrete element method (DEM) are explained in the first section and the results of simulated simple shear tests are summarized. They highlight rheological features which are discussed in the last part.

2. Micromechanical model of particles in a viscous fluid

We assume that the flowing material can be conceptualized as a granular material. This assumption is relevant for the grain size of silts and above. Clayey materials are very different and exhibit much more complex interactions between particles, thus this study does not consider pure clays. For the mixtures of clay and coarser particles, the present approach may be valid as soon as the clay content and the pore water can be accounted for altogether as an equivalent viscous suspension carrying the bigger particles. Hereafter we introduce a numerical coupling between the DEM and a hydrodynamic model.

2.1. Discrete Element Model

An explicit finite difference scheme is employed for updating the position of each particle in a time-marching algorithm. The particles move according to the Newton's second law. The interactions between particles are governed by elastic-frictional contact forces defined using a soft contact approach [4]. The contact parameters are the normal and shear stiffnesses k_n and k_s, and the angle of contact friction ϕ. The contact forces are supplemented hereafter with forces coming from the interstitial fluid. A three-dimensional implementation of the DEM as found in the open source software YADE is used herein. For more details about the implementation, please refer to [5].

2.2. Long range interactions

The DEM-PFV method is used to solve a pore-scale version of the mass balance equation which appears in the continuous theory of porous media and leads to the so-called poromechanical coupling - leading to long range interactions between the particles. Only the main steps of the method are outlined hereafter since the details can be found in previous papers. We assume incompressible phases as in [6, 7] (for compressible phases see [?]). A tetrahedral decomposition of the pore space is introduced based on regular triangulation (figure 1), where that part of a tetrahedron occupied by the fluid is called a *pore*. From now on V_i denotes the volume of pore i. It is uniquely defined by the positions $\mathbf{x}_i$ and sizes of the solid particles, while the rate of change $\dot{V}_i$ also depends on their velocities $\dot{\mathbf{x}}_i$.

An exchange of fluid between adjacent pores i and j is represented by the interface flux q_{ij}. The volume balance equation gives

$$\dot{V}_i = \sum_{j=1}^{j=4} q_{ij}. \tag{1}$$

A linear relationship between q_{ij} and the local pressure gradient $(p_i - p_j)/l_{ij}$ (where l_{ij} is a reference length [6]) leads to

$$\dot{V}_i = \sum_{j=1}^{j=4} k_{ij}\,(p_j - p_i)/l_{ij} = \sum_{j=1}^{j=4} K_{ij}\,(p_j - p_i). \tag{2}$$

In this equation K_{ij} is the local hydraulic conductivity. It reflects the small scale geometry of the packing. In details, the proposed expression of K_{ij} depends on a local hydraulic radius R^h_{ij}

(area of the fluid-solid interface divided by the fluid volume - again see [6]) as

$$K_{ij} = \alpha \frac{S^f_{ij} {R^h_{ij}}^2}{2\eta\, l_{ij}} \tag{3}$$

where S^f_{ij} is the cross-sectional area of the pore-throat, η is the viscosity of the fluid, and α can be interpreted as a calibration parameter. $\alpha = 1$ is known to give good estimates of the actual permeability of glass beads [8] but we used $\alpha < 1$ in this study. This is further discussed in section 3.

Substituting $\dot{V}_i$ by its expression in terms of particles velocity and writing equation 2 for every element gives a system of linear equations. At each time step in the motion integration, solving this system gives a discrete field of fluid pressure $\mathbf{P}$ as function of the particles velocity.

The drag forces are deduced from the pressure field. They are the integrals of the pressure p and the viscous stress τ on the surface of the particle, i.e.

$$F^f_k = \int_{\partial\Gamma_k} p\mathbf{n}ds + \int_{\partial\Gamma_k} \tau\mathbf{n}ds \tag{4}$$

The F^f_k are introduced in Newton's second law together with the forces coming from solid contacts ($\mathbf{F}^c$) and lubrication effects ($\mathbf{F}^L$ defined in the next section). I.e.

$$\mathbf{M}\ddot{\mathbf{x}} = \mathbf{F}^c + \mathbf{F}^L + \mathbf{F}^f, \tag{5}$$

The strong two-way coupling defined by equation 5 (remember that $\mathbf{F}^f$ depends on $\dot{\mathbf{x}}$) is the poromechanical coupling. It is integrated with an explicit scheme whose accuracy has been verified in [7].

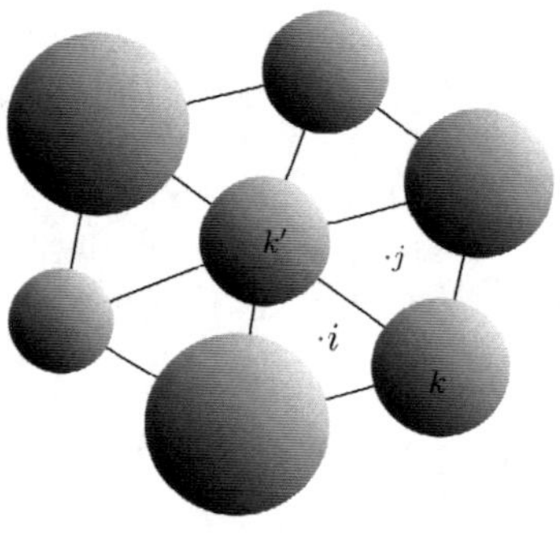

Figure 1. Regular triangulation in 2D.

2.3. Short range interactions

The lubrication forces are linked to the presence of the fluid and are only significant for nearly touching particles. We note particles k and k' with radii a_k and $a_{k'}$, linear velocities $\boldsymbol{v}_k$ and $\boldsymbol{v}_{k'}$ and angular velocities $\boldsymbol{\omega}_k$ and $\boldsymbol{\omega}_{k'}$, respectively. Their average radius is defined as $a = (a_k + a_{k'})/2$ and h denotes the inter-particle distance (surface to surface). An arbitrary relative motion between two particles can be decomposed in four elementary motions corresponding to normal displacement (subscript n), shear displacement (s), rolling (r) and twisting (t). For full details see [3]. Noting $\boldsymbol{\omega}_n = (\boldsymbol{v}_{k'} - \boldsymbol{v}_k) \times \mathbf{n}/(a_k + a_{k'} + h)$ the angular velocity of the local frame attached to the interacting pair, lubrication forces and torques are defined as follow:

$$\boldsymbol{F}_n^L = \frac{3}{2}\,\pi\,\eta\,\frac{a^2}{h}\,\mathbf{v}_n \tag{6}$$

$$\boldsymbol{F}_s^L = \frac{\pi\eta}{2}\left[-2a + (2a+h)\,ln\left(\frac{2a+h}{h}\right)\right]\,\mathbf{v}_t \tag{7}$$

$$\boldsymbol{C}_r^L = \pi\,\eta\,a^3\left(\frac{3}{2}\,ln\frac{a}{h} + \frac{63}{500}\,\frac{h}{a}\,ln\frac{a}{h}\right)\,[(\boldsymbol{\omega}_k - \boldsymbol{\omega}_{k'})\times\mathbf{n}] \tag{8}$$

$$\boldsymbol{C}_t^L = \pi\,\eta\,a^2\,\frac{h}{a}\,ln\frac{a}{h}\,[(\boldsymbol{\omega}_k - \boldsymbol{\omega}_{k'})\cdot\mathbf{n}]\,\mathbf{n} \tag{9}$$

where $\mathbf{v}_n = ((\mathbf{v}_{k'} - \mathbf{v}_k)\cdot\mathbf{n})\,\mathbf{n}$ is the normal relative velocity and $\mathbf{v}_t = (a_k(\boldsymbol{\omega}_k - \boldsymbol{\omega}_n) + a_{k'}(\boldsymbol{\omega}_{k'} - \boldsymbol{\omega}_n))\times\mathbf{n}$ is an objective expression of the tangential relative velocity. In this set of equations, the normal and shear forces, $\boldsymbol{F}_n$ and $\boldsymbol{F}_s$, are based on Frankel & Acrivos [9, 10] whereas C_r and C_t are based on Jeffrey & Onishi [11, 12]. The total lubrication force $\boldsymbol{F}_k^L$ (resp. $\boldsymbol{F}_{k'}^L$) applied by particle k' on particle k (resp. by particle k on particle k') and the total torque $\boldsymbol{C}_k^L$ (resp. $\boldsymbol{C}_{k'}^L$) applied by particle k' on particle k (resp. by particle k on particle k') relative to the particle center read:

$$\boldsymbol{F}_k^L = -\boldsymbol{F}_{k'}^L = \boldsymbol{F}_n + \boldsymbol{F}_s, \tag{10}$$

$$\boldsymbol{C}_k^L = (a_k + \frac{h}{2})\,\boldsymbol{F}_s + \boldsymbol{C}_r + \boldsymbol{C}_t, \tag{11}$$

$$\boldsymbol{C}_{k'}^L = (a_{k'} + \frac{h}{2})\,\boldsymbol{F}_s - \boldsymbol{C}_r - \boldsymbol{C}_t. \tag{12}$$

We account for the deformability of the particles near the contact region by combining the above normal lubrication model with the linear contact model via a Maxwell-type visco-elastic scheme ([3]. Lastly, we introduce a surface roughness such that the contact model and the lubrication model use slightly different values of h. The repulsive contact force appears even before $h = 0$ so that practically $h \leq 0$ never occurs in simulations.

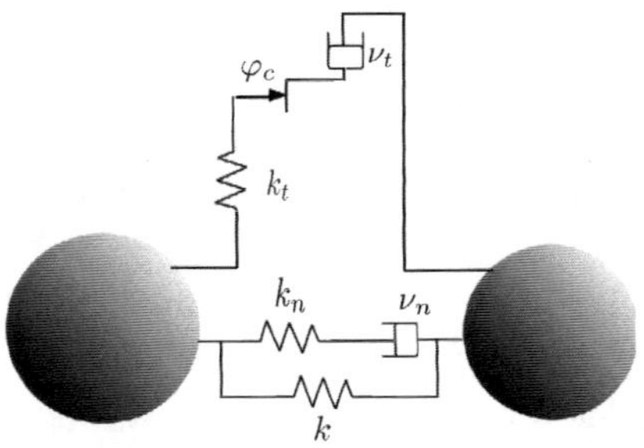

Figure 2. Visco-elastic scheme of the interaction between two elastic-like particles.

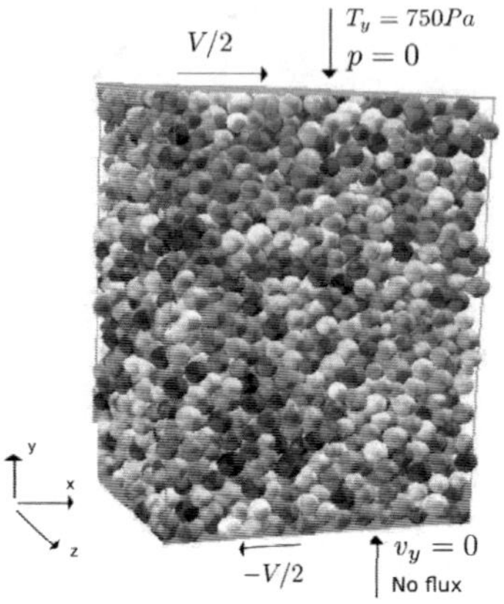

Figure 3. Simulation cell.

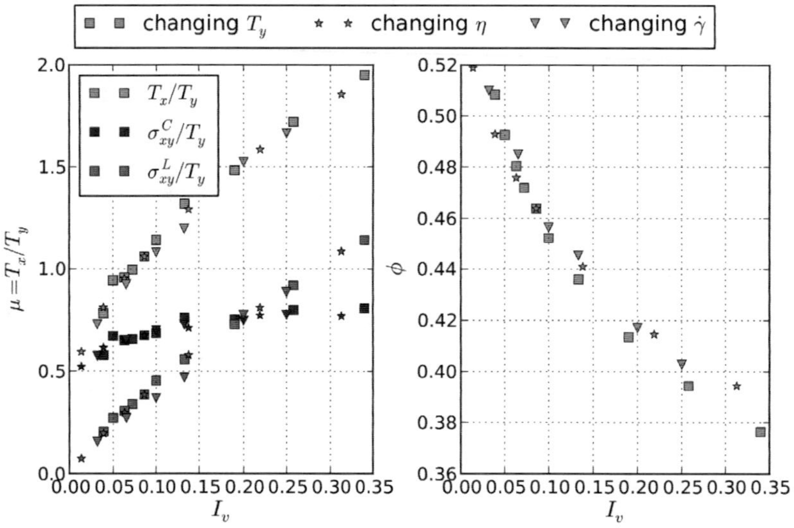

Figure 4. Normalized shear stress and solid fraction at steady state versus I_v. In each series the change of I_v is obtained by changing a different parameter: normal stress, viscosity, or shear rate.

3. Simple shear at imposed normal pressure

We simulate a saturated flowing material with bi-periodic boundary conditions, made of $N = 1000$ frictional spheres of average radius $a = 0.025 \pm 0.01$ m. The physical properties are roughness $\varepsilon = 0.035\,a$, density $\rho = 2500$ kg/m^3, normal contact stiffness $k_n/a = 5 \times 10^5$ Pa, shear stiffness $k_s = k_n/2$, and contact friction angle $\varphi = 30°$. There is no gravity. The numerical sample is first confined between two parallel plates then sheared by moving the top and the bottom plates at constant velocity $\pm V/2 = 1.5$ m/s. The boundary conditions for the top plate are the velocities $v_x = V/2$, $v_z = 0$, the total normal stress $T_y = 750$ Pa and the fluid pressure $p = 0$. At the bottom plate, $v_x = -V/2$, $v_z = 0$ and the fluid velocity along the y axis $v^f_y = 0$ (impermeable boundary). Periodic boundary conditions are defined along the horizontal axis for both the particles and the fluid. The external stress vector is $\boldsymbol{T} = \boldsymbol{F}/S$ where $\boldsymbol{F}$ is

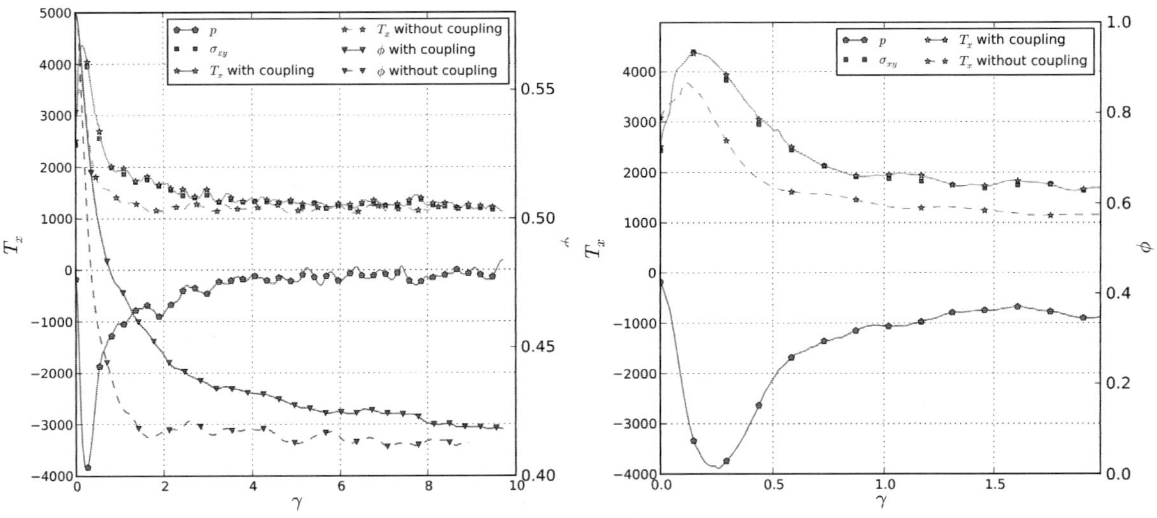

Figure 5. (left): The evolution of the shear stress and the solid fraction as a function of the deformation, with and without the poromechanical coupling. (right): Zoom on the transient regime.

the total force on the top plate and S is the horizontal cross sectional area. T_y is kept constant during the deformation, while T_x is a result of the imposed shear.

Figure 5 shows the evolution of the shear stress T_x, the pressure p and the solid fraction ϕ function of the deformation $\gamma(T) = \int_0^T \dot{\gamma}(t)dt$ where $\dot{\gamma}(t) = V/H(t)$ is the shear rate. The numerical results are presented for two cases: a first case where the poromechanical coupling (long range interactions) is included and another one where it is not (i.e. ignoring the last term in equation 5). In the early stage of deformation a transient regime is observed, reminiscent of the critical state theory. It is characterized by a peak stress, a decrease of the solid fraction and a negative pore pressure in the coupled case. This later effect entails a higher effective stress in the coupled problem, explaining why the shear stress reaches higher values. The system evolves toward a steady state for large deformations, in which the shear stress and the solid fraction are approximately constant and the pore pressure is nearly zero. The poromechanical coupling has no visible effect at steady state: the shear stress and the solid fraction reach similar values for both cases.

At steady state in non-inertial regimes the stress ratio $\mu = T_x/T_y$ and the solid fraction ϕ are entirely controlled by the viscous number I_v. I_v is a dimensionless form of the shear rate [13], reflecting the magnitude of viscous effects, and is defined as:

$$I_v = \frac{\eta\,|\dot{\gamma}|}{T_y}, \tag{13}$$

In other words, all possible combinations of confining pressure, fluid viscosity, and shear rate corresponding to a given value of I_v give the same result. It is easily confirmed by comparing simulations in which these control parameters are changed independently to produce different values of I_v (fig. 4).

Finally, a validation of our coupled model is obtained by comparing the apparent friction and solid fraction at steady state with the empirical equation of [13] (fig. 6, the empirical relation is plotted for convenience, it goes through data points not shown in the figure).

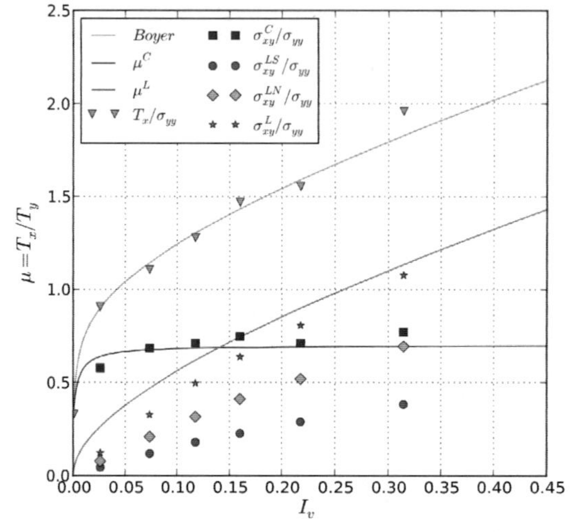

Figure 6. The stress ratio μ and the decomposition in contact stress, normal lubrication stress and shear lubrication stress. The solid line is the phenomenological law of [13].

4. Discussion

The previous section reveals a key feature of saturated geomaterials which has implications on debris flow: the shear stress and the solid fraction (or porosity) at large deformations are no longer uniquely defined by the confining stress as in critical state theory. They are both dependent on the shear rate. The effect is relatively well known as an experimental fact in the rheology of suspensions. It is attributed to the short range lubrication effects.

A consequence rarely discussed, however, is that the rate dependent volume changes are strongly coupled with long range poromechanical effects. In rheometer tests, the shear rate is usually applied for a sufficiently long time period so that the measurements are done at steady state. The transient regime is disregarded and the pore pressure is not recorded in most cases. For debris flow in natural conditions however, the shear rate is not constant since the the material is flowing down a slope where the slope angle is not constant (fig. 7). In such case the flowing material may never reach the steady state and it may be strongly influenced by the pore pressure changes associated to the poromechanical coupling.

Let us recall that the poromechanical coupling entails long range effects in the system and, ultimately, a dependency on the problem size (H in our case). It is known since Terzaghi that the characteristic time of such process scales with $\eta H^2/\kappa$ where κ is the intrinsic permeability. Since κ scales with a^2, the relaxation time of the transient regime is proportional to $\eta(H/a)^2$. A consequence is that the peak pore pressure and the peak stress in figure 5 scales with $(H/a)^2$. That is, they are much more significant in large scale in-situ conditions than in small scale lab tests. Arguably, the vision of non-colloidal solids suspended in water may not be always relevant to debris flow. Clay content can be reflected directly in our approach by considering an increased viscosity of the pore fluid. Possible non-Newtonian effects at the highest clay contents would need more developments, however. Either ways, more research is needed to validate the approach in practical cases.

Nevertheless, important consequences of the above findings for debris flow may be summarized as follow, most of these points need also further investigations:

- The bulk behavior of flowing geomaterials has a strong dependency on the shear rate and

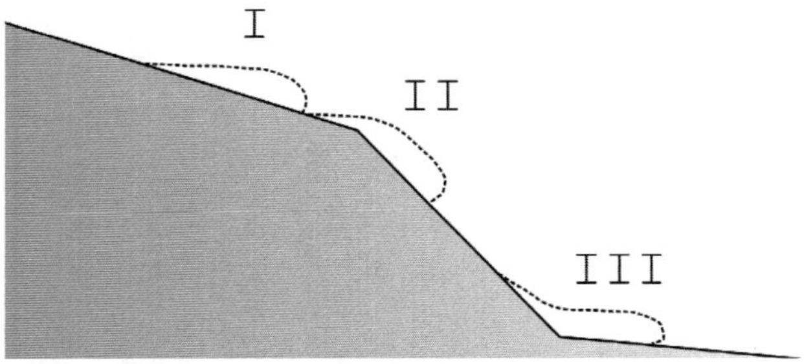

Figure 7. A typical scenario of debris flow: (I) steady state regime, pore pressure vanishes; (II) increasing slope angle, negative pore pressure and increased shear strength; (III) decreasing slope angle, excess pore pressure and transient liquefaction may accelerate the motion or increase the run-out distance.

pore pressure;

- a generalized critical state theory is possible by using the viscous number I_v, a normalized shear rate depending on the pore fluid viscosity;
- the changes of pore pressure are associated to the changes of slope angle;
- at a point of decreasing slope angle liquefaction is to be expected, associated with accelerated motion;
- when the slope angle is increasing negative pore pressure and high shear strength are expected, cavitation cannot be excluded;
- the effects are maximized for large problem sizes and low permeability

The changes of pore pressure result from a coupling and they have intricate consequences. Hence, a key conclusion of this discussion is that it is impossible to modelize a flowing geomaterials as an equivalent single-phase mixture. It is always necessary to solve a coupled problem. Promising steps in this direction can be found in [1], which provides an efficient computational techniques at the field scale. The micromechanical model we presented in this paper is complementary, in the sense that it hardly applies to large scale problems but it can provide insight into the rheology of the flowing materials and the coupling phenomena. It is freely available as part of Yade-DEM [14] and it can be used for further investigations.

References

[1] Pastor M, Blanc T, Haddad B, Drempetic V, Morles M, Dutto P, Stickle M, Mira P and Merodo J 2015 *Archives of Computational Methods in Engineering* **22** 67–104
[2] Cuomo S, Prime N, Iannone A, Dufour F, Cascini L and Darve F 2013 *Acta Geotechnica* **8** 125–136 ISSN 1861-1125
[3] Marzougui D, Chareyre B and Chauchat J 2015 *Granular Matter* DOI:10.1007/s10035-015-0560-6
[4] Cundall P and Strack O 1979 *Geotechnique* 4765
[5] Smilauer V and Chareyre B 2010 Yade DEM Formulation *Yade Documentation* ed Smilauer V (The Yade Project) 1st ed http://yade-dem.org/doc/
[6] Chareyre B, Cortis A, Catalano E and Barthélemy E 2012 *Transport in porous media* **94** 595–615
[7] Catalano E, Chareyre B and Barthélémy E 2014 *International Journal for Numerical and Analytical Methods in Geomechanics* **38** 51–71
[8] Tong A T, Catalano E and Chareyre B 2012 *Oil & Gas Science and Technology-Revue d'IFP Energies Nouvelles* **67** 743–752
[9] Frankel N and Acrivos A 1967 *Chemical Engineering Science* **22** 847853
[10] Van den Brule B and Jongschaap R 1991 *Journal of statistical physics* **62** 12251237

[11] Jeffrey D and Onishi Y 1984 *Zeitschrift für angewandte Mathematik und Physik ZAMP* **35** 634–641
[12] Jeffrey D and Onishi Y 1984 *Journal of Fluid Mechanics* **139** 261–290
[13] Boyer F, Guazzelli and Pouliquen O 2011 *Physical Review Letters* **107** 188301
[14] Smilauer V, Catalano E, Chareyre B, Dorofenko S, Duriez J, Gladky A, Kozicki J, Modenese C, Scholtès L, Sibille L, Stransky J and Thoeni K 2010 Yade Reference Documentation *Yade Documentation* ed Smilauer V (The Yade Project) 1st ed http://yade-dem.org/doc/

IOP Conf. Series: Earth and Environmental Science **26** (2015) 012024 doi:10.1088/1755-1315/26/1/012024

MPM dynamic simulation of a seismically induced sliding mass

C Moormann[1] **and F Hamad**
Stuttgart University, Pfaffenwaldring 35, 70569 Stuttgart, Germany

E-mail: christian.moormann@igs.uni-stuttgart.de

Abstract. In some geotechnical applications, material can undergo large displacement combined with excessive deformation; e.g. the sliding mass problem. Owing to the limitations of classical Lagrangian and Eulerian finite element methods to model these problems, the Material Point Method (MPM) has been developed about two decades ago to cope with the large deformation. In MPM, the continuum field is represented by Lagrangian material points (particles), which can move through a fixed background of a computational mesh. Therefore, it can be seen as a mesh-based method formulated in arbitrary Lagrangian–Eulerian description. Although MPM represents the continuum by material points, solution is performed on the computational mesh. Thus, imposing boundary conditions is not aligned with the material representation. In this paper, a non–zero kinematic condition is introduced where an additional set of particles is incorporated, which tracks the moving boundary by carrying the time–dependent boundary evolution. Furthermore, the material point method has been adopted to simulate the progressive failure of a sliding granular slope triggered by a seismic excitation. In order to represent the topographical bottom of the sliding mass, on which the seismic motion is applied, a rigid boundary is implemented by introducing an additional set of particles. A frictional contact algorithm is defined between the boundary and the descending mass, which allows sliding and rolling with friction. The traction due to contact is incorporated into the discretised momentum equation as an external force where the solution of this equation is performed separately for each body in contact. Defining the local coordinate system accurately in this algorithm is essential to avoid interpenetration. Thus, a two-dimensional triangular discretisation is utilised within the three-dimensional tetrahedral elements to track the surface progression of each body in contact. Complying with other continuum models findings performed on granular materials, the present model overpredicts the lateral deformations. Therefore, a local damping proportional to the out-of-balance nodal forces is included. In spite of the simple Mohr- Coulomb failure criteria being used, the results of the present numerical model are comparative to another continuum based model.

1. Introduction

The stability of slopes is one of the prominent problems in practical engineering. The major causes responsible for triggering a landslide includes heavy rainfall, imposed loads, strength degradation due to weathering and seismic excitation. The earthquake–induced landslides are among the most destructive slope movements, where the excessive failure might take place [11]. In some of the cases, it becomes inevitable to study the behaviour of the slopes even after the failure as the large scale movement over broad areas can result in serious damages and casualties. In these cases, the failures

[1] Corresponding author.

cannot be prevented, but it becomes a primary interest of research to know the extent and probability of an eventual slide to reduce the damage.
Numerical methods based on mesh deformation, e.g. Finite Element Method (FEM), have difficulties in modelling large deformations due to problems of mesh distortion and entanglement. In order to such type of large deformation problems, various methods have been proposed by coupling the two frame of references in a unified approach. Coupled Eulerian-Lagrangian (CEL) method is one of these methods, in which an updated Lagrangian finite element method (FEM) is coupled with the Eulerian description via interface models. The FE-software Abaqus utilised this feature and therefore it has been used in this paper. The Material Point Method (MPM), which is an innovative mesh-free particle method, performs the coupling procedure in an arbitrary form where the natural movement of the material is traced. MPM has been applied for many geotechnical applications involving large deformation of collapsing slopes and landslides (see for example [1,6]). In this paper, a strain-smoothening techniques based on nodal mixed discretisation has been implemented in MPM to relax the mesh locking problem, whereas the frictional contact algorithm is extended so that a prescribed velocity can be assigned directly to one of the two bodies in contact. Moreover in this paper, the two methods (MPM and CEL) have been applied to a landslide progression failure occurred during the 1999 Chi-Chi earthquake of Taiwan. To show the potential of applying the two continuum based models, some results are presented.

2. Coupled Eulerian-Lagrangian (CEL) method

Coupled Eulerian-Lagrangian (CEL) method is numerical framework to combine Lagrangian (has relatively small deformation) with Eulerian material (fluid behaviour) combined together via the interface of the Lagrangian object. By using this method, the Eulerian material is tracked as it flows through the mesh by computing its Eulerian Volume Fraction (EVF). Each Eulerian material is designated a percentage, which represents the portion of that element filled with a material. The coupling procedure between the two material descriptions inside CEL allows simulating the rock material as a flow running over the rigid base. The Eulerian mesh, where the sliding mass flows with large deformations, has no problems regarding mesh and element distortion as the Eulerian mesh remains fixed spatially.

2.1. Time integration

The CEL method is implemented in Abaqus/Explicit using explicit time integration [4]. The discretised equation of motion can be written in the form

$$\ddot{a} = M^{-1} \cdot \left(F^{ext} - F^{int}\right) \tag{1}$$

where, $\mathbf{M}$ is the mass matrix, $\ddot{\mathbf{a}}$ is the nodal acceleration vector with the superposed dot is a material derivative of the nodal displacement $\mathbf{a}$ with respect to time, $\mathbf{F}^{ext}$ and $\mathbf{F}^{int}$ are the external and internal nodal force vectors, respectively. Equation (1) is integrated explicitly using central difference integration as follows

$$\dot{a}^{(i+\frac{1}{2})} = \dot{a}^{(i-\frac{1}{2})} + \frac{\Delta t^{(i+1)} + \Delta t^{(i)}}{2} \cdot \ddot{a}^{(i)} \tag{2}$$

$$a^{(i+1)} = a^{(i)} + \Delta t^{(i+1)} \cdot \dot{a}^{(i+\frac{1}{2})} \tag{3}$$

where $\dot{\mathbf{a}}$ is velocity and Δt is the time increment. The superscript (i) refers to the increment number while (i-1/2) and (i+1/2) points to mid increment values. The central difference integration operator is explicit in that the kinematic state can be advanced using known values from previous increment. Explicit integration is conditionally stable, and is bounded with the following limits

$$\Delta t \leq \frac{2}{\omega_{max}} \tag{4}$$

in which ω_{max} is the maximum frequency of the system. Abaqus/Explicit uses an adaptive algorithm to determine conservative bounds for the highest element frequency. An estimate of the highest eigenvalue in the system can be obtained by determining the maximum element dilatational mode of the mesh. The stability limit based upon this highest element frequency is conservative in that it will give a smaller stable time increment than the true stability limit that is based upon the maximum frequency of the entire model. Abaqus/Explicit contains a global estimation algorithm, which determines the maximum frequency of the entire model. At the beginning of the analysis, the program evaluates the time step size based on element by element estimation. As the step proceeds, the stability limit will be determined from the global estimator once the algorithm determines that the accuracy of the global estimation is acceptable.

3. Material Point Method (MPM)

In MPM, the continuum body is discretised by Lagrangian material points. The momentum equation is solved on the background Eulerian mesh which provides a convenient means of calculating discrete derivatives.

3.1. Time integration

Euler-forward time integration scheme is adopted in MPM to solve Equation (1), which yields

$$\dot{\mathbf{a}}^{t+\Delta t} = \dot{\mathbf{a}}^{t} + \Delta t \ddot{\mathbf{a}}^{t}, \quad \ddot{\mathbf{a}}^{t} = [\mathbf{M}_{l}^{t}]^{-1}\mathbf{F}^{t} \tag{6}$$

where t is the current time increment, $\dot{\mathbf{a}}^{t}$ and $\dot{\mathbf{a}}^{t+ t}$ are the nodal velocities at time t and t+ t, respectively. The incremental nodal displacement is obtained by integrating the nodal velocity by the Euler-backward rule (e.g. Sulsky et al. [12], Wieckowski [14]).

$$\Delta\mathbf{a}^{t+\Delta t} = \Delta t \, \dot{\mathbf{a}}^{t+\Delta t} \tag{7}$$

The positions of the particles are subsequently updated from

$$\mathbf{x}_p{}^{t+\Delta t} = \mathbf{x}_p{}^{t} + N_p \Delta\mathbf{a}^{t+\Delta t} \tag{8}$$

in which $\mathbf{x}_p^{t}$ and $\mathbf{x}_p^{t+\Delta t}$ are the particle positions at time t and (t + t) respectively. After getting the grid node velocities, the strain increment of a material point p is calculated and based on the calculated strain increment, the stress increment at each material point is updated by applying the appropriate constitutive model. The constitutive models are applied at material points and this allows easy evaluation and tracking of history-dependent variables. At the end of time (t + t) all the material point variables are updated and a new cycle is begun.

3.2. Contact algorithm

The contact defined between the soil and rigid walls is a frictional contact. The contact algorithm proposed by Bardenhagen et al. [2] is used in this paper, which can be seen as a predictor-corrector scheme formulated in explicit manner, in which the velocity is predicted from the solution of each body separately and then corrected using the velocity of the coupled bodies following Coulomb friction. This solution scheme uses the concept of comparing the single and combined body velocities for a contact node and defining its behaviour accordingly. The algorithm is able to detect whether two bodies are approaching or separating from one another. In the case of approaching, the contact behaves according to the coulomb friction law. If the two bodies are separating, the contact algorithm allows free separation where each one moves according to its own equation of motion. Following this procedure, the bodies in contact would feel each other as soon as they contribute to the same computational grid node.

For the considered problem of the sliding mass, the bottom is defined as a rigid object being in contact with the sliding deformable mass. The calculation of the interface orientation is performed be tracking the configuration of the rigid body, then the deformable body is updated consequently so that the interpenetration of the two objects is prevented. As mentioned earlier, there is no need to specify the contact surface a priori as the algorithm detect it automatically, which is valid for explicit integration schemes where the time step size is rather small. For the bottom object, a prescribed velocity is assigned to the material points; therefore, Equation (1) is carried out only for the deformable object.

3.3. Prescribed velocity

In traditional dynamic FEM, the prescribed velocity is defined over nodes. These nodes always define part of the Lagrangian body boundary. On the other hand, in MPM the continuum is defined by Lagrangian particles which might change position from one element to another. Hence, there is no interface surface where prescribed velocity can be applied. Within the framework of MPM, prescribed velocity can be applied directly on the material points of a rigid body representing the moving boundary for simple one-dimensional problem. The active boundary nodes must be tracked. For applications with axial movement, part of the mesh can be displaced as a moving mesh having a prescribed value while the rest is stretched uniformly [9]. This becomes very complicated and inconvenient when applied to applications like imposing seismic motion for a slope problem. As an alternative, an additional set of particles is introduced which tracks the moving boundary by carrying the time-dependent boundary evolution [5]. At the beginning of a time step, a prescribed velocity is assigned to the boundary particle. Next, the prescribed velocity is mapped via the shape functions from the prescribed particles to the computational nodes, where the discrete equations are solved. Following the same methodology in this paper, prescribed velocities are applied as a boundary condition for the rigid walls and a contact is defined between the rigid walls and soil.

4. Earthquake induced landslide debris flow

The dynamic process of landslides induced by earthquakes is very complex in its nature. Various numerical methods are used to simulate different parts of the whole process of evolution of a landslide starting from the development of a critical sliding surface to initiation and triggering of failure to disintegration of the sliding mass to debris flow and finally deposition. Here, CEL and MPM are used to simulate the last part of the process i.e. progression of the sliding mass and deposition. For this, the Chiu-fen-erh-shan landslide of 1999 is used as a case study.

4.1. The Chiu-fen-erh-shan landslide

The Chiu-fen-erh-shan landslide is one of the major landslides caused by the disastrous M_w 7.6 Chi-Chi earthquake of 1999 in Taiwan. The landslide has been characterised as a translational rock-block slide on dip slope and known to be disintegrated into fragmented rock avalanche travelling long distance at high velocity [7]. The landslide debris travelled a distance of more than 1 km and covered an area of 1.95 km^2. Many studies have been conducted since then to understand the mechanism and movement of the landslide [8, 7, 15]. These studies concluded that the slope had undergone gravitational creep and was unstable even prior to the earthquake. A slip surface had been developed as a result of flexural folding. Observations by Wang et al [13] suggested the presence of clay seams between the alternating beds of shale and sandstone that composed the slope. The existence of the clay seam provides a very smooth surface for the slide to occur. The slope was retained in its position because of a sandstone bed that formed resistant ridges at the foot of the slope. The sandstone bed was probably damaged seriously during the earthquake giving way to the massive landslide. The topography and horizontal component of the velocity are depicted in Figure 1 [3].

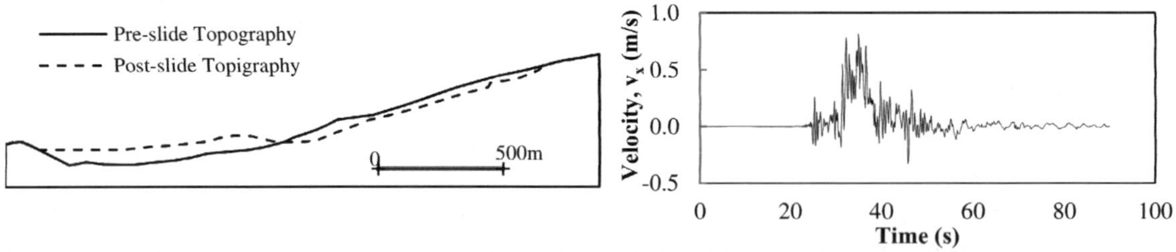

Figure 1. Chi-Chi landslide: (left) topography and (right) time history of the horizontal velocity

4.2. Numerical analyses

In this section, both CEL and MPM analyses are presented considering restrictions in each model. In CEL, the base block is assumed as a Lagrangian rigid body while the sliding mass is considered as an Eulerian material. The sliding mass, which is comprised, of a highly jointed rock mass is modelled as a continuum with a density of 2550 kg/m^3, Young's Modulus of 500 MPa and a Poisson's ratio of 0.19. The Mohr-Coulomb failure criterion is used in the analysis. The strength behavior of the sliding mass is mainly ruled by the strength parameters of the rock joints, hence, the strength parameters of joints are directly used to define the shear strength of the continuum. The values adopted for cohesion and friction are 20 kPa and 24° respectively [3]. The simulation is performed in two steps. In the first step, the quasi-static solution for the gravity stresses is obtained, while the seismic excitation is given in the second step to the base block. The numerical model of the slope is meshed with 3-dimensional, linear eight noded elements with reduced integration. The contact between the Lagrangian and Eulerian domain of the model is implemented by the general contact algorithm, which enforces the use of penalty contact method.

The same problem above has been repeated in MPM, where a three-dimensional program using tetrahedral elements has been used. Therefore, the plane-strain condition of the present two-dimensional problem is imposed using one element in depth while the out-of-plane boundary condition is restricted. To relax the numerical problem of the mesh locking associated with the low-order element, strain-smoothening technique based on nodal mixed discretisation has been integrated. Prescribed velocity is imposed to the particles of the sliding surface as a rigid block. Numerical damping has been added to the sliding material to imitate the material damping. The progression of the flow resulting from the landslide simulated by the two methods is depicted in Figure 2.

For the MPM model, an artificial damping has been added to approach the measurements (Figure 1). Therefore, the MPM deformation in Figure (2) shows the final steady-state whereas the sliding object in the Abaqus model flows further beyond the edge of the slope.

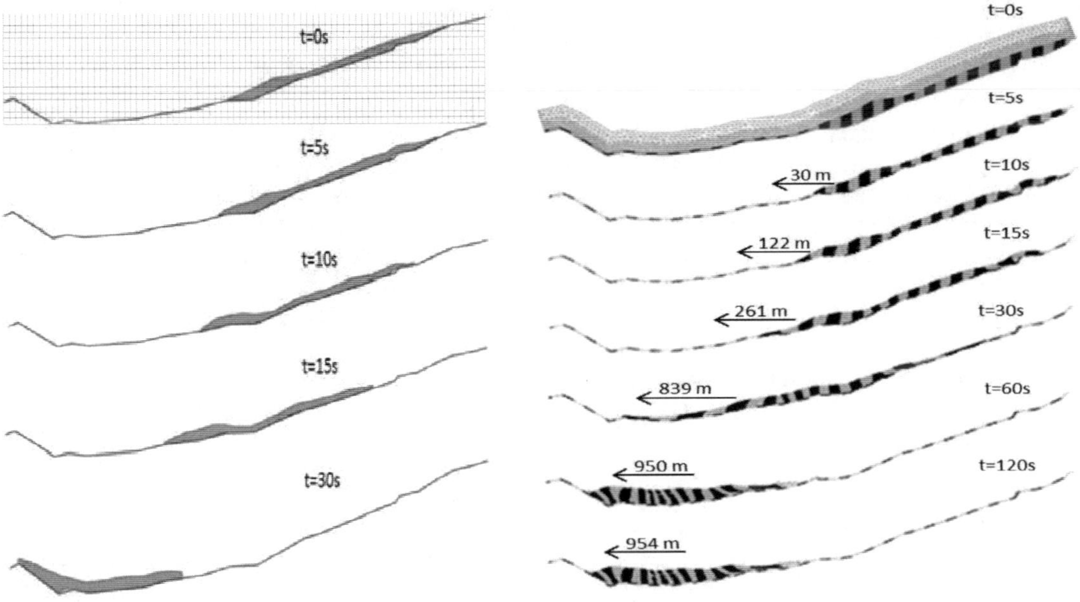

Figure 2. Progression of landslide: (left) CEL (right) MPM

5. Conclusion

The potential of simulating an important geotechnical application has been presented in this paper using CEL and MPM. Owing to the fact that both methods are based on continuum modelling, they overestimate the lateral movement of the sliding material when a simple constitutive model is used, (as the Mohr-Coulomb one implemented in this paper). Therefore, implementing more advanced constitutive model is essential to get better prediction or to add a numerical damping as an alternative. However, the amount of the required damping is difficult to evaluate. For the MPM model, the artificial damping has been shown to be efficient to predict the final configuration beyond failure, whereas the Rayleigh damping integrated in Abaqus was not sufficient to hold the sliding mass (see Figure (2)). Remembering that CEL treats the sliding material in Eulerian manner, therefore, tracking the local displacement is not possible. On the other hand, the Lagrangian nature of MPM allows obtaining the incremental displacement in more natural fashion (see the layering representation in Figure (2)). In order to proceed further with the continuum modelling of rock slopes, more advanced constitutive models are required, as well as updating the contact evolution basing on the mobilised resistance rather than the constant assumption in the current models.

6. Acknowledgment

The authors would also like to acknowledge "Deltares, The Netherlands" for providing access to their MPM source code, which was further developed in this paper.

References

[1] Andersen, S. & Andersen, L. 2010. Modelling of landslides with the material point method. *Computational Geosciences*, **14**(1):137-147.

[2] Bardenhagen, S.G., Brackbill, J.U. & Sulsky, D.L. 2000. The material point method for granular materials. *Computer methods in applied mechanics and engineering*, **187**(3-4):539-541.

[3] Chen, H. 2000. Engineering geological characteristics of Taiwan landslides. *Sino-Geotechnics*, **79**:59-70.

[4] Dassault Systems, ABAQUS. 2008. Version **6.8** EF documentation.

[5] Hamad, F.M., Vermeer, P.A, & Moormann, C. 2013. Development of a coupled FEM-MPM approach to model a 3D membrane with an application of releasing geocontainer from barge. In: *Int. Conf. Installation Effects in Geotechnical Engineering*, 176-183: Rotterdam, The Netherlands.

[6] Hamad, F.M., Vermeer, P.A. & Moormann, C. 2013. Failure of a geotextile-reinforced embankment using the material point method. In: Proc. 3rd Int. Conf. particle-based methods- Fundamentals and Applications, 498-509: Stuttgart, Germany.

[7] Huang, C-C., Lee, Y-H., Liu, H-P., Feefer, D.K., Jibson, R.W. 2001. Influence of surface-normal ground acceleration on the initiation of the Jih-feng-erh-shan landslide during the 1999 Chi-Chi, Taiwan earthquake. *Bull Seismol Soc Am*, **91**:953-958.

[8] Hung, J-J. 2000. Chi-Chi earthquake induced landslides in Taiwan. *Earthquake Eng Seismol,* **2**:25-33.

[9] Jassim, I.J., Hamad, F.M. & Vermeer, P.A. 27-29 April 2011. Dynamic material point method with applications in geomechanics. In: Proc. 2nd Int. *Symposium on Computational Geomechanics (COMGEO II)*, Cavtat-Dubrovnik, Croatia.

[11] Omidvar, B., Mousavi, S.M., Ghazban. F., Feyzi, R. 2011. Quantitative risk analysis for earthquake induced landslides- Emamzadeh Ali, Iran. *Engineering Geology*, **122** (2011) 191-203.

[12] Sulsky, D., Zhou, S.J. & Schreyer, H.L. 1995. Application of a particle-in-cell method to soil mechanics, **87**(1):236-252.

[13] Wang, W-N., Chigira, M., Furuya, T. 2003. Geological and geomorphological precursors of the Chiu-fen-erh-shan landslide triggered by the Chi-Chi earthquake in central Taiwan. *Engineering Geology*, **69**:1-13.

[14] Wieckowski, Z., Youn, S.K. & Yeon, J.H. 1999. A particle-in-cell solution to the silo discharging problem. *Int. J. Numerical Methods in Engineering*, **45**(9):1203-1226.

[15] Wu, J-H., Lin, J-S,, Chen, C-S. 2009. Dynamic discrete analysis of an earthquake-induced large-scale landslide. *Int J Rock Mech Mining Sci*, **46**:397-407.

IOP Conf. Series: Earth and Environmental Science **26** (2015) 012036 doi:10.1088/1755-1315/26/1/012036

Impact of dry granular masses on rigid barriers

F Calvetti[1], C di Prisco[1] and E Vairaktaris[2,3]

[1]Civil and Environmental Department, Politecnico di Milano, Italy
[2] School of Applied Mathematics and Physical Sciences, NTU of Athens, Greece

E-mail: mvairak@mail.ntua.gr

Abstract. This work concerns the impact of dry granular masses on rigid artificial obstacles. The authors approached the problem by performing an extensive campaign of numerical analyses with a commercial code based on the discrete element theory. The standard approaches employed to design sheltering structures are exclusively based on the assessment of the Maximum Impact Force (MIF) exerted by the soil mass on the obstacle, and the sheltering structure is usually designed according to simplified pseudo-static approaches. In a previous paper the authors considered the dependence of MIF on the Froude number and on a large series of both geometrical and mechanical parameters. Indeed, the impulsive nature of the force exerted by the soil onto the structure has to be considered in order to optimize the design of this type of structures. For this reason in this paper the evolution with time of the impact force and the mechanics of the phenomenon are investigated.

1. Introduction

Among the several aspects that have to be considered for landslide risk assessment, the work developed in the last years by the Authors is mainly focussed on the comprehension/modelling of the mechanical processes developing when the landslide interacts with either civil structures or protection works. At present, standard approaches employed to design sheltering structures are exclusively based on the assessment of the Maximum Impact Force (MIF) exerted by the soil mass on the obstacle, since the sheltering structure is usually designed according to simple pseudo-static approaches.

Existing classical approaches employed to design sheltering structures are inspired to:

(i) impacts of fluid masses onto rigid barriers [1], which can be subdivided in hydrostatic (HS) and hydrodynamic (HD);
(ii) impacts of boulders (BI) on elastic walls.

According to HS models ([2],[3], [4]), the MIF value is independent of the soil mass velocity and the pressure is linearly dependent on depth; according to HD models ([3], [4], [5], [6], [7]) the pressure is assumed to be constant with depth and dependent on the average mass velocity. On the other side, the most common BI models ([4], [5], [6], [8], [9]) consider elastic impacts and use the non linear Hertz's equation for the MIF evaluation. In general, hydraulic models provide lower bounds for the MIF, whereas solid impact models estimate upper bound values ([9], [10]).

The literature review clearly shows that numerical codes are not yet very commonly used for this kind of problems. Most of the works approaching this topic by employing numerical codes aim to simulate specific geometrical conditions/cases ([11], [12]). Nevertheless, with the exception of a few

[3] Corresponding author

contributions ([13], [14]), the role played by both geometrical and mechanical factors is not yet investigated in detail.

To analyse the impact of granular masses, the authors decided to employ a DEM code in which the soil mass is simulated as an assembly of rigid spheres with deformable contacts, and the initial conditions for the system are imposed at the instant of time just preceding the impact itself. In a previous work [15], the authors considered the effect of the energetic content, mean velocity, length of the flowing mass (ℓ), flow height (h), front inclination (α), inter-particle (f_c) and ground friction coefficients, mass porosity on the MIF value and they observed that:

1. the total mass impacting on the obstacle is not a representative factor influencing the MIF value. In contrast, the dominant factors are the flow height and the sliding mass porosity.
2. The quantitative dependence of the MIF value on the mean mass velocity seems to depend on the inter-particle friction coefficient,
3. The role of the inter-particle friction angle seems to become irrelevant for large porosities.
4. The inclination of the front plays a dominant role; the largest values of MIF are obtained in case of angles close to 90°.

Finally, in [15] the authors have shown that the Froude number is the main variable to be considered for the assessment of the MIF value and introduced a formula according to which the MIF value can be calculated as a function of all the geometrical and mechanical parameters cited above.

In the current work, in contrast to the previous reference, the authors try to qualitatively describe, from a micro-structural point of view, the mechanical process. The paper is thus organized as it follows. A brief description of the numerical model is given in section 2. For the sake of both clarity and brevity within the enormous number of numerical simulation results obtained by the authors, here below a reference numerical test is only taken into account, since it was considered to be representative. In section 3 the evolution of both force chains and velocity vectors within the soil mass is illustrated. To highlight the mechanical process taking place within the soil mass, three additional numerical tests characterised by different values of the inter-particle friction angle and front inclination are also discussed.

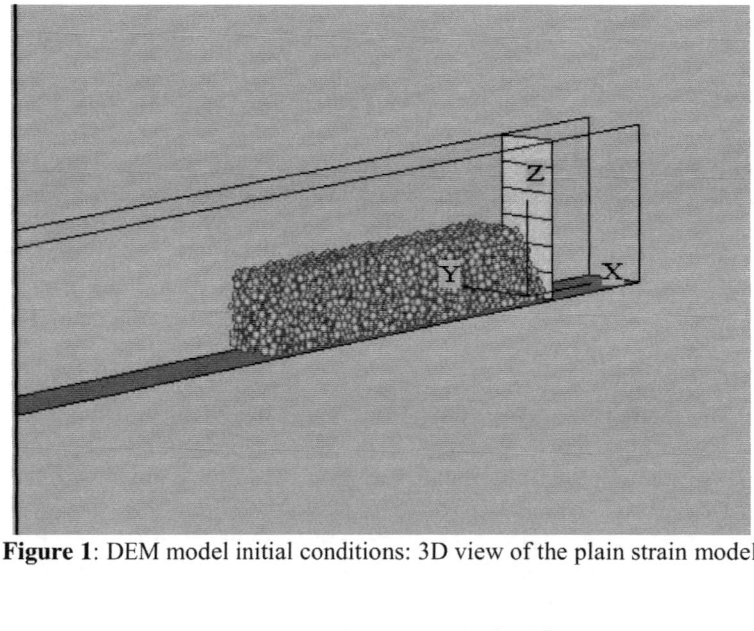

Figure 1: DEM model initial conditions: 3D view of the plain strain model

2. DEM model

The numerical simulations have been performed by employing the PFC 3D code ([16], [17]). For the sake of simplicity, plane strain conditions were externally imposed: the flow width is constant and, to avoid any dependence of numerical results on it, is equal to eight times the average grain diameter, whereas the friction among grains and side walls was taken equal to 0. The problem geometry taken into consideration is illustrated in Figure 1, where both the rigid barrier and the granular assembly of mass m and initial volume V, are represented. As was previously mentioned, the initial conditions (velocity) for the granular mass are assigned at the instant of time just preceding the impact.

Following the approach proposed in [18] for the quantitative modelling of the soil behaviour, particles are not allowed to rotate and a linear contact stiffness model, characterized by two parameters (k_n and k_s – Table 2) is chosen. Other micro-mechanical parameters that have to be assigned for the system to be modelled are: the inter-particle friction coefficient f_c, the friction coefficient between the grains and the ground and between the grains and the obstacle. The value of f_c=0.3 for the friction among particles was chosen as a typical value for sands/gravels [18], whereas for the grain – wall impact, by assuming an embankment being the "virtual" obstacle, a greater value was chosen (0.6).

In Table 1 the data concerning the initial conditions imposed for the reference test as well as for the model geometry are listed. In Table 2 the micro-mechanical properties are collected.

Table 1: Reference test: Initial conditions and model data

Number of grains	4201
Average grain diameter $\bar{D}$ (m)	0.30
Ratio between the largest and the smallest radius	2.4
Total Volume of grains, V, (m3)	59.4
Length of the grain assembly, ℓ (m)	15
Width of the grain assembly, b (m)	8 $\bar{D}$
Height of the grain assembly h (m)	3
Particle unit weight γ_s (kN/m3)	26
Porosity, *por*, (-)	0.45
Total Mass m (tons)	154
Initial Average Velocity v_0 (m/sec)	8.73
Initial Kinetic Energy, E_{kin}, (MJ)	5.9
Wall height (m)	6
Front inclination, α, [degrees]	60

Table 2: Reference Micromechanical Properties

$k_n/\bar{D}$ (MPa)	300
$k_s/\bar{D}$ (MPa)	75
Surface friction (ball-ball (f_c); ball-wall; ball-ground; ball-side wall)	0.3; 0.6; 0.3; 0.0

3. Discussion of the numerical results

To clarify the mechanical processes taking place during the impact of the soil mass onto the vertical obstacle, in this sub-section, some phenomenological information are discussed, not only with reference to macro-scopic but even to micro-scopic variables. In particular, in Figure 2 the evolution of the impact force with time for the reference test is illustrated. The dynamic nature of the process is revealed by the presence of numerous peaks until the attainment of the residual force value. The process is also very rapid; quasi-static conditions are reached in about 1 second.

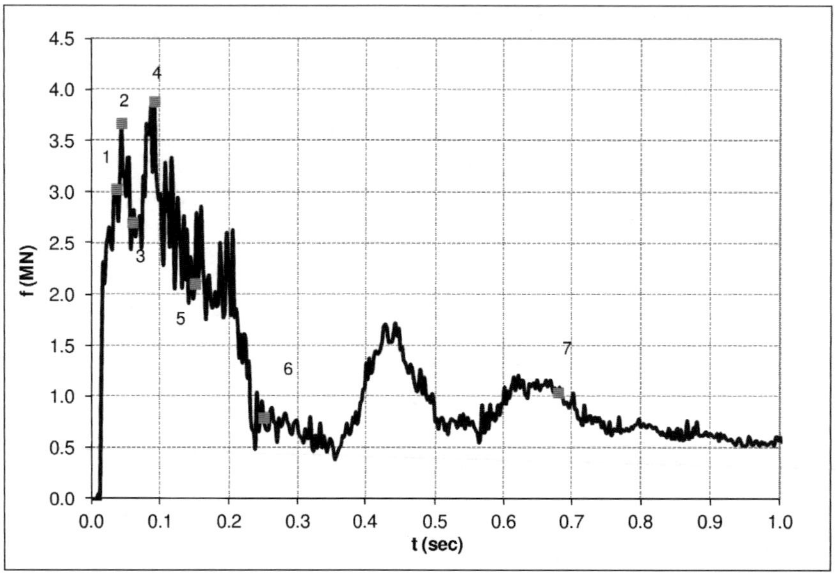

Figure 2: Evolution of the impact force with time for the reference test (Table 1)

In addition, in Figure 3 and Figure 4 the micro-mechanical response of the system is presented by showing the contact forces transmitted by the particles (left) and the velocity vectors (right). Each figure refers to the time instant i (with 1< i <7) of Figure 2. The evolution of the geometry of the soil mass is clearly shown: a marked increase of the effective impact height *H* seems to take place only after point 4 and the maximum height *H* is recorded for i = 7, when the energetic content of the soil mass has already almost totally dissipated. At the peak (point 4), the current *H* value is approximately half of *h,* i.e. the initial flow height of the soil mass.

The intensity of the inter-particle forces (force chains) is normalized with respect to 1 MN in all Figures. This implies that after time instant 4, the forces transmitted by the particles to the obstacle reduce progressively: in fact, owing to the initial geometry of the soil mass, the force concentration is maximum at the initial time instant.

From Figure 3 (d-f) and 4(e-h) it seems evident that in the sub-domain above the red solid lines velocity is characterised by values quite larger than those related to the soil particles positioned below that line. This implies that, below the red line, the soil can be assumed to behave under quasi-static conditions. In contrast, if we take into consideration Figure 3 (a-c) and Figure 4 (a-d), we derive that another frontier can be drawn: this is separating the material behaving like a sort of "granular fluid" from the other part of the mass. In fact, above the light blue straight lines the force chains seem are negligible. In particular, in Figure 3a-c and in Figure 4a,b the forces among particles are negligible in the zone above the light blue line, since that part of the soil mass seems not to be involved in the impact process. In Figure 4c and 4d, the soil mass inside the zone above the light blue line is fluidized, grain velocities are changing direction (most of them are pointing upwards as is evidenced from their relative movement; see also Figures 6, 8) and its agitation is quite relevant: energy is mainly dissipated for impacts and long lasting contacts among particles are almost absent.

The final condition, as is clearly suggested in Figure 4h, is characterized by only two sub-domains, since the blue and the red line are coincident: the material is there either approximately still, below both the drawn lines, or under fluidized conditions, above those lines.

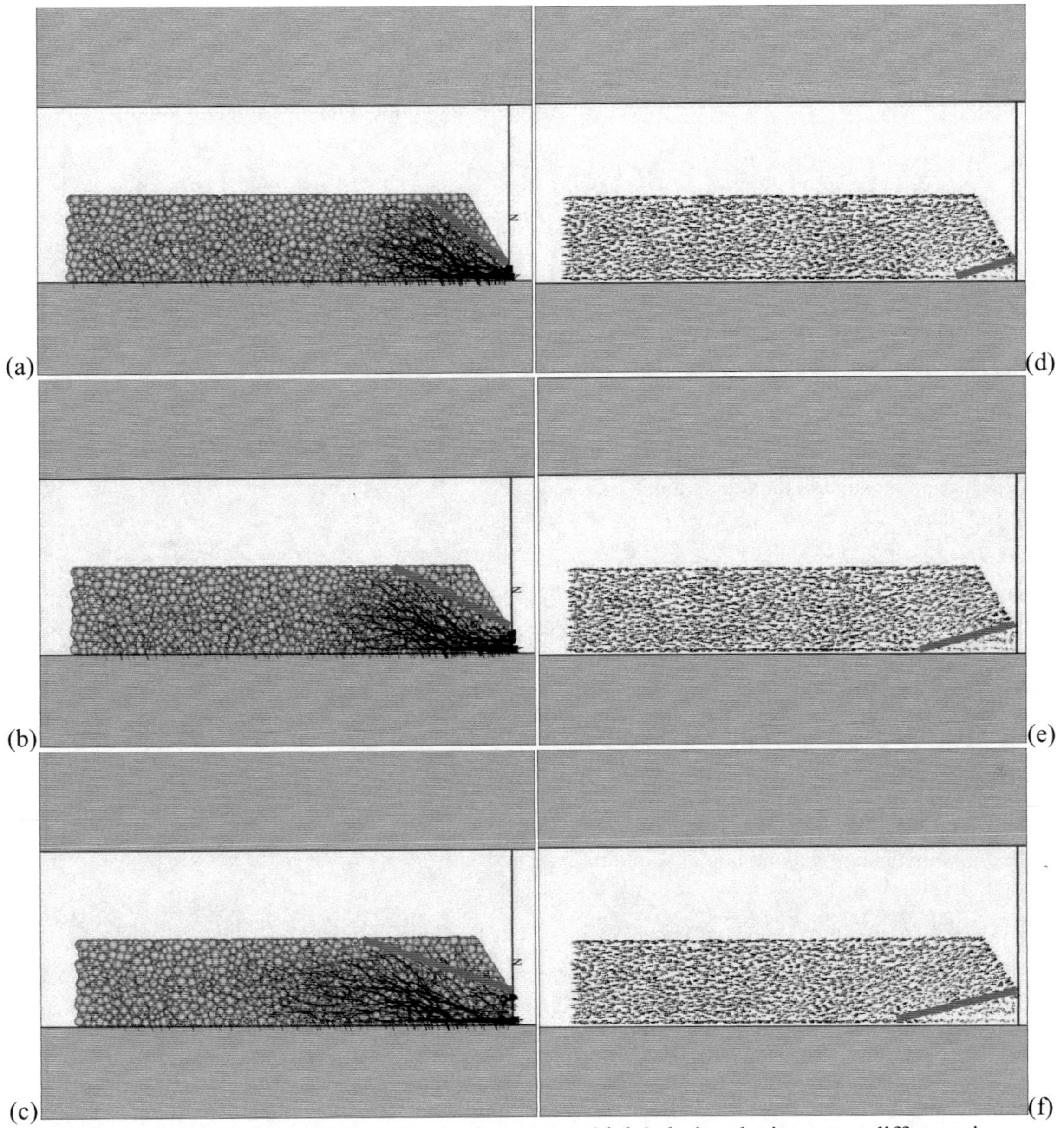

Figure 3: Force chains (left) and velocity vectors (right) during the impact at different time instants: point 1 of Figure 2 (a) and (d), point 2 (b) and (e), point 3 (c) and f).

By summarising, we can state that the impact mechanical process can be described by following the evolution of both the light blue and the red lines, since they subdivide the soil mass in three sub-domains: zone A, in which the soil mass is approximately still (mean velocity is negligible), zone B, in which the material is fluidized and impacts among grains dissipate most of the energy and zone C, which includes the rest of the material; inside this zone, a part of the material behaves like a solid and force chains dominate the mechanical response of the material, whereas some other part remains unaltered with respect to the initial conditions, i.e. this part has not yet participated in the impact.

Figure 4: Force chains (left) and velocity vectors (right) at different time instants: point 4 of Figure 2 (a) and (e), point 5 (b) and (f), point 6 (c) and (g), point 7 (d) and (h).

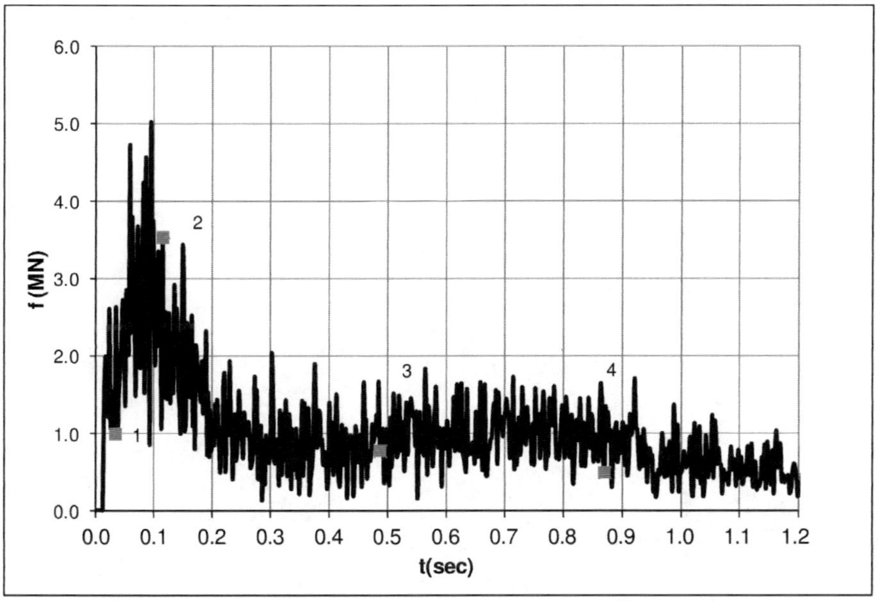

Figure 5: Evolution of the impact force with time in case $f_c = 0$ and $\alpha = 60°$ (all the other parameters are those of Table 1).

Figure 6: Force chains (black lines) within the soil mass during the impact at different time instants: point 1, 2, 3, 4 of Figure 5 correspond to (a), (b) (c) and (d) respectively.

To emphasize the role played by the parameters f_c and α in influencing the mechanical response of the soil mass, in Figures 5-10, analogous results concerning contact forces only (due to space limitations) are shown. Figures 5 and 6 present the tests with $f_c = 0$ and $\alpha = 60°$, while Figures 7 and 8 refer to the micromechanical evolution of the test with $f_c = 0$ and $\alpha = 90°$. Finally the results of the test concerning $f_c = 1$ and $\alpha = 90°$ are also shown in Figure 9 and Figure 10.

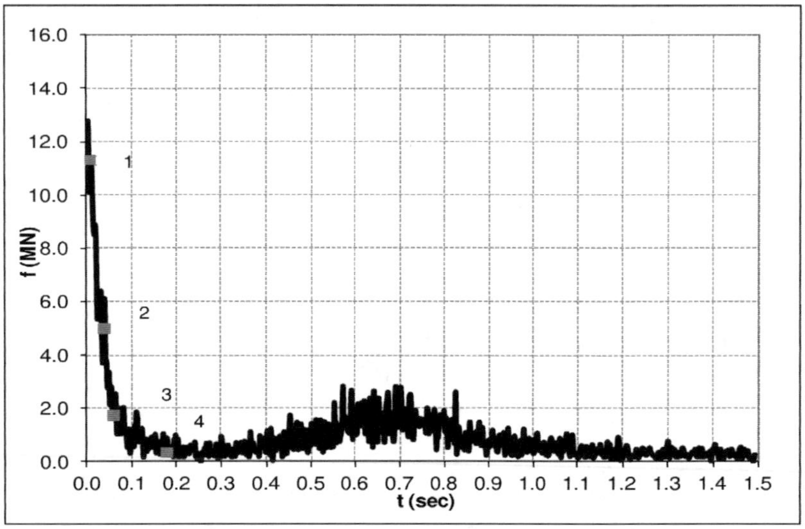

Figure 7: Evolution of the impact force with time in case $f_c = 0$ and $\alpha = 90°$ (all the other parameters are those of Table 1).

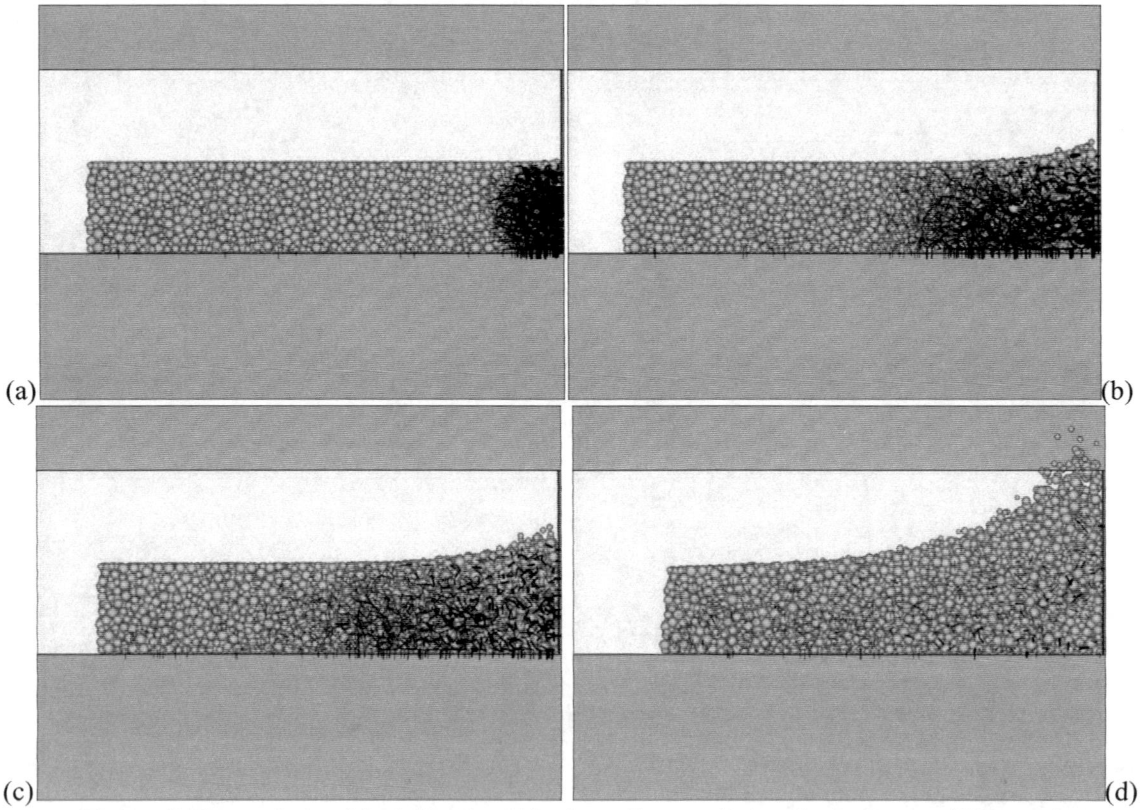

Figure 8: Force chains (black lines) within the soil mass during the impact at different time instants: point 1 of Figure 9 (a), point 2 (b), point 3 (c) and point 4 (d)).

For f_c = 0 and $\alpha = 60°$, (Figures 5, 6) the evolution of the impact force with time is more irregular than in the reference case. Moreover, part of the mass overtakes the obstacle, and force chains during the impact are practically absent and the mass appears to be totally fluidized (Figure 6). By changing only the front inclination, i.e. with f_c = 0 and $\alpha = 90°$, (Figures 7, 8) the force – time graph shows that the soil mass behaviour is more solid-like, initially; in fact the MIF is significantly increased and it develops very fast; then a force wave propagation starts spreading backwards, although it is vanishing very rapidly. After some time instants a fully fluidised behaviour is developed inside the whole mass, similar to the behaviour observed in a previous test (Figure 6c,d compared to Figure 8d).

In contrast, when f_c=1 and $\alpha = 90°$, the material seems to behave like a solid throughout the test: (a) the MIF is larger than the value corresponding to α=60° and (b) it is attained at a time instant very close to t=0, that is immediately after the impact (Figure 9). In this test the force propagation wave is more evident: in Figure 10b the force wave is travelling backward and has reached the backward boundary of the soil mass, in Figure 10c and d the force wave is travelling towards the barrier, until its nullification that is evident in Figure 10e (when impact force goes to zero). Again, in the most superficial part of the soil mass the force chains seem to be absent and the velocity fluctuations more pronounced.

By comparing the two last tests we can conclude that the front inclination and the inter-particle friction are influencing very much the overall behaviour of the soil mass. Nevertheless, the parameter which dominates the solid/fluid like behaviour of the mass is the inter-particle friction: when the inter-particle friction angle value is sufficiently large, the soil mass seems to prevalently behave like a solid.

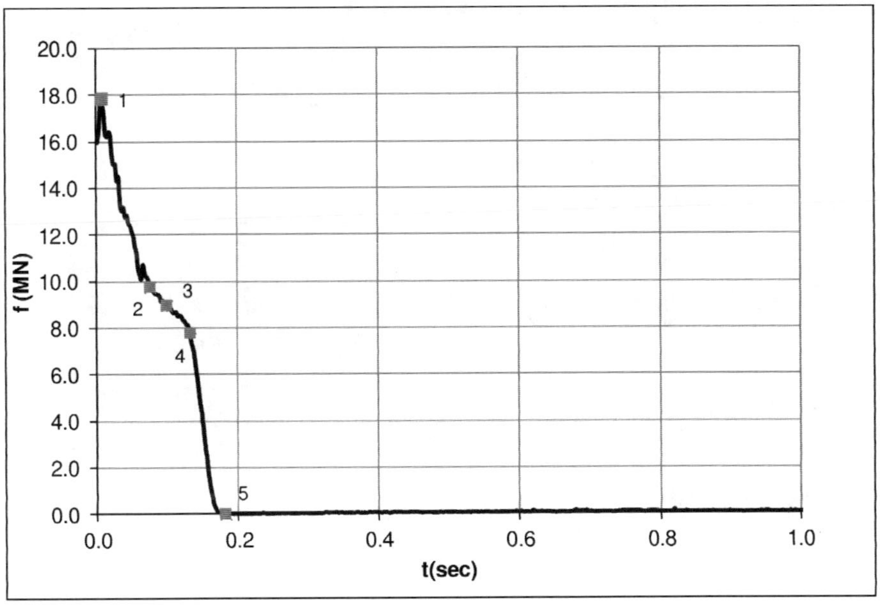

Figure 9: Evolution of the impact force with time in case f_c = 1 and $\alpha = 90°$ (all the other parameters are those of Table 1).

4. Conclusions

In this paper the impact of dry granular masses on rigid barriers is analysed. In particular, the authors illustrated some numerical results, putting in evidence from a qualitative point of view the nature of the mechanical process.

The numerical results were obtained by using a commercial DEM code. The soil mass impact has been described by analysing the evolution with time of the impact force, of the force chains and of the velocity field within the granular mass. Three different values of the inter-particle friction angle were taken into account to emphasize the double nature of the mechanical behaviour of the material that, during the impact, this seems to behave partially like a solid and partially like a granular fluid.

In particular, when the inter-particle friction angle is sufficiently small, the system tends to behave like a fluid, whereas when the inter-particle friction angle value is sufficiently large, the soil mass seems to prevalently behave like a solid. When the front inclination is large, some aspects of the solid-like behaviour are visible even for small values of inter-particle friction.

In contrast, when the inter-particle friction angle is intermediate, during the impact, part of the soil mass seems to behave like a solid and in the more superficial zone like a fluidized mass. To schematize the overall behaviour of the soil mass, three zones develop during impact: in one the material is approximately still, in another one the agitation governs the material response. The geometry of the three zones evolves with time: the first and the second one progressively increase, whereas the third one disappears at the end of the dynamic process.

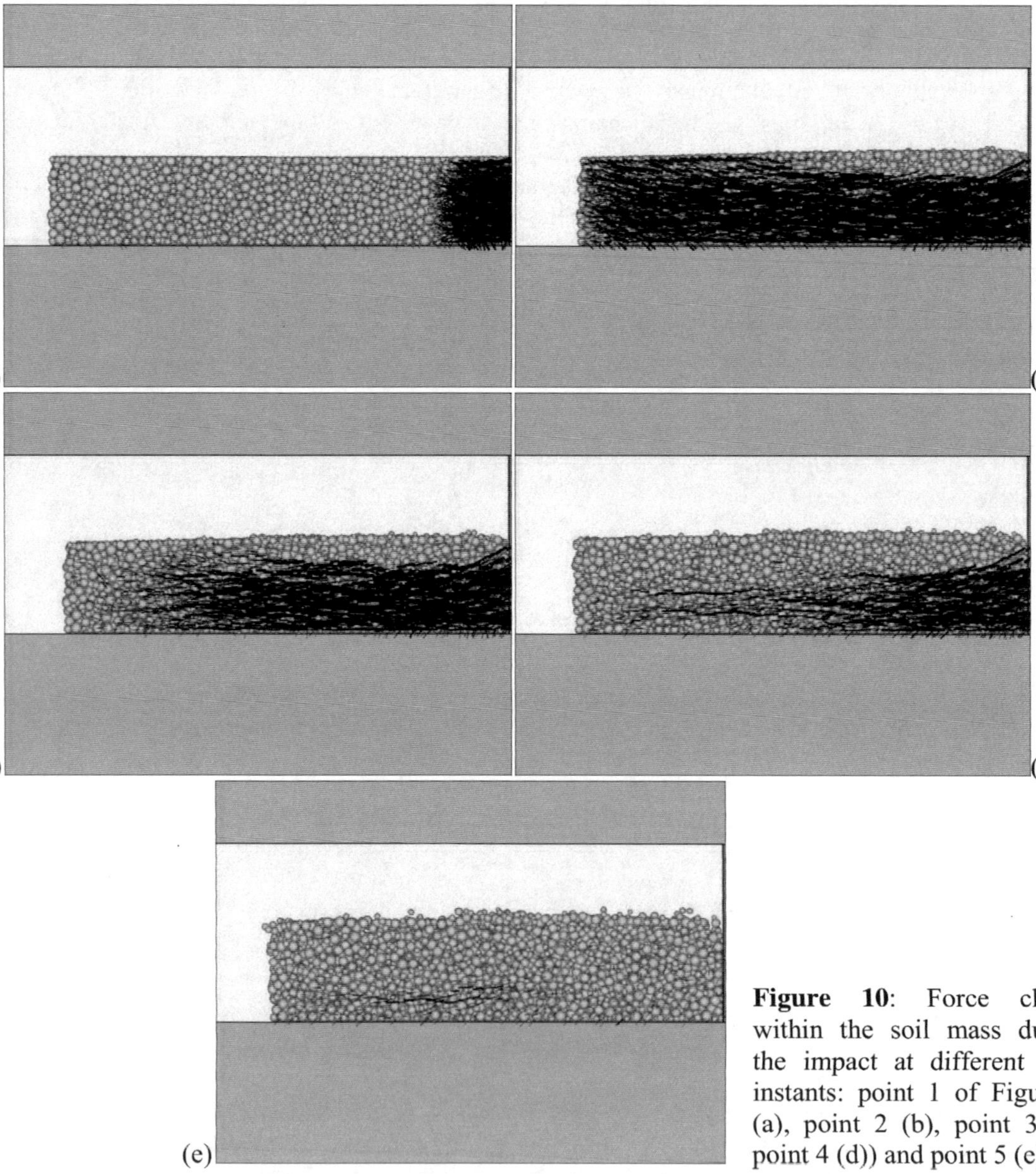

(a) (b) (c) (d) (e)

Figure 10: Force chains within the soil mass during the impact at different time instants: point 1 of Figure 9 (a), point 2 (b), point 3 (c), point 4 (d)) and point 5 (e).

5. Acknowledgments

This research was financially supported by the Italian government within the framework of PON01_01869 Project "Tecnologie e Materiali Innovativi per la Difesa del Territorio e la Tutela dell'ambiente" and of the PRIN 2011 Project titled "La mitigazione del rischio da frana mediante interventi sostenibili".

6. References

[1] Hübl J, Suda J, Proske D, Kaitna R, Scheidl C. Debris flow impact estimation. *Eleventh international symposium on water management and hydraulic Engineering* 2009 137–48.
[2] Ishikawa N, Inoue R, Hayashi K, Hasegawa Y and Mizuyama T Experimental Approach on measurement of impulsive fluid force using debris flow model *Conference proceedings interpraevent* 2008 343–54.
[3] Armanini A 1997 *On the dynamic impact of debris flows*. Recent Dev Debris Flows (Berlin: Springer)
[4] Suda J, Hübl J, Bergmeister K Design and construction of high stressed concrete structures as protection works for torrent control in the Austrian Alps *Proceedings of the third international fib congress* 2010 1–12.
[5] Dok L 2000 *Review of Natural Terrain Landslide Debris-resting Barrier Design* (Hong - Kong: Civil Engineering Department: Government of the Hong Kong Special Administrative Region), 123.
[6] Kwan JSH 2012 *Supplementary Technical Guidance on Design of Rigid Debris-resisting Barriers. GEO Technical Note No. TN 2/2012 (GEO Report No. 270),* (Hong Kong: The Government of the Hong Kong Special Administrative Region)
[7] Arattano M, Franzi L. On the evaluation of debris flows dynamics by means of mathematical models *Nat Hazards Earth Syst Sci* 2003 **3(6)** 539–44.
[8] Jóhannesson T, Gauer P, Issler P and K. Lied K (Eds) 2009 *The design of avalanche protection dams, Recent practical and theoretical developments*. (Brussels: European Comission) Climate Change and Natural Hazard Research - Series 2 Project report EUR 23339, DOI: 10.2777/12871.
[9] Huang, HP, Yang, KC, and Lai SW Impact force of debris flow on filter dam *Geophysical Research Abstracts* 2007 **9** 03218.
[10] Miyoshi I. Experimental Study of Impact Load on a Dam Due to Debris Flow. *Proceedings of the IUFRO Technical Session on Geomorphic Hazards in Managed Forests. Pacific Southwest Research Station* 1990 90.
[11] Salciarini D, Tamagnini C, Conversini P. Discrete element modeling of debris-avalanche impact on earthfill barriers *Phys Chem Earth, Parts A/B/C* Elsevier Ltd 2010 **35(3-5)** 172–81.
[12] Li X, He S, Luo Y, and Wu Y Discrete element modeling of debris avalanche impact on retaining walls *Journal of Mountain Science* 2010 **7(3)** 276-281.
[13] Spang R M 1998 Rockfall Barriers-Design and Practice in Europe *Proceedings of the Seminar on Planning, Design and Implementation of Debris Flow and Rockfall Hazards Mitigation Measures* 1-8.
[14] Yoshida H Recent experimental studies on rockfall control in Japan *Proc. of the Joint Japan-Swiss Scientific Seminar on Impact Load by Rock Falls and Design of Protection Structures* 1999 69-78.
[15] Calvetti F, di Prisco C and Vairaktaris E DEM numerical assessment of impact forces of dry granular masses on rigid barriers. *Acta Geotechnica* 2015 (submitted for publication)
[16] Cundall P A and Strack O D A discrete numerical model for granular assemblies. *Geotechnique*, **29(1)** 47-65.

[17] Cundall PA and Hart RD Numerical modelling of discontinua. *Engineering computations* 1992 **9(2)** 101-113.
[18] Calvetti F Discrete modelling of granular materials and geotechnical problems *European Journal of Environmental and Civil Engineering* 2008 **12(7-8)** 951-965.
[19] Calvetti F, di Prisco C and Vairaktaris E DEM modeling of soil/rock avalanche impact on barriers, *Int. J. Numer. Anal. Meth. Geomech.* 2015 (to be submitted for publication)

A deep seated movement in a marly-arenaceous formation: analysis of slope deformation and pore pressure influence

S. Assefa[1] A Graziani and A Lembo-Fazio

Roma Tre University, Engineering Department.
Civil Engineering Section, Via Vito Volterra 62, 00146, Rome, Italy

Abstract: The case history of a deep-seated slope movement in a complex rock formation (Marly-Arenaceous Formation) is analyzed. The movement, monitored for more than 20 years, was recognized after the discovery of intense cracking in the concrete lining of a hydraulic tunnel running across the slope. The time history of displacements shows that the ongoing deformation process is essentially a stationary creep phenomenon, also influenced by transient variations in pore pressure distribution. The slip surface is formed by a tectonized clay gouge layer and the mobilized shear strength is close to residual. The slope has been modelled (UDEC code) as a complex blocky structure defined by several joint sets: bedding joints, inclined and sub-vertical discontinuities. Different geometries of the slip surface, reasonably varied within the range of hypotheses compatible with field evidences, have limited influence on the limit friction angle of the slip surface. Joint patterns have influence on the deformation mode and minor impact on the mobilized friction angle. The model response is less sensitive to the water level at the slope toe as compared to the rise of groundwater table.

1. Introduction

The slope movement analyzed in this study was recognized in 1990 after the excavation of a hydraulic tunnel in the right bank of the Chiascio River (Italy) (figure 1). The movement affects a large portion of a ridge which has the longitudinal direction approximately perpendicular to the river. Just passed the sliding area, the river is crossed by a rockfill dam, which was intended to create a reservoir for water supply, by a diversion tunnel located on the right bank. So far, the diversion tunnel has been operated only once, in 1991-1992, not for regular water conveyance but for temporary deviation of the river flow.

The first evidence of ongoing slope deformation was the intense cracking of the concrete lining in a tunnel section located well inside the ridge, at a distance from the intake of some 250 m, under an average overburden of 50 m. The damaged tunnel stretch was soon interpreted as the place where the tunnel axis intersects a deep-seated slip surface.

The entire area of the dam reservoir, included the sliding slope and the dam foundation, is formed by an outcrop of the Marly-Arenaceous Formation, a Miocene flysch characterized by alternating sequence of marl, sandstone and calcarenite layers. As observed in many flysch formations of Italian Apennines [1], joints are often slickensided as a consequence of tectonic shearing and bending deformation of strata.

[1] Corresponding author (email: sirajm2000@gmail.com)
Tel.: +39 06 57333482, Fax: +39 06 57333469

The presence of major discontinuity planes of low strength represents a key factor for stability analysis in slope and dam engineering and has often required specific investigations [2], [3]. Weak layers of small thickness, such as clay gouge interbeds or thin shear bands, may be easily overlooked during ordinary borehole investigations, particularly in structurally complex formations [4], such as flysch and layered limestone formations [5], [6], [7].

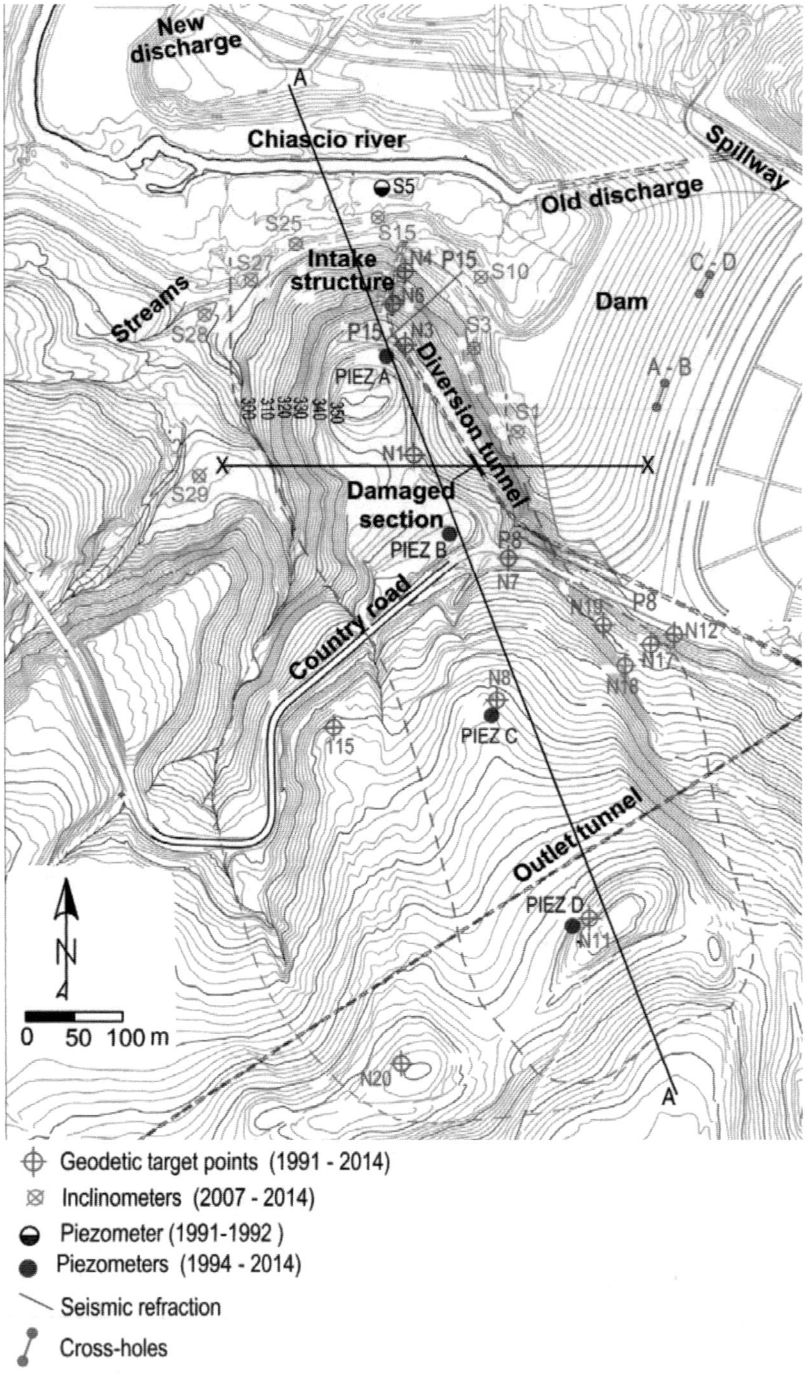

Figure 1. Study area.

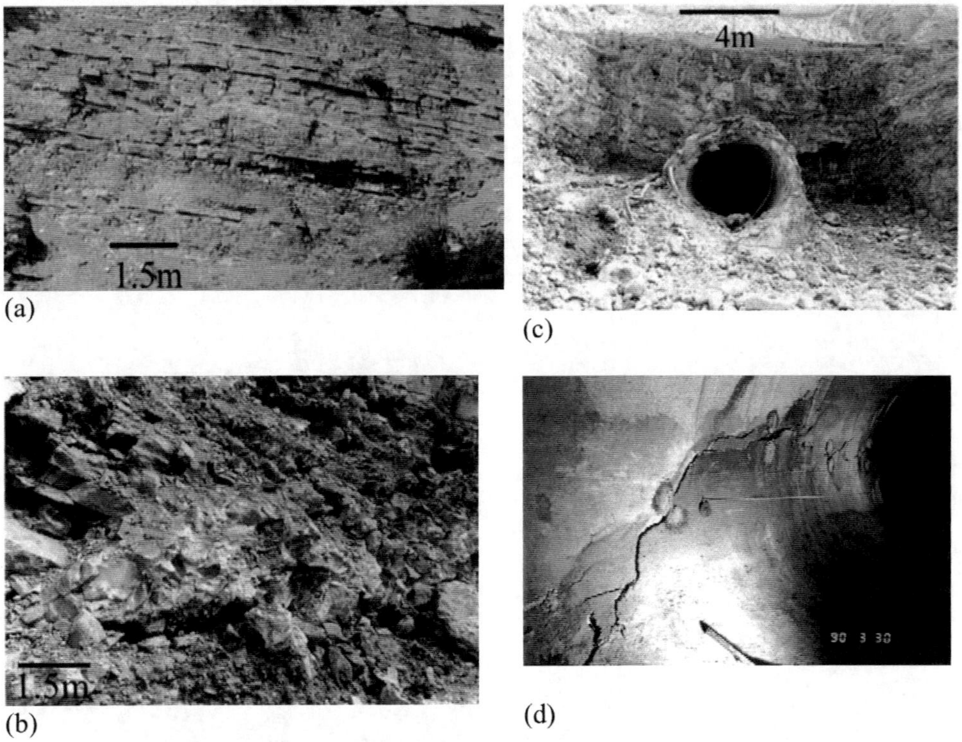

Figure 2. Regularly layered marly-arenaceous rock mass in the east side of the ridge (a); disturbed rock mass structure at the toe of slope (b); curved layers exposed during the excavation works for tunnel intake (c); main cracks and fissures seen in the concrete lining of the diversion tunnel (d).

In the present case, it is likely that most of the sliding surface is seated along an over-thrust plane gently dipping towards the Chiascio River valley. This circumstance adds further complexity to the structural setting of the slope and can explain some particular features of the sliding mass, whose overall volume has been estimated of some 18 Mm^3.

This paper focuses on the influence of rock mass structure (block shape, joint pattern) and pore pressure distribution on the mechanical response of the slope. First, the main results of geotechnical investigations and displacements measurements performed in more than 20 years are discussed in order to obtain a reference framework for slope modelling. The kinematic characters of the movement are essentially those of a planar-sliding (average velocity of 13-17mm/y) presumably localized along tectonized clay interbeds.

The structural setting of the slope was modelled by a discontinuum medium approach [8]. Therefore, the influence of rock mass structure (slip surface geometry, joint set orientation and continuity) on deformation mode and mobilized strength has been investigated. Finally, the influence of water table profile inside the slope and water level at the toe of the slope has been analysed.

2. Geological setting and morphology of the slope

The stretch of Chiascio River basin considered in this study is entirely based on a thick sequence of Miocene flysch deposits (figure 3). The main valley as well as the lateral gullies are carved in the same rock formation.

The Marly-Arenaceous Formation is characterized by a layered structure, at least in its less disturbed portions [9], [10]. The different layers can be grouped on the base of the prevailing lithological components:

a) Lower marls unit, with rare thin layers of clay shale,
b) Marly-arenaceous-calcarenitic unit,
c) Calcarenite unit, sparsely present as single layers,
d) Upper marls unit.

The lower marls are characterized by a more regular and continuous bedding. The a) unit was mostly exposed during the excavation for dam foundation, nearby the toe part of the sliding area (figure 3). The excavation works, started in 1984, required the removal of debris and loosened rock up to 15 m depth below the river bed elevation.

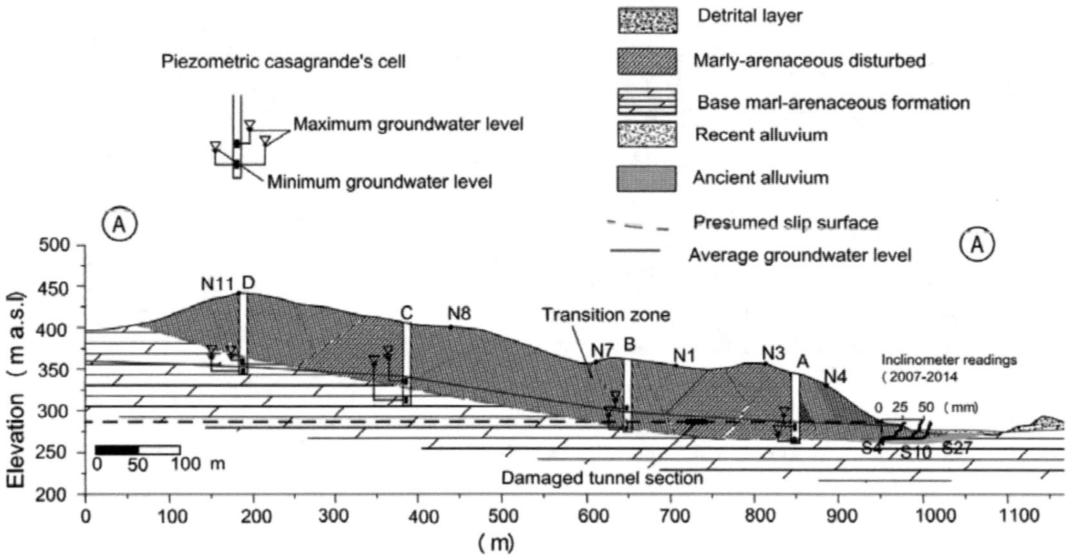

Figure 3. Longitudinal cross-section of the ridge with relevant geological data, location of piezometer boreholes and position of some geodetic targets.

The slope movement herein considered affects a large portion of a long ridge, upstream to the right abutment of the dam. The sliding mass seems mostly composed of alternating layers (thickness, 0.1–0.2 m) of unit b) lithotypes, with the upper marls unit and some debris deposits at the top. The Calcarenite unit consists of two separate strata (each one formed by single layers of 0.1–1 m for a total thickness of 3–4 m) intermingled with the thinner layers of unit b).

While the upper portion of the ridge is gently dipping (10°) towards the Chiascio valley, the frontal slope and lateral flanks are steep (25-40°) and mostly covered by a talus of rock debris and remolded clay. This situation hampers the visual detection of any trace of the slip surface at the toe of the slope (el. 280 m a.s.l.). The upper limit of the ridge (S sector) is characterized by a plateau, located at an average elevation of 395 m a.s.l.

In the east flank of the ridge a regular monocline structure (bedding dips to W, dip angle 20–40°) can be clearly identified. The same structure, with remarkably continuous and consistent orientations of the bedding joints, could be observed also on the foundation surface of the dam, as reported by the construction surveys. Conversely, the west flank of the ridge (bordered by the gully called "Fosso della Torre") is characterized by a less consistent orientation of bedding, especially in the lower part where bedding orientation is more dispersed, although dip direction from E to NE seems prevailing, with dip angle of 25-50°. Therefore, the overall pattern of the bedding planes suggests the hypothesis of a syncline structure, with axial plane having N-S direction.

The N sector of the ridge could be carefully inspected during the excavation works for the intake structure of the diversion tunnel (figure 2c) and, after the discovery of the ongoing movement, by extensive borehole investigations. In this area the rock mass exhibits a higher degree of fracturing and local disarrangement of the bedding joints.

A fundamental evidence acquired from borehole loggings in the N sector is the presence of quaternary alluvium under the Miocene flysch forming the toe of the slope (figure 3). This circumstance can be reasonably explained as the consequence of an overthrust deformation. By comparing the position of the contact between alluvium and marly-arenaceous formation at different boreholes, it can be argued that the basal plane of the overthrust is almost horizontal and is located at an elevation slightly lower than the river bed. Moreover, the cumulated amount of shearing displacement would be as high as 100-150 m. If such hypothesis holds true, it is likely that also the current process of slope deformation is governed by the same shear band, i.e., by a layer of strongly tectonized clay gouge.

3. Survey of tunnel damage and slope displacement

As already mentioned, the occurrence of a slow movement in the ridge crossed by the diversion tunnel of the dam, was recognized only after the completion of the tunnel lining. Nowadays, surface evidences of the ongoing deformation can also be detected, especially on the east boundary of the sliding area, where the pavement and curbs of some country roads appear locally displaced and fissured.

The diversion tunnel departs from the right flank of the valley, at an elevation of 287 m a.s.l., approximately 15 m higher than the toe of the slope, and runs through the long ridge with a direction almost perpendicular to the river (figure 1) for some 900 m, before turning to the W, by taking a N238° direction.

The clear evidence of a deep-seated movement came from the heavy damage of the tunnel lining, not yet in operation, which consists of a 4.5 m diameter, 0.5 m thick cast concrete ring. Cracking and opening of the construction joints were localized in a 23 m long stretch of tunnel, between chainage 235 and 258 m from the intake section. The limited width of the damaged stretch is strongly suggestive of a deep movement with localized shearing. The damaged section of the tunnel is under an average overburden of 50 m and is located, in plan, some 40 m inside the lateral boundary of the movement, as successively demonstrated by the geodetic survey of surface displacements.

The fissures in the lining were repeatedly sealed but formed again, then four 4 couples of crack-meters were installed across the main cracks (figure 2). A clear picture of the relative displacement trend was obtained in 1991-1992, during a short-time operation of the diversion tunnel as temporary discharge tunnel in order to allow the completion works of the dam bottom outlet. In that period of temporary operation, the reservoir water level reached the maximum elevation of 296.5 m. After 3 months of tunnel operation, with average water level at 287 m el, the maximum stroke (50 mm) of the displacement gauges was achieved. The relative displacement vectors measured across the cracks exhibit a negligible plunge and a direction close to the tunnel axis.

3.1. Geodetic survey of displacements

A first network of targets (N1-N11) for the geodetic survey of surface displacements, restricted to the frontal slope of the ridge (N sector), was implemented to integrate the crack-meters installed in the damaged section of the tunnel, before starting the temporary filling of the dam pool. The geodetic network was thereafter extended to the S and W areas, with targets (N12-N20) at greater distances from the N slope, in order to estimate areal distribution and limits of the movement (figure 1).

Since then, all the targets have been continuously monitored by a total station, installed on the opposite side of the valley, at the top of the concrete walls of the dam spillway. Measurements are performed regularly once a month. The prevailing components of displacement are in the horizontal plane; the vertical components are much smaller and usually are not processed.

3.2. Tunnel extensometer
In 1994, it was decided to install a steel lining within the damaged section of the tunnel. The annular gap between two steel pipes was not sealed but endowed with a "telescopic joint", in order to accommodate differential axial displacements between the two tunnel stretches separated by the cracked zone and, therefore, hosted, respectively, inside and outside the sliding rock mass. The telescopic joint was supplemented by an electric dial gauge, thereafter referred to as "tunnel extensometer".

3.3. Inclinometer measurements
The displacement data available after some years of geodetic monitoring were instrumental to trace the approximate boundaries of the slide surface. The lateral limit of the movement along the E flank of ridge were considered with particular attention due to proximity of the right abutment of the dam. Therefore, it was decided to refine the survey by installing a set of inclinometer tubes (figure 1). Taking advantage of the shallow depth of the slip surface in this peripheral zone of the movement, the length of the 8 boreholes (S1…S29 series in figure 1) could be limited to 20-45 m. The boreholes were drilled in 2006-2008, since then, readings have been carried every two months by removable sliding inclinometer probe.

4. Geotechnical investigations

The complex structure of the rock mass, as expected from the geological investigations, is clearly confirmed by a comprehensive analysis of borehole loggings (e.g., borehole S27 in figure 4). Borehole investigations have been focused on the N sector, particularly, in the zone of tunnel intake.

Bedding joints have discordant orientation in the upper and lower portion of the ridge, especially on the W flank, as already recognized from field surveys. This situation suggests that the entire body the ridge can be divided in different zones, characterized by a variable degree of fracturing and disarrangement of the rock mass.

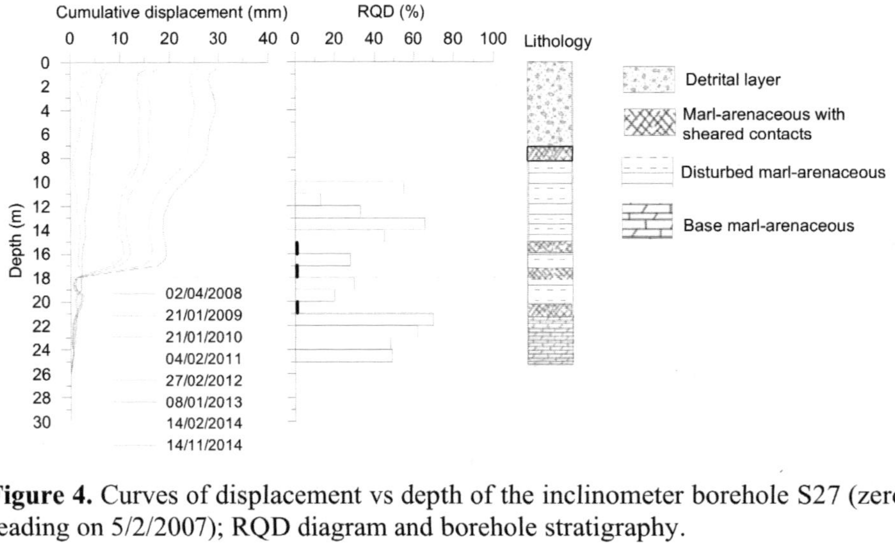

Figure 4. Curves of displacement vs depth of the inclinometer borehole S27 (zero reading on 5/2/2007); RQD diagram and borehole stratigraphy.

Though results obtained from borehole loggings are sparse, chaotic structure typically accounts for 10 to 40% of total borehole logging data. Comprehensive analysis of joint dips from borehole loggings shows that sub-horizontal strata and steep joints are inclined, respectively, between 5-20° and 30-60°.

For instance, the S3 borehole, drilled through the toe of the N slope of the ridge, shows that the base marly-arenacoues-calcarenitic unit has RQD value in the 40-70% range with discontinuities dipping between 18-20° while the upper layer of disturbed marl RQD value is from 15-40% with discontinuity inclined at 12-20°. Lower RQD values between 10-40 % and even null values were obtained in the more disturbed zones, presumably the zones of higher shearing deformation. Borehole S27 (figure 4), located in the NW side of the ridge, shows RQD values of 50-70% for the undisturbed marly-arenaceous-calcarenitic unit and RQD of 15-30% at the depth of the sheared zone.

4.1. Mechanical properties

The intense shearing deformation occurred along the basal zone of the slope suggests that the mobilized strength is likely to be close to residual conditions for most of the sliding surface.

The properties of the sheared material (clay gouge) were examined from the exploratory window opened at the damaged portion of the tunnel, where the slip surface intersects the tunnel axis. The core of the shear zone consists of a 0.15-0.20m thick band of clay gouge (figure 5) entrapping fragments formed by shear grinding of the rock materials (mainly marl and sandstone). Index properties of the clay gouge samples are: Clay = 46%, Silt = 45%, Sand = 9%, γ_s = 27.2 kN/m3, w_L = 72.6%, w_P = 40.2 %, I_p =32.4%, CaCO3 = 20.1%. This material possess high plasticity and its grain size composition is dominated by a high silt-clay proportion and less coarse materials. The results of shear box testing in drained condition are: peak, c_p =21.5 kPa and φ_p = 15° and residual, φ_r = 7.7° (figure 6).

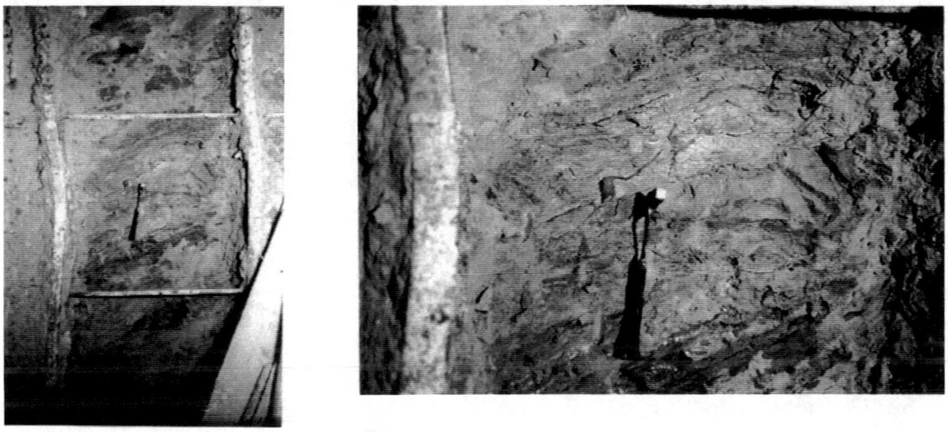

(a) (b)

Figure 5. Clay gouge interbed exposed at the damaged tunnel section (a); detail of photo (a) with the deformed steel set (b); the observed thickness of the clay layer is approximately 0.2 m.

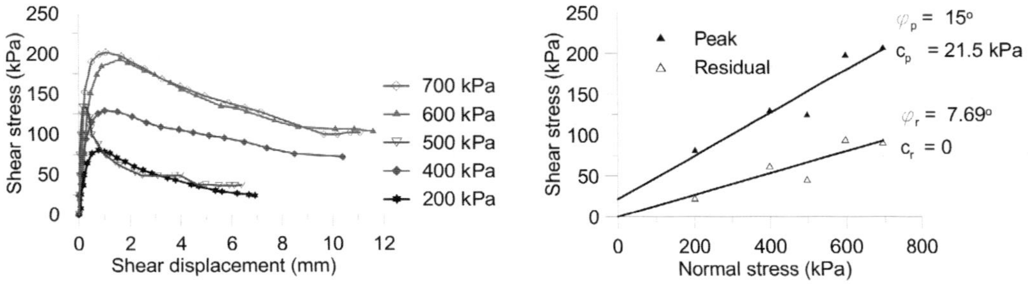

Figure 6. Results of direct shear tests on clay gouge sample.

The strength properties of the different rock materials have been obtained mainly by point loading tests. The unconfined compression strengths of the marly lithotypes and sandstone are 22-88 MPa and 73-140 MPa, respectively. For the disturbed zones of the rock mass, GSI value of 35-40 can be assumed, while for the marly-arenaceous formation forming the "bedrock", i.e.,below the shearing zone at the base of the slope, GSI values of 55-60 seem more appropriate. Rock joints can be characterized by JCS= 30-40 MPa and JRC= 3-5. Slickensided joints are frequently found in borehole cores (figure 7), mainly at the contact between marly and arenaceous layers but also within claystone beds.

Typical friction angles from in situ shear tests on slickensided and laminated bedding joints of the Marly-Arenaceous formation are in the range of 12-15° [11], [2]. Laboratory shear strength tests between clay filling and limestone contacts were also conducted by [6]; φ_r = 13.5-18° was obtained. In addition, from in situ and laboratory direct shear tests, [12] obtained peak and residual strength of 14° for sandstone – sandstone contacts and φ_p = 19° and φ_r = 13-18° for sandstone–claystone contacts.

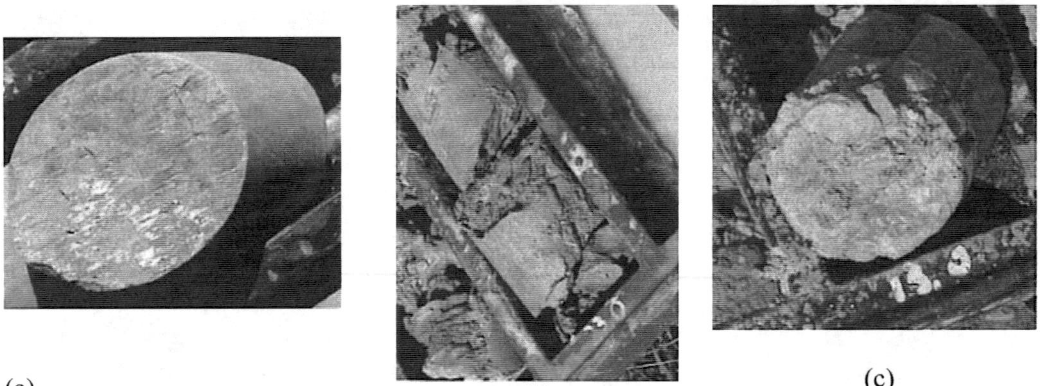

(a) (b) (c)

Figure 7. Typical aspect of joint surfaces: joint surface recovered from borehole S27 at depth of 7.4m (a), from borehole S3 at depth of 24.9m (b), presumably close to the depth of the slip surface, and from borehole S23 at depth 13m (c).

A report by [13] shows φ_r in the range of 10-26° after several ring shear test performed on specimens from an outcrop of the same Marly-Arenaceous in a nearby site, also containing thick clay interbeds. Ring shear tests from [14] show that φ_p = 13-27° and φ_r = 10-15° for different types of clay-rich interbeds.

Cross-hole tests up to 30m depth were performed to assess the elastic properties of the dam foundation, formed by the alternate layers of sandstones, marl-clay and calcarenite of unit a). Elastic properties (E_d, ν_d) were estimated from compression and shear waves velocities (figure 8). A comprehensive study by [15] on several dam sites in Italy shows good correlation between plate loading tests and seismic measurements. Oberti et al. [2] also performed several plate loading tests, perpendicular and parallel to the stratification in marly and sandstone units. Deformation moduli of 18-24 GPa were obtained for tests conducted parallel to the stratification and 16-18 GPa for tests perpendicular to the stratification.

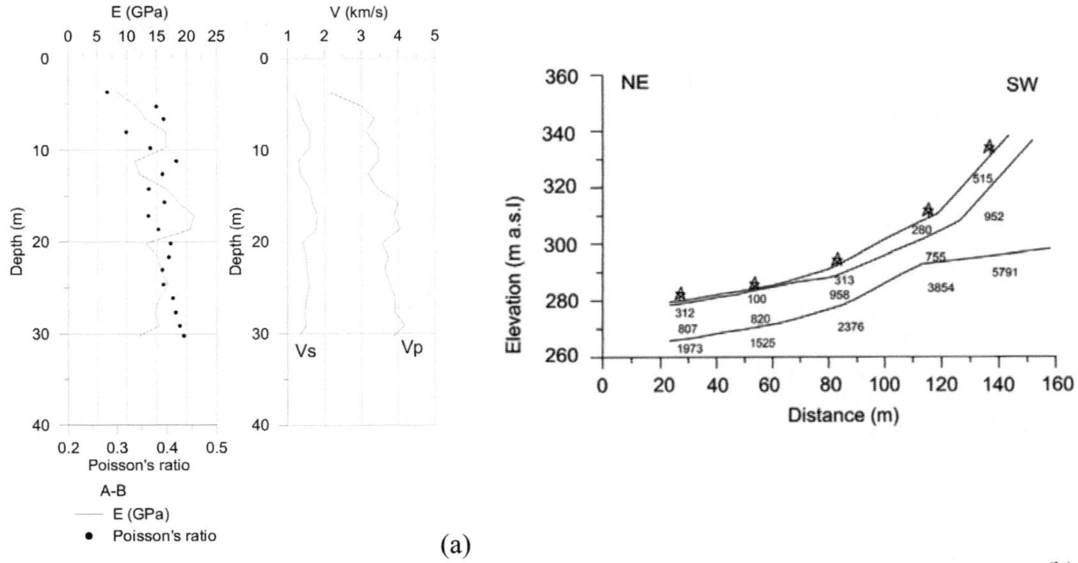

(a)

(b)

Figure 8. Results of cross-hole test for dam foundation, in boreholes A-B (a), and p-wave velocity from seismic refraction investigation P15 at the slope toe (b).

Investigations conducted on the slope body using down-hole and seismic refraction show that the first few meters, composed of loose detrital layers (i.e., silt and gravel deposits), have P- wave velocity of lower than 0.5 km/s while for the fractured marly-arenacous formation P-wave velocity increases to 0.5-0.8 km/s. Finally, P-wave velocity of 0.8-3 km/s can be assumed for massive sandstone and calcarenite layers (figure 8).

4.2. Permeability tests and piezometer measurements

Lugeon tests were conducted in the area of the dam foundation. The permeability of the rock mass lies between $1x10^{-7}$–$3x10^{-7}$ m/s (figure 9). Similar results were obtained by [2] in the area of Ridracoli dam, with average permeability of $1x10^{-7}$–$4x10^{-7}$ m/s at high and shallow depths, respectively.

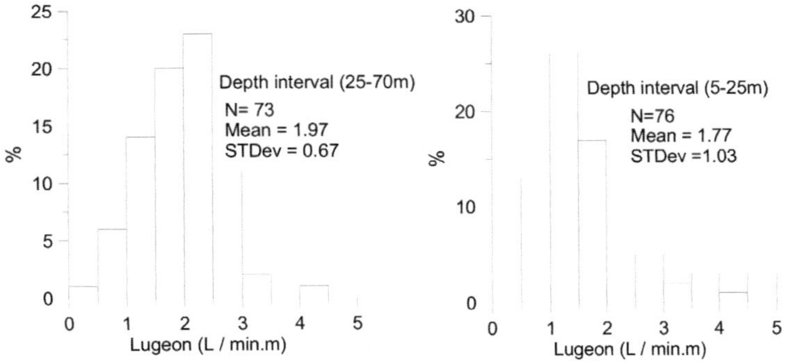

Figure 9. Frequency distribution of Lugeon values for a test conducted at the valley.

It is worthwhile to remark that the Marly-Arenaceous formation may exhibit higher hydraulic transmissivity in the direction of strata, with seepage paths limited to discontinuities planes and involving almost exclusively the more fractured sandstone strata, while the clay-rich layers represent low-permeability barriers.

To investigate the pore pressure distribution inside the slope, four deep boreholes were equipped with piezometer cells (figure 3). Each borehole hosts two cells installed, respectively, above and below the presumed elevation of the slip surface. During the monitoring period the water level of the dam pool never exceeded 282 m a.s.l. Therefore, the measured pore pressures have been mainly influenced by rainfall and infiltration process.

5. General hypotheses and kinematic characters of the movement

5.1. Geometry of the sliding mass

Borehole investigations and survey of the N sector of the ridge have shown that a basal slip plane can be reasonably assumed at the top of the buried alluvium layer. Inclinometer profiles and target displacement vectors can give further support to this assumption.

Although the density of measurement points decreases in the S sector, at greater distance from the river bed, magnitude and direction of horizontal displacements measured at different points are generally consistent (figure 11). The overall direction of displacement is approximately parallel to the longitudinal direction of the ridge, with some eastward rotation at the toe of the slope. The basic hypothesis of a planar sliding with some internal shearing of the moving body seems therefore reasonable.

The position of the damaged tunnel section represents a further restraint to trace a likely longitudinal profile of the slip surface (figure 10). Finally, the upper limit (S boundary) of the movement can be located mainly on the base of the morphology of the ridge, i.e., by observing the presence of a plateau with some deformation traces at elevation 400 m a.s.l. The distance from this plateau to the river bed, measured along the longitudinal axis of the ridge, is approximately 1000 m.

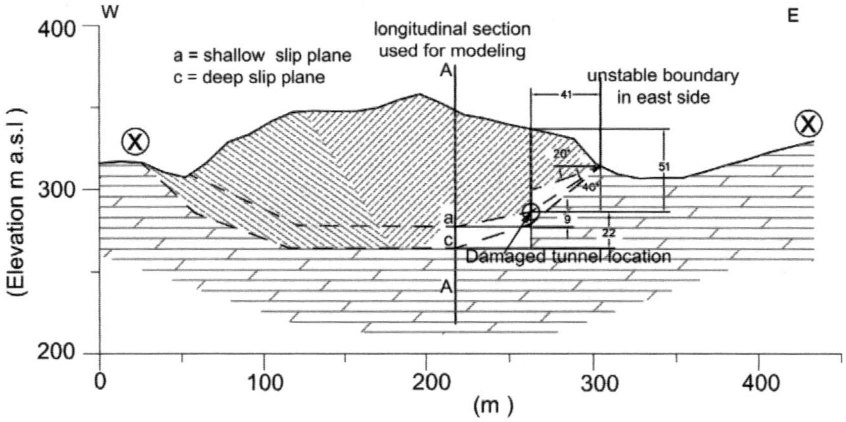

Figure 10. Slip surface reconstructed at section x-x considering the position of damaged tunnel section and typical dip angles (20-40°) of layers at the East side of the ridge, as evidenced from field investigations (all dimensions are in m).

The slip surface can be reasonably drawn as a near horizontal plane in the toe part. The slip surface is certainly more inclined in the upper part but the effective geometry cannot the defined in detail. This last issue will be further investigated by comparing different DEM models.

The lateral boundary of the movement can be traced more confidently on the E side, on the base of displacement measurements, while on the W side the separation line between stable and unstable zone fades away (figure 11). The slide boundary is close to the top of the dam but the movement does not affect the abutment zone, as demonstrated by the null displacement of targets N12 and N19.

IOP Conf. Series: Earth and Environmental Science **26** (2015) 012052 doi:10.1088/1755-1315/26/1/012052

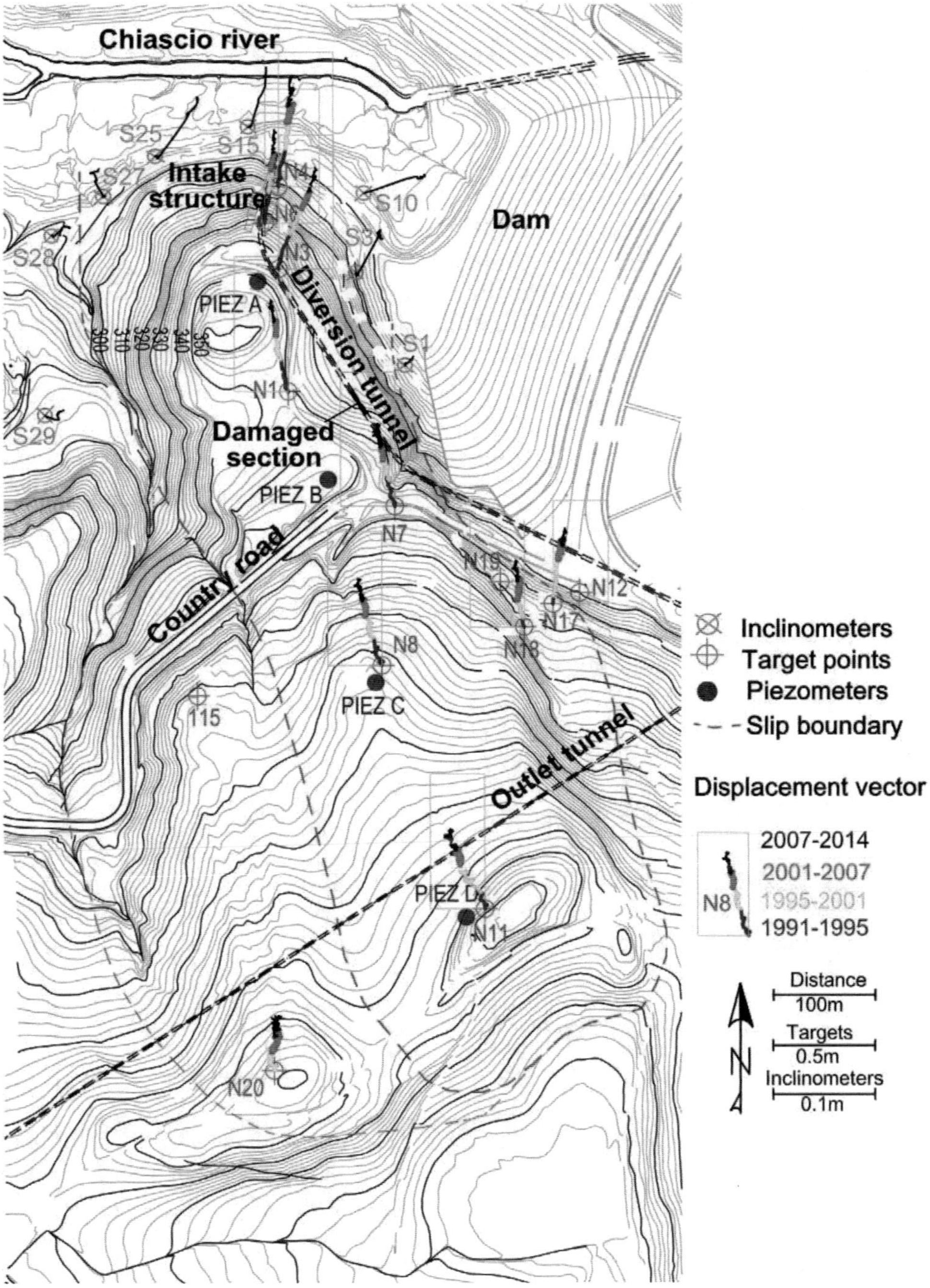

Figure 11. Plot of horizontal displacements and boundary of the movement.

The current kinematics of the movement can be outlined as a compound mechanism in which a blocky system is sliding on a low inclination basal plane. The depression along the upper profile of the ridge, between the targets N7 and N8 (figure 3), may represent a morphological evidence of past inter-block deformation. Inclinometers are located only at the toe of the slope: the deformation profiles generally show a rigid mass type of movement, with a clearly defined slip plane (figure 4), but in some cases a smeared deformation mode is also observed.

5.2. Time history of displacements

The time history of measured displacements (figure 12) show that the ongoing deformations are characterized by stationary velocity. Apart from the much higher velocities recorded in 1991, during the temporary operation of the diversion tunnel, the average velocity of the targets lies in the 13-17 mm/y range in the period 1992-2001 and tends to decrease in the following years. However, an increase in velocity was observed recently after the last measurements. The tunnel extensometer exhibits a similar trend and consistent magnitude of displacement.

A secondary fluctuation of velocity is also observed. This behavior is likely a consequence of transient seasonal variations in pore pressure distribution. Some targets also show alternated phases of negative and positive velocity, which may be tentatively explained as the combined effect of "stick-slip" phenomena and/or temperature variations. Additional difficulty in analyzing the velocity trend (figure 12) stems from the low frequency of measurements.

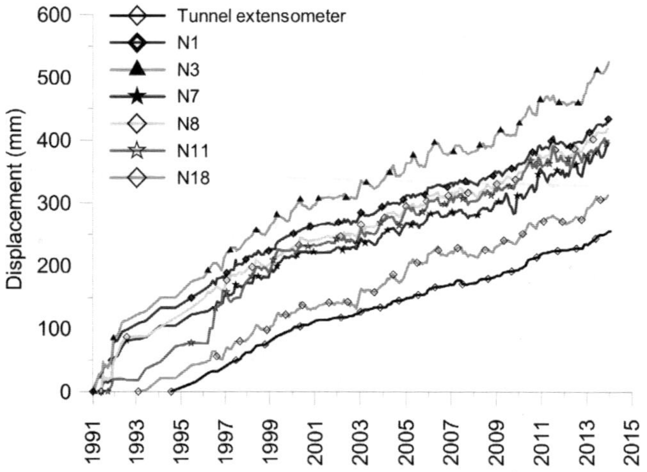

Figure 12. Displacement of the geodetic targets and deformation of the telescopic joint of the tunnel; the time of first measurement is not the same for all monitoring points.

6. Pore pressure regime and rainfall influence

The average groundwater table, as recorded by piezometer cells in borehole A, B, C and D, progressively decreases towards the slope toe (figure 3).

The cells in the uppermost borehole (D1 and D2), which are located at 80 m and 97 m below the ground surface, respectively, have given null pore pressures for most of the measurement period. The significant difference in piezometric head between the cells in the same vertical reflects a complex groundwater situation, where fractured rock aquifer are separated by the less permeable marly layers.

Readings from piezometer B2, located in the same zone of the damaged tunnel section, show a local falling of the groundwater table. This fact can be explained as the effect of a higher degree of loosening, and therefore higher permeability, of the rock mass. The steep flanks of the ridge in this sector may also facilitates the lateral drainage process.

The time-lag between pore pressure increases in piezometers and rainfall was also analyzed. The maximum fluctuation of groundwater levels around the average value is some 10 m. Peaks in piezometer levels seem best correlated with three months cumulated rainfall exceeding the average values (figure 13). Correlation analyses were also attempted between displacement velocities and rainfall, but results are unreliable due to the low frequency of measurement (figure 13).

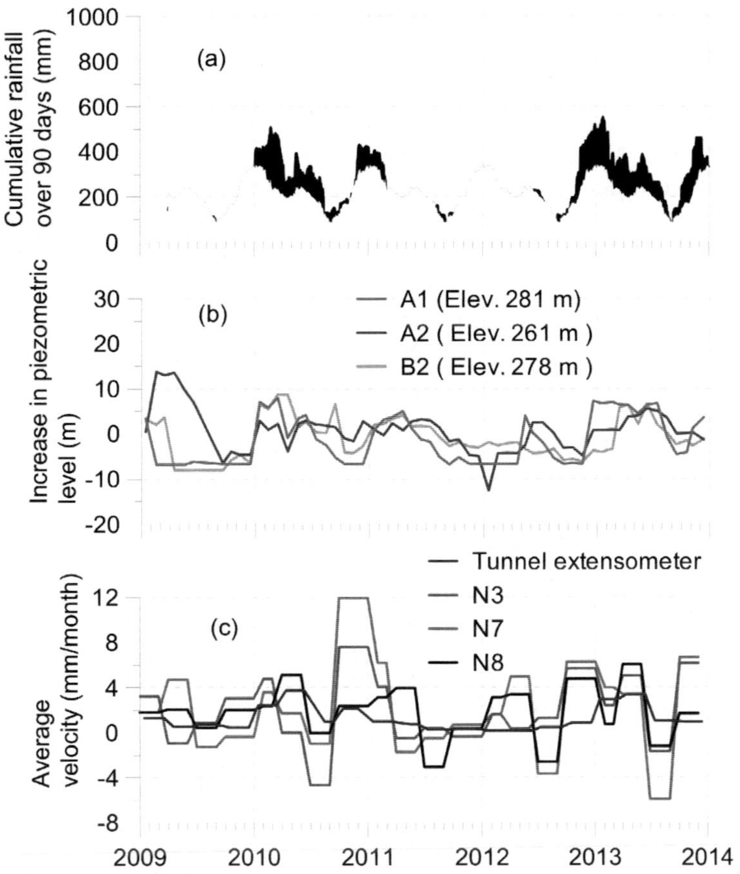

Figure 13. Correlation between 3-months cumulated excess rainfall (a), represented by the shaded area, variations in piezometer levels (b) and average displacement velocity of different measurement points (c).

7. DEM modelling of the slope

The aim of this first phase of modeling is to investigate the influence of the structure of the moving mass (i.e., geometry of slip surface and joint pattern) on slope displacements and mobilized friction angle of the slip surface. A 2D discontinuum modeling approach (DEM) was adopted via the UDEC code [8]. In the present modeling block materials are linearly elastic while joints are ideal elasto-plastic and obey the Mohr–Coulomb strength criterion. Block materials have been assumed linearly elastic while joints are ideal elasto-plastic and obey the Mohr–Coulomb strength criterion.

The following analyses are primarily a back analysis of the slope deformation observed in more than 20 years of field monitoring, which has allowed to categorize the movement as a "slow" process. Dynamics aspects often related to catastrophic slope failures (e.g. [16],[17]) are not essential for the present case. Therefore, local damping has been applied, as customary in UDEC code for static or quasi-static processes.

The slope has been modelled as a complex blocky structure defined by several joint sets. Bedding planes, gently inclined and markedly continuous are crossed by steeper and sub-vertical joints.

Frequency and number of joint sets have been increased in the more disturbed portions of the slope, i.e., at the toe and in the transition zone, i.e., where the dip angle of the slip surface changes. The plane section considered in the model is parallel to the longitudinal axis of the ridge (figure 14).

The mechanical properties of the joint sets are illustrated in table 1. All the joint sets have the same stiffness properties (k_n and $k_s = 0.1k_n$) and purely frictional strength. The normal stiffness of the joints has been calibrated so that the equivalent rock mass modulus is in agreement with the elastic modulus suggested by the GSI approach and in the same range of cross-hole data (see Section 4.1).

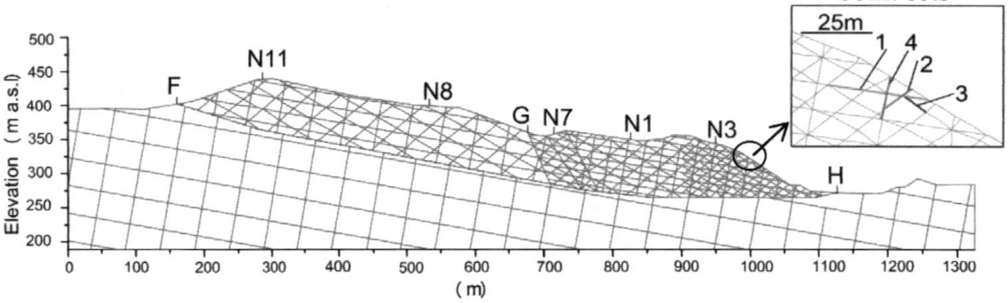

Figure 14. Model A with continuous joint pattern.

The present morphology of the slope has been formed by simulating an "erosion" process, starting from an ideal situation with horizontal ground surface and in situ stress condition with horizontal-to-vertical stress ratio $K_0 = 0.5$. The reference groundwater table (stationary conditions) considered in the following analyses represents the average profile obtained from piezometer measurements (vertical boreholes A, B, C and D).

The influence of rock mass structure on the mobilized friction angle of the basal slip surface has been investigated. The shear strength reduction method was employed to calculate the failure condition for a given joint pattern. Equilibrium conditions have been satisfied for each step of reduction in basal friction angle. To this aim, an unbalanced forces ratio [8] of 10^{-6} has been adopted as a convergence criterion. Normally, a progressively higher number of calculation steps is required as the model approaches the plastic collapse. Additionally, the displacement histories of some grid points were monitored. The selected control points correspond to the position of specific geodetic targets (N1-N11), located at the toe of the slope as well as in the transition and upper zone of the ridge.

As highlighted by previous studies [18], the problem of finding the strength corresponding to the onset of collapse involves computational difficulties and ambiguous results are likely to be obtained. The expedient of progressively decreasing the step-size of the strength reduction as the system approaches the stability limit proved to be useful but not decisive.

Finally, for the present analyses, it was decided to adopt the following empirical approach. The limit friction angle has been identified by inspecting the displacement vs mobilized friction curves, i.e., it corresponds to the point of maximum curvature, where the displacement exhibits a sharp increase (e.g., figures 15 and 17). The aforementioned criterion may not be fully satisfactory, in general, but seems effective in order to compare the responses of the different models.

Moreover, a factor of safety (F.S.) can be easily calculated as the ratio of the tangent of the actual friction angle to the tangent of the failure friction angle, as suggested in [19].

The shear strength assumed for the bedding planes corresponds to the typical properties determined for the contact planes between marly and arecaneous layers (see Section 4.1). The pattern of cross joints included in the model represents an idealized picture of the blocky structure of the marly-arenaceous formation. Persistent or discontinuous joints can be considered (Model A and Model B in figure 16, respectively), as shown in the following sections.

A particular remark concerns the shear strength of the cross joints (joint sets 3 and 4). A minimum friction angle of 30° and 18° is required for the cross joints of Model A and B, respectively, in order to mobilize a translational movement of the rock mass along the basal plane. For lower friction angles, a different collapse mechanism is obtained, i.e., the slip failure of a steep wedge of rock mass in the frontal zone of the slope (between points N3 and H in figure 14). This local failure mode was to be avoided since it is not in agreement with the observed characters of the real movement.

The aforementioned difference in the friction angle required for cross joints can represent a measure of the interlocking contribution to shear strength due to the non-persistent joints (figure 16). Conversely, it has been found that the friction angle of cross joint sets has limited influence on the mobilized friction angle of the basal slip surface.

Table 1. Joint properties adopted in UDEC simulations.

Description	Joint spacing[a] S (m)	Dip angle ψ (°)	Friction angle φ (°)	Normal stiffness k_n (GPa/m)	Shear stiffness k_s (GPa/m)
Bedding planes (1)	8-16	10	18	2.3	0.23
Joint set (2)	12-24	40	30	2.3	0.23
Joint set (3)	10-20	30	30	2.3	0.23
Joint set (4)	20-40	80	30	2.3	0.23
Slip surface	-	0-10.5[b]	Varies[c]	2.3	0.23

[a]Spacing is different for the upper and lower (transition and toe) portions of the slope.
[b]Range of dip angles of the slip surface (c) assumed for the model in figure 14.
[c]Friction angle of slip surface is varied according to the strength reduction method.

7.1. Influence of blocky structure

7.1.1. Block size. Since the joint spacing of the model must be increased with respect to the real situation, to get a reasonable number of blocks for modelling [18], [20], it is worthwhile to analyze the influence of block scaling. The effect of block size has been preliminary investigated by comparing the mechanical response of two simplified models in which only two joint sets, (1) and (4), are implemented with spacing S and 0.5*S, respectively (figure 15).

The results of figure 15 show that block scaling affects the overall stiffness of the rock mass [21] but does not change the mobilized friction at collapse. The block size was further decreased (1/3 of block size shown in figure 15a), again, the model results show an increase in displacement magnitude but the effect on the mobilized friction angle of the slip surface is still negligible. Similar results were obtained by [18], where the failure friction angle was found to be progressively less sensitive to block size as the number of blocks becomes sufficiently higher.

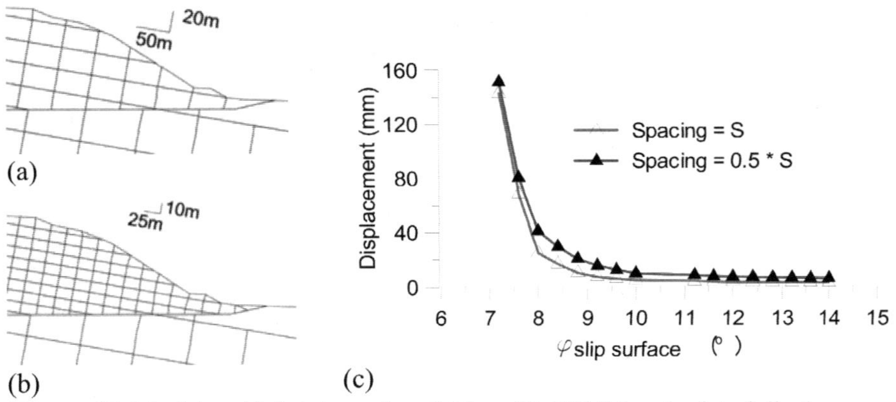

Figure 15. Models with joint spacing S (a) and 0.5*S (b); calculated displacement of point N3 vs mobilized friction angle of the slip surface (c).

For the models shown in figure 15, as the number of blocks increases the model exhibits not only a basal sliding but also significant shearing along the sub-vertical joints and toppling deformation in the frontal part of the slope. These features are not in agreement with the observed deformation style.

7.1.2. Joint pattern. A set of analyses were performed with the model of figure 14 (Model A), in which all the joint sets are continuous. On the base of field survey and borehole data, a staggered joint pattern has been also implemented (Model B in figure 16), in order to analyze the influence of non-persistent joints and interlocking.

The model with staggered joints exhibits less internal shearing of the rock mass and a more uniform distribution of horizontal displacements. The deformation mode of Model (A) is more complex: the lower portion of the slope, particularly N1 and N3 points, moves more than the upper portion (figure 17), where shearing deformations along the joints dipping towards S (joint set 2) are also activated. Similar influences of joint pattern on the deformation mechanism of rock slopes have been reported by [18], [22], and [23].

The mobilized friction angle at collapse is 7.6° and 7.2° (figure 17), respectively, for Model (A) and Model (B). The limit strength is therefore close to the residual friction obtained from shear box tests. For the staggered joint model (Model B) the limit condition is anticipated by a marked increase of displacement since the friction angle is reduced to 8°.

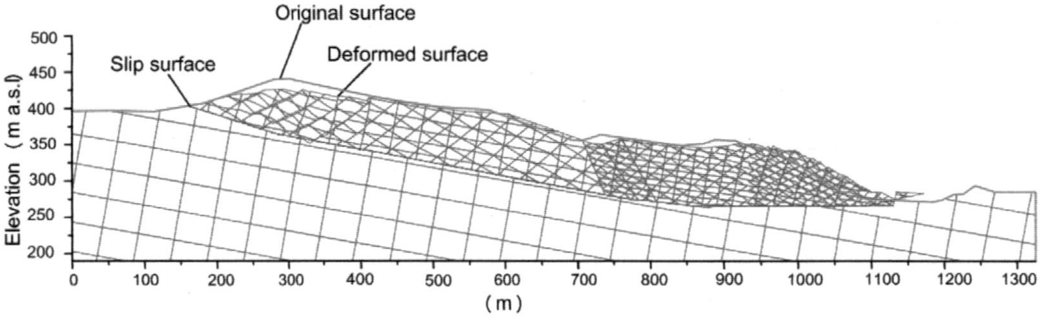

(a) Model A

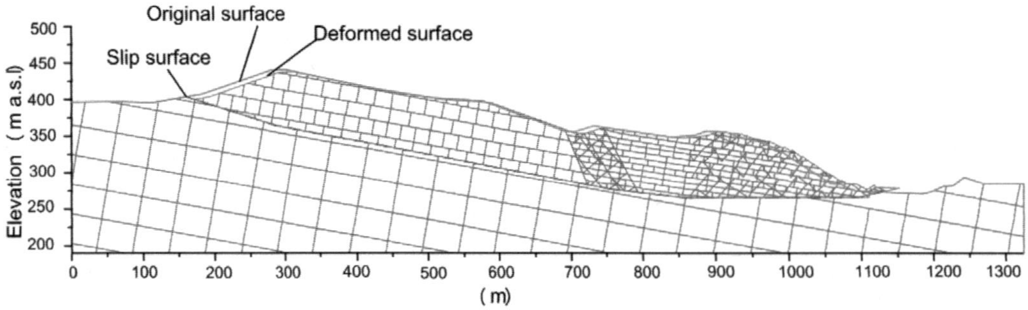

(b) Model B

Figure 16. Deformed shape (x200) of Model A (a) and Model B (b) for a mobilized friction angle φ =7.6° on the basal slip surface.

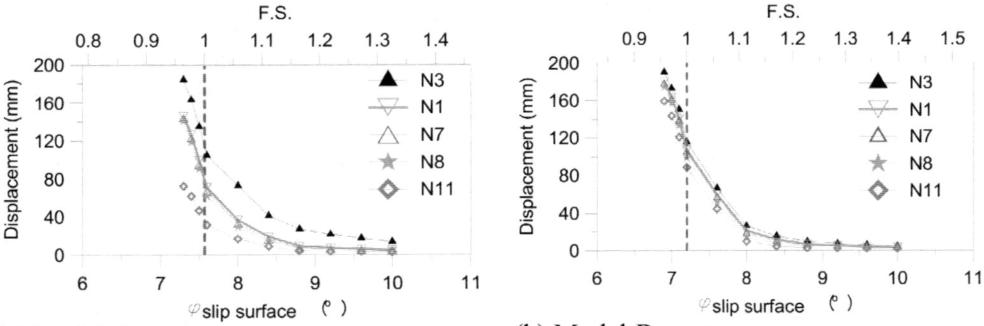

(a) Model-A (b) Model-B

Figure 17. Calculated displacement vs mobilized friction angle of basal slip surface for Model A (a) and Model B. (b).

7.1.3. Depth and shape of the slip surface. As already discussed in Section 5, depth and shape of the slip surface can be unambiguously defined only in the toe portion of the slope. However for the upper part of the slope some reasonable hypotheses can been formulated, taking into account the location of the damaged tunnel section and the slope morphology. Figure 18 shows the set of four different hypotheses (a, b, c and d) which have been considered and implemented in Model A (model with continuous joints). Dip angle and depth of the upper portion of the slip surface vary reasonably between 7–13° and 54–100 m, respectively. As figure 18 shows, the slip surface (c) gives a minimum friction angle very close to the residual value of shear box tests. The slip surfaces a and b, for which the limit friction angle is in the 9-9.5° range, can be considered realistic too.

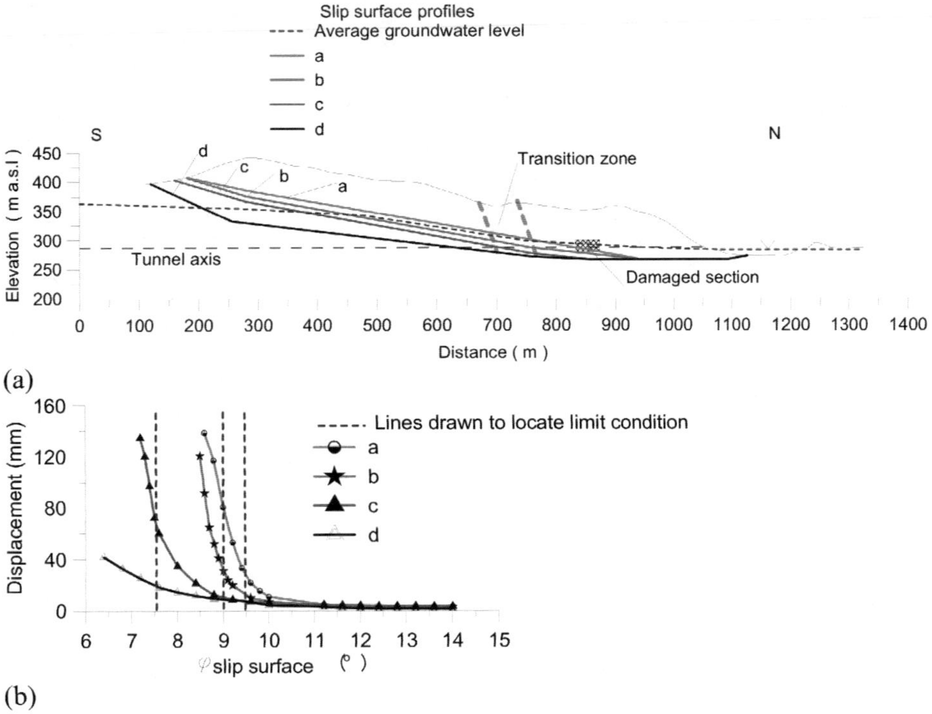

Figure 18. Considered slip surface profiles (a) and their limit condition (b).

7.2. Influence of groundwater table rise and reservoir filling

In a first set of analyses, the water level of the reservoir was kept constant at 282 m a.s.l., while the groundwater table inside the slope rises from the average conditions (AGW) to the maximum level recorded throughout the monitoring period (AGW + 6 m, AGW + 10 m and finally AGW + 15 m). The limit friction angle of the slip surface increases from 7.6 to 8.8° (figure 19). Figure 20 shows the displacement contours when the limit friction angle of slip surface is 7.6° for the case with maximum groundwater level (AGW +15 m).

Then the influence of reservoir filling has been considered. In these analyses a horizontal water table is introduced inside the slope till the intersection point with the average groundwater table profile. The effect of reservoir level increase, up to the maximum project level of 330 m a.s.l, has been analyzed by keeping constant the friction angle of the slip surface (7.6°). Figure 19 shows the "rate" of increase in displacement (displacement increment divided by the increase in water level) obtained for different water pool elevation.

The reservoir filling alters pore pressure only in the lower portion of the slope. The main effect consists in buoyancy forces induced in the rock mass.

As it is shown from the displacement of target point N3 in (figure 19) the rise of reservoir level from 305 to 310 m a.s.l. gives the maximum rate of displacement. The upper portion of the slope, N8 point, experiences a slightly higher displacement as compared to the lower zone.

The slope deforms differently in the upper part, between points F and G, where effective stress is less affected by filling. Vertical displacements are directed upwards between G and H and downwards in the upper zone of the ridge. The surcharge load from the reservoir seems to provide also some restraint to the toe of the slope, even when it is mostly submerged.

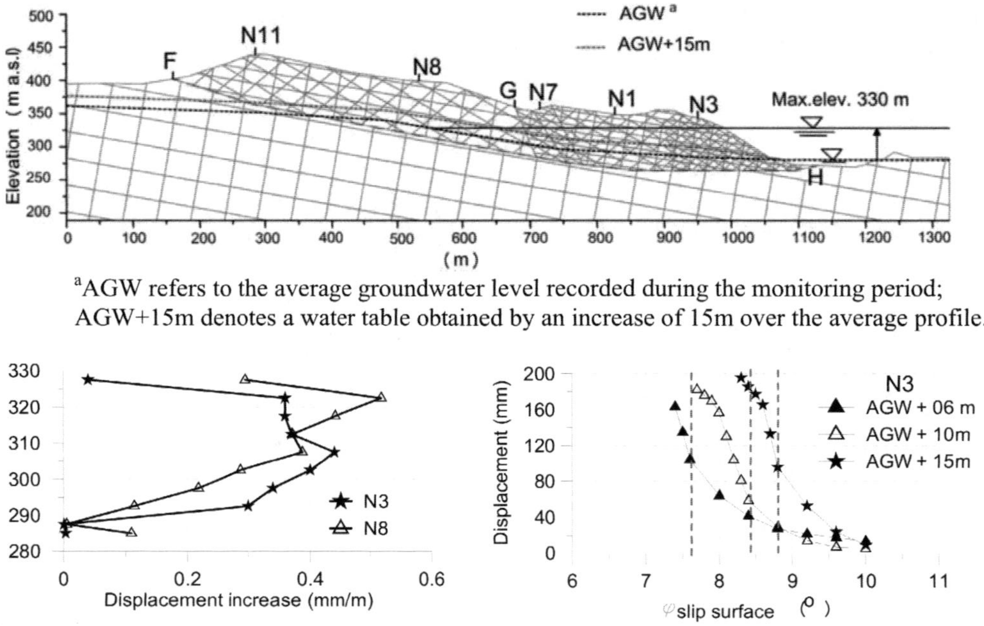

(a) [a]AGW refers to the average groundwater level recorded during the monitoring period; AGW+15m denotes a water table obtained by an increase of 15m over the average profile.

(b) (c)

Figure 19. Average and maximum elevation of groundwater table, maximum level of reservoir considered in DEM modelling (a); horizontal displacement increases calculated for progressive reservoir filling (b); horizontal displacements vs mobilized friction angle curves calculated for different groundwater profiles (c).

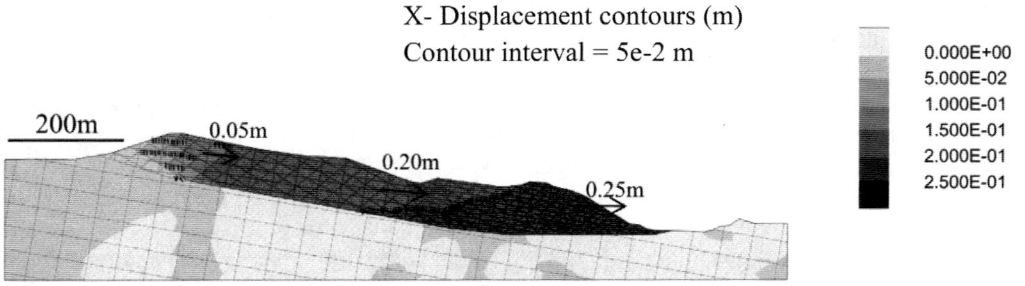

(a)

Figure 20. Horizontal displacement contours when maximum groundwater level (i.e. AGW+15 m) is applied to the model keeping φ-mobilized = 7.6° along the slip surface.

8. Conclusions

The analysis of monitoring data has shown that the slope movement exhibits the typical features of a stationary creep with seasonal reactivation due to change in pore pressure and rainfall regime. Three months cumulative excess rainfall gives better correlation with pore pressure increase and peaks in displacement velocity. The limit strength of the slip surface obtained from DEM modelling is close to the residual friction of the clay gouge.

The influence of the slip surface geometry on the mobilized friction angle has been examined by considering different reasonable hypotheses. It was found the limit friction angle can vary as much as 2°. The influence of block size on model response was investigated particularly for the case of persistent joints. A change in block scale affects deformation and displacement magnitude but has negligible influence on the mobilized friction angle. Also, the joint pattern has limited influence on mobilized strength, but the deformation mode is different for persistent and staggered joint systems. The model with staggered joints exhibits less shearing inside the rock mass and more uniform deformation. Both models can be considered meaningful and instrumental for a better understanding of slope behavior.

The influence of groundwater level rise was investigated with respect to the average groundwater table derived from piezometer measurements within the slope. The minimum friction angle required for stability increases from 7.6 to 8.8° when the groundwater table rises from the average conditions to the maximum expected level. Further analyses were focused on the response of the slope to reservoir filling. The increase in reservoir level determines small displacements and negligible changes of the mobilized friction angle. Vertical displacements are directed upwards in the lower part of the slope, directly affected by buoyancy forces, and downwards in the upper part.

Acknowledgments

Special thanks to Prof. S. Olivero and Engineer S. Cola of EAUT for sharing their data and invaluable experience.

References

[1] Angelucci A, De rosa E, Fierro G, Gnaccolini M, La monica G B, Martinis B, Parea G C, Pescatore T,Rizzini A and Wezel F C 1967 Sedimentological characteristics of some italian turbidites *Geol. Rom.* **VI** pp 345-420

[2] Oberti G, Bavestrello F, Rossi P and Flamigni F 1986 Rock Mechanics Investigations, Design and Construction of the Ridracoli Dam *Rock Mechanics and Rock Engineering* **19** 113-142

[3] Graziani A, Rossini C, Rotonda T 2012 Characterization and DEM modelling of shear zones at a large dam foundation *International Journal of Geomechanics* **6** 648-664

[4] D'Elia B, Picarelli L, Leroueil S and Vaunat J 1998 Geotechnical characterisation of slope movements in structurally complex clay soils and stiff jointed clays *Rivista Italiana Geotecnica* **3** 5-47

[5] Alonso E, Gens A and Lloret A 1993 The landslide of Cortes de Pallas, Spain; a case study *Gèotechnique* **43** 507-521

[6] Hatzor YH & Levin M 1997 The shear strength of clay-filled bedding planes in limestones back analysis of a slope failure on a phosphate mine, Israel *Geotechnical and Geological Engineering* **15** 263-282

[7] Graziani A, Rotonda T and Tommasi P 2009 Stability and deformation mode of a rock slide along interbeds reactivated by rainfall *1st Italian Workshop on Landslides Napoli* **1** 62-71

[8] Itasca 2011 UDEC-Universal Distinct Element Code Version 5 User's Manual *Itasca Consulting Group Inc Minneapolis*

[9] Celluzzi E, Graziani A, Lembo-Fazio A 2014 Analysis of a deep-seated slope movement in a marly- arenaceous formation *Int. Conf. Eurock 2014 Vigo paper* **643** 1-6

[10] Assefa S, Graziani A and Lembo-Fazio A 2014 Analisi di uno scivolamento planare in una formazione marnoso-arenacea. *Incontro Annuale dei Ricercatori di Geotecnica 2014-IARG 2014 Chieti e Pescara,* (in italian)

[11] Baldovin G 1968 La resistenza a taglio delle rocce con alternanza di strati litologicamente eterogenei *Rivista Italiana Geotecnica* **4** 170-181(in italian)

[12] Alonso E and Pinyol N M 2014 Slope stability in slightly fissured claystones and marls *Landslides* 10.1007/s10346-014-0526-5

[13] Grana V and Tommasi P 2013 A deep-seated slow movement controlled by structural setting in marly formations of Central Italy *Landslides* 10.1007/s10346-013-0384-6

[14] Tommasi P, Verrucci L, Campedel P, Veronese L, Pettinelli E and Ribacchi R 2009 Buckling of high natural slopes: The case of Lavini di Marco (Trento-Italy) *Engineering Geology* **109** 93–108

[15] Ribacchi R 1987 Rock mass deformability. Insitu tests, their interpretation and typical results in Italy *2nd international symposium. Field measurements in Geomechanics* **1** 171-192. Kobe

[16] Alonso E E and Pinyol, N M 2010 Criteria for rapid sliding I A review of Vaiont case. *Engng Geol.* **114** No. 3–4 198–210

[17] Kveldsvik V Kaynia A M Nadim F Bhasin R Nilsen B and Einstein H H 2009 Dynamic distinct-element analysis of the 800 m high Åknes rock slope. *Int. J. Rock Mech. Mining Sci.***46** No. 4 686–698

[18] Boon C W, Houlsby G T and Utili S 2014 New insights into the 1963 Vajont slide using 2D and 3D distinct-element method analyses *Gèotechnique* **64** 800-816

[19] Dawson EM, Roth WH and Drescher A 1999 Slope stability analysis by strength reduction. *Gèotechnique* **49**(6) 835–840

[20] Zangerl C, Eberhardt E and Loew S 2003 Ground settlements above tunnels in fractured crystalline rock numerical analysis of coupled hydromechanical mechanisms *Hydrogeology Journal* **11** 162–173

[21] Goodman R E 1989 *Introduction to rock mechanics*, 2nd edn. (New York, NY, USA: Wiley)

[22] Hammah, RE Yacoub T and Curran JH 2009 Variation of failure mechanisms of slopes in jointed rock masses with changing scale. In: Diederichs MS, Grasselli G (eds) *Third Canada-U.S. rock mechanics symposium, Toronto, Canada*, Paper 3956

[23] Brideau M A and Stead D 2012 Evaluating Kinematic Controls on Planar Translational Slope Failure Mechanisms Using Three-Dimensional Distinct Element Modelling *Geotech Geol Eng* **30** 991–1011

Coupled DEM-CFD Investigation of Granular Transport in a Fluid Channel

T. Zhao[1*], F. Dai[1], N. W. Xu[1]

[1]State Key Laboratory of Hydraulics and Mountain River Engineering, College of Water Resource and Hydropower, Sichuan University, 610065, China

[*] E-mail: zhaotao@scu.edu.cn

Abstract This paper presents three dimensional numerical investigations of granular transport in fluids, analysed by the Discrete Element Method (DEM) coupled with Computational Fluid Mechanics (CFD). By employing this model, the relevance of flow velocity and granular depositional morphology has been clarified. The larger the flow velocity is, the further distance the grains can be transported to. In this process, the segregation of solid grains has been clearly identified. This research reveals that coarse grains normally accumulate near the grain source region, while the fine grains can be transported to the flow front. Regardless of the different flow velocities used in these simulations, the intensity of grains segregation remains almost unchanged. The results obtained from the DEM-CFD coupled simulations can reasonably explain the grain transport process occurred in natural environments, such as river scouring, evolution of river/ocean floor, deserts and submarine landslides.

Keywords granular transport, fluid channel, DEM-CFD, segregation

1. Introduction

Granular transportation and deposition in fluids are relevant to subjects of public safety, water research management, and environmental sustainability [1]. Actually, this topic is at the centre of a wide range of practical and fundamental subjects. For example, this process is very important in understanding how beaches, rivers and deltas evolve, mountains erode and landscapes form. The related research topics can be the sediment transport in rivers, along lake and ocean shores, transport of desert sand, soil erosion, drift of snow and so on. Based on previous researches, the coarse and fine grains behave differently regarding their motions in fluid [2]. In this process, the segregation of grains along the flow path and the depositional depth is dominant. This phenomenon has been widely observed in river scouring, evolving of ocean costal line and submarine landslides [3]. For instance, in typical submarine landslides, the fine grains can be transported to the sliding front, while the majority of coarse grains normally deposit near the slope source region [4]. In this process, the displaced fine grains are entrained by fluid currents and can normally travel to very long distances, representing significant hazards to the offshore and seabed infrastructures, such as the telecommunication cables, pipelines, wind and tidal turbines [3,5-8]. In addition, the grain sizes can characterize different river patterns, and by bridging the regional hydrology and geomorphology, it can also reflect the sediment transport process [9]. For example, the downstream variation of riverbed grain sizes would reflect the morphological evolution of the river channel [10]. Therefore, it is important to understand the characteristics of granular transport in fluid and depositional morphology of sediments.

The current research investigates the transport of solid grains driven by open channel flows via a coupled discrete-continuous method, aiming to provide new insights into the relationship between the flow velocity and granular depositional morphology.

2. Model configuration

The numerical investigation of granular transport reported in this paper is based on the idea that the motion of particles is completely governed by the Newtonian equations of motion, in which the inter-particle collisions are modelled by the soft particle approach and the fluid flow is calculated by the

Navier-Stokes equations [11-13]. Consequently, the Discrete Element Method (DEM) [14] and Computational Fluid Dynamics (CFD) [15] techniques can be used to study the mechanical and hydraulic behaviour of particles and fluid flow, respectively. By coupling these two methods, a complete analysis of the granular transport in a fluid channel can be made [16]. In this study, the DEM and CFD open source codes ESyS-Particle [17,18] and OpenFOAM [19] were employed for the simulations presented herein. The governing equations of particle motion are expressed as below:

$$m_i \frac{d^2}{dt^2}\vec{x_i} = m_i\vec{g} + \vec{F_c} + \overrightarrow{f_{fluid}} \tag{1}$$

$$I_i \frac{d}{dt}\vec{\omega_i} = \sum_c \vec{r_c} \times \vec{F_c} + \overrightarrow{M_r} \tag{2}$$

where m_i is the mass of a single particle i; $\vec{x_i}$ is the position of its centroid; $\vec{g}$ is the gravitational acceleration; $\vec{F_c}$ is the contact force exerted by the neighbouring particles, which is calculated using a linear-spring model [14]; $\overrightarrow{f_{fluid}}$ is the force exerted by fluid flow on the particle; $I_i = 2m_i r^2/5$ is the moment of inertia about the grain geometric centroid, with r being the particle radius; $\vec{\omega_i}$ is the angular velocity; $\vec{r_c}$ is the vector from the particle mass center to the contact point; $\overrightarrow{M_r}$ is the rolling resistance moment induced by the non-spherical particle shape and interlocking between particles.

The force exerted by the fluid flow on the particle $\left(\overrightarrow{f_{fluid}}\right)$ consists of two components: the buoyant $\left(\vec{f_b^i}\right)$ and drag forces $\left(F_{di}\right)$. The buoyant force is calculated as follows:

$$\vec{f_b^i} = -v_{pi}\nabla p \tag{3}$$

where v_{pi} is the volume of particle i; p is the pressure of fluid flow.

The drag force is induced by the relative motion between particle and fluid [20]. It is calculated by the Di Felice [21] empirical correlation as:

$$F_{di} = \frac{1}{2} C_d \rho_f \frac{\pi d_p^2}{4} n^2 |U - V| (U - V) n^{-(\chi+1)} \tag{4}$$

where $C_d = \frac{24}{\text{Re}}\left(1 + 0.150\,\text{Re}^{0.681}\right) + \frac{0.407}{1 + 8710/\text{Re}}$ is the drag force coefficient [22]; ρ_f and U are the fluid density and velocity; d_p and V are diameter and velocity of a particle, respectively; $\chi = 3.7 - 0.65\exp\left[-\frac{(1.5 - \log_{10}\text{Re})^2}{2}\right]$, with $\text{Re} = \rho_f dn|U - V|/\mu$ the Reynolds number defined at the particle size scale, with n being the fluid volume fraction in CFD mesh cells, μ being the fluid viscosity. In the current analyses, χ ranges from 3.4 to 3.7.

The governing equations of fluid flow are shown as follows:

$$\frac{\partial(\rho_f n)}{\partial t} + \nabla \cdot (\rho_f n\mathbf{U}) = 0 \tag{5}$$

$$\frac{\partial(\rho_f n\mathbf{U})}{\partial t} + \nabla \cdot (\rho_f n\mathbf{U}\mathbf{U}) - n\nabla \cdot \boldsymbol{\tau} = -n\nabla p + n\rho_f \vec{g} + \overline{\boldsymbol{f}_d} \tag{6}$$

where $\overline{f_d} = \sum_{i=1}^{N} F_{di} \Big/ V_{mesh}$ is the drag force per unit fluid volume, with V_{mesh} being the volume of a fluid mesh cell and N being the number of particles within the cell; $\boldsymbol{\tau}$ is the fluid viscous stress tensor.

The input parameters are listed in Table 1. The values of these input parameters have been calibrated by research reported in Utili *et al.* [18] and Zhao *et al.* [2].

Table 1. Input parameters for the simulations presented herein.

DEM Parameters	**Values**	**CFD Parameters**	**Values**
Particle diameter, d (mm)	[1.8, 3.8]	Fluid density, ρ_f (kg/m^3)	1000
Granular density, ρ_s (kg/m^3)	2650	Viscosity, μ (Pa·s)	0.001
Normal stiffness, K_n (N/m)	3.0×10^7	Flow velocity, U (m/s)	0.1, 0.5, 1.0
Shear stiffness, K_s (N/m)	2.7×10^7	**Simulation Parameters**	**Values**
Particle friction angle, ϕ_μ (°)	30	Gravity, g (m/s^2)	-9.81
Coefficient of rolling stiffness, β	1.0	DEM Time step size, Δt_s (s)	10^{-7}
Coefficient of plastic moment, η	0.1	CFD Time step size, Δt_f (s)	10^{-5}
		Coupling frequency*, α	100

* The coupling frequency is the number of iteration steps used in the DEM in one coupling interval.

In the DEM model, the linear-elastic and rolling resistant contact model has been used. Here, the rolling law is employed with the only aim of accounting for the shape effect of non-spherical particles (as quantified by the two parameters, β and η). The law accounts for moments arising from the fact that the line of action of the normal contact force in the case of non-spherical particles no longer passes through the centre of mass of the particles and hence generating rotational moments [23,24].

The model configuration of the granular transport is shown in Figure 1.

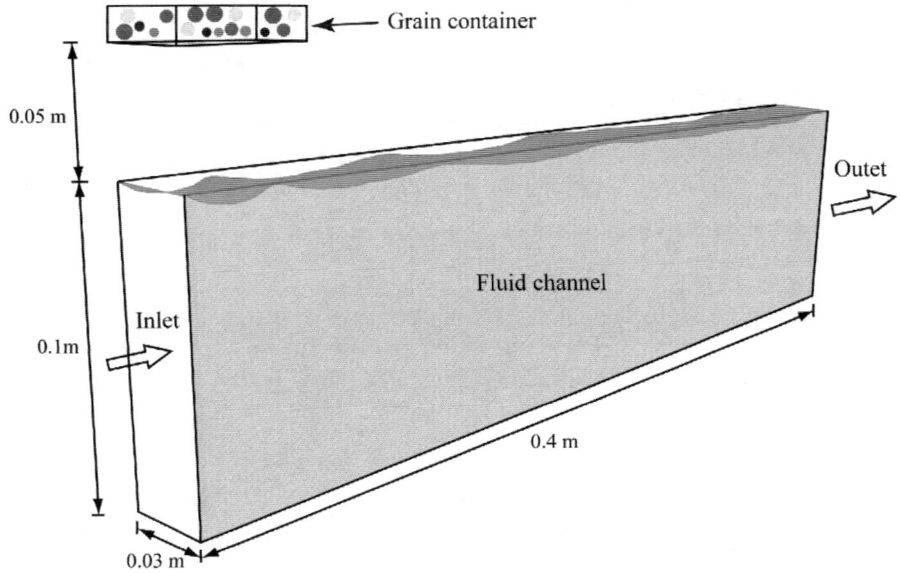

Figure 1. Schematic view of the grain transport model

As shown in Figure 1, the granular transport model consists of the solid phase (DEM component) and the fluid phase (CFD component). During the simulation, the two components work as follows:

IOP Conf. Series: Earth and Environmental Science **26** (2015) 012016 doi:10.1088/1755-1315/26/1/012016

1) The DEM component

(1) Grain container: a DEM engine (coded as a C++ class) can continuously generate polydispersed particles within a cubic box. Particles are coloured as red, pink, yellow, blue and cyan as their diameters decrease gradually. The grain container is placed 0.05 meter above the fluid channel.

(2) An initial downward velocity (2 m/s) has been set to all grains, so that they can drop quickly into the open water channel and thus speed up the simulation.

(3) Coarse granular bed: a layer of polydispersed coarse grains are paved on the bottom of the water tank to simulate the frictional floor.

(4) In this research, he height of the grain container and the initial granular downward velocity would determine the grain velocities near the upper surface of the fluid channel. The choice of these two parameters turns out to be crucial for this type of simulations. However, we also understand the granular velocity would reach a constant value in fluid, as determined by the grain diameter and fluid viscosity [2]. Thus, the height of grain container and initial downward velocity are chosen without any detailed parametric analyses.

2) The CFD component:

(1) An open fluid channel has been placed horizontally on the floor. The dimension of the fluid channel is 0.03 m × 0.4 m × 0.1 m. The upper boundary has been set as open air condition. Left and right side walls and the bottom floor have slip boundary conditions. The back and front boundaries are inlet and outlet of the flow channel.

(2) A constant fluid flux (Q) is maintained by pumping water into the tank from the inlet at a constant velocity.

(3) The inlet flow velocity has been set as 0.1 m/s, 0.5 m/s and 1.0 m/s in a series of simulations.

3. Results

The granular transport for fluid flow velocities of 0.1 m/s, 0.5 m/s and 1.0 m/s are illustrated as a series of snapshots, as shown in Figure 2. According to Figure 2, the solid grains are continuously generated in the grain container and dropped downwards into the flow channel. Grains falling into the channel are entrained in the flow, as the fluid flows from the left to the right sides. As the flow velocity increases from 0.1m/s to 1.0 m/s, the majority of grains are transported to a long distance away from their source region. This is especially true for fluid velocity equals to 1.0 m/s, in which all grains accumulate in the outlet region of the channel (the DEM model has been set as an enclosed parallelepiped, such that the grains cannot move out of the DEM domain.). After a distance of transport, the flow velocity might slow down, or other mechanisms such as the gravity of grains, friction forces acting on grains, will cause the grains to deposit along the base of the channel. The coarse grains will settle quickly and deposit near the flow inlet region, while the fine grains can suspend in the fluid and be transported to the flow front. Finally, all grains will settle continuously onto the base of the channel. In Figure 2 (a) (b), the segregation of solid grains is evident, that most of the coarse grains (coloured as red) are deposited near the flow inlet region, while finer grains can be transported further to locations near the flow outlet region. In the meantime, the fluid velocity increases near the deposit surface. The distribution of flow velocity field changes with the shape of grain deposits. However, the grain segregation phenomenon is not evident for simulations shown in Figure 2 (c), in which fast moving flow can transport all grains to the outlet region and grains with different sizes are mixed in the deposit.

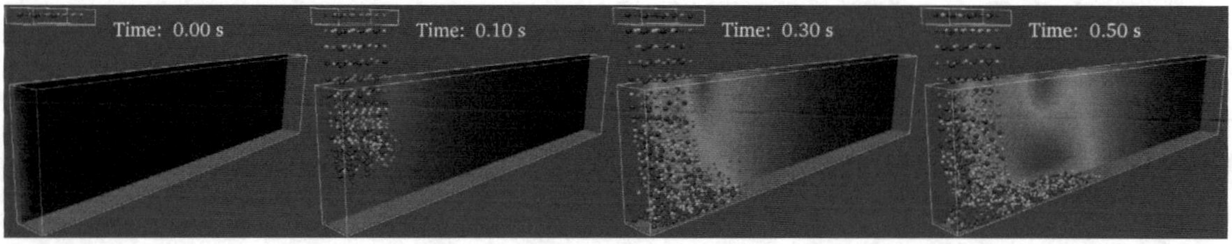

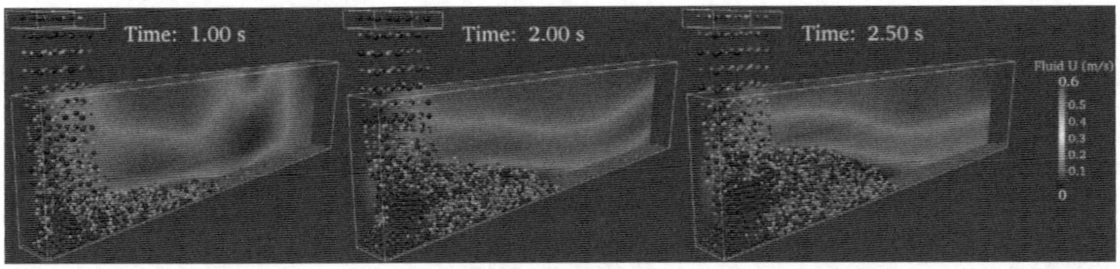

(a) transport of grains for U = 0.1 m/s

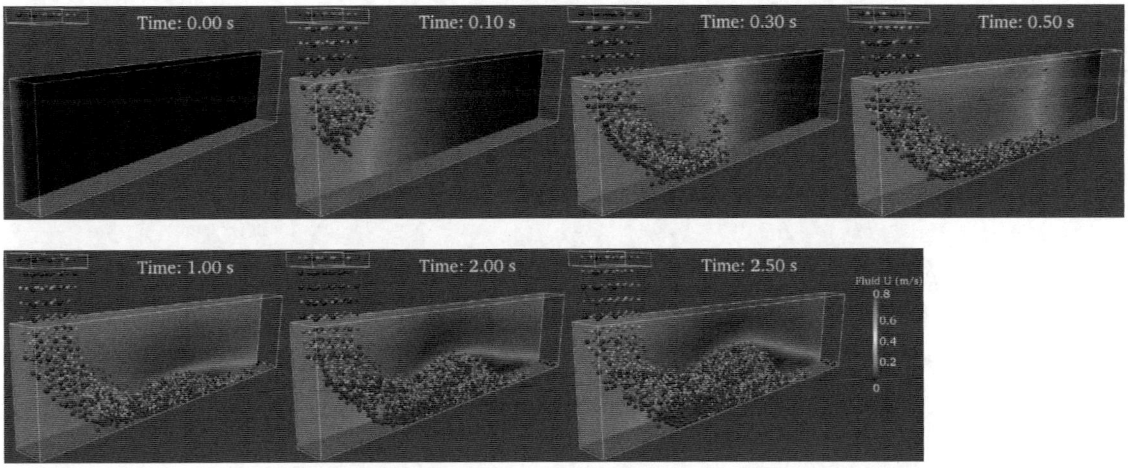

(b) transport of grains for U = 0.5 m/s

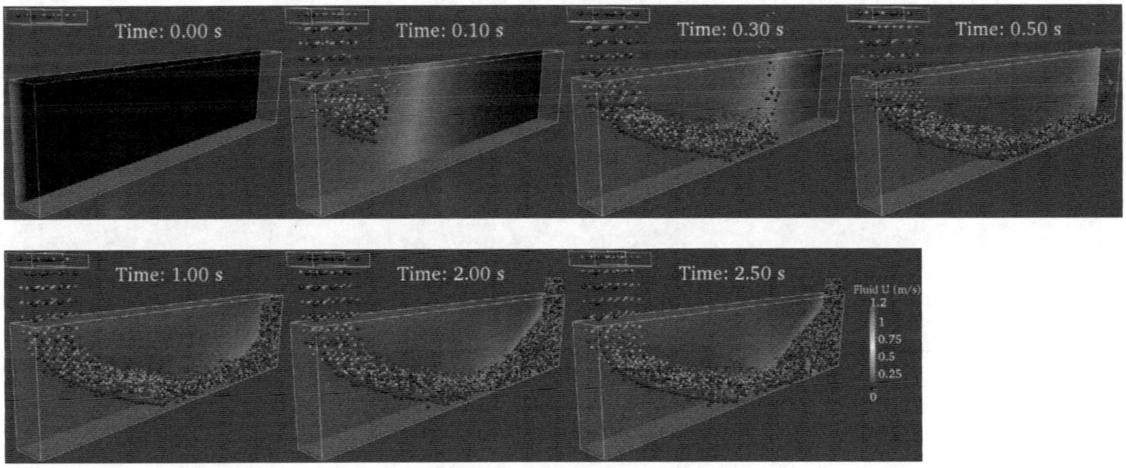

(c) transport of grains for U = 1.0 m/s

Figure 2. Snapshots of grain transport in a fluid channel (the background is the flow velocity contour)

In order to quantify the grain segregation phenomenon shown in Figure 2, a series of particle size distribution (PSD) curves of sediments at the time 2.5 seconds after the initiation of flows are plotted below in Figure 3. In this section, the stable sediments layer has been divided into three parts (*i.e.* left, middle and right), with each part containing approximately the same amount of grains. According to Figure 3, it can be observed that regardless of the flow velocity in these three different tests, the PSD curves have similar shapes. The majority of coarse grains accumulate at the left section, as represented by a PSD curve dramatically different from the initial PSD curve. For grains in the middle section, the PSD curve deviates below the initial PSD curve, indicating that the portion of accumulated coarse grains in this region is larger than the initial value at generation. In the right section, the PSD curve locates slightly above the initial PSD curve, as the majority of fine grains have been transported there.

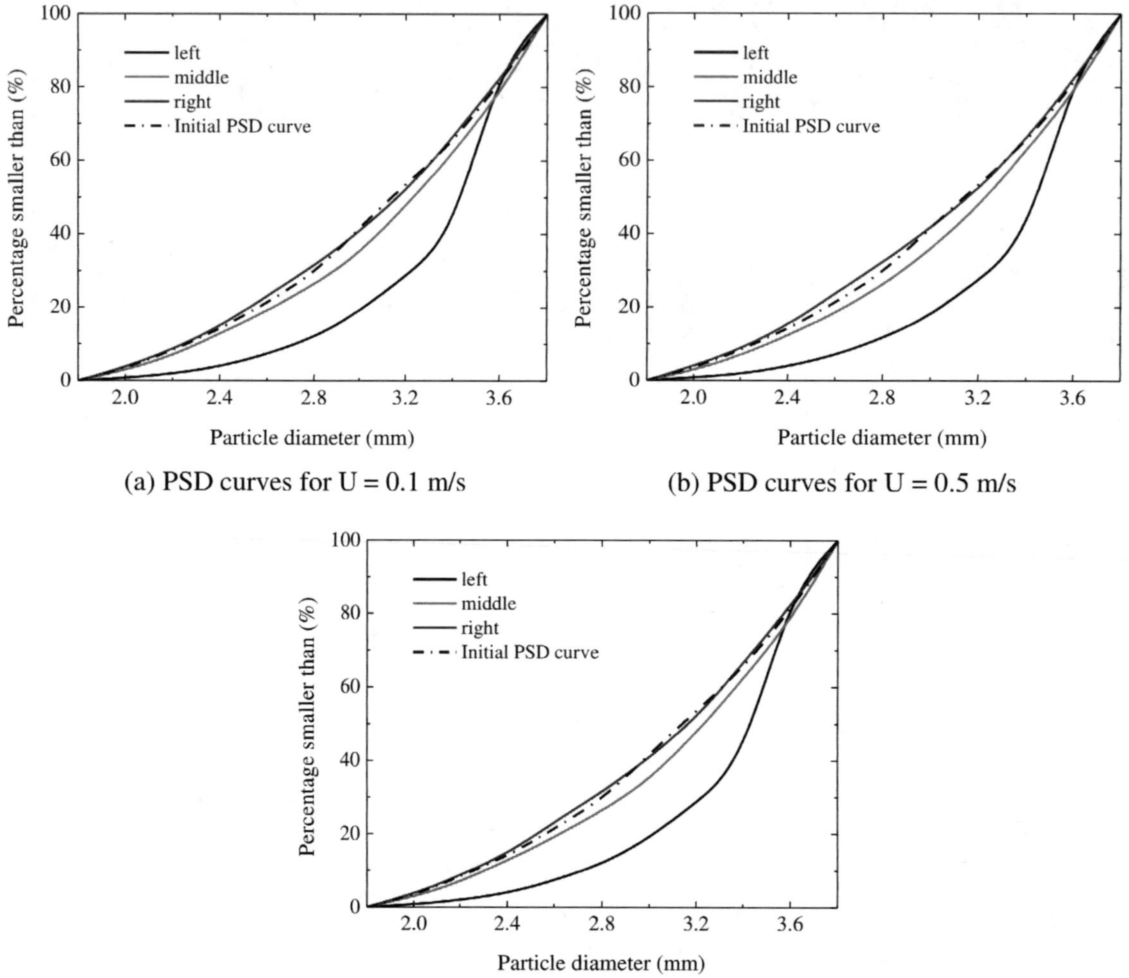

(a) PSD curves for U = 0.1 m/s

(b) PSD curves for U = 0.5 m/s

(c) PSD curves for U = 1.0 m/s

Figure 3. Particle size distribution of sediments in different tests

4. Conclusions

This paper presents a numerical investigation of the behaviour of granular transport in a fluid channel via the DEM-CFD coupling simulations. By varying the flow velocity in the channel, three distinctive types of granular transport phenomena has been observed. The main concluding points are summarized as below:

(1) The locations of the deposited grains are closely related to the flow velocity. The larger the flow velocity is, the longer distance the grains can be transported to.
(2) The segregation of solid grains is evident during the granular transport, in which the majority of coarse grains would accumulate near the source region, while the fine grains can be transported to the flow front. Though the flow velocity is dramatically different for a series of simulations, the PSD curves of the sediments are almost the same.
(3) The discussions presented above can reasonably explain the mechanisms of grain segregation occurred in river scouring, evolution of river/ocean floor, deserts and submarine landslides.

5. Acknowledgements

This work is supported by the research grant from the 'Program of Young scholars from Renowned Universities or Institutes' at Sichuan University, China， Program for New Century Excellent Talents in University (NCET-13-0382) and the Youth Science and Technology Fund of Sichuan Province (2014JQ0004).

6. References

[1] P. Frey and M. Church, Science **325**, 1509 (2009).
[2] T. Zhao, G. T. Houlsby, and S. Utili, Granular Matter **16**, 921 (2014).
[3] M. A. Hampton, H. J. Lee, and J. Locat, Reviews of Geophysics **34**, 33 (1996).
[4] V. Topin, Y. Monerie, F. Perales, and F. Radjai, Phys Rev Lett. **109** (2012).
[5] F. Legros, Engineering Geology **63**, 301 (2002).
[6] H. J. Lee, W. C. Schwab, and J. S. Booth, in *Submarine Landslides: Selected Studies in the U.S. Exclusive Economic Zone* (U.S. Geological Survey Bulletin 2002, 1991).
[7] J. Locat and H. J. Lee, Canadian Geotechnical Journal **39**, 193 (2002).
[8] D. G. Moore, Journal of Sedimentary Research **31**, 343 (1961).
[9] O. P. Harmar and N. J. Clifford, Geomorphology **84**, 222 (2007).
[10] Z. Wang, Z. Chen, M. Li, J. Chen, and Y. Zhao, Geomorphology **113**, 217 (2009).
[11] W. Kalthoff, S. Schwarzer, G. Ristow, and H. J. Herrmann, International Journal of Modern Physics C **07**, 543 (1996).
[12] B. Wachmann, W. Kalthoff, S. Schwarzer, and H. J. Herrmann, Granular Matter **1**, 75 (1998).
[13] V. Komiwes, P. Mege, Y. Meimon, and H. Herrmann, Granular Matter **8**, 41 (2006).
[14] P. A. Cundall and O. D. L. Strack, Géotechnique **29**, 47 (1979).
[15] J. D. Anderson, *Computational fluid dynamics: the basics with applications* (McGraw-Hill, 1995), p.^pp. 1-547.
[16] F. Chen, E. C. Drumm, and G. Guiochon, Computers and Geotechnics **38**, 638 (2011).
[17] S. Abe, D. Place, and P. Mora, Pure appl. geophys. **161**, 2265 (2004).
[18] S. Utili, T. Zhao, and G. T. Houlsby, Engineering Geology (2014).
[19] OpenFOAM - The open source CFD toolbox, http://www.openfoam.com/.
[20] K. D. Kafui, C. Thornton, and M. J. Adams, Chemical Engineering Science **57**, 2395 (2002).
[21] R. Di Felice, International Journal of Multiphase Flow **20**, 153 (1994).
[22] P. Brown and D. Lawler, Journal of Environmental Engineering **129**, 222 (2003).
[23] M. J. Jiang, H. S. Yu, and D. Harris, Computers and Geotechnics **32**, 340 (2005).
[24] N. Belheine, J. P. Plassiard, F. V. Donzé, F. Darve, and A. Seridi, Computers and Geotechnics **36**, 320 (2009).

Runout analysis of landslides using material point method

Y SUN, J YANG and E SONG
Department of Civil Engineering, Tsinghua University, Beijing 100084, China

E-mail: sunyj12@mails.tsinghua.edu.cn

Abstract. Dynamic simulation of runout landslides is essential for assessment, prevention and mitigation of landslide-induced disasters, but difficult for traditional numerical methods. This paper use the material point method (MPM) to simulate the long runout landslides. The basic theories and the solution procedure of MPM are presented first. Then a landside experiment with the soil modelled by aluminium bars is simulated to validate this method for landslide analysis. The numerical results of the final configuration and the failure surface of the landslide agree well with the experiment results. Finally, the landslide of Erman Mountain in China Sichuan province is simulated. The velocities, affected areas, flow intensity and impact forces on structures are investigated.

1. Introduction

The simulation of long runout landslides such as debris flows, avalanches and large-scale liquefaction failures is very important for determining affected areas and flow intensity, which are essential for assessment, prevention and mitigation of these disasters[1]. Traditional numerical methods such as finite element method, finite volume method, finite difference method and boundary element method are hard to deal with problems in these fields due to mesh tangling, free surface tracking, dissipation problem associated with convection and inherent problems such as modelling slipping, contact and separation. Some meshfree methods such as smoothed particle hydrodynamics and discrete element method suffer from lower accuracy and higher computational cost. The material point method (MPM), developed recently, has enormous advantages in managing problems involving very large deformation and dynamic process[2,3].

MPM developed by Sulsky et al.[4,5] is an extension of the particle-in-cell method [6] to solid mechanics problems. A solid is discretized by a collection of particles (or material points), which carry all the material information. The particles are followed throughout the deformation of the solid and provide a Lagrangian description. A background grid, which can be held fixed or adapted as needed, is used to determine spatial gradients, formulate and solve the governing equations of updated Lagrangian form. The MPM combines the advantages of both Eulerian and Lagrangian schemes, the problem of mesh distortion, which leads to numerical difficulties in the Lagrangian frame, is avoided. The MPM is most suitable for the simulation of large deformation problems.

This paper presents the simulation of long runout landslides by the MPM. The simulation procedure is as follows: (1) The initial stress field within the stable slope is generated with K_0 procedure or gravity loading. (2) A landslide is triggered by decreasing the strength of the soil[7]. (3) The dynamic evolution of the long runout landslides is simulated. At first, a simple numerical example of a 2D landslide experiment conducted by Bui et al.[8] is presented to investigate the accuracy of the MPM in runout analysis. The numerical results of the final configuration and the failure surface of the landslide agree well with the experiment results. Then, the landslide of Erman Mountain in China Sichuan province, which caused 20 casualties and the destruction of many buildings with about one million cubic meters of debris on July 27, 2010, is simulated by the MPM. The velocities, affected areas, flow intensity and impact forces on the structures are investigated.

2. Material point method

The material point method can be regarded as a special updated Lagrangian finite element method with material point discretization of material and finite element discretization of space. Within the MPM, large deformation is handled by particles moving through a computational grid, which is fixed in space. The followings are the basic theories of MPM[9].

2.1. Governing equations

The governing equations on the current configuration can be formulated in updated Lagrangian form as follows,

$$\sigma_{ij,j} + \rho b_i = \rho \ddot{u}_i, \qquad \text{in } \Omega \tag{1}$$

with the boundary conditions

$$\begin{cases} n_j \sigma_{ij} = \overline{T}_i, & \text{on } S_\sigma \\ u_i = \overline{u}_i, & \text{on } S_u \end{cases} \tag{2}$$

where i,j=1,2,3 are spatial coordinate indices and obey Einstein summation convention; σ_{ij} is the Cauchy stress tensor; ρ is the current material density; b_i is the body force per unit mass; u_i is the displacement, a superposed dot denotes material differentiation with time; Ω, S_σ and S_u are the material domain, the traction boundary and the displacement boundary, respectively; n_j is the unit outward normal vector of S_σ; $\overline{T}_i$ is the surface traction; $\overline{u}_i$ is the displacement of S_u.

The weak form of the governing equations is

$$\int_\Omega \delta u_{i,j} \rho \sigma^s_{ij} dV + \int_\Omega \delta u_i \rho \ddot{u}_i dV - \int_\Omega \delta u_i \rho b_i dV - \int_{S_\sigma} \delta u_i \rho \overline{T}^s_i dA = 0 \tag{3}$$

where $\sigma^s_{ij} = \sigma_{ij}/\rho$ and $\overline{T}^s_i = \overline{T}_i/\rho$.

2.2. Material point discretization of material

In MPM, solid bodies are discretized into particles (see Figure 1), termed material points, which carry all the state information such as masses, velocities and stresses.

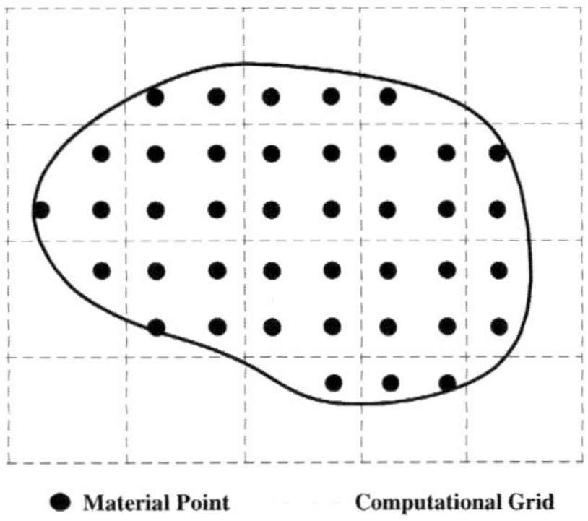

Figure 1. Material point discretization.

The density of an arbitrary point of the soil body is approximated by

$$\rho(x_i) = \sum_{p=1}^{n_p} m_p \delta(x_i - x_{ip}) \tag{4}$$

where the subscript p denotes the material point number, x_i (i=1,2,3) is the coordinate of the point, n_p is the number of material points, m_p is the mass of material point p, $\delta(x)$ is Dirac Delta function, x_{ip} is the coordinate of material point p.

2.3. Finite Element Discretization of space

The computational space is subdivided into finite elements e=1 to n_e with n_I nodes. Inside each computational step, the computational grid is binded to the material points and deform with the material points. Any variables on the material point can be interpolated from the computational grid nodes by interpolation functions $N_I = N_I(\boldsymbol{x})$. Linear B-spline interpolation function is adopted in standard MPM.

The finite element trial function u_i is

$$u_i = N_I u_{iI} \tag{5}$$

The test function δu_i is

$$\delta u_i = N_I \delta u_{iI} \tag{6}$$

where the subscript I denotes the computational grid node number. u_{iI} and δu_{iI} are the nodal value of the trial function and test function, respectively. δu_{iI} is zero on displacement boundary S_u and arbitrary otherwise.

2.4. Global equations

Substituting Equation (4),(5) and (6) into Equation (3) leads to

$$m_{IJ} \ddot{u}_{iJ} = f_{iI}^{\mathrm{ext}} - f_{iI}^{\mathrm{int}}, \quad x_I \notin S_{\mathrm{u}} \tag{7}$$

where mass

$$m_{IJ} = \sum_{p=1}^{n_p} m_p N_{Ip} N_{Jp} \tag{8}$$

In MPM, the lumped mass matrix

$$m_{II} = \sum_{J=1}^{n_I} \sum_{p=1}^{n_p} m_p N_{Ip} N_{Jp} = \sum_{p=1}^{n_p} m_p N_{Ip} \tag{9}$$

is usually adopted to reduce the cost of calculation, and the Equation (7) becomes

$$m_{II} \ddot{u}_{iI} = f_{iI}^{\mathrm{ext}} - f_{iI}^{\mathrm{int}}, \quad x_I \notin S_{\mathrm{u}} \tag{10}$$

The external force

$$f_{iI}^{\mathrm{ext}} = \sum_{p=1}^{n_p} N_{Ip} m_p b_{ip} + \sum_{p=1}^{n_p} N_{Ip} \frac{m_p}{\rho_p} \bar{T}_{ip} h^{-1} \tag{11}$$

where h is the thickness of the traction boundary surface, and the internal force

$$f_{iI}^{\mathrm{int}} = \sum_{p=1}^{n_p} N_{Ip,j} \frac{m_p}{\rho_p} \sigma_{ijp} \tag{12}$$

Adopting an explicit scheme over time increment Δt, we get the updated velocities of the background grid nodes from Equation (10).

$$\dot{u}_{iI}^{k+1} = \frac{f_{iI}^{\text{ext}} - f_{iI}^{\text{int}}}{m_{II}} \Delta t + \dot{u}_{iI}^{k} \tag{13}$$

with the subscript k being a time step counter.

The maximum time step is estimated using the CFL stability condition[10]

$$\Delta t_{\text{cr}} \approx \min(\frac{\Delta x}{c_p}) \tag{14}$$

where Δx is the cell length, c_p is the longitudinal wave speed. The time step size $\Delta t = \alpha_c \Delta t_{\text{cr}}$, α_c (≤ 1) is the time step scale factor called Courant number.

2.5. Solution procedure

Each calculation step of MPM can be classified in three phases[11]: Initial phase where the information of the material points is mapped to the computational grid node to provide initial values for the solution of global equations; Lagrangian phase where the global equations are setup within updated Lagrangian framework and solved to get the nodal velocities and displacements, then the state information, such as locations, velocities and stresses, of the material points are updated; and the convective phase where the computational grid is reset, while keeping the location of the material points constant.

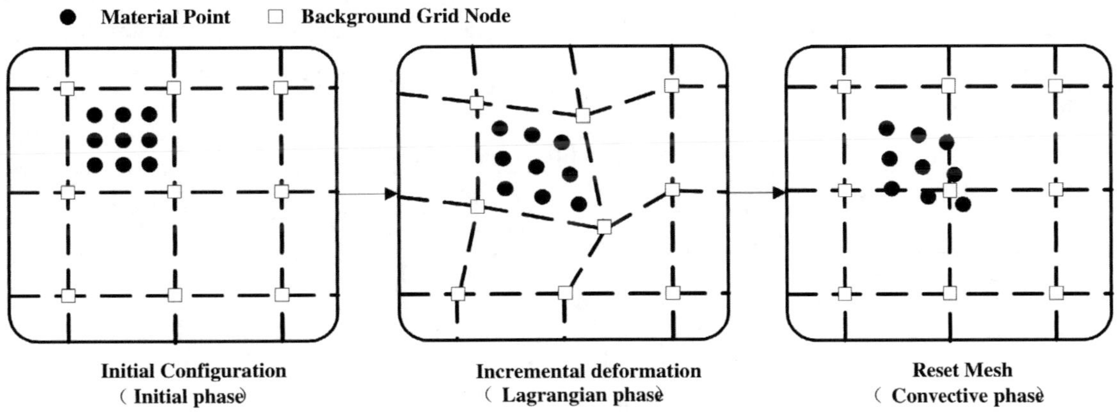

Figure 2. Three phases of one computational step with MPM.

The numeric procedure of one MPM step is as follows.

(1) Map the momenta, external forces and internal forces of material points to the regular background grid nodes

(2) Impose the essential boundary conditions on the background grid nodes.

(3) Solve Equation (13) to update the nodal velocities.

(4) Update the velocities and positions of the material points using interpolation of the nodal accelerations and updated nodal velocities, respectively.

(5) Remap the velocities of the material points to the background grid nodes to calculate the incremental strain tensor and the incremental vorticity tensor of the material points. Then update the stress tensor by an objective constitutive model, as well as the density of the material points. In this paper, we adopt an incremental elastoplastic constitutive model

$$\sigma_{ij}^{\nabla} = C_{ijkl}^{ep} \dot{\varepsilon}_{kl} \tag{15}$$

with Jaumann stress rate:

$$\sigma_{ij}^{\nabla} = \dot{\sigma}_{ij} - W_{jk}\sigma_{ki} - W_{ik}\sigma_{kj} \tag{16}$$

where, W_{ij} is vorticity tensor

$$W_{ij} = \frac{1}{2}\left(\frac{\partial v_i}{\partial x_j} - \frac{\partial v_j}{\partial x_i}\right) \tag{17}$$

and the strain tensor

$$\dot{\varepsilon}_{kl} = \frac{1}{2}\left(\frac{\partial v_i}{\partial x_j} + \frac{\partial v_j}{\partial x_i}\right) \tag{18}$$

(6) Reset the deformed grid and use the initial regular background grid in the next step.

3. Numerical examples

Two examples are presented to demonstrate the capabilities of the MPM in capturing the behaviour of long runout landslide, that is: a landslide experiment simulation; Erman Mountain landslide simulation.

3.1. Landslide experiment simulation

A plane strain experiment of soil collapse was conducted by Bui et al.[8] using small aluminum bars to model the soil. To determine the parameters for the Mohr Coulomb constitutive model of the aluminum bars, they conducted a shear box test. The material parameters are shown in Table 1. This soil model was initially arranged into a rectangular area (200mm×100mm), which is generated by standing two flat solid walls on a flat surface. The simulated slide was triggered by quickly removing the right wall horizontally to the right.

Table 1. Material constants of the aluminum bars

Density(ρ)	Young's modulus(E)	Poisson's ratio(ν)	Cohesion(c)	Friction angle(φ)	Dilatancy angle(ψ)
2650 kg/m^3	0.84 MPa	0.3	0 kPa	19.8°	0°

The numerical model is constructed in plane strain condition with symmetric boundary condition and frictional boundary condition at the left and bottom side, respectively. The friction angle δ between the aluminum bars and the flat surface at the bottom equal to the friction angle of aluminum bars, that is $\delta=\varphi$. The calculation is performed using square elements with the side length of Δx=2 mm and a time step size of Δt= 48.41μs. Initially, There are 4 material points per element. The initial stress field is generated by K_0 procedure with K_0=1-sin(φ).

a.

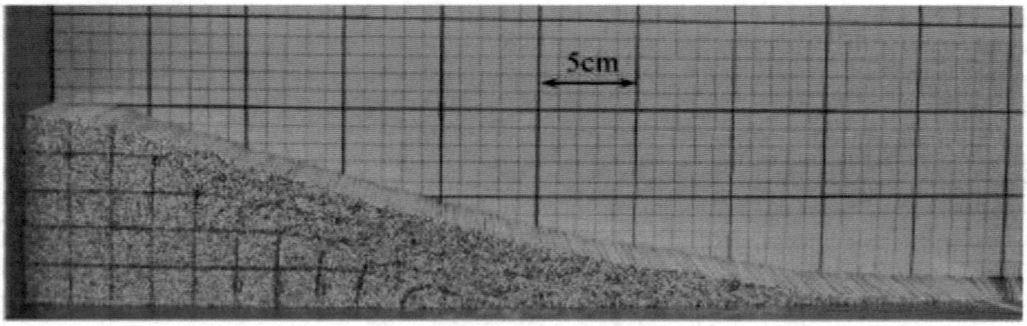

b.

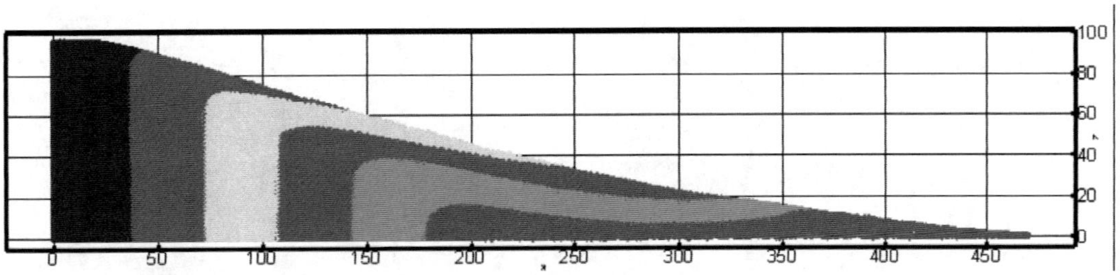

c.

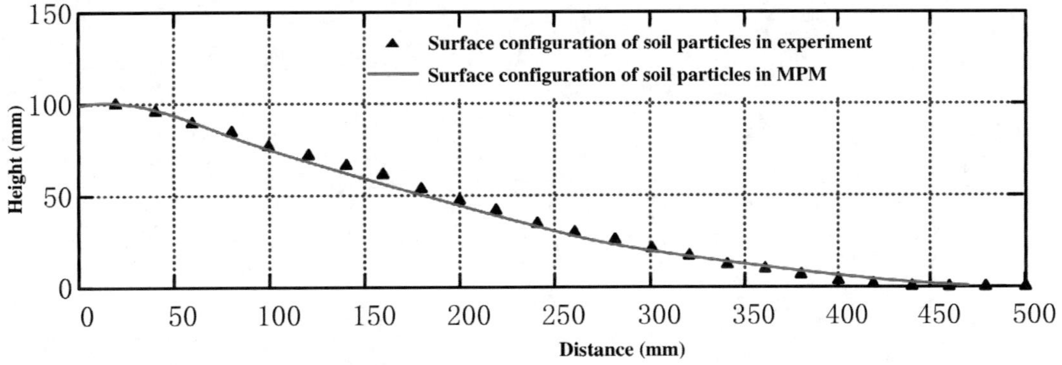

Figure 3. Comparison between experiment and MPM simulation of soil failure.

(a) Experimental result; (b) MPM simulation result; (c) Comparison of surface configurations.

Figure 3 shows the results of the experiment and the computation. Figure 3a is the failure configuration of the experimental result. Figure 3b shows the failure configuration of the MPM simulation result. Figure 3c is the comparison of the free surfaces of the final configuration by the experimental and MPM simulation. From the comparison, we can find that the results of the MPM simulation agree well with those of the experiment. So MPM has the capacity to simulate problems involving very large deformation.

3.2. Erman Mountain landslide simulation

The Erman Mountain lies in China Sichuan province and the landslide took place on July 27, 2010 affecting approximately 48 $\times 10^4 m^3$ of weathered basalt. The landslide was caused by long time rainfall and consisted of two sliding phases. The first sliding caused 20 casualties and destroyed 5 buildings. After 30mins of the first sliding, the secondary landslide started and ran into the Wangong town at foot of the mountain. The secondary landslide damaged 92 buildings and the evacuation 1500 people[12]. The images of the landslide are shown in Figure 4 and the geological longitudinal section (A-A') is shown in Figure 5.

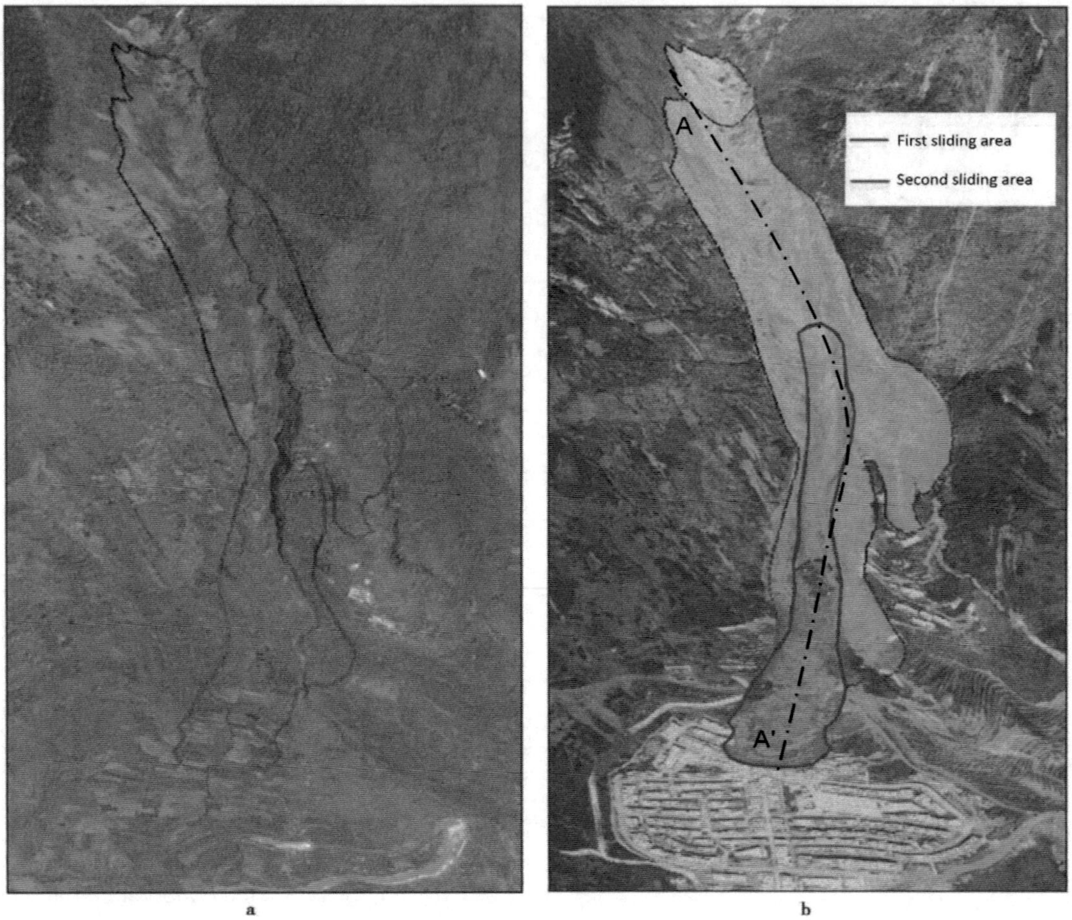

Figure 4. The remote sensing images before and after the Erman Mountain landslide[12].
a. before landslide; b. after landslide

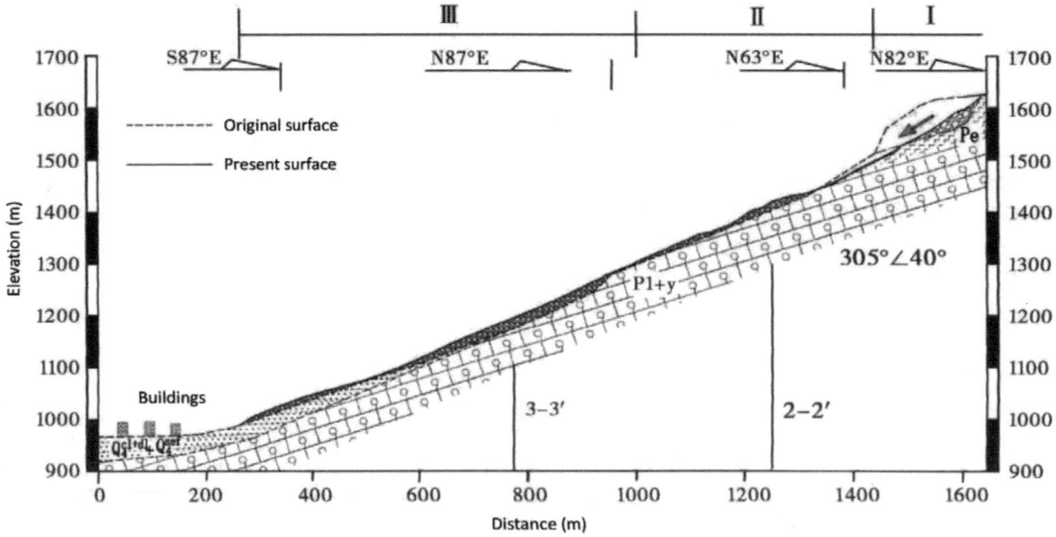

Figure 5. Longitudinal cross section (A-A') of Erman Mountain landslide [12].

In order to illustrate the capacity of the MPM in managing the long runout analysis, we simplified the profile of the section (A-A') to a constant slope model with the slope angle θ=25°, as shown in Figure 6. The model consists of two types of soil: the base soil and the weak soil. The material constants of the two types of soil are shown in Table 2. The weak soil is modelled with pure cohesive material to approximate undrained shearing behaviour.

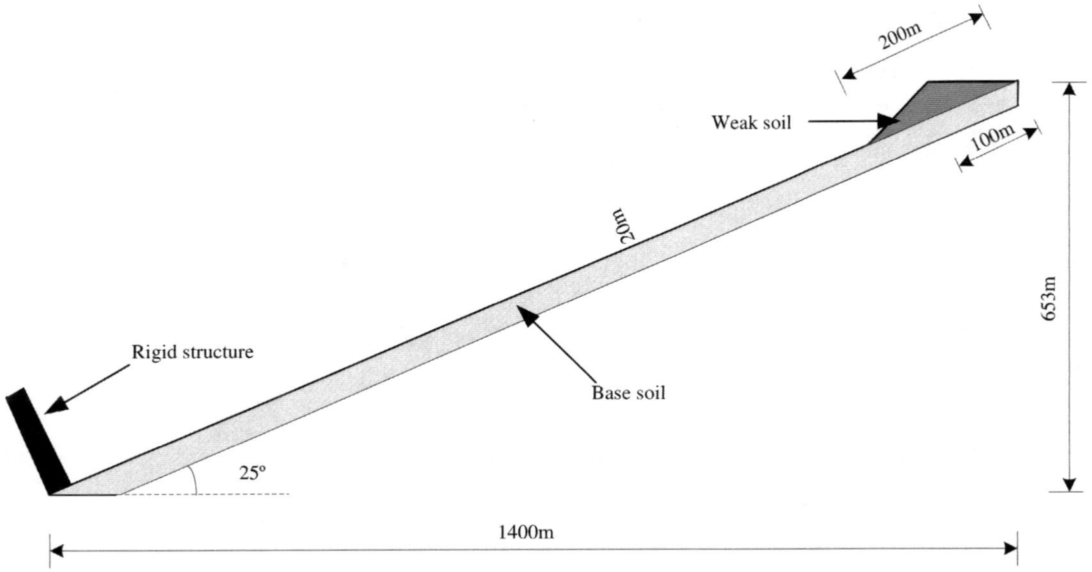

Figure 6. Simplified section (A-A') of Erman Mountain.

Table 2. Material constants of the soil

Soil type	ρ (kg/m^3)	E(MPa)	v	c(kPa)	φ(°)	ψ(°)
Weak soil	2200	10	0.3	10	0	0
Base soil	2200	10	0.3	10	30	0

The numerical model is constructed in plane strain condition with fixed boundary conditions. The calculation is performed using square elements with the side length of Δx=2m and a time step size of Δt= 2.557×10^{-4}s. Initially, there are 4 material points per element. The initial stress field is generated using gravity loading by increasing the strength of the sliding soil.

Figure 7 shows the velocities of material points during the dynamic process of the landslide. The simulation of landslide take place at t=0s. Then the weak soil begins to slide down the slope with the velocity increasing rapidly. At t=20s the front of the sliding soil impacts on the rigid structure with the maximum velocity of about 130 m/s. At this time, the kinetic energy reaches the maximum value of 1.7×10^{7}kJ. From this time, the kinetic energy begins to decrease and the deposit becomes larger and larger. At t=30s the deposit reaches the maximum height and the base soil far behind the front of the deposit is also sliding down with velocity decreasing with time. At about t=50s the velocity of the soil is small and the deposit of the landslide reaches the maximum volume. The simulation lasted 75s and the final configuration of the computation is shown in Figure 8.

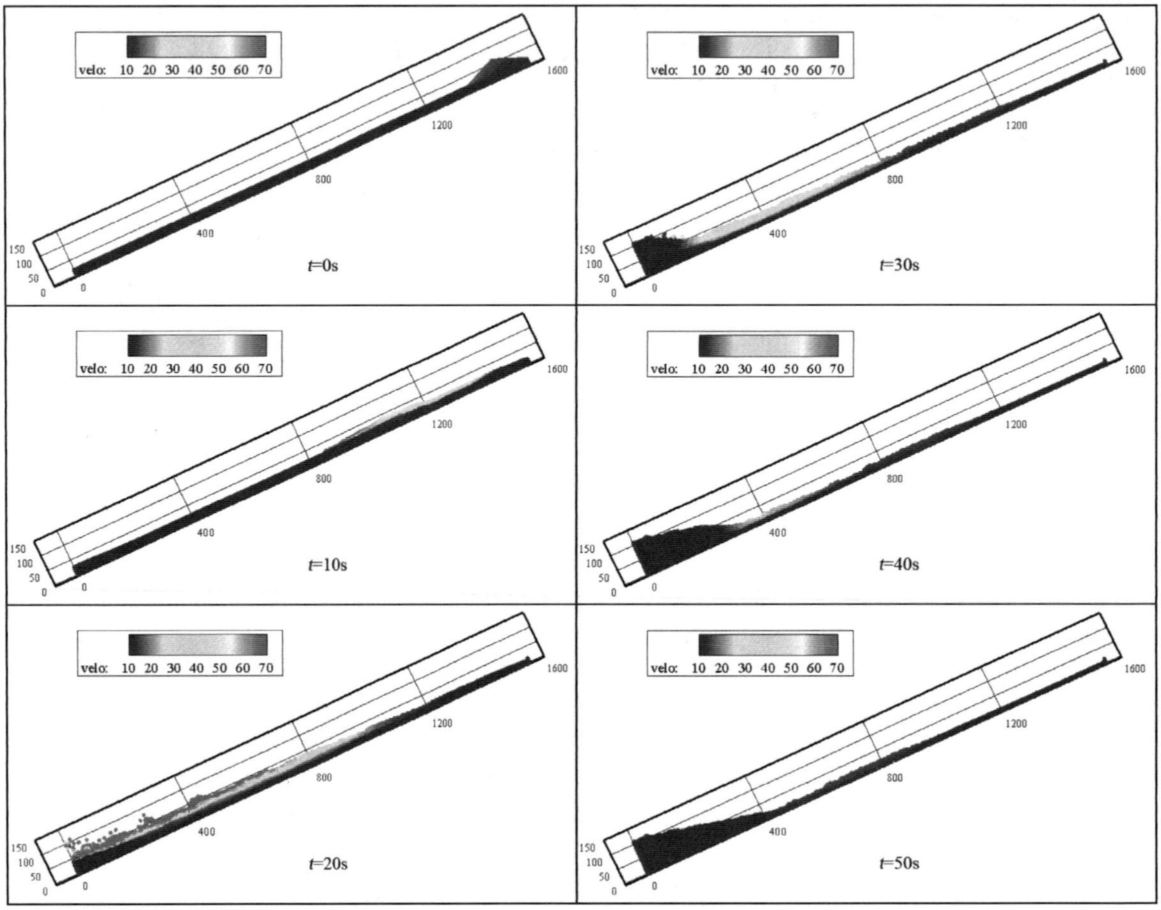

Figure 7. Velocity of sliding soil in Erman Mountain landslide (m/s)

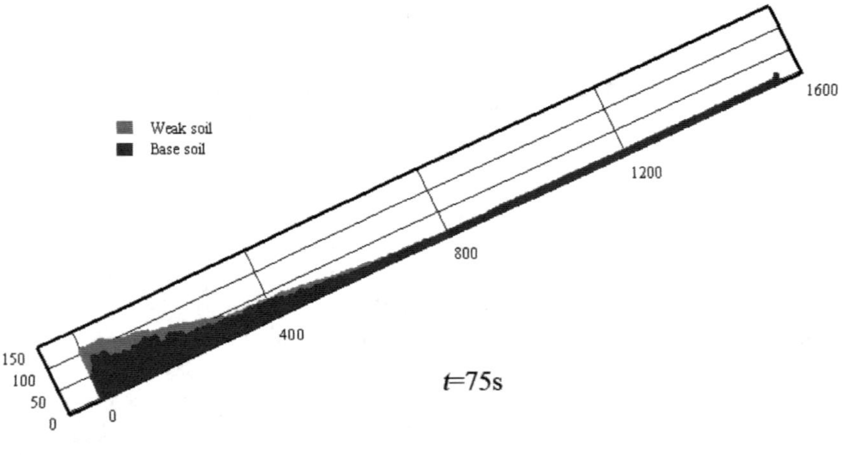

Figure 8. Final simulation configuration of Erman Mountain landslide

Figure 9 shows the impact force on the rigid structure. Before t=20s, the impact force increases linearly with lower rate, because of the effect of the stress wave. From t=20s to t=30s the impact force oscillates frequently due to the intense impact of the sliding soil. After t=30s the impact force increases stably and reaches the peak value at t=40s then decreases to the asymptotic value of 100MN/m.

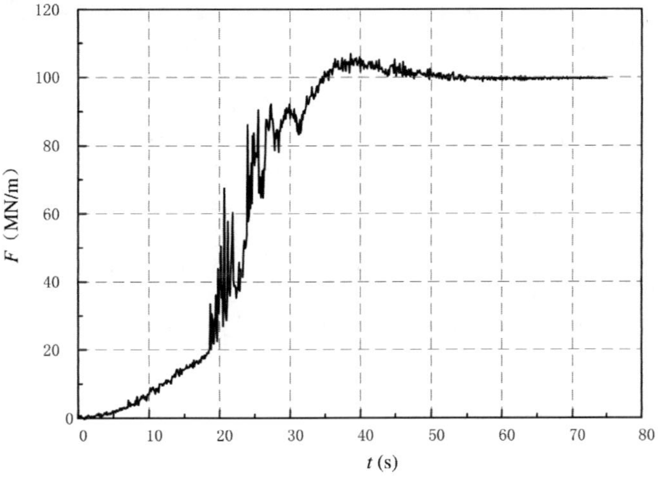

Figure 9. Impact force on the rigid structure of the landslide

4. Conclusions

Dynamic simulation of runout landslides by the MPM is presented. The MPM uses two representations of the continuum: Lagrangian representation of the material points and Eulerian description of the computational grid. Lagrangian scheme, which is convenient to record history-dependent variables in material constitutive model, is wildly used in solid mechanics. Eulerian scheme, which avoids mesh distortion, is usually applied in fluid mechanics. The MPM combines the advantages of both Eulerian and Lagrangian descriptions and is very suitable for simulating long runout landslide.

Two numerical examples illustrate the capability of the MPM in runout analysis. It is inherent for MPM to model no-slip contact between deferent soil layers and contact. It is convenient to model the accumulation effect of the sliding soil and impact force on the structure. The velocities, affected areas, flow intensity are investigated.

Acknowledgements

This work was supported by the National Key Fundamental Research and Development Program of China (Project No. 2014CB047004).

References

[1] Hungr O 1995 A model for the runout analysis of rapid flow slides, debris flows, and avalanches *Can. Geotech. J.* **32** 610-23
[2] Coetzee CJ 2014 Discrete and continuum modelling of soil cutting *Comput. Part. Mech.* **1** 409-23
[3] Mast CM, Arduino P, Miller GR and Mackenzie-Helnwein P 2014 Avalanche and landslide simulation using the material point method: flow dynamics and force interaction with structures *Comput. Geosci.* **18** 817-30
[4] Sulsky D, Chen Z and Schreyer HL 1994 A particle method for history-dependent materials *Comput. Method. Appl. Mech. Eng.* **118** 179-196
[5] Sulsky D, Zhou S and Schreyer HL 1995 Application of a particle-in-cell method to solid mechanics *Comput. Phys. Commun.* **87** 236-252
[6] Harlow FH 1964 The particle-in-cell computing method for fluid dynamics *Methods Comput. Phys.* **3** 319-343

[7] Andersen S, Andersen L 2010 Modelling of landslides with the material-point method *Comput. Geosci.* **14** 137-147
[8] Bui HH, Fukagawa R, Sako K and Ohno S 2008 Lagrangian meshfree particles method (SPH) for large deformation and failure flows of geomaterial using elastic-plastic soil constitutive model *Int. J. Numer. Anal. Methods Geomech.* **32** 1537-1570
[9] Liu P, Liu Y, Zhang X and Guan Y 2015 Investigation on high-velocity impact of micron particles using material point method *Int. J. Impact. Eng.* **75** 241-254
[10] Belytschko T, Liu WK and Moran B 2000 Nonlinear finite elements for continua and structures *New York: John Wiley & Sons*
[11] Coetzee CJ, Vermeer PA and Basson AH 2005 The modelling of anchors using the material point method *Int. J. Numer. Anal. Methods Geomech.* **29** 879-895
[12] Xu Q Dong X Deng M Chen L and Hu Z 2010 The Ermanshan rock slide-debris flow of Junly 27, 2010 in Hanyuan, Sichuan: characteristics and failure mechanism *J. Eng. Geol.* **18** 609-622 (in Chinese)

IOP Conf. Series: Earth and Environmental Science **26** (2015) 012015 doi:10.1088/1755-1315/26/1/012015

The effects of the coordination on the fragmentation of a single grain

Salami Y., Dano C., Hicher P.-Y., Colombo G. and Denain P.

Research Institute in Civil and Mechanical Engineering (GeM, UMR CNRS 6183), École Centrale Nantes, Nantes, France.

Email: younes.salami@ec-nantes.fr

Abstract. The main objective of this study is to understand the effects of the coordination number on the behavior of a single grain, before and during fracture. A new apparatus is designed, with the purpose of conducting multipoint crushing tests, and to investigate the effects of the contacts number, type and position on the failure of a particle. Some of the tests were monitored using digital image correlation (DIC), in order to get an in-depth view into the mechanics of fracture. The fragmentation of the grains was studied, and the cracks were shown to follow specific paths. The failure forces of multiple contact configurations were compared to show the effect of the number and position of contact forces on the primary and secondary cracks. It was shown that the position and magnitude of the contact forces plays a significant role in the fragmentation of the grain.

1. Introduction

Investigating particle crushing in soils is crucial in understanding its behavior, because, soil behavior could be explained from a grain level analysis, where the grain interactions aren't limited to sliding and rotating, but also to fragmenting. It is widely accepted that the probability of grain breakage is a combination of the distribution of local forces on the grains as well as the distribution of the particle strength. The larger particles are usually weaker, but can see their survival probability significantly increased if their coordination number is high. Smaller particles however, have a higher resistance to crushing, but their coordination number is limited by their size. In addition, the strength of particles, generally represented by a Weibull distribution as done by McDowell & Amon [2] is usually identified through point load tests or splitting tests between stiff plates, inducing a diametral compression of the tested particle [1]. This methodology is not necessarily representative of the map of contacts into a stressed granular packing. Finally, the confining effect provided by the contacts on the particle crushing force is only seldom taken into account whereas many authors have assumed its existence (Auvinet & Marsal, [2]; Tsoungui, [3]). Surprisingly, no experimental systematic investigation has been conducted to precise this.

Numerically, the multi-axial compression of round particles was investigated from a fracture mechanics point of view by Tsoungui *et al.* [3]–[5], Sukumaran *et al.* [6] and Ben-nun & Einav [7]. They proposed two failure criteria for a two dimensional grain subjected to a random distribution of contact forces. In Tsoungui's criterion, the complex distribution of forces is replaced by an equivalent distribution (see Figure 1-(a)) of orthogonal forces F_{max} and F_{min}, orientated by the tilt angle Φ, in the form of an inclined cross (Figure 1 –(b)). This model considers that fracture occurs according to a tensile splitting mode (mode I), in a manner similar to that of a grain between two diametrically opposed forces. A series of finite element simulations were then performed on disks of radius R under diametrical compression, and disks subjected to multiple contact forces. These simulations confirmed the idea that the crack originates from the center of the disk where the tensile stresses are concentrated, before propagating through the continuum. The stresses perpendicular to the loading axis near the grain center σ_{xx}^0 for the two simulations (for pure diametral compression and for the equivalent set of contact forces F_{min} and F_{max}) can become positive (tension), while the other stresses are always in compression for both cases, and cannot possibly produce a mode I fracture. σ_{xx}^0 (positive for tension and negative for compression) is expressed by (1) for the diametric compression, and (2) for the equivalent set of contact forces:

$$\sigma_{xx}^0 = \sigma_{xx}(x \to 0, y \to 0) \approx \frac{F}{\pi R} \tag{1}$$

$$\sigma_{xx}^0 = \frac{F_{min} - 3F_{max}}{\pi R} \tag{2}$$

The stresses near the center of the grain appear to be independent from the contact and loading conditions. When the tensile stresses inside the disk σ_{xx}^0 exceed a critical value σ_c, assumed to be the same for every contact configuration, a central crack that moves along the direction of F_{max} causes the specimen to break. The failure criterion is expressed in terms of the hydrostatic stress p and the shearing stress τ in Eq. (3):

$$\sigma_{xx}^0 = 2\tau - p \geq \sigma_c \tag{3}$$

This formula suggests that a grain subjected to a random distribution of contact forces has the highest probability of breakage when $2\tau \gg p$, meaning a force configuration close to a diametral compression. In the case when $2\tau \ll p$, the grain has a low probability of breakage since its stress state tends to a hydrostatic pressure.

Ben-Nun & Einav [7] proposed an alternative criterion, supposed to overcome the principal limitation of Tsoungui's model: the inability of a grain to crush under isotropic compression, or when the set of forces applied is symmetrical. This criterion supposes that the cracks result from an in-plane shear fracture mode (mode II) (Figure 1–(c)). The average of the normal components of the contact forces is compared to a critical threshold force. This critical force is based on Sukumaran's DEM simulations [6], and the resulting formula for the failure force is (4):

$$F_c = d\sigma_c f_w f_D f_{CN} \tag{4}$$

σ_c being the tensile stress at failure of a grain of size d, f_w, f_D and f_{CN} are reduction factors taking account respectively the variability in strength, curvature and coordination number. The first factor f_w is deduced from a statistical law (Weibull [8]) describing the distribution of the flaws or imperfections directly affecting the strength. The two other factors are defined empirically, and the one accounting for the effect of the coordination number C is expressed as a function of the crushed grain size d and the average size of the contacting grains D (5):

$$f_{CN} = (C-1)\exp\left(\left(\frac{D}{d}\right)\frac{(C-2)(C-3)}{4C}\right) \tag{5}$$

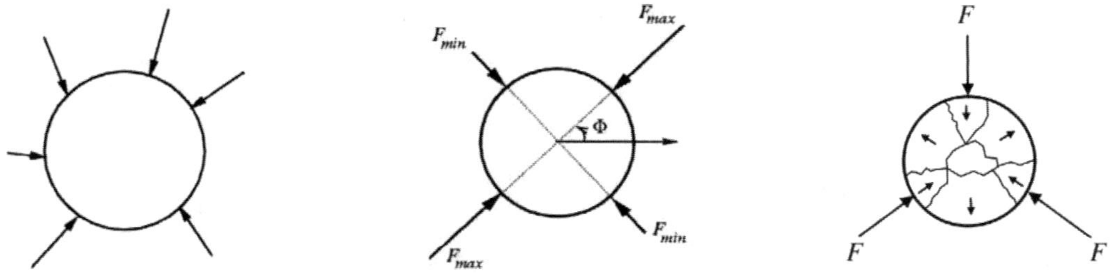

(a) (b) (c)

Figure 1. The two failure criteria for a grain under an arbitrary set of contact forces (a), according to Tsoungui *et al.* (b), and to Ben-Nun & Einav (c). (after[3], [7]).

2. The experimental procedure

A new device, used to simulate the effects of the surrounding particles on the fragmentation behavior of a specimen, was designed. It is worth noting that the results presented here are only preliminary since further improvements are planned on the device. The apparatus consists of a round metal mounting frame, equipped with multiple clamping bars intended to lock the displacements of a cylindrical specimen from various points on its lateral surface. The bars can be moved freely around the frame, and then be locked in a desired position. This apparatus is then placed on a loading frame, and the specimen is crushed under a particular configuration of contacts. The contacts imposed on the specimen can either be linear or along a surface. The loading frame has a capacity of 50 kN, and the force and displacements are measured using a load cell and a displacement transducer, with a resolution of 0.06 N and 0.01 mm respectively. The results presented here correspond to a displacement velocity of 0.6 mm/min.

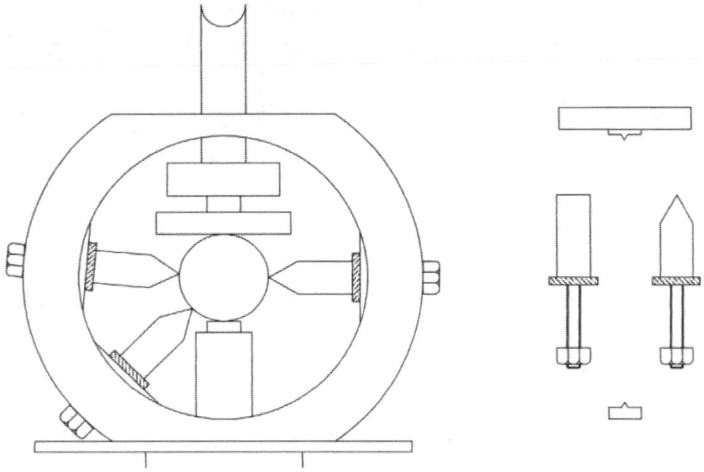

Figure 2. Schematic view of the apparatus and the clamping bars.

The tests are conducted on cylindrical disks of standardized mortar with a diameter of 52 mm and a width of 22 mm, prepared in accordance with the French standard NF EN 196 –1 in batches of 15 units. A CEM1 52.5R cement and a Hostun sand HN 0.6/1.6 are used for this formulation. The water/cement ratio is 0.5. The disks are then cured for 7 days in a humidity chamber (100% relative humidity and a temperature of 18°C). The preparation mode was perfected during the first weeks, as we tried to minimize all sources of discrepancy by controlling every step of the process. From the mixture of the paste, to the vibration of the specimens, to the storage conditions, everything was handled with extreme care, to ensure a standard deviation less than 0.5 for the diametric compression tests, and less than 0.6 for the other configurations.

The cylindrical (disk-like) shape was chosen to represent a two dimensional system, and to eventually compare the results with the theoretical 2D crushing models. Besides, this configuration provides us with an unobstructed view into the specimen's base as the contacts are along the lateral surface, allowing direct access to the surface strains through Digital Image Correlation.

3. Digital image correlation

Digital image correlation (DIC) is an optical experimental technique allowing the measurement of surface displacements in the case of a 2 dimensional analysis. This technique has been widely validated theoretically and experimentally, and has recently appeared in many research papers in material science as a robust and accurate method. The technique is quite easy to carry out as it doesn't require any specific technology other than a digital camera and DIC processing software. The strains are calculated from the computed displacements using the continuum mechanics theory.

The digital camera used has a resolution of 1040x1392 pixels and delivers up to 256 levels of gray output. The camera is programmed to take one picture each second (maximum capability), and is placed at 50 cm from the specimen (see Figure 3). The data is stored on a computer, and the image correlation is performed later using the software VIC2D. The configurations tested are: 5 Brazilian tests (diametral compression), 5 tests in the 4-90 configuration, 5 in the 6-45 configuration, and 2 tests in a random configuration (See Figure 4).

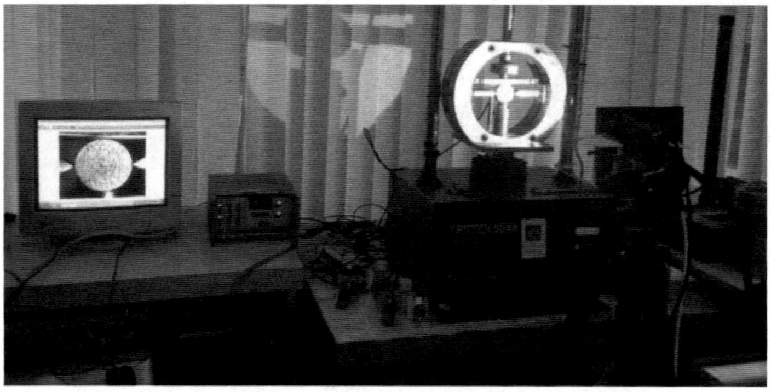

Figure 3. The camera setting for the DIC

4. Results

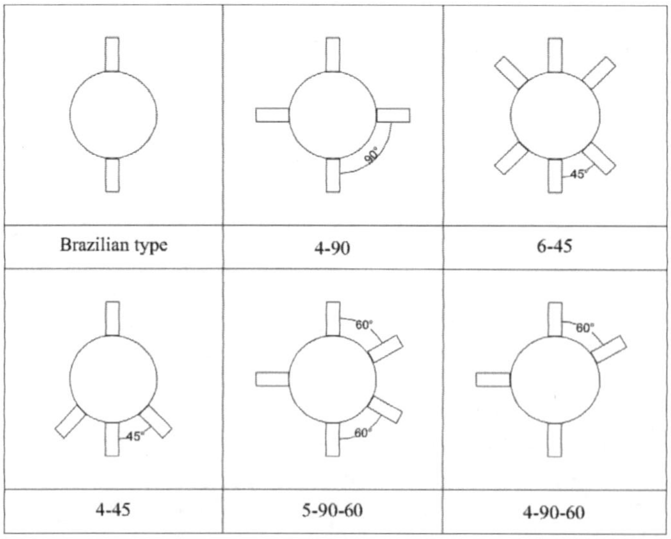

Figure 4. The different loading configurations tested

IOP Conf. Series: Earth and Environmental Science **26** (2015) 012015 doi:10.1088/1755-1315/26/1/012015

4.1. The fragmentation of the disk

For all of the tests conducted, a central crack appears along the loading axis. Other cracks develop later, following specific paths corresponding to each configuration, and apparently originating from the contact area. Since the displacement imposed is along the diametric axis, a volume fracture model [5] should explain the appearance of the central crack, although the digital image correlation suggests otherwise. The secondary cracks could be explained using a contact fracture model.

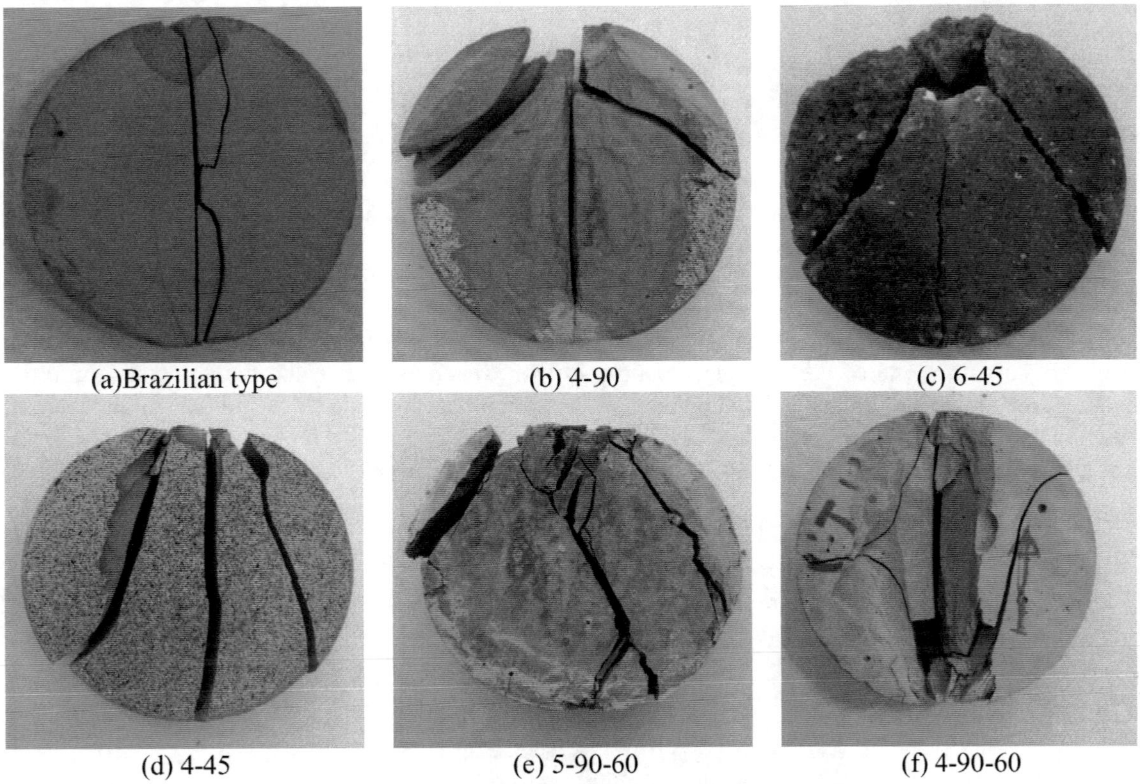

Figure 5. Typical fragments resulting from the crushing of cement disks in the configurations studied Figure 4

The predominance of the effect of the two contacts along the vertical axis, since they are the ones carrying the load, leads to the appearance of the loading axis crack first. This was explained in the literature by the formation of a traction zone far away from the contacts (generally near the center of the disk) and ultimately splitting the specimen in half. The digital image correlation of a diametral compression test shows that the extension strain is highly concentrated near the contacts (Figure 6), and this traction zone advances from the top and the bottom, before merging in the form of a first diametral crack. Because of the small area of contact, both the horizontal and vertical compressive strains are highly concentrated near the contacts. Wang & Xing [9] determined a minimum loading angle for the crack initiation to be at the center of the cylindrical specimen, and proposed using a specimen with flattened edges to better reproduce the theoretical behavior. The contact area is then large enough to spread the load along a greater area, and the flat contact geometry is stable and prevents the specimen from moving during the test. It was also suggested that a longer duration of the test could result in a flattening of the contact area, and might lead to a tensile zone near the center of the disk.

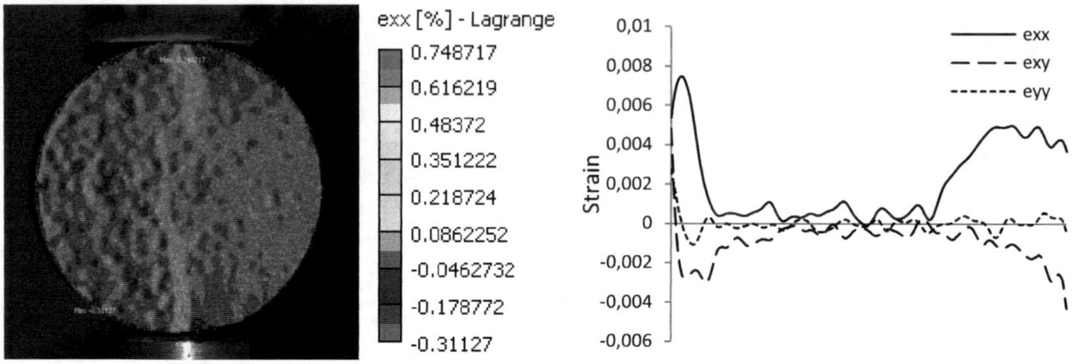

(a) (b)

Figure 6. (a) Distribution of the strains in the x direction prior to fracture (b) the strains along a vertical line

In multipoint configurations, secondary cracks result from another fracture mechanism. The cracks appear to start from one of the secondary contacts before reaching one of the two contacts along the loading axis. These cracks appear to be caused by a shear fracture mode II, as the shear strains are at their lowest negative value along this second crack (see Figure 7). The high shear strains along the first diametric crack are simply caused by the friction of the two initial fragments

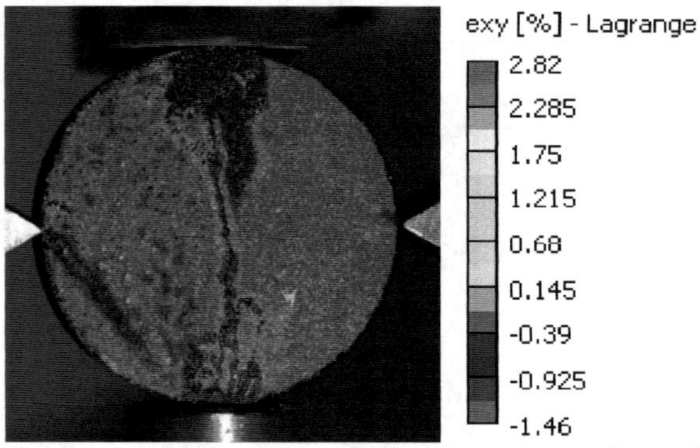

Figure 7. Distribution of the shear strains along the surface of the disk

To sum up, in almost all of the tests conducted in the different configurations in Figure 4, two initial cracks appear near the contacts, and propagate along the diametral loading axis before joining and splitting the specimen in half. Secondly, one or more cracks start from near the lateral contacts, and reach the top or bottom contact. These cracks seem to follow a shearing fracture mode. In addition, a crack that starts from one of the lateral contacts will most likely join the farthest loading contact.

4.2. The effects of the contacts number

For all the tests performed, the failure forces corresponding to the first and second crack appearances, marked by a sudden drop in the force, were recorded. The mean value for each configuration is presented in Table 1. We only considered the tests with surface contacts. The first column F_1 corresponds to the force for the first crack (the one along the loading axis), and F_2 to the force for a second crack. We chose not to take into account other cracks that appear later in the loading, as the solid is highly damaged. It should be noted that we were able to keep the standard deviation below 0.6 kN for all of the configurations. From a first look into the table, we can immediately see that a higher

IOP Conf. Series: Earth and Environmental Science **26** (2015) 012015 doi:10.1088/1755-1315/26/1/012015

coordination number means a higher resistance to crushing. Another important consideration is that whenever a set of two contact forces are diametrically opposed, the resistance is slightly higher. This is apparent in the comparison between the 4-90 and the 4-90-60 configurations. This raises the issue of the importance of the contact position on the crushing behavior of a grain. Until now, only the number of contacts was thought to affect the critical stress of a grain.

The lateral contacts allow the grain to sustain greater forces until they finally fail.

Table 1. Mean value of the critical force for the first two cracks.

Contact configuration	F_1(kN)	F_2(kN)
Brazilian	5.95	-
4-90-60	6.45	9.70
4-90	7.12	13.70
4-45	7.28	9.88
5-90-60	8.84	13.75
Random (5contacts)	8.44	12.76

The secondary cracks appear to be equally affected by the contacts positions.

4.3. The effects of the contact type

The effect of the contact type is investigated through a series of tests on two sample groups composed of 15 identical mortar disks each. The tests on the first group consist of 3 diametral compression tests (Brazilian type) and 12 tests in the 4-90 configuration (Figure 4), with a surface contact type for both the clamping bars and the compression axis contacts. The same tests were performed on the second group, the only difference being the contact type, which is now linear. Typical force-displacement curves are presented in Figure 8:

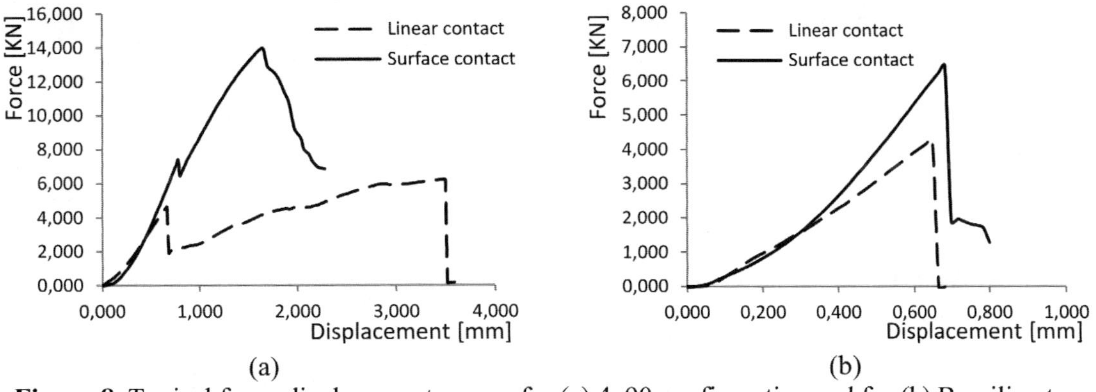

Figure 8. Typical force-displacement curves for (a) 4x90 configuration and for (b) Brazilian type crushing tests

For each of the diametral compression type tests, a mean value of the critical force for the linear contacts is 4.28kN, and 6.25kN for the surface contacts. This force is associated with a diametral crack originating near the contacts along the loading axis, and propagating diametrically through the specimen, breaking it in half. Both the contact types are non-conforming, so the stresses are highly concentrated near the contacts independently from the shape of the body, with an even higher concentration in the case of the linear contacts than the surface ones. This generates localized strains around the contacts favoring the crushing of the grain at a lower stresses.

In the 4-90 configurations, two force peaks are observed for both contact types, the first corresponding to the diametral crack, and the second to the secondary cracks. We can see that the first peaks in the 4-90 configurations are slightly higher than the ones in the Brazilian configuration. The failure post second peak is sudden in the case of the linear contact, but more ductile like in the case of the surface contact. The surface contact provides a higher adherence area blocking the specimen from moving after failure, explaining the post peak behavior of the surface contact: the two fragments are held together by the surface contacts, and the specimen continues to carry the load applied, causing multiple little cracks to appear, and leading to an increasing damage, especially near the contacts. In the case of a linear contact, the tip of the contact penetrates in the crack after failure, leading to a quick crack opening, and a sudden failure.

4.4. The effects of the contact force

A preliminary investigation on the effect of the contact force was conducted through 12 compression tests in the 4-90 configuration. The lateral contacts were tightened to simulate the effect of a pre-stress. No force was measured, as this experiment goal was just to get an oversight of the problem, but a specific tightening of the clamping bolts ensured reasonably comparable conditions for the tests.

The first crack along the loading diameter appears, but no other crack followed, instead, many little cracks developed near the contacts, causing the specimen to break into a multitude of small fragments. A ductile-like behavior like the one in Figure 8 (a) is then observed in the force displacement curves. The mean critical forces associated with the crack in this case is 11.36 kN, which is higher than the 7.12 kN at which a 4-90 specimen without any pre-stress breaks.

5. Conclusion

The importance of the coordination number was investigated through a series of crushing tests on mortar cylinders, with varying contacts numbers and position. 6 configurations of contacts positions were tested, and monitored using DIC analysis. It was found that each configuration results in a specific set of fragments. Digital image correlation allowed us to track each crack and determine its point of initiation, and its defined path. It was confirmed that the number of contacts plays a determining role in the crushing of a single grain, as the grains with the higher coordination number are harder to break. Our new device allowed us to control the contact's position, and to show the importance of the contact's position on the failure of the grain.

References

[1] J. C. Jaeger, "Failure of Rocks Under Tensile Conditions," *Int. J. Rock Mech. Min. Sci.*, vol. 4, pp. 219–227, 1967.

[2] G. Auvinet and R. Marsal, "Statistical model of grain breakage," *Proceedings, Fifth Pan-American Conf. Soil Found. Eng. Buenos Aires*, vol. 1, pp. 193–204, 1975.

[3] O. Tsoungui, D. Vallet, and J. C. Charmet, "Numerical model of crushing of grains inside two-dimensional granular materials," *Powder Technol.*, vol. 105, pp. 190–198, 1999.

[4] O. Tsoungui, D. Vallet, and J.-C. Charmet, "Use of contact area trace to study the force distributions inside 2D granular systems," *Granul. Matter*, vol. 1, pp. 65–69, 1998.

[5] O. Tsoungui, D. Vallet, J.-C. Charmet, and S. Roux, "Size effects in single grain fragmentation," *Granul. Matter*, vol. 2, pp. 19–27, 1999.

[6] B. Sukumaran, I. Einav, and Arcady Dyskin, "Qualitative Assessment of the Influence of Coordination Number on Crushing Strength using DEM," *5th World Congr. Part. Technol.*, pp. 1–8, 2006.

[7] O. Ben-Nun and I. Einav, "The role of self-organization during confined comminution of granular materials.," *Philos. Trans. A. Math. Phys. Eng. Sci.*, vol. 368, pp. 231–247, 2010.

[8] W. Weibull, "A statistical distribution function of wide applicability," *Journal of applied mechanics*, vol. 18. pp. 293–297, 1951.

[9] Q.-Z. Wang and L. Xing, "Determination of fracture toughness KIC by using the flattened Brazilian disk specimen for rocks," *Eng. Fract. Mech.*, vol. 64, no. 2, pp. 193–201, Sep. 1999.

DEM analyses of the whole failure process of shallow foundation in plate load test on dense sand

L Li[1,2,3,5], M J Jiang[1,2,3], T Li[1,2,3] and S L Chen[4]
[1] State Key Laboratory for Disaster Reduction in Civil Engineering, Tongji University, Shanghai 200092, China
[2] Key Laboratory of Geotechnical and Underground Engineering of Ministry of Education, Tongji University, Shanghai 200092, China
[3] Department of Geotechnical Engineering, Tongji University, Shanghai 200092, China
[4] Transportation and Municipal Mngineering Institute, Hydrochina Huadong Engineering Corporation, Hangzhou 310014, China

E-mail: lilei_tongji@126.com

Abstract. Shallow foundations are widely used in civil engineering practice, but the instability mechanism is still unclear yet. Previously, the Finite Element Method (FEM) was commonly used to analyze the failure process of shallow foundations, but it meets difficulty in properly simulating the whole failure process of shallow foundation on the strain-softening material. Hence, the Discrete Element Method (DEM) is employed in this paper to study the instability mechanism of the shallow foundation via numerical plate load test with focus on the microscopic features evolution during vertical loading. In the simulation, an amplified gravity was applied to a dense granular ground to reproduce a gravity stress state at a large scale. Then, a plate was put on the granular ground to simulate the plate load test. Deformation pattern, particle velocity and distribution of void ratio in the ground were examined to illustrate the microscopic features in the whole failure process of the granular ground. The results show that: 1) There are a marked peak value and a settlement softening branch in the stress-settlement relationship. 2) The grids close to the edge of the plate are peculiarly extended and twisted. 3) Four particle motion patterns were observed in the velocity fields and the percentage of each motion pattern changes during loading. 4) The void ratio field varies during loading, and the distinguishing interface tends to be similar to Terzaghi's shear failure surface.

1. Introduction

The shallow foundation, one of the two classic foundation categories - shallow foundation and deep foundation, has been widely used in geotechnical engineering practice. In the traditional research, some macroscopic phenomena such as the relationship between stress and settlement were studied by the laboratory and in-situ plate load tests. However, the microscopic features in the ground, such as particle velocity and distributions of void ratio, cannot be observed. In recent decades, by simulating plate load test using the Finite Element Method (FEM) and finite-difference method [1-5], some microscopic features in the ground were observed. Nevertheless, there are some shortcomings using the continuum numerical method ascribe to some limitations including constitutive model and the

[5] To whom any correspondence should be addressed.

assumption of small deformation. For example, when the bearing capacity neither increases gradually nor keeps stable with the increase in settlement, the features cannot continually reflect the ground's status correctly.

An alternative tool, the Discrete Element Method (DEM) developed by Cundall and Strack [6], has been used more and more widely in geotechnical engineering to obtain mechanical behaviors of soil samples [7-11], and to investigate the responses of soils under different boundary conditions [12-16]. More importantly, this numerical analysis technique can provide not only observations on the macroscopic mechanical response of a particle assemblage but also insights into the microscopic mechanical information.

The main aim of this paper is to study on the plane-strain instability mechanism in granular material with focus on the microscopic features in the whole failure process of the shallow foundation on dense sand. For the purpose, the plate load test was carried out on the granular ground with K_0 boundary. Relationship between stress and settlement, deformation pattern, velocity field and void ratio field are analyzed in the failure process.

2. Sand ground specimen

The material used in the ground is composed of 15 types of discs. The maximum and minimum diameter of the granular particles is 0.5 mm and 2.0 mm, respectively. The mean particle diameter is d_{50}=1.3 mm and the uniformity coefficient is $C_u=d_{60}/d_{10}$=2.4. Due to the geometric symmetry of the boundary-value problem, only a half of the ground and plate was generated using the DEM in the simulation. The target ground consists of 160,000 particles with planar void ratio of about 0.18, and has a width and depth of W=0.48 m and D=0.3 m, respectively. The multi-layer under-compaction method (UCM) proposed by Jiang et al.[17] was used to generate the ground. In order to improve the homogeneity, the wall-particle frictional coefficient is set to be zero, and inter-particle frictional coefficient (μ) is also set to be zero in the ground generation in order to produce a dense assembly of particles. After the ground was generated, the inter-particle frictional coefficient is chosen to to be 1.0, and then the granular ground was consolidated under an amplified gravity field of 10 g. After the ground was consolidated, a loading plate with a width of B/2=0.06 m is formed at the central line as shown in Figure. 1. The loading plate represented by a rigid wall is then pushed downward at a velocity of 5.0×10^{-3} m/s. The frictional coefficient between the loading plate and particle is set to be 1.0 in the process of loading.

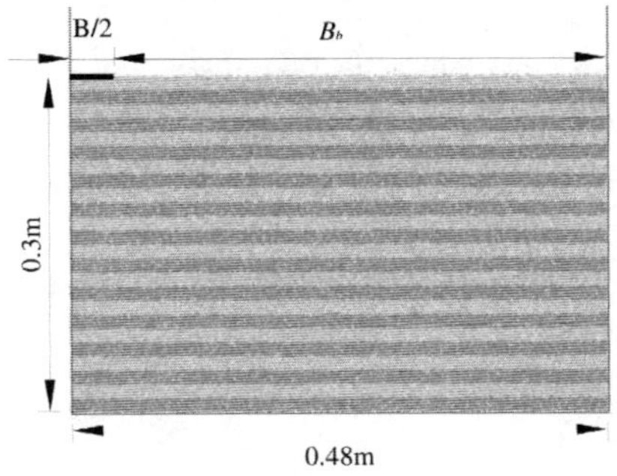

Figure 1. Boundary conditions of the ground in the DEM analyses.

3. Macroscopic strength features of the ground material

The macro mechanical behavior of the ground specimen was studied by conducting DEM simulation of biaxial tests. The DEM specimen for the biaxial tests consisting of 6,000 particles was compressed at a strain rate of 5%/min under the confining stresses of 50 kPa, 100 kPa, 200 kPa and 400 kPa, respectively. These biaxial tests indicate that the peak and residual internal friction angle of the material is 31° and 12°, respectively.

4. Simulation results

4.1. Relationship between stress and settlement

Figure 2 presents that the vertical stress on the rough foundation exists a peak value which decreases sharply to a residual value with the increase in the foundation settlement. The residual value decreases slightly with the increase in the foundation settlement. In order to analyze the instability mechanism of the ground in the whole process of the plate load test, the relationship curve between stress and settlement can be divided into four sections. As shown in Figure 2, the first section from the point of origin to point A can be named as compression stage, the second section from point A to point C can be named as shear stage, the third section from point C to point D can be named as softening stage, and the last section from point D to the end can be named as residual stage. At the points (A, B, C and D) on the curve, deformation pattern, particle velocity and distributions of void ratio in the ground were observed to investigate the evolution of the granular ground.

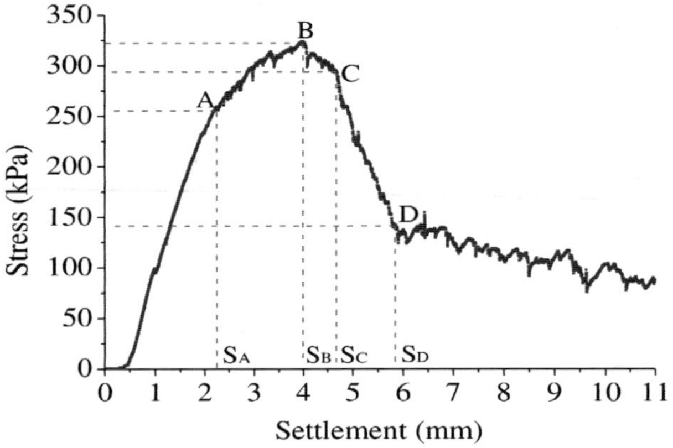

Figure 2. Relationship between stress and settlement in the plate load test.

4.2. Deformation fields

Figure 3 presents the painted grids observed in the plate load test at different settlements. The painted grid method proposed by Jiang et al. [18] was adopted to show the deformation pattern of the granular ground.

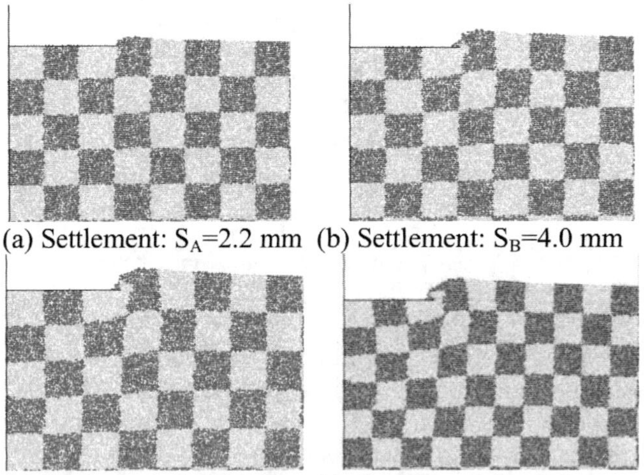

(a) Settlement: S_A=2.2 mm (b) Settlement: S_B=4.0 mm

(c) Settlement: S_C=4.7 mm (d) Settlement: S_D=5.8 mm

Figure 3. Deformation fields in granular ground

At the end of the compression stage, see Figure 3(a), the plate has slight influence on the painted grids which basically keeps the original shape. When the vertical stress comes to the peak value of the ultimate bearing capacity as shown in Figure 3(b), the top-right grid under the plate is pushed sideward and upward while the top-left grids under the plate are pushed down. In contrast, the top grid to the right of the plate is "dragged" downward and sideward. At the end of the shear stage and the end of the softening stage, see Figure 3(c) and (d), the top-right grid under the plate is pushed downward, sideward and upward, and the grid to the right of the top-right grid is squeezed sideward, downward and upward, hence these grids are peculiarly extended and twisted. The deformation of these grids is very difficult to be described properly with the "elements" used in conventional finite element analysis.

4.3. Velocity fields

Figure 4 presents the velocity vectors of all particles which were colored based on their velocity magnitudes. These velocity vectors were divided into seven intervals with their velocities from zero to the maximum velocity v_{max} and presented by vectors with seven colors.

Figure 4 provides velocity fields observed at different settlements: S_A=2.2 mm (a), S_B=4.0 mm (b), S_C=4.7 mm (c) and S_D=5.8 mm (d). Figure 4 shows that velocity field is different at different settlements. In Figure 4(a), i.e. in the compression stage, most of the particles under the plate move downward while some particles beneath the edge of the plate downward and sideward. The particles to the right of the plate and under the plate are affected by the plate movement. In the granular ground, four particle motion patterns are observed: downward, downward and sideward, sideward, sideward and upward. Figure 4(b) shows that when the vertical stress comes to the peak value, the number of particles to the right of the plate moving sideward and upward increases significantly compared with that of the compression stage while the number of particles under the plate moving downward decreases obviously. In Figure 4(c) and (d), the aforementioned four motion patterns of particles still exist in the granular ground, but the number of the particles moving downward decreases gradually. In addition, the shear interface tends to be similar to Terzaghi's shear failure surface [6].

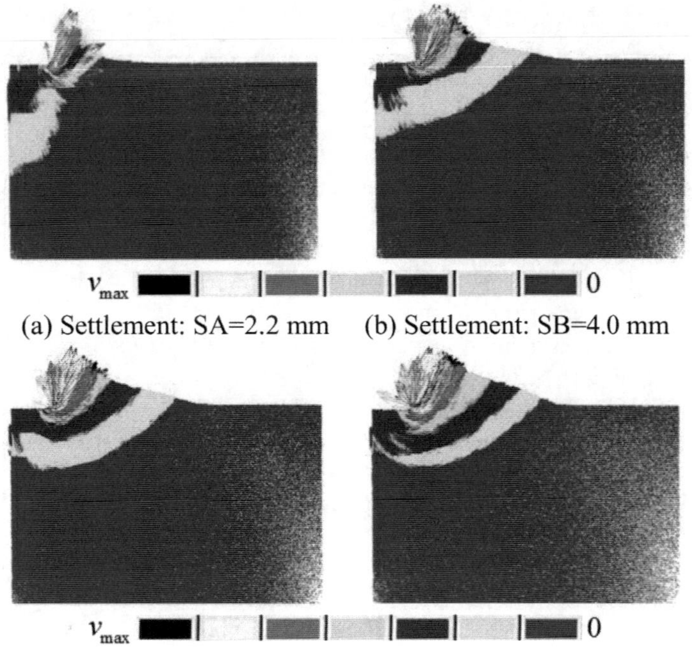

(a) Settlement: SA=2.2 mm (b) Settlement: SB=4.0 mm

(c) Settlement: SC=4.7 mm (d) Settlement: SD=5.8 mm

Figure 4. Distributions of velocity vectors in the DEM analyses

4.4. Void ratio fields

The void ratio is a classical variable in soil mechanics, but it is very difficult to describe the variable properly by the FEM when dilation appears in the soil. In contrast, the void ratio field in the granular ground can be easily measured by measurement circle in the DEM analysis.

Figure 5 provides the distribution of the void ratios observed at settlements of S_A=2.2 mm (a), S_B=4.0 mm (b), S_C=4.7 mm (c), and S_D=5.8 mm (d). Figure 5(a) shows that the expansion area (the void ratio increases within the area) is significantly located at the region close to the edge of the plate. When the stress comes to the peak value in Figure 5(b), the expansion occurs in a larger area, and a through-interface is formed preliminarily. In Figure 5(c) and (d), it is obvious that the expansion only occurs in a specific area at which the void ratio increases during loading, and the distinguishing interface tends to be similar to Terzaghi's shear failure surface [6].

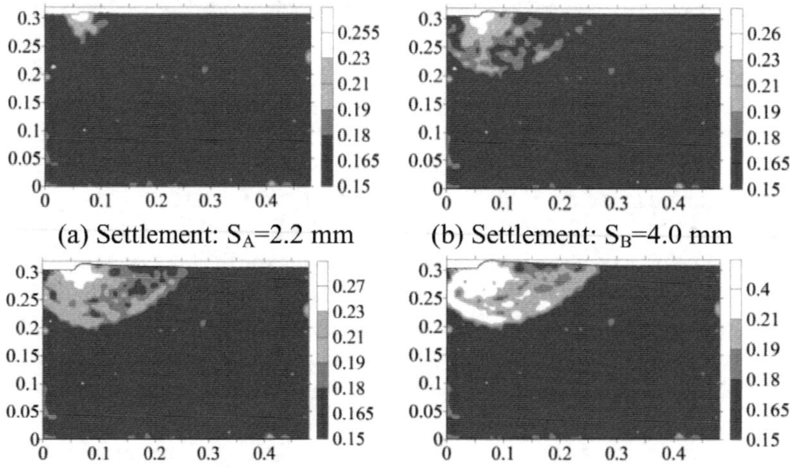

(a) Settlement: S_A=2.2 mm (b) Settlement: S_B=4.0 mm

(c) Settlement: S_C=4.7 mm (d) Settlement: S_D=5.8 mm

Figure 5. Distributions of void ratio in the granular ground

5. onclusions

The discrete element method numerical technique was used to investigate the failure process of the shallow foundation on dense sand ground by simulating plate load test. Deformation pattern, velocity field and void ratio field are analyzed to study on the plane-strain instability mechanism with focus on the microscopic features in the granular ground. The main conclusions are as follows:

1) The relationship curve between stress and settlement can be divided into four stages: compression, shear, softening and residual stages.

2) The grids are pushed towards many directions so that some grids close to the edge of the plate are peculiarly extended and twisted.

3) In the velocity fields, four particle motion patterns are observed: downward, downward and sideward, sideward, sideward and upward. The percentage of each motion patterns changes during loading. In addition, the shear interface tends to be similar to Terzaghi's shear failure surface.

4) The void ratio field varies during loading as well. The distinguishing interface tends to be similar to Terzaghi's shear failure surface.

Acknowledgements

The research was funded by the National Basic Research Program of China (973 Program) with grant Nos. 2014CB046901 and 2011CB013504, the State Key Laboratory of Disaster Reduction in Civil Engineering with Grant No. SLDRCE14-A-04, and the China National Funds for Distinguished Young Scientists with grant No. 51025932. All of these supports are greatly appreciated.

References

[1] Yin J H, Wang Y J, and Selvadurai A P S 2001 Influence of nonassociativity on the bearing

capacity of a strip footing *J. Geotech. Geoenviron. Eng.* **127** 985-989

[2] Erickson H L, and Drescher A 2002 Bearing capacity of circular footings *J. Geotech. Geoenviron. Eng.* **128** 38-43

[3] Hjiaj M, Lyamin A V, and Sloan S W 2005 Numerical limit analysis solutions for the bearing capacity factor Nγ *Int. J. Solids Struct.* **42** 1681-1704

[4] Lyamin A V, Salgado R, Sloan S W, and Prezzi M 2007 Two-and three-dimensional bearing capacity of footings in sand *Géotechnique* **57** 647-662

[5] Loukidis D, and Salgado R 2009 Bearing capacity of strip and circular footings in sand using finite elements *Comput. Geotech.* **36** 871-879

[6] Cundall P A, Strack O D L 1979 A discrete numerical model for granular assemblies *Géotechnique* **29** 47-65

[7] Jiang M J, Zhang F G, Hu H J, Cui Y J, and Peng J B 2014 Structural characterization of natural loess and remolded loess under triaxial tests *Eng. Geol.* **181** 249-260.

[8] Jiang M J, Chen H, Tapias M, Attoyo M, and Fang L 2014 Study of mechanical behaviour and strain localization of methane hydrate bearing sediments with different saturations by a new DEM model *Comput. Geotech.* **57** 122-138

[9] Jiang M J, Yan H B, Zhu H H, and Utili S 2011 Modeling shear behaviour and train localization in cemented sands by two-dimensional distinct element method analyses *Comput. Geotech.* **38** 14-29

[10] Jiang M J, Li T, Hu H J, and Thornton C 2014 DEM analyses of one-dimensional compression and collapse behaviour of unsaturated structural loess *Comput. Geotech.* **60** 47-60

[11] Jiang M J, Liu J D, Sun Y G, and Yin Z Y 2013 Investigation into macroscopic and microscopic behaviors of bonded sands using distinct element method *Soils Found.***53** 804-819

[12] Jiang M J, Dai Y S, Cui L, Shen Z F, and Wang X X 2014 Investigating mechanism of inclined CPT in granular ground using DEM *Granul. Matter* **16** 785-796

[13] Jiang M J, He J, Wang J F, Liu F, and Zhang W C 2014 Distinct simulation of earth pressure against a rigid retaining wall considering inter-particle rolling resistance in sandy backfill *Granul. Matter* **16** 797-814

[14] Jiang M J, and Yin Z Y 2012 Analysis of stress redistribution in soil and earth pressure on tunnel lining using the discrete element method *Tunn. Undergr. Sp. Tech.* **32** 251-259

[15] Jiang M J, Shen Z F, and Zhu F Y 2013 Numerical analyses of braced excavation in granular grounds: continuum and discrete element approaches *Granul. Matter* **15** 195-208

[16] Jiang M J, and Murakami A 2012 Distinct element method analyses of idealized bonded-granulate cut slope *Granul. Matter* **14** 393-410

[17] Jiang M J, Konrad J M, and Leroueil S 2003 An efficient technique for generating homogeneous specimens for DEM studies *Comput. Geotech.* **30** 579-597

[18] Jiang M J, Yu H S, and Harris D 2006 Discrete element modelling of deep penetration in granular soils *Int. J. Numer. Anal. Met.* **30** 335-361

Inter-relationship between joint dilatancy and frictional resistance: impact on fracture behaviour

Kenneth Imo-Imo Eshiet[1] **and Yong Sheng**[1]

1 School of Civil Engineering,
University of Leeds, LS2 9JT, Leeds, United Kingdom
E-mail: kenieshiet@yahoo.com

Abstract. Rock joints possess features that alter both the characteristics of rock formations and the fracturing process. Two important properties that govern the shear behaviour and dilatancy behaviour of discontinuities have been analysed in this paper by the discrete element method (DEM). The ability of a frictional joint to suppress fracture growth decreases as the frictional resistance increases; however, the rate and extent of fracturing increases with joint dilatancy. The influence of joint frictional resistance is more dominant at high values and in this range effects of small magnitudes of dilatancy are correspondingly insignificant. At low joint friction the occurrence of even a small amount of dilation increases the severity of the fracturing process. This study highlights the interactions between two main joint properties with the anticipation that the concepts derived here will be useful for predicting fracture behaviour at the subsurface.

Keywords: Discrete Element Method (DEM), hydraulic fracture, rock joint, rock joint friction, rock joint dilatancy

1. Introduction

The presence of discontinuities affects the structure and behaviour of rock masses. The magnitude of the impact is very pronounced when viewed at the macro level. Thus, the macro-structure and macro-behaviour are significantly affected by the existence of discontinuities which create areas of weaknesses. Investigations to ascertain this process have taken numerous forms, carried out by a number of studies. An important subject includes the impact of discontinuities on certain aspects of the fracturing process [1-16]. For instance, Casas *et al.* [10] determined the effect of interfacial behaviour on fracture development by analysing the influence of varying interfacial physical and material properties, Chuprakov *et al.* [16] examined contributions from key variables including rock frictional resistance, fracture inclination, net pressure at the fracture/fault and differential stresses, and Blair, *et al.*[8] introduced an alternative method, using pressure history records, to monitor fracture interactions at interfaces.

Some earlier works carried out by Bandis *et al.* [19], Barton and Choubey [20], Barton [21], Ladany and Archambault [22], Lee *et al.* [23], and Park and Song [24] have explored issues that control the shear behaviour of joints, which include dilatancy, the impact of asperities and contact of joint planes. This work investigates the relationship between conditions that instigate hydraulic

Corresponding Author:

[1] Kenneth Imo-Imo Eshiet.

fracturing, fracture growth and joint features. Tensile strength, shear strength and cohesion are key properties that govern the behaviour of bonded joints. Joint strength is also affected by joint roughness, stiffness of the contacting planes and the presence of infills. The characteristics and effect of infills are described in Indraratna *et al.* [25], Trivedi [26] and Zare *et al.* [27].

An association between hydraulic fracturing processes and rock joint behaviour is investigated using the Discrete Element Method (DEM) and comprises such phenomena as the intensity and pattern of fracture growth, fracture interaction, properties of joints and the characteristics of the rock mass. The inter-dependency of key joint properties (e.g. shearing resistance and dilation), their impact on the strength and displacement of joints, and the corresponding effect on fracturing behaviour is the primary focus of the paper.

The outcome of this investigation is particularly relevant in the evaluation of zones prone to landslides as well as the assessment of slope stability. This also include the evolution of natural slopes [28, 29]. The role of active fault behaviour to cause landslide processes is an emerging research area [30, 31]. The results are also applicable in scenarios such as opencast mine slopes and reservoir embankments; for instance, in Biringen et al. [32] the growth of faults has been shown to cause a reduction in strength and slope stability of a water reservoir embankment.

2. Test and calibration

The behaviour of a jointed rock mass is to a great extent influenced by the joint characteristics. The shear strength of joints therefore plays an important role. One of the major influences of the shear strength is the cohesive strength and angle of internal friction. The angle of friction is also affected by the dilatancy which, in the case of joints, is controlled by the joint roughness. The joint roughness coefficient is a measure of its smoothness [21, 33] and could be used for the assessment of non-planer joints [34]. The effective normal stress across the joint also contributes to its shear strength. Direct shear tests were carried out mainly to determine joint properties such as the cohesive strength, frictional resistance, joint wall strength (JCS) and joint roughness coefficient (JRC). The shear strength can be derived from known values of cohesive strengths and frictional resistance.

In order to ascertain as well as calibrate the joint properties, a joint with specified values of properties was created in samples of the rock mass and direct shear tests conducted under varying effective normal stress conditions (ranging between 1e6 MPa to 5e6 MPa). For the first series of tests the micro-property representing the joint friction angle was varied with each successive test but the joint cohesive strength was kept constant. The values of the microscopic properties of the synthetic rock sample and assigned joint properties are given in Table 1 and Table 2.

Table 1. Micro-Properties of rock sample.

Parameter Description	Values
Contact-bond normal strength (mean)	5.0 MPa
Contact-bond normal strength (std deviation)	1.25 MPa
Contact-bond shear strength	5.0 MPa
Contact-bond normal strength (std deviation)	1.25 MPa
Particle size (radius)	0.002 m – 0.004 m
Particle friction coefficient	1.0
Particle normal stiffness, k_n	29.0 MPa
Particle shear stiffness, k_s	10.36 MPa
Particle density	2650 kg/ m^3
Porosity	0.16
Particle-particle contact modulus	14.5 GPa
Particle stiffness ratio	2.8

Parameter Description	Values
Joint Properties	
Normal stiffness, k^n	1.583 x 10^{12} Pa/m
Shear stiffness, k^s	0.565 x 10^{12} Pa/m
Friction coefficient, ϕ	Varied accordingly (btw 0.0 & 1.0)
Dilation angle, φ	Varied accordingly (btw 0.0 & 40)
Cohesive strength, c	0.0

Table 2. Rock mechanical properties.

Parameter Description	Values
Mechanical Properties	
Contact-bond normal strength, $\hat{q}$	11.7 MPa
Elastic modulus, E	9.7 GPa
Poisson ratio	0.19

Using samples with dimensions of height=0.3 m and width 0.6 m, a single planar longitudinal joint was created along the centre of the intact sample (Figure 1). The joints were made to be smooth and as such have negligible joint roughness coefficients ($JCR \approx 0$). Barton and Choubey [20] provides standard curves that state values of JCR corresponding to the roughness of joints. A value of JCR ranging between 0 and 2 is suggested for smooth planar joints. Particles bordering the two planar surfaces of the joint that meet the criterion for selection for the smooth-joint contact were identified and a zero bond strength allocated to those contacts (Figure 2b) dividing the two joint surfaces.

a) Position and orientation of joint joint

b) Collection of smooth joint contacts that form the

Figure 2. Joint configuration and position of contacts

The layout of the shear tests is shown in Figure 3. Vertical stresses representing effective normal stresses were applied via the horizontal walls. The bottom left wall was fixed in the horizontal direction but particles in contact with it were allowed to slide vertically. Similarly, particles were able to slide laterally along the top and bottom horizontal walls. To apply a steady load, a constant horizontal velocity was applied to the top left vertical wall in the W-E direction. The horizontal velocity was set to 0.003 m/s to guarantee quasi-static equilibrium during the test.

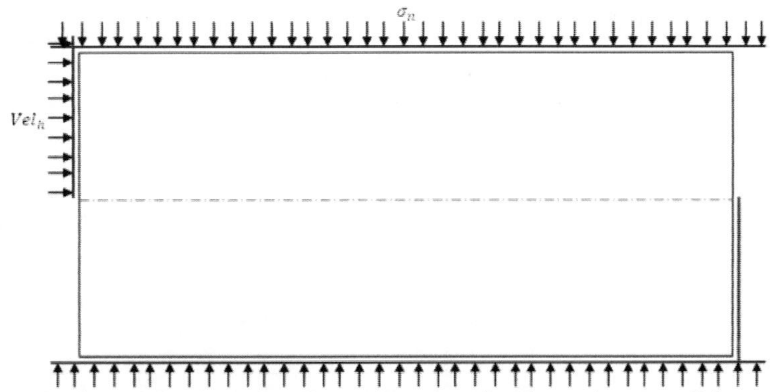

(a) Schematic of test configuration showing boundary conditions.

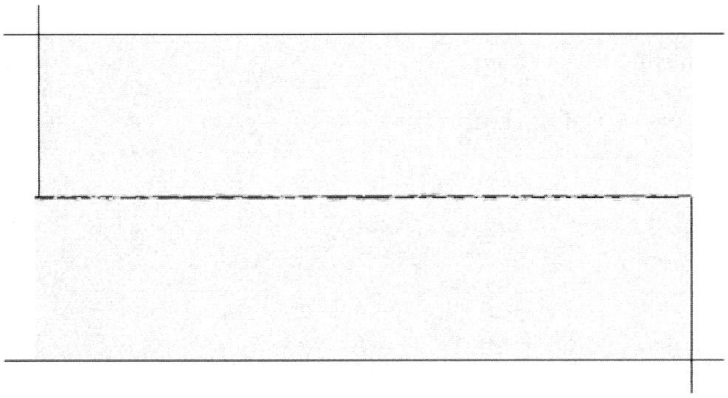

(b) Alignment of walls with respect to the synthetic sample.

Figure 3 Configuration of shear test and boundary conditions.

Figures 4 and 5 depict the shearing behaviour of different frictional joints under different normal loading conditions. The shear stress reaches a peak strength value before gradually reducing to a residual strength.

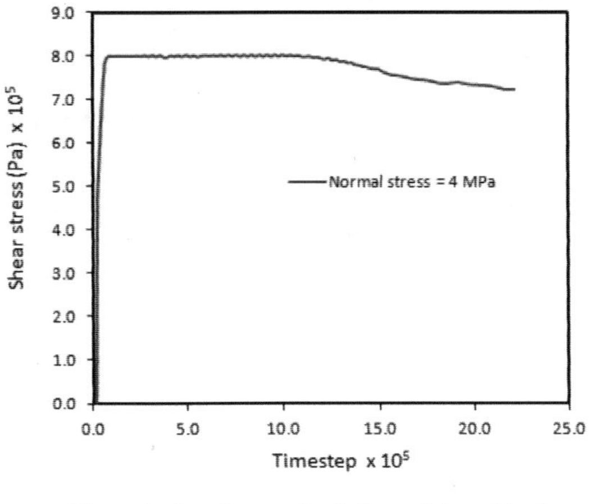

(a) Shear behaviour of a joint with a friction coefficient of 0.2.

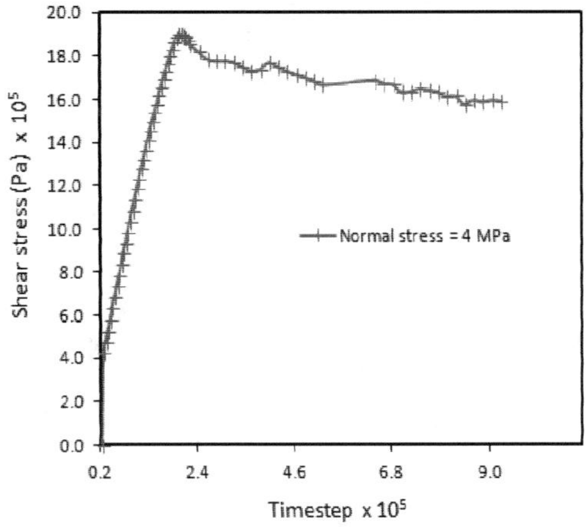

(b) Shear behaviour of a joint with a friction coefficient of 0.5.

Figure 4. Shear behaviour of the joint during the test.

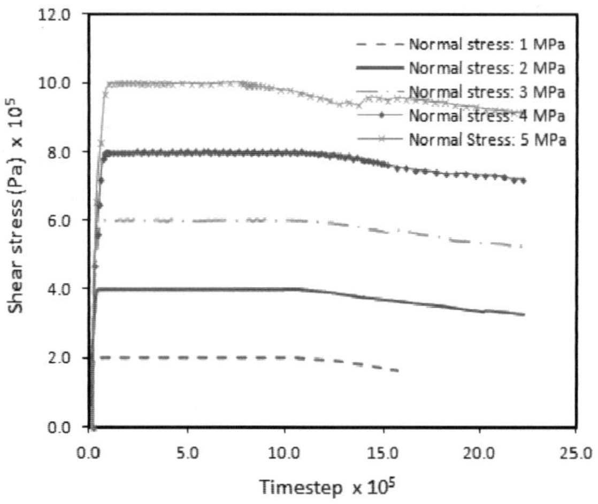

(a) Comparison of shear behaviours of a joint with a friction coefficient of 0.2.

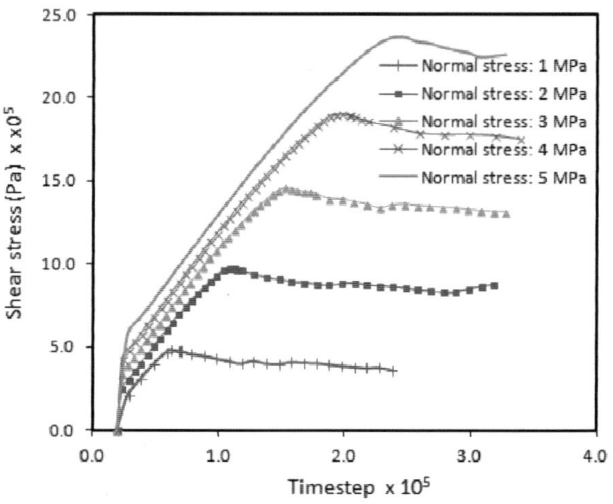

(b) Comparison of shear behaviours of a joint with a friction coefficient of 0.5.

Figure 5. Shear behaviour under different normal stress conditions.

To complete the calibration of the joint friction property failure envelopes were constructed for various assigned friction coefficients, using peak shear strength values and effective stresses normal to the joint plane as shown in Figures 6. In the figure, the limit of the shear strength when a friction coefficient of 0.2 is specified is shown. The equation describing the curve in terms of the peak shear stress and effective normal stress is

$$\tau = 0.0045 + 0.1993\sigma_n \quad (4a)$$

From Equation 5 the cohesive strength (c) is 0.0045 MPa, which is relatively negligible $c \approx 0$. The friction coefficient ϕ is 0.1993(≈ 0.2) and the corresponding friction angle is given as

$$\phi = tan^{-1}(0.1993) = 11.27^o \quad (4b)$$

$$\phi \approx 11.3^o \text{ and } c \approx 0.0\ MPa$$

Likewise, when a friction coefficient of 0.5 is set as an input parameter value to define the joint characteristics, the shear envelope derived is describe by:

$$\tau = 0.0261 + 0.4729\sigma_n \quad (5a)$$

The cohesive strength (c) is 0.0261 MPa, which is also relatively negligible. The friction coefficient (ϕ) is 0.4729 (≈ 0.5) and the corresponding friction angle is given as

$$\phi = tan^{-1}(0.4729) = 25.31^o \quad (5b)$$

$$\phi \approx 25.3^o \text{ and } c \approx 0.0\ MPa$$

From Equations 4-5, the numerical experimentally derived joint strength properties match the values of the input micro-parameters (e.g. friction angle and cohesive strength). Shear tests using other values of inputted friction coefficient and cohesive strength show matching results. Table 3 shows a comparison of some of the results.

Table 3. Comparison between inputted and derived joint properties.

Parameters	*Batch*	*Micro-property*		*Derived value*	
		Coefficient	Angle	Coefficient	Angle
Friction	Test 1	0.20	11.30 °C	0.199	11.27 °C
	Test 2	0.50	26.57 °C	0.473	25.30 °C
Cohesive strength (MPa)	Test 1	0.0		0.0261	
	Test 2	0.0		0.0045	

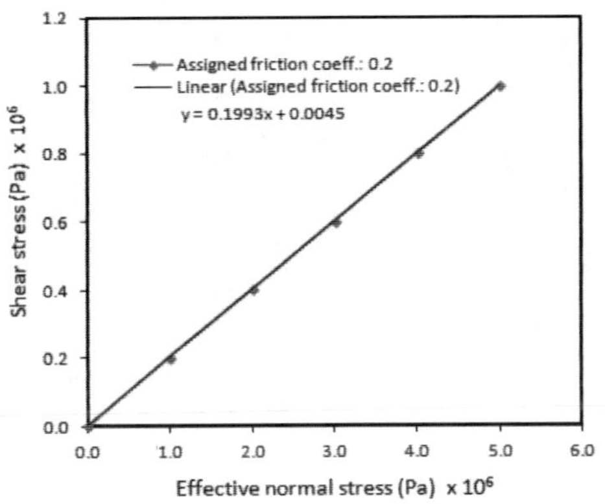

Figure 6. Joint failure envelope for an assigned friction coefficient of 0.2.

The contact force distribution for joints with friction coefficients of 0.2 is shown in Figure 7. Localised concentration of contact forces occur at the joint surface and the distribution of such spots along the joint increases with the normal stress. This is attributed to the increase in contact area between the two planes of each joint as the stress acting normally to it increases. Higher concentrations of contact forces also exist at the top left and bottom right sections of the rock mass, mainly because of the pressure/loading on the top left wall as well as the lateral restrictive support of the bottom right wall. Concentration and distribution of contact forces increases with the normal stress as localisation is more distinct at lower normal stresses. At higher normal stresses, dilation of the joint is restrained. Figure 8 shows the micro tensile and shear cracks, mostly initiated near the joint planes. The prevalence of micro cracks increases with normal stress and joint frictional resistance.

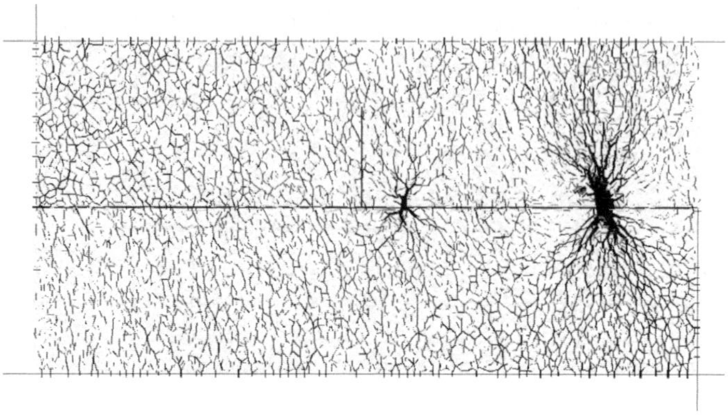

(a) Contact force distribution (Normal stress, 1 MPa)

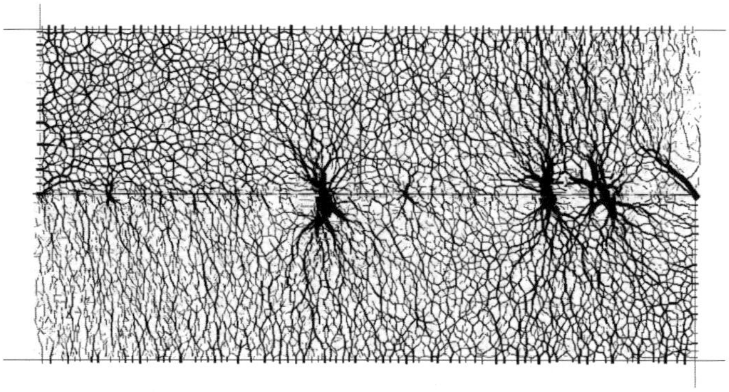

(b) Contact force distribution (Normal stress, 3 MPa)

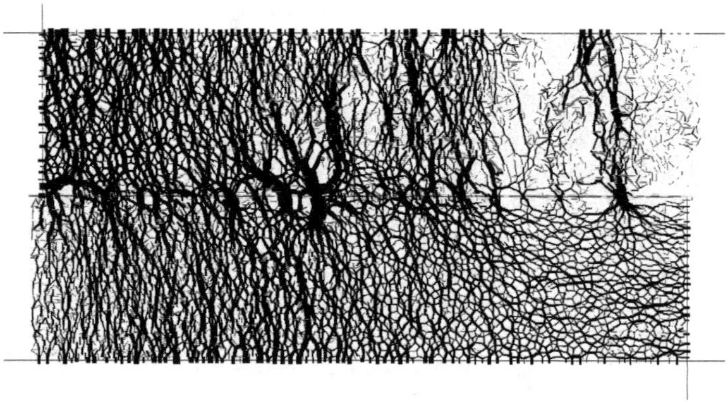

(c) Contact force distribution (Normal stress, 5 MPa)

Figure 7. Contact force distribution at various normal stress conditions (Friction coefficient = 0.2)
[Tensile: red, Compressive: black]

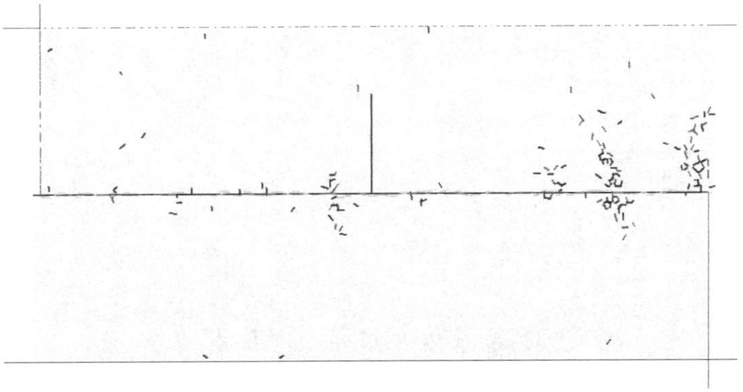

(a) Micro-crack distribution
(Normal stress, 4 MPa; Friction coefficient, 0.2)

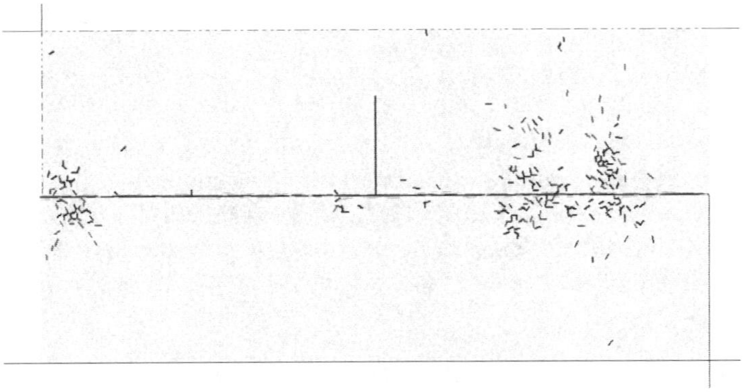

(b) Micro-crack distribution
(Normal stress, 3 MPa; Friction coefficient, 0.5)

Figure 8. Micro-crack distribution at different conditions

3. Simulation methodology

3.1. Features of the model domain

Samples of synthetic rock materials with properties similar to that tested (e.g. using biaxial and direct shear tests) were adopted. The synthetic rock material is representative of sandstone. Whereas the micro properties and mechanical properties are given in Table 1-2, the sample dimensions as shown in Figure 9 is width = 2 m and height = 1 m. Also the rock domain is deliberated jointed within by placing joints at strategic locations. The number, positions and spacing of the joints make up for different scenarios of the model set-up. The effect of the various layouts as well as the features of the network of joints is considered. The joint network comprises the size, position, spacing between each joint and connectivity if applicable. The first round of tests was conducted on a rock mass consisting of two parallel through joints inserted at locations of equal distance (0.15 m) from the centre of the domain. Both joints are lateral spanning in the *XY* direction in 2D (Figure 9-10). It is expected that in a 3D domain the joints will cut across the out of plane direction (*Z* plane).

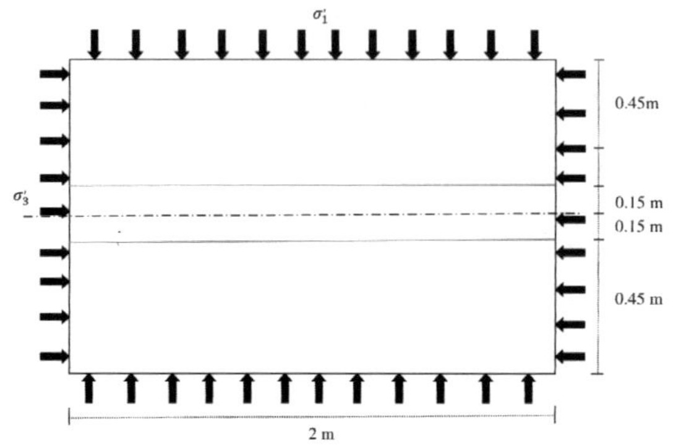

Figure 9. Layout of rock mass including two parallel lateral joints.

3.2. Boundary conditions and loading

Top and bottom vertical stresses, representing an overburden effect, were applied in addition to lateral confining stresses. The combination of these generates in-situ stress conditions. The maximum principal stress (σ_1') and minimum principal stresses σ_3' act in the vertical and horizontal directions respectively, where σ_1' = 2.5 MPa and σ_3' = 2.0 MPa. Fluid is introduced at the centre of the rock mass (Figure 10) at a final injection pressure of 35 MPa, maintained for the entire duration of the simulation. The loading is intended to cause a perturbation of fluid pressure as a result of the flow of fluid from a remote and singular location (e.g. an injection well). The main fluid properties are presented in Table 4.

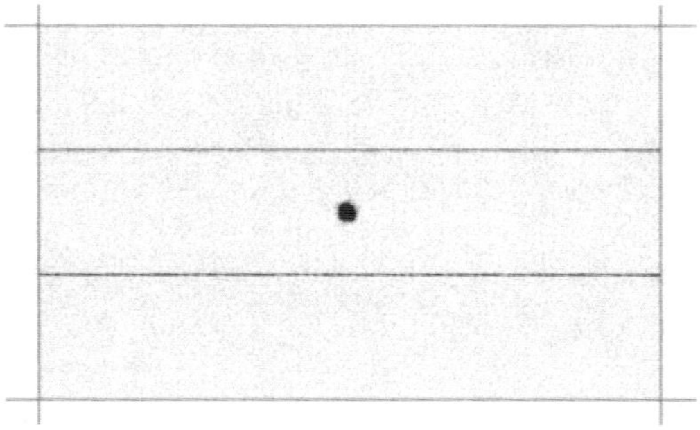

Figure 10. Fluid injected at the centre of the rock mass in between two parallel joints

Table 4. Fluid properties.

Parameter Description	Values
Viscosity, μ	3.95 x 102 Pa-s
Density, ρ_f	479 Kg/m^3
Bulk modulus, K_f	0.035 GN/m^2

IOP Conf. Series: Earth and Environmental Science **26** (2015) 012053 doi:10.1088/1755-1315/26/1/012053

4. Results and discussion

The key properties controlling joint behaviour include the shearing resistance, dilatancy, surface roughness and joint wall compressive strength. The direct impact of these properties on the overall joint performance and the corresponding role of the affected joints in association with subsurface events are of interest. Joint shearing resistance is described by its friction angle (or friction coefficient) which contributes to the shear strength and is in fact considered a measure of the joint shear strength. Variations in the joint friction angle will therefore affect its responses to natural and induced phenomena as well as alter the way in which the joint impacts on surrounding activities. In previous studies the extent of dependence of subsurface events on the frictional behaviour of joints within the proximity of such occurrences was determined. A typical case of a subsurface activity is the perturbation caused by fluid pressure (hydrostatic or via fluid flow) and the resulting onset and proliferation of fractures. Propagation and interconnectivity of fractures is an important phenomenon that could have an enormous impact. As a product of fluid pressure perturbation, the pattern and intensity of fractures as a function of both changes in the joint frictional resistance and changes in dilatancy is appraised. The friction coefficient was varied according to the following values: 0.0, 0.2, 0.5, 0.7 and 1.0, which correspond to friction angles of 0.0°, 11.3°, 26.6°, 35.0° and 45.0° respectively.

4.1. The relationship between joint dilatancy and frictional behaviour

The effect of the friction resistance of non-dilatant ($\varphi = 0$) rock joints on the fracturing process has been illustrated in a previous study. A similar investigation to determine the influence of frictional resistance on dilatant ($\varphi \neq 0$) rock joints invariably requires that its dilatant behaviour be considered. Consequently, a combination of two properties is essential: rock joint friction angle and dilation angle. The joint frictional resistance is a dependent variable that is also influenced by its dilation. In order to establish a link between the properties with respect to their joint effect on the fracturing process the two properties are simultaneously varied. The following presentation examines the characteristics of dilatant rock joints in an attempt to highlight the conformity of certain fracturing processes to peculiar combinations of frictional and dilatant behaviour. A summary of values representing some combinations of the joint dilatancy and frictional resistance is presented in Table 5.

Table 5. Combination of properties for dilatant rock joints

Case 1		***Case 2***		***Case 3***		***Case 4***	
Friction	**Dilatancy**	**Friction**	**Dilatancy**	**Friction**	**Dilatancy**	**Friction**	**Dilatancy**
0.2 (11.3°)	2.5° 5.0° 10.0° 20.0° 30.0° 40.0°	0.5 (26.6°)	2.5° 5.0° 10.0° 20.0° 30.0° 40.0°	0.7 (35.0°)	2.5° 5.0° 10.0° 20.0° 30.0° 40.0°	1.0 (45.0°)	2.5° 5.0° 10.0° 20.0° 30.0° 40.0°

The range of friction coefficient is 0.2 – 1.0 while the range of dilation angle is 2.5° – 40° and for some cases the frictional resistance is fixed whereas the degree of dilatancy is adjusted. For the combination of a low friction coefficient and a very low dilation angle (0.2, 2.5°), fracture propagation is constrained between the joints with only a few isolated cracks occurring at the outer regions away from the perimeter of the joint plane. Although, the intensity of fracturing is similar to cases involving non frictional non-dilatant rock joints and low frictional non-dilatant rock joints, the pattern of fracturing is somewhat different. The fracturing due to the presence of non-dilatant non-frictional/low frictional joints is more widely spread. For the same non/low frictional rock joint, a small increment in the dilatancy; for instance, an increase to φ = 5.0°, results in an upward and perpendicular propagation of fractures across the top joint and the generation of pockets of crack initiation at several remote locations along the plane of the top and bottom joints (Figure 11a). The fracture growth at the outer section becomes relatively rapid once it crosses the joint plane. The low frictional resistance of the joints means that the joints are more sensitive and as such the impact of dilation is more pronounced causing a greater severity in the movement and shearing of the joint planes.

Due to the lateral movement of the joints the occurrence of cracks initiated along the joint plane is preponderantly shear induced. Nevertheless, propagation of the resulting fractures is mainly caused by tensile failure of the rock material. The increment in dilatancy not only creates additional spots of crack initiation, it intensifies propagation in directions away from the joint plane. With a further increase in joint dilatancy (φ = 20.0°), the intensity of fracturing increases correspondingly (Figure 11b). The fractures created along the joint plane, as shown in this case, propagate away from the plane as well as in directions parallel and adjacent to the joint plane. There is a preference for fractures to grow along the joint surface than away from it. Isolated spots of fractures initially created are readily linked to form a coalescence of cracks before further proliferation away from the joint plane. A more intense state of fracturing occurs when the dilatancy is within the high range (φ = 40.0°) (Figure 11c). The fractures spread along many sections of the joint planes and then grow extensively into other regions outside the enclosure of the two joints.

Another effect of joint dilatancy is the possibility of cavity initiation along the joint surfaces. This phenomenon is more noticeable when the dilatancy is in the medium to high range with the number and sizes of cavities increasing as dilatancy increases. To illustrate, a greater number of cavities are created and a greater number grow into considerably sizes when the joint dilation angle is 40° (Figure 11c) as compared with a dilation angle of 20° (Figure 11b). Apart from the initial location above the point of fluid injection where the fracture propagating directly from the point of fluid injection crosses the upper joint, the location of individual cavities is inconsistent with the degree of dilatancy. A comparison between joints of dilation angle 10°, 20°, and 40° indicates that besides the cavity created just above the point of fluid injection the number and location of other cavities is different for dissimilar joint dilatancies.

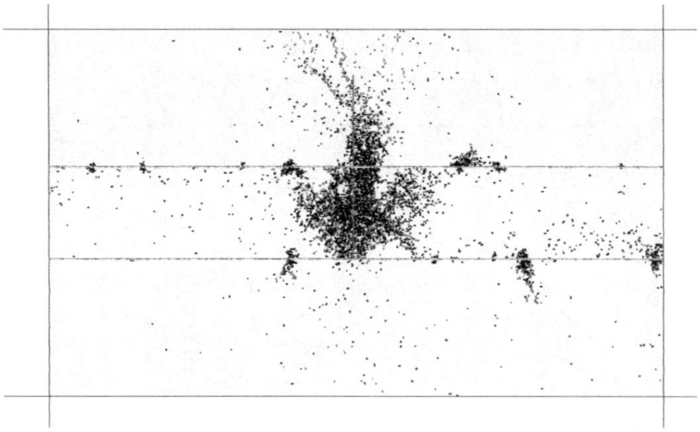

(a) Fracture development (dilation angle: 5.0).

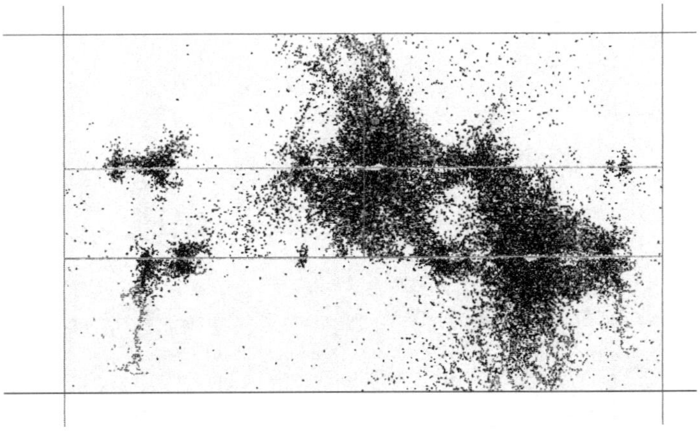

(b) Fracture development (dilation angle: 20.0).

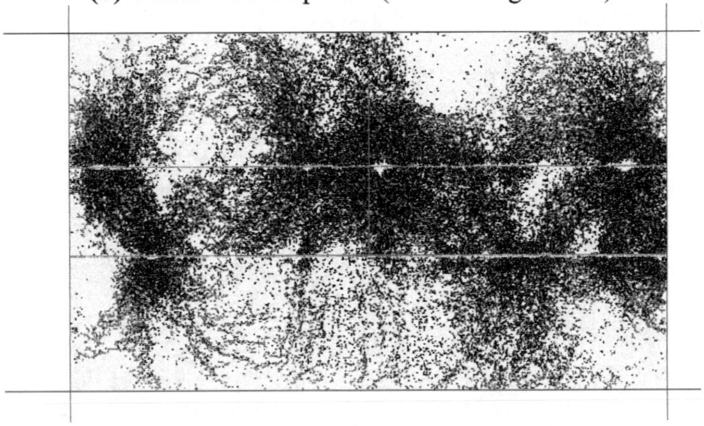

(c) Fracture development (dilation angle: 40.0).

Figure 11. Influence of dilatancy on fracturing behaviour (friction coefficient: 0.2)

The rate of generation and population of cracks indicates an intensity of shear induced cracks significantly higher than tensile induced cracks (Figure 12); the disparity in the rate and number of cracks increases progressively with time such that at the later stages the extent of shear cracks is so much greater than tensile cracks. The preponderance of shear cracks is attributed to the sliding performance of the joint surfaces which drives its shearing behaviour. The less frictional a joint is the greater the tendency for the surfaces to slide against each other that may result in the onset and propagation of cracks caused by shear stresses. Highly frictional joints allow relatively fewer cases of shearing events.

The effect of rock joint dilatancy on the fracturing process becomes apparent when results for the different magnitudes of dilatancies are juxtaposed. The intensity of total fracturing increases with the dilation angle (Figure 13a). The rate of increment between the two variables is disproportionate especially within the range of high dilation angles (Figure 14) and a similar trend occurs throughout the elapse period. For instance, at a timestep of 2.48e2, the total number of cracks is as follows: 22, 51, 100, 254 and 641, for a joint dilation angle of 2.5, 5.0°, 10.0 °, 20.0 ° and 40.0 °, respectively. The same pattern is observed for both shear and tensile cracks (Figure 13 b-c).

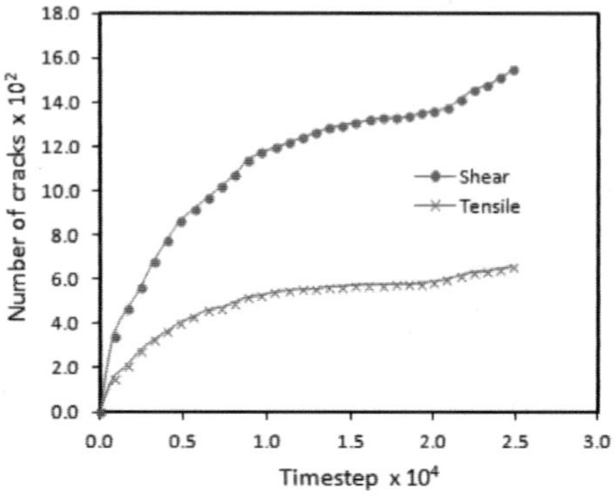

(a) Fracture development (dilation angle: 2.5).

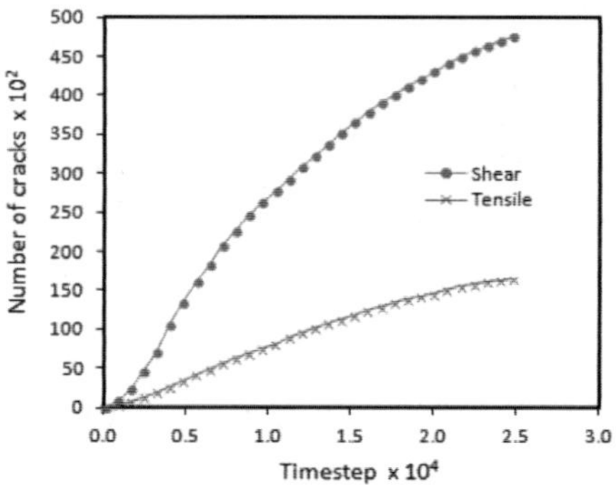

(b) Fracture development (dilation angle: 40.0).

Figure 12. Tensile and shear fracture development at various joint dilatancies (Friction coefficient: 0.2).

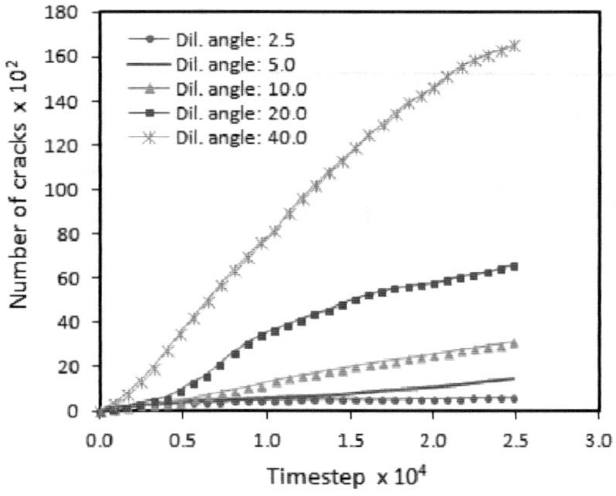

(a) Tensile fracture development

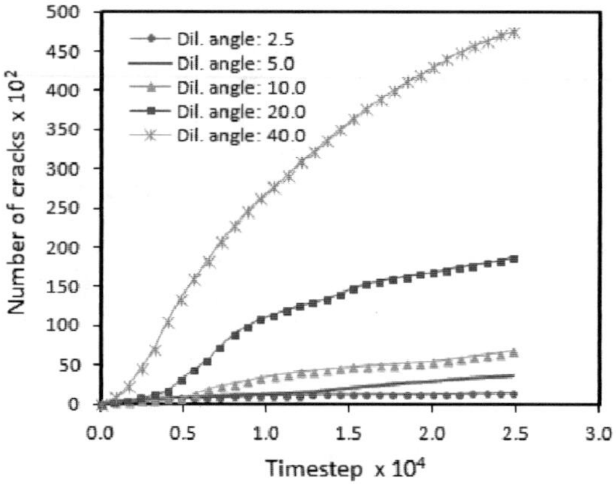

(b) Shear fracture development.

Figure 13. Effect of dilatancy on fracture development (Friction coefficient: 0.2)

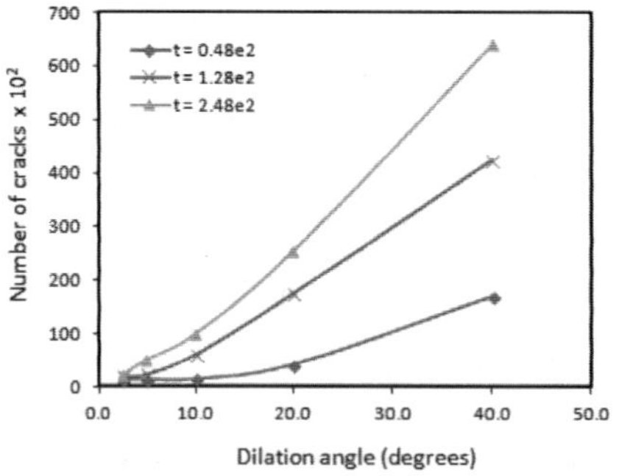

Figure 14. Magnitude of total fracture development at different rock joint dilatancies (Friction coefficient: 0.2).

6. Conclusions

The existence of discontinuities (joints) in rock masses affects its strength and response to both in-situ and external forcings. They major elements that determine the attributes of discontinuities are often contained in their mechanical and geo-chemical properties, which makes the study of their characteristics essential, especially if the joint interface is bonded. The behaviour of bonded discontinuities is determined by the nature of the fill, which may be permeable or impermeable; on the other hand, the behaviour of unbounded discontinuities is mainly governed by its mechanical and physical properties. The compressive strength (JCS), shear strength and dilatancy of the interface are some prevailing factors. This is especially so in cases where they dominate the over-burden pressure and in-situ stresses (e.g. at shallow depths).

Prior investigations conducted by, for instance, Hanson *et al.* [4], Hanson *et al.* [39] and Anderson [40] are limited in scope. Unlike previous studies [4, 39, 40], a more detailed analysis of the fracturing phenomenon which elaborately highlights the role of the key properties that control the behaviour of unbonded interfaces and the interaction between approaching fractures, the interface and fluid pressure perturbation has been presented in this research. The interplay between selected interface properties was also discussed and the numerical methodology adopted provided the means for a more rigorous investigation of the processes at the inter-particle level as well as enabled the flexible control of certain features of the interface. Two important and related phenomena were investigated: the shear behaviour of interfaces represented by the frictional resistance (the cohesive strength is negligible for unbounded interfaces), interface dilatancy and the interplay between the two properties. Although the numerical experiments involved the interface between rock in formations (also referred herein as joints), the concepts applied and results obtained are applicable to analogous forms of discontinuities. The following outlines the outcome of this research:

The dilatancy of rock joints contribute to its performance and impacts on its frictional behaviour. At low joint friction the occurrence of even a small amount of dilation increases the severity of the fracturing process. In addition to facilitating extra zones of fracture initiation at the joint plane, increments in dilatancy intensify fracture propagation in directions away from the joint plane, albeit there is a preference for fractures to grow along the joint plane. Dilation enhances the tendency for cavity initiation along the joint plane and the number and size of cavity increases as dilation increases.

The onset of cracks at joint planes is predominantly instigated by shear failure and tensile failure of the rock material induces propagation of the resulting fractures. The intensity of shear induced fractures is considerably greater than tensile induced fractures and the predominance of shear fractures is attributed to the sliding performance of the joint planes. Generally, the rate and magnitude of shear and tensile fracturing increases with joint dilatancy. For a constant magnitude of joint dilatancy the intensity of fracturing reduces appreciably as the joint frictional resistance increases. The effect of the frictional resistance becomes more dominant as it tends towards high values and within this range the effect of small magnitudes of dilatancy is trivial. At the vicinity of highly frictional joints, fracture development at low joint dilatancy is greater than fracture development at medium joint dilatancy. Unless joint dilatancy is low, the rate and magnitude of fracturing decreases as joint friction increases. Further studies are essential to include the impact of variations in overburden pressure conditions as well as differences in rock material properties due to layering.

Nomenclature

c	Cohesive strength
E	Young's Modulus
E^*	Young's Modulus in plain strain

IOP Conf. Series: Earth and Environmental Science **26** (2015) 012053 doi:10.1088/1755-1315/26/1/012053

e_y	Axial strain
e_x	Lateral strain
ε_v	Volumetric strain
ε_1	Major principal strain
ε_3	Minor principal strain
JCR	Joint roughness coefficient
JCS	Joint wall compressive strength
k_n	Particle normal stiffness
k_s	Particle shear stiffness
k^n	Normal stiffness
k^s	Shear stiffness
K_f	Bulk modulus
$\hat{n}_j$	Unit normal vector defining the joint plane
N_{tc}	Estimated rate of development of tensile cracks
N_{sc}	Estimated rate of development of shear cracks
$\hat{q}$	Compressive strength
q_u	Unconfined compressive strength
T	Tensile strength
t	Elapsed time
τ_p	Peak shear strength
τ_r	Residual value of shear strength
τ_θ	The shear stress required to overcome the volumetric expansion
ρ_f	Density
υ	Poisson's ratio
υ^*	Poisson's ratio in plain strain
γ	Plastic shear strain
γ_{max}	Maximum plastic shear strain
θ	Dip angle
θ_d	Average angle of deviation of the joint plane/ joint surface particles from the direction of applied shear stress
ϑ	Dip direction
σ_D	Differential stress
σ_n	Normal stress
σ_y	Axial stress
σ_x	Lateral stress
σ_1'	Major effective principal stress
σ_3'	Minor effective principal stress
ϕ	Angle of internal friction (friction angle)

ϕ_r	Residual value of friction angle
ϕ_b	Basic friction angle
ϕ_{crit}	Critical friction angle
ϕ_f	Interparticle friction angle corrected for work done or energy dissipated due to expansion
ϕ_t	The true angle of friction between the mineral surfaces of the particles
ϕ_{cv}	Angle of friction under constant volume
φ	Dilation of a material, joint or discontinuity
φ_p	Peak dilation, which is the same as the maximum dilation
μ	Viscosity

References

[1] Daneshy A A 1978 Hydraulic fracture propagation in layered formations *SPE Journal* **18**: p. 33-41

[2] Simonson E R, Abou-Sayed A S and Clifton R J 1978 Containment of massive hydraulic fractures *SPE Journal* **18** (1): p. 27-32.

[3] Blanton T L 1982 An experimental study of interaction between hydraulically induced and pre-existing fractures *SPE Unconventional Gas Recovery Symposium*, Pittsburgh, May 16-18 1982.

[4] Hanson M E, et al. 1982 Some effects of stress, friction, and fluid flow on hydraulic fracturing *SPE Journal* **22**: p. 321-332.

[5] van Eekelen H A 1982 Hydraulic fracture geometry: fracture containment in layered formations *SPE Journal* **22** (3): p. 341-349.

[6] Warpinski N R and Teufel L W 1987 Influence of geologic discontinuities on hydraulic fracture propagation *Journal of Petroleum Technology* **39**: p. 209-220.

[7] Settari A 1988 Quantitative analysis of factors influencing vertical and lateral fracture growth *SPE Production Engineering* **3** (3): p. 310-322.

[8] Blair S C, et al. 1989 Laboratory observations of the effect of geologic discontinuities on hydrofracture propagation *Rock Mechanics as a Guide for Efficient Utilization of Natural Resources* Khair (ed.) Balkema, Rotterdam

[9] Renshaw C E and Pollard D D 1995 An experimentally verified criterion for propagation across unbounded frictional interfaces in brittle, linear elastic materials *International Journal of Rock Mechanics and Mining Science and Geomechanics Abstracts* **32**: p. 237-249

[10] Casas L, et al. 2006 Laboratory hydraulic fracturing test on a rock with artificial discontinuities *SPE Annual Technical Conference and Exhibition,* San Antonio, Texas, USA, 24-27 September 2006

[11] Zhang X and Jeffrey R G 2006 The role of friction and secondary flaws on deflection and reinitiation of hydraulic fractures at orthogonal pre-existing fractures *Geophysical Journal International* **166**: p. 1454-1465

[12] Thiercelin M and Makkhyu E 2007 Stress field in the vicinity of a natural fault activated by the propagation of an induced hydraulic fracture *Proc., First Canada-US Rock Mechanics Symposium-Rock Mechanics Meeting Society's Challenge and Demands* **2** p. 1617-1624

[13] Athavale A S and Miskimins J L 2008 Laboratory hydraulic fracturing on small homogeneous and laminated blocks *42nd US Rock Mechanics Symposium*, San Francisco June 29-July 2, 2008

[14] Zhang X and Jeffrey R G 2008 Reinitiation or termination of fluid-driven fractures at frictional bedding interfaces *Journal of Geophysical Research* **113**: p. B08416, 16 PP

[15] Philipp S, et al. 2009 Field studies and numerical models of hydrofracture propagation layered fractures reservoirs *Geophysical Research Abstracts* **11**.

[16] Chuprakov D A, et al. 2010 Hydraulic fracture propagation in a naturally fractured reservoir *SPE Oil and Gas India Conference and Exhibition* Mumbai, India, 20-22 January, 2010

[17] Cai M and Horii H 1992 A constitutive model of highly jointed rock masses *Mechanics of Materials* **13**: p. 217-246

[18] Kulatilake P H S W, Malama B and Wang J 2001 Physical and particle flow modeling of jointed rock block behaviour under uniaxial loading *International Journal of Rock Mechanics and Mining Sciences* **38**: p. 641-657

[19] Bandis S, Lumsden A and Barton N 1983 Fundamentals of rock joint deformation *International Journal of Rock Mechanics and Mining Sciences and Geomechanics Abstracts* **20** (6): p. 249-268

[20] Barton N and Choubey V 1977 The shear strength of rock joints in theory and practice *Rock Mechanics* **10**: p. 1-54

[21] Barton N 1976 The shear strength of rock and rock joints *International Journal of Rock Mechanics and Mining Sciences and Geomechanics Abstracts* **19** (9): p. 255-279

[22] Ladany B and Archambault G 1970 Simulation of shear behaviour of a jointed rock mass *Proceedings of the 11th symposium on rock mechanics*: New York, p.105-125. p. 105--125

[23] Lee H et al. 2001 Influence of asperity degradation on the mechanical behaviour of rough rock joints under cyclic shear loading *International Journal of Rock Mechanics and Mining Sciences* **38**: p. 967-980

[24] Park J.-W and Song J-J 2009 Numerical simulation of a direct shear test on a rock joint using a bonded-particle model *International Journal of Rock Mechanics and Mining Sciences* **46**: p. 1315-1328

[25] Indraratna B, Oliveira D and Jayanathan, M 2008 Revised shear strength model for infilled rock joints considering overconsolidation effect *Proceedings of the 1st Southern Hemisphere International Rock Mechanics Symposium SHIRMS,* 16-19 September, Perth. 2008

[26] Trivedi A 2010 Strength and dilatancy of jointed rocks with granular fill *Acta Geotechnica* **5** (1): p. 15-31

[27] Zare M, Kakaie S R and Jalali S M E 2008 A new empirical criterion for prediction of the shear strength of natural infilled rock joints under constant normal load (CNL) conditions, *Proceeedings of ISRM International Symposium - 5th Asian Rock Mechanics Symposium (ARMS5)* 24-26 November, Tehran, Iran. 2008

[28] Utili S., Crosta G.B. (2011). Modelling the evolution of natural slopes subject to weathering: Part I. Limit analysis approach. Journal of Geophysical Research – Earth Surface, 116, F01016, doi:10.1029/2009JF001557.

[29] Utili S., Crosta G.B. (2011) Modelling the evolution of natural slopes subject to weathering: Part II. Discrete element approach Journal of Geophysical Research – Earth Surface, 116, F01017, doi:10.1029/2009JF001559.

[30] Khorsandi A and Ghoreishi S H 2013 Studying the Interaction Between Active Faults and Landslide Phenomenon: Case Study of Landslide in Latian, Northeast Tehran, Iran *Geotech Geol Eng* **31**: p. 617-625

[31] Jibson R W, et al. 2004 Landslides Triggered by the 2002 Denali Fault, Alaska, Earthquake and the Inferred Nature of the Strong Shaking *Earthquake Spectra* **20** (3): p. 669-691

[32] Biringen E, Davie J and Fenster D 2013 Impacts of Growth Faults on Slope Stability *Geo-Congress* 2013 p. 1359-1368

[33] Byerlee J 1978 Friction of Rocks, in Rock Friction and Earthquake Prediction, J Byerlee and M. Wyss, Editors *Birkhäuser Basel* p. 615-626

[34] Barton N 1973 Review of a new shear-strength criterion for rock joints *Engineering Geology* **17**: p. 287-332

[35] De Josselin de Jong G 1976 Rowe's stress-dilatancy relation based on friction *Geotechnique* **26** (1): p. 527-534

[36] Rowe P W 1969 The relation between the shear strength of sands in triaxial compression, plane strain and direct shear *Geotechnique* **19** (1): p. 75-86

[37] Rowe P W 1962 The stress-dilatancy relation for static equilibrium of an assembly of particles in contact *Proceedings of the Royal Society of London, Series A - Mathematical and Physical Sciences* **269** (1339): p. 500-527

[38] Newland P L and Allely B H 1957 Volume changes in drained triaxial tests on granular materials *Geotechnique* **7**: p. 17-34

[39] Hanson M E, Shaffer R J and Anderson G D 1981 Effects of various parameters on hydraulic fracturing geometry SPE Journal **21** (4): p. 435-443

[40] Anderson G D 1981 Effects of Friction on Hydraulic Fracture Growth Near Unbonded Interfaces in Rocks *SPE Journal*, **21** (1): p. 21-29

IOP Conf. Series: Earth and Environmental Science **26** (2015) 012044 doi:10.1088/1755-1315/26/1/012044

A simplified DEM numerical simulation of vibroflotation without backfill

M J Jiang[1,2,3,4], W W Liu[1,2,3], J He[1,2,3] and Y Sun[1,2,3]
[1] State Key Laboratory for Disaster Reduction in Civil Engineering, Tongji University, Shanghai, China, 200092
[2] Key Laboratory of Geotechnical and Underground Enigineering of Ministry of Education, Tongji University, Shanghai, China, 200092
[3] Department of Geotechnical Engineering, College of Civil Engineering, Tongji University, Shanghai, China, 200092

E-mail: mingjing.jiang@tongji.edu.cn

Abstract. Vibroflotation is one of the deep vibratory compaction techniques for ground reinforcement. This method densifies the soil and improves its mechanical properties, thus helps to protect people's lives and property from geological disasters. The macro reinforcement mechanisms of vibroflotation method have been investigated by numerical simulations, laboratory and in-situ experiments. However, little attention has been paid on its micro-mechanism, which is essential to fully understand the principle of the ground reinforcement. Discrete element method (DEM), based on discrete mechanics, is more powerful to solve large deformation and failure problems. This paper investigated the macro-micro mechanism of vibroflotation without backfill under two conditions, i.e., whether or not the ground water was considered, by incorporating inter-particle rolling resistance model in the DEM simulations. Conclusions obtained are as follows: The DEM simulations incorporating rolling resistance well replicate the mechanical response of the soil assemblages and are in line with practical observations. The void ratio of the granular soil fluctuates up and down in the process of vibroflotation, and finally reduces to a lower value. It is more efficient to densify the ground without water compared to the ground with water.

1. Introduction

Vibroflotation is one of soil improvement techniques to increase soil strength and bearing capacity of ground using vibration and the jetting of water. It is an efficient method to protect the ground from geohazards such as liquefaction under seismic. It was firstly proposed by Steuerman in 1936, and well developed in America.

The applicability and mechanism of vibroflotation have been studied by different scholars. Webb and Ian Hall [1] thought the densification zone and efficiency will significantly reduce with the increase of silt content, according to vibroflotation tests with backfill. Slocombe et al. [2] pointed out that 15% is the maximum fines content to permit the performance of vibroflotation method without the addition of stone aggregate. The vibroflotation method is also used to improve the dynamic property such as liquefaction-resistant behavior from Baez [3] and Dobson [4]. Bement and Selby [5] took a laboratory tests to investigate the compaction settlement of granular soils during pile driving. Arnold et al. [6]

[4] To whom any correspondence should be addressed.

presented an elastic finite element analysis of vibrocompaction, comparing horizontally circling deep vibrators with vertically oscillating top vibrators. Results show better performance of deep vibrators with multidirectional shearing mode. However, owing to several simplifications (i.e., without considering the influence of pore water), the observations between numerical simulations and in-situ data were not good. Susana et al. [7] also utilized a new finite element model to simulate the mechanical response of sandy soils using the vibroflotation technique.

Although meaningful results have been obtained, the micro-mechanism in the process of vibroflotation remains poorly understood. On the other hand, finite element method (FEM) has inherent disadvantages to analyze geohazard problems which related to large deformation (e.g., the study of shear band in cemented sands [8]). It's also a challenge to obtain a proper constitutive model of complex soils. In contrast, the discrete element method (DEM), firstly proposed by Cundall and Strack [9], has obvious advantages to solve large deformation and failure problems. It treats the soil as non-continuous aggregates to better simulate the macro-micro mechanical behavior by choosing appropriate contact models and corresponding parameters. The paper analyzes the compaction mechanism using the 2D DEM to simulate the vibroflotation process without backfill under two conditions (i.e., below and above ground water).

2. The contact model considering inter-particle rolling resistance

The inter-particle rolling resistance model proposed by Jiang et al. [10] was used in this paper. The model only introduces one additional parameter β to consider the influence of irregular shapes of soil particles. Comparing to other standard DEM models (e.g., Iwashita et al. [11]), this model thus solves the drawback of using a lot of parameters to define rolling resistance, which needs to be determined through trial and error. The influence of different rolling resistance coefficient on passive earth pressure was also examined from the micro-macro scale [12].

The contact models of two rigid disks illustrated in Figure 1 are composed of normal, tangential and rolling contact components. The mechanical responses of each component are presented in Figure 1. The normal contact model resists normal pressure which is proportional to the normal stiffness k_n and the amount of particle overlap u_n, as shown in Eq. (1).

$$F_n = k_n \times u_n \tag{1}$$

The tangential force is proportional to the tangential displacement u_s and the tangential stiffness k_s, as illustrated by Eq. (2), maintains constant when it reaches to the maximum value that is calculated by friction coefficient multiplied by the normal contact force.

$$F_s = k_s \times u_s \tag{2}$$

The rolling contact model is characterized by the stiffness parameter k_m, which is dependent on rolling resistance coefficient β, particle contact normal stiffness k_n, and the common radius r. The rolling resistance maintains at a value in Model 1 or abruptly reduces to zero in Model 2 after reaching a critical relative rotation angle θ_r. Model 1 is adopted in this paper, thus the moment is calculated by Eq. (3) as follows:

$$M = \begin{cases} K_m\theta = \frac{\theta \times k_n \times r^2 \times \beta^2}{12}, & \theta \le \theta_r, \\ M_0 = \frac{1}{6}\left(F_n \times r \times \beta\right), & \theta > \theta_r. \end{cases} \tag{3}$$

where r is the average radius of the two contact particles, namely:

$$r = \frac{2r_1r_2}{r_1 + r_2}. \tag{4}$$

where r1 and r2 are the radius of two contact particles respectively.

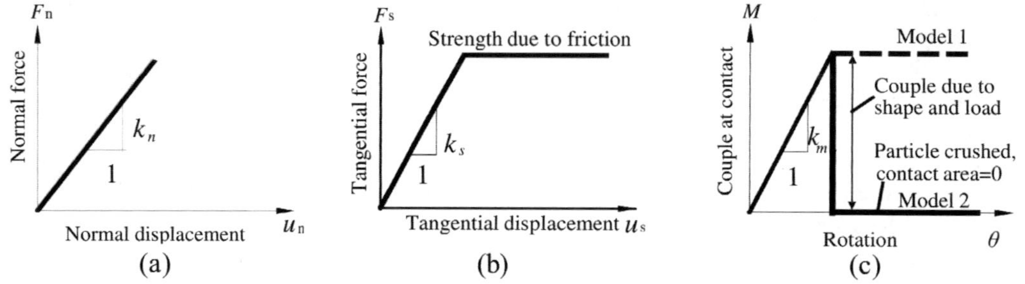

Figure 1. Schematic illustration of the contact model proposed for grains and its mechanical response: (a) Normal direction; (b) Tangential direction; (c) Rolling direction [10]

3. Biaxial compression tests

Twelve numerical simulations were conducted to study the frictional behavior of the granular materials incorporating the rolling resistance model. Each specimen has a height of 80mm and a width of 40mm, and is composed of 6,000 particles. The materials are all composed of discs with a maximum diameter of 9.0mm, minimum diameter of 6.0mm, and an average grain diameter d_{50} =7.6mm. The initial void ratio is 0.30. The friction coefficient between particles is chosen to be 0.5. The normal and tangential stiffness of particles are 1.5×10^8 N·m^{-1} and 1.0×10^8 N·m^{-1}, respectively. The biaxial compression tests under different confining pressures (i.e., 100kPa, 200kPa, and 400kPa) and different rolling resistance coefficients (i.e., 0, 0.4, 0.8 and 1.2) are conducted to obtain mechanical parameters of specimens. Figure 2 shows the deviator stress axial strain and volumetric strain responses obtained from the DEM biaxial compression tests with different β under confining pressure of 200kPa. The sample with $\beta = 0$ shows strain hardening and shear contraction while other specimens (i.e., β = 0.4, 0.8 and 1.2) shows a behavior of strain softening and shear dilatant. With the increases of β , the strain softening and shear dilatant response are more distinct. Figure 3 presents that the peak and residual frictional angle of the soil with the variations of β . The peak and residual internal frictional angles both increase with the increases of β .

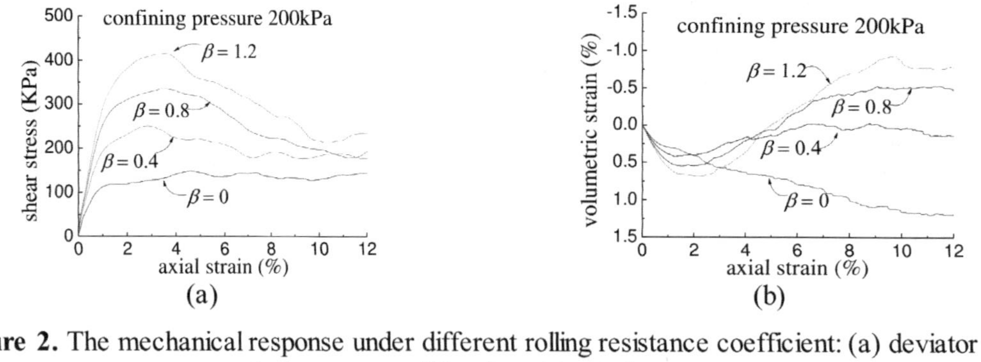

Figure 2. The mechanical response under different rolling resistance coefficient: (a) deviator stress strain response; (b) volumetric strain response

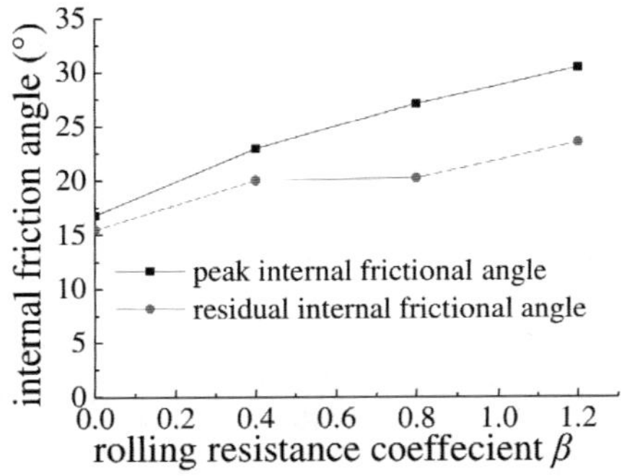

Figure 3. Relationship between the peak and residual internal friction angle and rolling resistance coefficient β

4. DEM simulation of vibroflotation

The Multi-layer under-compaction technique (UCM) [13] is chosen to produce the ground. The original ground size is 0.5 m×0.14 m. Under the action of 20 g gravitational field, the real size of ground is 10 m ×2.8 m. It is composed of 500,000 particles.

This paper aims to simulate the process of vibroflotation without backfill. Note that the stage of the penetration of the vibrator into the ground is ignored. First, a square with 0.2m long is generated by command clump in the commercial software PFC to represent the vibrator, and the center is located in the point of (5.0, 1.6). Then, as shown in Figure 4, 14 measurement circles (M1-M14) are arranged around the vibrator to monitor the variations of each variable. The vibrator vibrates in the horizontal direction, i.e., the direction from M3 to M6, in accordance with actual circumstances. Figure 5 shows the variation of the loading velocity chosen. Positive velocity represents the vibrator moves left, and vice versa. The amplitude of vibration is similar to that of actual vibrator. In addition, corresponding to soils above the ground water, the density of granular soils below the ground water is set to effective unit weight without changes of any other parameters. The influence of excess pore water pressure, which couldoccur in the process of vibration, is not considered here in the DEM simulations.

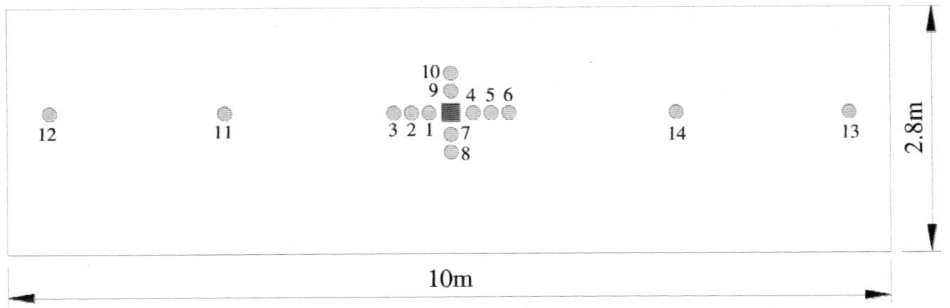

Figure 4. The ground sample and layout of measurement circles

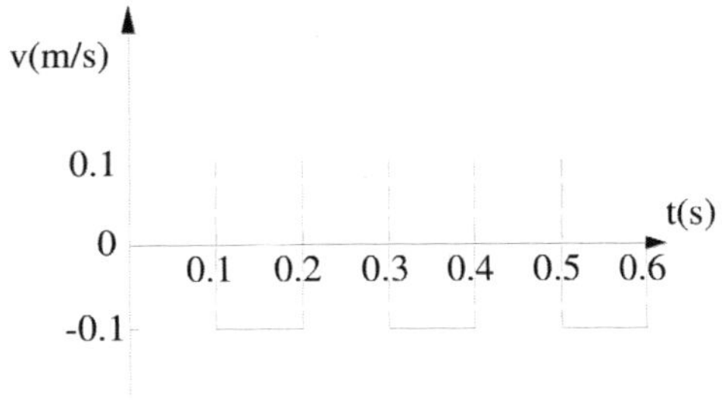

Figure 5. The variation of the loading velocity

5. DEM simulation results

5.1. Force chains

The force chains of the sample under different circumstances are similar, thus we only present the force chains above the ground water at different times under the rolling resistance coefficient of β = 0.8, as shown in Figure 6. The black line presents the direction of the pressure and the width presents its size. The color of the force chains in the ground sample are deepen from up to down, which shows the increase of pressure in accordance with actual ground under gravity. As demonstrated in Figure 6, the pressure acts on the left side of the specimen and the value of the force chains is larger than that of the other side when the vibrator moves to the left, i.e., t = 0.1 s, 0.3 s, and 0.5 s. Thus the soil on the left side is compacted. Correspondingly, when the vibrator moves to the right, i.e., t = 0.2 s, 0.4 s, and 0.6 s, the force chains concentrate and deepen due to the effect of the vibrator on the right soil.

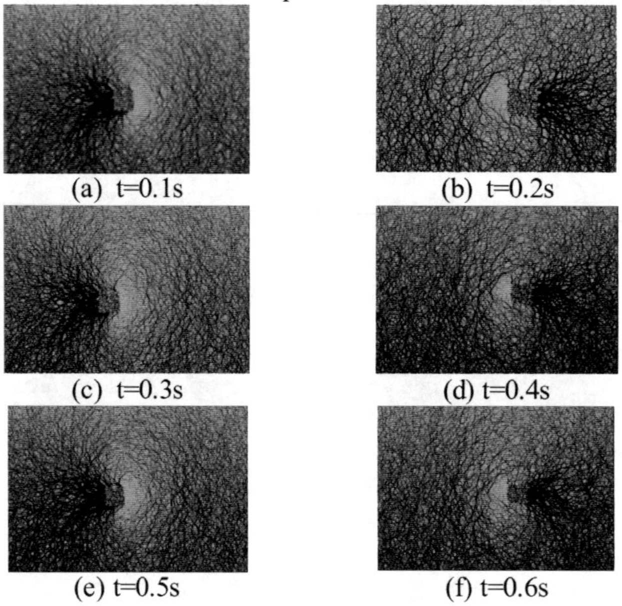

Figure 6. Force chains in the ground without water during the vibroflotation process (β =0.8).

5.2. Void ratio distributions

Figure 7 presents the variations of the void ratio monitored by 14 measurement circles with different values of β under two conditions, i.e., below or above the ground water. Typical measurement circles (i.e., M1, M2, M4, M7, M9, and M13) that represent the near, down, up, and far fields around the vibrator, respectively, are selected to analyse the changing void ratio. As illustrated in Figure 7, the void ratios in all of the measurement circles reduce in the vibroflotation process. Note that the initial void ratios in each measurements are less than the initial void ratio of the sample, and this is due to consolidation process under gravity.

As shown in Figure 7 (a), the variation of void ratio in M1 shows fluctuation up and down. During the first 0.1 s, the reduction of void ratio is due to the vibrator is moving to the left side and the soil in the left part is strengthened. Then the void ratio of M1 increases when the vibrator takes part in the soil compaction of the right side from the time of 0.1 s to 0.2 s. With the vibrator moves in the reciprocating way, accordingly, the void ratio of M1 fluctuates. Figure 7 (b) shows that the void ratio in M2 reduces when the vibrator moves to the left in the first 0.1 s and then rebounds to a relatively lower value in the next 0.1 s. In the next several loops, the void ratio in M2 fluctuates to a lower level and the particles in that area are thus compacted.

Figure 7 (c) presents the void ratio in M4 keeps constant at first, due to the vibrator moves to the left initially, and then reduces,similar to the variation of void ration in M1 or M2. The void ratios of M7 and M9 in Figure 7 (d)-(e) show that the soil is strengthened below and above the vibrator.

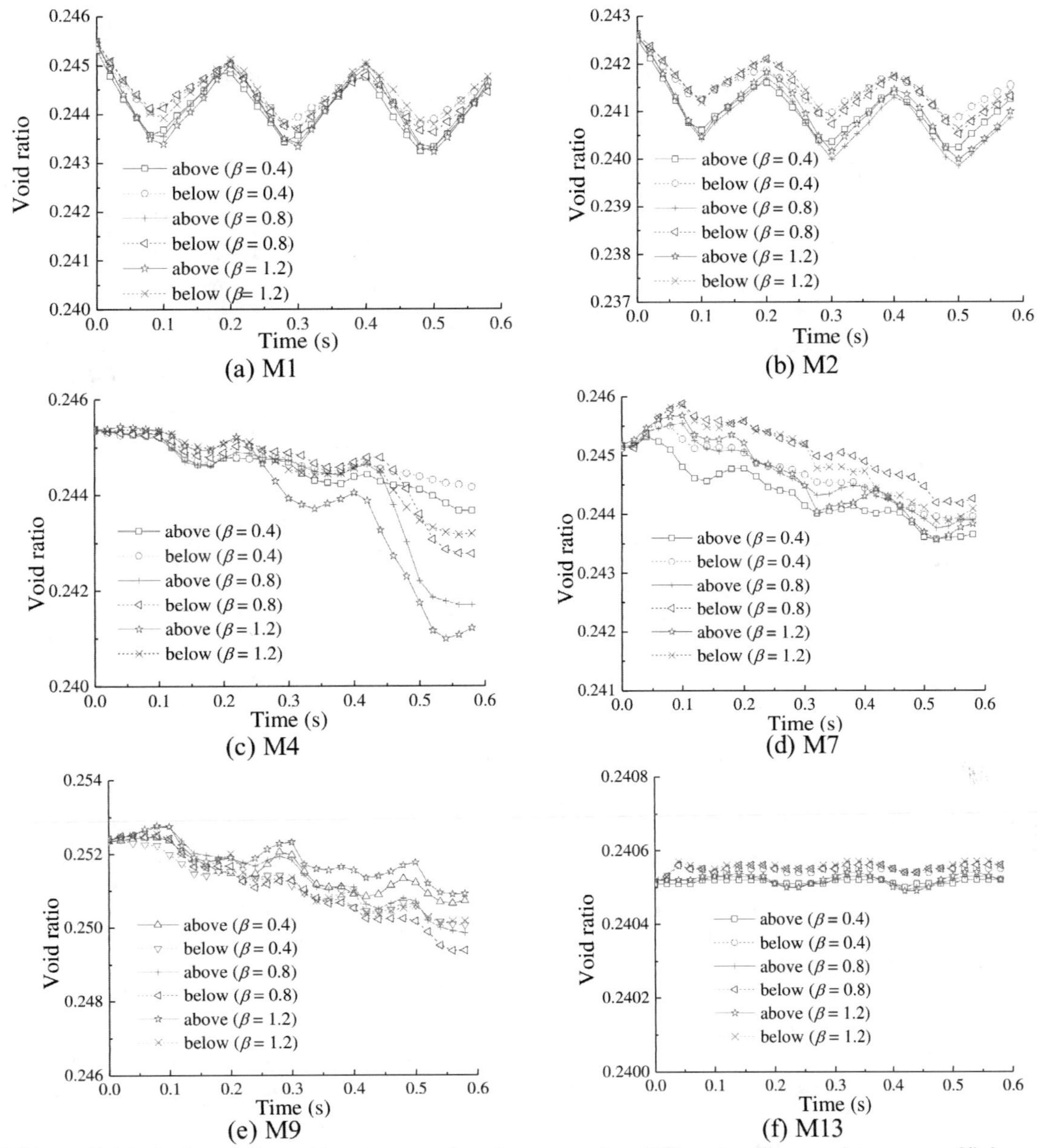

Figure 7. Variation of the void ratio near the vibrator under different rolling resistance coefficients

However, in M13, little fluctuations and reductions of void ratio are observed in both two conditions with different rolling resistance coefficients. Therefore, we can conclude that in that area far from the vibrator, it is of no significance to improve the compaction of the ground. Note that the slight difference of initial void ratio is owing to the different position of measurement circles.

The variation of the void ratio above the ground water are greater than that of below the ground water due to the better reinforcement effect. Note that the wet vibration modeled by using effective unit weight considers the buoyancy factor that could be used as a comparison. This is a simplified way to initially consider the effect of water.

6. Discussion on main limitations

The purpose of this paper is to initially analyse the macro-micro mechanics of soil during vibroflotation using DEM for two cases (i.e., the dry vibrations and wet vibrations). For wet vibrations, using effective unit weight is a simplified way to consider the influence of water. However, under dynamic vibration, granular soil is easily liquefied due to built-up excess pore water pressure. In order to more truly simulate the vibroflotation and analyze the possible arising liquefaction, CFD-DEM could be used to analyze the coupling effect between water and soil particles. The limitations stated above need to overcome in the near future.

7. Conclusion

This paper simulates the vibroflotation process without backfill using the 2D DEM. Rolling resistance model is introduced in the DEM to better reflect the actual soil characteristics. Macro and Micro mechanisms of the vibration process are investigated. The main results are as follows: (a) The force chains show that the reinforcement effect on the ground using the vibroflotation method result from the compaction of soil particles around the vibrator. (c) The horizontal vibration of the vibrator has different reinforcement effect on the surrounding soils. The further the measurement are located from the vibrator, the smaller the variation of void ratio is. (d) Generally speaking, considering the effect unit weight of granular soil will reduce the vibroflotation reinforcement effect and the real interaction between excess pore water and soil particles is not considered in this paper.

Acknowledgement
The research was funded by the Major Project of Chinese National Programs for Fundamental Research and Development (973 Program) (No. 2014CB046901), China National Funds for Distinguished Young Scientists (No. 51025932), and State Key Lab. of Disaster Reduction in Civil Engineering (No. SLDRCE14-A-04), which are greatly appreciated.

References

[1] Webb D L, Hall R I 1900 Effects of vibroflotation on clayey sands *J. Soil Mech. and Found. Div.* **95** 1365-78
[2] Slocombe B C, Bell A L and Baez H U 2000 The densification of granular soils using vibro methods *Géotechnique* **50** 715-725
[3] Baez J I 1995 A design model for the reduction of soil liquefaction by vibro-stone columns *Los Angeles: University of Southern California*
[4] Dobson T 1987 *Case histories of the vibro systems to minimize the risk of liquefaction Soil Improvement-A Ten-Year Update* (New Jersey: ASCE Specialty Publication) pp 167-183
[5] Bement R A P, Selby A R 1997 Compaction of granular soils by uniform vibration equivalent to vibrodriving of piles *Geotech. Geol. Eng.* **15** 121-143
[6] Arnold M, Herle I 2009 Comparison of vibrocompaction methods by numerical simulations *Int. J. Numer. Anal. Meth. Geomech.* **33** 1823-38
[7] López-Querol S, Peco J and Arias-Trujillo J 2014 Numerical modeling on vibroflotation soil improvement techniques using a densification constitutive law *Soil Dyn. Earthquake Eng.* **65** 1-10
[8] Jiang M J, Yan H B, Zhu H H and Utili S 2011 Modelling shear behaviour and strain localosation in cemented sands by two demensional Distinct Element Method Analyses *Comput. Geotech.* **38** 14-29
[9] Cundall P A, Strack O D L 1979 A discrete numerical model for granular assemblies *Geotechnique* **29** 47-65
[10] Jiang M J, Yu H S and Harris D 2005 A novel discrete model for granular material incorporating rolling resistance *Comput. Geotech.* **32** 340-357
[11] Iwashita K, Oda M 2000 Micro-deformation mechanism of shear banding process based on modified distinct element *Powder Technol.* **109** 192-205
[12] Jiang M J, He J, Liu F and Wang H N 2013 Effect of inter-particle rolling resistance on passive earth pressure against a translating rigid retaining wall *AIP Conf. Proc.* **1542** 281-284
[13] Jiang M J, Konrad J and Leroueil S 2003 An efficient technique for generating homogeneous specimens for DEM studies *Comput. Geotech.* **30** 579-597

3D numerical analyses for the quantitative risk assessment of subsidence and water flood due to the partial collapse of an abandoned gypsum mine.

R. Castellanza[1], G.M. Orlandi[2], C.di Prisco[3], G. Frigerio[3], L. Flessati[3], J.A. Fernandez Merodo[4], F. Agliardi[1], S. Grisi[1], G. B. Crosta[1]

[1]DISAT – Università di Milano Bicocca, riccardo.castellanza@unimib.it
[2]Studio Associato di Geologia Spada, Ranica (BG) info@studiogeospada.it
[3]DICA – Politecnico di Milano, claudio.diprisco@polimi.it
[4]Universidad Politécnica de Madrid – IGME Madrid, jose.fernandez@igme.es

Abstract. After the abandonment occurred in the '70s, the mining system (rooms and pillars) located in S. Lazzaro di Savena (BO, Italy) , grown on three levels with the method rooms and pillars, has been progressively more and more affected by degradation processes due to water infiltration. The mine is located underneath a residential area causing significant concern to the local municipality. On the basis of in situ surveys, laboratory and in situ geomechanical tests, some critical scenarios were adopted in the analyses to simulate the progressive collapse of pillars and of roofs in the most critical sectors of the mine. A first set of numerical analyses using 3D geotechnical FEM codes were performed to predict the extension of the subsidence area and its interaction with buildings. Secondly 3D CFD analyses were used to evaluated the amount of water that could be eventually ejected outside the mine and eventually flooding the downstream village. The predicted extension of the subsidence area together with the predicted amount of the ejected water have been used to design possible remedial measurements.

1. Introduction

Aim of this study is the assessment of stability conditions for an abandoned gypsum mine located in San Lazzaro di Savena (Bologna, Italy). Mining was carried out until the end of the 70s by room and pillar method (Figure 1). During mining, a karst cave was crossed bringing karstic water to infiltrate into the mine. As a consequence, the lower level of the mining is now completely flooded and portions of the mining levels show critical conditions and are structurally prone to instability. Buildings and infrastructures are located above the first (L1) and second (L2) levels and a large portion of the area downstream the mine, just above Savena river, is urbanised.

Gypsum geomechanical properties change over time; water, or even air humidity, dissolves or weakens gypsum pillars, leading progressively to collapse. The mine is located in macro-crystalline gypsum beds belonging to the Messinian Gessoso Solfifera Formation. Selenitic gypsum beds are interlayered by centimetre to meter thick shales layers. In order to evaluate the risk related to the collapse of the flooded level 3 (L3) a deterministic approach based on 3D numerical analyses has been considered. The entire abandoned mine system up to the ground surface has been generated in a 3D model. The considered critical scenario implies the collapse of the pillars and roof of the flooded level 3.

In a first step, a sequential collapse starting from the most critical pillar has been simulated by means of a 3D Finite Element code. This allowed the definition of the subsidence basin at the ground surface and the interaction with the buildings in terms of ground displacements. 3D numerical analyses have been performed with an elasto-perfectly plastic constitutive model.

In a second step, the effect of a simultaneous collapse of the entire L3 has been considered in order to evaluate the risk of a flooding due to water outflow from the mine system. Using a 3D CFD (Continuum Fluid Dynamics) finite element code the collapse of L3 level has been simulated and the volume of outflowing water has been quantified together with water velocity and pressure. On the basis of the results, remediation measures have been defined and the risk has been evaluated with a deterministic approach.

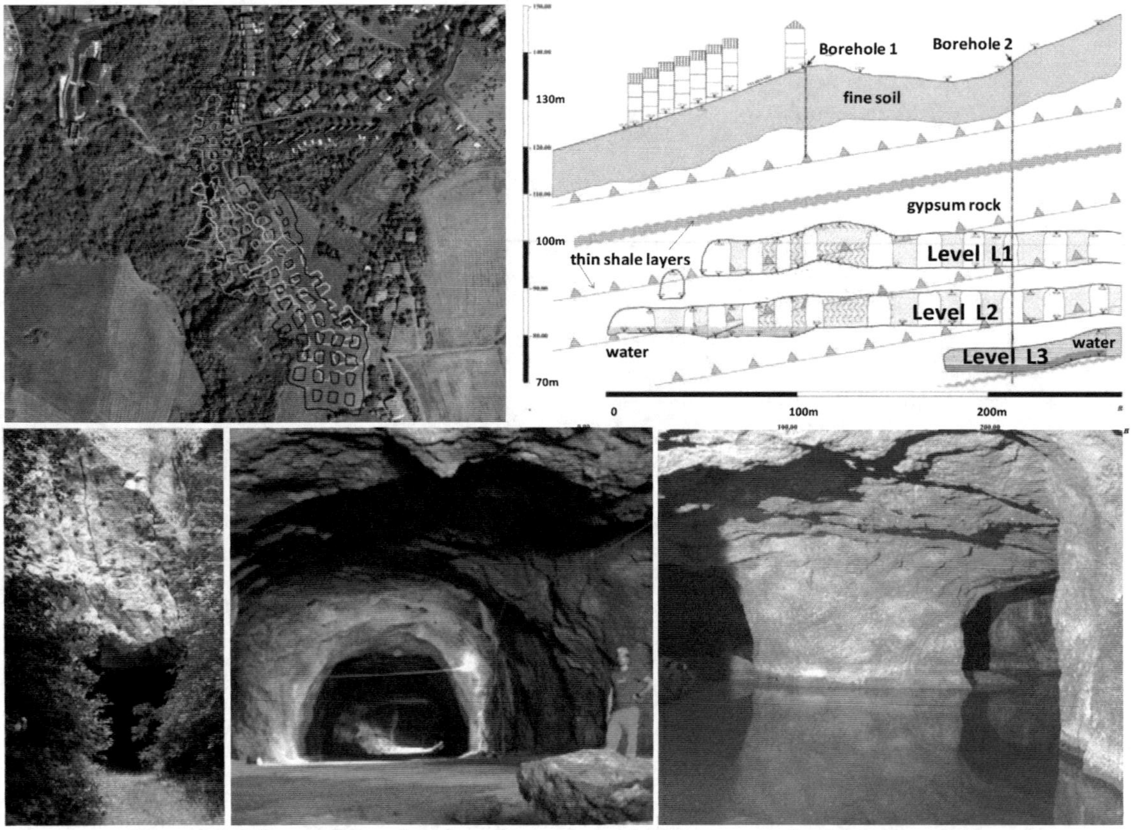

Figure 1. Views of the abandoned gypsum mine system and schematic section.

2. Geomechanical characterization of gypsum including weathering effects

To evaluate the effects of weathering processes caused by water on the mechanical behavior of gypsum a specific experimental campaign was designated. Performing a series of uniaxial compression tests on cores extracted at different depth, it was possible to measure a value of the compressive strength (UCS). The material in dry conditions exhibited a compressive strength of about 12 MPa; the strength reduced to 7 MPa when the material was taken from the roofs of L1 and L2 levels as it was affected by conditions of saturate humid air (Hr $\cong$ 100%) and it resulted 5 MPa for the gypsum of L3 level, as it was immersed in water.

In addition to UCS tests indirect tensile and triaxial tests have been performed. Subsequently a series of small scale experimental tests on cylindrical pillars simultaneously loaded and immersed in a water flow were performed. In Figure 3a, similarly to what presented in Castellanza et al. (2010), it is shown a test carried out on a sample with an axial load of 4 MPa (representative of the conditions of the site of pillars) that reaches failure after 9 days from the immersion in a water flow.

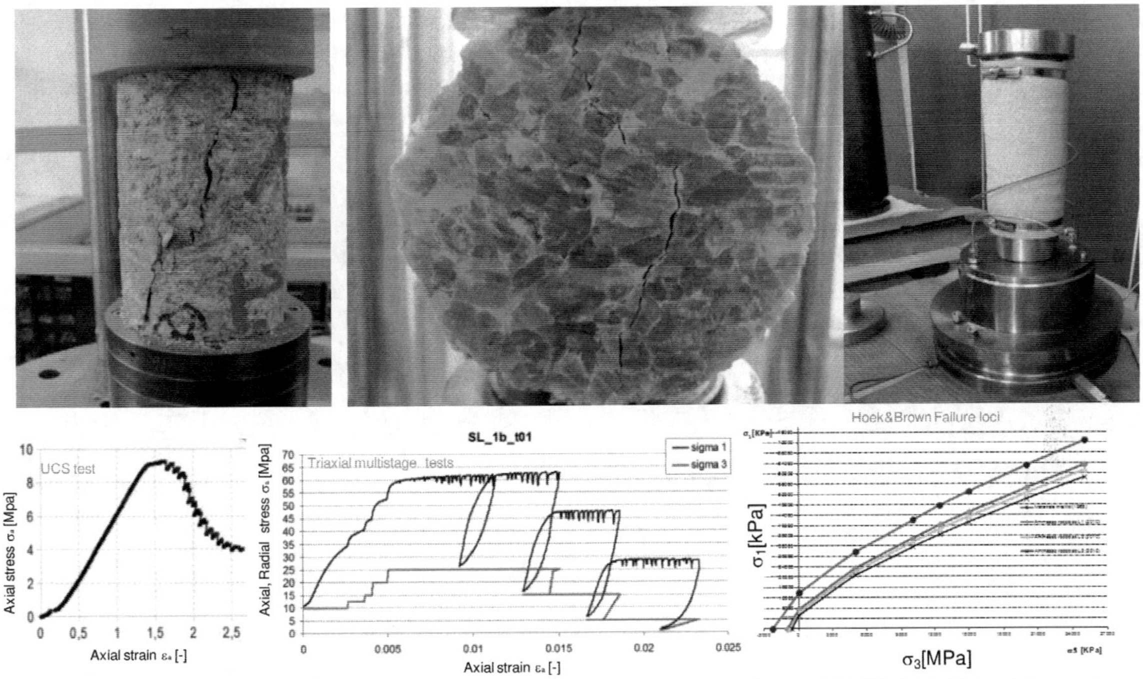

Figure 2 Geomechanical characterization in the laboratory: standard tests (UCS, BT, Triaxial tests).

Finally a series of uniaxial compression tests on specimens degraded by immersion in water at different time both in laboratory and in situ is shown in Figure 3b. It shows the reduction in UCS strength for specimens immerged in situ in the water at L3 level for times longer than 300 days.

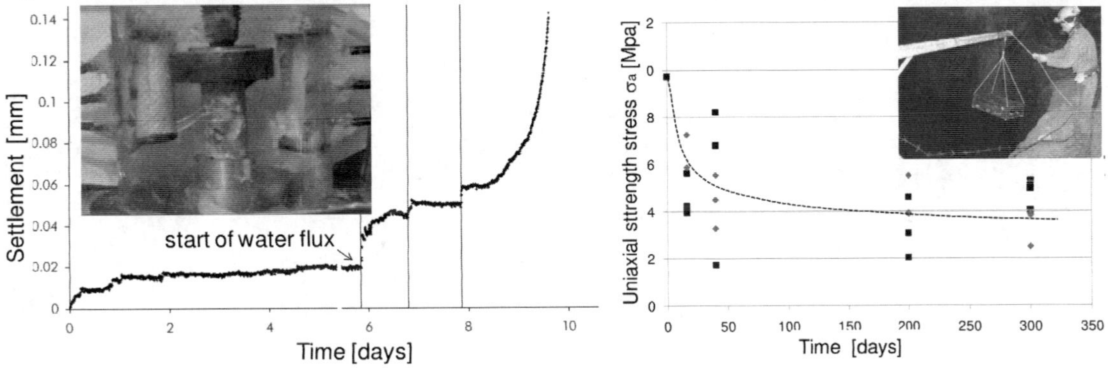

a) b)

Figure 3. a) Small-scale simulation of collapse of a pillar in a water flux , b) Decay resistance to uniaxial specimens immersed in situ (the flooded quarry).

The experimental tests carried out allowed to define the parameter of the Hoek&Brown criteria for rock masses (1997). As visible in Figure 2 the Hoek and Brown yield surface shrinks as weathering proceeds in the different levels of the abandoned mine. The minimum strength is associated to the flooded L3 level.

3. Estimation of the subsidence area with 3D FEM non linear analyses

The next step was to create a detailed three-dimensional model of the entire system, including the geometric elements (external topography, location buildings and infrastructure, mining plan levels) and those related to the geomechanical characterization of materials and their distribution (figure. 4).
Numerical modelling has been developed with non-linear FEM code Midas GTS. Initially a series of preliminary assessments to verify the pillar stability were performed using the method proposed by Obert & Duvall (1967) . The in situ stresses of each pillar were evaluated by the 3D numerical analyses.

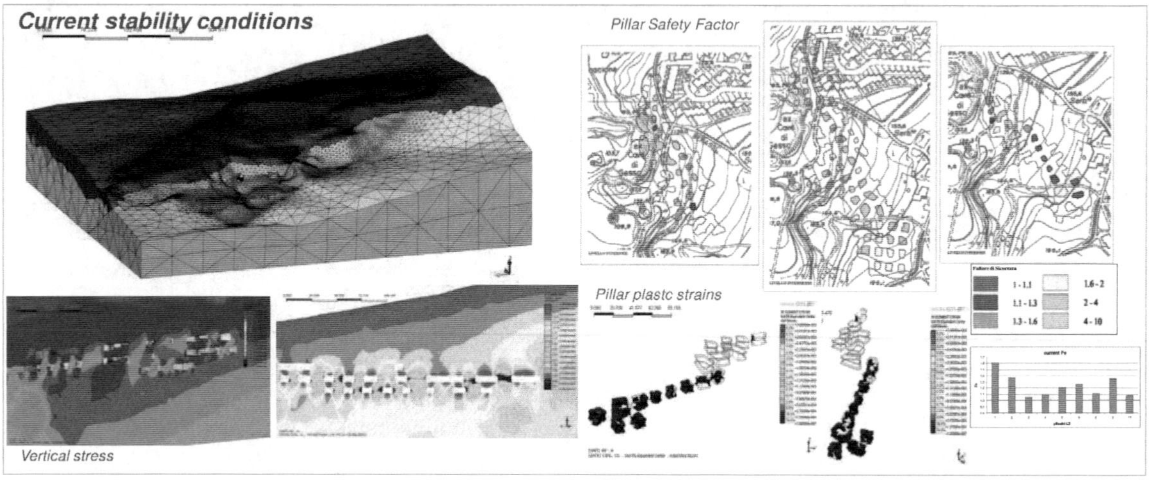

Figure 4. Assessment of current conditions of stability of the pillars by the method of Obert&Duvall and non-linear FEM analysis.

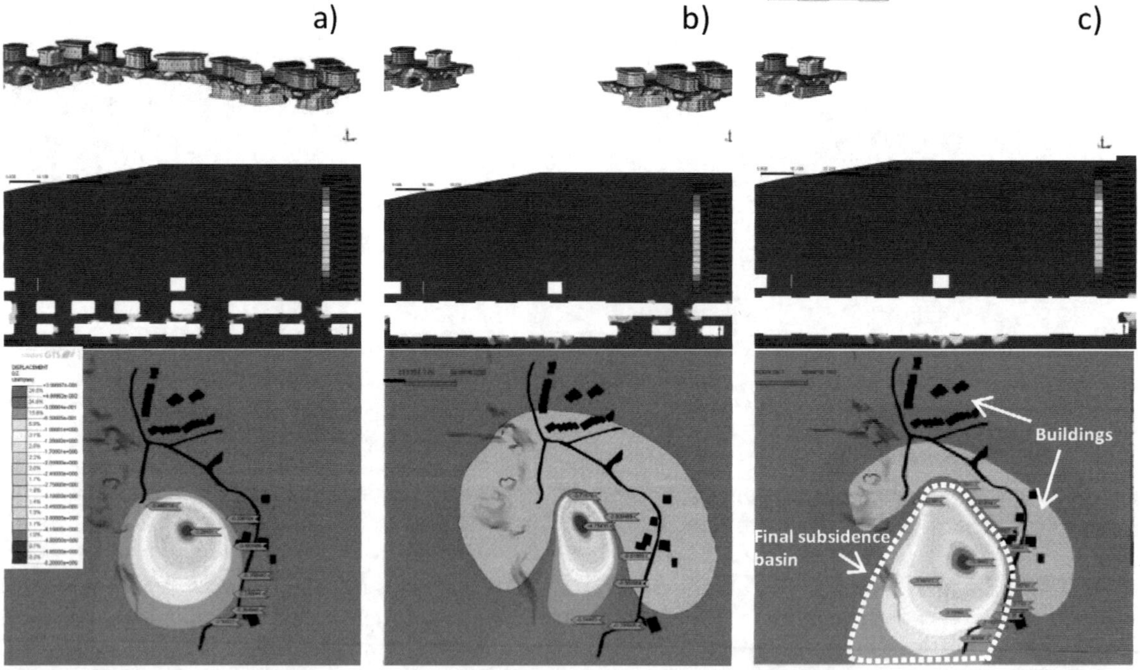

Figure 5. Simulation of the progressive collapse (a, b, c) of the pillars of the lower flooded level (L3) and the corresponding intermediate level (L2) with prediction of subsidence induced at the surface in relation to the existing buildings.

Using this approach the most critical situations were identified. They are represented by a series of pillars in the lower L3 level, characterized by a safety factor close to unity (Figure 4) and by significant plastic strains. Next, using a procedure of excavation for phases, the pillars of L3 level and the roof of the intermediate L3/L2 level were progressively removed (Figure 5a,b,c). In this way the numerical model was used to estimate the basin of subsidence at the surface (Figure 5). The numerical analyses reveal that even in case of collapse of all the L3 level pillars, the system is able to maintain equilibrium and convergence. A sort of large underground tunnel with a diameter of about 20m will be created when the pillar of L3 will be remove. Nevertheless the upper levels (L2 and L1) does not fails.

In this scenario of progressive removal of pillars and portion of roof, a subsidence basin is predicted in terms of vertical settlements. Since it not affect directly the buildings and the infrastructure (Figure 5c) it has been concluded that the risk related to the failure of all the flooded pillars at L3 level is very low.

4. 3D Numerical estimation of flooding due to impulsive water outflow from the mine.

In this section the 3D numerical results of water flooding induced by the collapse of critical pillar and roof portions inside L3 and L2 levels is presented. More than 80.000m^3 are now flooding L3 level; consequently the goal of this analysis is to quantify the outflow water that in case of catastrophic scenario can rise from L3 level to L1 level. This water could proceed towards the urbanized area around the gypsum mine inducing flooding.

3D CFD analyses have been performed with FLOW 3D. In Figure 6 a preliminary numerical test is reported. The main aspects required in the simulation are: i) Advanced Surface Modelling Fluid TruVOF (technical volumes of fluids), ii) viscous flow of water subject to the gravitative action, iii) turbulence model laminar; iv) friction coefficient for shear wall boundary conditions.

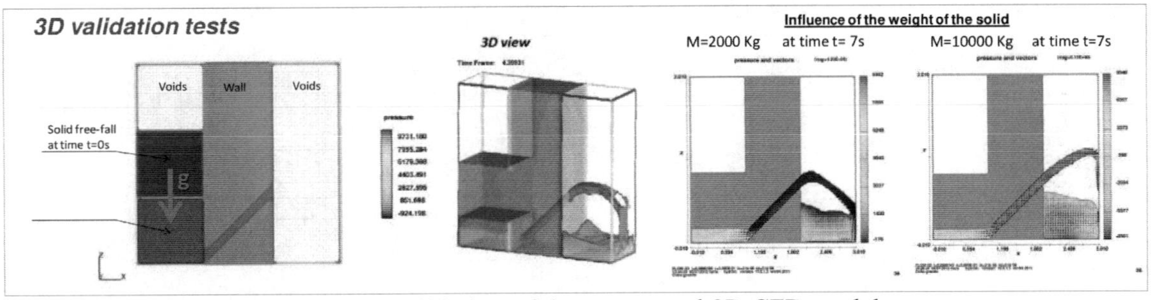

Figure 6. Preliminary tests for the validation of the conceptual 3D CFD model.

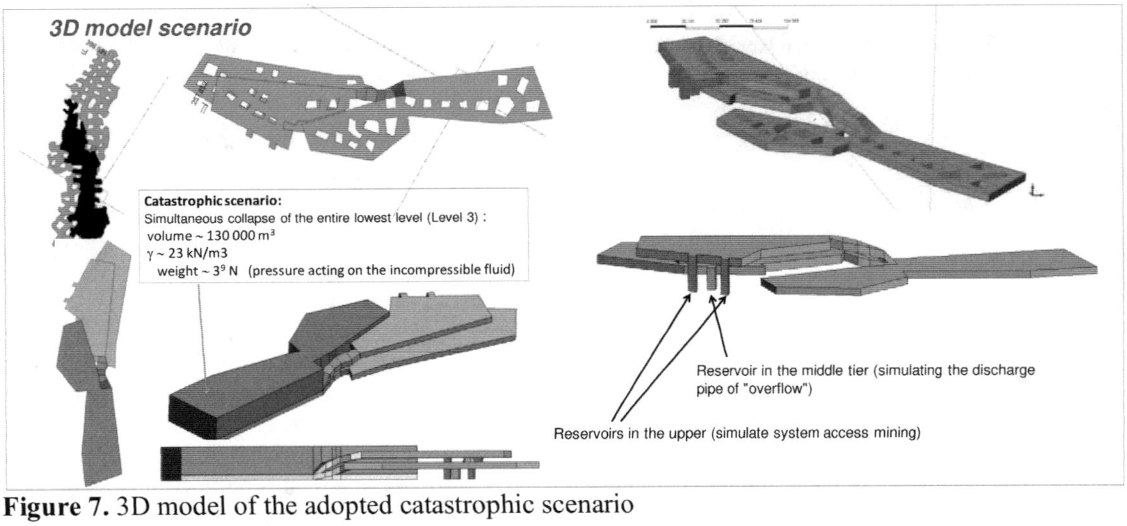

Figure 7. 3D model of the adopted catastrophic scenario

Concerning the Fluid-Structure Interaction a moving objects has been adopted to simulate the collapse of the slab between the second and third levels. The modelling of the hydraulic behaviour of a fluid volume (volume of water stored in the lower level of the quarry) subjected to a pressure (derived from the collapse of a certain portion of the mining system) was initially sketched and validated through simple models two and three dimensional (figure 6). In a finite domain, a volume of incompressible fluid comes to lie below a solid characterized by a certain density, initially bound (in equilibrium). Leaving the solid free from the constraints, it tends to move in a free fall due to gravity, causing a pressure increase in the fluid volume, which in turn has the possibility to move thanks to the presence of connections with expansion chambers (tunnels and wells that connect the mining flooded level with the level above, not flooded). The catastrophic scenario built in 3D, for this mine is shown in Figure 7. It was decided to consider a further unfavourable condition which is the simultaneous collapse the entire roof of L3 level to maximize the impact of the falling down gypsum mass on the volume of water present in mine.

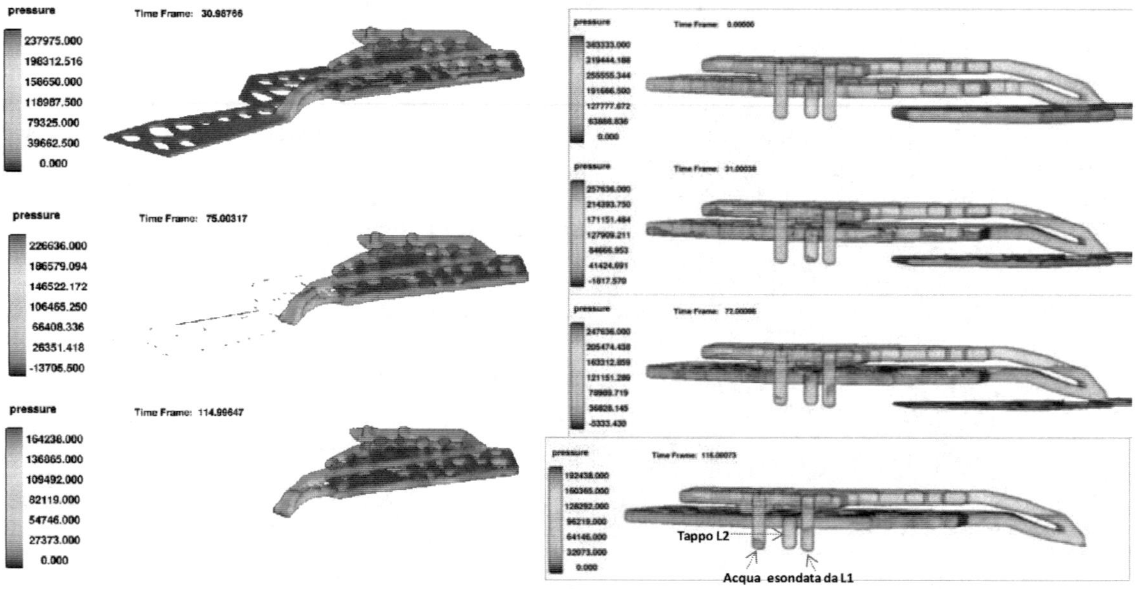

Figure 8. Forecast flooding volume of water (instant collapse - catastrophic scenario)

The results are shown for different time intervals in Figure 8. It could be shown that the analysis revealed the internal buffering capacity of the intact portion of the L2 (figure 8). Furthermore, the analysis allowed to assess the volume and the velocity of the water that can flood out from L1 level towards the urbanized area. The estimated values of the flooding water is about 2000 m^3. It represent approximately the 2% of the water that flood the mine. This amount is compatible with small channelling works to bring this water in the Savena River located just outside the mine. The forecasting of flooding water reveal again a low risk for the urbanized area

5. Conclusion

Numerical analysis performed, characterized by an accurate three-dimensional modelling, enable to assess the effects of the collapse of L3 level both in terms of subsidence basin and volume of flooding water. These effects were considered negligible in terms of interaction with urban areas resting below and above the considered abandoned mine. Therefore, these results were used for a better planning that lies in the absence of relevant remediational measures for L3 level. In fact, even in case of extremely critical scenarios for L3 level no significant risk of citizens living close this abandoned mine. The decommissioning of the mine was consequently efficiently focused on L1 and L2 levels.

The main efforts are now concentrated to increase the strength of the pillars located at L1 and L2 levels that directly interact with buildings.

6. References

[1] Castellanza R., Nova R., Orlandi G. (2010), Flooded gypsum mine remedial by chamber filling, *Journal of Geotechnical and Geoenvironmental Engineering ASCE* Vol. **136**(4), pp. 629-639, DOI: 10.1061• ASCE GT.1943-5606.0000249.

[2] Hoek, E., and E. T. Brown (1997). Practical estimates of rock mass strength. *International Journal of Rock Mechanics and Mining Sciences* **34**.8: 1165-1186.

[3] Obert L, Duvall W.I. (1967).Rocks mechanics and design of structures in rocks, Wiley, New York

Experiences from the small historical dams failures during heavy floods

I Vaníček[1], M Vaníček[2], D Jirásko[1] and T Pecival[3]

[1] Czech Technical University in Prague, Faculty of Civil Engineering, Geotechnical Dept. Thákurova 7, 166 29 Prague 6, Czech Republic

[2] Geosyntetika s.r.o., Nikoly Tesly 3, 160 00 Prague 6, Czech Republic

[3] Vodní díla TBD – TBD a.s., Hybernská 1617/40, 110 00 Prague 1, Czech Republic

E-mail:daniel.jirasko@fsv.cvut.cz

Abstract. Roughly between 400 and 600 years ago many small earth dams were constructed mainly in the south part of the Czech Republic. They were used for fish production and flood protection. To our days roughly one third survived, which means about 25 000 of them. During catastrophic floods in 2002 many of them had some problems but less than 0.3% failed. Experiences gained from the failure evaluation are presented. Firstly from the view of limit states of failures, when limit states of internal erosion and surface erosion played most important role and were the main reason of failures. Secondly, from the view of so called domino effect of failure, when the most important dam on the catchment basin failed and after that the other ones, situated below, had limited chance to survive. The failures are described for catchment basin of the small river Lomnice in south part of the Czech Republic close to the town Blatna. The experiences obtained there led to the evaluation of other catchment basins where domino effect of failure can play also very important role. For the evaluation of potential risk, the numerical modelling was used to study the flood wave propagation below the critical dam, especially at the moment when this wave is reaching the dam situated below the critical one. Finally, the recommendations are specified, not only for individual dams but also for catchment basin, where the risk of domino effect failure is very high.

1. Introduction

During last two decades, floods caused many problems in the Czech Republic. Therefore, more attention is devoted to this problem, either with respect to their frequency, as to the environment protection with the help of different flood protection measures. During the last period the emphasis is directed on the information system, on the first signal about potential dike failure, on the recommended behaviour during floods, and on the actions which can play positive role from the safety point of view, etc.

In the Czech Republic, there are roughly 3 different types of floods:

- Regional – affecting one or more river-basins, coming usually in summer and is caused by long term heavy rainfall;
- Local – affecting small area. Caused by extremely high local rainfall (storm) – most difficult from the forecast point of view;
- Floods in foothills – during spring, when the soil is still frozen – caused by combination of snow melting and rain.

Very often during the last period, we are speaking about frequency of floods, as for example in the Czech Republic different floods were recorded, especially the regional ones, in 1997, 1998, 2002, 2006 and 2010. Figure 1 is showing the situation in Prague during the heavy floods in 2002. However, from the figure 2 – showing flood events in Prague during the period of 1825-2002 – it is clear that the flood in 2002 was strongest during the last 2 centuries; however, flood frequency during the last decade of the 19th century was also exceptional. The 2002 flood also heavily affected the Prague metro system [1].

Figure 1. View on River Vltava in Prague during heavy floods in 2002 (August 14, August 22).

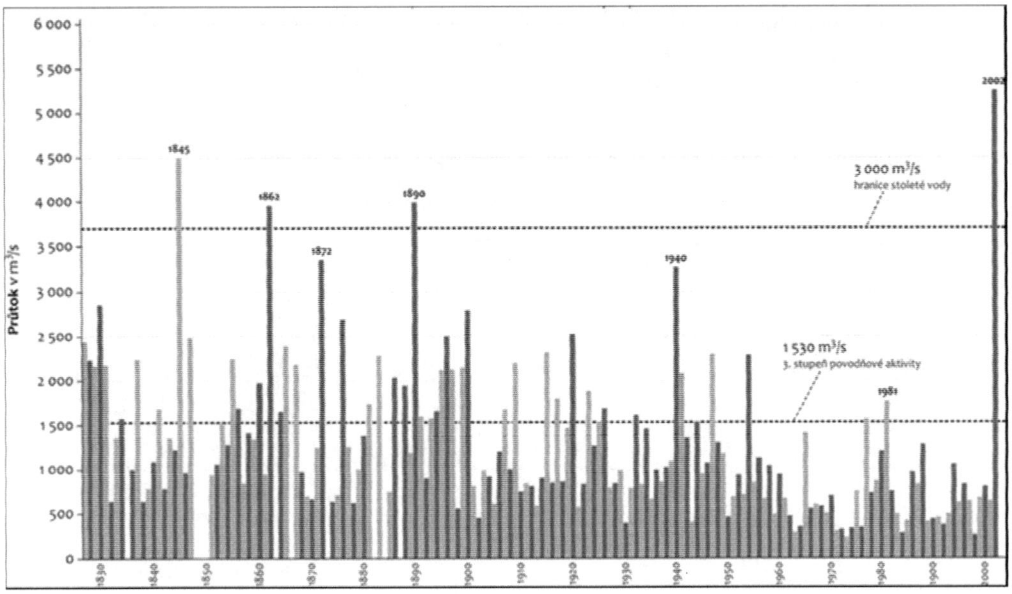

Figure 2. Prague – floods during the period of 1825 – 2002.

Uniqueness of the flood in 2002 was that the heavy rainfalls causing the flood came in two peaks very close to each other. Nearly whole country was strongly affected by this flood with maximum rainfall in south part of the country. Since nearly all rivers are flowing in a northerly direction the highest flood wave reached on the river Labe (Elbe) boundary with Germany close to Dresden.

The highest concentration of failures of small earth historical dams during the 2002 flood were connected with small river Lomnice, (figure 3), where five dams failed and two others were damaged, [2] or [3].

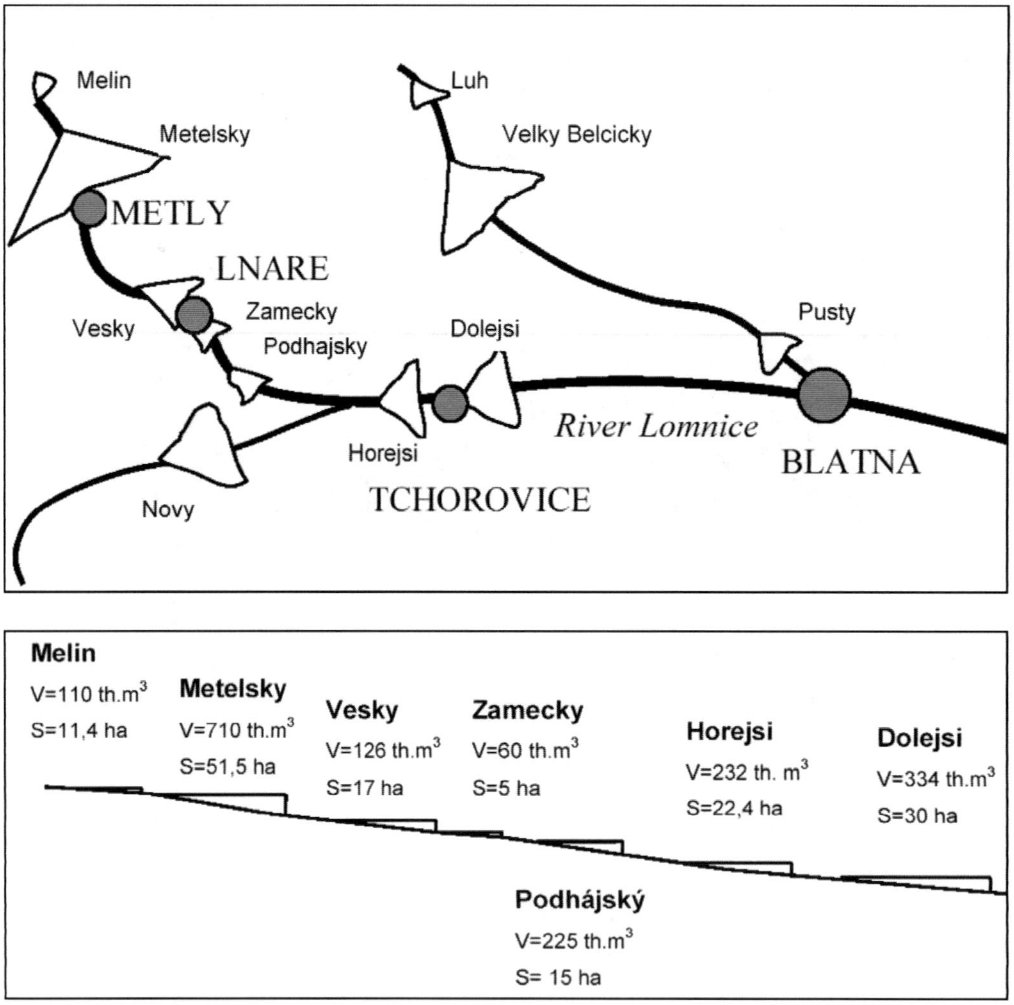

Figure 3. Little River Lomnice – situation and scheme of dams.

Heavy rainfalls during 7th and 9th August 2002 did not cause any problems; pond reservoirs were nearly full and fulfilled their flood protection role. The situation was improved during Saturday and Sunday (10-11th August) and with the help of directed outlets, a certain reservoir volume was restored, for example in three cases water level was 0.3 – 0.6 m below normal level.

New heavy rainfalls started on Monday (12th August), which had very quick negative effect due to the full saturation of surrounding land. On Monday evening, the flow volume reached 100 years flow rate. But still all dams were able to catch all these volumes.

The critical situation started at around 4 am of the next day (Tuesday) when - due to overflowing - the first dam (Melín) on this river basin failed. Additional wave reached the crest of the next dam

"Metelský" in few minutes and very quickly, this dam failed as well. Relatively large volume from this reservoir (about 1 mil. m^3), with the estimated (and recalculated) outlet more than 500 $m^3 .s^{-1}$ (which was much higher than 100 years flow rate – about 20 $m^3.s^{-1}$), affected the villages Metly and Předmíř, where some houses failed and one man died. The embankment of the next pond "Veský" survived due to the very wide crest with asphalt pavement but was strongly eroded on the downstream side. The same situation was observed on the next pond "Zámecký", the crest of which is also protected by asphalt pavement – European road E 49 is passing on top of the crest, figure 4.

Figure 4. Crest of the dam Zamecky - European road E 49.

Figure 5. Town Blatná – on 13th August 2002 morning.

IOP Conf. Series: Earth and Environmental Science **26** (2015) 012042 doi:10.1088/1755-1315/26/1/012042

Nevertheless all the other three dams (figure 3) situated below failed, and the village Tchořovice was strongly affected, many houses were destroyed. After the failure of the last one (Dolejší), the flood wave was getting flatter in much wider and flatter valley so that the impact on the town Blatná was diminished. Nevertheless, even there the damages were very high, figure 5, because on the brook, joining River Lomnice in Blatná, another three dams (Luh, Velky Belcicky and Pusty) failed as well.

2. Character of failure

In all observed cases, the surface erosion played the most important role, but in two cases, the internal erosion probably also participated in the final collapse.

With respect to the internal erosion two different reasons were recorded. For example in one place of failure – gash – of the embankment of the pond Metly, the part of the old wooden outlet was found. From the historical records, this old outlet was left there roughly in seventeenth century, because from these days a new place for the new outlet was selected. In this profile on the downstream side, the place with higher humectation (moisture content) was observed for last decades, probably as the result of higher permeability in this profile. Due to the higher hydraulic gradient, the clogging of this old outlet failed and internal erosion started, leading to the complete collapse of the dam. The second case was observed for the pond Podhajsky, the embankment of which was reconstructed only 13 years before the event. Sealing wall formed by jet grouting was applied in one part of the dam embankment. At the same place dam completely collapsed, with high probability as the result of the combination of high hydraulic gradient and bad connection of this wall with dam bedrock, where large granite boulders were disclosed.

With respect to the surface erosion, the significant role played the dam crest arrangement. For the asphalt pavement, the erosion was strongly limited, protecting against total collapse – see also figure 4. The uniformity of the crest level played also important role, as in this case water overflow and the crest erosion happened throughout the whole length (figure 6) and dam collapsed in the place with minimum erosion resistivity, in one case two water gaps occurred at the same moment at two different places.

Figure 6. Dam Dolejsi – Downstream slope eroded along the entire crest before the dam failed in two places.

3. Reconstruction

The basic question connected with reconstruction was connected with the type of soil, which has to be used. In Middle Ages the granite eluvium was used, as the bedrock in this region is composed by granite rock. However, grain size distribution curve was on the boundary, which is now accepted from the permeability point of view. Nevertheless, as this eluvium contains high percentage of mica the measured properties were sufficient:

- Percentage of fine particles – between 10 and 35%,
- Liquid limit 30-37%
- Plasticity limit 23-28% (but in some cases not detectable due to high sandy content);
- Optimum moisture content (Proctor standard) 12.5 – 17.0 %
- Coefficient of permeability after compaction in the range of $10^{-8} – 10^{-10}$ m.s^{-1}

Special care was devoted to the connection of the new part of the dam body with the older one. Contact was done with the help of cut-off, the same method was applied for the connection of the new part of the dam with subsoil. In the case of direct connection of the reconstructed part with boulders in the bedrock, either in the bottom or on the side slopes, bentonite slurry was applied on these boulders to obtain good connection and to restrict the development of preferential paths for the seepage.

As the old dams had no toe drain on the downstream side, the great care during reconstruction was devoted also to this drain to affect flow net is the positive way.

Nevertheless during the reconstruction the main attention was devoted to the protection against potential surface erosion in the future. For long crest of the earth dam or for dikes the first step is connected with unification of the crest height along all of its length, as surface overflowing starts in lowest point followed by erosion. The second step is coming out from the same principle. We recommend to select the place where overflowing below the dam (dike) can cause minimum problems and at this place to construct new additional spillway, the crest of which is few decimetres below of the main crest. This principle was proposed and realized during the reconstruction of small dams on the River Lomnice, however the crest of this additional spillway was reinforced. Firstly by concrete slab, later on the principle of the reinforced embankment was recommended. As reinforcing layer the new material was used, so called brick-fiber-concrete. This material is similar to concrete, however it is using demolition waste, mostly old bricks, instead of aggregates with addition of short synthetic fibers, randomly distributed in the mix. The high erosion resistivity of such reinforced embankment was approved by field test in the scale close to 1:1. The advantage of such additional spillway is not only in reduction of amount of water which overflow the main crest, but also as first signal, that the critical situation can start. This approach, which can be called as robustness approach, is now preferred in civil engineering generally.

For shorter length of the crest the reinforcement can be applied along the whole length. Cheaper and environmentally friendlier is solution recommended for dry dam, the main aim of which is to decrease flood wave during floods. Bottom outlet has limited capacity enabling to pass through only the flow equal let say to $Q_{10} – Q_{30}$. For higher inflow rates the water level in the reservoir is increasing followed by overflowing along the whole crest length. In this case we recommended only crest reinforcement with the help of anti-erosion geosynthetic mattress which is very well connected with the subsoil, not only at the crest, but also with the help of long spins on the downstream face of the dam. After seeding with grass the root system help to improve this connection. Practical experience with this method is very good, as after two crest overflowing no damage was observed.

Some aspects influencing internal and surface erosion can be found in rather large range of literature, e.g. Sherard [5], Vaníček [6], Briaud et al [7], Dyer, Utili and Zielinski [8] or Utili et al [9].

4. Catchment basin evaluation – potential of the domino effect of failure

The reason for higher concentration of small dam failures on the small River Lomnice we attributed to the so-called domino effect of failure [2]. This effect can be typical for the small catchment basin, where the most important dam is situated on its beginning. In our case, it was pond Metelsky with highest volume of water in the reservoir. After the failure of this dam the others situated below had limited chance to survive. Therefore the main experience obtained from the described case is connected with closer specification of this domino effect of failure.

With the help of water-resources maps (in the scale 1: 50 000 covering all country), it was possible to specify other catchment basins in the Czech Republic, which are sensitive to this effect. Very quickly, it was recognized that the River Lomnice is not an exceptional case. Therefore, it was recommended to increase the care about dams, which can play initial role to this domino effect of failure. This recommendation is in accordance with Water law as this law specifies demands on technical-security control for different category of dams. In addition, for selected dams this category can be after the evaluation increased. Nevertheless, before the final specification of the dam category we recommend the numerical modelling of the propagation of this additional flood wave below this critical dam.

5. Flood wave propagation modelling

The numerical modelling, with which the specificity for selected catchment basin and dam can be defined, is based on two phases. For the first step, we have used software BossBreach, version 2.0 – [4] which is able to model the size of the breach of the dam and outlet from the failed dam as function of time. This software of the erosion of the earth dam is based on hydraulic, erosion and transportation equations, which take into account the geometry and material characteristics of the dam. Although this software is simple, can be useful as is giving the first estimation about the peak of the wave leaving the profile of the failed dam.

For the simulation of the range of flooded area below failed dam and for the gradual decrease of the flood wave peak (flood wave transformation) in individual profiles and namely for the profile of the next dam, software HEC-RAS 4.0 (Hydrologic Engineering Canters River Analysis System) was used. Detailed parametric study was done and very good experiences were obtained for specific dam Brevnice. Cao et al (2014) discuss a similar problem – [10]. Morphological map for the river basin with dam Brevnice in digital form was the first condition of the study, followed by the information about vegetation, about different interaction with transport infrastructure in this inundation area. Subsequent parametric study helped us very much to estimate the impact of the potential engineered measures to flatten flood wave peak. Nevertheless the final result for the profile of the dam situated below the critical one is most important for further evaluation and recommendation.

6. Conclusion

Very careful evaluation of the reasons of great concentration of failed small historic dams during the heavy floods in 2002 helped to improve our knowledge about individual aspects having strongest negative impact. Limit state of surface erosion followed by limit state of internal erosion are playing most important role during the phase of dam crest overflowing. The reconstruction (and newly constructed dams or dikes) should fully exploit this knowledge and therefore some of them were presented. Nevertheless, the main reason for higher concentration of small dam's failures was attributed to the domino effect of failure, especially for the case where the dam with highest reservoir volume is situated at the start of the catchment basin. After the failure of such critical dam the other dams situated below have limited chance to survive. After the specification of this type of domino effect of failure the subsequent study showed that similar situation can happen for many other catchment basins, as there is still a very high number of old historical small earth dams – about 25 000. Therefore, new conditions for monitoring of these dams with higher potential risk for the individual catchment basins were specified in laws updated following the described event evaluation. Numerical modelling can be very useful tool for more detailed evaluation of the potential risk of failure or about potential safety measures.

References

[1] Soga K, Vaníček I, Gens 2011 *Micro-Measurement and Monitoring System for Ageing Underground Infrastructures* (Prague: CTU Press)

[2] Vaníček I, Vaníček J 2004 *Rehabilitation of old earth dams failed during heavy floods in 2002 Proc. New Developments in Dam Engineering* (Balkema) pp 889-898

[3] Vaníček I, Vaníček M 2008 *Earth Structures in Transpoort, Water and Environmental Engineering* (Berlin: Springer)

[4] Pecival T, 2015 *Research of frequency and types of failures for small earth dams – domino effect* PhD thesis, (Czech Technical University in Prague, in Czech)

[5] Sherard JL, 1973 *Embankment Dam Cracking. Embankment Dam Engineering* Casagrande Volume (New York: Willey) pp 271-353

[6] Vanicek I, 1988 *Creation and behaviour of cracks in clay core of earth and rockfill dams* (Praha: SNTL – Publishing house of technical literature in Czech) 168 p.

[7] Briaud JL, Chen HC, Govindasamy AV, Storesund R, 2008 *Levee Erosion by Overtopping in New Orleans during the Katrina Hurricane*, Vol. 134, No. 5, May 2008, *Journal of Geotechnical and Geoenvironmental Engineering,* (Reston Virginia USA: ASCE)

[8] Dyer MR, Utili S, Zielinski M, 2009 *Field survey of desiccation fissuring of flood embankments*. (London: ICE Proc. Water Management, 162(3)) pp 221-232

[9] Utili S, Castellanza R, Galli A, Sentenac P, 2015 *Novel approach for health monitoring of earthen embankments*. Journal of Geotechnical and Geoenvironmental Engineering (ASCE, 141(3): 04014111)

[10] Cao Z, Huang W, Pender G, Liu X, 2014 *Cascade dam break floods*, J. Flood Risk Management, 7(4) pp 357-373

DEM analyses of shear behaviour of rock joints by a novel bond contact model

M J Jiang[1,2,3,5], J Liu [1,2,3], C Sun [1,2,3] and H Chen [4]
[1] State Key Laboratory for Disaster Reduction in Civil Engineering, Tongji University, Shanghai, China, 200092
[2] Key Laboratory of Geotechnical and Underground Enigineering of Ministry of Education, Tongji University, Shanghai, China, 200092
[3] Department of Geotechnical Engineering, College of Civil Engineering, Tongji University, Shanghai, China, 200092
[4] Yunnan traffic planning design and research institute, Kunming, China

E-mail: mingjing.jiang@tongji.edu.cn

Abstract. The failure of rock joints is one of the potential causes for the local and general rock instability, which may trigger devastating geohazards such as landslide. In this paper, the Distinct Element Method (DEM) featured by a novel bond contact model was utilized to simulate shear behaviour of centre/non-coplanar rock joints. The DEM results show that the complete shear behaviour of jointed rock includes four stages: elastic shearing phase, crack propagation, the failure of rock bridges and the through-going discontinuity. The peak shear strength of centre joint increases as the joint connectivity rate decreases. For intermittent non-coplanar rock joints, as the inclination of the rock joints increases, its shear capacity decreases when the inclination angle is negative while increase when positive. Comparison with the experimental results proves the capability of this DEM model in capturing the mechanical properties of the jointed rocks.

1. Introduction

Shear failure of rock joints can induce the instability of rock mass, a common medium in the field of mining, oil exploitation, water conservancy and hydropower. Thus, the shear behavior of rock joints becomes a topic of the geological engineering. Lajtai [1-2] conducted direct shear tests to examine and analyze the strength of discontinuous rocks, and concluded that the total shear strength is composed of two parts, namely, the joint friction along the separated parts of the weakness plane and the fundamental shear strength and internal friction in solid bridges. Gehle and Kutter [3] investigated the shear behavior of intermittent rock joints, arguing that the failure pattern and the shear resistance of jointed rock are significantly affected by both the orientation of the joints and the normal stress. Cao et al. [4] further found that the rock specimen takes a turn from wing crack propagation failure to crack coalescence failure with the increase of the angle of the rock bridge.

Although theoretical and experimental studies have given some promising results of the shearing of rock joints, some very important microscopic information controlling the failure patterns still remain unknown. More and more researchers have used numerical methods to further investigate this problem [5-11]. Among these methods, the Distinct Element Method (DEM) proposed by Cundall and Strack

[5] To whom any correspondence should be addressed.

[8] is a feasible method to study the shear behavior and failure mechanism of sand and rocks [9-12]. Park et al. [9] examined the effects of the geometrical features and the micro-properties of a joint on its shear behavior by using a bond model. Potyondy and Cundall [10] proposed a numerical model for rock and investigated the sensitivity of the results to micro-properties. Jiang et al. [11] investigated the shear behavior and strain localization in cemented sands by 2D DEM. More recently, Bewick et al. [12] investigated mechanism of rupture by comparing DEM direct shearing result with experimental one under different normal pressures.

In this study, a new DEM bond model proposed by one of the authors has been adopted to simulate mechanical response of jointed rock mass under direct shearing condition. The shear curves and shearing capacities of the DEM virtual rock joints are qualitatively compared with the experimental results. Also, the force chains inside the jointed rock sample are presented to provide a view in the shear behaiviour microscopically.

2. DEM bond contact model

The DEM bond contact model is proposed by Jiang et al. [13] for the purpose of simulating heavily cemented materials such as granite. The bond contact model was first derived theoretically [14-16] and then further verified by laboratory tests on bonded aluminum rods [17-18].

The total bond contact is assumed to be composed by two paralleled parts: inter-particle and inter-bond contacts. Contact forces are assumed to be transferred through not only the bond but also the inter-particle contact:

$$F_n = F_{bn} + F_{pn}, F_s = F_{bs} + F_{ps}, M = M_b + M_p \quad (1)$$

where F_{bn}, F_{bs} and M_b are the normal force, the shear force and the moment transferred through the bond, respectively, while F_{pn}, F_{ps} and M_p are the corresponding components transferred through the inter-particle contact. The inter-particle contact laws can be approached in reference [15] and the mechanical response of inter-bond contact can be expressed as:

$$F_{bn} = \begin{cases} \min\left[k_{bn}u_n, R_{bc}\right], (u_n \geq 0) \\ \min\left[k_{bn}u_n, R_{bt}\right], (u_n < 0) \end{cases} \quad (2)$$

$$F_{bs} = \min\left[k_{bs}u_s, R_{bs}\right] \quad (3)$$

$$M_b = \min\left[k_{bm}\theta, R_{br}\right] \quad (4)$$

where min[•] is the operator taking the minimum; u_n, u_s and θ are the overlap, the shear displacement and the relative rotation angle, respectively; k_{bn}, k_{bs}, k_{bm} are the normal, shear and rolling contact stiffness of bonds, respectively; R_{bc}, R_{bt}, R_{bs} and R_{br} are the compressive, tensile, shear and rolling bond strengths, respectively.

The projections of the strength envelope in F_s-F_n, M-F_n planes and critical normal force R_{cr} can be expressed as:

$$R_s = \mu_b f R_{bc}[1 + g_s(\ln\frac{1}{f})^{f_s}] + \mu_p F_{pn} \quad (5)$$

$$R_r = \frac{1}{6}\beta_b \bar{r} f R_{bc}[1 + g_r(\ln\frac{1}{f})^{f_r}] + \frac{1}{6}\beta_p \bar{r} F_{pn} \; F_{bs} = \min\left[k_{bs}u_s, R_{bs}\right] \quad (6)$$

$$R_{cr} = \frac{k_{bn} + k_{pn}}{k_{bn}} R_{bc} \quad (7)$$

where f=$(F_{bn}$+$R_{bt})/(R_{bt}$+$R_{bc})$ is the stress ratio, which can be used to define the types of bond failures: when $f = 0$, compressive failure occurs; when $0 < f < 1$, tensile-shear-torsional failure ($F_n < 0$) or compressive-shear-torsional failure ($F_n > 0$) occurs; when $f = 1$, tensile failure occurs; μ_b and μ_p are friction coefficients of bonds and particles; β_b and β_p are rolling resistance coefficients of bonds and particles; f_s, f_r, g_s and g_s are the envelope shape factors; $\bar{r}$ is the common radius of two particles in contact and k_{pn} is inter-particle normal contact stiffness. It should be noted that the strength in Eqs. (5)

IOP Conf. Series: Earth and Environmental Science **26** (2015) 012021 doi:10.1088/1755-1315/26/1/012021

and (6) includes both of the components of bond and inter-particle contacts when $F_n > 0$, which means that when $F_n < 0$ the second term disappears as inter-particle contact cannot transfer tensile force.

If forces reach strength envelope/critical values, they will fall to the residual state. The residual shear strength R_{sr} and rolling strength R_{rr} can be expressed by:

$$R_{sr} = \mu_b F_{bn} + \mu_p F_{pn} \tag{8}$$

$$R_{rr} = \frac{1}{6}\beta_b \bar{r} F_{bn} + \frac{1}{6}\beta_p \bar{r} F_{pn} \tag{9}$$

which also contain two terms corresponding to bond and inter-particle components.

3. DEM modelling

The DEM rock sample containing 10,000 particles was generated by Under-compaction Method (UCM) proposed by Jiang et al. [18], with maximum, minimum and average diameters being 2.0, 0.5 and 1.3 mm respectively. The size of the specimen is 12.23 cm×7.82 cm with planar void ratio being 0.20. The synthetic intact rock material has already been calibrated by comparison with uniaxial compression and Brazilian test results of granite [19]. A set of joints in the given arrangement are then generated within the rectangular DEM sample, and the joints were identified by as a 2 mm thick layer of particles including no bond effect. The assigned parameters for both intact rock and joints are listed in Table 1.

Table 1 Microscopic material parameters used in DEM simulations

Parameters	Values for intact rock	Value for joint
Particle density ρ_s (kg/m^3)	2700	2700
Planar void ratio e	0.20	0.20
Normal stiffness of particles k_{np} (N/m)	1.8×10^{11}	1.8×10^{11}
Tangential stiffness of particles k_{sp} (N/m)	9.47×10^{10}	9.47×10^{10}
Normal stiffness of bonds k_{nb} (N/m)	9.0×10^{10}	9.0×10^{10}
Tangential stiffness of bonds k_{sb} (N/m)	4.735×10^{10}	4.735×10^{10}
Tensile strength of bonds R_{bt} (N)	6.5×10^4	0.0
Compressive strength of bonds R_{bc} (N)	8.0×10^7	0.0
Interparticle friction coefficient μ_p	1.0	0.2
Interparticle rolling resistance coefficient β_p	1.5	0.0
Friction coefficient of bonds μ_b	0.5	0.1
Rolling resistance coefficient of bonds β_b	0.5	0.0

Figure 1 and Figure 3 present the schematics and the DEM samples with the centre and non-coplanar rock joints respectively. The parameter g in Fig. 2 is the joint connectivity rate and is determined by Eq. (10):

$$g = \frac{L_t - L_j}{L_j} \tag{10}$$

During the direct shear test, shear load was applied to the DEM jointed rock sample by moving walls #2 and #3 towards each other at a quasi-static velocity while maintaining walls #1 and #4 under a constant normal pressure. For the centre joints, the influence of joint connectivity rate (e.g., 0.6, 0.8, 0.9) is investigated with normal stress being 1.0 MPa, while for intermittent non-coplanar joints with a given joint length l and spacing e, the effect of the inclination of joints i (varying from -75° to 90°) is explored.

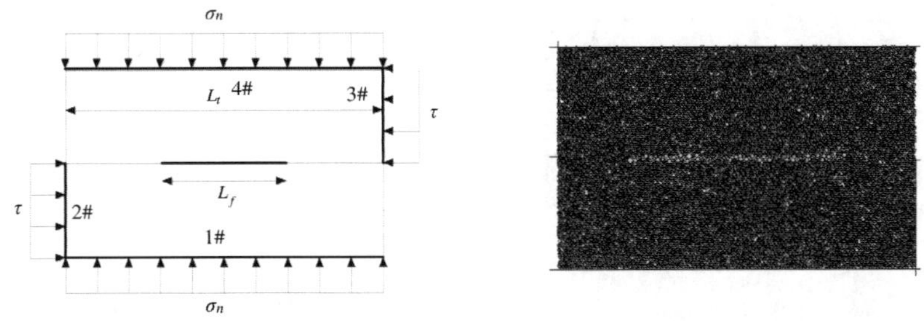

(a) Schematic (b) DEM Specimen

Figure 1 Schematic and DEM model of rock sample containing centre joint

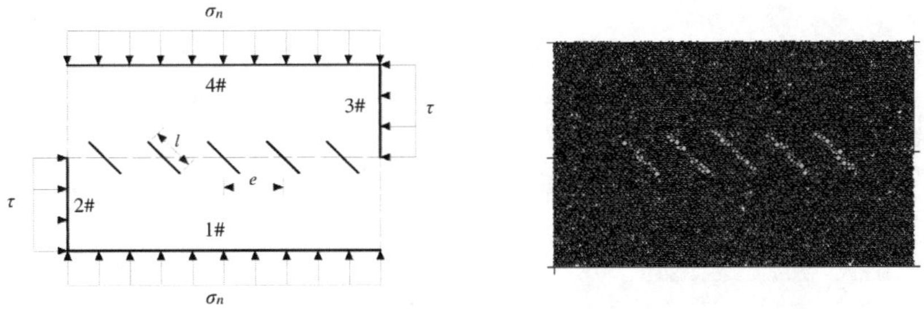

(a) Schematic (b) DEM Specimen

Figure 2 Schematic and DEM model of rock sample containing intermittent non-coplanar joints used in direct shear test.

4. Results

4.1 Centre joint

Figure 3 illustrates the relationships between the shear stress/number of bond breakages and shear displacements for the specimen with $g = 0.6$, it shows that the deformation and failure process of DEM specimens can be divided into four stages: (1) elastic shearing phase: shear stress increases almost linearly with the increase of shear displacement, while the no bond breakages happen during this shearing phase; (2) crack propagation: as the shear displacement increases, shear stress fluctuates until reaching the peak stress, and a large number of bond breakages occur with tensile-shear-torsional failures dominating in number; (3) the failure of rock bridges: the shear stress quickly drops to a residual value after arriving the peak; (4) through-going discontinuity: the shear stress remains at a relatively low residual value in spite of some fluctuations, during this period of shearing, the number of the microscopic bond breakages consistently grows. The shear behaviour above is quite typical compared with that normally obtained from experimental results [20].

Figure 4 shows the relationships between the peak shear strength and joint connectivity rate, the shear strength in numerical and experimental results are normalized by their own peak stresses for the case with $g = 0.6$ separately. It can be observed that both of the peak shear stresses significantly decrease with the increase of normal stress, and the trends of the DEM and experimental results are similar. This phenomenon can be reasoned that growth of the proportion of joint weakens the shearing capacity within the shear axis, causing the decrease of the shear strength of the jointed samples.

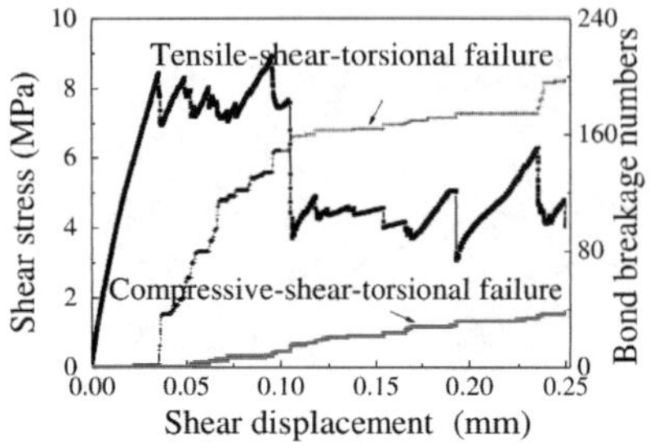

Figure 3 Relationship between shear stress/broken bond number and shear displacement by DEM simulation

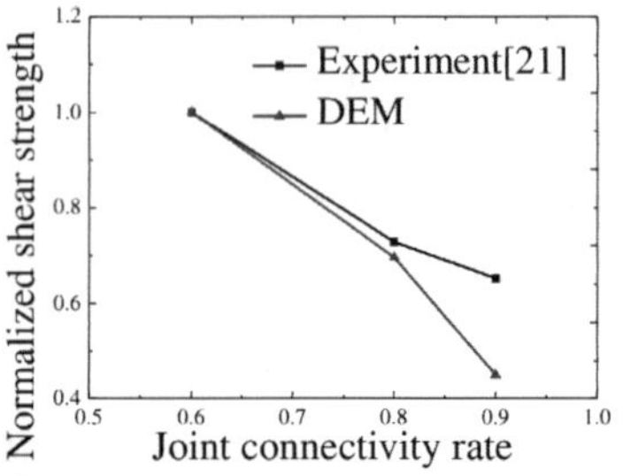

Figure 4 Relationship between the normalized peak shear stress and the joint connectivity rate

4.2 Intermittent non-coplanar joints

Considering that all the cases with different inclinations share almost the same trend, only the shear curve of non-coplanar intermittent rock joints with inclination i = 45° is showed in Fig. 6. The shear curve of the non-coplanar joints is quite similar to that of the centre ones, except that no apparent yielding when approaching the peak stress. The reason for this might be that the number of the bond breakages rises exponentially when the jointed sample arrives yielding state, so the stress drops sharply simultaneously, but not the case for the centre joint.

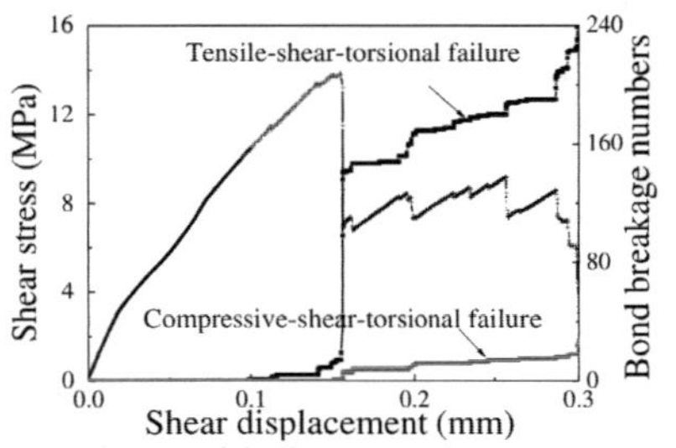

Figure 5 Relationship between shear stress/broken bond number and shear displacement by DEM simulation with joint inclination angle of 45°

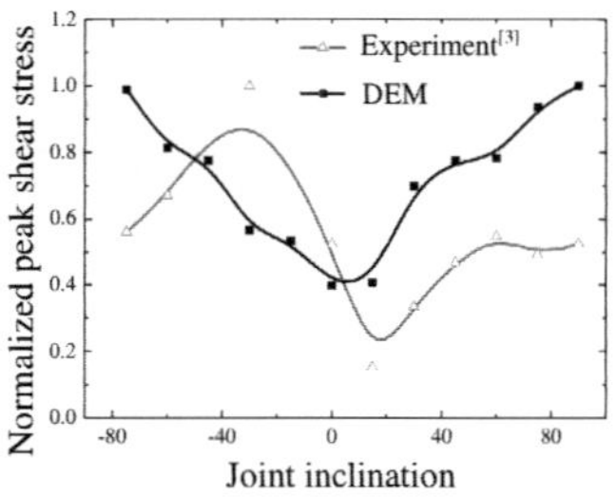

Figure 6 The relationship between the joint angle and the normalized peak value of shear stress

Figure 6 illustrates the shear loading capacities of non-coplanar intermittent joint assemblages with different inclinations. The numerical and experimental results are both normalized by the maximum peak stresses obtained in experimental tests (i.e., -30° for experimental results). The results show that in the DEM simulation, the peak stress decrease as the inclination angle changes from -75° to 15°. It reaches its lower limit value at around 15° and then increase till 75°. The experimental results show almost the same trend for inclination angles varying from -30° to 75° except for the values near -45°. The difference can be partially explained that in the DEM simulation, the joints are filled with particles, whereas in the laboratory, the man-made joints are just open cracks with nothing inside.

5. Conclusions

The numerical simulation results indicate that the complete shear behaviour of jointed rock includes elastic shearing phase, crack propagation, the failure of rock bridges and the through-going discontinuity. For centre rock joints, the decrease of the joint connectivity rate will enhance the peak shear stress. As to intermittent non-coplanar rock joints, the relationship between shear stress and shear displacement presents the elastic-brittle-plastic behavior. The shear capacity drops when the joint inclination angle is negative but rises when positive with the increases of rock joints inclination. The failure process of intermittent non-coplanar rock joints begins with the transmission of compressive forces through the tips of rock joints and ends with the breakage of rock bridges. The comparison between DEM and experimental results proves that this bond contact model is capable of describing the shear behaviour of jointed rocks during direct shearing.

Acknowledgements
This research is supported by the Major Project of Chinese National Programs for Fundamental Research and Development (973 Program) (Nos. 2011CB013504 and 2014CB046901), China National Funds for Distinguished Young Scientists (No. 51025932), State Key Lab. of Disaster Reduction in Civil Engineering (No. SLDRCE14-A-04). These supports are greatly appreciated.

References

[1] Lajai E Z 1969 Shear strength of weakness planes in rock *Int. J. Rock Mech. Min.* **6** 499-515
[2] Lajai E Z 1969 Strength of discontinuous rocks in direct shear *Géotechnique* **19** 218-233
[3] Gehle C, Kutter H K 2003 Breakage and shear behavior of intermittent rock joints *Int. J. Rock Mech. Min.* **40** 687-700
[4] Cao P, Liu T Y, Pu C Z and Lin H 2015 Crack propagation and coalescence of brittle rock-like specimens with pre-existing cracks in compression *Eng. Geol.* **187** 113-121
[5] Zhang S, Leech C 1986 FEM analysis on mixed-mode fracture of CSM-GRP *Eng. Fract. Mech.* **23** 521-535
[6] Hayashi K, Ono A, Abe H 1989 BEM Analysis of a Cylindrical Three-Point Bend Specimen with a Chevron Crack for Fracture Toughness Test of Rock *JSME Int. J. A-Solid M.* **32** 427-431
[7] Hatzor Y, Arzi A, Zaslavsky Y and Shapira A 2004 Dynamic stability analysis of jointed rock slopes using the DDA method: King Herod's Palace, Masada, Israel *Int. J. Rock. Mech. Min.* **41** 813-832
[8] Cundall P A, Strack O D L 1979 A discrete numerical model for granular assemblies. *Géotechnique* **29** 47-65
[9] Park J W, Song J J 2009 Numerical simulation of a direct shear test on a rock joint using a bonded-particle model *Int. J. Rock Mech. Min.* **46** 1315-1328
[10] Potyondy D O, Cundall P A 2004 A bonded-particle model for rock *Int. J. Rock Mech. Min.* **41** 1329-1364
[11] Jiang M J, Yan H B, Zhu H H and Utili S 2011 Modelling shear behaviour and strain localisation in cemented sands by two dimensional Distinct Element Method Analyses *Comput. Geotech.* **38** 14-29
[12] Bewick R P, Kaiser P K, Bawden W F and Bahrani N 2014 DEM simulation of direct shear: 1. Rupture under constant normal stress boundary conditions *Rock Mech. Rock Eng.* **47** 1647-1671
[13] Jiang M J, Chen H, Crosta G B 2015 Numerical modeling of rock mechanical behavior and fracture propagation by a new bond contact model *Int. J. Rock Mech. Min.* (accepted for publication)
[14] Jiang M J, Yu H S, Harris D 2006 Bond rolling resistance and its effect on yielding of bonded granulates by DEM analyses *Int. J. Numer. Anal. Met.* **30** 723-761
[15] Jiang M J, Yu H S, Harris D 2005 A novel discrete model for granular material incorporating rolling resistance *Comput. Geotech.* **32** 340-357
[16] Jiang M J, Yu H S, Leroueil S 2007 A simple and efficient approach to capturing bonding effect in naturally micro-structured sands by discrete element method *Int. J. Numer. Meth. Eng.* **69** 1158-1193
[17] Jiang M J, Sun Y G, Li L Q and Zhu H H 2012 Contact behavior of idealized granules bonded in two different interparticle distances: An experimental investigation *Mech. Mater.* **55** 1-15
[18] Jiang M J, Sun Y G, Xiao Y 2012 An experimental investigation on the mechanical behavior between cemented granule *Geotech. Test J.* **35** 678-690
[19] Jiang M J, Konrad J M, Leroueil S 2003 An efficient technique for generating homogeneous specimens for DEM studies *Comput. Geotech.* **30** 579-597
[20] Martin C D 1993 *The strength of massive Lac du Bonnet granite around underground openings: Winnipeg* (Canada: University of Manitoba)
[21] Bai S W, Ren W Z, Feng D X and Zhou S H 1999 Research on the strength behavior of rock containing coplanar close intermittent joints by direct shear test *Rock Soil Mech.* 20(2): 10-16

IOP Conf. Series: Earth and Environmental Science **26** (2015) 012010 doi:10.1088/1755-1315/26/1/012010

Stabilization Columns for Embankment Support – Investigation, Verification and Further Development of Analytical Analyses

H Pankrath[1,3], H Kaya[2] and R Thiele[1]
[1] Leipzig University of Applied Sciences, G² Group Geotechnics, Karl-Liebknecht-Str. 132, Leipzig 04277, Germany
[2] Hamburg University of Technology, Harburger Schlossstraße 20, Hamburg 21079, Germany

E-mail: holger.pankrath@htwk-leipzig.de

Abstract. As a technical and economical alternative to foundations on piles, but also to shallow foundations on improved soil, in recent decades a high number of soil improvement methods have been developed and established. Many of these methods use non-reinforced, cylindrical load bearing elements. A very common application of stabilizing columns is the improvement of a few meters thick soft soils below dams and embankments. But especially for this application, many failure cases are documented worldwide. In the contribution the substantial content and results are presented for investigation, testing and further development of methods for evaluating the slope stability. After a description of the problem and consequential tasks the contribution contains main results of the investigations of international sources with the stepwise development of analytical solutions. Next to the in practice well-known approaches for gravel columns, less common approaches from Scandinavia are explained. The contribution is completed with a presentation and discussion of an illustrative example, taking into account a number of different failure modes of the columns and the surrounding soil. The example was compared and validated with a 3D Model using the Finite Element Method.

1. Introduction

Within the German Geotechnical Society (*DGGT*) the working group 2.8 is working on the revision of recommendations for manufacturing, design and quality control of non-reinforced, cylindrical load bearing elements, further referred as stabilization columns. In the working group the column types are distinguished due to their manufacturing processes and the nature of the contribution of the column materials. A very common yet highly sensitive application of stabilizing columns is the improvement of several meters thick, low load-bearing soils below dams and traffic embankments. Analytical design programs (based on the the state of equilibrium) are currently not able to consider the stability columns in the safety analyses in terms of a slope stability failure. Therefore, the effect of the columns must be incorporated indirect into the programs, so that the effect can be taken into account in the slope stability evaluation.

[3] To whom any correspondence should be addressed.

Figure 1. Failure of an embankment, supported by stabilization columns [17]

Motivation for the authors was to investigate, compare and further develop approaches for evaluating the slope stability using software based on the state of equilibrium. This may lead for arbitrary constellations to safe and economical dimensioning and arrangement of the columns. In this paper, main content and results are presented for the development of the analytical concepts, discussed and tested on a concrete example using a 3D numerical model.

2. Research and Analysis

General information on the design, analysis and design for applications in geotechnical engineering can be find currently in DIN EN 1997-1: 2009-09 (Eurocode 7). The guideline is part of a series of standards for influence and design, which only makes sense in the package. For national application also individual parameters are defined in Annexes (in Germany - NAD: DIN EN 1997-1 / NA: 2010-12). Additional regulations also can be found in DIN 1054: 2010-12. In addition to the European standards for specifications to ground and ground failure calculations reference is made to DIN 4084: 2009-01.

For a consideration of stability columns in an analytical calculation, using an equilibrium analysis, two basic approaches are very common:

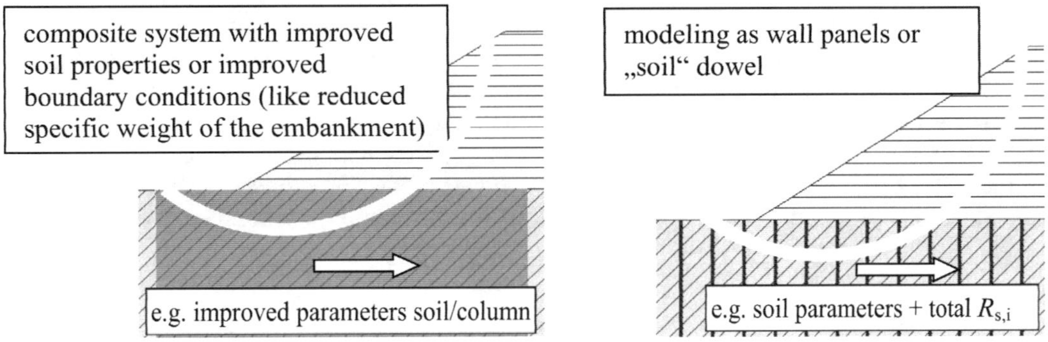

Figure 2. Basic approaches for consideration of stability columns in a calculation: "global" by improved soil properties of a composite system or by improved boundary conditions (left figure); by modelling the columns as wall panels or "soil" dowel (right figure)

For estimating of improved soil properties of the composite system many approaches exist which take into account the columns according to a load-component and an area-ratio e.g. according to [7], [12] and [13].

Especially for the use of stiff column materials with binder the advices of the authors [4], [9], [10] and [11] have to be taken into account:

- The load transfer behavior is decisively influenced by the column stiffness.
- Columns with very low stiffness transfer loads in interaction with the surrounding soil better than columns with high stiffness. The failure modes can be approximated to an improved/composite system.
- Columns with high stiffness behave towards the surrounding soil with low bearing capacity as a pile-like element. A column failure is assessed as more critical due to the brittle material behavior (progressive failure or "zipper" principle possible).

It is therefore necessary to distinguish between two basic modes of operation:

- behavior as a block or as a composite system for granular columns without binder and for "soft" columns with binder
- behavior as a pile-like element for "rigid" columns with binder

3. Approaches und Separation

3.1. Granular Columns (gravel columns)

For granular columns, the approach of [12], supplemented by [13] is recommended. In this approach improved soil properties of a composite system are calculated considering the load-component and the area-proportion of column and surrounding soil. Compared with a just area proportional determination of the improved soil properties, this may easily lead to an overestimation of the improvement in slope stability considerations, especially for deep slip circles. In [13] approaches are explained which deal with this problem, e.g. via calculating improved soil properties with consideration of the ratio between the respective load above the improved layer (Q_{top}) and the total load (Q_{total}).

3.2. "Soft" Columns with Binder (deep mixing method)

For "soft" stabilization columns the specifications according to [7] are recommended. Basis of this approach is the determination of improved soil properties according to the area ratio of column and the surrounding soil (see (1) and (2)) and the limitation of shear strengths of the columns of 150 kN/m² or max. 200 kN/m² depending on the arrangement of the columns below an embankment or a dam. A distinction is made between the location of the column between "passive zone", "shear zone" and "active zone". Furthermore, the approach provides a separate consideration of undrained and drained conditions (less favorable is relevant). Other failure mechanisms of the columns are not considered. Therefore an effective interconnection between column and surrounding soil as well as a shear failure along a slip surface through soil and columns is assumed, analogous to the approach for granular columns. This is considered to be realistic when strength or stiffness between columns and soil does not vary too much or if the ratio of the improved area is greater than 50%. As a result of numerical benchmark calculations, [11] indicates 15 psi (100 kN/m) for a column compression strength as an upper limit suitability for a calculation with improved soil properties.

$$\varphi_{avg} = A_S/A \cdot \tan\varphi_S + (1 - A_S/A) \cdot \tan\varphi_B \quad (1)$$

$$c_{avg} = A_S/A \cdot c_S + (1 - A_S/A) \cdot c_B \quad (2)$$

Especially in northern Europe groups of stabilization columns are used for settlement reduction and stabilization of dams. In this case, the area ratio of the pillars from the surrounding soil is usually well below 30%. In contrast to this in the Japanese region, also with frequent applications of similar improvement methods, the area ratio is from 30% to 50% and higher [11]. For this column arrangements with high area ratio separate recommendations have been published [15].

3.3. „Rigid" Columns with Binder

The approach for "soft" columns with binder with a determination of improved soil properties and evidence of equilibrium conditions along a sliding body assumes an uniform deformation behavior of columns and surrounding soil (similar strengths of columns and soil). This approach leads to an overestimation of the improvement for the use of groups of "rigid" columns with binder. [9] and [4] evaluated analyses of experimental studies on the failure behavior in centrifuge tests and numerical analyses. They pointed out that for "rigid" columns with binder in addition to shear failure along a sliding surface other failure modes of the columns and the surrounding soil will occur. Particularly in areas below or next to embankments an arrangement of the columns in overlapping form is much safer. Groups of individual columns should not be used in next to embankments [15] and [10].

4. Application Example for "Rigid" Columns with Binder

In this chapter an application example for the approach for groups of individual "rigid" columns with binder according to [9] and [4] is illustrated. In modified approach, six major failure modes of the columns were evaluated (see also [1]). For the traffic dam, the overall stability has to be proofed in the limit state GEO-3 for the design situation BS-P. The applied load on the dam crest is composed of $q_{Q,k}$ = 60.0 kN/m with a width of 2 x 2.6 m and $q_{G,k}$ = 12.5 kN/m with a width of 8.5 m (see Figure 3). The water table is 20 m below the dam crest.

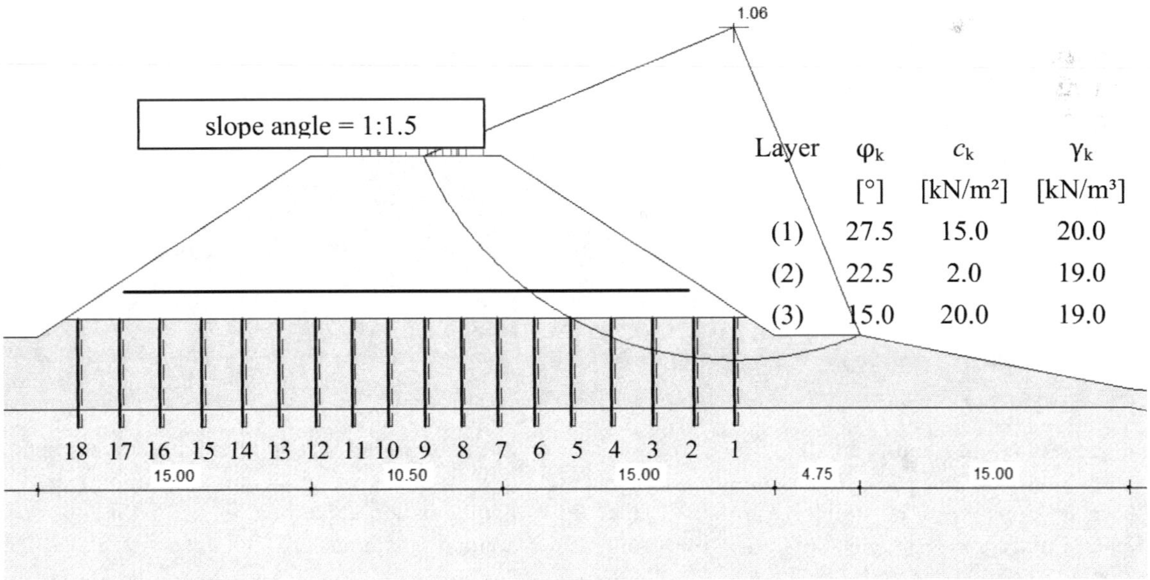

Figure 3. Geometry and basic details of the embankment, decisive slip circle and utilization factor μ_{max} without consideration of stability columns (only illustration of the proposed column arrangement)

For the initial situation (without columns) a utilization factor μ_{max} <1.0 was determined for the limit state GEO-3 in the design situation BS-A. Using the partial safety factors obtained for a design situation BS-P leads to an utilization factor μ_{max} = 1.06 (an increase of the overall stability is required, s. Figure 3).

In Table 1 the required geometric data which is necessary to calculate the applied load Q_E according to (3) is collected:

Table 1. Determination of the vertical load per unit cell (dead load dam).

number of column	**1**	**2**	**3**	**4**	**5**	**6**	**7**	**8**
height of embankment h_i (m)	0.5	2.0	3.5	5	6.5	7.8	9.0	9.0
length of column above the slip surface H_1 (m)	2.4	2.2	1.8	1.1	0.0	0.0	0.0	0.0
length of column below the slip surface H_2 (m)	3.6	3.8	4.2	4.9	6.0	6.0	6.0	6.0
load per unit cell Q_E (kN)	48.1	192.4	336.7	480.9	559.6	595.2	384.7	384.7

$$Q_E = \gamma \cdot h_i \cdot A_E \tag{3}$$

Decisive for the moment capacity of the stability columns M_u is the vertical stress in the columns (see also Table 2). Depending on the type and design of the load distribution layer an increased load component of the columns *m*' has to be estimated using an appropriate method for example according to [6], [16] oder [8]. For this example it is assumed that an increased load component of the columns *m* is 0.8 (80%) for the entire column length.

$$Q_S = Q_E \cdot m' \tag{4}$$

$$\sigma_{v,S} = Q_S / A_S \tag{5}$$

$$M_u = f_{c,d} \cdot A_{pl} \cdot e_{pl} \tag{6}$$

$$A_{pl} = \sigma_{v,S} / f_{c,d} \cdot A_S \tag{7}$$

$$e_{pl} = \frac{d_S}{2} \cdot \left(1{,}65\left(\frac{A_{pl}}{A_S}\right)^4 - 4{,}05\left(\frac{A_{pl}}{A_S}\right)^3 + 3{,}49\left(\frac{A_{pl}}{A_S}\right)^2 - 2{,}08\left(\frac{A_{pl}}{A_S}\right) + 1\right) \tag{8}$$

Table 2. Determination of the moment capacity M_u per column according to [9].

number of column	**1**	**2**	**3**	**4**	**5**	**6**	**7**	**8**
Q_S (kN)	38.48	153.90	269.33	384.75	447.65	476.17	307.74	307.74
$\sigma_{v,S}$ (kN/m^2)	136.08	544.31	952.54	1360.77	1583.23	1684.11	1088.40	1088.40
A_{pl} (m^2)	0.0030	0.0121	0.0212	0.0302	0.0352	0.0374	0.0242	0.0242
e_{pl} (m)	0.2934	0.2751	0.2587	0.2438	0.2364	0.2331	0.2536	0.2536
M_u (kNm)	11.29	42.34	69.67	93.81	105.81	111.00	78.03	78.03

In addition to the failure modes shown in Figure 3, columns with less strength can fail by shearing, compressing or buckling. For columns with medium strength or as in this example, high strength, these failure modes are unlikely and only expected in areas in the centre of a dam ("active zone"). Other failure modes according to [9] for groups of individual columns next to the slope ("passive zone") are not considered.

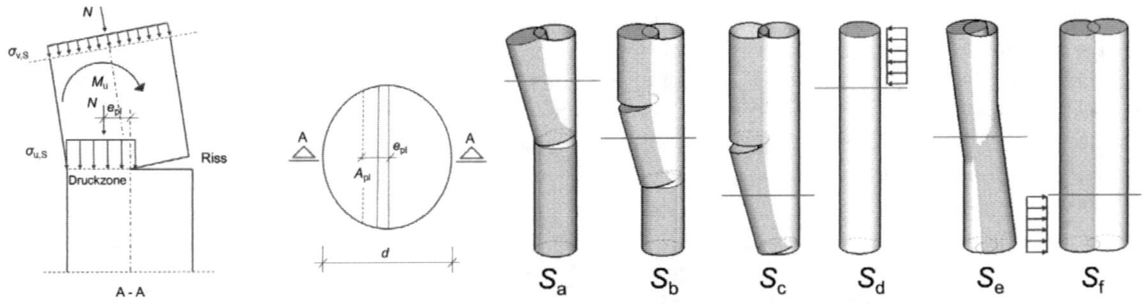

Figure 4. Review of the moment capacity M_U of the columns (left) and identification of the column forces $R_{S,i}$ depending on different failure modes (pictured right)

According to [9], the column resistance S (Figure 4) is defined as the resistance acting parallel to the slip surface and is made up of units of normal force and column strength. The column force T corresponds to the horizontal resistance of the column and is hereinafter referred as $R_{S,i}$. With (9) to (14) the resistance values are determined for the failure modes a to f.

$$S = N \sin\alpha_a + T \cos\alpha_a$$

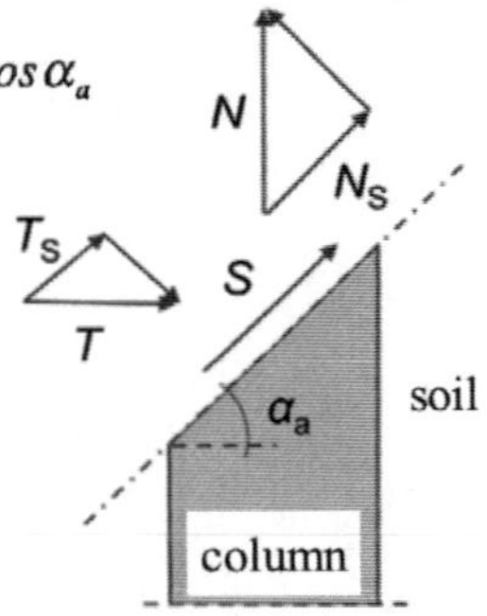

Figure 5. Approach to calculate the column resistance as a function of normal force and column strength according to [9]

$$R_{S,a} = kc_u d_S{}^2 \left(\sqrt{\frac{4}{9}\left(\frac{H_1}{d}\right)^2 + \frac{4}{3}\frac{M_u}{kc_u d_S^3}} - \frac{1}{3}\left(\frac{H_1}{d_S}\right)\right) \tag{9}$$

$$R_{S,b} = \sqrt{2M_u kc_u d_S} \tag{10}$$

$$R_{S,c} = kc_u d_S{}^2 \left(\sqrt{\frac{4}{9}\left(\frac{L-H_1}{d}\right)^2 + \frac{4}{3}\frac{M_u}{kc_u d_S^3}} - \frac{1}{3}\left(\frac{L-H_1}{d_S}\right)\right) \tag{11}$$

$$R_{S,d} = kc_u d_S H_1 \tag{12}$$

only if $M_{max} = kc_u d_S \frac{H_1{}^2}{2} < M_u$

$$R_{S,e} = \frac{kc_u d_S}{2}\left(\sqrt{3L^2 - 4LH_1 + 4H_1{}^2} - L\right) \tag{13}$$

only if $H_1 \approx H_2$

$$R_{S,f} = kc_u d_S (L - H_1) \tag{14}$$

only if $M_{max} = kc_u d_S \frac{(L-H_1)^2}{2} < M_u$

The load capacity factor k is for large deformations above or below a sliding surface at up to 9 [9]. Particularly for large differences in strength between column and the surrounding soil, a failure of the columns is assumed, just before the soil can mobilize its full shear strength. For small deformations in the soil k should be set to 2.0 according to [4]. This approach takes into account the very limited ductility of the columns and is used in the further calculation example. In recognition of k> 2, the deformation compatibility of columns and soil should be considered and proofed separately.
For determining the column resistance in accordance to the cutting angle α_a proportions for normal force and column strength have to be considered (Figure 4). The approach of the normal force leads, in particular for large cutting angles α_a to computationally higher column resistances. Weaknesses by imperfections and/or by tilting of the columns are not considered. Subsequently, the column resistance $R_{s,i}$ was simplistically equated with the column strength T. This corresponds to columns in which the angle of cutting angle α_a is small as in this example. Subsequently, the column forces $R_{S,i}$ are determined as depending on the failure modes a to f. The minimum value (min $R_{S,i}$) is decisive.

Table 3. Determination of the relevant column forces $R_{S,i}$.

number of column	**1**	**2**	**3**	**4**	**5**	**6**	**7**	**8**
$R_{S,a}$ (kN)	20,24	29,77	37,15	44,58	52,85	54,13	45,39	45,39
$R_{S,b}$ (kN)	21,14	40,95	52,52	60,95	64,73	66,30	55,59	55,59
$R_{S,c}$ (kN)	26,80	35,20	42,37	49,27	55,62	56,33	51,68	51,68
$R_{S,d}$ (kN)	47,52	43,56	35,64	21,78	0,00	0,00	0,00	0,00
$R_{S,e}$ (kN)	25,44	26,08	27,90	32,64	43,48	43,48	43,48	43,48
$R_{S,f}$ (kN)	71,28	75,24	83,16	97,02	118,80	118,80	118,80	118,80
min $R_{S,i}$ (kN)	20,24	29,77	35,64	21,78	0,00	0,00	0,00	0,00

light grey: condition $M_{max} < M_u$ (failure mode d and f) or/and $H_1 = H_2$ (failure mode e) not satisfied

The sum of the forces in the relevant slip circle and the exceeding of the stresses (ΔE) may be determined as followed (see Figure 3):

$\sum$ min $R_{S,i}$ = 107.43 kN
ΔE = ΔM / radius (r) with $\Delta M = E \cdot M - r \cdot M$
= (19,601.5 kNm – 18,413.6 kNm) / 20.66 m
= 57.5 kN

For this example calculation the review of the column forces Σ min $R_{S,i}$ to ΔE leads to sufficient capacity of the improvement (107.4 kN > 57.5 kN). In the analytical model, soil resistances and the corresponding column strength $R_{s,i}$ can be added analogous to Figure 3 for further iteration. For the evidence of the overall stability it has to be proofed whether a reassessment is required on a different relevant slip circle. In this example, the column resistance for column 1 to 5 and a limited soil resistance were used in the analytical model in limit state GEO-3 and the design situation BS-P. The result of the calculation no significantly changed the position of the slip circle.

5. Comparative Calculation with Finite Element Method (FEM)

The results obtained with the approach of "rigid" binder columns according to [9] and [4] has to be recalculated and verified with a numerical model. For this purpose, the FEM is used with the algorithm of the software Plaxis3D [3].

5.1. Numerical Model

For the numerical analysis, a three-dimensional model of the dam is used. In addition to the formation of slip surfaces and the determination of the stability, the interaction between the columns is considered. Taking advantage of the symmetries, a limited section of the dam is modeled for a spacing of one column. The geometry used, groundwater conditions and the material properties are illustrated in Figure 3. At the bottom and at the lateral edges of the model the displacement degrees of freedom are kept in horizontal and vertical directions. Figure 6 shows the discretization of the FE model. The mesh consists of about a hundred thousand 10-node tetrahedral elements.

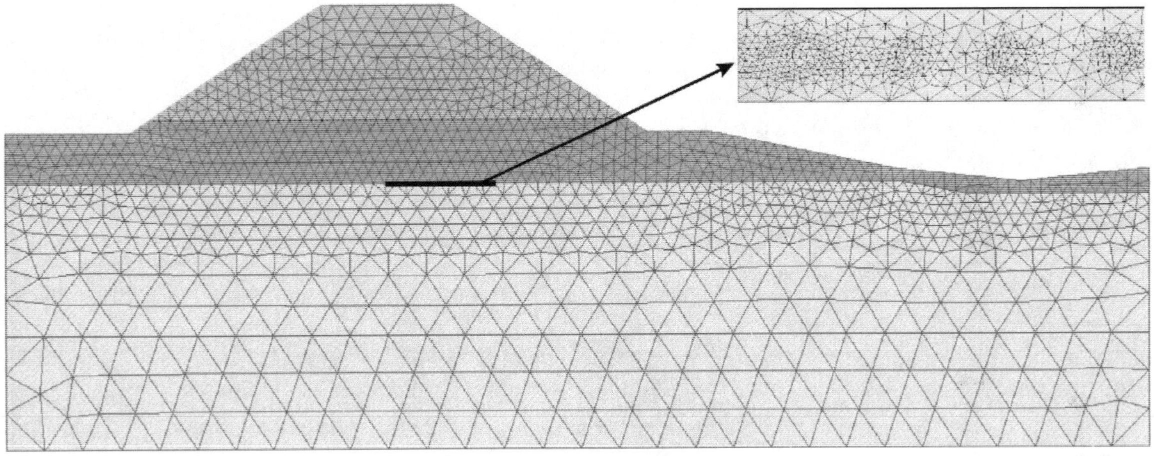

Figure 6. FE model with the used discretization

5.1.1. Load History

At the beginning of the analysis the initial stress state is calculated before modeling the dam in two phases. Another plastic step is to stabilize the equilibrium state in a so-called "zero phase" without further design changes. Now, the stability of the dam is determined without use of stabilizing columns under the previously applied loads. Alternatively the columns are considered wished-in-place, in the next step the loads are applied and the stability of this condition is calculated.

5.1.2. Modeling of the Soil

The global stability of the slope is determined by an φ-c reduction in Plaxis3D. According to the recommendations of the Working Group "Numerical Analysis in Geotechnical Engineering" [5] the safety of numerical stability analyzes is less dependent on the choice of material model. Thus, for the FE analysis also as in the above calculation, the linear elastic, perfectly plastic material model with boundary condition according to Mohr-Coulomb (MCM) and the material parameters illustrated in figure 3 were used.

5.1.3. Modeling of the Columns

The columns are modeled as volume elements with MCM using fitted parameters for the material (C30/37), to take into account the resistance reduction of the column. To map the yield surface of the concrete, the material parameters listed in Table 4 are used. This allows an approximate consideration of the plastic behavior of the column and the automatic φ-c-reduction contained in Plaxis3D also for the columns.

Table 4. Material parameters for mapping the stabilization column (C30/37) to the linear elastic, perfectly plastic material model with the boundary condition according to Mohr-Coulomb.

Parameter	Unit	C30/37	Note
γ/γ_r	kN/m^3	24/24	
E	MN/m^2	33,000	according to EC2-1-1, section 3.1.3
υ	-	0.2	
φ'	°	50.7	$\tan\varphi' = (f_{cm} - 5/3\, f_{ct})/\, (f_{cm} + 5/3\, f_{ct})$ *
c'	kN/m^2	6,776	$c' = (0{,}5\, f_{cm}(1 - \sin\varphi'))/\sqrt{1-(\sin\varphi')^2}$ *
$\sigma_{\text{Tension cut off}}$	kN/m^2	2,900	

* according to [14]

A direct force transmission between soil and columns is assumed. To determine the internal forces and moments, as well as the deformation of the columns, a beam element is added in the axis of the column. The column stiffness of the volume elements and of the beams has to be equal to the overall stiffness. In this case, the stiffness of the beam is taken as one thousandth of the overall stiffness, whereas the stiffness of the volume elements corresponds to the overall stiffness. The resulting difference of 0.1% for the overall stiffness is tolerable. The ratio of the cutting forces corresponds to (16) and (17).

$$u_{column} = u_{beam} \tag{15}$$

$$M_{column} = \frac{(EI)_{column} + (EI)_{beam}}{(EI)_{beam}} M_{beam} = 1001\, M_{beam} \tag{16}$$

$$N_{column} = \frac{(EA)_{column} + (EA)_{beam}}{(EA)_{beam}} N_{beam} = 1001\, N_{beam} \tag{17}$$

5.2. Results

5.2.1. Increase of Stability

The essential for a proof of the slope is the stability of the observed slope. In the FEM φ-c-reduction the utilization factor can be increased with considering the stability columns by approximately 15%. This result is comparable to the 12% increase in the analytical calculation.

5.2.2. Failure Mechanism

In order to assess the influence of the columns on the failure mechanism for the depth of the dam, a section in the plane of the column axis (A) and between the columns (B) is evaluated.

Figure 7 (top) shows the shear plane of the non-stabilized embankment by illustrating the incremental shear strain $\Delta\gamma$. As expected, due to the symmetrical loading of the embankment, an identical failure arises on both slopes. The failure surface is constant over the depth (in the direction of dam axis) and is almost equivalent to the failure image of the calculations with the equilibrium software (see Section 4, Figure 3), a single monolithic failure body. The radius and the center of the slip circle differ slightly from each other.

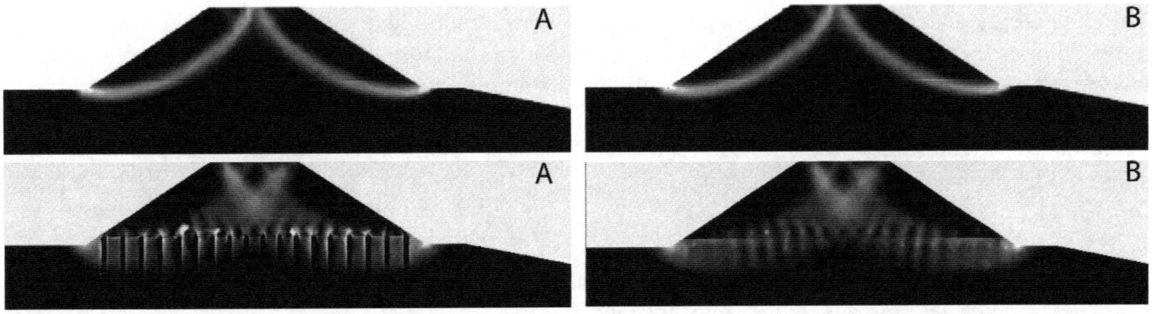

Figure 7. Failure mode of the dam after strength reduction in section A and section B without stabilizing columns (top) and with stabilization column (below)

In Figure 7 (below) the failure body in the ultimate limit state for the dam is illustrated with stabilizing columns. The left picture shows the positions of the columns are clearly visible due to the very low shear strains. The columns lead to a shifting of the shear plane and a strong dispersion of the shear plane towards a shear band. The shear strains are also significantly reduced. Section B also illustrates that the failure body is almost constant in the depth. Thus, the selected distance is sufficient to ensure interaction between the columns and to prevent a flow of soil between the columns. In this case, a two-dimensional calculation of the system is justified as in the approach of "rigid" binder columns according to [9] and [4].

5.2.3. Comparison with the Approach of "Rigid" Columns with Binder

For the verification of the method for "rigid" binder columns according to [9] and [4] in particular the attachable lateral forces at the level of the slip surface are considered. The lateral force curves calculated in the FEM are shown for the right slope in Figure 8.

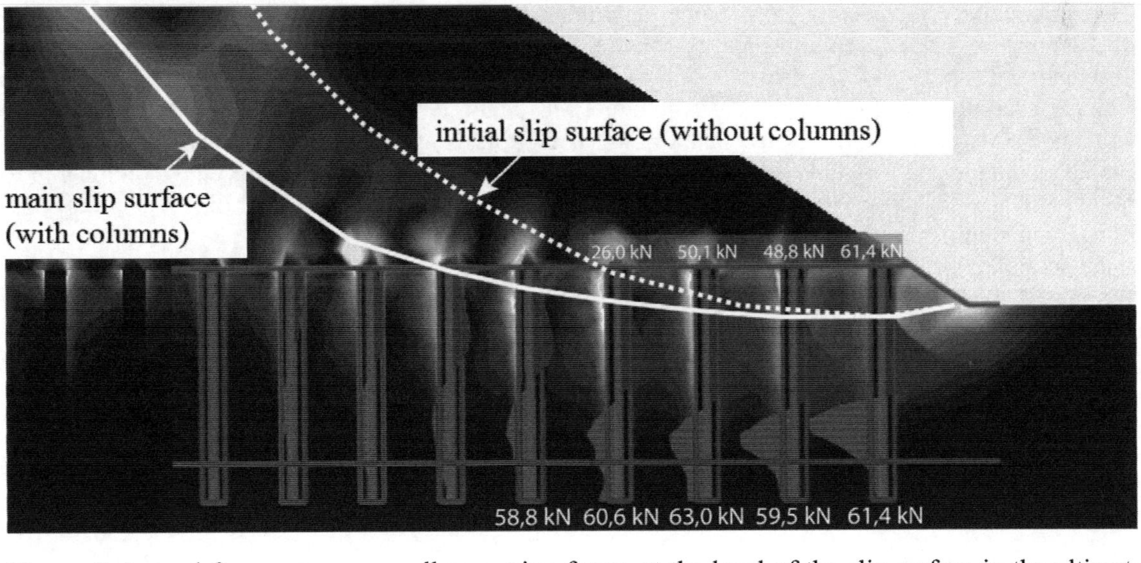

Figure 8. Lateral force curves as well as cutting forces at the level of the slip surface in the ultimate limit state

The shear surface is influenced by all columns, but only the outermost five stability columns are in the initial slip surface and therefore taken into account for the analytical calculation. The procedure for "rigid" binder columns is based on the initial slip surface without columns. This leads to attachable transverse forces R_s in the order of 20 to 40 kN. The transverse forces calculated with the FEM are based on the slip surface of the embankment stabilized with columns and are between 59 to 63 kN. By shifting the slip surface more columns are considered. Tab. 5 shows a comparison of the calculated forces.

Table 5. Summary of attachable lateral forces of the approach for "rigid" binder columns (BMS) and the stability analysis with FEM.

number of column	**6-8**	**5**	**4**	**3**	**2**	**1**
min $R_{S,i}$ (“rigid" columns) [kN]	0	0	21.8	35.6	39.8	20.2
Q_s (FEM) [kN]	0	58.8	60.6	63.0	59.5	61.4
Q_s (FEM; initial slipe surface) [kN]	0	0	26.0	50.1	48.8	61.4

In addition, the shear forces at the level of the original slip surface are indicated in the table. Due to the more accurate replica of the approach, this leads to a significantly better agreement with the forces min $R_{s,i}$ ("rigid" columns). However, these are also smaller. By shifting the slip surface in the range of stability columns always occurs an increase in the attachable columns or shear forces compared to the initial slip surface.

6. Summary

As a result of the conducted research and analysis of analytical approaches for estimating the slope stability with consideration of stabilizing columns, essential and appropriate approaches were identified. The current types of non-reinforced cylindrical load bearing elements have to be separated between granular columns and "soft" binder columns with a consideration via improved soil properties of a composite system of columns and surrounding soil. For "rigid" binder columns, an approach was presented and discussed with consideration of different failure modes of the individual columns in a concrete example using the analytical software GGU Stability. In addition, a comparative analysis with the FEM has been illustrated and discussed on this example. The analytical approach is in principle suitable and plausible in the form shown and it has reserves in the safety level compared to the numerical calculation. The comparative calculation with FEM using "rigid" binder columns also illustrates a significant widening / scattering of the failure zone.

Thus, with the presented approaches to computational consideration and assessment of stability columns, the current types of stability columns can be suitable considered and measured in a slope stability analyses in limit state GEO-3.

7. References

[1] Borchert K.-M, Kirsch F. (2008): Bemessung der Erweiterung von Bestandsdämmen auf gering tragfähigem Untergrund mit Rüttelstopf- oder Betonsäulen. *VSVI-Seminar "Bautechnik"*. Berlin, Dt. Technikmuseum, 02.04.2008

[2] Buß J. (2012): *Handbuch GGU-STABILITY*. Böschungsbruchberechnungen und Berechnungen von Bodenvernagelungen und Bewehrte-Erde-Wänden nach DIN 4084 und EC 7. Hg. v. Civilserve GmbH.

[3] Brinkgreve R.B.J., Engin E., Swolfs W.M. (2014): *Plaxis 2014 manual*.

[4] Broms B.B. (2004): Lime and lime/cement columns. Chapter 8 in: M. P. Moseley and K. Kirsch (Hg.): *Ground improvement*, **2**. Aufl., London, New York, Spon Press, S.254 – 330.

[5] EANG (2014): *Empfehlungen des Arbeitskreises Numerik in der Geotechnik*. **1**. Aufl., DGGT, Wilhelm Ernst & Sohn, Berlin.

[6] EBGEO (2010): *Empfehlungen für den Entwurf und die Berechnung von Erdkörpern mit Bewehrungseinlagen aus Geokunststoffen*. **2**. Aufl. Berlin, Ernst.

[7] EuroSoilStab (2002): *Design Guide Soft Soil Stabilisation* - Development of design and construction methods to stabilise soft organic soils. ESS (CT97-0351, Project No.: BE 96-3177).

[8] Filz G., Sloan J., McGuire M., Collin J. and Smith M. (2012): Column-Supported Embankments. Settlement and Load Transfer. *Geotechnical Engineering State of the Art and Practice*, pp. 54-77.

[9] Kivelö M. (1998): *Stabilisation of embankments on soft soil with lime/cement columns*. Royal Institut of Technology, Stockholm.

[10] Kitazume, M. & Terashi, M. (2013): *The deep mixing method*. Leiden, The Netherlands: CRC Press/Balkema.

[11] Navin M. P. (2005): *Stability of Embankments Founded on Soft Soil Improved with Deep-Mixing-Method Columns*. Virginia Polytechnic Institute State University, Blacksburg, USA.

[12] Priebe H. J. (1995): Die Bemessung von Rüttelstopfverdichtungen. Fachaufsatz GT 07-13 D. In: *Ground Engineering*.

[13] Priebe H J. (2003): Zur Bemessung von Rüttelstopfverdichtungen - Anwendung des Verfahrens bei extrem weichen Böden, bei schwimmenden Gründungen und beim Nachweis der Sicherheit gegen Gelände- und Böschungsbruch. In: *Bautechnik* (**80**), S. 380–384.

[14] Wolffersdorff P.-A. von, Koletzka A., Rosner, S. (2010): Scherkondetalbrücke – Untersuchungen zum Interaktionsverhalten zwischen Brückenwiderlager und Anschlussdamm. Baugrund Dresden Ingenieurgesellschaft; S. Marx, TU Dresden, Institut für Massivbau.

[15] Public Works Research Center (2004): *Technical Manual on Deep Mixing Method for On Land Work*. Japan, 334p

[16] Zaeske D. (2001): Zur Wirkungsweise von unbewehrten und bewehrten mineralischen Tragschichten über pfahlartigen Gründungselementen, *Schriftenreihe Universität Kassel*, **10**, Februar 2001.

[17] Raithel, M. / Leusink, E. / Kollmer, D. (2005): Sanierung eines Straßendammes mit EPS-Hartschaum. *Symposium Geotechnics*. **18**, University of Kassel.